Microscopic organic-walled fossils are found in most sedimentary rocks. The organic particles – spores, pollen and other land- and marine-derived microfossils representing animals, plants, fungi and protists – can be extracted and used to date the rock, reveal details of the original sedimentary environment and provide information on the climate of the time of deposition. The mix within a sediment of these discrete organic particles – palynomorphs and palynodebris – form palynofacies. This book presents research work on the sedimentation of components of palynofacies and details their importance for sequence stratigraphy and the interpretation of ancient biologic and geologic environments.

A comprehensive introduction to the subject is presented in the first chapter. Palynosedimentation in modern environments, the reconstruction of terrestrial vegetation and the application of the data to sequence stratigraphy are then considered. Later chapters detail the various quantitative methods and their specific applications in the subject.

This is a valuable reference work for palynologists, sedimentologists and paleobiologists, and for professionals working in the hydrocarbons industries.

Sedimentation of organic particles

Sedimentation of organic particles

edited by ALFRED TRAVERSE
Department of Geosciences, The Pennsylvania State University

CAMBRIDGE
UNIVERSITY PRESS

CAMBRIDGE UNIVERSITY PRESS
Cambridge, New York, Melbourne, Madrid, Cape Town, Singapore, São Paulo

Cambridge University Press
The Edinburgh Building, Cambridge CB2 2RU, UK

Published in the United States of America by Cambridge University Press, New York

www.cambridge.org
Information on this title: www.cambridge.org/9780521384360

© Cambridge University Press 1994

First published 1994
This digitally printed first paperback version 2005

A catalogue record for this publication is available from the British Library

Library of Congress Cataloguing in Publication data
Sedimentation of organic particles / edited by Alfred Traverse.

   p. cm.
ISBN 0 521 38436 2
1. Palynology.   2. Sedimentation and deposition.
3. Micropaleontology.   I. Traverse, Alfred, 1925–
QE993.S43   1994
561´.13–dc20   92-39221 CIP

ISBN-13 978-0-521-38436-0 hardback
ISBN-10 0-521-38436-2 hardback

ISBN-13 978-0-521-67550-5 paperback
ISBN-10 0-521-67550-2 paperback

Contents

Preface

ALFRED TRAVERSE

My first job as a palynologist was with Shell Development Company in Houston, Texas. I was hired in 1955, and my assignment was to study pollen and spores in recent sediments in whatever way or ways I wished. For the next seven years I did this for Shell, and have, since then, been fascinated by the relationship between pollen, spores and other organic particles to the organisms from which they derive and to the sediments in which they occur as fossils. When Dr. Mary Dettmann invited me to organize a symposium on this subject for the 6th International Palynological Congress in Brisbane, Australia (August–September, 1988), I accepted enthusiastically. After the Congress I invited the participants in the symposium to produce chapters on this subject for a proposed book. However, as it worked out, there is very slight overlap between this volume and that symposium. For one reason or another, some of the participants in Brisbane did not prepare chapters. Others (including me) felt that the papers they had published in the Proceedings volume for the Congress (*Review of Palaeobotany and Palynology*, **64**, 1990) need not be duplicated, and they produced something quite different for this book. It is also very significant for the shape of the present volume that I invited a number of people to write chapters who were not involved in the symposium at all, in order to assure a wide-ranging coverage of the subject. The new project *was* inspired by the symposium, but publishes a much different and broader array of material than was presented in Australia in 1988.

The logic of the organization of this book seems right to me, but I realize from discussions with some of the authors that others might organize it differently. One of an editor's relatively few real privileges is to arrange the chapters! It seemed appropriate to me to begin with an introduction to the subject as I see it, including, where appropriate, references to individual chapters as they contribute to understanding organic particle sedimentation. The introduction is followed by chapters on palyno-sedimentation in modern environments, including discussion of the relatively new concepts of palynodebris and palynofacies. These matters seem to me basic to the whole field. Then come two chapters on the reconstruction of terrestrial vegetation from palynological data. These chapters were added to the project on the basis of reviews of the original book proposal, which noted that this specialized application of palynosedimentation is important to a sizable group of researchers specializing in Cenozoic biology and geology. Next come chapters on the application of studies of palynosedimentation data to a number of currently important geologic problems, such as sedimentary cycles and sequence stratigraphy. Several concluding chapters on various quantitative methods and their specific applications seemed to me to provide a good counterbalance at the end to the general information at the very beginning of the volume.

An appendix presents a previously unpublished annotated bibliography on the literature of palynomorph transport and sedimentation. The bibliography has been regularly updated for years, and this most recent version should be a very valuable tool for anybody interested in palynosedimentation.

Because of the varied backgrounds of authors who contributed to this volume, there is some heterogeneity of usage. For example, in the bibliographies to chapters by George Hart and co-authors, references are made to privately circulated company reports. Citation of these is somewhat controversial, but Dr Hart indicates that he will provide photocopies of these reports to interested persons who write him.

All of the chapters by my colleagues have been reviewed by at least one (usually two) highly qualified expert(s), and in every case this has resulted in considerable to extensive change. Some manuscripts unfortunately had to be rejected on the basis of review. The reviewers were mostly authors of other chapters. Practically *all* authors were involved as reviewers (generally anonymously), most with more than one manuscript, and they are acknowledged only by being listed as authors here. However, the help in evaluating manuscripts of the following 'external' reviewers is gratefully acknowledged: Jan Jansonius, Ronald J. Litwin, D. Colin McGregor and Cathy Whitlock. Two of my Penn State colleagues, Roy J. Greenfield and S. J. Mackwell, offered extensive computer assistance in the course of preparation of the final manuscript. Other Penn State colleagues, Lee R. Kump and Rudy L. Slingerland, were very helpful in editorial work with several chapters. Discussions with Martin B. Farley were central to initiation and continuation of this project. He has been very generous with help throughout the long time it has taken to complete the work. At the very end of the task I was a guest as Fulbright Professor at the Senckenberg Research Institute and Natural History Museum in Frankfurt am Main, Germany. Friedemann Schaarschmidt, the leader of the Paleobotanical Section, and his co-workers, Karin Schmidt, Birgit Nickel, Kurt Goth and Volker Wilde, have all helped with the project. I am also grateful to Prof. Dr Willi Ziegler, Senckenberg's Director, for inviting me to work at the museum. Elizabeth I. Traverse has word-processed every letter in this book, mostly many times over, served as general editorial assistant and dealt with countless other problems, large and small, beginning years ago and not ending until the index was prepared and all proof read. Nobody has ever had a better assistant.

Contributors

David J. Batten
  Palynological Research Centre
  Institute of Earth Studies
  The University College of Wales
  Aberystwyth, Wales SY23 3DB, UK

Michael C. Boulter
  Palaeobiology Research Unit
  University of East London
  Romford Road
  London E15 4LZ, UK

Richard Bradshaw
  Faculty of Forestry
  Swedish University of Agricultural Sciences
  S-230 53 Alnarp, Sweden

W. A. Brugman
  Royal Dutch/Shell
  8 IPM-EPX 36
  P.O. Box 162
  2501 den Haag, Netherlands

Grace S. Brush
  Department of Geography and Environmental
  Engineering
  The Johns Hopkins University
  Baltimore, MD 21218, USA

Lucien M. Brush
  Department of Geography and Environmental
  Engineering
  The Johns Hopkins University
  Baltimore, MD 21218, USA

Vaughn M. Bryant, Jr.
  Department of Anthropology
  Palynology Laboratory
  Texas A&M University
  College Station, TX 77843, USA

Claude Caratini
  French Institute
  BP 33
  Pondicherry, India 605001

David L. Carlson
  Archaeological Research Laboratory
  Texas A&M University
  College Station, TX 77843, USA

Jason D. Darby
  United States Environmental Protection Agency
  Region IV
  345 Courtland St.
  Atlanta, GA 30365, USA

Yoram Eshet
  Geological Survey of Israel
  30 Malkhei Israel Street
  Jerusalem 95501, Israel

Martin B. Farley
  Exxon Production Research Co.
  P.O. Box 2189
  Houston, TX 77252, USA

R. Farley Fleming
  United States Geological Survey
  Federal Center, MS 919
  Box 25046
  Denver, CO 80225, USA

Robert A. Gastaldo
  Department of Geology
  Auburn University
  Auburn, AL 36849-5305, USA

William A. Gregory
  Exxon Exploration Company
  P.O. Box 4279
  Houston, TX 77210-4279, USA

Daniel Habib
  Department of Geology
  Queens College
  City University of New York
  Flushing, NY 11367, USA

George F. Hart
  Department of Geology and Geophysics
  Louisiana State University
  Baton Rouge, LA 70803, USA

Richard G. Holloway
  Biology Department
  University of New Mexico
  Albuquerque, NM 87131, USA

Phillip L. Holmes
  243 Christchurch Ave
  Wealdstone, Harrow
  Middlesex, HA3 5BA, UK

Stephen T. Jackson
  Department of Biological Sciences
  Northern Arizona University
  Flagstaff, AZ 86011, USA

John G. Jones
  Palynology Laboratory
  Texas A&M University
  College Station, TX 77843, USA

J. H. F. Kerp
  Westfälische Wilhelms-Universität
  Abt. Paläobotanik
  Geol.-Paläont. Institut und Museum
  Hindenburgplatz 57-59
  D-48143 Münster, Germany

Warren L. Kovach
  Kovach Computing Services
  85 Nant-y-Felin
  Pentraeth, Anglesey
  LL75 8UY, Wales, UK

Douglas J. Nichols
  United States Geological Survey
  Federal Center, MS 919
  Box 25046
  Denver, CO 80225, USA

Mark A. Pasley
  Amoco Production Co.
  P.O. Box 3092
  Houston, TX 77253, USA

Fredric L. Pirkle
  E. I. Du Pont de Nemours and Co.
  Starke, FL 32091, USA

David T. Pocknall
  Amoco Production Co.
  P.O. Box 3092
  Houston, TX 77253, USA

Fredrick J. Rich
  Department of Geology and Geography
  Georgia Southern University
  Statesboro, GA 30460, USA

Claire Richelot
  Paleontology, The University
  7, pl. Vingt-Août
  B-4000 Liège, Belgium

Maurice Streel
  Paleontology, The University
  7, pl. Vingt-Août
  B-4000 Liège, Belgium

Paul K. Strother
  Weston Observatory
  381 Concord Road
  Weston, MA 02193, USA

Arthur R. Sweet
  Geological Survey of Canada
  Institute of Sedimentary and Petroleum Geology
  3303 33rd St. NW
  Calgary, Alberta, Canada T2L 2A7

Alfred Traverse
  Palynological Laboratories
  Department of Geosciences
  Pennsylvania State University
  University Park, PA 16802, USA

P. F. Van Bergen
  Netherlands Institute of Sea Research
  Division of Marine Biogeochemistry
  P.O. Box 59
  1790 AB Den Burg, Texel
  The Netherlands

Robert Van Pelt
  Westinghouse Savannah River
  P.O. Box 616
  Merrill Lynch Bldg.
  Aiken, SC 29801, USA

1 Sedimentation of palynomorphs and palynodebris: an introduction

ALFRED TRAVERSE

Introduction

The title above is really the subject of this whole book. Some time ago, Egon Degens (1965) estimated that about 2% of the total volume of sedimentary rocks on Earth is organic matter, that is, 20 m of a total of 1000 m of sediment in the whole of Earth history. All but 5 cm ($=$coal and oil) of the 20 m ($=19.95$ m) is finely disseminated organic matter in shales, limestones and sandstones. Much of it is amorphous, much is from marine animals, algae and protists, either degraded or in recognizable particles. Since at least the Late Devonian, however, terrestrial plant biomass has been a (perhaps the) major source of the organic matter in sedimentary rocks (see discussion in Traverse, 1992). At present, approximately 150×10^6 metric tons of chemically resistant, particulate organic matter (POC) reaches the continental shelves from the major rivers of the world (Deuser, 1988; Ittekot, 1988). That can be compared with estimates of total sedimentation of 70×10^8 metric tons annually (Holland, 1978).

As a rather rash guess, I would say that modern sedimentary rocks probably contain something like 10 times as much resistant-walled pollen-and-spore-like material as do 375×10^6-year-old Devonian shales and sandstones (Schuyler & Traverse, 1990). Nevertheless, land-derived plant debris in microscopic particles has been a very significant factor in building the organic bankroll of the Earth for some hundreds of millions of years, and much of it is recognizable as to source. In addition, marine-derived organic matter has been and is making a sizable contribution. Robust organic-walled, microscopic animal remains (chitinozoans, scolecodonts, 'microforams'), protists (for example, dinoflagellate cysts and some acritarchs), algae (various acritarchs, tasmanitids, among others) are obvious examples, but there is a much larger component of amorphous matter, as well as difficult to identify organic structures. It is clear from several chapters in Huc (1990) that organic matter in marine sediments includes important contributions from both the indigenous marine realm itself (autochthonous) and from terrigenous material derived from the continents (allochthonous) (see Fig. 1.1).

Palynologists have a keen interest in all of the fossil and potentially fossil organic matter just mentioned, but naturally are especially interested in identifiable particles, because their recognizability permits derivation of information about patterns of events on the continents and their shelves, such as climatic alteration, orogeny and other tectonic events, sedimentation patterns, and world oceanic current alterations. In addition, because of the relationship of organic particles to coal and oil formation, their study obviously is connected to investigations of coal and oil genesis and oil exploration. Hence, studies of the origin and eventual sedimentation of organic particles are currently an exciting frontier of both theoretical and applied paleontology.

As is so often true in science, developments far removed from palynology-proper have had, and continue to have, obviously unpredictable impact on this area of study. Computer applications, including the automatic electronic recognition of particles, searching of records about them, and multivariate statistical programs are already in use. New fields of study such as the science and technology of particle systems (Hogg, 1989) may well have impact soon. Studies with their roots in well established basic geological and biological research are producing significant results. For example, studies of the origin and taphonomy of organic matter generally (Allison & Briggs, 1991) tell us much about what to expect in the sedimentation of organic particles. Conversely, the results of our research provide information that students of taphonomy of organic matter need to consider. Chapter 7 (Gastaldo, this book) is enlightening in this respect. Petrologists, paleobotanists, and other students of organic matter have for decades given order to our understanding of such diverse matters as the constitution of kerogen generally, and of coal in particular (Murchison, 1987; Scott, 1987). Their work has, for example, helped to clarify the lacustrine algal contribution to some oil shales (Goth

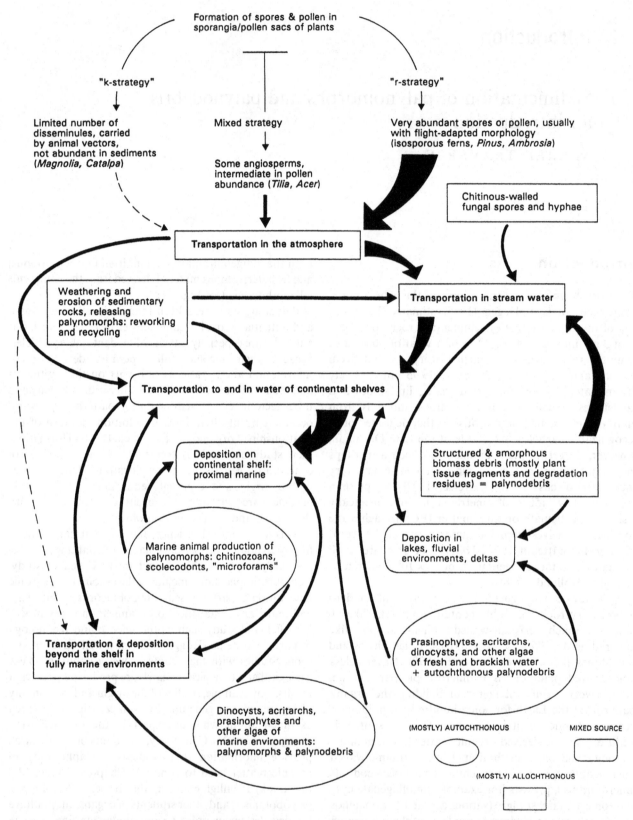

Figure 1.1. The complex origin, transportation and deposition as sediment, of robust organic (carbonaceous) particles. The shape of the boxes indicates whether particles concerned are produced mostly *in* the given fluid environment (autochthonous), or delivered mostly to that environment from outside of it (allochthonous), or have a mixed source. (See legend at bottom of figure.) The size of the arrows indicates very roughly the relative magnitude of particle movements. More detail is provided for tracheophyte spores and pollen, because information derived from fossil sporomorphs is widely used for reconstruction of terrestrial climate and events.

et al., 1988). In this book, Chapters 9 and 17 (Hart and colleagues) present examples of the significance of various categories of particulate organic matter, not only for understanding the origin of sedimentary rocks, but also for correctly interpreting a great variety of sedimentary environments and evaluating them, for example, for sequence stratigraphy. The purpose of the entire book is to present recent contributions to the general study of the sedimentology of organic matter, the development of which subject was heralded several years ago by Pelet and Deroo (1983).

Definition of the particle types

While it has no force of 'law,' the preliminary *Amsterdam Palynological Organic Matter Classification*, drawn up on 27–28 June, 1991, by a committee which included David J. Batten, a co-author for this book (Batten, personal communication, 16 Dec., 1991), goes a long way toward explaining much of what we are talking about (underlined items, including the numbers, were added or slightly changed from the original; other modifications are explained):

I.   PALYNOMORPHS
     sporomorphs
        pollen
        spores
           megaspores (>200 *μm*)
           small spores (≤200 *μm*)
     'algae'
        dinoflagellate cysts
        prasinophytes
        chlorococcales
        (cyanobacteria also included in original)
     acritarchs
     zoomorphs
        foram linings (='microforams')
        chitinozoans
        scolecodonts
        (tintinnids and rhizopods also
        included in original)
     fungal spores
        spores
        sclerotia
     indeterminate; opaque; microscopic seeds

II. (PALYNODEBRIS): STRUCTURED DEBRIS
     woody
     plant epidermis/cuticle
     plant tissue (other)
     animal
     fungal
     opaque, indeterminate

III. (PALYNODEBRIS): AMORPHOUS
     finely dispersed
     homogeneous
     heterogeneous
     opaque

IV. INDETERMINATE (=not assignable to I–III)
     opaque
     other

The Amsterdam classification includes cyanobacteria under algal palynomorphs, and tintinnids and rhizopods under zoomorph palynomorphs, all of which I regard as inappropriate in the list and have removed. Scolecodonts were presented with a question-mark under zoomorphs, which seems unnecessary. 'Small spores,' which the new classification uses under sporomorphs, *is* a more felicitous word than 'miospore,' which has the disadvantage of being confusable with both meiospore and microspore. Sections II, III and perhaps IV comprise what some now call 'palynodebris,' an expression around long enough to have picked up contradictory attributes. Highton *et al.* (1991), for example, classify palynodebris particles as >200 *μm*, whereas Boulter, in his classification of organic particles (Chap. 11, this book), describes palynodebris as not size-related. It is interesting that Combaz (1964) had, in his classification of the organic constituents of rock, a three-fold subdivision: (1) (what we now call) palynomorphs; (2) (what we now call) palynodebris; (3) amorphous and indeterminate organic matter. Gastaldo provides (Chap. 7, this book) a summary of how plant organics become phytoclasts and thus palynodebris. It is clear that the Amsterdam classification is strongly influenced by the detailed and profusely illustrated classification of Van Bergen *et al.* (1990). Brugman *et al.* (Chap. 19, this book) employ this classification as a basic assumption. Boulter's related but different classification in Chapter 11, and that of Hart, Hart *et al.*, and Darby and Hart (Chaps. 9, 17 & 10, respectively) are important additional contributions to this subject. For completeness one should also study the botanically-based classification of Masran and Pocock (1981) and Habib's organic petrologic classification, which is referenced and used in Chapter 16 (Habib *et al.*, this book).

Natural history of organic particles as clasts

Elements of the Earth's biomass with resistant cell walls join other clastic particles in the atmosphere and hydrosphere, as suggested in Figure 1.1. Pieces of organic matter in the fluids of the Earth's surface are more than just passive fellow travellers of mineral clastic particles.

It is known that living organic particles such as diatoms may tend to affect directly the stability of a stream bed (Heinzelmann *et al.*, 1991). As generators of methanol and as substrate for sulfur-bacteria, organic particles undergoing taphonomic processes directly affect the properties of the co-sediment. In their final resting place they also provide raw material for hydrocarbon generation. From the point of view of palynology, these particles most significantly act as indicators or clues for the combinations of circumstances that put them where they are found in sediments, as suggested in Figure 1.1. As Robbins *et al.* (1991) point out, the biological information provided by palynomorphs/palynodebris must be coupled with their geological provenance to maximize the utility of the information their presence provides. Similarly, in Chapter 16 of this volume, Habib *et al.* show that the geochemistry of sediments is related to the suite of palynomorphs they contain. In Figure 1.1, I have indicated eventual sites of deposition, and thus incorporation in the sediments that become sedimentary rocks. The natural history of organic particles before they reach an eventual site of deposition is much more complicated than such a diagram can display.

Palynofacies

It is now well known that when a particular assortment of palynomorphs/palynodebris in rocks is considered as a whole, it can yield information about (1) the biosphere segment producing the fossil particles, and (2) the environment of deposition in which the rock was produced, hence, the rock type produced. Combaz (1964) coined the term 'palynofacies' for the organic content of sedimentary rock, indicating both sedimentary environment of deposition and information about the producing biosphere elements.

Palynolithofacies

I note from my company reports at Shell Development Company in 1960 that I was at that time using the term 'palynofacies' to refer to a more or less local concentration of particular palynomorphs, indicating a sort of *biofacies*. It is clear that much of the application of the word since then has been geologically oriented, and 'palynofacies' is used primarily to indicate information about the enclosing *rock*, especially vis-à-vis its environment of deposition. I would call this a study of 'palynolithofacies.'

LeNoir and Hart (1988), for example, described seven 'palynofacies' in Miocene sands of Louisiana. The palynofacies were characterized by dinoflagellate assemblages, but the meaning is *litho*facies, not *bio*facies. The dinocysts by themselves are not used to extract biologic information. Streel and Scheckler (1990) and Schuyler and Traverse (1990) similarly used Devonian miospore assemblages to characterize environments of deposition, from the point of view of the sedimentary rocks produced. In Chapter 19 of this book, Brugman *et al.* present detailed information about precise subdivisions of palynolithofacies, and in Chapter 18, Kovach and Batten describe computer-based quantitative methods for handling such data. Caratini (Chap. 8, this book) illustrates a helpful graphical technique for presenting such palynolithofacies.

Palynobiofacies

The application of palynological studies to determine relationships between palynomorph concentrations and biosphere associations such as vegetation types I would designate with the term 'palynobiofacies.' Pleistocene palynologists use fossil pollen and spore assemblages to reconstruct plant associations. They are thus at pains to correct for over- and under-representation, and have developed very sophisticated quantitative methods to accomplish this end. Bradshaw and Jackson (Chaps. 13 & 14, respectively, this book) deal with aspects of such research. This is, in a sense, a way of eliminating biofacies concentrations from the data, to 'restore the norm.' All biofacies are to some degree counter to the norm, or average. That is why they are different 'faces' or facies. In Chapter 12 of this book Nichols and Pocknall describe the use of palynomorph-based plant association biofacies to determine the sorts of environments in which extensive ancient peats were formed. Traverse's Chapter 6 on palynomorph sedimentation in Trinity–Galveston Bay, Texas and vicinity, shows that in that set of environments, local sedimentary factors affect local total palynomorph concentrations, and it is clear that cores of sediment would yield quite different local palynofloras, depending on the location sampled. However, Brush and Brush's Chapter 3 on sedimentation in Chesapeake Bay shows that the overall pollen and spore flora of such a region at various stages in the recent past, determined from cored sediment, can be successfully used for interpreting climatic trends. Strother's Chapter 23 shows that, even in the early Paleozoic and before, organisms played an important role in sedimentation, either directly, or indirectly, as indicators of environment.

Palyno-biolithofacies

Many studies of palynomorph/palynodebris-based facies are 'bi-facial.' That is, they have as goals both elucidation of the enclosing rock, and the biosphere association from which the palynological fossils were derived. This was the emphasis in Farley and Traverse's (1990) study of Paleogene environments in Wyoming, and is further elucidated in Farley's Chapter 20 in the present volume. Obviously, palyno-biolithofacies studies tend to shift toward one or the other emphasis. Fleming and Nichols (Chap. 21, this book) show how multivariate analytical methods can be used to determine both sedimentary and biological environments. Rich and Pirkle (Chap. 15, this book) present a study of sedimentation in the southeastern USA, in which the emphasis leans toward understanding the source vegetation for palynofloras, but the plant associations indicated by the pollen assemblages are also used to help interpret the origin of various sedimentary types. Darby and Hart's study of carbonate deposition (Chap. 10, this book) uses palynologic data, especially on palynodebris, to determine depositional environment in the absence of clastic mineral particles. The goal is lithofacies analysis, but the means are very biological. Sweet (Chap. 22, this book) emphasizes floral changes across the K/T boundary, as they relate to lithofacies.

Palynosedimentation

It is now quite well known that palynomorphs and palynodebris particles behave in the same way as do clastic particles of their size and specific gravity (see Traverse, 1988: Chap. 17; Spicer, 1991, makes a very significant contribution to understanding plant taphonomic processes that produce, among other things, most kinds of palynodebris). In general, palynomorphs/palynodebris are transported to sites of deposition in water currents (see Fig. 1.1). An exception is found in arid regions, where all significant deposition is from the atmosphere. Bryant *et al.* (Chap. 4, this book) present a study of palynomorphs so sedimented. Also exceptions are some significant parts of the continental shelf, such as off northwest Africa (Hooghiemstra, 1988a,b; Dupont *et al.*, 1989), where wind transportation overwhelms the insignificant stream contributions. On the other hand, many studies of streams have demonstrated the dominance of water transportation of palynomorphs/palynodebris to continental shelves (Muller, 1959, for the Orinoco; Darrell, 1973, for the Mississippi; Federova, 1952, for the Volga, and many others). Holmes (1990, Chap. 2 this volume) has given a theoretical basis to the expectation of such transport by laboratory experiment. Darby and Hart's chapter on Florida (Chap. 10, this volume), and Traverse and Ginsburg's (1966) study of the Bahamas, have shown that 'pollen rain' delivered by either wind (virtually the only source in the Bahamas) or by wind and stream (as in Florida) is moved in non-clastic environments by water currents.

Habib *et al.* and Hart *et al.* (Chaps. 16 & 17, respectively, this volume) both show that such large scale phenomena/events at the marine/nonmarine boundary as transgression/regression can be traced by following the associated palynomorph/palynodebris sedimentation, both as to deposition of marine-derived palynomorphs (dinoflagellate cysts, for example) and as to sedimentation of different mixes of land-derived palynomorphs/palynodebris.

Transportation of palynomorphs/palynodebris in water

Although study of the transportation of pollen and spores in the atmosphere has a long history, relatively little has been done regarding the water transportation of palynomorphs/palynodebris. I have discussed the status of such studies recently (Traverse, 1990, 1992), as well as in Chapter 17 of Traverse (1988). Not surprisingly, as Koreneva (1964) and Chmura and Liu (1990) have pointed out, palynomorphs behave in the water as sedimentary particles, and their concentration is correlated strongly with mineral sediment concentration. Studies by Vronskiy (1975) in the Aral Sea and by Dupont *et al.* (1989), Hooghiemstra (1988a,b) and Hooghiemstra *et al.* (1986) in the Atlantic Ocean off northwest Africa, and by Brush and Brush (Chap. 3, this volume), have shown, however, that in some situations airborne pollen exceeds or even dominates that brought by streams. For the most part, it seems clear that stream action and current movements on the continental shelf play the leading role in sedimentation of the palynomorphs/palynodebris found in shales and sandstones. Thus the experimental work reported by Holmes (Chap. 2, this book; see also Holmes, 1990) on the behavior of palynomorphs under laboratory conditions is an important contribution to basic understanding of the phenomena governing pollen suspension in and sedimentation out of water. Farley's annotated bibliography of the literature relating to all aspects of pollen/spore transportation in atmosphere and hydrosphere (Appendix, this book) should prove an invaluable resource to researchers in this field.

Sorting of palynomorphs/palynodebris during sedimentation

Holmes (1990, Chap. 2 this book) has shown that sorting, mostly by size, can occur in the water transport of palynomorphs. As size sorting is possible in watch-glasses in the laboratory by 'swirling' (see discussion in Traverse, 1988, Chap. 6 this book), this is not surprising. Pleistocene palynologists have noticed sorting, or concentration, of certain forms and call it 'focusing.' Boulter and Riddick (1986) have noted that sorting of palynodebris elements occurs – and preferential deposition in different environments. Catto (1985) observed sorting into silts and sands in the somewhat unusual setting of a sub-arctic braided stream. Muller (1959), of course observed in his classic paper on the Orinoco drainage that smaller pollen becomes a larger percentage of total palynomorphs as one moves offshore. It must be mentioned that the 1:4 ratio of large to small palynomorphs on which the Hoffmeister (1954) method for finding ancient shorelines depended, was based on palynomorph sorting. However, there is little evidence that such sorting is efficient enough either to make the Hoffmeister method work reliably, or to prevent use of palynofloral analysis, for example, for stratigraphy based on joint occurrence of populations of many taxa of different pollen size.

Reworking, recycling and related matters

The resilience of the walls of most sorts of palynomorphs and some kinds of palynodebris permits them not surprisingly to be resuspended and redeposited during depositional processes (Starling & Crowder, 1981). In Chapter 5 (this book), Streel and Richelot describe a situation in which pollen originally deposited in upland locations is carried downstream and out to sea during large floods. Palynomorphs also are frequently weathered and eroded out of rocks representing original sites of deposition, and recycled and redeposited. I am currently investigating cores of present–interglacial sediment from the Meadowlands complex, New Jersey. The 'native' pollen and spores are only poor to fair in quality of preservation, but (brown) Triassic (*Corollina*) and (yellow-green) late Cretaceous (*Rugubivesiculites*, Normapolles, *Cicatricosisporites*) spores and pollen, obviously eroded from nearby outcrops when they were above sea level several thousands of years ago are beautifully preserved (Traverse, in press). Eshet *et al.* (1988) have shown that reworked palynomorphs in the fossil record, derived from previously exposed surfaces, have potential for study of source rocks and of depositional cycles that might not otherwise be observed. Reworked pollen found in the water of Trinity River, Texas, comes from all stratigraphic levels, Carboniferous to latest Cenozoic, intercepted by the river in its course toward the Gulf of Mexico (Traverse, 1990, 1992, Chap. 6 this book).

Sequence stratigraphy and organic sedimentation

The relatively new development of sequence stratigraphy represents, among other things, acceptance of sedimentological insights and information into stratigraphic interpretation. The sedimentation of organic particles is ideally suited for a role in this new approach to stratigraphy, because many of the organic particles are age-specific and thus directly useful in stratigraphic interpretation, but all categories of organic particles can also be used in interpreting sedimentary environments.

Chapter 17 (Hart *et al.*, this book) deals specifically with the contribution to sequence stratigraphy of palynofacies determined from the content of suites of particulate organic matter. Chapter 19 (Brugman *et al.*, this book) describes use of an integrated palynofacies study in the Triassic of the Germanic Basin, revealing the sort of regressive, transgressive and mosaic environmental sequences that are useful for sequence stratigraphy. It seems clear, at least to this editor, that the future of paleopalynology is secure for some time, since it deals not only with fossil palynomorphs and their direct applicability to solution of stratigraphic problems, but also with resistant organic particles in general and their great potential for interpreting the genesis and subsequent history of sediments.

Suggestions for the Future

Some of the blocks in Figure 1.1 indicate parts of the total subject that are especially open for research – areas about which relatively little is known:

Fungal spores, from sporulation to sedimentation and subsequent history in sedimentary rock

The only fungal spores that are of interest palynologically are those with resistant walls, presumably all made of chitin (melanin probably plays a role as well; see Traverse, in press). Which super-taxa of Fungi this involves, what fraction of Fungi as a whole this comprises, the chemistry, the sedimentation, the geologic history, are all largely open questions.

Zoo-palynomorph sedimentation

Sedimentation of such particles, especially of chitinozoans, scolecodonts and chitinous foraminiferal linings ('micro-forams'), has been relatively unstudied, although the stratigraphic range, especially for chitinozoans, has been extensively researched. The chemistry of these fossils is also very much an open question in need of solution.

Algal palynomorphs

The sedimentation of prasinophytes, zygnemataceous zygospores, and various other categories of algal bodies, including most acritarchs, will be fruitful subjects for further investigation.

Organic particle transportation in the marine realm

Movement of palynomorphs/palynodebris by various current systems, by turbidity currents, and by tidal action would be well worth more study. The investigation of palynomorphs in ocean water is wide open. Farley's (1987) research in the Caribbean is practically alone as a systematic survey of pollen and spore transportation in the water of a significant segment of the oceans.

Reworking, recycling, redeposition

These phenomena comprise a complex, fascinating and important subject that could constitute the basis of many very significant projects. The ability of resistant-walled palynomorphs to survive repeated attack on the substrate in which they are found and subsequent transportation should not be an embarrassment to palynology, but a source of very significant information about source rocks, cycles of erosion and deposition, differential diagenetic responses, and the courses of paleo-streams.

Sedimentation of palynodebris

Discussed in a number of chapters in this book, this subject is still very new. We don't begin to know what the very diverse assemblages of such organic particles mean, except in the limited number of cases in which the materials have been carefully studied. This is clearly a very large subject which is now opening up ahead of us, and I hope this book stimulates study of it.

Acknowledgments

As always, I acknowledge with gratitude the assistance of E. I. Traverse in preparation of the manuscript, and for many helpful suggestions. Discussions with Martin B. Farley at the beginning of this project were useful in designing the chapter. F. Schaarschmidt and members of his staff in the Paleobotanical Section of the Senckenberg Research Institute and Natural History Museum, Frankfurt am Main, Germany, were very cooperative in many ways during the final stage of this project, when I was a Fulbright Professor at that institution. B. Nickel was especially helpful in the preparation of Figure 1.1.

References

Allison, P. A. & Briggs, D. E. G. (1991). Taphonomy of nonmineralized tissues. In *Taphonomy: Releasing the Data Locked in the Fossil Record*, ed. P. A. L. & D. E. G. B. Vol. **9** of *Topics in Geobiology*. New York: Plenum Press, 25–70.

Boulter, M. C. & Riddick, A. (1986). Classification and analysis of palynodebris from the Palaeocene sediments of the Forties Field. *Sedimentology*, **33**, 871–86.

Catto, N. R. (1985). Hydrodynamic distribution of palynomorphs in a fluvial succession, Yukon. *Canadian Journal of Earth Science*, **22**, 1552–6.

Chmura, G. L. & Liu, K-b (1990). Pollen in the lower Mississippi River. *Review of Palaeobotany and Palynology*, **64**, 253–61.

Combaz, A. (1964). Les Palynofaciès. *Revue de Micropaléontologie*, **7**(3), 205–18.

Darrell, J. H., II (1973). Statistical evaluation of palynomorph distribution in the sedimentary environments of the modern Mississippi River delta. Ph.D. diss. Louisiana State University, Baton Rouge.

Degens, E. T. (1965). *Geochemistry of Sediments: A Brief Survey*. Englewood Cliffs, New Jersey: Prentice-Hall.

Deuser, W. G. (1988). Whither organic carbon? *Nature*, **332**, 396–7.

Dupont, L. M., Beug, H.-J., Stalling, H. & Tiedemann, R. (1989). 6. First palynological results from site 658 at 21° N off northwest Africa: pollen as climate indicators. In *Proceedings of the Ocean Drilling Program, Scientific Results* **108**, ed. W. Ruddiman, M. Sarnthein, *et al.*, 93–111.

Eshet, Y., Druckman, Y., Cousminer, H., Habib, D. & Drugg, W. S. (1988). Reworked palynomorphs and their use in the determination of sedimentary cycles. *Geology*, **16**, 662–5.

Farley, M. B. (1987). Palynomorphs from surface water of the eastern and central Caribbean Sea. *Micropaleontology*, **33**, 270–9.

Farley, M. B. & Traverse, A. (1990). Usefulness of palynomorph concentrations in distinguishing Paleogene depositional environments in Wyoming (U.S.A.). *Review of Palaeobotany and Palynology*, **64**, 325–9.

Federova, R. V. (1952). Dispersal of spores and pollen by water currents. *Trudy Instituta Geografii Moskva, Akademiya Nauk SSSR*, **52**, 46–72. (In Russian).

Goth, K., de Leeuw, J. W., Püttmann, W. & Tegelaar, E. W. (1988). Origin of Messel oil shale kerogen. *Nature*, **336**(6201), 759–61.

Heinzelmann, C., Baumgartner, B. & Rehse, C. (1991). Algal films stabilise the river bed: studying the effects of benthos settlement on bed erosion. *German Research*, **2/91**, 12–14.

Highton, P. J. C., Pearson, A. & Scott, A. C. (1991). Palynofacies and palynodebris and their use in Coal Measure palaeoecology and palaeoenvironmental analysis. *Neues Jahrbuch für Geologie und Paläontologie, Abhandlungen*, **183**(1–3), 135–69.

Hoffmeister, W. S. (1954). Microfossil prospecting for petroleum. US Patent 2,686,108.

Hogg, R. (1989). The science and technology of particle systems. *Earth and Mineral Sciences* (Pennsylvania State University, College of Earth and Mineral Sciences), **58**(1), 1–5.

Holland, H. E. (1978). *The Chemistry of the Atmosphere and Oceans*. New York: John Wiley & Sons.

Holmes, P. L. (1990). Differential transport of spores and pollen: a laboratory study. *Review of Palaeobotany and Palynology*, **64**, 289–96.

Hooghiemstra, H. (1988a). Palynological records from northwest African marine sediments: a general outline of the interpretation of the pollen signal. *Philosophical Transactions of the Royal Society of London*, **B 318**, 431–49.

Hooghiemstra, H. (1988b). Changes of major wind belts and vegetation zones in NW Africa 20,000 – 5,000 yr B.P., as deduced from a marine pollen record near Cap Blanc. *Review of Palaeobotany and Palynology*, **55**, 101–40.

Hooghiemstra, H., Agwu, C. O. C. & Beug, H.-J. (1986). Pollen and spore distribution in recent marine sediments: a record of NW-African seasonal wind patterns and vegetation belts. *'Meteor' Forschung-Ergebnisse*, **C40**, 87–135.

Huc, A. Y., ed. (1990). *Deposition of Organic Facies*. AAPG Studies in Geology **30**. Tulsa, Oklahoma: American Association of Petroleum Geologists.

Ittekot, V. (1988). Global trends in the nature of organic matter in river suspensions. *Nature*, **332**, 436–8.

Koreneva, E. V. (1964). Distribution and preservation of pollen in the western part of the Pacific Ocean. *Geological Bulletin* (Department of Geology, Queens College, City University of New York), **2**, 1–17.

LeNoir, E. A. & Hart, G. F. (1988). Palynofacies of some Miocene sands from the Gulf of Mexico, offshore Louisiana, U.S.A. *Palynology*, **12**, 151–65.

Masran, T. C. & Pocock, S. A. J. (1981). The classification of plant-derived particulate organic matter in sedimentary rocks. In *Organic Maturation Studies and Fossil Fuel Exploration*, ed. J. Brooks. London: Academic Press, 145–75.

Muller, J. (1959). Palynology of Recent Orinoco Delta and shelf sediments. *Micropaleontology*, **5**, 1–32.

Murchison, D. G. (1987). Recent advances in organic petrology and organic geochemistry: an overview with some reference to 'oil from coal'. In *Coal and Coal-bearing Strata: Recent Advances*, ed. A. C. Scott. *Geological Society Special Publication* **32**, 257–302.

Pelet, R. & Deroo, G. (1983). Vers une sédimentologie de la matière organique. *Bulletin de la Societé geologique de France*, **25**(4), 483–93.

Robbins, E. I., Zhou, Zi & Zhou, Zhi. (1991). Organic tissues in Tertiary lacustrine and palustrine rocks from the Jiyang and Pingyi rift depressions, Shandong Province, eastern China. *Special Publications International Association of Sedimentologists*, **13**, 291–311.

Schuyler, A. & Traverse, A. (1990). Sedimentology of miospores in the middle to Upper Devonian Oneonta Formation, Catskill Magnafacies, New York. *Review of Palaeobotany and Palynology*, **64**, 305–13.

Scott, A. C., ed. (1987). Coal and coal-bearing strata: recent advances and future prospects. In *Coal and Coal-bearing Strata: Recent Advances*, ed. A. C. S. *Geological Society Special Publication* **32**, 1–6.

Spicer, R. A. (1991). Plant taphonomic processes. In *Taphonomy: Releasing the Data Locked in the Fossil Record*, ed. P. A. Allison & D. E. G. Briggs. Vol. **9** of *Topics in Geobiology*. New York: Plenum Press, 71–113.

Starling, R. N. & Crowder, A. (1981). Pollen in the Salmon River system, Ontario, Canada. *Review of Palaeobotany and Palynology*, **31**, 311–34.

Streel, M. & Scheckler, S. E. (1990). Miospore lateral distribution in upper Famennian alluvial lagoonal to tidal facies from eastern United States and Belgium. *Review of Palaeobotany and Palynology*, **64**, 315–24.

Traverse, A. (1988). *Paleopalynology*. London: Unwin Hyman.

Traverse, A. (1990). Studies of pollen and spores in rivers and other bodies of water, in terms of source-vegetation and sedimentation, with special reference to Trinity River and Bay, Texas. *Review of Palaeobotany and Palynology*, **64**, 297–303.

Traverse, A. (1992). Organic fluvial sediment: palynomorphs and 'palynodebris' in the lower Trinity River, Texas. *Annals of the Missouri Botanical Garden*, **79**, 110–25.

Traverse, A. (in press). Manifestations of sporopollenin, chitin and other 'non-degradable plastics' in the geologic record, as evidence for major biologic events. In *Proceedings Birbal Sahni Birth Centenary Palaeobotanical Conference*, ed. B. S. Venkatachala, K. P. Jain & N. Awasthi. *Geophytology*, **22**.

Traverse, A. & Ginsburg, R. N. (1966). Palynology of the surface sediments of Great Bahama Bank, as related to water movement and sedimentation. *Marine Geology*, **4**, 417–59.

Van Bergen, P. F., Janssen, N. M. M., Alferink, M. & Kerp, J. H. F. (1990). Recognition of organic matter types in standard palynological slides. In *Proceedings: International Symposium on Organic Petrology*, ed. W. J. J. Fermont & J. W. Weegink. *Mededelingen Rijks Geologische Dienst* **45**, 9–21.

Vronskiy, V. A. (1975). On modern 'pollen rain' above the surface of the Aral Sea. *Doklady Akademii Nauk SSSR*, **222**(1), 167–70. (In Russian).

A Palynomorph sedimentation

2 The sorting of spores and pollen by water: experimental and field evidence

PHILLIP L. HOLMES

Introduction

Palynologists extract fossil spores and pollen (collectively 'sporomorphs') from rocks representing many different sedimentary environments; these are used as indicators of the surrounding vegetation and climate as well as for stratigraphy. They are of particular use to stratigraphers, as they may be found in both marine and terrestrial sediments and can therefore be used as a link between the two environments.

In the past it was common to believe that most of the pollen was transported to the sediment via the atmosphere as a 'pollen rain.' This pollen rain was considered to be relatively uniform over a wide area.

The distribution of pollen through the atmosphere has been studied by many authors (for example, Lanner, 1966; Janssen, 1973). Numerous experiments have been carried out on the dispersion of pollen from both point sources (Colwell, 1951; Chamberlain, 1966) and from forests (Lanner, 1966). Some of these studies were initiated by investigators in other fields of study. May (1958), for example, was attempting to model the dispersion of radioactive particles in the atmosphere and she used *Lycopodium* spores to do this.

As palynologists recognized the need to be able to differentiate between pollen that had travelled a few meters from that which had arrived at the site from several kilometers away, the pollen rain was interpreted as having three components with arbitrary boundaries: (1) 'local' (sporomorphs deposited within a few tens of meters of their source plant), (2) 'extra-local' (deposited within a few hundred meters of the source), and (3) 'regional' (deposited more than a few hundred meters from their source). The boundaries between these groups are vague, and many authors also use a fourth term, 'extra-regional,' for the long-distance transport of sporomorphs.

As pollen analysis was extended to sediments other than peats (and their fossil derivative, coal) it became apparent that some of the pollen was transported to the site of deposition by water (e.g., Muller, 1959; Traverse & Ginsburg, 1966), rather than by air. Quaternary investigators in particular, who have been using lake sediments as an alternative to moss polsters, have become increasingly aware of a substantial fluvial input of pollen to lakes. Among the more compelling evidence for this is the increase in absolute numbers of pollen found in sediments adjacent to river mouths in both lacustrine (Fletcher, 1979) and marine settings (Heusser & Florer, 1973; Heusser, 1988). Palynology is used in an increasing number of research fields, and a finer resolution time scale is demanded. Hence there is an increasing need to understand the taphonomic processes that act upon the sporomorphs.

Nearly all of the work in this area has concentrated on field studies (e.g., Peck, 1973; Bonny, 1978; Crowder & Starling, 1980). There have been very few controlled laboratory experiments studying the behavior of sporomorphs in water, despite the unanswered questions raised by the field studies. For example, are the differences between sporomorphs great enough for physical processes to act upon? If they are, under what circumstances will sorting occur and is it possible to detect these effects? We have very little direct observation of the factors that affect sporomorph transport by water.

In this chapter I review this limited experimental work and add new data based on my own studies. This will be followed by a discussion on the transport routes that sporomorphs may follow on their way to lake sediments. The processes that I describe, however, could be used in connection with any sedimentary system.

The pros and cons of experimental palynology

Before we can interpret pollen records in taphonomic terms we need to know what factors affect their deposition rates. Of these processes we need to know which, if any, affect sporomorphs differentially (i.e., which processes can 'sort' them).

Among the problems of making field observations is the lack of control of variables such as water velocity and the suspended load of the flow. Sporomorphs may also reach the water via different routes, so the physical condition of the exine and protoplasmic contents may differ. Much of the sporomorph input to a stream or river is from surface runoff. Some will also be deposited directly from the atmosphere. The latter is more likely to be dry and uncorroded. The sporomorphs brought in by surface runoff may be partially degraded, having spent some time exposed on various substrates such as leaves, twigs and the soil, where they are open to microbial attack. These are also much more likely to be waterlogged. Rainfall is known to clear the atmosphere of much of its sporomorph load, by capture in raindrops (McCully *et al.*, 1956; McDonald, 1962). These grains should be in good physical condition but will also be waterlogged. A whole range of preservation states and degrees of saturation will occur, and these may behave differently when they are incorporated into the sporomorph load of the river.

By running laboratory experiments, it is possible to overcome these problems. Variables such as water velocity and depth can be controlled. The bed material over which the water flows is easily changed. The state of the sporomorphs used is also known. It is even possible to test for differences in behavior between fresh and partially degraded spores and pollen.

Laboratory experiments also have disadvantages. The first of these is scale. The natural environment cannot be recreated in the laboratory. It is therefore only possible to test sporomorph behavior on a small scale, testing nature a piece at a time. The results from what appears to be a large flume in the laboratory should not be used to predict particle behavior in a river of varying width and depth, without careful consideration. One of the objectives of laboratory experiments such as those described in this chapter is to remove as many unknown quantities as possible from the scenario. Such simplifications should be remembered when trying to interpret the results in a meaningful and helpful manner.

Laboratory experiments should not be run in isolation from the field, and all results should also be tested there. In common with many subjects, the results from experiments often raise as many questions as answers, especially in the early stages of the work. Even so, the results presented should prove to be of value, for example, in showing where future work may be directed.

The data base

When sporomorphs are studied from a taxonomic viewpoint, the size of the grain and the thickness of the exine are frequently used for diagnostic purposes. These sorts of data are therefore relatively easy to come by. When looking at the behavior of sporomorphs in water, other variables such as the specific gravity (or density) and volume of the grain are also of interest. Such characters are not routinely measured. Some data are available from the literature, but this is relatively sparse. Measurement of such characters is extremely time-consuming.

Table 2.1 summarizes our knowledge of some of the more important variables of pollen and spores. Not all sources are in agreement, because of differences in the techniques used for measurement. Values for grain density will vary with the water content of the grain. Most grains rapidly equilibrate with the external humidity (Harrington & Metzger, 1963). In this case, all of the measurements have been placed in the table, but more attention should be been given to the highest value, as this is most likely to represent the exine density. The density of a ragweed (*Ambrosia*) grain as a whole approaches that of the exine at 100% humidity (Harrington & Metzger, 1963).

Incorporation of sporomorphs into a water body

The transport of sporomorphs through the atmosphere is adequately covered in the literature elsewhere (Tauber, 1967a,b; Bradshaw & Webb, 1985). Similarly, the processes involved in the degradation of sporomorphs have been studied (Havinga, 1964, 1984; Sangster & Dale, 1964). In the following discussion, I therefore concentrate on the behavior of the sporomorphs beginning when they come into contact with the water surface. Spores and pollen that enter a water body via surface runoff may be considered to be saturated and will therefore be readily incorporated into the water body. I therefore ignore this input for now, concentrating instead on the input of sporomorphs from the atmosphere.

Once a spore or pollen grain has landed on a water surface, the time taken for it to become incorporated into the flow depends on both the time taken to saturate the grain and on the turbulent nature of the water.

Saturation of sporomorphs

Saturation has been studied by Hopkins (1950) who experimented with arboreal pollen. He tapped pollen from male inflorescences into a beaker filled with water and noted the time required for each to sink. Most deciduous pollen types sank within five to ten minutes, while bisaccate types tended to float. Bisaccate grains float because the sacci are sealed and thereby keep the grain buoyant even when the rest of the grain is saturated.

The buoyancy of bisaccates is a well known phenomenon. Pohl (1933) noted that pine pollen could retain its buoyancy for up to four years, others have reiterated this

Table 2.1. *A compilation of pertinent spore/pollen characteristics.*

An asterisk (*) denotes measurements made on wet grains. It should be noted that difference authors use different techniques to obtain their results, hence the wide variation in, for example, sporomorph densities. Much of the variation in density measurements is probably due to differences in the level of saturation of the grains. In order to conserve space the information from different authors has not been segregated; thus the fall velocity quoted for a particular species does not necessarily relate to the other data on the same line. Data sources: [1]Heathcote (1978); [2]Fletcher (1979); [3]Niklas & Paw (1982); [4]Gregory (1961); [5]Chamberlain (1966); [6]Dyakowska (1937); [7]Brush & Brush (1972); [8]Durham (1946); [9]Harrington & Metzger (1963); [10]Davis & Brubaker (1973).

| Species | Size (μm) | Specific Gravity | Density (g/cc) | Fall velocity (V_t) (cm/sec) | | | |
| | | | | In Air | | In Water | |
				Calc.	Obs.	Calc.	Obs.
Acer sp.	39.9[1]		1.236*[1]			0.0202[1]	0.00949[1]
Abies sp.	86.2[1]		1.270*[1]			0.110[1]	0.0285[1]
							0.0518[7]
Alnus sp.	21.22[1]	0.75[8]	1.17[1]			0.00347[1]	0.00438[1]
	22[7]	0.97[8]	1.16[7]			0.0037[7]	
A. glutinosa		0.97[8]			1.46[8]		
	26.0[8]			1.98[8]			
	30.0[8]			2.59[8]			
A. viridis	30.0[8]			2.59[8]	1.7[4]		
Ambrosia sp.	19.52[1]		1.2*[2]			0.0037[1]	0.0035[1]
	18[7]	1.5[7]	1.28*[9]				0.0078[7]
Artemisia sp.	20.4[8]	1.02[8]		1.28[8]	1.006[8]		
	25.4[1]		1.14[1]			0.0048[1]	0.0047[1]
Betula sp.	27.5[7]	1.55*[7]				0.019[7]	
			1.4[1]				
B. lutea	31.22[1]		1.375[1]			0.0197[1]	0.0104
B. nigra	24.6[8]	0.94[8]		1.7[8]	1.68[8]		
B. papyrifera	28.79[1]		1.368[1]			0.0102[1]	0.0165[1]
B. pubescens	27.5[6]			2.13[8]			
B. verrucosa	24.5[8]			1.78[8]			
Corylus sp.	22[7]	1.1*[7]				0.0023[7]	
C. americana	23.6[8]	1.09[8]		1.83[8]	1.31[8]		
C. avellana	24.2[6]			2.96[8]	2.5[4]		
Fagus sp.	47.0[1]	0.71[6]	1.08[1]			0.00955[1]	0.00903[1]
		0.94[8]					
F. grandifolia	44.0[8]	0.94[8]		5.52[8]	3.54[8]		
F. sylvatica	38.4[6]	0.713[4]			5.5[4]		
Fraxinus sp.	24.0[1]		1.17*[1]			0.00452[1]	0.00417[1]
Lycopodium sp.	31.6[5]	1.175[4]		1.76[5]			
	32.4[5]			1.90[5]			
	34[5]			2.14[5]			
Pinus banksiana	41[3]			2.0[3]			
P. echinata	38[3]			2.0[3]			
P. resinosa	52[3]		1.315[1]	2.3[3]		0.037[1]	0.0111[1]
P. strobus	49[3]			2.1[3]		0.0175[1]	0.0105[1]
P. sylvestris		0.391[4]		2.5[3]	2.49[3]		
	54[3]						
	57[7]	0.8[7]				0.0316[7]	
	52[8]	1.2*[7]			2.49[8]		
POACEAE							
Capriola dactylon	28.5[8]	1.01[8]		2.50[8]	1.86[8]		
Dactylis glomerata	34.0[8]	0.98[4]		3.17[8]	3.10[4]		
		0.91[8]			2.78[8]		

Table 2.1. *Continued*

Species	Sixe (μm)	Specific Gravity	Density (g/cc)	In Air Calc.	In Air Obs.	In Water Calc.	In Water Obs.
				Fall velocity (V_t) (cm/sec)			
Phleum pratense	34.0[8]	0.9[8]		3.17[8]	2.8[8]		
Poa pratensis	28.0[8]	0.9[8]		1.46[8]			
	30.0[8]			2.44[8]	1.74[8]		
Secale cereale	60[8]			7.25[8]	4.3[4]		
	49.5[5]				6.0[8]		
					8.8[4]		
Zea sp.	81.6[1]	1.42[1]					
Z. mays		0.35[8]		24.52[8]	18.0[8]		
	90[8]	1.0[8]					
Populus	30.13[1]		1.161*[1]			0.00789[1]	0.00451[1]
	24[7]	1.02*[7]				0.0056[7]	
Quercus sp.	32.66[1]	1.04[8]	1.31[1]			0.00178[1]	
	29[7]	1.2[7]	1.16[10]			0.0081[10]	
Q. imbricaria	33.1[8]	1.04[8]		3.42[8]	2.2[8]		
Q. macrocarpa	32.3[8]	1.04[8]		3.3[8]	1.83[8]		
Q. robur	27.7[6]			2.41[8]	2.9[4]		
Salix sp.	17.35[1]		1.113[1]			0.00184[1]	0.00278[1]
Ulmus sp.	27.5[1]		1.175*[1]			0.00715[1]	0.00567[1]
						0.0093[7]	
URTICACEAE							
Cannabis sativa	30[8]			2.22[8]	1.04[8]		
	25[8]	0.82[8]		1.56[8]	2.55[8]		
Urtica sp.	14.46[1]		1.126*[1]			0.00134[1]	0.000289[1]
U. gracilis	14.0[8]		0.77[8]	0.4[8]	0.34[8]		

Notes: * = measurements on saturated grains.
Further results for the observed V_t of spores in air may be found in Gregory (1961).

point (e.g., Erdtman, 1943; Traverse & Ginsburg, 1966; Traverse, 1988). Hopkins' (1950) results show that there is some differential buoyancy between *Pinus* species. The grains that sank are those with smaller or deformed/broken air sacs. When placed on a water surface in a large water tank, over which a breeze of 13 km/h was blown, conifer pollen drifts at a rate of 0.16 to 0.32 km/h. When oak pollen (*Quercus palustris*) was put through a similar process most of it sank within the first meter (experiments by Hopkins, 1950). In Hopkins' experiments the surface of the water was calm. In nature the surface of streams and lakes are usually broken by wave action or turbulence within the flow. Most sporomorph types will therefore be rapidly incorporated into the water body.

Bisaccate grains will only sink if the bladders are pierced either by physical, chemical or microbial processes. They are therefore more likely to be saturated if they enter turbulent flow, or if they have been resident in a position (such as the soil surface) where they are open to microbial attack prior to introduction to the water.

All of the pollen types tested by Hopkins had at least one pore or colpus. These provide an easy route for entry of water into the grain. Many pollen grains have mechanisms to prevent dessication but cannot prevent water uptake. Sporomorphs, many of which have no openings in the exine, have no inlet for water. Some, such as *Lycopodium*, also have an oily coat (Balick & Beitel, 1988). Attempts to saturate *Lycopodium* in the laboratory have shown that such grains are extremely resistant to water uptake. Extreme methods have to be invoked, such as placing the sporomorphs in a beaker in a vacuum for up to eight hours (own method) or boiling in water (Reynolds, 1979). This means that they are likely to remain afloat in the water, in nature, for long periods of time. This is consistent with results from the Volga River where Federova (1952) found *Lycopodium* spores over 700 km from their source.

Transport of sporomorphs in a water body

Once they have been incorporated into the flow most sporomorph types are transported as part of the suspended load. They are normally associated with the fine silt/clay fraction (Muller, 1959; Cross et al., 1966; Chen, 1987). This is the size class below that of the sporomorph size due to differences in their density compared to mineral sediment. The distribution of sporomorphs within a river channel will depend on the flow characteristics of the river. Peck (1973) assumed an even distribution occurred in Oakdale Beck (Yorkshire) while Starling & Crowder (1980/81) found a concentration of sporomorphs in the zone of maximum flow in the Salmon River (Ontario). They also found secondary areas with high counts in the bedload of the river. The Salmon River is much larger than Oakdale Beck and is also in an area of lower relief; it is therefore likely to be less turbulent, allowing such patterns to persist.

On a smaller scale Brush and Brush (1972) found a slight increase in the total numbers of sporomorphs with depth at the downstream end of the flume used in their experiments. The flume used for their study measured 4 m long by 0.23 m wide, water depth was 0.14 m and flow velocity set at 33 cm/sec. Similar experiments run by the author in a larger flume (7.5 m × 0.3 m, water depth 0.26 m) also show a slight increase of sporomorph numbers with depth over a distance of seven meters and at a range of low flow velocities (3 cm/sec–12.5 cm/sec).

Several authors have found that water currents can sort sporomorphs. This was first seen in the field by Muller (1959) who was looking at the palynomorph distribution in recent sediments in the Orinoco delta area. He noted that the pollen of *Rhizophora*, which is relatively small, was carried further out to sea than larger sporomorphs from similar source areas. Cross et al. (1966) also found evidence of sorting in a marine setting as lighter, more buoyant, grains were carried further toward the southern end of the Gulf of California.

Crowder and Cuddy (1973), looking at pollen spectra from Hay Bay (Lake Ontario), found that ragweed (*Ambrosia*) pollen was carried further from the mouth of Wilton Creek than was pollen from the grasses (Poaceae).

They put this difference down to the differing specific gravities of the two pollen types. There is also a significant size difference.

Peck (1974b) found down-lake decreases in many sporomorph types in two Yorkshire reservoirs. The larger sporomorph types were deposited in slightly higher numbers close to the stream mouth. Fletcher (1979) studied the sporomorph distribution in the surface sediments of Lake Michigan and found that lake currents carried small grains further than large grains, *Ambrosia* being particularly buoyant.

Davis et al. (1971) and Davis and Brubaker (1973) explained differences in the distributions of oak (*Quercus*) and ragweed (*Ambrosia*) in the sediments of several small North American lakes as resulting from sorting. The smaller ragweed grains remain in suspension for longer periods of time, allowing them to be transported to the littoral areas of the lake by wind generated currents.

Sorting was also used to explain results presented by Chen (1987) from the sediments of Lake Barine in Queensland, Australia. Many of the grains are transported to the lake by surface runoff. On entering the lake the runoff forms a current running from the lake's edge toward the center. Larger grains are deposited in the littoral areas, while smaller grains are carried to the lake center by these currents.

Clearly, sorting of sporomorphs can occur in many depositional environments. The behavior of sporomorphs as sedimentary particles is, however, poorly understood. To date there has only been one other attempt at studying their transport in the laboratory, that of Brush and Brush (1972). Experiments were also run by Peck (1972); these, however, were designed to test the efficiency of the Tauber Trap in water. Some of the data presented by Peck can be used to provide useful information on sporomorph transport.

The experiments undertaken by Brush and Brush (1972) were run in a small flume over a sand bed (grain size 0.63–2.0 mm) with a flow velocity of 33 cm/sec. A bed regime of dunes and ripples was established prior to the introduction of the sporomorphs. Total experiment time was 15 minutes in each experiment. They found that sporomorphs were preferentially deposited with the finest sediments in the bottomset beds of the ripples. They also noticed that there was some selectivity in the transport of the sporomorphs and of their inclusion in the bed. No overall pattern for selectivity emerged from this study, and the fate of many of the sporomorph types in the experiments could not be determined. They emphasized that a larger flume and longer experiment times would be needed for further investigation.

Peck's (1972) investigation of the Tauber Trap revealed that at low velocities the trap would preferentially remove larger sporomorphs from the water flow. Experiments with higher flow velocities showed no such differentiation.

Here the selection was thought to occur within the trap. It is possible, however, for the selection to have occurred at the trap orifice.

In a more recent study of alluvial sediments from a canyon stream in the Chuska mountains of Arizona, Fall (1987) tested the assumption that the pollen in the alluvium was from local vegetation. She found that pollen concentrations in general were highest in the fine grained sediments. Non-arboreal sporomorph types however were found in higher numbers in the sandy sediments. Similar results have been reported from glacio-lacustrine sediments in Alaska (Goodwin, 1988) and braided stream deposits adjacent to the Caribou River in Yukon (Catto, 1985). These results may be due to differential transport (as suggested by Fall, 1987; Goodwin, 1988; Catto, 1985) or they may be due to the differing source areas of the tributaries involved (Hall, 1989).

These results suggest that the deposition of fine grained sediments, sporomorphs in particular, is related both to the flow conditions of the water and the bed material over which the water flows. I undertook a series of experiments to investigate these relationships. The basic aims were as follows: (1) to elucidate those sporomorph characters (size, density) which control transport; (2) to determine at what flow velocities a significant level of differential transport ceases to occur; (3) to investigate the effect of bed material on sporomorph deposition.

Experimental methods

Treatment of sporomorphs

After collection (from living plants), all sporomorph types (except *Lycopodium*) were air dried for a period of up to ten days. *Lycopodium* spores were purchased from a chemical supplier. A few days prior to each experiment the sporomorphs were saturated with water by placing the required amount in a beaker of water in a partial vacuum. Pine pollen was saturated by putting the grains through an alcohol series as the vacuum treatment caused the air sacs to collapse. Some sporomorph types were also acetolysed, and the effect of this treatment was investigated in separate experiments.

Flume conditions

The flume used for this study was an Armfield tilting flume. This has a sediment loop fitted so that it may continuously recycle the water and any suspended sediment. The main section of the flume measured 7.5 m × 0.3 m, the water depth varied between experiments from 0.23 to 0.26 m. The flow velocities used were in the range 0.03–0.3 m/sec, with Reynolds (Re) numbers in the range 14 400–95 000 and Froude numbers in the range 0.1628–0.2067. Experiments were run over two different sand beds as well as over the smooth base of the flume.

Basic experimental method

Once the flume was set in the configuration required for the experiment the sporomorphs were added to the water over a period of several minutes. This, combined with the turbulence encountered on passage through the motor, sufficed to give a fairly homogeneous distribution of sporomorphs in the water. Once in the flume three things could happen to the sporomorphs: (1) they could be deposited, (2) they could remain in suspension, or (3) they could be lost, either by lodging in the pipes or on the glass sides of the flume. Very few sporomorphs are found in the water after the flume has been cleaned, which suggests that a negligible amount lodge there during the experiments. The sporomorphs that stick to the sides of the flume form a visible line at the downstream end of the working section. Even so their number is small in comparison to the total numbers in the water. Such loss may therefore be ignored when considering the fate of the sporomorphs in each experiment.

The total number of sporomorphs added to the flow at the beginning of each experiment can be calculated. By taking samples of the water throughout each experiment and counting the number of sporomorphs in one milliliter it is possible to estimate how many remain in suspension. This means that we can also calculate the numbers that have been deposited. A graph of the number of sporomorphs in suspension plotted against time can then be used to calculate a rate of deposition (see Fig. 2.1).

The gradient of the regression line shown in Figure 2.1 describes the rate of loss of sporomorphs from suspension with time. As the only alternative site for the sporomorphs is on the flume bed, this can be translated as the rate of deposition to the bed. If the flow velocity were reduced we would expect sporomorph deposition to occur at a faster rate. This would be manifest as a steeper line on a graph of the type shown in Figure 2.1. This procedure has been used in Figure 2.1b where the inverse rate of deposition (taken as the reciprocal of the gradient) is plotted against flow velocity.

Statistical methods

The data from each experiment were put through a least-squares regression analysis to give the best fit line. The gradient of each line was tested against the null hypothesis that it equalled zero (i.e., H_0: $\beta = 0$) using analysis of variance and an F-test. The null hypothesis was rejected at the 95% confidence level for all of the data presented here. For experiments run at the same velocity, the gradients were tested against each other using the Student's t test, with the null hypothesis that all gradients were equal (H_0: $\beta_1 = \beta_2$).

Data from experiments with *Lycopodium* (ave. diameter 32μm), flow over medium sand bed. r^2 =coefficient of determination

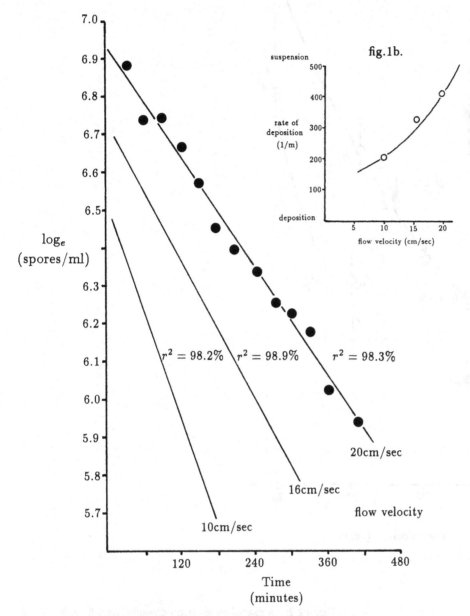

Figure 2.1. Calculation of a rate of deposition from the flume results. The main graph shows the relationship between sporomorph concentration in the flow ($\log_e$(spores/ml)) with time. The gradient of the least-squares regression line (m) in each case is equal to the rate at which sporomorphs are lost from the flow. Data points are given for one line only to retain clarity. The second graph (1b – inset) uses this relationship with the inverse of the gradient (m^{-1}) plotted versus flow velocity. This illustrates clearly the changes in the deposition rate under different flow conditions; in this example the decreasing flow velocity results in a decrease in the inverse deposition rate (i.e., an increase in the deposition rate). The line drawn through these points is curved as there is a fourth data point (not plotted) from an experiment run at 25cm/sec with an inverse rate of deposition of 8000 (approx.). The data on this figure are derived from experiments with acetolysed *Lycopodium* spores and thus differ from the data points for *Lycopodium* in Figure 2.2.

Evidence for the sorting of sporomorphs

Figure 2.2 displays a set of results I obtained from recent flume experiments, along with results recalculated from data given in Peck (1972). The inverse deposition rates of *Lycopodium* spores at several different flow velocities are shown. This provides a line for a 32 μm spore against which deposition rates for sporomorphs of various sizes may be compared. Peck's data for *Lycopodium* show a very close correspondence to this line, as do *Dactylis* and *Quercus* pollen, both of which are similar in size (33 μm and 28 μm respectively). *Quercus* also has a specific gravity similar to that of *Lycopodium* (approximately 1.2); however, *Dactylis* is lighter, with a specific gravity of 0.98

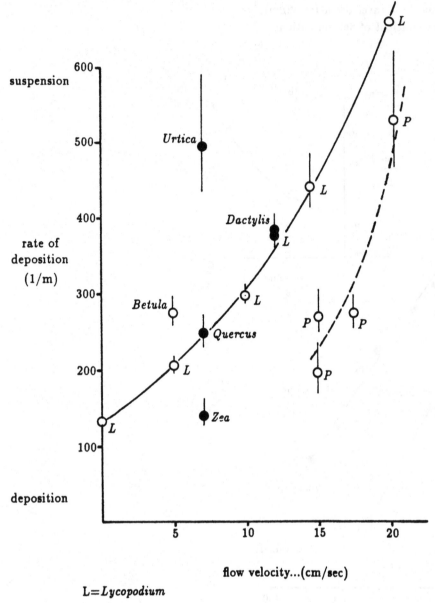

suspension

rate of
deposition
(1/m)

deposition

flow velocity...(cm/sec)

L = *Lycopodium*
P = *Pinus*
● Peck's data
○ Own data

Figure 2.2. A comparison of inverse deposition rates for different sporomorph types under a variety of flow conditions. These show a general relationship between the sporomorph size and the inverse deposition rate, with larger sporomorphs deposited at a faster rate than smaller forms. The error bars give the 95% confidence limits for each point. The line for *Pinus* was fitted by eye; the *Lycopodium* line fits the least-squares regression equation:

$$\log_e y = 4.85 + 0.0885x$$

where $p = 0.002$, $r^2 = 99.6\%$.

(see Table 2.1). This may explain why *Dactylis* plots above the *Lycopodium* line, although the difference is slight (the regression lines for the data from *Dactylis* and *Lycopodium* are not statistically separable at the 95% confidence level).

Betula pollen is smaller than *Lycopodium* spores and plots above the line, even though it has a higher specific gravity (see Table 2.1). *Urtica* pollen, the smallest and lightest grain for which data are available, shows very slow rates of deposition. Two larger grains are also plotted. *Zea* pollen is relatively light (sg 1.0), but it is much larger than *Lycopodium* and therefore plots in the lower part of the graph. The points plotted for *Pinus*

pollen are for completely saturated grains (i.e., the air sacs are waterlogged). The specific gravity of *Pinus* is similar to that of *Lycopodium*, but as it is nearly twice as large, it also plots beneath the line.

Discussion

The rate at which a grain falls through a liquid will depend on the difference in density between the grain and the liquid. If the grain is more dense it will sink. The rate at which this occurs will depend on the weight force acting upon the grain. In turbulent flow the relative velocity of the grain to any upcurrents in the flow is also important. If the upcurrent velocity is greater than the sporomorph's terminal velocity the grain will remain in suspension.

Sorting will clearly occur when the velocity of the upcurrents is of a similar magnitude to the particles' terminal velocity (Heathcote, 1978). This is recreated in these experiments. The relative rates of deposition in this situation depend on the sporomorph's terminal velocity. This is, in turn, related to the density and size of the grain.

The terminal velocity (V_t) of a sphere may be calculated using Stokes' Law. Sporomorphs, however, are not solid spheres, and calculation of terminal velocities by this method therefore tend to be inaccurate when compared to observed fall velocities (Brush & Brush, 1972; Heathcote, 1978). It may still be used, however, to give a range of hypothetical values to illustrate the effects of changes in the size and specific gravity on a grain's fall velocity. A series of such calculations were derived using a version of Stokes' Law given in Brush and Brush (1972):

$$V_t = \frac{d^2 g (sg - 1)}{18v}$$

where d = nominal diameter of grain, g = gravitational force constant (9.8 m/sec/sec), v = kinematic viscosity of water (approximately 0.01 cm^2/sec), sg = specific gravity of the particle.

A change in specific gravity from 1.1 to 1.3 (i.e., a relative change of $\times 3$, when considered as a difference from the specific gravity of water) gives a three-fold increase in fall velocity from 1.095×10^{-3} to 3.285×10^{-3} cm/sec. A size change, however, from 15 μm to 45 μm (i.e., three-fold) gives a nine-fold increase in V_t from 1.095×10^{-3} to 9.854×10^{-3} cm/sec. So, while sorting of sporomorphs depends on both parameters, the size of a grain appears to be more important. This is borne out by the results in Figure 2.2, where *Betula* ($d = 25$ μm, $sg = 1.5$) plots above *Lycopodium* ($d = 32$ μm, $sg = 1.2$).

The shape of a grain is thought to be unimportant for most sporomorphs. The exceptions are those grains, such as bisaccate types, with large departures from sphericity (Heathcote, 1978; Winkelmolen, 1982). In such instances, the fall velocity of a grain will also depend upon its orientation within the water flow.

Velocities at which sorting occurs

It can be seen from Figure 2.2 that, as the flow velocity is increased, the inverse deposition rate of the sporomorphs also increases, as more grains remain in suspension for longer periods of time. At flow velocities of 20 cm/sec all grains smaller than *Lycopodium* will remain in suspension. The curve for *Pinus* is very similar, with most grains remaining in suspension at flow velocities of 25 cm/sec. Most sporomorph types have sizes below that of *Pinus*. Larger types will probably remain in suspension once a flow velocity of 35 cm/sec is reached. At velocities greater than this differential transport should cease to occur.

Most natural rivers flow at rates greater than 35 cm/sec. Sorting will therefore only occur where the current is slowed. This occurs where rivers flow into either a marine or a lacustrine environment. In such situations a river may rapidly deposit much of its sediment. Many sporomorphs may also be deposited as they are trapped by the descending sediment. The addition of sediment to a liquid will aid the settling of sporomorphs during centrifugation (Jemmett & Owen, 1990). Such deposits should contain a sporomorph assemblage that has not been sorted.

When sporomorphs are carried away from the river mouth and deposited with other fine grained sediments, sorting is likely to have occurred. The degree of sorting will be dependent on the distance covered by the water body at velocities below 35 cm/sec. In the flume experiments, sorting could not be detected in the water over the length of the working section (7.5 m). Heathcote (1978) suggests that sorting will not be noticeable over distances below 30 m. Field evidence from Silwood Lake suggests that this estimate is of the right order of magnitude for small lakes. The differentiation of sporomorph types will be blurred by variations of current strength. Faster flow velocities that occur after periods of heavy rainfall will carry larger sporomorphs over greater distances and this should lead to a more homogeneous spectrum in the lake sediments (see Peck, 1974a,b).

The figure of 35 cm/sec derived from these experiments as the upper limit for sorting appears to contradict data collected by other authors. Brush and Brush (1972) ran their experiments at 33 cm/sec and found evidence for significant amounts of sorting. Similarly, Fall (1987) found evidence of sorting in a canyon stream where flow velocities are likely to have been in excess of 40 cm/sec. These results require some explanation.

The effect of bed material on sporomorph deposition

The roughness of a bed will disrupt the flow pattern of the water adjacent to it. Particles carried by the flow will also be affected by this disruption. On a large scale this can cause the formation of particle clusters with large protuberances from the bed 'collecting' smaller particles in their wake (Brayshaw *et al.*, 1983).

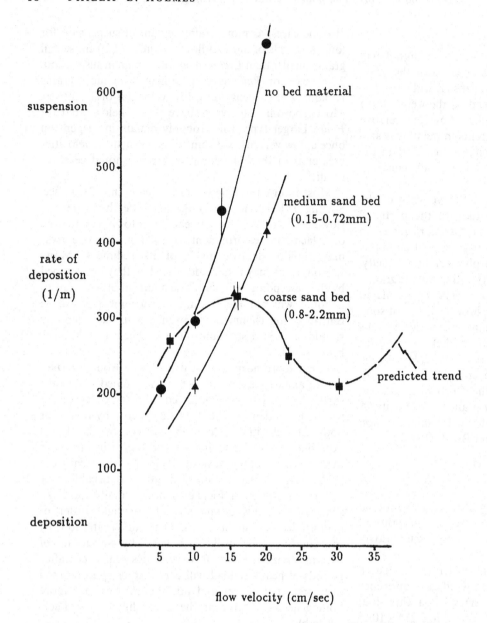

suspension

600

500

rate of
deposition
(1/m)

400

300

200

100

deposition

no bed material

medium sand bed
(0.15-0.72mm)

coarse sand bed
(0.8-2.2mm)

predicted trend

5   10   15   20   25   30   35

flow velocity (cm/sec)

Figure 2.3. Results from experiments run over different bed materials. All experiments used *Lycopodium* spores. Note how the coarse sand bed decreases the inverse deposition rate at the higher velocities. The error bars give the 95% confidence limits for each data point; lines are fitted by eye.

The velocity profile of a current will show a gradual decline in flow velocity close to the bed. At the base of this is a thin layer in which the flow is said to be laminar as viscous forces dominate (in any flow the degree of turbulence may be described by the ratio between viscous and inertial forces in the flow: Leeder, 1983). This layer is known as the laminar sub-layer (or viscous sub-layer). Flow in this layer will be disrupted by any grain that lies within it and has a diameter greater than one fifth of the layer thickness (Raudkivi, 1967).

Experiments have been run with two different sand beds. The minimum clast size exceeded the thickness of the laminar sub-layer in all of these experiments. Clasts will therefore disrupt the flow pattern adjacent to the bed and may entrain particles in their wake. The results from these experiments are displayed in Figure 2.3. When the flow is over a coarse-grained sand bed, the inverse rate of deposition can be seen to decrease with flow velocity once a critical point is reached. Experiments could not be run at higher velocities, because of the limitations of the equipment used. I suggest, however, that once the threshold for grain movement is reached, the inverse rate of deposition of the sporomorphs will begin to rise as sporomorphs are resuspended.

At flow velocities of up to 15 cm/sec, sporomorphs are deposited at a rate similar to that calculated from experiments where no bed was used. This suggests that they drop through the flow under their own weight. The

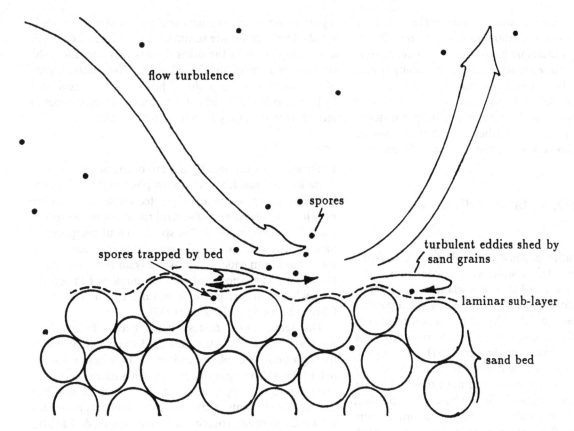

Figure 2.4. The mechanism by which sand beds increase the rate of sporomorph deposition. At low velocities there is very little turbulence, and the grains settle out at a rate determined by their individual fall velocity. At higher velocities turbulence carries sporomorphs to the bed at a much faster rate. Once near the bed they may be caught in turbulent eddies shed by prominent sand grains. Once in the bed they are trapped between grains.

flow conditions at these velocities are not highly turbulent ($Re = 14\,401$ at a flow velocity of 5 cm/sec). As the flow velocity increases the degree of turbulence also increases ($Re = 95\,047$ at 33 cm/sec). Turbulent eddies may then bring sporomorphs down to the bed at a rate faster than they could achieve before. Once in the vicinity of the bed they are trapped by turbulent eddies shed by the sand grains (see Fig. 2.4) and cannot be resuspended.

When a sand of medium grain-size was used, a bed regime of ripples was established. These migrated along the flume at a rate commensurate with flow velocity. The results from these experiments also show an increase in the deposition rates of sporomorphs, when compared to results from experiments with no bed material (not shown). The increase is not as drastic as that seen in experiments with a coarse sand bed, which is probably a result of the mobility of the smaller sand grains. Sporomorphs may be released as the sand grains around them are moved. The deposition rates in this case would then be a reflection of the residence time of the sporomorphs in the bed.

These results at least partially explain why Brush and Brush (1972) and Fall (1987) found sporomorphs in sand beds. To date no explanation has been found as to why this filtering effect might affect the sporomorphs differentially. It is possible that the relationship is connected to the ratio of sporomorph size to the pore diameter between sand grains. This has not been tested. Reid and Frostick (1985) have suggested that the inclusion of fine-grained heavy minerals in coarse sediments (to form placer deposits) is controlled by the relative sizes of the bed sediment pores and the ingressing particles.

Conclusions from laboratory study

From the evidence presented here it is suggested that, where water currents drop below 35 cm/sec, sporomorphs may be sorted. This sorting is primarily due to size, but sporomorph density is also important.

The bed material over which the water flows also affects the deposition rate of the sporomorphs. Both of the beds tested here (medium and coarse grain sizes) increased the rate of deposition at flow velocities between 15 and 35 cm/sec (i.e., a decrease in the inverse deposition rate on the graphs). The effects of higher velocities are not known.

These experiments cannot be applied directly to field results. Average transport distances based on the rates of deposition given here would be meaningless. This is because many flows are considerably deeper than 26 cm. Sporomorphs will therefore have a longer residence time in suspension, as they are less likely to be brought into contact with the bed material by flow turbulence.

It is therefore desirable to look at the effects of differential transport in the field, where deeper water is found. By looking at different field situations it may be possible to calculate transport distances for different flow depths.

Sporomorphs in lake sediments

Silwood Lake

This is a small anthropogenic lake in Berkshire, England. It is a little over 100 m wide with a maximum water depth of two meters. It lies in a lowland area of gentle relief and is nearly completely enclosed by small patches of woodland (see Fig. 2.5). The lake was constructed in 1815 and partially emptied in 1958 to allow the construction of a new weir (Spicer, 1981). The current strength in the lake was measured using an Ott velocity meter. The maximum flow rate did not exceed 5 cm/sec. In other words, flow velocities in the lake are of the right order of magnitude to allow sorting to occur. During periods of rainfall, flow rates may exceed those measured. Wind generated currents may also occur, but the strength of these has not been measured.

The sediments from the open part of the lake were sampled on a grid pattern with a Jenkin Corer and 46 samples were taken (see Fig. 2.5). The top five centimeters of each sample were removed and homogenized, and a sub-sample of known weight was taken from the homogenate. The sediment was acetolysed and then put through a heavy liquid separation to concentrate the sporomorphs. A known quantity of exotic grains (*Lycopodium*) were added after the acetolysis, so that the exotic grains would be easily recognizable as such (Bonny, 1972). This allowed the calculation of absolute numbers of sporomorphs per gram (wet weight) of sediment in a manner similar to that of Matthews (1969). This method for calculating absolute numbers is one of the least time-consuming (Bonny, 1972), and losses due to processing can be ignored, assuming that they are non-selective (Peck, 1973). Sporomorph counts have been made for all of the common types found in the lake sediments. These have been plotted on distribution maps. Before these maps can be interpreted, it is necessary to have a basic understanding of the processes that affect sporomorph distribution.

Sporomorph input to a lake

Much has been written on the various factors affecting the sporomorph spectra in a lake. I concentrate on the inputs to an open lake (i.e., one with a stream or river input). These inputs are summarized in Figure 2.6, which also shows some of the other factors that can affect the sporomorph spectra in lake sediments. The relative merits of the use of open and closed lakes has been reviewed by Pennington (1979), who illustrates the different types of information that may be gained from each.

THE AERIAL INPUT

For many years this was thought to be the main transport route for the transfer of sporomorphs to lake sediments. The sporomorphs that are deposited on the lake surface may have followed one of several routes from the source plant. The aerial input may be split into four components: local, extra-local, regional, and extra-regional. These terms have been widely used by many authors. The limits of these zones are not fixed, and terminology has become somewhat confused in the literature. In this chapter, I follow the usage of Janssen (1973).

The aerial input creates certain biases. Pollen from trees will be carried over greater distances than pollen from ground-cover plants, as it is released from a greater height. This also means that tree pollen is more likely to be carried into the upper atmosphere (Lanner, 1966). For ground-cover plants in general, most pollen is deposited within 25 m of the source plant (Kozumplic & Christie, 1972). This depends, of course, on atmospheric conditions. Moseholm et al. (1987), when modelling the dispersion of grass pollen, have said that attention should be paid to air temperature, humidity, wind velocity and precipitation. They calculated that 99% of grass pollen is deposited on the ground within a distance of 800 m. Precipitation is particularly important, as it has been shown to rapidly clear the atmosphere of its pollen content (McDonald, 1962). Sporomorphs from plants under the forest canopy will be deposited very close to their source, as wind velocities are reduced in the trunk space (Tauber, 1967a; Andersen, 1973).

The relative importance of each of these pollen sources will depend on the size of the lake. For small lakes, such as Silwood, local and extra-local plant pollen will dominate the aerial input. The regional component will become progressively more important with increasing lake size (Prentice, 1985), and in areas with little or no local sporomorph production the extra-regional component may form a significant proportion of the total spectra (Van der Knaap, 1987). The aerial component will also be affected by other factors such as local airflow conditions around the lake (Tauber, 1967a,b). Such conditions will depend on the surrounding vegetation. If the lake is surrounded by woodland, the trees would have a sheltering effect, and the main air currents may never reach the lake surface. In open areas, air currents will have a much greater effect (McLennan & Mathewes, 1984).

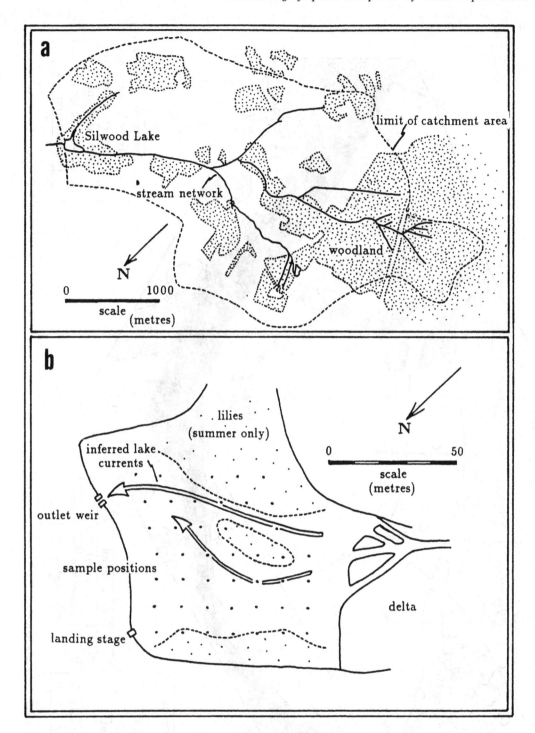

Figure 2.5. Silwood Lake, Berkshire, England, showing (a) catchment area, drainage network; (b) lake currents and sample grid. Note that the lake is surrounded by woodland, as is much of the stream network. The shaded areas represent parts of the lake seasonally covered by water lilies.

THE RIVER INPUT

This source can provide up to 97% of the sporomorph input to a moderate-size lake (Peck, 1974a,b; in the Oakdale catchment, lower reservoir 226 × 58 m, maximum depth of 9 m). This percentage will be dependent on many factors, such as the ratio of lake area to river discharge, the figure for Blelham Tarn in the English Lake District being 87% (Bonny, 1978; lake area 10.2 ha, max. depth 14.5 m), although a proportion of this figure may represent grains resuspended from the lake bottom. Most of the sporomorphs in a river will have arrived via surface runoff. The total numbers of sporomorphs in a river will therefore also be related to climate (amount and type of

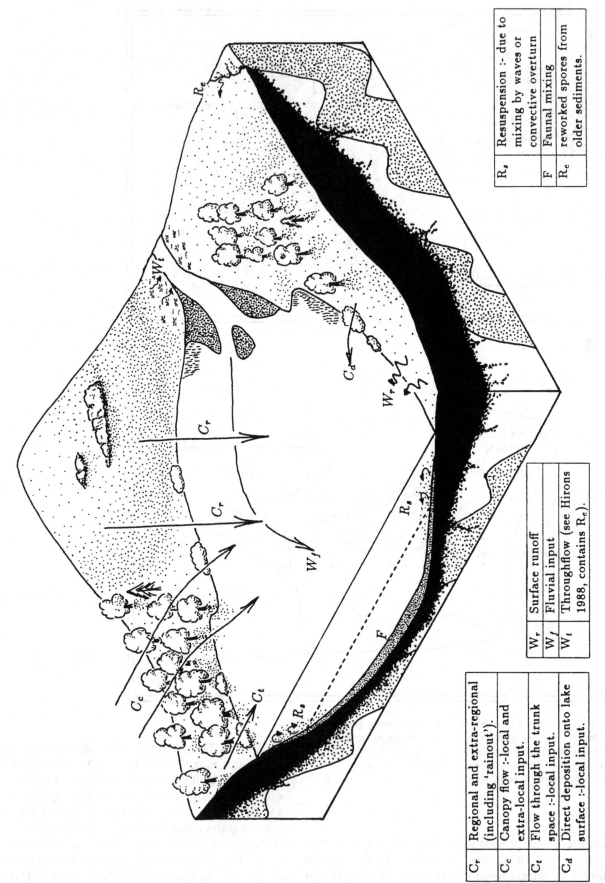

C_r	Regional and extra-regional (including 'rainout').
C_c	Canopy flow :- local and extra-local input.
C_t	Flow through the trunk space :- local input.
C_d	Direct deposition onto lake surface :- local input.

W_r	Surface runoff
W_f	Fluvial input
W_t	Throughflow (see Hirons 1988, contains R_e).

R_s	Resuspension :- due to mixing by waves or convective overturn
F	Faunal mixing
R_e	reworked spores from older sediments.

Figure 2.6. A summary of the sporomorph inputs to a lake.

rainfall), topography and geology, all of which affect the amount of surface runoff in an area. Large numbers of sporomorphs are commonly found in rivers at times of peak flow (Federova, 1952; Peck, 1974b; Brown, 1985). The relative importance of these is reviewed by Peck (1974b).

In the previous section it was noted that much of the pollen and spores from ground-cover plants is not carried very far by air. A large proportion of it is deposited on leaves, stems and the ground around the source plant. Much of the pollen produced by trees is also captured in a similar manner. These grains provide the source of sporomorphs for surface runoff. Pollen production per unit area is roughly equal for arboreal and non-arboreal pollen (Faegri & Iversen, 1975; this assumes that entomophilous plants are not dominant). If a larger proportion of the non-arboreal pollen is deposited closer to the source then these pollen types should dominate in the spectra for runoff and hence also in rivers. Crowder and Cuddy (1973), in a study of Wilton Creek, Ontario, found that samples from Hay Bay, where the creek empties into the lake, gave much higher non-arboreal pollen counts than samples from adjacent peat bogs.

SURFACE RUNOFF

In most areas direct surface runoff into a lake will be relatively small in comparison to the water inputs from streams and groundwater seepage. Lake Barine in northeastern Queensland, however, is an example of a lake where a majority of the water input is via surface runoff (Chen, 1987), because of the physical conditions of the area. This also occurs elsewhere, especially after periods of heavy rainfall (Birks, 1969; Battarbee & Flower, 1984); in such instances this input may disrupt the sporomorph spectra obtained from the lake sediments.

Homogenization of lake sediments

After initial deposition, the sediments in a lake may be resuspended by physical processes, or they could be mixed by burrowing lake benthos.

RESUSPENSION

Resuspension may be facilitated by a number of processes. In lakes that develop a thermocline, convective overturn will resuspend sediment in all areas of the lake. The resuspension generally affects the sediment in littoral areas more than that in deeper areas. Littoral parts of lakes will also be affected by wave action, a process that affects a lake throughout the year. In well-mixed lakes, the resuspended material will be dispersed uniformly throughout the water column. Subsequent deposition will therefore concentrate sediment in the deeper parts of the lake, resulting in an overall movement of material from the shallow to the deeper areas (Davis, 1968; Hilton

et al., 1986). This also leads to a more homogenized pattern of sporomorph deposition within the lake (e.g., Sayles Lake, area 6 ha, max. depth 3 m; see Davis et al., 1971).

The extent of homogenization is linked, in part, to the bottom topography and circulation patterns of the lake. Blelham Tarn in the English Lake District (area 10.2 ha, average depth 6.5 m, maximum depth 14.5 m) has highly variable spectra in its littoral zone, despite frequent resuspension of the sediments (Bonny, 1978).

In addition to lateral mixing, resuspension mixes the accumulated input from several years of deposition. In Frains Lake (size 500 m × 200 m, max. depth 10 m), Davis (1968) calculated that most pollen grains were redeposited two to four times before they were finally buried deeply enough to escape disturbance.

Silwood Lake is too shallow for thermal stratification and is also fairly sheltered from the wind. Resuspension is therefore thought to be relatively uncommon across most of the lake. The delta, however, is frequently flooded after periods of high rainfall. Sediments from the delta top may be resuspended at these times and carried into the lake.

FAUNAL MIXING

R. B. Davis (1967) studied the benthos of four lakes in Maine, USA, and the effects that it had on the sediment. He noted that the depth of penetration was related to the degree of compaction of the sediment, with deeper burrows occurring in softer sediments. The overall effect of the benthic fauna is to mix the sediments from several years' deposition. Webb (1974) observed that burrowing organisms kept tree pollen in surface sediments for several years after the trees had 'left' the area. In some experiments in soils with earthworms, Ray (1959) found that the worms rapidly redistributed the pollen throughout the soil. The passage of the pollen through the gut of the worms did not affect the exine.

Overall, if the mixing is restricted to the top few centimeters of sediment it may be beneficial, averaging out several years input (pollen production being highly variable from year to year as it is dependent on weather conditions). Deep mixing is generally considered to be detrimental (M. B. Davis, 1967) as it obscures long-term changes in the pollen spectra.

The faunal content of Silwood Lake is not known with any precision. None of the cores removed from the lake during my sampling periods (in the summer months) contained benthic organisms.

Losses from the system

In short, sporomorphs may be removed from a lake in two ways. First, they may be chemically or biologically degraded. This process affects sporomorphs differentially, as shown by Havinga (1964) and Sangster and Dale (1964). Some sporomorphs, such as *Populus* and *Acer* pollen, may be completely removed from the fossil record by degradation, while others, e.g., *Lycopodium* spores and *Pinus* pollen, are almost completely unaffected. Both studies showed that degradation is related to the oxygen content and the pH of the enclosing sediments and hence the depositional environment. Peats appear to be the better preserver of sporomorphs when compared to results from pond sediments (Sangster & Dale, 1964).

Sporomorphs may also be removed from the lake by an exiting current. Obviously the more buoyant types are the most prone to this, as they are the most likely to remain in suspension. The extent to which this occurs will depend on both current velocity and lake morphology. Elongate lakes with a single input and outlet will have a greater proportion of direct throughflow from one end of the lake to the other (Pennington *et al.*, 1972). Losses will therefore be correspondingly higher.

Anthropogenic lakes such as Silwood have a different flow pattern at their outlet in comparison with natural lakes. This is because the lake bathymetry is asymmetrical, with depth increasing from the inlet to the dam. At Silwood the outflow is over a weir, and currents approaching it will be split, with a backflow occurring at depth; therefore only the surface waters are lost from the lake. This will accentuate the differential loss from the lake, as only the grains in the top part of the water column are removed. The back current may also result in a large number of sporomorphs being deposited in front of the outlet. Rowley and Walch (1972) found such distributions from their experiments in Yosemite National Park, California. They released 1400 g of exotic pollen into a glacial stream network and later sampled both stream and lake sediments. They had a very low recovery rate, because of the large dilution factors involved. Transects across several of the lakes revealed high numbers of exotic grains near the stream inlets and outlets, with lower values in the lake center. The lakes that they sampled were created by natural glacial dams, so the outflow pattern may be similar to that suggested for Silwood Lake.

It is also possible for sporomorphs to be lost from the air–water interface, albeit under exceptional conditions. In a series of experiments run by Valencia (1967), droplets of water created by two interfering wave patterns were found to contain significant amounts of pollen. This is more likely to be important in large lakes, and in the marine environment, where strong winds create whitecaps which release small droplets of water. This may also explain some of the results obtained by Rowley and Walch (1972), who found some exotic grains in lakes which had no connection to the network in which they had been released.

Theoretical sporomorph distributions in a lake

Experiments suggest a possible link between the size of sporomorphs and their sorting by water currents. It should therefore be possible to turn to a field situation to see if a similar relationship exists there. The closest natural analogue for the flume would be a small stream of similar proportions. In such a setting, however, sporomorphs would enter the stream along its entire length. It would not be possible to calculate transport distances, as the point of entry would be unknown. For this reason a lake has been used, the stream input providing a point source for many of the spores/pollen.

Having established which processes affect sporomorph distributions in lakes and the biases of the different transport routes for certain types, it should be possible to predict distributions under specific sets of conditions. This has been attempted for Silwood Lake. The basic assumptions behind these predictions are as follows:

(1) sporomorph input is via two routes, the air and the stream;
(2) direct surface runoff into the lake is negligible;
(3) there is only one lake current, this being provided by the throughflow from the inlet to the outlet;
(4) resuspension does not alter the original deposition pattern.

Of these, (1) and (2) are thought to be reasonable, considering the nature of the area. The third assumption may create difficulties, as the lake may also be affected by wind generated currents which are more variable. Resuspension may or may not be important in Silwood Lake, but it has not been looked for. The sporomorph distributions suggest that there is very little resuspension, but this is a circular argument.

Suggested distributions under these conditions are given in Figure 2.7. The effect of changing two variables is shown. The two distributions at the top of the diagram assume that all sporomorphs reach the lake from the atmosphere. Figure 2.7b assumes that the source plants are found on the lake shore and release their pollen close to the ground surface. In this case most of the pollen is deposited close to the shoreline, very little reaching the lake center. If the source plants were taller, or the sporomorphs were carried to the lake by wind currents, from further afield (i.e., an extra-local source), then higher numbers of sporomorphs will be deposited in the lake center, as shown (Fig. 2.7a). This is because the air currents that carry the sporomorphs to the lake will not reach the lake surface until they have already travelled part way across the lake. This 'skip distance' will vary according to factors such as tree height or spacing.

The bottom of Figure 2.7 shows the suggested distribution of sporomorphs if they entered the lake exclusively via the stream, with highest deposition close

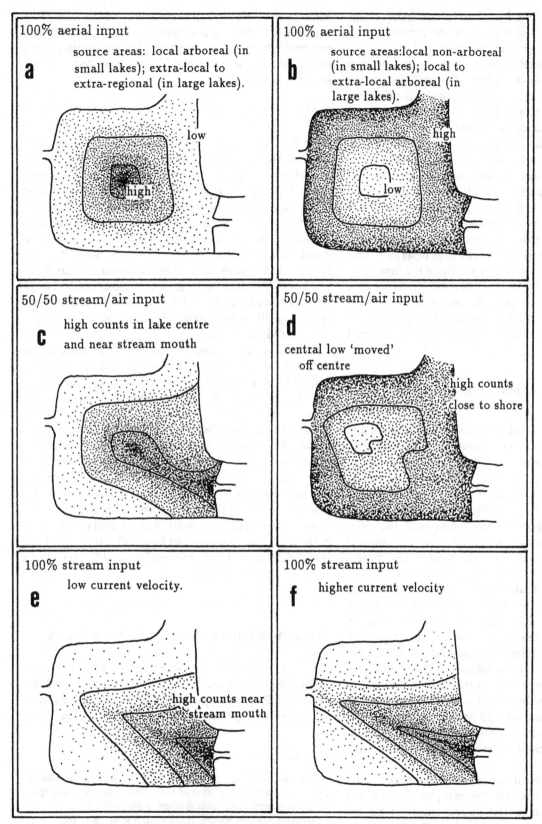

Figure 2.7. Theoretical sporomorph distributions in lake sediments: (a) and (b) represent patterns created in lakes completely dominated by an aerial input. In (a) the source area is regional, with maximum deposition in the lake center; in (b) the source area is local with maximum deposition close to the lake shore. (c) and (d) are examples of patterns formed by mixed air/stream inputs. (e) and (f) show the pattern that would be created in the case of an overwhelming stream input, with maximum deposition close to the stream mouth; (f) illustrates the effect of increasing stream velocity, the sporomorphs being carried further into the lake before deposition.

to the stream mouth (Fig. 2.7e). The pattern is dependent upon the flow velocity of the stream and corresponding lake current. Higher current velocities will stretch the contours, giving a band of high values across the lake center (Fig. 2.7f). If the current strength were high enough, it might exclude sporomorph deposition completely. This pattern will also change for the different types, if their transport distance is related to their size and hence current velocity. Larger sporomorphs would be concentrated closer to the stream mouth, while smaller types are carried further out into the lake.

Figure 2.7 (c,d) represents an attempt to integrate the patterns from above and below. This pattern obviously depends on the relative importance of the two inputs. If this is correct, then actual distributions will be comparable to one of the theoretical cases that I have described. This will then give an idea of the proportion of the input that arrives via the stream. We can then use the types that show closest correspondence to the theoretical distribution for a stream input to look for evidence of sorting, as these distributions are less likely to be distorted by an aerial input.

Sporomorph distribution patterns in Silwood sediments

Figure 2.8 illustrates the distributions of the arboreal and non-arboreal sporomorphs within the sediments of Silwood Lake. The numbers used are for absolute counts. Both arboreal and non-arboreal pollen maps show very similar patterns. There are, however, important differences which may be seen by comparison with Figure 2.7. The high numbers of sporomorphs found near the delta suggest that the stream is a major contributor to the total sporomorph influx of the lake (it shares many characteristics with Figs. 2.7e and 7f) for both groups.

The arboreal distribution however also shows a region of higher values in the lake center. The theoretical interpretation (Fig. 2.7) indicates a higher input of sporomorphs from the atmosphere for this region. A majority of the arboreal pollen that enters Silwood Lake is from *Alnus* and *Betula* (average 57.2% of the total arboreal pollen). These two genera occupy 49% of the total crown cover within a radius of 400m around the lake (Spicer, 1981), and are also among the most prolific of pollen producers compared to the other genera present (Gregory, 1961). There is probably a substantial airborne input into the lake.

There is no corresponding high value in the lake center on the map for non-arboreal sporomorphs. This suggests that very few of the non-arboreal sporomorph types are transferred to the lake directly by fallout from the atmosphere. It should be noted that neither diagram includes counts made for *Corylus*, as the source area is not known. The only other anomalous 'high' on the maps

lies adjacent to the southeastern arm of the lake. This suggests that there is a secondary input of waterborne sporomorphs from this direction. The southeastern arm of the lake has a stream inlet at its head but was not sampled in this study.

STATISTICAL SIGNIFICANCE OF THE SPOROMORPH DISTRIBUTIONS

With data such as these it is desirable to obtain some level of statistical significance which would give the probability of the patterns observed occurring by chance alone. Initially, the range of observed values was tested by calculating the mean for the data set and the 95% confidence limits about the mean. Over half of the data points lay outside this range. No level of significance could be attached to this, but it does show that the sporomorph distribution in the sediments is far from uniform. This means that a sample taken at random from the lake has only a 50:50 chance of giving a count which is statistically indistinguishable from the mean (at the 95% confidence level).

To test the significance of the contoured pattern, the data were subjected to a spatial autocorrelation analysis, using Moran's *I* as the coefficient. This tests whether 'a variable at one locality is independent of values of the variable at neighbouring localities' (Sokal and Oden, 1978a). The test may be applied at several levels, initially using the 'nearest neighbours' and then involving progressively more of the data points. The values for '*I*' thus obtained may then be plotted on a correlogram. The method for this is given by Sokal and Oden (1978a,b), who also show how the correlogram may be interpreted.

All of the sporomorph distributions tested to date show patterns that are calculated to be significant at the 99.9% confidence level. They are therefore highly unlikely to have occurred by chance alone.

Evidence for sorting in Silwood Lake

There is a need to be selective in deciding which sporomorph types are used when looking for evidence of sorting. Preferred types would be those that are almost exclusively transported to the lake via the stream. These

Figure 2.8. The distribution of total arboreal and total non-arboreal pollen in the sediments of Silwood Lake. These maps are based upon the absolute sporomorph concentrations in the sediment — sporomorphs/gm (wet weight) of sediment. A minimum of 250 sporomorphs/sample were counted to obtain the relative proportions of the taxa, with a minimum of 500 sporomorphs (including the 'spike' of *Lycopodium* spores added to the sample for calculation purposes) counted to determine the absolute numbers.

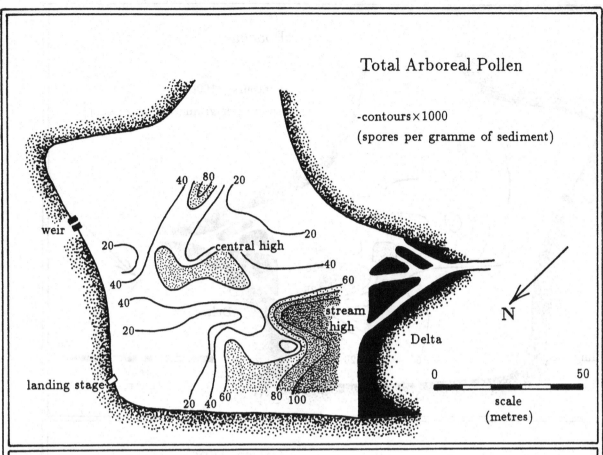

Total Arboreal Pollen

-contours×1000
(spores per gramme of sediment)

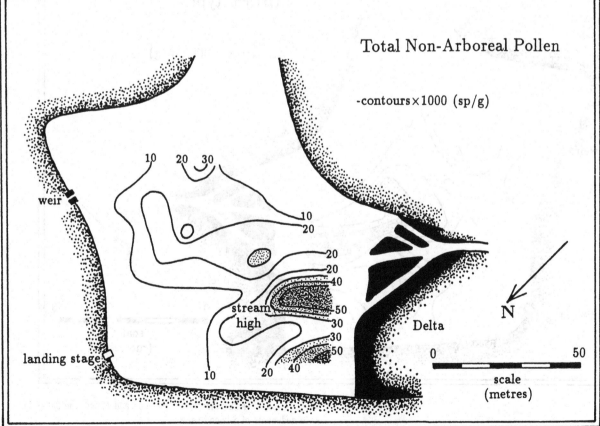

Total Non-Arboreal Pollen

-contours×1000 (sp/g)

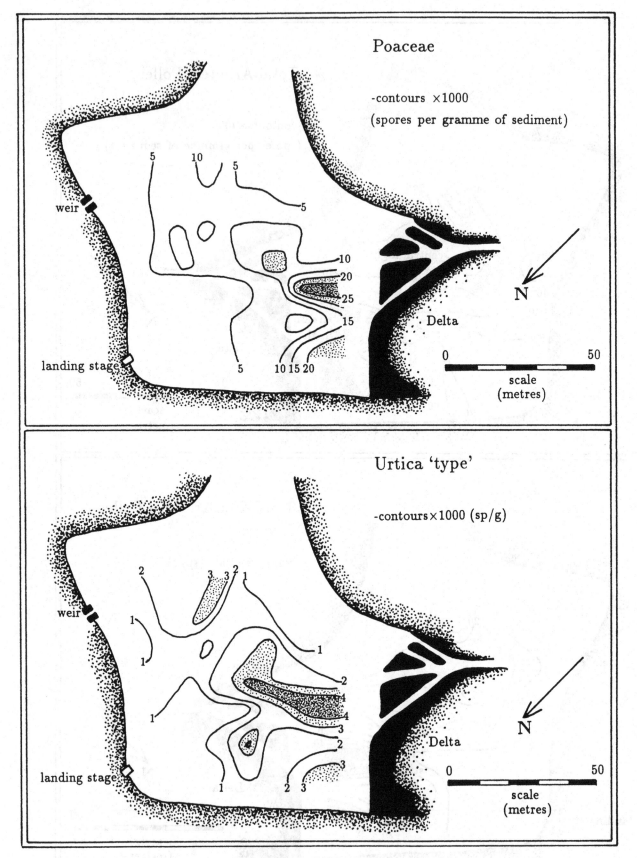

Figure 2.9. The distribution of Poaceae and *Urtica*-type pollen in the sediments of Silwood Lake.

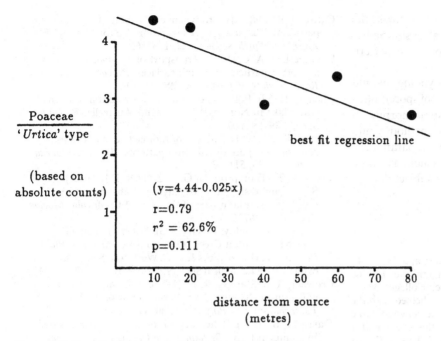

Poaceae
—————
'*Urtica*' type

(based on
absolute counts)

$(y=4.44-0.025x)$

$r=0.79$

$r^2 = 62.6\%$

$p=0.111$

best fit regression line

distance from source
(metres)

Figure 2.10. The ratio of Poaceae to *Urtica*-type pollen across Silwood Lake.

are most likely to occur within the non-arboreal component. Very few spores/pollen in the non-arboreal component are present in large enough numbers to be of significance, because of the low numbers counted (usually 250 grains per slide, excluding exotic grains). Of those that are present in large numbers, the largest grains (Poaceae) and the smallest grains (cf. *Urtica*) show some evidence of having been sorted (Fig. 2.9).

The patterns displayed are similar, both showing high concentrations of sporomorphs proximal to the delta. Precise correspondence to the location of the stream mouth is not important, since the streams are transitory and rapidly migrate over the delta surface. The position of the contours on the distribution map for *Urtica*-type suggests that grains of this type are carried farther out into the lake than grains of Poaceae.

To test this relationship, the ratio of Poaceae to *Urtica*-type has been plotted in Figure 2.10 for a transect across the lake. The line of the transect follows the axis of the 'high' from the delta and then cuts contours perpendicularly, heading toward the lake outlet. A simple linear regression was applied to the five points plotted; these have a coefficient of determination of (r^2) 62.6%. This was tested with the same tests described previously for the flume experiment results. The null hypothesis however could only be rejected at the 89% confidence level. While this is not statistically conclusive, it does suggest that sorting has occurred along the current in Silwood Lake.

Conclusions

Evidence from both the field and laboratory studies has led to the suggestion that water currents can, and do, sort sporomorphs when the current velocity drops below 35 cm/sec. Most sorting occurs at current velocities below 20 cm/sec. At this velocity, it is thought that the effects of sorting will be noticeable within a distance of 30 m (Heathcote, 1978). Flume experiments give a minimum 'detection distance' of seven meters for current velocities of 7 cm/sec. The implications of this are clear. In a stratigraphic sequence, relative changes of different sporomorph types may be due to changes in the hydrology and sedimentology of the sample site. In Silwood Lake, for example, a core taken from the lake center should show a gradual rise in the numbers of Poaceae grains, as the stream mouth has been prograding toward the sample site.

The sorting process appears to be size-selective, although sporomorph density is also significant. This fact should enable us to detect the effects of sorting. If the core is from an area where sorting is thought to have occurred, the counts should be rearranged so that pollen diagrams are produced for sporomorphs within certain size ranges instead of by species. The curves for the largest and the smallest types should give an indication of the degree of the problem. If non-arboreal sporomorphs are used, a clearer picture should emerge. Obviously, if it were possible to collect two cores over 30 m apart, the situation could be better appraised.

The nature of the bed material is also important. Experiments have shown that the effect of a sand bed (grain sizes used; 0.8–2.2 mm, 0.15–0.72) is to increase the rate of deposition while the bed material is stationary. This filtering effect is also selective although no pattern has yet been found.

Study in this area of experimental palynology would benefit greatly from an increased data base of spore/pollen parameters (size, density etc. – for sporomorphs in equilibrium with water). If a uniform method could be found for the measurement of these parameters, interpretation of experimental results (such as those presented here) would also be much less subjective.

Acknowledgments

I would like to thank Prof. W. G. Chaloner and Dr L. E. Frostick for their enthusiasm and support. Thanks are also due to Dr I. Reid and Mr C. Hawkins, of Birkbeck College, for the use of the flume. Mr S. Janes and Mr G. Prior helped with the collection and processing (respectively) of the Silwood Lake samples. Editorial work on the manuscript by the general editor and anonymous reviewers is much appreciated. This work was carried out during the tenure of an NERC studentship which is gratefully acknowledged.

References

Andersen, S. (1973). The differential pollen productivity of trees and its significance for the interpretation of a pollen diagram from a forested region. In *Quaternary Plant Ecology*, ed. H. J. B. Birks & R. G. West. 14th Symposium of the British Ecological Society, Oxford: Blackwells.

Balick, M. J. & Beitel, J. M. (1988). *Lycopodium* spores found in condom dusting agent. *Nature*, **332**, 591.

Battarbee, R. W. & Flower, R. J. (1984). The inwash of catchment diatoms as a source of error in the sediment-based reconstruction of pH in an acid lake. *Limnology and Oceanography*, **29**, 1325–9.

Birks, H. J. B. (1969). Inwashed pollen spectra at Loch Fada, Isle of Skye. *New Phytology*, **69**, 807–20.

Bonny, A. P. (1972). A method for determining absolute pollen frequencies in lake sediments. *New Phytology*, **71**, 393–405.

Bonny, A. P. (1978). The effect of pollen recruitment processes on pollen distribution over the sediment surface of a small lake in Cumbria. *Ecology*, **66**, 385–416.

Bradshaw, R. H. W. & Webb, T. (1985). Relationships between contemporary pollen and vegetation data from Wisconsin and Michigan, USA. *Ecology*, **66**, 721–37.

Brayshaw, A. C., Frostick, L. E. & Reid, I. (1983). The hydrodynamics of particle clusters and sediment entrainment in coarse alluvial channels. *Sedimentology*, **30**, 137–43.

Brown, A.G. (1985). The potential use of pollen in the identification of suspended sediment sources. *Earth Surface Processes and Landforms*, **10**, 27–32.

Brush, G. S. & Brush, L. M. (1972). Transport of pollen in a sediment laden channel: a laboratory study. *American Journal of Science*, **272**, 359–81.

Catto, N. R. (1985). Hydrodynamic distribution of palynomorphs in a fluvial succession, Yukon. *Canadian Journal of Earth Sciences*, **22**, 1552–6.

Chamberlain, A. C. (1966). Transport of *Lycopodium* spores and other particles to rough surfaces. *Proceedings of the Royal Society of London A*, **296**, 45–70.

Chen, Y. (1987). Pollen and sediment distribution in a small crater lake in Northeast Queensland, Australia. *Pollen et Spores*, **29**, 89–110.

Colwell, R. N. (1951). The use of radioactive isotopes in determining spore distribution patterns. *American Journal of Botany*, **38**, 511–23.

Cross, A. T., Thompson, G. G. & Zaitzeff, J. B. (1966). Source and distribution of palynomorphs in bottom sediments, southern part of the Gulf of California. *Marine Geology*, **4**, 467–524.

Crowder, A. & Cuddy, D. G. (1973). Pollen in a small river basin: Wilton Creek, Ontario. In *Quaternary Plant Ecology*, ed. H. J. B. Birks & R. G. West. 14th Symposium of the British Ecological Society, Oxford: Blackwells, 61–78.

Crowder, A. & Starling, R. N. (1980). Contemporary pollen in the Salmon River basin, Ontario. *Review of Palaeobotany and Palynology*, **30**, 11–26.

Davis, M. B. (1967). Pollen deposition in lakes as measured by sediment traps. *Bulletin of the Geological Society of America*, **78**, 849–58.

Davis, M. B. (1968). Pollen grains in lake sediments: redeposition caused by seasonal water circulation. *Science*, **162**, 796–9.

Davis, M. B. & Brubaker, L. B. (1973). Differential sedimentation of pollen grains in lakes. *Limnology and Oceanography*, **18**, 635–46.

Davis, M. B., Brubaker, L. B. & Beiswenger, J. M. (1971). Pollen grains in lake sediments: pollen percentages in surface sediments from southern Michigan. *Quaternary Research*, **1**, 450–67.

Davis, R. B. (1967). Pollen studies of near surface sediments in Maine lakes. In *Quaternary Paleoecology*, ed. E. J. Cushing & H. E. Wright, Jr., Proceedings 7th Congress INQUA. New Haven: Yale University Press, 143–73.

Durham, O. C. (1946). The volumetric incidence of atmospheric allergens III – the rate of fall of pollen grains in still air. *Journal of Allergy*, **17**, 70–8.

Dyakowska, J. (1937). Researches on the rapidity of the falling down pollen of some trees. *Akademija Umiejetnosci, Kraków, Series B. Natural Sciences (Bulletin International de L'Académie Polonaise des Sciences et des Lettres. Série B: Sciences Naturelles)*, 155–68.

Erdtman, G. (1943). *Introduction to Pollen Analysis*. Waltham, Massachusetts: Chronica Botanica.

Faegri, K. & Iversen, J. (1975). *Textbook of Pollen Analysis*. 3rd ed. Copenhagen: Munksgaard.

Fall, P.L. (1987). Pollen taphonomy in a canyon stream. *Quaternary Research*, **28**, 393–406.

Federova, R. V. (1952). Distribution of pollen and spores by flowing water. *Works of the Geographical Institute of the Academy of Sciences, USSR*, **52**, 46–72. (In Russian).

Fletcher, M. R. (1979) Distribution patterns of Holocene and reworked Paleozoic palynomorphs in sediments from southeastern Lake Michigan. Ph.D. diss. Case Western Reserve University, Cleveland, OH.

Goodwin, R. G. (1988). Pollen taphonomy in Holocene glaciolacustrine sediments: a cautionary note. *Palaios*, **3**, 606–11.

Gregory, P. H. (1961). *Microbiology of the Atmosphere*. London: Leonard Hill.

Hall, S. A. (1989). Pollen analysis and paleoecology of alluvium. *Quaternary Research*, **31**, 435–8.

Harrington, J. B. & Metzger, K. (1963). Ragweed pollen density. *American Journal of Botany*, **50**, 532–9.

Havinga, A. J. (1964). Corrosion of pollen and spores. *Pollen et Spores*, **6**, 621–35.

Havinga, A. J. (1984). A 20-year experimental investigation into the differential corrosion susceptibility of pollen and spores in various soil types. *Pollen et Spores*, **26**:3/4, 541–58.

Heathcote, I. W. (1978). Differential pollen deposition and water circulation in small Minnesota lakes. Ph.D. diss. Yale University, New Haven, CT.

Heusser, C. J. & Florer, L. E. (1973). Correlation of marine and continental Quaternary pollen records from the NE Pacific and Western Washington. *Quaternary Research*, **3**, 661–70.

Heusser, L.E. (1988). Pollen distribution in marine sediments on the continental margin off Northern California. *Marine Geology*, **80**, 131–47.

Hilton, J., Lishman, J. P. & Allen, P. V. (1986). The dominant processes of sediment distribution in a small, eutrophic, monomictic lake. *Limnology and Oceanography*, **31**, 125–33.

Hopkins, J. S. (1950). Differential pollen flotation and deposition of conifers and deciduous trees. *Ecology*, **31**, 633–41.

Janssen, C. R. (1973). Local and regional pollen deposition. In *Quaternary Plant Ecology*, ed. H. J. B. Birks & R. G. West. 14th Symposium of the British Ecological Society. Oxford: Blackwells, 31–42.

Jemmett, G. & Owen, J. A. K. (1990). Where has all the pollen gone? *Review of Palaeobotany and Palynology*, **64**, 20511.

Kozumplik, V. & Christie, B. R. (1972). Dissemination of orchard-grass pollen. *Canadian Journal of Plant Science*, **52**, 997–1002.

Lanner, R. M. (1966). Needed: a new approach to the study of pollen dispersion. *Silvae Genetica*, **20**, 50–2.

Leeder, M.R. (1983). *Sedimentology: Process and Product*. London: Allen & Unwin.

Matthews, J. (1969). The assessment of a method for the determination of absolute pollen frequencies. *New Phytology*, **68**, 161–6.

May, K. R. (1958). The washout of *Lycopodium* spores in rainfall. *Quaternary Journal of the Royal Meteorological Society*, **84**, 451–8.

McCully, C. R., Fisher, M., Langer, G., Rosinski, J., Glaess, H. & Werle, D. (1956). The scavenging action of rain on airborne particulate matter. *Industrial and Engineering Chemistry*, **48**, 1512–6.

McDonald, J. E. (1962). Collection and washout of airborne pollen and spores by raindrops. *Science*, **135**, 435–6.

McLennan, D. S. & Mathewes, R. W. (1984). Pollen transport and representation in the coast mountains of British Columbia: I. Flowering phenology and aerial deposition. *Canadian Journal of Botany*, **62**, 2154–64.

Moseholm, L., Weeke, E. R. & Petersen, B. N. (1987). Grass pollen dispersion. *Pollen et Spores*, **29**, 305–21.

Muller, J. (1959). Palynology of Recent Orinoco Delta and shelf sediments. *Micropaleontology*, **5**, 1–32.

Niklas, K. J. & Paw, U. K. T. (1983). Conifer ovulate cone morphology: implications on pollen impaction patterns. *American Journal of Botany*, **70**, 568–77.

Peck, R. (1972). Efficiency tests on the Tauber Trap used as a pollen sampler in turbulent water flow. *New Phytologist*, **71**, 187–98.

Peck, R. (1973). Pollen budget studies in a small Yorkshire Catchment. In *Quaternary Plant Ecology*, ed. H. J. B. Birks & R. G. West. 14th Symposium of the British Ecological Society. Oxford: Blackwells, 43–60.

Peck, R. (1974a). A comparison of four absolute pollen preparation techniques. *New Phytologist*, **73**, 567–87.

Peck, R. (1974b). Studies of pollen distribution in the Oakdale catchment. Ph.D. diss. Cambridge University, UK.

Pennington, W. (1979). The origin of pollen in lake sediments: An enclosed lake compared with one receiving streams. *New Phytology*, **83**, 189–213.

Pennington, W., Haworth, E. Y., Bonny, A. P. & Lishman, J. P. (1972). Lake sediments in Northern Scotland. *Philosophical Transactions of the Royal Society of London*, **B264**, 191–294.

Pohl, F. (1933). Untersuchungen über Bestäubungsverhältnisse der Traubereiche. *Beihefte zum botanischen Zentralblatt*, Abt.1, **51**, 693–6.

Prentice, I. C. (1985). Pollen representation, source area and basin size; toward a uniform theory of pollen analysis. *Quaternary Research*, **23**, 76–86.

Raudkivi, A. J. (1967). *Loose Boundary Hydraulics*. Elmsford, NY: Pergamon Press.

Ray, A. (1959). The effect of earthworms on soil pollen distribution. *Journal of the Oxford University Forestry Society*, **7**, 16–21.

Reid, I. & Frostick, L. E. (1985). Role of settling, entrainment and dispersive equivalence and of interstice trapping in placer formation. *Journal of the Geological Society of London*, **142**, 739–46.

Reynolds, C. S. (1979). Seston sedimentation: experiments with *Lycopodium* spores in a closed system. *Freshwater Biology*, **9**, 55–76.

Rowley, J. R. & Walch, K. M. (1972). Recovery of introduced pollen from a glacier mountain stream. *Grana*, **12**, 146–52.

Sangster, A. G. & Dale, H. M. (1964). Pollen grain preservation of under-represented species in fossil spectra. *Canadian Journal of Botany*, **42**, 437–49.

Sokal, R. R. & Oden, F. M. (1978a). Spatial autocorrelation in biology: 1. Methodology. *Biological Journal of the Linnean Society*, **10**, 199–228.

Sokal, R. R. & Oden, F. M. (1978b). Spatial autocorrelation in biology: 2. Some biological implications and four applications of evolutionary and ecological interest. *Biological Journal of the Linnean Society*, **10**, 229–49.

Spicer, R. A. (1981). The sorting and deposition of allochthonous plant material in a modern environment at Silwood Lake, Silwood Park, Berkshire, England. *Geological Survey Professional Paper* **1143**.

Starling, R. M. & Crowder, A. D. (1980/81). Pollen in the Salmon River system, Ontario. *Review of Palaeobotany and Palynology*, **31**, 311–34.

Tauber, H. (1967a). Differential pollen dispersion and filtration. In *Quaternary Paleoecology*, ed. E. J. Cushing & H. E. Wright, Jr. New Haven: Yale University Press, 131–42.

Tauber, H. (1967b). Investigations of the mode of pollen transfer in forested areas. *Review of Palaeobotany and Palynology*, **3**, 277–86.

Traverse, A. (1988). *Paleopalynology*. London: Unwin Hyman.

Traverse, A. & Ginsburg, R. N. (1966). Palynology of the surface sediments of the Great Bahama Bank, as related to water movement and sedimentation. *Marine Geology*, **4**, 417–59.

Valencia, M. J. (1967). Recycling of pollen from an air-water interface. *American Journal of Science*, **265**, 843–7.

Van der Knaap, W. O. (1988). Deposition of long-distance transported pollen and spores since 7900 B.P. studied in peat deposits from Spitsbergen. *Pollen et Spores*, **30**, 449–53.

Webb, T. (1974). Corresponding patterns of pollen and vegetation in Lower Michigan: a comparison of quantitative data. *Ecology*, **55**, 17–28.

Winkelmolen, A. M. (1982). Critical remarks on grain parameters, with special emphasis on shape. *Sedimentology*, **29**, 255–65.

3 Transport and deposition of pollen in an estuary: signature of the landscape

GRACE S. BRUSH and LUCIEN M. BRUSH

Introduction

Pollen preserved in lake and peat sediments has been used to reconstruct changes in vegetation induced by climatic events (Davis, Spear, & Shane, 1980; Webb, Cushing, & Wright, 1983; Davis & Jacobson, 1985) and by anthropogenic alterations to the landscape (Brugam, 1978; Burden *et al.*, 1986). Only recently has the record from estuarine sediments been similarly exploited (Brush, 1986; McGlone, 1988). In this chapter, we describe the general pathways of pollen in an estuary, and show from laboratory experiments and field observations how the transport and deposition of pollen in the estuary results in distributions in the sediments that record regional vegetation, land use and the effect of land use on rates and patterns of estuarine sedimentation. Our area of study is the Chesapeake Bay estuary extending between latitudes 37° and 39° 30′ N in the mid-Atlantic region of the USA (Fig. 3.1).

Estuarine circulation

First, we examine the fluid motion, sediment motion, and salinity characteristics of Chesapeake Bay. Chesapeake Bay and many of its tributaries are partially mixed estuaries (Dyer, 1973), characterized by vertical, lateral and longitudinal gradients in salinity, but with no abrupt change from fresh to salt water as is found, for example, in the salt wedge environments of many fjords. On the other hand, Chesapeake Bay estuary is neither completely mixed nor homogeneous, as are estuaries characterized by large tidal flows and small river inflows. An example of an idealized, partially mixed estuary is shown in Figure 3.2A. When river water enters the estuary and meets ocean water, partial mixing occurs and causes lines of equal salinity to deform. Salinity also changes across the estuary in response to the Earth's rotation and as a result of lateral fresh water inflows. Tides tend to form a zone of sediment resuspension at the bottom of the estuary,

causing sediment concentration to vary through a tidal cycle (Schubel, 1968). A turbidity maximum generally occurs near the point where the toe of the lower, saltier water reaches fresh water (Fig. 3.2A). The turbidity maximum, where sediment concentration is usually much higher than in adjacent waters, may move upstream or downstream seasonally or during storms in response to the amount of fresh water entering the estuary.

Typical velocity distributions at a section in the middle of a partially mixed estuary and at the river section are shown in Figure 3.2B. The distributions shown are typical for turbulent flow in a partially mixed estuary. Within a tidal cycle, the longitudinal velocities at various points along a vertical line vary in magnitude and direction. Because the river (fresh) water ultimately flows toward the ocean, a net nontidal flow exists equal to river flow (Fig. 3.2C; see also Pritchard & Kent, 1953). While the tides oscillate, turbulent diffusion occurs in all three coordinate directions. Weak vertical velocities, many orders of magnitude smaller than the longitudinal net non-tidal velocities, also occur in the estuary. The line of zero net tidal velocity is shallower toward the river and rises toward the ocean, causing the estuary to be divided into two major zones. The upper zone has net non-tidal motion toward the ocean, and the lower, saltier zone has net non-tidal flow up-estuary. Nevertheless, there is some diffusion and advection across the line of zero net non-tidal motion (Pritchard, 1967).

Figure 3.2D is a schematic view of the fate of pollen grains entering an estuary, assuming no aggregation of particles. A pollen grain entering at point 3, once wetted, will travel up and down the estuary through many tidal excursions as it slowly settles. Unlike ponds or small lakes, Chesapeake Bay almost always has waves large enough to insure rapid wetting of the pollen. There is a net down-estuary movement of the grain as it passes through the upper zone of net non-tidal flow. In the lower zone, a similar back and forth motion will occur with each tidal cycle, but the grain will slowly move up-estuary as it settles toward the bed. As the particle nears the bed,

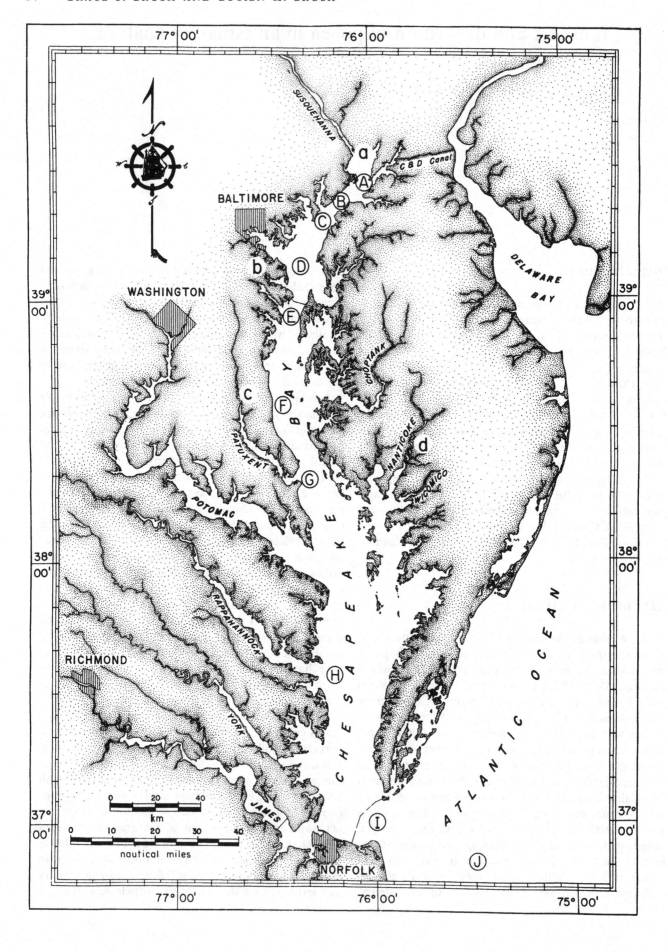

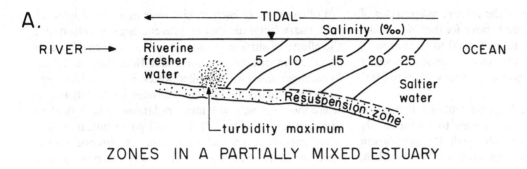

ZONES IN A PARTIALLY MIXED ESTUARY

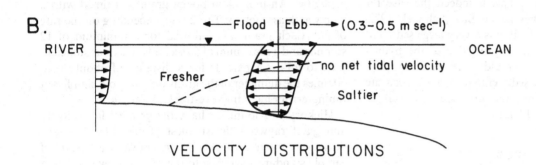

VELOCITY DISTRIBUTIONS

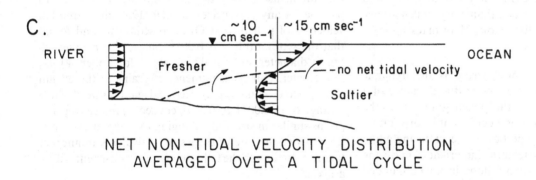

NET NON-TIDAL VELOCITY DISTRIBUTION
AVERAGED OVER A TIDAL CYCLE

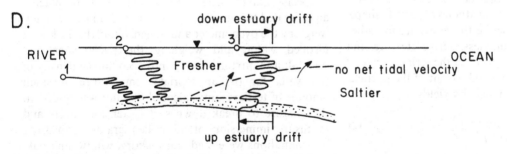

Figure 3.2. Diagram of some characteristics of an idealized partially mixed estuary (patterns in part after Pritchard, 1967).

it will participate in any resuspension occurring there, and may possibly be caught up in aggregations of particles, should they exist. With time and subsequent deposition of other particles, the pollen grain will reside

Figure 3.1. Map of Chesapeake Bay, showing core sites (a, b, c, d) and water samples (A through J).

in the bed and remain there unless moved by a huge storm. A pollen grain entering at point 2 in the fresh water part of the estuary or at point 1 from river flow may never reach the saline water. Because of the large surface area of the part of the estuary characterized by a two-zoned division of net non-tidal motion, most atmospherically derived pollen grains are expected to follow a path similar to the hypothetical pollen grain entering at point 3. Pollen grains will also make up some small fraction of the turbidity maximum.

To give some indication of the relative magnitudes of the various velocities mentioned above, for the Chesapeake Bay typical tidal velocities range from 0 to 50 cm sec^{-1}, net non-tidal velocities 0 to 15 cm sec^{-1}, vertical velocities 0 to 0.001 cm sec^{-1}, and pollen settling velocities 0.002 to 0.008 cm sec^{-1}.

The above description of the movement of pollen grains in the estuarine system also applies to hydraulically equivalent fine silt particles, although the proportion arriving from river water is much greater than for pollen. The model suggests that very few pollen or sediment particles leave the estuary. This is indeed the case for sediment particles in Chesapeake Bay (Schubel, 1972, p. 5), and indicates that the Bay is a very large sediment trap or settling basin. Some knowledge of the hydrodynamics of pollen grains should therefore allow their distributions in estuarine sediments to provide accurate documentation of sedimentary processes as well as changes in vegetation and land use.

Pollen hydrodynamics

Because of their small size (10–100 μm in diameter) and specific gravities slightly greater than unity, pollen grains, when wet, follow a generalized Stokes' Law of resistance:

$$F = 3\mu\pi d_n V_t K \qquad (1)$$

where F is the Stokes drag force, μ is the dynamic viscosity, d_n is the nominal diameter (the diameter of a sphere of equal volume to the pollen grain), V_t is the terminal fall velocity, and K is a coefficient incorporating a number of physical properties of the grain including shape, roundness, surface texture, and orientation during fall. $K = 1$ for a sphere falling alone in an unbounded fluid and is usually >1 for ordinary particles. McNown *et al.* (1951) introduced K in their studies of particle shape, but it has been used to describe the resistance for other characteristics such as roundness, surface texture and protuberances. The resisting force at the terminal rate of fall is balanced by the submerged weight of the particle. Solving for the terminal fall velocity yields

$$V_t = \frac{d_n^2 g}{18 K v}\left(\frac{\rho_s}{\rho} - 1\right) \qquad (2)$$

where ρ_s and ρ are are the particle and fluid densities respectively, g is the acceleration due to gravity, and v is the kinematic viscosity.

Experiments were run to determine actual fall velocities, and measurements were made of size and specific gravity of pollen grains of five species common in the Chesapeake Bay area. The species are *Carya tomentosa* Nutt. (mockernut hickory), *Quercus alba* L. (white oak), *Salix nigra* Marsh (black willow), *Ambrosia artemisifolia* L. (common ragweed), and *Ambrosia trifida* L. (giant ragweed) (Fig. 3.3).

Pollen grains of each species were first mixed in ≈ 0.1 ml tertiary butyl alcohol to prevent agglomeration and the effects of surface tension. A small amount of pollen was then introduced into a test tube 18 cm long and 3 cm in diameter, filled with distilled water. Fall velocities were observed for individual particles through a 25 × reticulated telescopic lens with 0.1 mm gradations, and a field of vision of 4.5 mm (Fig. 3.4). The pollen concentration was small enough to prevent significant interference among particles. Spacing between particles was observed to be at least 15 times the diameter, and boundary influence was negligible. An individual pollen grain was timed with a stop watch as it fell for 1 to 2 mm, depending on the rate of fall. Each grain was followed for a minimum of 12 seconds and a maximum of 50 seconds. This was sufficient time to follow a particle for a distance of about 20 to 50 times the diameter of the grain, and to avoid significant timing errors including eye-tracking fatigue.

Hickory was found to have the greatest fall velocity and giant ragweed the smallest (Table 3.1). Standard deviations and resulting coefficients of variation (ratio of sample standard deviation to the mean) show consistent measurements with little variation from particle to particle for any one species. Muller (1959) measured fall velocities of pollen from Orinoco sediments and found that the settling rate was 0.0019 cm sec^{-1} for pollen 20 μm in diameter and 0.0047 cm sec^{-1} for pollen 40 μm in diameter. Our measurements of grains in the 20 μm category (oak, common ragweed and giant ragweed) show a range from 0.0021 to 0.0041 cm sec^{-1}, indicating that grains similar in size and belonging to different taxa may fall at twice the rate of their size equivalents in another taxon, even though the fall velocities are consistent within a taxon.

Specific gravity measurements were made by placing an estimated two grams of pollen in a 15 ml pycnometer, weighing the pycnometer and pollen, filling the flask with distilled water, and weighing the pycnometer and pollen-liquid mixture, taking into account the density of the water for the appropriate temperature. A small amount of tertiary butyl alcohol (0.1 ml) was added to the mixture to break down surface tension effects and to insure immersion of all pollen grains. Consistent determinations were made for hickory, willow, and oak, but severe problems were encountered with both types of ragweed. Giant ragweed did not disperse properly, and common ragweed caused a foam to form in the pycnometer, possibly due to release of air from the pollen during the wetting procedure. After several failed attempts, measurements of ragweed were abandoned, and determinations made by Harrington and Metzger (1963) used.

Size determinations were made by measuring the two axes of the pollen grain seen on a microscope slide under 400 × magnification. For subsequent computations, it is

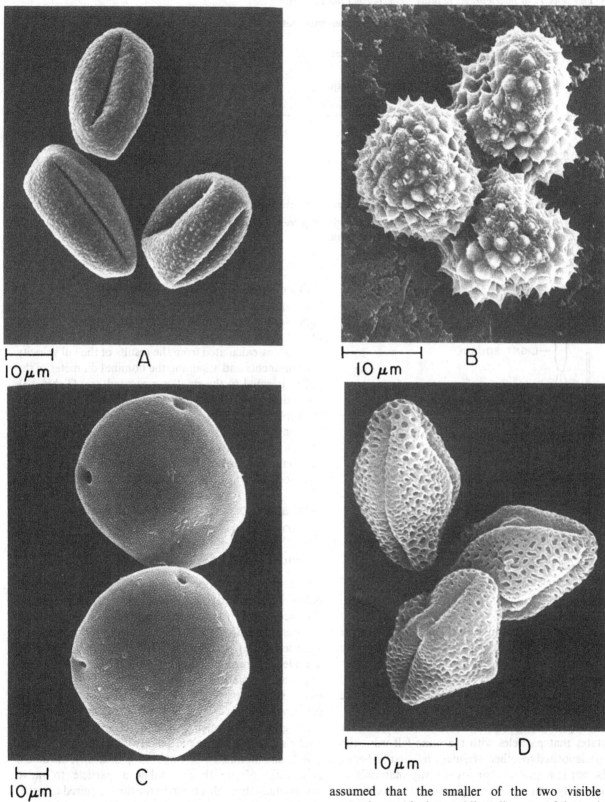

Figure 3.3. Scanning electron micrographs of pollen grains used in laboratory experiments. (A) *Quercus alba* (white oak); (B) *Ambrosia trifida* (giant ragweed); (C) *Carya tomentosa* (mockernut hickory); (D) *Salix nigra* (black willow).

assumed that the smaller of the two visible axes approximates the intermediate diameter of the grain, and is considered a close approximation to the nominal diameter of the particle (Cui & Komar, 1984). The third axis aligned parallel to the viewer's eye, is the smallest dimension due to the orientation of the grain on the slide.

Table 3.1. *Fall velocity of pollen grains in order of decreasing fall velocity.*

Temperature ranged from 24.2 to 26 °C during these measurements, but for any one pollen type the temperature did not range more than 0.5 °C.

Pollen grain[a]	Fall velocity V_T (cm sec^{-1})	Sample standard deviation (cm sec^{-1})	Coefficient of variation
Hickory	0.007 51	0.001 32	0.176
Oak	0.004 07	0.000 493	0.121
Common ragweed	0.002 80	0.000 372	0.133
Willow	0.002 15	0.000 268	0.125
Giant ragweed	0.002 12	0.000 123	0.058

[a] Hickory (*Carya tomentosa*), oak (*Quercus alba*), common ragweed (*Ambrosia artemisiifolia*), willow (*Salix nigra*), giant ragweed (*Ambrosia trifida*).

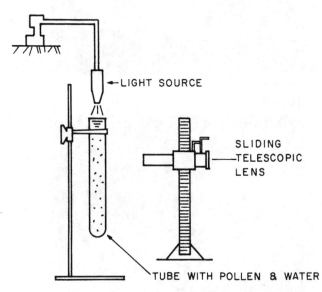

Figure 3.4. Schematic of the apparatus for measuring fall velocities of individual pollen grains.

LIGHT SOURCE

SLIDING TELESCOPIC LENS

TUBE WITH POLLEN & WATER

The coefficient of variation for oak and willow is larger than for the other grains (Table 3.2). These grains have the largest axial ratios, and hence greatest departures from a sphere. The high density of willow compensates for its small size to give a fall velocity similar to that of the larger but less dense ragweed (Table 3.1).

The concept of hydraulic equivalence (Rubey, 1933), which states that particles with the same fall velocities tend to be deposited together, originally focused on heavy minerals, but is applicable for low-density materials as well, particularly with regard to depositional environments. For example, it is possible to determine the size of a spherical quartz grain (or any other mineral) having the same fall velocity as a particular pollen grain (Table 3.3). Using Stokes' Law for particles with Reynolds numbers < 0.1, the equivalent sizes of pollen and quartz spheres

with similar settling velocities are computed (Table 3.3). Although quartz grains are not generally spherical, an approximate size can be determined. Each pollen type can be assigned a particular quartz equivalent.

K values, calculated from the results of the fall velocity measurements and assuming the nominal diameter of the pollen is equal to the smaller measured axis (Table 3.3), show that pollen grains depart significantly from values for a sphere, indicating that shape, spines, protuberances, etc. contribute a significant effect on the drag of the falling grain. Consequently, observed fall velocities are preferable to computed values for consideration of transport distances.

Distance of transport of a pollen grain in the estuary

A model for the distance of pollen transport

The above observations indicating certain similarities between pollen and fine sediment allow us to predict the distance of pollen transport using a settling basin model. As was noted previously, the Chesapeake Bay acts as a huge sediment trap. Therefore, it seems reasonable to use a model, designed for estimating the dimensions of a tank required for the removal of particles in a waste water treatment system, to estimate the distance a pollen grain (or any small particle) will travel in the estuary, before settling out. One such model described by Cordoba-Molina *et al.* (1978), was originally formulated by Hazen (1904), and modified by Camp (1946). Basically, the model is a ratio of the time it takes a particle to move horizontally through a basin to the time required to settle, modified by the effects of turbulence. The model assumes no erosion or resuspension as a grain reaches the bed. This assumption may not be true for an estuary, but it is a good first approximation, and as shown by Schubel (1972), small sediment particles in Chesapeake Bay do

Table 3.2. *Sizes and specific gravities of pollen used in the fall velocity experiments.*

Specific gravity measurements for both species of ragweed are values given by Harrington and Metzger (1963) and entered with parentheses on the table.

Pollen	Smaller axis (μm)	Standard deviation (μm)	Coefficient of variation	Ratio of long to small axis	Specific gravity
Hickory	45.36	4.375	0.096	1.09	1.21
Oak	21.43	3.239	0.151	1.39	1.27
Giant ragweed	18.62	1.288	0.069	1.07	(1.28)
Common ragweed	18.60	1.815	0.098	1.09	(1.28)
Willow	13.66	1.784	0.131	1.53	1.37

Table 3.3. *Computations based on the fall velocities of pollen grains*

Pollen	Quartz equivalent (μm)	Pollen diameter (μm)	Reynolds number $\dfrac{V_T d}{v}$	McNown's K factor (equation 3)
Hickory	9.19	45.36	0.003 4	3.10
Oak	6.76	21.43	0.000 86	1.77
Giant ragweed	4.88	18.62	0.000 39	2.46
Common ragweed	5.61	18.60	0.000 52	1.86
Willow	4.92	13.66	0.000 29	1.75

not travel very far following resuspension. Expressed mathematically, the fraction of total particles removed is given by the following equation

$$r = 1 - \exp\left(\frac{-V_t}{V_o}\right) \qquad (3)$$

where r is the fraction removed, V_t is the settling velocity and V_o is the so-called overflow rate. The overflow rate is the throughflow, in this case the fresh water discharge divided by the surface area of the basin. Note that $r = 1 - c$, where c is the vertically averaged concentration of material left in suspension. Depth does not enter the equation directly and is important only if significant erosion occurs at the bed.

The first example is closely tied to the actual parameters found in a wastewater treatment plant, designed for the removal of particles by settling. Assume that pollen is introduced into the influent of a continuous sedimentation tank, such as found in typical wastewater treatment plants, where the overflow rate is 0.0259 cm sec^{-1} (550 gal day^{-1} ft^{-2}) with a discharge of 131.4 l sec^{-1} (3×10^6 gal day^{-1}). The length of the tank is 50 m and the width 10.2 m. Using standard removal curves (Cordoba-Molina *et al.*, 1978) for turbulent flow in the absence of scour, 25% of hickory pollen and 8% of giant ragweed pollen would be removed after one pass through the tank. Table

Table 3.4. *Removal of pollen in sedimentation tanks using the curves found in Cordoba-Molina* et al. *(1978)*

% pollen removed	No. of tanks or passes required	
	Hickory	Giant ragweed
50	3	9
75	5	17
95	11	36
99	16	56

3.4 gives the integer number of tanks or passes required to accomplish various percentages of removal. Note that remixing is assumed to occur at the entrance to each tank in the series. Pollen grains have settling velocities slightly lower than particles found in wastewater effluent and therefore the removal efficiency is not as large. However, despite the rather low estimate for a single pass, the resulting distances of travel are not long. For the example given, 99% of hickory pollen would be removed over a distance of 800 m and 99% of giant ragweed pollen over 2850 m. Another approach would be to ask how long a settling tank need be to trap 99% of the pollen with all other conditions equal. The length would be 343

Table 3.5. *Concentrations of pollen (grains liter⁻¹) in the water column of the Chesapeake Bay in 1971.*

Station locations are marked on Fig. 1. (n.d. = no data)

Station		March	May	July	October
Surface waters					
A		213	1740	595	359
B		705	2120	403	0
C		331	1080	50	41
D		142	270	0	0
E		0	780	0	0
F		0	260	0	0
G		0	610	0	0
H		129	0	0	0
I		0	600	0	0
J		158	220	0	256
Mid-depth	Depth in meters				
A	(4.0)	n.d.	n.d.	n.d.	540
B	(6.0)	348	4930	653	386
C	(8.0)	538	890	112	106
D	(6.0)	78	1490	32	0
E	(16.0)	0	460	0	0
F	(16.0)	0	1870	0	0
G	(16.0)	103	620	302	0
H	(8.0)	230	600	265	253
I	(10.0)	0	200	0	0
J	(10.0)	546	600	334	191

m for hickory and 1215 m for giant ragweed. Regardless of which model is chosen, the distances are quite short in terms of a natural system, such as Chesapeake Bay. Settling velocities for the pollen studied range from 1.83 m day⁻¹ to 6.49 m day⁻¹.

The main parameter in the previous illustration is the ratio of the pollen fall velocity to the overflow rate (discharge divided by surface area). In estuaries like Chesapeake Bay, analogous to settling basins, overflow rates can be determined. For example, the Potomac River estuary (Fig. 3.1) during average inflows has an overflow rate of 3.26×10^{-5} cm sec⁻¹. Over 99.9% of the pollen of hickory and giant ragweed entering this estuary, once wetted, would be removed over its length. Only during intense storms would the removal efficiency be decreased appreciably. In fact, for the maximum known discharge, the removal efficiency of hickory pollen for the Potomac River would be 98.7%, and 80.3% for pollen of giant ragweed. All of the other estuaries of the Chesapeake Bay system, including the main stem of the Bay have similar removal efficiencies. Not only are the removal efficiencies large, but the total distance of travel is small because

trap efficiencies would remain above 99% even for short segments of the estuary, e.g., 1/10 of the length. Other estimates of the trap efficiencies of estuaries show similar large removal rates (Biggs & Howell, 1985). Thus, even most of the airborne, or for that matter waterborne, pollen entering partially mixed estuaries would be trapped after a short distance of travel.

Observed distributions of pollen in water

Water samples collected in the surface and mid-depth of the water column throughout Chesapeake Bay (Fig. 3.1) at four different times of the same year, show the majority of pollen present in the water column in early May, the period of maximum pollination (Table 3.5). There is a small amount of pollen in the surface water at several locations in March when some early-flowering trees shed pollen. Pollen occurs in the surface water in July only in the area of the turbidity maximum in the upper Bay, and in October in the area of the turbidity maximum and where the estuary empties into the ocean, both regions of high resuspension. The distributions indicate that the majority of pollen is transported relatively short distances before being deposited, and are in agreement with calculated settling velocities of 1.83 to 6.49 m day⁻¹ based on the experimental fall velocity data. Pollen found at mid-depth in the water column outside of pollination times, is in areas of high resuspension, i.e., the shallow upper Bay, the mouths of tributaries and the mouth of the estuary proper.

Observed distributions of pollen in the sediment

The concentration of *Chenopodium* pollen in surface sediments of cores taken in Port Tobacco, a small tributary of the Potomac River, provides field corroboration of the simulated distances of transport (DeFries, 1986). Surface sediments 1 km from the source area of *Chenopodium* pollen contain 5057 grains g⁻¹ of sediment. Concentrations of *Chenopodium* pollen become progressively less in the downstream direction away from the source (Table 3.6), and are rarely seen in surface sediments in the Potomac River estuary proper. Within 4.2 km, 94.5% of the pollen is removed from the water column.

Trees of *Tsuga canadensis* (L.) Carr. (Eastern hemlock) comprise approximately 38% of stands of the hemlock–birch forest association of the Appalachian physiographic province in Maryland, 60 km northwest of Washington, DC (Brush *et al.*, 1980). Hemlock trees produce large amounts of pollen, but hemlock pollen constitutes only 0.3% of the total pollen in surface sediments of the Potomac River estuary at Washington, DC (Brush & DeFries, 1981), indicating that the pollen is deposited

Table 3.6. *Concentrations (number of pollen grains/gram of sediment) of* Chenopodium *pollen in Port Tobacco estuary*

Station A (1 km from source of pollen)	Station B (1.8 km downstream from Stn A)	Loss between A and B (%)	Station C (1.4 km downstream from Stn B)	Loss between B and C (%)	Loss between A and C (%)
5057	1240	75	276	77.7	94.5

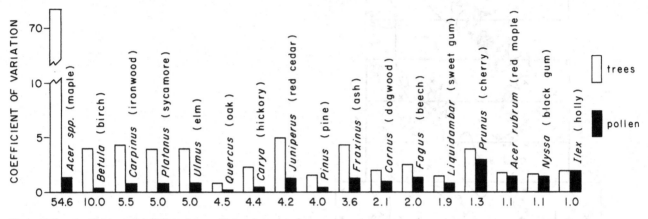

Figure 3.5. Coefficients of variation of trees (% basal area) and pollen (% frequency) along the Potomac River estuary. Ratios of tree c.v. to pollen c.v. are below each pair of histograms.

before it reaches Washington. Our observations are in agreement with those of Bradshaw and Webb (1985), who found in Wisconsin and Michigan, hemlock pollen present in large numbers in surface sediments of lakes and bogs of various sizes within 30 km of the source trees; there is much less dispersion at farther distances. Groot (1966) reported high percentages of hemlock pollen in Delaware estuary water collected in August, and interpreted this as evidence of the long distance transport of hemlock pollen, based on a regional vegetation map which shows the *Acer–Tsuga* (maple–hemlock) association occurring in New York and northern Pennsylvania. However, hemlock pollen in the Delaware estuary could also originate from stands of hemlock trees in other vegetation associations closer to the estuary. Sedimentation rates determined throughout 18 tributaries in Chesapeake Bay show at least a twofold increase in the upstream and midstream sections since European settlement, but no significant increase in the downstream reaches (Brush 1984). These historical patterns of sediment accumulation also indicate that fine particles travel short distances in the estuary prior to deposition.

Pollen–vegetation relationships

The following analyses support the hypothesis that pollen distributions are related to vegetation distributions in

accordance with the pollen transport and deposition mechanisms in the Chesapeake Bay estuary as described above.

Effects of transport on pollen distributions in sediment were identified by comparing the coefficient of variation (c.v.) (standard deviation divided by the mean) of the frequency of pollen in surface sediment samples with the c.v. of frequency of the respective tree taxa in plots adjacent to the estuary (Fig. 3.5). The comparison was made between vegetation in a broad strip adjacent to the Potomac River estuary and pollen concentrations in the surface sediment of the estuary (Brush & DeFries, 1981). The ratio of tree c.v. to pollen c.v. was used to evaluate similarity in distributions, with a ratio of 1 indicating perfect similarity. Maple and birch, with ratios of 55 and 10 respectively, are characterized by greater homogeneity in pollen than tree distributions. Ratios for all other species ranged from 5.5 to 1, indicating similar distributional patterns among the pollen and vegetation.

Mean percentage basal area of pine, oak, and hickory trees and mean percent of pine, oak, and hickory pollen in surface sediments of the Chesapeake Bay and tributaries were measured in 14 (for trees) and 11 (for pollen), 7.5' latitudinal strips from 38° to 39°45' north latitude (Fig. 3.6). The strips spanned 75°30' to 77°15' west longitude. Pine, oak and hickory are the taxa with the greatest number of species in the Chesapeake Bay region, and they form the most important components of the existing vegetation. Both pine trees and pine pollen increase southward (Fig. 3.7). The slope of the regression line is similar for both, with an increase in basal area of

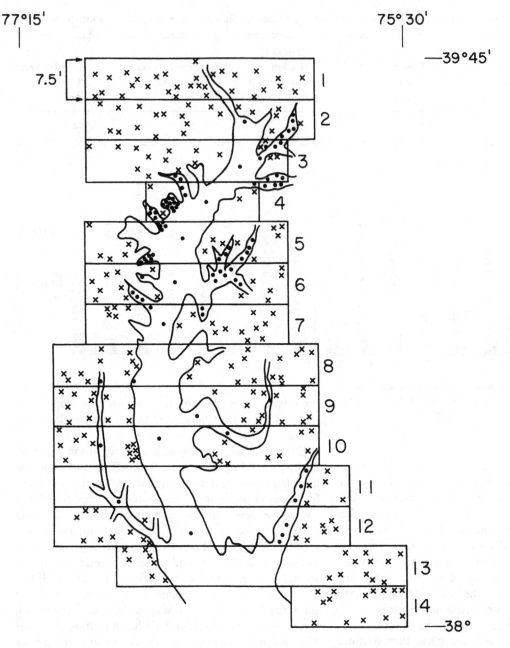

Figure 3.6. Map of the Chesapeake Bay region showing tree samples (×) and pollen samples (●) in 7.5′ latitudinal strips. Comparisons were made of tree (mean % basal area in 400 m² plots) in 14 strips and pollen (mean % of total pollen concentration in surface sediments) in 11 strips (absent from 1, 13 and 14) along a latitudinal gradient from 38° to 39°45′ N. Taxa analyzed were pine, oak and hickory.

trees of 2.55% per 7.5′ of latitude southward and an increase in pollen concentrations of 2.59% per 7.5′ latitude, also southward. There are no latitudinal trends in either oak or hickory, reflecting the random distribution of both oak and hickory trees on the landscape and oak and hickory pollen in the sediments. Groot (1966) recognized a similar relationship between pollen of pine, oak and hickory in the Delaware estuary and pine, oak and hickory trees in the surrounding vegetation, but his observations do not include data from measured plots.

Similar comparisons performed for 28 taxa of trees in a band adjacent to the Potomac River estuary and their respective pollen in the surface sediments showed that the most significant gradient was for pine trees and pollen,

which increased from north to south and downstream (Brush & DeFries, 1981). A similar gradient, though much smaller, was observed for both trees and pollen of sweet gum. A few other gradients were observed in either trees or pollen of the same taxa, but they were too small to indicate any significant difference in the distributions between the two.

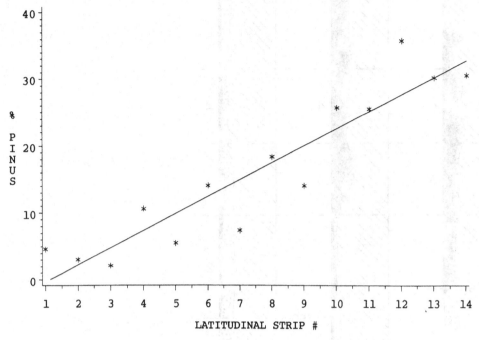

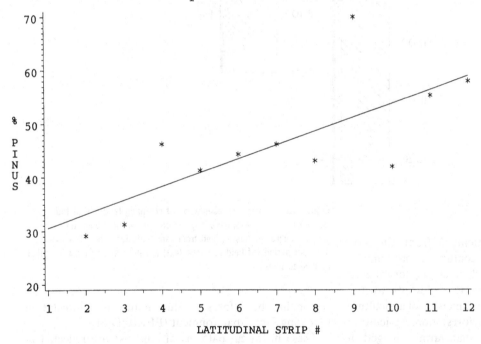

Figure 3.7. Regressions of pine trees and pollen in surface sediments against latitude.

Pollen–land use relationships

The close relationship between vegetation on the landscape and pollen in estuarine sediments allows reconstruction of land use from fossil pollen distributions.

On a regional scale, the history of land use in the Chesapeake Bay region can be divided into five major periods. The periods are not synchronous as settlement began at different times in different places, and different kinds of agriculture were practiced at different localities. Prior to European settlement, the entire region was forested except for tidal wetlands and small Indian clearings (Bruce, 1896; Stetson, 1956) (PERIOD 1).

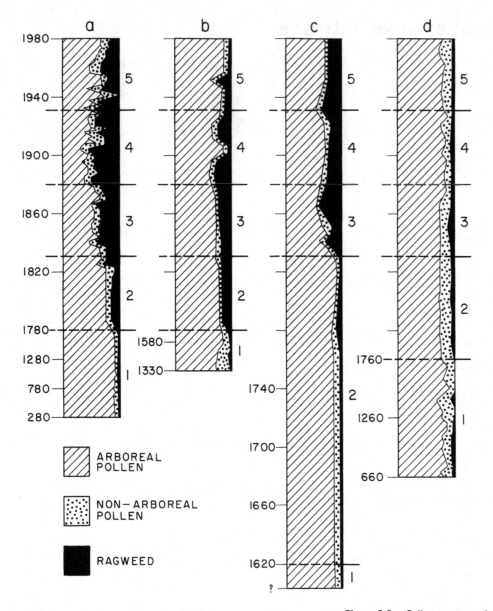

Figure 3.8. Pollen representation of changing land use at four locations in the Chesapeake Bay region. The vertical axis, time, is based on chronologies derived from sedimentation rates. The five different periods of land use (see text) are delineated from each other by dashed lines.

During initial European settlement from the late seventeenth to mid eighteenth century, approximately 20 to 30% of the land was cleared, much of it for tobacco farming (PERIOD 2). From the late eighteenth to mid nineteenth century, the continued expansion of agriculture resulted in approximately 40% deforestation. Agriculture consisted mainly of individual small farms engaged in both tobacco and grain farming (PERIOD 3). Beginning in the late nineteenth century, small farms were combined into predominantly large commercial operations. From then until the 1930s, when farms were abandoned because of depressed economic conditions, the amount of land under cultivation ranged from 60 to 80%, and included the added impact of widespread use of heavy machinery and a deep plow zone (PERIOD 4). Since the 1930s, approximately 40% of the land in the Chesapeake Bay

region has been forested, due mainly to afforestation following farm abandonment (PERIOD 5).

These historical patterns of land use are reflected in pollen distributions of arboreal and non-arboreal plants and ragweed found in sediment cores. Ragweed, a prolific pollen producer, colonizes freshly disturbed ground (Bazzaz, 1974). Consequently, an increase of ragweed pollen in the sediments is a reliable marker of deforestation followed by cultivation. Stratigraphic profiles of the proportion of arboreal pollen, non-arboreal pollen excluding ragweed, and ragweed pollen are shown for four locations (Fig. 3.8). These are an embayment in the

Upper Chesapeake Bay severely impacted by regional and local land use, including in addition to agriculture, eighteenth century charcoal production and mid twentieth century stone quarrying (Fig. 3.1a), a bog at the head of a tributary, also severely impacted by deforestation and agriculture (Fig. 3.1b), a tidal fresh water marsh downstream from mid twentieth century urban development, surrounded by an area of intensive agriculture since the mid nineteenth century, preceded by a long period of tobacco agriculture beginning in the mid seventeenth century; (Fig. 3.1c), and a tidal fresh water marsh with a relatively undisturbed local history (Fig. 3.1d).

In order to establish a chronology for the cores, the bottom layer of all cores, and up to three horizons in some cores were radiocarbon-dated. The agricultural horizon, recognized by increased ragweed concentrations, was identified and dated based on historical records of agricultural land use (Brush, 1984). Sedimentation rates were calculated for each 1- or 2-cm interval of the cores by adjusting the average sedimentation rate between dated horizons according to the pollen concentrations in the interval (Brush, 1989). The method assumes uniform pollen influx during relatively short time periods. Since pollen grains are Stokesian particles, hydrodynamically similar to fine sediment, and their transport in water is also similar to fine sediment, rates of sediment accumulation correspond inversely to pollen concentration. A chronology for each core was established by dividing the sedimentation rate by the depth of the interval, giving the number of years represented by that interval of sediment. The vertical time axes (Fig. 3.8) are based on these chronologies.

Pollen representation of regional land use is affected by local land use at each location, but at all locations, the demarcation between pre- and post-European settlement is clearly represented by increases in ragweed pollen (Fig. 3.8). The post-European patterns of land use are reflected by changes in ragweed and other non-arboreal pollen more or less proportional to the percentage of land deforested (Fig. 3.8). Tobacco farming (PERIOD 2) which extended for over 200 years at location c, is reflected by a uniform ragweed profile in the sediment core for the period from 1620 to 1840. The pollen profile for the more or less pristine tidal marsh (Fig. 3.8d) shows very little change, reflecting minimal land use following European settlement. The amount of non-arboreal pollen other than ragweed present in the pre-European profiles is higher in the bog and marsh environments than in the embayment, reflecting large populations of non-arboreal wetland species native to those areas.

Conclusions and discussion

Pollen grains, preserved in sediments deposited in partially mixed estuaries, provide an accurate record of the composition and abundance of vegetation on the landscape. Considerations of estuarine processes, hydrodynamics of pollen, laboratory measurements and field observations of transport and deposition of pollen provide the basis for this conclusion. This coupled with the fact that geographically estuaries span different vegetation and geologic gradients, provides a spatial framework for paleoecological studies not so readily available in other depositional environments.

As noted at the outset of this chapter, relatively few researchers have utilized the wealth of information stored in estuarine sediments. Nevertheless, the stratigraphic record of many estuaries is on a par with records found in lake and bog deposits. This may be due to the shallow nature of the estuary, and the suspected atmospheric source of the majority of the pollen. In these respects the estuary resembles small lakes more than large deep lakes. McAndrews and Power (1973), in their study of pollen distributions in Lake Ontario, found that pollen was less well preserved than in small lakes and attribute this degradation to the importance of stream rather than atmospheric input of pollen into the lake. Pollen is well preserved in estuarine sediments. Despite occasional storms and the constant twice daily flood and ebb of the tides (at least in the region of Mid-Atlantic USA) spatial and historical patterns of vegetation and land use are preserved in the pollen record, as is the case also in Lake Ontario (McAndrews & Power, 1973). Some deposits may be disturbed by bioturbation, but we have collected many undisturbed cores in the Chesapeake Bay. Pollen may be caught up in flocs in certain parts of the estuary, but this process, where it occurs, helps to keep pollen in place, thus preventing homogenization and preserving the spatial patterns of vegetation and land use.

The scenario is fairly simple. Our data suggest that most pollen enters the estuary from the atmosphere, with some possibly entering from rivers and streams. Other studies suggest a larger population originating from river and stream input (see Brush, 1989, for a discussion of differences in data and interpretations). Either way, once in the water, the pollen slowly settles as it is moved up and down the estuary by tides. Resuspension tends to disperse the pollen somewhat, but the ultimate fate of pollen more dense than water (the majority) is to settle to the bed within several kilometers of entry into the estuary.

The simplicity of the process is illustrated by field observations showing the close relationship between local vegetation and pollen in the sediment. The history of land use is documented by shifts in arboreal to non-arboreal pollen reflecting forested conditions on the landscape. If land use changes result in high rates of erosion and sedimentation, this is also captured in the sediment record by a lower concentration of pollen in the sediment (Brush, 1989). And finally, the observations reported here provide the basis for using the pollen record in estuaries like Chesapeake Bay for deciphering the long-term vegetation record in terms of climatic change.

While Chesapeake Bay and its tributaries are used as examples in this chapter, similar results could be expected in partially mixed estuaries throughout the world. The rates of transport and deposition may differ somewhat from estuary to estuary, but the basic processes are the same. With proper modification, the results can be used for studying relationships between pollen in the sediment and local vegetation in both highly stratified and well mixed estuaries.

Acknowledgments

We thank W. Fastie and R. Pelton of the Physics Department, The Johns Hopkins University, for help in setting up the telescopic lens for measuring fall velocities of pollen; L. Kim for measuring specific gravities; P. Thornton for computing the vegetation and pollen regressions, S. Cooper for the SEMs of pollen; J. Clark, S. Cooper, C. ReVelle and P. Wilcock for commenting on parts of earlier versions of the manuscript; and M. Farley and A. Traverse for many helpful comments and suggestions in the final version.

References

Bazzaz, F. A. (1974). Ecophysiology of *Ambrosia artemisifolia*: a successful dominant. *Ecology*, **55**, 112–9.

Biggs, R. B. & Howell, B. A. (1984). The estuary as a sediment trap. In *The Estuary as a Filter*, ed. V. S. Kennedy. New York: Academic Press, 107–29.

Bradshaw, R. H. W. & Webb III, T. (1985). Relationships between contemporary pollen and vegetation data from Wisconsin and Michigan, USA. *Ecology*, **66**, 721–37.

Bruce, P. A. (1896). *Economic History of Virginia in the Seventeenth Century*, Vols. 1 and 2. New York: Macmillan.

Brugam, R. B. (1978). Human disturbance and the historical development of Linsley Pond. *Ecology*, **59**, 19–36.

Brush, G. S. (1984). Patterns of recent sedimentation in Chesapeake Bay (Virginia–Maryland, USA) tributaries. *Chemical Geology*, **44**, 227–42.

Brush, G. S. (1986). Geology and ecology of Chesapeake Bay: a long-term monitoring tool for management. *Journal of the Washington Academy of Sciences*, **76** (3), 146–60.

Brush, G. S. (1989). Rates and patterns of estuarine sedimentation. *Limnology and Oceanography*, **34** (7), 1235–46.

Brush, G. S. & DeFries, R. S. (1981). Spatial distributions of pollen in surface sediments of the Potomac estuary. *Limnology and Oceanography*, **26** (2), 295–309.

Brush, G. S., Lenk, C. & Smith, J. (1980). The natural forests of Maryland: an explanation of the vegetation map of Maryland (with 1:250,000 map). *Ecological Monographs*, **50**, 77–92.

Burden, E. T., McAndrews, J. & Norris, G. (1986). Palynology of Indian and European forest clearance and farming in lake sediment cores from Awenda Provincial Park, Ontario. *Canadian Journal of Earth Sciences*, **23**, 55–65.

Camp, T. R. (1946). Sedimentation and the design of settling tanks. *Transactions, American Society of Civil Engineers*, **111**, 895–923.

Cordoba–Molina, J. F., Hudgins, R. R. & Silveston, P. L. (1978). Settling in continuous sedimentation tanks. *Journal of the Environmental Engineering Division, American Society of Civil Engineers*, **104** (EE6), 1263–75.

Cui, B. & Komar, P. D. (1984). Size measures and the ellipsoidal form of clastic sediment particles. *Journal of Sedimentary Petrology*, **54** (3), 783–97.

Davis, M. B., Spear, R. W. & Shane, L. C. K. (1980). Holocene climate of New England. *Quaternary Research*, **14**, 240–50.

Davis, R. B. & Jacobson, G. L. Jr. (1985). Late Glacial and Early Holocene landscapes in northern New England and adjacent areas of Canada. *Quaternary Research*, **23**, 341–68.

DeFries, R. S. (1986). Effects of land use history on sedimentation in the Potomac Estuary, Maryland. *US Geological Survey Water Supply Paper*, **2234–K**, 1–23.

Dyer, K. (1973). *Estuaries: A Physical Introduction*. London: John Wiley & Sons.

Groot, J. J. (1966). Some observations on pollen grains in suspension in the estuary of the Delaware River. *Marine Geology*, **4**, 409–16.

Harrington, J. B. & Metzger, K. (1963). Ragweed pollen density. *American Journal of Botany*, **50**, 532–9.

Hazen, A. (1904). On sedimentation. *Transactions, American Society of Civil Engineers*, **53**, 45–71.

McAndrews, J. H. & Power, D. M. (1973). Palynology of the Great Lakes: the surface sediments of Lake Ontario. *Canadian Journal of Earth Science*, **10**, 777–92.

McGlone, M. S. (1988). Report on the pollen analysis of estuarine cores from Whangapoua and Whitianga Harbours, Coromandel Peninsula. *Botany Division Report*, Canterbury Agricultural and Science Center, Christchurch, New Zealand.

McNown, J. S., Malaika, J. & Pramanik, H. R. (1951). Particle shape and settling velocity. *Proceedings, International Association for Hydraulic Research*, 4th Meeting, Bombay, 511–22.

Muller, J. (1959). Palynology of recent Orinoco delta and shelf sediments: reports of the Orinoco Shelf expedition, Vol. 5. *Micropaleontology*, **5** (1), 1–32.

Pritchard, D. W. (1967). Observations of circulation in coastal plain estuaries. In *Estuaries*, ed. G. H. Lauff. Washington, DC: American Association for the Advancement of Science, 37–44.

Pritchard, D. W. & Kent, R. E. (1953). The reduction and analysis of data from the James River Operation Oyster Spot. *Technical Report*, **6**, Chesapeake Bay Institute, The Johns Hopkins University.

Rubey, W. W. (1933). The size distribution of heavy minerals within a water-laid sandstone. *Journal of Sedimentary Petrology*, **3**, 3–29.

Schubel, J. R. (1968). Suspended sediment of the northern Chesapeake Bay. *Technical Report*, **35**, Chesapeake Bay Institute, The Johns Hopkins University.

Schubel, J. R. (1972). The physical and chemical conditions of Chesapeake Bay: an evaluation. *Special Report*, **21**, Chesapeake Bay Institute, The Johns Hopkins University.

Stetson, C. W. (1956). *Washington and His Neighbors*. Garrett & Massie.

Webb, T., III, Cushing, E. J. & Wright, H. E. Jr. (1983). Holocene changes in the vegetation of the Midwest. In *Late Quaternary Environments of the United States*, vol. II, *The Holocene*, ed. H. E. Wright, Jr. University of Minnesota Press, 142–65.

4 Pollen preservation in alkaline soils of the American Southwest

VAUGHN M. BRYANT, JR., RICHARD G. HOLLOWAY,
JOHN G. JONES and DAVID L. CARLSON

Introduction

Pollen analyses form the data base for many types of inferences ranging from sequential changes in past environments to lifestyles and diets of prehistoric human populations. In all cases, interpretation of pollen data must account for those factors that may have influenced the composition of the original pollen rain, and also for those factors that may have altered the composition of the pollen assemblage following deposition.

During the last 50 years palynologists have learned that there are many complex factors that determine the original composition of the pollen rain in an arid region. These include mode of pollination, differences in pollen production, differential dispersion patterns, and the size, weight, and aerodynamic ability of pollen types to remain airborne (see Chaps. 3, 13, 14, and other related chapters in this volume). Following deposition, other factors influence eventual loss or recovery of specific pollen types. These factors include pollen redeposition, the chemical composition of a pollen grain's exine, its morphological shape and types of surface ornamentation, and its susceptibility to various types of degradation processes including those from mechanical, chemical, or biological agents (Bryant, 1978, 1988; Bryant & Holloway, 1983; Holloway, 1989; O'Rourke, 1990). In this chapter we focus on the post depositional degradation processes.

One of the first agents that can affect pollen grains is mechanical degradation. After pollen is released from its source, it can become abraded or broken during the transportation phase. These alterations can result from impact or from changes in the natural environment. Studies by Duhoux (1982), for example, have shown that changes in atmospheric moisture levels can result in high numbers of exine ruptures in taxoid-type pollen taxa such as *Taxodium*, *Juniperus*, and *Thuja*. Later, after being deposited, taxoid-type pollen, as well as other types, can become further abraded by various causes in the natural environment. These include impact against objects, water and wind erosion, changes in temperature, changes in atmospheric or soil moisture contents, volcanic eruptions, and soil movement. Other factors that can cause damage to deposited pollen include the cultural activities of humans, such as land surface modifications, construction activities, and plowing of agricultural fields.

Morphological structure and ornamentation of pollen walls seem to be important factors in determining their potential susceptibility to mechanical degradation. For example, pollen grains having protruding structures, like the bladders of many conifer species or the spines of some Malvaceae grains, have a tendency for their projections to break off or erode through a variety of mechanical processes. In some cases, the actual appearance of a pollen grain may become so altered after the loss of an appended structure or structures, that accurate identification is no longer possible. In addition, structural alteration by mechanical means can also cause severe exine weakening, thereby facilitating further abrasion or even the destruction of the entire grain through other processes. Brooks and Shaw (1968), Shaw (1971), Rowley and Prijanto (1977), and Rowley (1990) have examined various aspects of pollen grain chemistry and exine composition and have found that differences in sporopollenin composition and molecular structure can make pollen grains either more or less resistant to chemical deterioration.

Using the effects of pH as an example, Dimbleby (1957) was the first to chart differences in pollen preservation caused by soil chemistry. His research demonstrated that soils with a low, acidic pH are ideal vectors for pollen preservation, while fossil pollen in sediments with a pH above 6.0 is often degraded or completely destroyed. Since Dimbleby's original study (1957), other studies conducted in arid regions of the American Southwest by Martin (1963), Bryant (1969), and Hall (1981, 1991) have demonstrated that fossil pollen can be recovered from alkaline soils with a pH as high as 8.9. Even when this is possible, however, the recovered pollen has often deteriorated, making accurate pollen analyses difficult, or, in some cases, nearly impossible.

Related to Dimbleby's (1957) original work on pH is

Tschudy's (1969) research on the Eh (oxidation potential) of sediments. Tschudy asserts that Eh actually may be a more important guide to the eventual preservation or destruction of palynomorphs than pH. Low Eh reflects a reducing, anaerobic environment where carbon dioxide and hydrogen sulfide are the byproducts of microbe respiration and combine to decrease the pH values. Thus, in some sediments the creation of a negative Eh potential results in the formation of a strongly reducing environment (Tschudy, 1969). Because a reducing environment retards oxygen retention, low Eh environment becomes ideal for pollen preservation. On the other hand, an oxidizing sediment with high Eh speeds the destruction of pollen.

The chemical composition of pollen walls and pollen wall structural morphology also play important roles in determining whether pollen grains will remain preserved in various sediments. In a 20-year study beginning in 1964 and ending in 1984, Havinga (1964, 1984) reported that the relationship between amounts of sporopollenin and cellulose in the pollen exines seems to affect their susceptibility to eventual destruction through oxidation. He found, for example, that pollen grains having high percentages of sporopollenin in their walls tend to remain preserved longer, even in soils with high pH and Eh values, than do pollen grains with walls composed mostly of cellulose.

Biological agents, such as fungi and bacteria, can cause pollen grain degradation. Recent studies (Holloway, 1981) show that some taxa of phycomycete fungi seek out and feed on the nutrient materials in the cytoplasm of pollen grains. His experimental studies show that fungal mycelia often enter a pollen grain through natural apertures, although they are also capable of dissolving areas of the exine in order to enter the grain (Fig. 4.1). Both types of attack contribute to the eventual destruction of pollen grains by creating new holes or enlarging tiny cracks in the exine, thus weakening the overall grain and making it more susceptible to other forms of degradation.

Some years earlier, phycomycete fungi were investigated by Goldstein (1960) and Elsik (1966), who found they were a causative factor in the destruction of pollen. Data from Goldstein's initial study show that some taxa of Phycomycetes are selective in their preference for pollen types and will infect certain pollen taxa at a much faster rate than others. For example, he found that pollen grains from certain species of coniferous trees, especially *Pseudotsuga*, were attacked much more frequently by phycomycete fungi than were types of angiosperm pollen. Unlike Holloway's (1981) study, Goldstein did not focus on how fungi actually damage pollen grains. Instead, his data concluded only that pollen from many conifer taxa are the most susceptible types to fungal infection and, thus, by inference, eventual destruction.

Elsik (1971) noted that bacterial degradation of pollen grains also occurs. He found that certain bacteria, especially types of Actinomycetes, attack pollen walls in a definite pattern. He found that in some cases this type of bacterial destruction can continue to occur long after pollen grains have lost their cytoplasm and have become preserved in sediments for thousands, or even millions, of years.

Data base

Fossil pollen samples

Between 1983 and 1988, we carried out a survey of geological and archaeological sediments along a proposed oil and gas pipeline route in areas just north of the United States–Mexico border (Fagan, 1989). The proposed route began on the Pacific Coast near Santa Barbara, California, and extended eastward to the town of McCamey, in west Texas, a distance of 7338 km (Fig. 4.2). The initial purpose for collecting and examining these sediments, which ranged in age from 1000 to 5000 years old, was to provide a paleoenvironmental record and to recover culturally important information about the prehistoric inhabitants who once lived along the proposed pipeline route. The 509 fossil samples we examined from this transect form one of the more extensive sampling records of arid land sediments yet attempted in the United States and provided us with an opportunity to examine differences in pollen preservation over a wide geographical region of the American Southwest.

All fossil pollen samples were processed in the following manner. First, soil samples of 20–25 cm^3 were spiked with $22\,600 \pm 800$ *Lycopodium* spores to enable us to determine pollen concentration values. We used *Lycopodium* spores because the genus is rarely found naturally in arid regions in the southwestern part of North America. Previous palynological studies in this region, as noted by Hall (1985), reveal no evidence that *Lycopodium* ever grew in these arid regions during the last 5000 years.

All of the fossil sediment samples were collected from environments having a pH value higher than 6.0. Thus, anhydrous carbonates (e.g., $CaCO_3$, $MgCO_3$) were among the most common compounds found in these samples. The first step of processing removed these with concentrated hydrochloric acid. The second step focused on removing small rocks and coarse-grained silicates by decanting. Each sample was placed in a 4000 ml beaker filled with distilled water and then stirred in all directions to keep pollen suspended. The liquid fraction was then quickly poured into another beaker and saved. This process was repeated three times for each sample. The samples were then treated with hydrofluoric acid for 24 hours to remove residual finer-grained silicates. Samples were then deflocculated in a weak, non-foaming detergent,

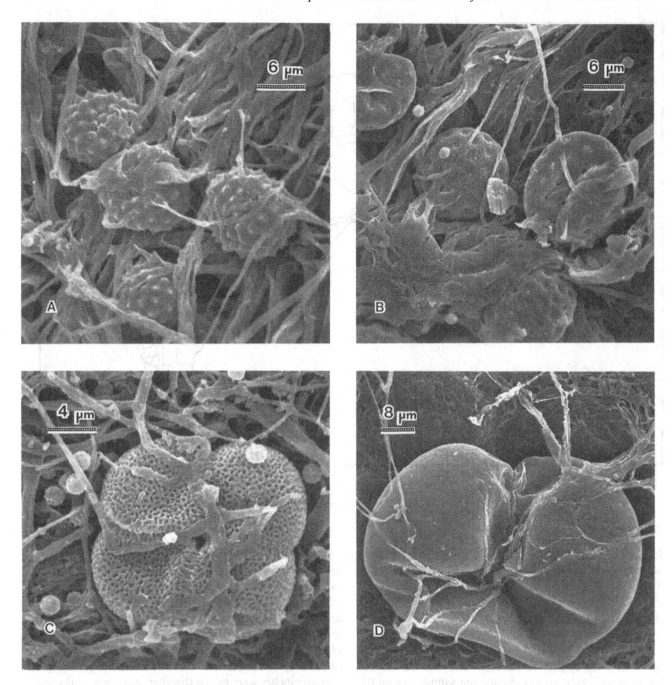

Figure 4.1. Scanning electron micrographs of pollen infected with fungi: (A) *Iva angustifolia* D.C., infected with *Aspergillus niger*; (B) *Chenopodium album* L., infected with *Fusarium* spp.; (C) *Typha latifolia* L., infected with *Aspergillus niger*; (D) *Picea pungens* Engelm., infected with *Fusarium* spp.

then sonicated for two minutes in a Delta D-5 sonicator. This was followed by density separation with zinc bromide (s.g. 2.0), a process that separated most of the remaining detritus from the pollen. Because most of these soils had a high Eh potential, little organic debris was present, and it was not necessary to remove non-polliniferous plant material. Processed materials were stained in safranin and mounted in glycerin. Fossil pollen counts of 200+ fossil grains were attempted for each sample. Each separated pine bladder was counted as half a grain. Grains recognizable as being pollen, but too badly broken or degraded to be identified, were included in the counts as indeterminate. Cryptogamic spores and the *Lycopodium* spores used to spike each sample were counted, but not included as part of the pollen sum.

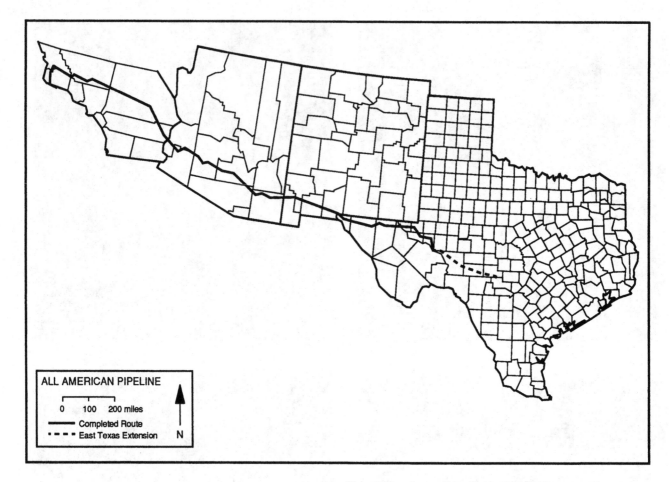

Figure 4.2. Route of the All American Pipeline across the southwestern portion of the US.

Modern pollen surface samples

In 1987, as part of this study, we analyzed 90 surface samples collected from locations in west Texas (Fig. 4.3). These served as modern, control samples and formed a data base for comparative studies with the information derived from the fossil samples. We selected arid regions of west Texas as the site for our modern pollen surface samples because no surface samples were taken by those who made the original fossil sediment collections. Later, we were unable to collect surface soil samples along the pipeline route because of on-going construction and lack of access to the original sampling sites. Climatic and edaphic studies of the pipeline route (Ponczynski, 1989) suggest that these conditions in the areas of west Texas, from which we collected surface samples, are very similar, although not identical in all cases, to those of the original 509 samples we examined. Based on these data, we feel pollen comparisons between the studies are warranted.

Surface samples were collected along east–west and north–south transects (Fig. 4.3). Because much of the land in this region is under private ownership, most of the collection sites were located near major highways in the region. At intervals of approximately 15 km, we sampled areas no closer than 200 m from the highway. Each surface sample was collected using the 'pinch' method (Adam & Mehringer, 1975). This was done by selecting a sampling area approximately 50 m wide by 100 m long and then walking back and forth within the area collecting 'pinches' of surface soil. Each pinch of soil was combined with the other 20 or more pinches from the same sampled locale and all pinches were mixed together in a single collecting bag, to prevent the possibility of over-representation of a single pollen type from any one of the pinch samples. Bags were labeled, sealed, and then frozen to prevent microbe activity. Tests using the pinch method (Adam & Mehringer, 1975) reveal that more than eight pinches are needed from each surface sample to ensure a reliable pollen assemblage of the regional vegetation.

The same laboratory extraction and counting procedures used for the fossil sediments were used for the modern pollen surface samples. The only difference was an added step of screening each sample through a screen with mesh openings of 250 μm after the hydrofluoric acid (HF) treatment to remove small fragments of organic material.

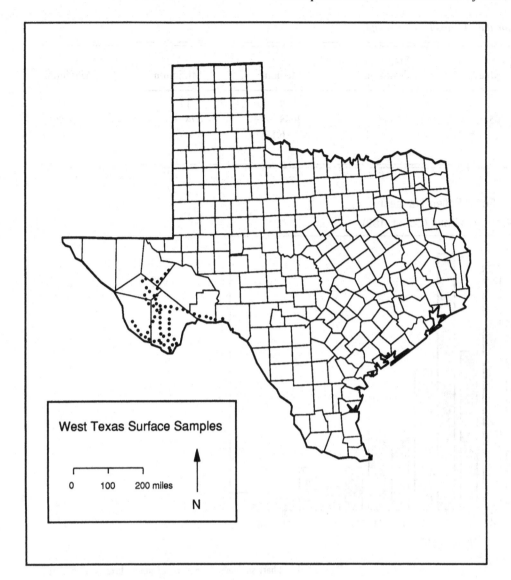

Figure 4.3. Location of collection sites for each of the 90 modern pollen surface samples.

Results

Fossil pollen samples

Of the 509 samples collected for analysis, only 243 (48%) contained sufficient fossil pollen to conduct statistically valid counts in excess of 200 grains, as recommended by Barkley (1934), Martin (1963), and others. We found the remaining 266 samples (52%) were almost entirely devoid of fossil pollen. The small amounts of pollen found in the productive samples represented: (1) taxa known to be highly resistant to various agents of destruction, (2) types that could still be identified even though they were severely degraded, or (3) highly degraded pollen grains that were no longer identifiable. This type of limited pollen recovery phenomenon is common in soils of the American southwest and has been previously reported by Hall (1981), Holloway (1981), Bryant and Schoenwetter (1987) and Bryant and Hall (1993).

The average number of pollen taxa per fossil sample was only 7.5, and the maximum of 17 was observed in only three samples. Two of the samples contained only one pollen type each (Fig. 4.4, Table 4.1). The five most frequent types in these samples were (1) *Pinus*, (2) cheno-ams [a combined term used to represent pollen taxa in both the family Chenopodiaceae and the genus *Amaranthus*], (3) genera of the Asteraceae family (including high and low spine composites, *Artemisia*, and Liguliflorae types), (4) *Ephedra*, and (5) various genera of the Poaceae family. Although these five pollen types represent plants common in the floral communities of the region sampled, they also represent distinctive pollen types that can be recognized easily even after being severely degraded.

Pollen concentration values for each of the 243 fossil

Table 4.1. *Descriptive statistics for the 243 fossil pollen samples*

	Mean	Standard deviation	Median	Minimum	Maximum
Number of taxa	7.5	3.01	7	1	17
Pollen concentration	6545	13 302	3688	674	165 828
Percent Indeterminable	7.6	6.51	6.1	0	34.1

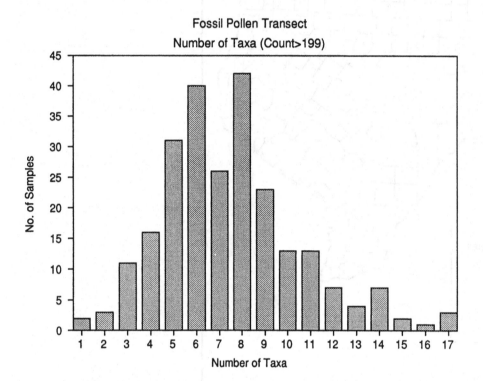

Figure 4.4. Chart showing the number of pollen taxa in each of the 243 fossil pollen samples.

pollen samples were generally low (Table 4.1). That the distribution is skewed to the right is indicated by the difference between the mean and median statistics. While the mean concentration is 6545 grains/cm^3, half the samples have fewer than 3688 grains/cm^3. Table 4.1 also shows that the mean percentage of indeterminable pollen was 7.6 and that the distribution of these values is skewed as well. As we will show later, there is a tendency for the highest percentages of indeterminate pollen to come from samples with the lowest concentration values.

Pollen listed in the indeterminable category included grains that were broken, corroded, degraded or folded in ways that made their identification impossible. This category included grains that were so broken or fragmented that accurate identification was impossible, grains that had so many areas missing from corrosion that accurate identification was no longer possible, grains that were so thinned and degraded that their accurate identity remained in question, and pollen that was so folded their identity remained uncertain. If pollen grains

were broken, corroded, degraded, or folded, yet still identifiable, they were counted and included in the taxa to which they belonged and were not added to the indeterminable category. Although we did not split the indeterminable category into the groups listed above for counting purposes, almost all of the indeterminable grains in our study were those in the degraded category.

Modern pollen samples

All 90 of the surface pollen samples collected for this study contained sufficient pollen to count more than 200 grains per sample. We have excluded one of these 90 samples from our analyses because it is unlike the rest and contained an extremely high pollen concentration value (sample 22, concentration 238 430). For the remaining 89 samples, the mean number of taxa per sample was 17.4

Table 4.2. *Descriptive statistics for the 89 surface pollen samples*

	Mean	Standard deviation	Median	Minimum	Maximum
Number of taxa	17.4	3.04	18	11	24
Pollen concentration	21 311	13 854	17 374	2517	70 625
Percent indeterminable	6.6	3.23	6.2	0.8	15.9

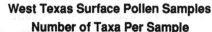

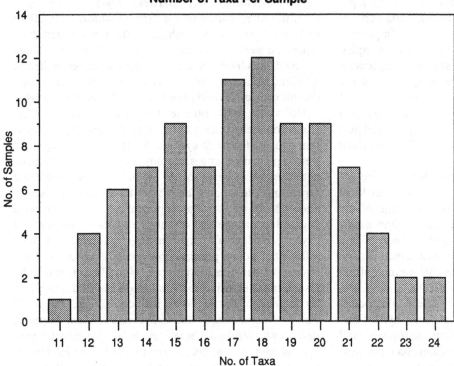

Figure 4.5. Chart showing the number of pollen taxa in each of the 90 modern pollen samples.

(Fig. 4.5, Table 4.2). A difference of means test comparing the average number of taxa from the fossil and surface samples shows a significant difference between the two groups ($t = 26.53$, $p < 0.0001$).

Pollen concentration values are significantly higher for the surface samples ($t = 8.86$, $p < 0.0001$), with mean values roughly three times those found among the fossil samples. Although the sample sizes are sufficiently large that the skewed distributions are not likely to affect the results of the difference of means test, a nonparametric difference of medians test (the Wilcoxon 2-sample test) was also used to compare the samples. Again the difference between the samples was significant ($p < 0.0001$).

Even though the modern samples represented pollen that had been deposited fairly recently, some of the grains already showed evidence of severe degradation. This is why the average percentage of indeterminable pollen is only slightly smaller for the surface samples. The difference is significant according to the Student's t test ($t = 2.03$, $p < 0.05$), but not according to the more conservative Wilcoxon test ($p < 0.84$).

Discussion

For over five decades palynologists have been seeking to understand pollen composition and the causes of pollen degradation. However, pollen wall morphology and chemical composition are still being debated, and we do not yet understand fully the elements that determine pollen destruction or preservation in different environmental settings. Our present study of fossil and modern pollen samples from arid regions in the American West and Southwest is a case in point. We suspect many factors (i.e., pH, Eh, bacteria, fungi, soil chemistry, mechanical agents, soil moisture, temperature, etc.) may have contributed to the destruction of fossil pollen in the samples we examined for this study.

We believe the various agents of pollen destruction may play different roles in different geographic regions and depositional settings. However, of these, we suspect that oxidation, soil chemistry, and mechanical degradation are most important in causing pollen loss in soil sediments of arid regions.

Soils in the arid regions of the American southwest are generally alkaline, with pH above 6.0. Many have a pH higher than 8.5. We suspect that the high alkalinity of the soils, and the resulting high pH values, are important causative factors that lead to pollen loss in fossil samples. To test this hypothesis, we conducted a linear regression analysis of our fossil pollen samples comparing the pH and pollen concentration values. However, this test revealed no apparent correlation. A similar regression test comparing the pH and the percentage of indeterminable grains in each fossil sample also failed to show any apparent correlation.

We feel that the lack of correlation between the pH values and the pollen concentration values of our fossil pollen samples is a result of the narrow range of pH values represented by these samples rather than a true test of the role of pH as a destructive agent. Demonstration of a correlation between pH and pollen concentration will require samples which span the entire range of pH values, as was noted by Dimbleby (1957).

We believe a more important factor determining pollen preservation is the frequency of soil hydration–dehydration. Various forms of pollen destruction and deterioration seem to be linked to phenomena associated with the evaporative process of changing a liquid to a gas (Holloway, 1989). Experiments conducted by Burstyn and Bartlett (1975) showed that significant pressure is exerted on a curved surface of an organic-walled, hollow sphere at the instant when water is transformed, by evaporation, from a liquid to a gas. This pressure phenomenon would be especially critical for water-filled, tiny, spherical structures like pollen grains. We believe that these forces could cause major structural damage to the exine of pollen grains. Each time the soil hydration–dehydration process occurs, pollen in the soil is subjected to two potentially destructive processes: (1) the expansion–contraction process caused by being wetted or dried, and (2) the pressure phenomenon described by Burstyn and Bartlett (1975). The more frequently this cycle occurs in a soil, the greater the potential for fossil pollen to be destroyed or to deteriorate, first by developing stress cracks in the exine, and next by breakdown of the pollen wall through mechanical processes.

Many soils in the temperate regions of the world are subjected to periodic wetting and drying and yet are able to retain enough moisture throughout each cycle to prevent complete dessication. When this water retention in soils occurs, the hydration–dehydration effect upon buried pollen is not severe. As long as some moisture is present in soils, capillary action keeps buried pollen grains from fully drying out, and thus prevents them from undergoing the shock and stress associated with the change from hydration to dehydration. Likewise, soils that remain dry are also ideal for pollen preservation, because they also do not subject fossil pollen to the destructive processes of the hydration–dehydration cycle.

In most of the semi-arid and desert regions of the world, moisture often comes in the form of sudden rainfall or early morning dew. In both cases, the hot sun generally bakes the wetted soil completely dry within a few hours or few days. Under these circumstances, buried pollen in arid land soils would be subjected to repeated hydration–dehydration cycles and the associated destructive stresses.

Holloway (1989) examined the destructive effects of the hydration–dehydration cycle on pollen grains. He noted that, after as few as 25 cycles of wetting and drying, there were significant changes and noticeable amounts of exine deterioration in many of the 14 pollen taxa he selected for testing. He found that 76% of the fresh pollen tested and 86% of the acetylated pollen tested showed exine degradation at the end of only 25 cycles of wetting and drying. Holloway's experiments also indicated how differential pollen preservation occurs as a result of the hydration–dehydration process. Of the 14 pollen taxa he tested, those showing the greatest degree of degradation by the end of the 25 cycles included *Carya*, *Juniperus*, *Populus*, *Pseudotsuga*, *Salix*, *Typha*, and *Zea*. Types of deterioration included breakage, surface corrosion, severe folding, and general decomposition. Other pollen taxa in the same experiment, such as *Iva* and *Amaranthus*, also showed some degradation but were still recognizable.

These experimental results are consistent with observations of fossil pollen assemblages. In the early 1970s a fossil pollen study of the Hershop Bog, central Texas, revealed excellent pollen preservation covering a time span of more than 10 000 years (Larson *et al.*, 1972). The only exception was the absence of well-preserved pollen in the uppermost 80 cm of peat, representing the last 2000–3000 years of deposition. As explained by the authors, trenching of the bog's margins to drain the area was begun in the early 1940s. That effort failed but lowered the water table from the surface down to the 80 cm level. During the 50 years since the bog surface was first lowered, periodic rains have flooded the bog many times, subjecting the upper 80 cm to repeated cycles of hydration–dehydration. It was presumably these cycles of hydration–dehydration, combined with subsequent processes such as mechanical breakdown and oxidation, that led to the destruction of fossil pollen in the uppermost

80 cm section of the bog. Hall (1981) noted the same phenomenon in deposits of the Hughes peat bed in Iowa.

The sediments in many dry rockshelters are often excellent sources of fossil pollen, yet at the Levi rockshelter in central Texas, the soils proved barren of fossil pollen (Bryant, 1969). Later, it was discovered that there had been a seep spring along the back wall of the rockshelter. Although inactive today, travertine deposits around the seep indicate it had been active over a long period of time. We suspect that during each rainy season the seep spring probably wet the rockshelter's soils with oxygenated water, resulting in cycles of hydration–dehydration and oxidation, which led to eventual pollen destruction.

In another rockshelter, the Meadowcroft Rockshelter in Pennsylvania, pollen and other organic materials were reportedly preserved in the dry and protected soils of the site, but were destroyed in the shelter's soils located outside the drip line (Adovasio & Carlisle, 1984). The difference in preservation between fossil pollen in the soils inside and outside the rockshelter's drip line is attributed mainly to differences in cycles of hydration–dehydration followed by accompanying destruction processes such as mechanical breakdown and oxidation. Whereas the soils outside the drip line were subjected to cycles of hydration–dehydration, the soils inside the drip line remained dry, thus preserving their fossil pollen.

Data from Holloway's (1989) experimental study were also useful in understanding the results of our fossil pollen transect study. As mentioned earlier, Holloway noted that *Iva* and *Amaranthus* were still recognizable even when badly degraded. Two of the five most common pollen taxa found in our fossil pollen transect were types that included composites similar to *Iva* and taxa in the group termed cheno-ams, which includes *Amaranthus*.

As discussed in our results section, we found that the average number of pollen taxa and the average pollen concentration value for all fossil pollen samples were lower than those of our modern samples. Because both sets of samples were collected from similar environmental settings, we believe the differences in pollen taxa number and lower concentration values are the result of differential pollen degradation. This assumption is supported by the percentage of indeterminable pollen grains found in each of these two groups.

To examine the possible correlations between the percentage of indeterminable pollen and pollen concentration values, we conducted a regression analysis using pollen concentration as the independent variable. Because the data distributions were skewed (Tables 4.1 and 4.2), the logarithms of each value were used instead of the raw values. Four samples with zero percentage indeterminable values were replaced with a value of 0.39 percent. The linear regression of the logarithms was then converted into a nonlinear equation. For the fossil pollen group,

Table 4.3. *Expected percentage of indeterminable pollen*

Concentration values	Fossil pollen[a]	Surface pollen[b]
1 000	9.4	11.5
2 000	6.8	9.7
4 000	4.9	8.2
8 000	3.5	6.9
12 000	2.9	6.3
16 000	2.6	5.9
20 000	2.3	5.5
25 000	2.1	5.3
30 000	1.9	5.0
40 000	1.7	4.7
50 000	1.5	4.4
75 000	1.2	4.0
100 000	1.1	3.7

[a] Percentage indeterminable $= 242.87 \times \text{Concentration}^{-0.470\,56}$.
[b] Percentage indeterminable $= 62.64 \times \text{Concentration}^{-0.244\,76}$.

the correlation between log (Percent Indeterminable) and log (Concentration) is 0.39, which is statistically significant ($p < 0.0001$). Although the correlation is not as strong as we expected, it does demonstrate that as concentration values decrease, the percentage of indeterminable pollen increases. Figure 4.6 shows the observed values and the regression line for the concentration values between 0 and 30 000 grains/cm^3. Concentration values above 4000 grains/cm^3 should have less than 5% indeterminable pollen, although the data are quite variable until concentration values of 12 000 grains/cm^3 are reached. The same correlation for the surface pollen samples is 0.27, which is also statistically significant ($p < 0.0101$). Figure 4.7 shows the observed values and the regression line. Table 4.3 shows the predictions for each equation for a range of concentration values. Although the fossil pollen equation gives lower percentage indeterminable predictions than the surface pollen equation, the differences are not statistically significant.

Our correlations between the percentage of indeterminate pollen and the pollen concentration values in fossil pollen samples from arid lands are supported by data noted by Hall (1981). In his initial study, Hall examined pollen preservation from a number of soil profiles in the southwestern USA, and found a link between low concentration values and high percentages of indeterminate pollen grains. Although Hall did not conduct regression tests as part of his study, he did note that there was a trend towards increasing pollen destruction and lower pollen concentration values when the percentages of indeterminant pollen increased.

Thus, as Hall (1981) pointed out, and as we have confirmed, fossil pollen assemblages in arid land regions

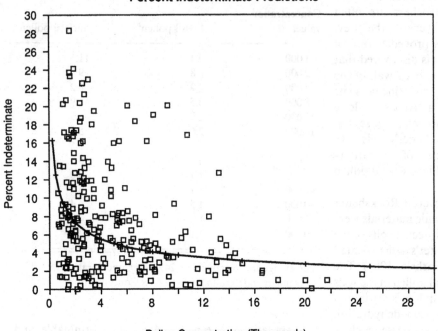

Figure 4.6. Regression analysis of fossil pollen data testing the percentage of indeterminate pollen against the pollen concentration values.

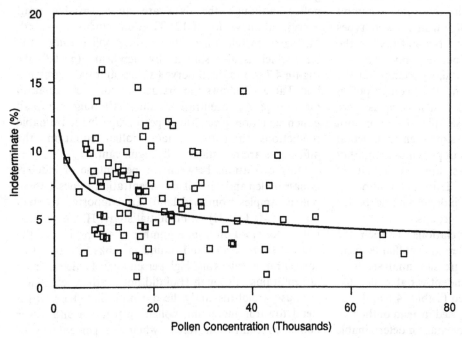

Figure 4.7. Regression analysis of modern pollen data testing the percentage of indeterminate pollen against the pollen concentration values.

are altered by progressive deterioration. In other words, we believe that palynologists working with arid land sediments should be concerned when the percentages of indeterminate pollen types increase rapidly in samples to levels approaching 50%. Furthermore, as both Hall's study and ours have demonstrated, the presence of low fossil pollen concentration values in arid land soils, combined with high percentages of indeterminate grains, generally indicates a loss of pollen by deterioration. In most cases those losses will be differentially distributed among the various fossil pollen taxa, and will show the greatest effect on the frequency of the most abundant taxa.

Summary

For years archaeologists and palynologists have been examining soils from both geological and archaeological sites in the arid regions of southern North America. Hundreds of research papers now exist that report on fossil pollen analyses from this region (Hall, 1985), and these papers vary widely in the extent to which the possibility of differential pollen preservation is acknowledged. Our results are not the first time these concerns have been expressed about pollen studies from arid land soils (Hall, 1981; Holloway, 1981; Bryant & Schoenwetter, 1987; Horowitz, 1992). However, our present study is different from some of the earlier ones, because we examined a broad range of fossil pollen samples covering a wide geographical region.

We do not propose that all fossil pollen data collected from arid land regions of the world must be discarded. We suggest, instead, that while conducting routine analyses, it is essential to obtain estimates of pollen concentration and to record the percentage of indeterminate grains present in each sample. Once these data are collected, it is easier to determine if a pollen assemblage from a particular lithological setting has been severely altered. Once that is known, palynologists should then be able to assess the reliability of the pollen data and identify which data sets are inaccurate and which can be effectively interpreted as being reliable.

In conclusion, our examination of pollen samples from a wide area of the arid American Southwest reveals:

(1) significant pollen deterioration occurs in most buried sediments;
(2) fossil pollen deterioration occurs differentially, thus distorting the preserved record;
(3) severe pollen deterioration in sediments can be recognized by the low number of taxa present, the types of taxa that remain identifiable, the percentage of indeterminant grains, and the low pollen concentration values of the sediments.

References

Adam, D. P. & Mehringer, P. J., Jr. (1975). Modern pollen surface samples – an analysis of subsamples. *Journal of Research of the US Geological Survey*, **3**, 733–6.

Adovasio, J. M. & Carlisle, R. C. (1984). An Indian hunter's camp for 20,000 years. *Scientific American*, **25**, 130–6.

Barkley, F. A. (1934). The statistical theory of pollen analysis. *Ecology*, **47**, 439–47.

Brooks, J. & Shaw, G. (1968). Chemical structure of the exine of pollen walls and a new function for carotenoids in nature. *Nature*, **219**, 523–4.

Bryant, V. M., Jr. (1969). Late full-glacial and post-glacial pollen analysis of Texas sediments. Ph.D. dissertation (Dept. of Botany), The University of Texas, Austin, TX.

Bryant, V. M., Jr. (1978). Palynology: A useful method for determining paleoenvironments. *Texas Journal of Science*, **45**, 1–45.

Bryant, V. M., Jr. (1988). Preservation of biological remains from archaeological sites. In *Interdisciplinary Workshop on the Physical–Chemical–Biological Processes Affecting Archaeological Sites*, ed. C. Mathewson. US Army Corp of Engineers Waterways Experiment Station, Vicksburg, MS, 85–115.

Bryant, V. M., Jr. & Hall, S. A. (1993). Archaeological palynology in the United States: a critique. *American Antiquity*, **58**, 277–86.

Bryant, V. M., Jr. & Holloway, R. G. (1983). The role of palynology in archaeology. In *Advances in Archaeological Method and Theory #6*, ed. M. Schiffer. New York: Academic Press, 191–224.

Bryant, V. M., Jr. & Schoenwetter, J. (1987). Pollen records from Lubbock Lake. In *Lubbock Lake Late Quaternary Studies on the Southern High Plains*, ed. E. Johnson. College Station, TX: Texas A & M University Press, 36–40.

Burstyn, H. P. & Bartlett, A. A. (1975). Critical point drying: application of the physics of the PVT surface to electron microscopy. *American Journal of Physics*, **43**, 414–9.

Dimbleby, G. W. (1957). Pollen analysis of terrestrial soils. *New Phytologist*, **56**, 12–28.

Duhoux, E. (1982). Mechanism of exine rupture in hydrated taxoid type of pollen. *Grana*, **21**, 1–7.

Elsik, W. C. (1966). Biologic degradation of fossil pollen grains and spores. *Micropaleontology*, **12**, 515–8.

Elsik, W. C. (1971). Microbiological degradation of sporopollenin. In *Sporopollenin*, ed. J. Brooks, P. Grant, M. Muir, P. van Gijzel, and G. Shaw. New York: Academic Press, 480–511.

Fagan, B. (1989). Introduction. In *Cultural Resources Report for the All American Pipeline Project*, ed. All American Pipeline Company. Las Cruces, NM: New Mexico State University, 1–9.

Goldstein, S. (1960). Destruction of pollen by Phycomycetes. *Ecology*, **41**, 543–5.

Hall, S. A. (1981). Deteriorated pollen grains and the interpretation of Quaternary pollen diagrams. *Review of Paleobotany and Palynology*, **32**, 193–206.

Hall, S. A. (1985). Bibliography of Quaternary palynology in Arizona, Colorado, New Mexico and Utah. In *Pollen Records of Late-Quaternary North American Sediments*, ed. V. Bryant and R. Holloway. Dallas: American Association of Stratigraphic Palynologists, 407–23.

Hall, S. A. (1991). Progressive deterioration of pollen grains in south-central U.S. rockshelters. *Journal of Palynology* (1990–1991), 159–64.

Havinga, A. J. (1964). Investigations into the differential corrosion susceptibility of pollen and spores. *Pollen et Spores*, **6**, 621–35.

Havinga, A. J. (1984). A 20-year experimental investigation into the differential corrosion susceptibility of pollen and spores in various soil types. *Pollen et Spores*, **26**, 541–58.

Holloway, R. G. (1981). Preservation and experimental diagenesis of the pollen exine. Ph.D. dissertation, Texas A & M University (Botany), College Station, TX.

Holloway, R. G. (1989). Experimental mechanical pollen degradation and its application to Quaternary age deposits. *Texas Journal of Science*, **41**, 131–45.

Horowitz, A. (1992). *Palynology of Arid Lands.* London: Elsevier.

Larson, D. A., Bryant, V. M., Jr. & Patty, T. S. (1972). Pollen analysis of a central Texas bog. *American Midland Naturalist*, **88**, 358–67.

Martin, Paul S. (1963). *The Last 10,000 Years: A Fossil Pollen Record of the American Southwest.* Tucson, AZ: University of Arizona Press.

O'Rourke, M. K. (1990). Pollen reentrainment: contributions to the pollen rain in an arid environment. *Grana*, **29**, 147–52.

Ponczynski, J. (1989). The environmental setting. In *Cultural Resources Report for the All American Pipeline Project*, ed. All American Pipeline Company. Las Cruces, NM: New Mexico State University, 150–81.

Rowley, J. R. (1990). The fundamental structure of the pollen exine. In *Morphology, Development, and Systematic Relevance of Pollen and Spores*, ed. M. Hesse & F. Ehrendorfer. *Plant Systematics and Evolution-Supplementa*, **5**, 13–9.

Rowley, J. R. & Prijanto, B. (1977). Selective destruction of the exine of pollen grains. *Geophytology*, **7**, 1–23.

Shaw, G. (1971). The chemistry of sporopollenin. In *Sporopollenin*, ed. J. Brooks, P. Grant, M. Muir, P. van Gijzel & G. Shaw. New York: Academic Press, 305–50.

Tschudy, R. H. (1969). Relationship of palynomorphs to sedimentation. *Aspects of Palynology*, ed. R. H. T. & R. Scott. New York: John Wiley & Sons, 79–96.

5 Wind and water transport and sedimentation of miospores along two rivers subject to major floods and entering the Mediterranean Sea at Calvi (Corsica, France)

MAURICE STREEL and CLAIRE RICHELOT

Introduction

In the Mediterranean region, several studies have been devoted to the distribution of pollen in coastal marine areas (Rossignol, 1961; Koreneva, 1971; Rossignol & Pastouret, 1971; Belfiore et al., 1981; Brun, 1983). Other studies have tried to understand the pollen rain in mountain areas where peat bogs are excellent pollen recorders (Reille, 1975; Beaulieu, 1977). However, little is known about how much of the pollen produced in high and middle altitudes is transported to the area of sedimentation near the shoreline, and how it is transported.

The region of Calvi in Corsica is typical of Mediterranean climatic conditions, where mountains rise to 2000 m a short distance (20 km) from the coast, with rather humid climate at the top (near 2000 mm precipitation, concentrated during the cold season; Simi, 1964), and rather dry climate (near 700 mm precipitation) near the coast. Autumn, winter, or spring high rainfall on the top and on steep slopes (Fig. 5.1) produces huge and violent floods (Guilcher, 1979), which carry large stones that accumulate on the narrow coastal plain and which also deliver fine material to the sea. During the dry season (summer) the rivers are at minimum flow and often vanish in the coarse sediments of the coastal plain. Two small rivers enter the sea near Calvi, the Fiume Secco and the Ficarella (Fig. 5.1). Both cross a succession of distinctive vegetation zones (Richelot & Streel, 1985) present on the slopes (Fig. 5.1):

(1) *at the lower level*, a xerophytic 'maquis' of Cistaceae and Asteraceae with a few riverside forests of alder (*Alnus glutinosa*); there are also cultivated olive trees (*Olea europaea*), more in the Fiume Secco than in the Ficarella areas;

(2) *at the intermediate level*, forests of pine (*Pinus pinaster* subsp. *hamiltonii*) and oak (*Quercus ilex*), with shrubs of heath (*Erica arborea*); this forest is more widespread in the Ficarella than in the Fiume Secco areas;

(3) *at a higher level*, forests of pine (*Pinus nigra* subsp. *laricio*), a forest more widespread in the Fiume Secco than in the Ficarella areas. Alpine moor-heaths ('Landes') with peat bogs top the mountains.

Sampling

Samples were taken from various sediments on the bottom of the sea in Calvi Bay and on land, in the rivers and between them. The nearshore bottom of the sea is too sandy to have accumulated pollen grains except where *Posidonia oceanica*, a marine angiosperm grass, is developed in thick mats (Bay, 1984). The dense vertical growth of its axes allows this grass to avoid being covered by sediment, and it traps fine sediment which otherwise would be carried offshore by the marine currents. Twelve samples were taken in the uppermost 5-cm-thick layer of the *Posidonia* mats (Fig. 5.2).

Two samples of muddy sands were available from the center (Fig. 5.2) of Calvi Bay (Burhenne, 1981): sample 19 at 70 m depth, 2 km offshore, and sample 75 at 45 m depth, 500 m offshore. These two samples probably represent more time than any shallow samples taken nearshore or on land, because they correspond to several cm of sediment, but they were helpful to check the relative proportions of pollen of the selected taxa in offshore sediments. On land, samples were taken between the rivers at eight localities and sub-localities (Fig. 5.3). All are mosses covering granitic rocks at levels never reached by any flood. Each locality is a mean sample averaged from several small samples in an area of about 5 m². Twenty-five samples were taken from the bottom, or more commonly, near the sides of the two rivers (Fig. 5.2).

Maceration and counting methods and display of results

All samples were sieved with water on a 200 μm sieve, the fine material being macerated with 10% KOH for 15 minutes and then with cold 38% HF, continuously stirred

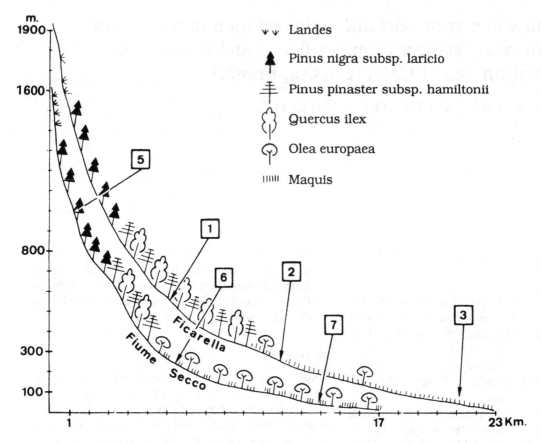

Figure 5.1. Sample localities and vegetation along the two rivers entering the Mediterranean Sea near Calvi. 'Landes' refers to moor-heaths. (From Richelot & Streel, 1985.)

for 12 hours and again sieved at 61 μm to eliminate the abundant organic matter. Material retained on this sieve was checked for any significant amount of pollen and was found almost barren. The fine material was then acetolysed and the residue filtered on a 11 μm sieve. Three hundred pollen grains were counted in each sample, more when one of the taxa was over-represented. Forty taxa were identified, 20 being present in most samples. The complete record (Richelot, 1984) is available from the Palynological Laboratory of the University of Liège, Belgium. In this chapter we will be concerned with only eight taxa: (A) *Pinus nigra*, (B) *Pinus pinaster*, (C) *Quercus ilex*, (D) Oleaceae, (E) *Alnus*, (F) Poaceae, (G) *Erica arborea*, and (H) Cistaceae and Asteraceae. These are the most abundant kinds of pollen recorded and represent the different kinds of land environments described above. The relative value of each is given in percentage of the total. The results from the two marine muddy sands are given individually (Fig. 5.4), but the data from the *Posidonia* mats, which happened to be quite homogeneous, are given as mean values (Figs. 5.2, 5.4). On land, the eight localities of samples taken between the rivers and the 25 samples from the river bottoms are grouped in three altitude classes for each of the rivers (Figs. 5.2, 5.3). We made one exception, however, grouping in the same class (Fig. 5.2, locality 4) river samples from short, lateral tributaries of the two main

rivers, because these short tributaries do not originate from high altitude in the mountains.

Local pollen rain and transport by winds

Winds blow from the mountains during the night and from the sea during the day, and one might wonder whether these quite regular winds have the effect of making the pollen rain uniform throughout the region from the coast to the mountain. The analysis of the samples taken away from the rivers (Fig. 5.3, localities 8–12) is most helpful in finding an answer to this question. Transport of pollen by winds is indeed obvious. Small amounts of pollen (less than 20%) of Oleaceae are present at localities 8, 9 and 11, although the producing plants live only at lower levels. Smaller amounts were noted by Reille (1975) in the peat bogs on the top of the mountains and were explained by wind transport.

Small amounts of *Pinus nigra* are present in the lowest localities, 10 and 12, although the producing plants live only higher in the mountains. Because these localities are

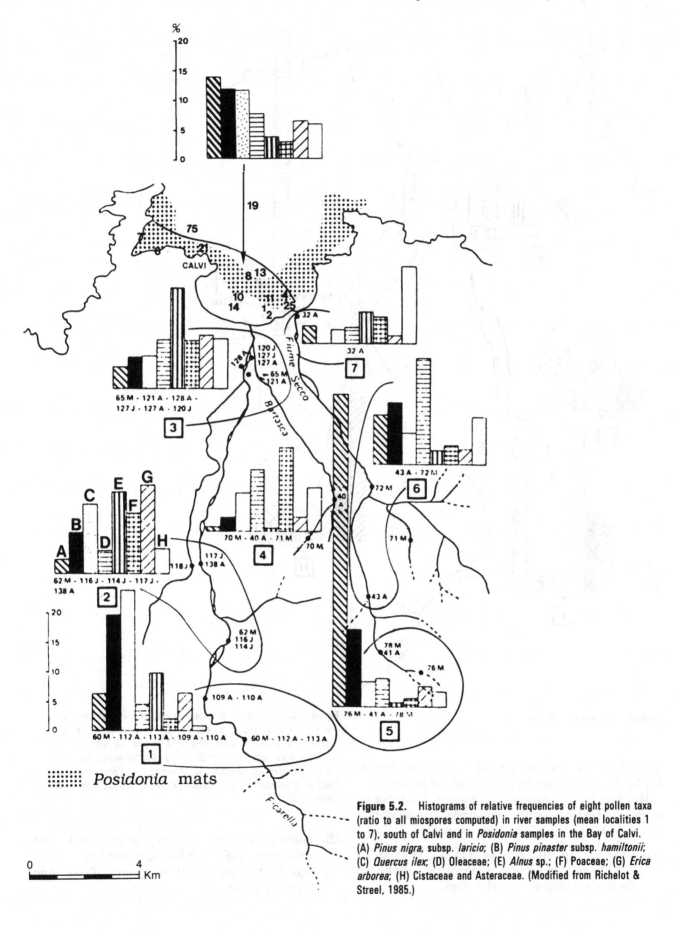

Figure 5.2. Histograms of relative frequencies of eight pollen taxa (ratio to all miospores computed) in river samples (mean localities 1 to 7), south of Calvi and in *Posidonia* samples in the Bay of Calvi. (A) *Pinus nigra*, subsp. *laricio*; (B) *Pinus pinaster* subsp. *hamiltonii*; (C) *Quercus ilex*; (D) Oleaceae; (E) *Alnus* sp.; (F) Poaceae; (G) *Erica arborea*; (H) Cistaceae and Asteraceae. (Modified from Richelot & Streel, 1985.)

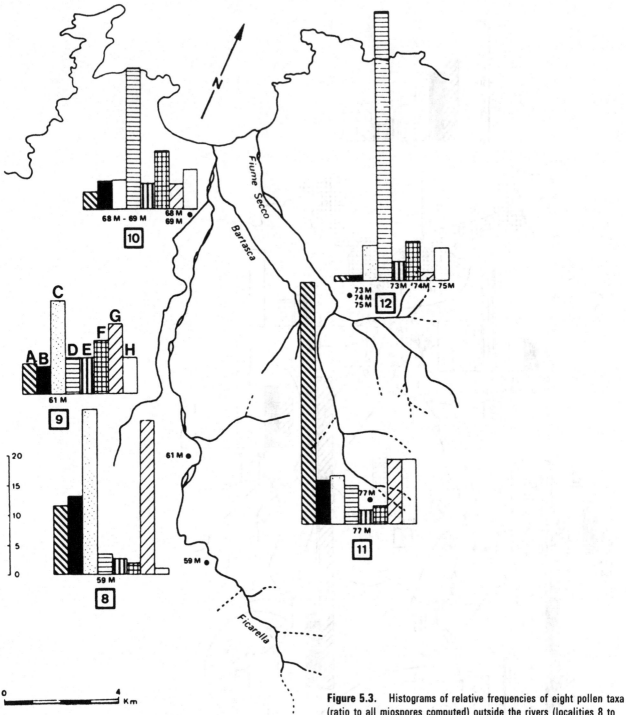

Figure 5.3. Histograms of relative frequencies of eight pollen taxa (ratio to all miospores computed) outside the rivers (localities 8 to 12), south of Calvi. Localities 10 and 12 include sub-localities. (A) *Pinus nigra*, subsp. *laricio*; (B) *Pinus pinaster* subsp. *hamiltonii*; (C) *Quercus ilex*; (D) Oleaceae; (E) *Alnus* sp.; (F) Poaceae; (G) *Erica arborea*; (H) Cistaceae and Asteraceae. (Modified from Richelot & Streel, 1985.)

never flooded by the rivers, wind transport is obvious. So, both day and night winds carry pollen grains.

However, the dominant pollen types are not transported by these rather long-ranging winds. Instead, such pollen is produced by the locally dominant vegetation:

(1) *lower level* (*plain*), Oleaceae (Fig. 5.3, localities 10 and 12), with a larger proportion where (in the Fiume Secco area) *Olea* is more abundant;

(2) *intermediate level of Ficarella area*, *Quercus ilex* and *Erica arborea* (Fig. 5.3, localities 8 and 9);

(3) *high level of the Fiume Secco area*, *Pinus nigra* (Fig. 5.3, locality 11).

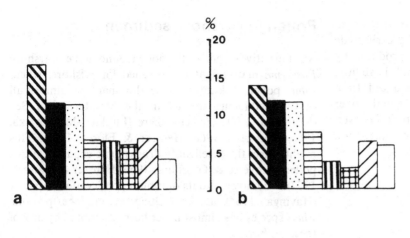

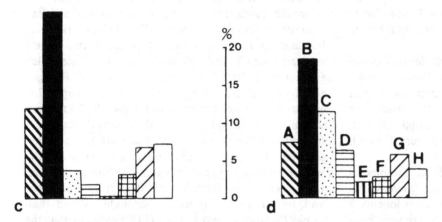

Figure 5.4. Histograms of relative frequencies of eight pollen taxa (ratio to all miospores computed): (a) mean of four river sample localities (1, 2, 5, 6) in the intermediate and high level; (b) mean of all samples in the *Posidonia* mats. (c) sample 19, at 70 m depth and 2 km offshore. (d) sample 75, at 45 m depth and 500 m offshore. (See Fig. 5.2 for localities of samples 19 and 75.) (A) *Pinus nigra*, subsp. *laricio;* (B) *Pinus pinaster* subsp. *hamiltonii;* (C) *Quercus ilex;* (D) Oleaceae; (E) *Alnus* sp.; (F) Poaceae; (G) *Erica arborea;* (H) Cistaceae and Asteraceae.

Pollen transport and sedimentation in the rivers

Pollen transport by the rivers depends on the energy level of the flow. We have noted that large floods, under the climatic–topographic conditions of this region, carry the fine sediments directly to the sea. Most of the coastal onshore deposits are too coarse to contain any pollen. The deposition of pollen with muddy sediments on the bottom and sides of steep rivers is possible only under conditions of low energy current, when the rivers are near minimum flow. The data shown in Figure 5.2 reflect this condition.

Analyses of sediment from locality 5 (Figs. 5.1 and 5.2) are typical for the results obtained from sediments taken

in the Fiume Secco at high altitude. Percentages are rather similar to those described away from the river (Fig. 5.3, locality 11), except that the proportion of conifers is higher in the river sediment than away from the river, probably reflecting the well-known capacity of bisaccate pollen to be easily carried by water. This is confirmed by the data obtained from the intermediate level in the same river. Indeed, locality 6 (Figs. 5.1 and 5.2) has significant proportions of bisaccate pollen when compared to the nearby localities 4 (Fig. 5.2) and 12 (Fig. 5.3). Locality 4 combines the data obtained from sediments taken in short, lateral tributaries to the main river, which do not originate high in the mountain and therefore were not fed by the montane conifer forest pollen rain, where *Pinus nigra* is dominant. Locality 12 demonstrates that the pollen rain there is very poor in bisaccate pollen. The bisaccate pollen found at locality 6, therefore, must have been transported downstream by the Fiume Secco River.

Along the Ficarella River, two groups of localities at the intermediate level were analyzed (Fig. 5.2, localities 1 and 2). When compared with the corresponding sites away from the river (Fig. 5.3, localities 8 and 9), larger proportions of *Alnus* pollen are noted in the river sediments. These proportions increase downstream, as confirmed by locality 3, the nearest to the coast. *Alnus* pollen is from a dominant tree of the riverside forest, and the local pollen rain is mainly produced by this tree and accumulates downstream. Winds have little influence in this transport, as can be deduced from comparison of the data at locality 3 (Fig. 5.2), in the river sediment, and locality 10 (Fig. 5.3), away from the river, where *Alnus* pollen is largely overwhelmed by other taxa.

Pollen delivery into the sea

In low-energy conditions, when the streams contain little water, only the Ficarella has a permanent, very slow delivery of surface water to the sea. The Fiume Secco, most often, vanishes into its own coarse alluvium before reaching the sea.

Should this low but almost permanent delivery of the Ficarella feed the pollen content of the *Posidonia* mats immediately nearshore, we should expect the pollen spectrum of the coastal locality 3 (Fig. 5.2) to match the pollen spectrum found in the *Posidonia* samples. In fact, it does not. The relative values (Fig. 5.2) are very different, almost complementary, rather than similar.

The pollen spectrum in the *Posidonia* samples is dominated by *Pinus nigra*, *P. pinaster* and *Quercus ilex*, which are represented by the lowest values at locality 3 onshore. On the contrary, *Alnus* pollen is rare nearshore but dominates in the onshore distal alluvium of the Ficarella. A sorting effect at such short distance might explain the bisaccate abundance offshore, but not the dominance of *P. nigra* over *P. pinaster* pollen and the high frequency of *Quercus ilex* in the *Posidonia* samples. Most probably, this pollen is delivered to the sea by the irregular large floods occuring in autumn, winter or spring. The *Posidonia* pollen spectrum is indeed more similar to the intermediate and upstream river samples than to any samples from the low level. We suggest that these intermediate and high level sediments are recycled by the floods. The contribution of the Fiume Secco to these floods is certainly important because of the dominance of *P. nigra* pollen in the marine nearshore sediments, the related forest being mostly developed upstream of this river. Locality 10 (Fig. 5.3), 3 km from the coast, has a pollen spectrum very different from that of *Posidonia*, which rules out direct wind transport from high and middle altitude to the sea. If wind were a factor, we would expect the pollen rain to be similar at nearby localities on both sides of the shoreline. By mixing in equal parts the pollen contribution found in the four intermediate and high level localities 1, 2, 5 and 6 of both rivers, one would obtain a mean spectrum very similar to the *Posidonia* spectrum (Fig. 5.4).

Pollen in offshore sediments

An alternative origin for the pollen found in the nearshore *Posidonia* mats might be searched for offshore, where *Pinus* pollen is known to be dominant in almost all bottom sediments studied in the Mediterranean Sea (Koreneva, 1971) and elsewhere (Lubliner-Mianowska, 1962; Koreneva, 1968; Heusser & Florer, 1973). This might reflect the excellent hydrodynamic capacity of this pollen (Traverse & Ginsburg, 1966; Heusser & Balsam, 1977) and maybe also its optimum preservation capacity (Havinga, 1964). Distinguishing among the possible *Pinus* species has almost never been attempted by any of these authors.

Could offshore water currents introduce to Calvi Bay the pollen found trapped in the *Posidonia* mats? The two muddy sands that we have analyzed from Calvi Bay demonstrate that *P. pinaster* is (Fig. 5.4) two times as abundant as *P. nigra*, bisaccate pollen being largely dominant over all other pollen (14.3% of bisaccate pollen of size intermediate between *P. pinaster* and *P. nigra* should be added to sample 19, 9.5% to sample 75). Bisaccates are more common in sample 19 (2 km from the coast) than in sample 75 (500 m from the coast). If one excludes the *P. pinaster*/*P. nigra* ratio, the pollen spectrum of sample 75 (500 m) is very close to the *Posidonia* mat pollen spectrum and therefore might have, in part, a common origin. Sample 19 might, on the other hand, be interpreted as more thoroughly sorted than sample 75. Heusser and Balsam (1977) consider that the decrease in proportion of *Alnus* pollen in the offshore direction might be the result of poor adaptation to water transport and of poor resistance to sea water. Brun (1983) also observed that the proportions of Oleaceae pollen decreased in the offshore direction. Therefore, we interpret the muddy-sand pollen spectrum as originating from the coastal discharge of the two rivers. In these muddy sands, bisaccate pollen was probably selectively sorted and also mixed to some extent with *P. pinaster*-type pollen carried by Mediterranean Sea currents.

Discussion and conclusion

Our results support the conclusion that most of the pollen trapped in the mixed mineral and organic littoral deposits formed by the *Posidonia* mats originates from the intermediate and high altitude coastal vegetation. It is mostly delivered to the sea by river systems during large floods. The Fiume Secco, which drains a large, high altitude, *P. nigra* forest, plays a major part in this delivery process. The very slow but permanent delivery of surface water to the sea by the Ficarella should carry, by

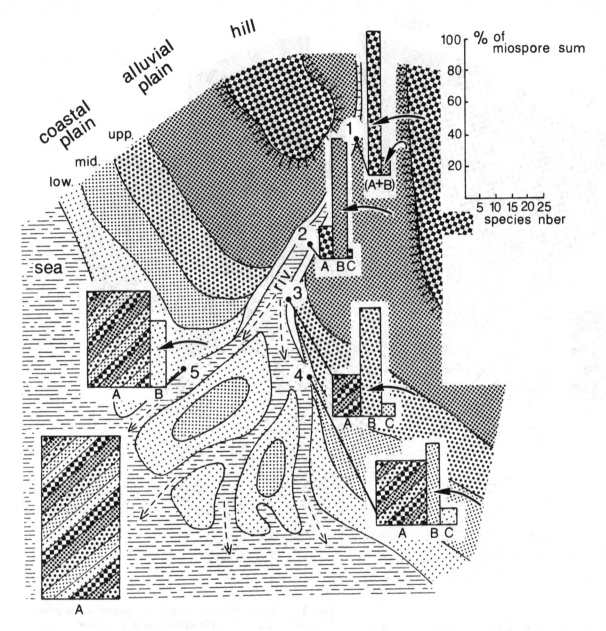

Figure 5.5. Reconstruction of wind and river transport of miospores to the sea in a deltaic environment with well developed coastal and alluvial plains. (A) wind and river transport from upstream; (B) local wind transport; (C) wind transport from downstream. (See also Hopping, 1967).

comparison, only a small quantity of pollen, and the proportions of taxa of this small supply are very different from those observed in the *Posidonia* mats.

Delivery to the sea of miospores through the fluviatile system has been studied by Hopping (1967), who developed a model for the distribution pattern of Tertiary pollen in a deltaic environment. By combining the data provided by Hopping (1967), we have tried to determine the progressive downstream enrichment of species (x-axis of the histograms in Fig. 5.5, as well as the changing

proportion of the miospores originating upstream (y-axis of the histograms in Fig. 5.5). This model shows extended coastal and alluvial plains, features which are not matched near Calvi (Fig. 5.6), where the supply of sediments to the sea has no true deltaic character.

Among the similarities between the models, the local vegetation provides most of the miospore content in each locality. Winds carrying spores and pollen for long distances are obviously less important than local pollen rain. We also note that the miospore spectra are distorted by fluviatile supply originating upstream. Among the differences, it should be emphasized that we have not observed near Calvi the progressive downstream enrichment of species noted in the ancient deltaic environment. Moreover, the sporomorph proportions

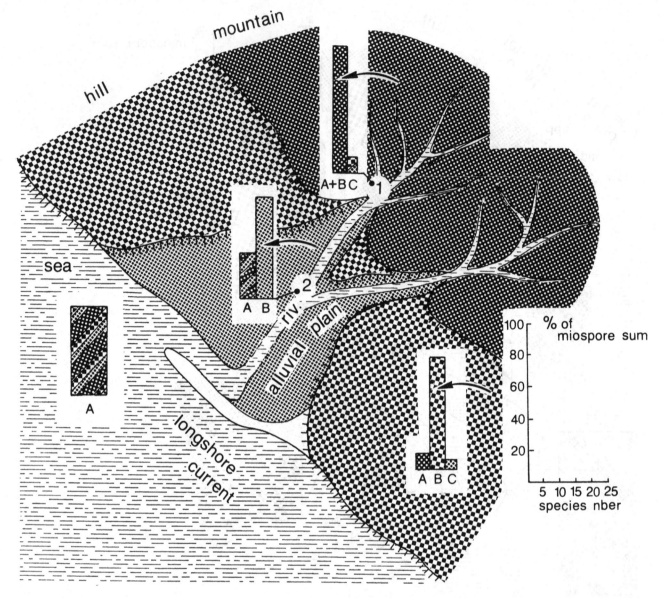

Figure 5.6. Reconstruction of wind and river transport of miospores to the sea in a coastal mountainous environment with poorly developed coastal and alluvial plains, such as that in the Calvi area. (A) wind and river transport from upstream; (B) local wind transport; (C) wind transport from downstream.

seen nearshore do not show, in this environment, equivalent supply from all upstream vegetation, but, near Calvi, a more important supply from upland mountainous vegetation.

Important and direct supplies to the sea of miospores originating upstream and/or upland have been discussed by several authors (even for the Paleozoic), and these data are well summarized by Chaloner and Muir (1964). These authors refer to the 'Neves effect,' a mechanism of miospore dispersion, 'in which miospore assemblages change successively in response to varying proximity and extent of the parent communities brought about by changing base-level.'

This 'Neves effect' mechanism is based on the assumption that the relative total quantity of miospore production by communities within the region concerned, carried by the wind, would dictate the composition of

the miospore assemblage in the nearshore marine sediments. It does not emphasize the possible importance of water transportation of miospores, and, moreover, does not take into account the possibility that relatively remote upland vegetation might directly feed the nearshore marine sediments through the river system, as it does indeed at the present time near Calvi.

A study of spore assemblages in Jurassic rocks of various lithologies (Muir, in Chaloner & Muir, 1964) has given results more consistent with the present situation near Calvi than with the 'Neves effect' *sensu stricto*. In this Jurassic delta, the marine rocks and the coarse fluviatile sandstones are rich in miospores of plants that were apparently growing in a habitat remote from the marine environment. They represent water-transported miospores from the hinterland. We want to exclude this kind of mechanism of miospore dispersion in the Jurassic from the 'Neves effect' *sensu stricto*, and propose two new concepts where miospore assemblages are more dependent on water transportation than on wind transportation:

(1) The 'Hopping effect' refers to a mechanism resulting in assemblages which are dominated by water-transported miospores mainly derived from nearby coastal and alluvial plains, as well as from remote 'upland.'

(2) The 'Muir effect' refers to a mechanism resulting in assemblages which contain dominant water-transported miospores mainly derived from the remote 'upland.'

Being dependent on the regime of the river systems, both mechanisms are probably more closely related to climatic changes than to base-level changes, which mainly explain the 'Neves effect.'

References

Bay, D. (1984). A field study of the growth dynamics and productivity of *Posidonia oceanica* (L.) Delile in Calvi Bay, Corsica. *Aquatic Botany*, **10**, 43–64.

Beaulieu, J.-L. de (1977). Contribution pollenanalytique à l'histoire tardiglaciaire et holocène de la végétation des Alpes méridionales françaises. Thèse ès Sciences, Univ. Aix-Marseille III.

Belfiore, A., Dambion, F., Moncharmont, M., Ozer, A., Pescatore, T., Streel, M. & Thorez, J. (1981). La sédimentation récente du golfe de Tarente (Italie méridionale). Aspects minéralogiques et micropaléontologiques. *Bulletin de la Société Royal des Sciences de Liège*, **50** (11–12), 373–83.

Brun, A. (1983). Etude palynologique des sédiments marins holocènes de 5000 BP à l'actuel dans le golfe de Gabès (mer pélagienne). *Pollen et Spores*, **3–4**, 437–60.

Burhenne, M., (1981). Faciès sédimentaires du précontinent calvais. Notes introductives. *Bulletin de la Société Royal des Sciences de Liège*, **50** (11–12), 387–404.

Chaloner, W. & Muir, M. (1964). Spores and floras. In *Coal and Coal-bearing Strata*, ed. D. G. Murchison & T. S. Westoll. Edinburgh: Oliver & Boyd, 127–46.

Guilcher, A. (1979). *Précis d'hydrologie marine et continentale*. Paris: Masson.

Havinga, A. J. (1964). Investigation into the differential corrosion susceptibility of pollen and spores. *Pollen et Spores*, **4**, 621–635.

Heusser, C. J. & Florer, L. E. (1973). Correlation of marine and continental Quaternary pollen records from the northeast Pacific and western Washington. *Quaternary Research*, **3**, 661–70.

Heusser, L. E. & Balsam, W. L. (1977). Pollen distribution in the northeast Pacific Ocean. *Quaternary Research*, **7**, 45–62.

Hopping, C. A. (1967). Palynology and the oil industry. *Review of Palaeobotany and Palynology*, **2**, 23–48.

Koreneva, E. V. (1968). Distribution and preservation of pollen in sediments in the western part of the Pacific Ocean. *Geological Bulletin* (Department of Geology, Queens College, Flushing, NY), **2**, 1–17.

Koreneva, E. V. (1971). Spores and pollen in Mediterranean bottom sediments. In *The Micropaleontology of Oceans*, ed. B. M. Funnell & W. R. Reidel. New York: Cambridge University Press, 361–71.

Lubliner-Mianowska, K. (1962). Pollen analysis of the surface samples of bottom sediments in the Bay of Gdansk. *Acta Societatis Botanicorum Poloniae*, **31**, 305–12.

Reille, M. (1975). Contribution pollenanalytique à l'histoire tardiglaciaire et holocène de la végétation de la montagne corse. Thèse ès Sciences, Univ. Aix-Marseille III.

Richelot, C. (1984). Le transport du pollen par les courants aériens, fluviatiles et marins dans la région de Calvi (Corse). *Mémoire de Licence en Sciences Géographiques, Université de Liège*.

Richelot, C. & Streel, M. (1985). Transport et sédimentation du pollen par les courants aériens, fluviatiles et marins à Calvi (Corse). *Pollen et Spores*, **3–4**, 349–64.

Rossignol, M. (1961). Analyse pollinique de sédiments marins quaternaires en Israël, 1. Sédiments récents. *Pollen et Spores*, **3**, 303–24.

Rossignol, M. & Pastouret, L. (1971). Analyse pollinique de niveaux sapropéliques post-glaciaires dans une carotte en Méditerranée orientale. *Review of Palaeobotany and Palynology*, **11**, 227–38.

Simi, P. (1964). Le climat de la Corse. *Bulletin de la Section de Geographie/Comité des Travaux Historiques et Scientifiques* **76**. Paris: Imprimerie nationale.

Traverse, A. & Ginsburg, R. (1966). Palynology of the surface sediments of Great Bahama Bank, as related to water movement and sedimentation. *Marine Geology*, **4**, 417–59.

6 Sedimentation of land-derived palynomorphs in the Trinity–Galveston Bay area, Texas

ALFRED TRAVERSE

Introduction

The research presented here was originally done in 1958–1962, when I was working for Shell Development Company in Houston, Texas. It represents the final segment of investigations I carried out for Shell on the palynology of modern sediments. The basic information for this last sub-project was submitted to Shell as an internal report after I left the company in 1962. The overall project included studies of a carbonate platform (the Great Bahama Bank; Traverse & Ginsburg, 1966), and the Trinity River – Trinity Bay – Galveston Bay clastic depositional complex of Texas (Traverse, 1990, 1992). The lower Trinity river system was selected for study because: (1) the river system was at that time relatively natural, undisturbed by dams, (2) Shell Development Company geologists had studied sedimentation in the area, and (3) the lower river and associated bays were easily accessible from my place of employment.

That these research results are presented so long after they were obtained can be justified only in that they represent a previously unpublished, unique data-set for a medium-sized relatively undisturbed river and associated bays, smaller streams, and adjacent nearshore parts of the Gulf of Mexico. Inasmuch as I have published results of studies of palynomorphs in the water of part of the area (Traverse, 1990, 1992), it also seems desirable to complete the story by presenting information available for the palynomorph content of surface sediment of the region. It should be emphasized that throughout this chapter I deal with the status of palynomorph sedimentation in the area as it was about 1960. How it may have changed by 1992 I could only speculate, although I have shown (Traverse, 1990, 1992) that the completion in 1968 of a very large dammed lake, Lake Livingston, about 200 km upstream from the mouth of the Trinity River, has significantly affected the palynomorph content of river water downstream from the dam. This chapter describes distribution of pollen and pteridophyte spores and associated microfossils (i.e., those found in the same preparations) in the uppermost layer of sediment in the Galveston–Trinity Bay area, including fungal spores and some algal remains, fragments of wood and bark ('tracheary fragments'), and leaf epidermal fragments ('leaf cuticles'). The pollen, pteridophyte spores, fungal spores, and algal cysts would now be included in the category 'palynomorph,' and the tissue fragments would now usually be called 'palynodebris'.

General information about the study area

The location of the area studied with reference to North America in general is displayed in Figure 6.1. Figure 6.2 shows the layout of the area in some detail, emphasizing the drainage features and other bodies of water from which samples were obtained. (Detailed modern information on sedimentation, vegetation, and other general features of the area has been published by White et al., 1985.)

Trinity–Galveston Bay is a very shallow body of water, mostly less than 3 m. It was formed by the drowning of the Trinity and San Jacinto River valleys, which had entrenched during the previous low sea level (Bernard et al., 1959). Trinity Bay has very little tidal action, mean tide being less than 30 cm. In 1962, the depth of East Bay averaged 1.0–1.5 m, and that of West Bay about 1.2–2.0 m. As has been pointed out by Shepard (1953), most parts of this bay system were filling with sediment. West Bay and southernmost Galveston Bay were exceptions, as they were being sharply flushed by currents (partly tidal), and West Bay was actually deepening. According to Lohse (1955), occasional, enormously powerful, intrusions of air ('northers') in the winter were the principal agents of flushing. As shown in Figure 6.2, the shipping channels and Intracoastal Waterway were

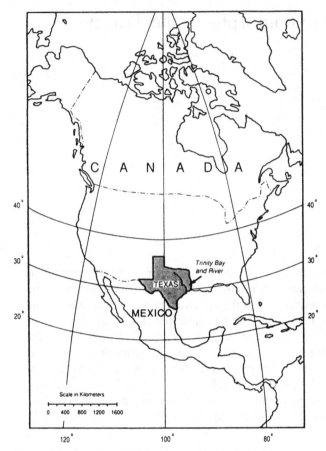

Figure 6.1.   Location of Trinity River and Bay in North America.

in 1962 being kept clear by frequent dredging operations. The main channel of the Trinity River in its delta had also been regularly dredged for over a century.

The Trinity River in this region was in 1960 a moderately fast-moving stream, becoming quite swift after heavy rains. The annual average runoff at Oakwood, a little more than halfway from Anahuac (near the mouth of the river; see Fig. 6.2) to Dallas, was 3 702 000 acre-feet, about the same as the Brazos River at a corresponding distance from the Gulf of Mexico. Trinity Bay, into which the river empties, forms one large bay with Galveston Bay, into which the San Jacinto River and Buffalo Bayou (Houston Ship Channel) empty. This combined bay is almost sealed off from the Gulf by Galveston Island and the Bolivar Peninsula. In the 1960s, the water of Trinity Bay was only slightly salty, but there was a salt-wedge tendency, with salty water from the Gulf protruding below the fresh water brought down by the river. The silt load of the Trinity was relatively light, estimated at 4000 acre-feet per year at Romayor in 1962. This can be compared with 22 000 acre-feet in the Brazos River at Richmond, a roughly equivalent station on a stream, the mouth of which, at Freeport, is about 70 km southwest of Galveston.

The San Jacinto River, the only other sizable stream, and the other streams ('bayous') of the area have relatively small discharge rates and silt loads compared to the Trinity River. The Trinity River has been one of the major conduits for sediment to the East Texas–Louisiana shelf in the late Pleistocene (Mazzullo, 1986).

East Bay (see Fig. 6.2) was partially shut off from the main bay system by Hanna Reef, an oyster bank, and was, therefore, filling up quite rapidly in 1962. West Bay, as we have seen, was flushed by tidal and storm action and was not shoaling at the time of this study.

Trinity Bay is an extension of the Galveston Bay complex and is the widened estuary (drowned river valley) of the Trinity River. Development of the Trinity River delta began within the last 1000 years (LeBlanc & Hodgson, 1959). This has enclosed the head of the bay, producing the lake labelled Turtle Bay in Figure 6.2. At the time of this study, this lake was being converted to a freshwater reservoir, which would later be called Lake Anahuac, and it was already no longer natural. The main channel of the Trinity River had been dredged in the delta, so that water flowed primarily in this 6-km-long channel with spoil-bank sides, rather than also in the natural distributaries, which were filling rapidly (Wantland, 1964; McEwan, 1963). A detailed description of the geology of the Galveston–Trinity Bay area, particularly of its sedimentation characteristics as it was a few years after the present study was completed, was presented in a compilation by Lankford and Rogers (1969).

I have already mentioned Lohse's (1955) observation that, while the prevailing wind is south to southeast, the occasional dominant winds from the north during the winter ('northers') had a greater effect on sedimentation, serving as powerful agents in keeping passes open between the bays and the Gulf that would otherwise have silted up. A study by Matty et al. (1987) of the nearby artificial reservoir, Lake Houston, demonstrates that silt-size particles are readily resuspended by wind-driven currents and waves. These seem to be the only ways in which winds could be said to have played a major role in pollen distribution in the sediments investigated here. The final pollen sedimentation was prevailingly a result of water action, and there appeared to be very little if any direct relationship between wind pattern and pollen distribution in the sediments, although much pollen obviously reached the various bodies of water directly from the atmosphere.

Climate

The climate of the area is somewhat monsoonal. From May through September, the weather is rather uniform. There is usually a southeast wind, 3–9 km/h. Temperatures May–September are almost always 24 ±3 °C at night and 35 ±3 °C in the day, with a July mean of 28 °C.

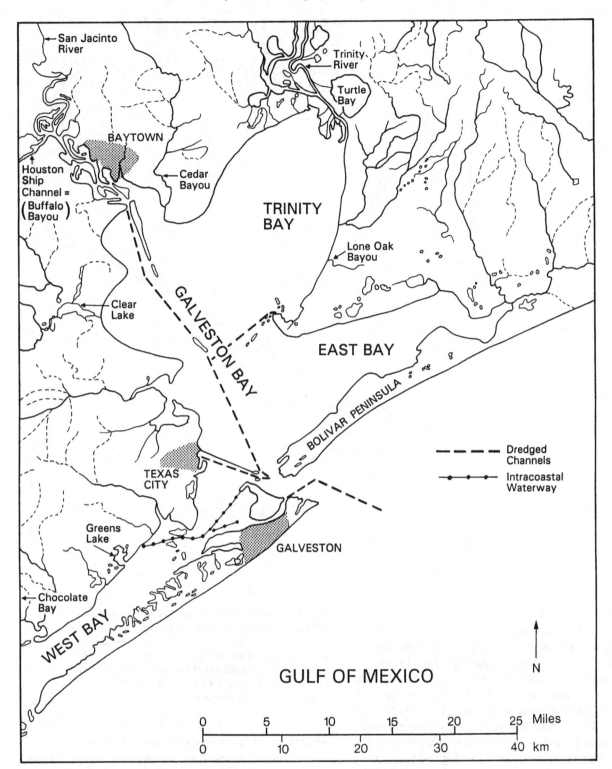

Figure 6.2. Study area as it was in 1960. Compare with Fig. 6.1 for orientation in North America. The lake near the Trinity River delta labelled 'Turtle Bay' is now known as Lake Anahuac. Only the portion of the Intracoastal Waterway that was a dredged channel in West Bay is shown. Coastline areas are somewhat simplified.

Severe tropical storms occur occasionally during this season. From October through April, on the other hand, weather in the area is much more variable. During this period, instead of the constant southeast winds of summer, frequent cold air masses from the north ('northers') bring either northwest to north winds and

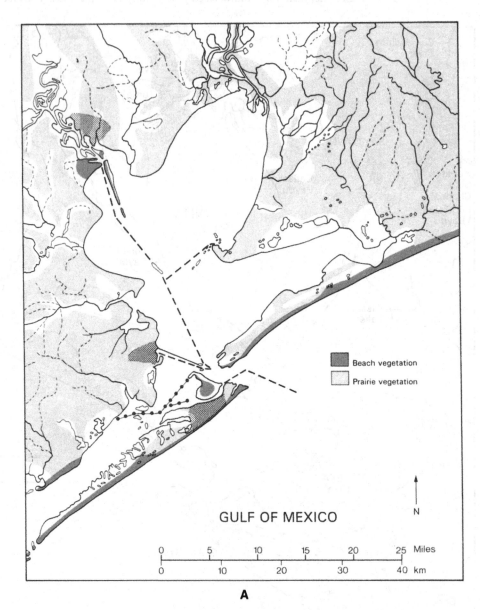

A

Figure 6.3. Reconnaissance vegetation map of study area.
(A) beach and prairie vegetation; (B) general vegetation areas
traversed by Trinity River; (C) forested areas and extensive
marsh vegetation. (B was published in Traverse, 1990, and is
reproduced here by permission of Elsevier Science Publishers.)
Compare with Figs. 6.1 and 6.2 for orientation, and with
Fig. 6.4 for relation, to sample sites.

cold, dry air (high pressure), or north to northeast winds
with cold, wet air (low pressure). During the high-pressure
'northers' the temperature frequently drops to near or
below freezing, occasionally as low as −10 °C. The
January mean temperature is 13 °C. There are intervals
of warm, equable weather with southerly winds between
'northers.' Thus, over the year as a whole, the prevailing
wind is from northeast to south–southeast (Shepard &
Moore, 1960). Humidity is generally high all year, except
when dry northern (for example, during cold, dry
'northers') or western air invades the area. Annual average
precipitation in the Galveston Bay area is about 130 cm
(declining to about 65 cm in the area of the headwaters
of the Trinity River, northwest of Dallas), but there are
great deviations from this mean. An August tropical
storm may bring 50 cm of rain to Galveston one year,
while the next August may have very little precipitation.

The amount of rain received in summer thundershowers
is very erratic. However, the mean precipitation, on the
basis of a long record, does not vary much from month
to month. Because summer precipitation comes from
thundershowers and occasional tropical storms, rainfall
is more regular in any given year during the winter
months, when frontal storms provide the moisture.

In its course from northwest of Dallas–Ft. Worth to
Trinity–Galveston Bay (about 750 km), the river flows
through Carboniferous, Cretaceous and progressively

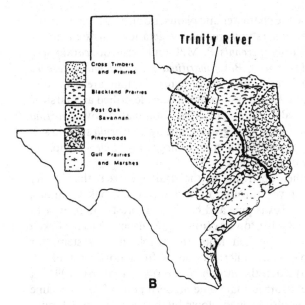

B

younger Cenozoic rocks, at times picking up reworked palynomorphs as it goes (Traverse, 1990). In terms of numbers, these do not play a significant role in the modern palynomorph depositional picture presented here.

Vegetation

Figure 6.3 presents a reconnaissance vegetation map of the area made in 1963, based on my field study notes and observations, plus study at that time of aerial photographs, with appropriate extrapolations. Figure 6.3B shows the general vegetation types through which the Trinity River flows from its headwaters to the Gulf area. The area lies near the intersection of three 'vegetation formations' per Barbour and Billings (1988): (1) central prairies and plains, (2) piedmont oak–pine forests, (3) coastal plain forests,

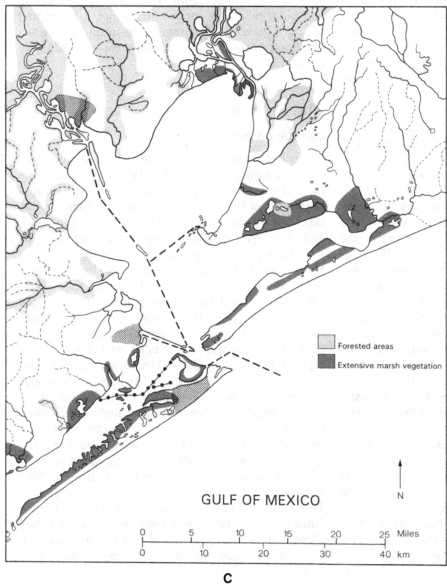

C

bogs, swamps, marshes and strand. As a guide to approximate location and general proportions of the vegetation types, the map is useful for indication of possible relationship of vegetation types to pollen sedimentation. A very detailed map of the present situation has been published by White, et al. (1985). The forested areas indicated in the western half of the map represented gallery forests along the streams, typical of prairie country. The dominant species were *Quercus nigra* L. (water oak), *Quercus phellos* L. (willow oak), *Ulmus crassifolia* Nutt. (cedar elm), *Carya illinoensis* (Wangenheim) K. Koch (pecan), other *Carya* spp. (hickory), *Taxodium distichum* (L.) Richard (swamp cypress), and *Fraxinus pennsylvanica* Marsh (ash), and a variety of other, less important, species of trees. In the north-central part of the area (east of Houston), several tongues of East Texas forest occurred. These patches of forest included the species just named in the damper and less well-drained portions, plus woods on the sandier, better-drained portions, dominated by *Pinus taeda* L. (loblolly pine), *Liquidambar styraciflua* L. (sweet gum), and *Quercus stellata* Wangenheim (post oak).

The prairies indicated were dominated by a variety of grasses, especially species of the genera *Spartina* and *Andropogon*. In moist places, thickets of trees and shrubs also were found. *Acacia farnesiana* (L.) Willd. (huisache) was especially abundant as a shrub or small tree in the dampest of such spots. *Baccharis halimifolia* L. (groundsel, a large shrub belonging to the Asteraceae = Compositae) and *Myrica cerifera* L. (wax myrtle) were also very common shrubs in the prairie.

Next to the water, narrow rims of aquatic plants occurred that are not shown in Figure 6.3, except for the more extensive stands of marsh plants. Fringing the Trinity River delta, the marshes consisted of the giant grass (called 'reed'), *Arundo donax* L., various rushes (*Juncus* spp.), sedges such as *Scirpus californicus* (C.A. Meyer) Steudel and *Scirpus validus* Vahl, *Typha* spp. (cattails), and other water-loving plants. Along the edges of East and West Bays, extensive marshes of grasses, sedges, and rushes, occurred, and these were rimmed by prairies of the tufted coastal grass, *Spartina spartinae* (Trin.) Hitchcock. The marsh localities of the area were similar to, but much smaller than, some of those of coastal Louisiana described in the classic paper of Penfound and Hathaway (1938).

True salt marshes were limited in extent in 1962 in this area. One large and typical salt marsh occurred on the western tip of Bolivar Peninsula, around the tidal ponds there. Others were found on the north side of Galveston Island and fringing West Bay. The bays have been described as surrounded by salt marshes (Reid, 1955), but, west of Trinity–Galveston Bay, water was too fresh and tidal flat area insufficient for true salt marshes to develop. Salt marshes depend on extensive tidal flooding, because the characteristic plants are all salt-loving. At the latitude of the study area such vegetation consisted mostly of *Salicornia perennis* P. Mill. and other *Salicornia* spp. (the glassworts), *Batis maritima* L., *Borrichia frutescens* (L.) DC., and other halophytes. Certain areas around East and West Bays that have been described as 'marshes' were really damp prairies, dominated by *Spartina spartinae* and other grasses, not by *Salicornia* spp. (in salt marshes), or *Typha* spp., *Juncus* spp., *Scirpus* spp. (in freshwater marshes).

Farther upstream in the drainage area, the Trinity River flowed through extensive prairies, oak savannas, and east Texas pine and hardwood forests, on its way to Trinity–Galveston Bay (Fig. 6.3B). Inasmuch as reworked spores and pollen from the whole river system (for example, Carboniferous spores from northwest of Ft. Worth) reach the mouth of the river (Traverse, 1992), it can be assumed that some modern pollen from the same area was also brought down by the river. But, as Chmura and Liu (1990) found for the Mississippi, it is doubtful that, say, *Quercus* pollen from far upstream, comprises a significant fraction of the oak pollen encountered in the lower river and bay.

Materials and methods

Sample collections

The surface sediment samples used in this study were collected during the years 1958–1962. Figure 6.4 shows the location and number for each sample. In each case uppermost surface sediment was sampled. Samples were taken from most parts of Galveston and Trinity Bays, and from West and East Bays, which are generally included under the comprehensive heading, Galveston Bay, but are really large lagoons protected by the barriers of Galveston Island and Bolivar Peninsula, respectively. Samples were also taken from many of the principal streams of the area, from backswamps (for example, sample 786), from ponds and lakes, and from tidal channels. Two groups of samples were taken in the adjacent portion of the Gulf of Mexico, to a depth of 14 m. The samples consisted of clays, sands, silts, shells, and organic matter, in varying proportions. Samples from streams, ponds, swamps, and other shallow, nearshore locations were collected by hand from a canoe or other small boat, or by wading into the water from shore. The sampling was accomplished by scooping up the sediment with a glass jar or, in deeper water, with a half-liter, weighted, scissor-levered, hand-operated grab sampler. The samples from deeper water were collected from a 9-m-long shrimp boat, using the grab sampler described above. All samples were stored in screw-top jars with polyethylene liners in the caps, using ethanol as a preservative.

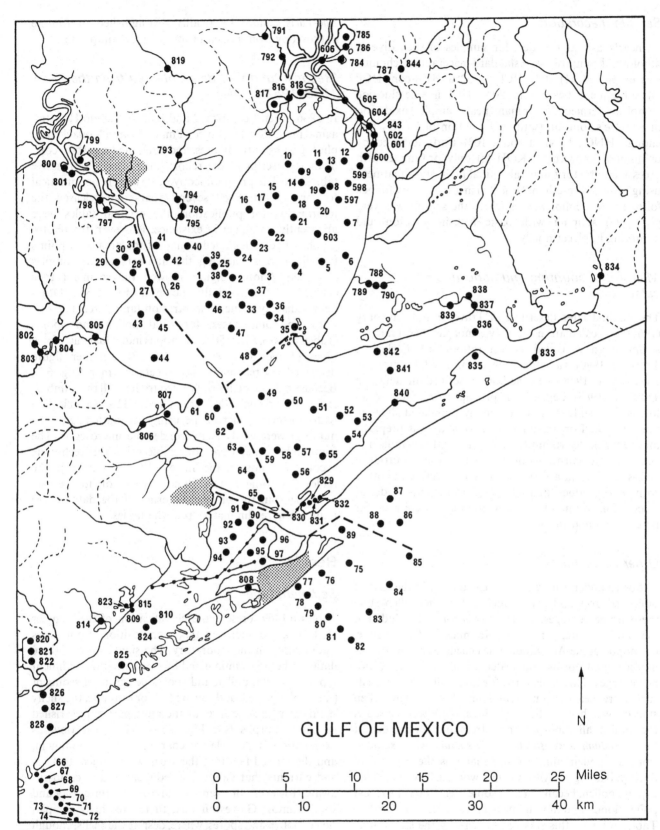

Figure 6.4. Map of study area area showing location and numbers for samples studied. All samples were surface samples collected as described in text. For sediment-type descriptions, see Tables. The bracketed samples 820, 821, 822, were collected in Chocolate Bay, just off the map to the west.

Sample maceration

As nearly as was practical for samples of such diverse lithological composition, a standard maceration technique was employed: (1) 10% HCl; (2) 52% HF + con. HCl; (3) subsequent heating in 20% HCl if silicofluoride complexes formed; (4) (for many peaty samples) oxidation in Schulze's solution (saturated solution of $KClO_3$ and enough HNO_3 to yield about 10% acid), followed by dehumification with 5% KOH; (5) (only for samples with excessive refractory mineral) a final float–sink treatment using bromoform-ethanol (s.g. 2.2) in a separatory funnel, followed by alcohol washes where the sample was very peaty; (6) staining with basic fuchsin solution and mounting in glycerin jelly.

Microfossil counting and concentration calculation

The method later described in Traverse and Ginsburg (1966) for calculation of palynomorphs per gram of sediment (now called concentration; see Farley & Traverse, 1990) was already in use in my laboratory in Houston in 1958. The method, described in detail in Traverse (1988), depends, in principle, on counting all fossils on a slide. For some densely populated slides in this study half or one-quarter of the slide was prepared for counting, by ruling the coverslip with an India ink pen from one corner to the opposite corner to produce 'half-slides,' or, in a very few cases, connecting the other corners to produce 'quarter-slides.' The counts made for calculation of microfossils per gram also were used for percentages and ratios.

Other calculations

Ratios to pollen sum were also calculated. In the case of pollen of trees and shrubs included in the pollen sum, these are percentages, but for *Taxodium*, not included in the pollen sum, and for Gramineae, Compositae, Chenopodiaceae-Amaranthaceae ('cheno-ams'), and other non-arboreal forms, the calculations are ratios, not percentages. This is also true for non-pollen microfossils such as tracheary fragments and fungal spores. The pollen sum on which the ratios and percentages were based is the total of all wind-pollinated trees and shrubs, except for *Taxodium* and *Baccharis*. *Taxodium* was excluded from the pollen sum, because that was the practice of Shell palynologists with whom I worked. They excluded cypress pollen, because it is in some areas subject to local pollen floods, and thus, to over-representation. In fact (see Tables 6.1–6.8), this was not the case in the study area, where dense stands of *Taxodium* were relatively small and not widespread. *Baccharis* was excluded, because its family, Compositae (=Asteraceae), was mostly herbaceous in this area, and the echinate pollen of

Baccharis cannot be readily distinguished in routine microscopy from pollen of many other composites.

'Reconstitution' of data obtained from the 1958–1962 studies

In 1966, when I completed and sent to Shell in Houston a final report for this project, I made 35-mm Kodachrome photographs of the figures, and I retained a carbon copy of the typescript of the report, but not of the massive tables, which in pre-photocopying days were not practical to copy. When Shell several years ago released the information for publication, the original tables were apparently no longer in existence. However, the photographed figures included a map of the area with points for all samples, showing the numbers of the samples (Fig. 6.4). Another figure showed the sediment type (Tables 6.1–6.8; partially shown in Fig. 6.5A). Other figures displayed the concentration (and percentages or ratios in some cases) for total pollen and spores (Tables 6.1–6.8; Fig. 6.5B) and various individual microfossil types (Tables 6.1–6.8; Figs. 6.6–6.8). Another figure displayed my reconnaissance vegetation map (Fig. 6.3). Because enlargement prints failed to show all the numbers legibly, the original slides were studied at 32 × magnification with a dissecting microscope, using transmitted light. The numbers were easily legible under the microscope. The few numbers that were not legible using this technique account for most of the 'no data' (= +) entries in the tables. The others are instances where obvious errors were made in the original entering of the data. These numbers were eliminated from the tables.

Results

Sediment type

Tables 6.1–6.8 present the classification of the samples studied, as to sediment-type composition, from gross examination in the laboratory. The significance of these data can be best summarized by comparing the sediment types with total pollen and spores per gram of sediment (see Tables 6.1–6.8 and Fig. 6.5B). For example, the shelly sediment-type zones are for the most part impoverished in palynomorphs (see Fig. 6.5A). The most obvious explanation is the relative coarseness of the sediment, and the related fact that the areas were subject to tides and currents that swept fine sediment and microfossils away. However, for example in parts of Trinity Bay and southernmost Galveston Bay, there are shelly samples rich in pollen and spores, where, despite the shells, enough palynomorph-rich silt was present (see discussion in caption to Fig. 6.5A). It is clear that shells alone do not indicate palynomorph poverty of a sediment. This is only the case if the shells are part of a total silt-poor picture.

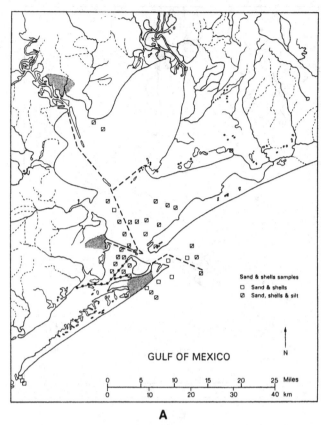

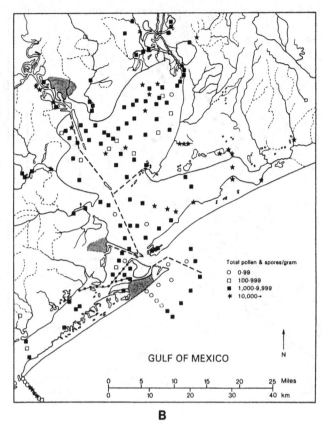

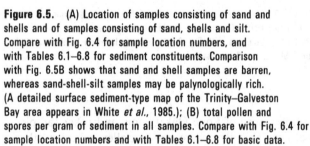

A   B

Figure 6.5. (A) Location of samples consisting of sand and shells and of samples consisting of sand, shells and silt. Compare with Fig. 6.4 for sample location numbers, and with Tables 6.1–6.8 for sediment constituents. Comparison with Fig. 6.5B shows that sand and shell samples are barren, whereas sand-shell-silt samples may be palynologically rich. (A detailed surface sediment-type map of the Trinity–Galveston Bay area appears in White *et al.*, 1985.); (B) total pollen and spores per gram of sediment in all samples. Compare with Fig. 6.4 for sample location numbers and with Tables 6.1–6.8 for basic data.

Concentration of total pollen and spores per gram of sediment (Tables 6.1–6.8; Fig. 6.5B)

The different environments and sediment types encountered in various parts of the study area should be emphasized.

EAST BAY (TABLE 6.3)

This relatively quiescent, very shallow body of water, in 1962 was sedimenting (shoaling) rapidly. The sediment had a relatively high pollen concentration, reaching nearly 4×10^4/g. The sluggish bayous, marshes and small lakes near East Bay also had high concen-trations. Where rapid sedimentation of silt and clays was occurring, palynomorphs (fine silt-sized) were also accumulating. This agrees exactly with findings of Darrell (1973) for the Mississippi River delta area.

WEST BAY AND THE SOUTHERN PART OF CENTRAL GALVESTON BAY, NEAR THE WESTERN TIP OF BOLIVAR PENINSULA (TABLE 6.2)

These areas were relatively low in palynomorph concentration. The most important factor was probably flushing resulting from tidal action and from storms. Sand and shells prevailed in the sediments, reflecting the same conditions.

GULF OF MEXICO, OFFSHORE (TABLE 6.1)

An area of high concentration existed in the Gulf of Mexico beyond the shelly zone swept by alongshore currents. This was especially marked offshore from the southwest end of Galveston Island (samples 69–72). Apparently this was a relatively low-energy environment, where silt sedimented out. That such silt sedimentation occurred in the Gulf in this manner, as a result of alongshore currents with local drops in energy level, was discussed by Price (1954).

SMALL STREAM SYSTEMS AND SMALL LAKES (TABLES 6.3, 6.5)

Most of these areas, such as Clear Lake and its bayou, Lone Oak Bayou, as well as the bayous and lakes around East Bay already mentioned, had high concentrations in many samples, reflecting both silty deposition and large amounts of locally derived pollen.

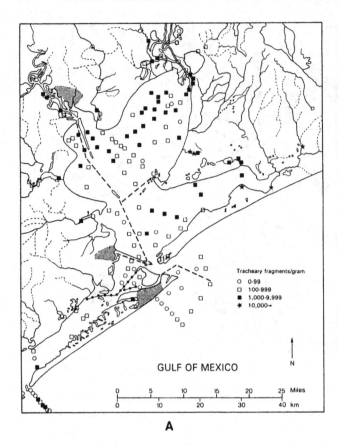

A

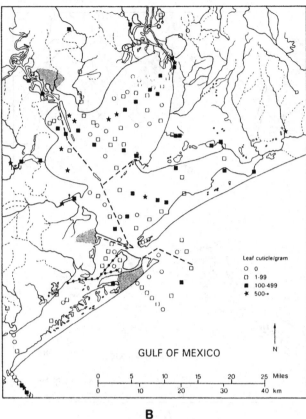

B

Figure 6.6. (A) Distribution of tracheary fragments per gram of sediment; (B) leaf cuticle (+epidermis) fragments per gram of sediment. Compare with Fig. 6.4 for sample location numbers and with Tables 6.1–6.8 for basic data.

SAN JACINTO RIVER, BUFFALO BAYOU, CEDAR BAYOU, AND THE LOWER TRINITY RIVER SYSTEM ITSELF (TABLES 6.7, 6.8)

Sediment from these areas yielded mixed results. Some samples were quite rich in palynomorphs, while others had moderate or low concentrations. This apparently reflects local sedimentation conditions and environment of deposition, with, for example, overbank samples yielding lower concentrations than channel fills (see Schuyler & Traverse, 1990, for Devonian riverine sediments).

GENERAL OBSERVATIONS

Sediment type and environment of deposition were obviously the most limiting factors. For example, the low pollen samples from southern Galveston Bay consisted prevailingly of sand-shell sediment. Pollen supply is not a limiting factor. The studies I have previously published on palynomorphs in the water in this area (Traverse, 1990, 1992) demonstrate that abundant pollen was available in the water at all times of the year. Environment of deposition is of utmost importance, and it is not a simple matter. Schuyler and Traverse (1990), for example, showed for Devonian fluviatile sediments that coarse breccias can be rich in palynomorphs if they are channel breccias, not if they are breccias basal to a whole fluviatile unit. In the present study area, some peaty samples, for

example, those of Lone Oak Bayou, and other samples from near East Bay, were rich in pollen, whereas other peat-rich samples, for example, those from marshes at the tip of Bolivar Peninsula, and from the Trinity River delta, were low in pollen. This result does not surprise a paleopalynologist, who often encounters carbonaceous shales or even coals that are poor in palynomorphs. On the other hand, samples from many parts of the study area contain sediment poorly sorted enough that even a grossly shelly sample can contain enough fine silt-sized particles to be relatively pollen-rich.

Distribution of selected microfossil groups

Tables 6.1–6.8 present the data for all the major types of palynomorphs, plus tracheary and cuticular palynodebris counted in this study. For some important forms, in terms of possible interest in connection with palynofacies studies, maps of the distributions are also presented (Figs. 6.6–6.8). Representatives of the forms discussed were illustrated photomicrographically in Traverse (1992).

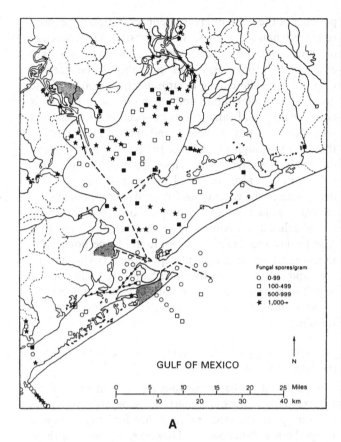

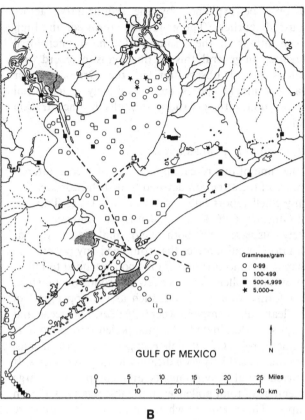

Figure 6.7. (A) Fungal spores per gram of sediment; (B) Gramineae (=Poaceae) pollen per gram of sediment. Compare with Fig. 6.4 for sample location numbers and with Tables 6.1–6.8 for basic data.

TRACHEARY FRAGMENTS (TABLES 6.1–6.8; FIG. 6.6A)

For the distribution of microscopic wood fragments, only particles larger than 15 μm with recognizable, lignified, cellular structure of wood or bark were counted. This was a very abundant microfossil type in the residues studied. As can been seen in Figure 6.6A, distribution was rather similar to that for total pollen and spores (Fig. 6.5B), except that areas of high concentration were geographically more limited. The area of high concentration in East Bay reflects the high rate of organic accumulation in the sediments from this area, while those of northern Galveston Bay and of Trinity Bay reflect the high rate of discharge of degraded vegetable matter from the streams of the area, especially the Trinity and San Jacinto Rivers. The very low concentrations in West Bay and western Bolivar Peninsula depended partly on low supply; the samples came mostly from marshy areas which did not produce much woody tissue, even though the sediment could be peaty.

LEAF CUTICLES (TABLES 6.1–6.8; FIG. 6.6B)

Fragments of leaf epidermis with cutinized cell outline are usually called 'leaf cuticles.' It appears that the explanation for the geographic distribution seen in Figure 6.6B is the same as that for wood fragments – discharge of cuticular matter by the larger streams into Trinity and Galveston Bays, plus the relatively undisturbed accumulation of organic matter in the East Bay 'settling basin' area. One unusual feature of this distribution is the high number of samples with no cuticles at all. Leaf cuticle fragments are prevailingly much larger than pollen and spores; several hundreds of microns across and larger are common. Their sedimentation is probably sensitive to energy levels as well as to supply. The barren samples in the salt-marsh areas of western Bolivar Peninsula and West Bay were a direct result of no supply. Most salt-marsh plants apparently do not produce resistant cuticles.

FUNGAL SPORES (TABLES 6.1–6.8; FIG. 6.7A)

Spores of a variety of fungi were encountered in these samples. Most of the forms (see illustrations in Traverse, 1992) are spores of saprophytic ascomycete fungi, and their supply in the area seems to have depended mostly on the supply of organic matter to the streams. The highest readings were in the Trinity River delta and adjacent northern Trinity Bay, plus a station in Chocolate Bay, just off the map to the west. However, there were sizable concentrations also in the samples from the southwest end of Galveston Island. Such high readings can only be explained by the sedimentary factors discussed earlier. These factors resulted in high concentrations in these samples for most microfossil categories. In general, this distribution is in accord with Muller's (1959) observation in the Orinoco River area that fungal spores are prevailingly a terrestrial phenomenon, with very sharp drop-offs in concentration offshore.

PINUS POLLEN (TABLES 6.1–6.8)

At the time of this research, pine was of special interest because of the attention devoted to its behavior in water in my Shell report later published in part as Traverse and Ginsburg (1966). Pine pollen is very buoyant and is widely dispersed by both wind and water. It was ubiquitous in all samples that contained any pollen. For this reason, concentration of pine pollen tended to reflect that of total pollen. Only in a few environments in which other forms were dominant, such as the upper part of the Clear Lake complex (where *Quercus*, Ulmaceae and Compositae dominated), did pine pollen retreat from its dominant role. Of all pollen types, pine pollen was, therefore, least likely to indicate a local palynofacies. It occurred in all environments and comprised a large percentage of total pollen, even in most relatively barren samples. This is evident when stations in sandy offshore and marsh regions of the West Bay area and western Bolivar Peninsula are compared.

QUERCUS POLLEN (TABLES 6.1–6.8)

Oak pollen distribution was similar to that of pine, although not as abundant, and therefore, a number of samples that were poor in pine pollen were barren of oak. Oak pollen had more of a tendency to predominate along streams than was true of pine pollen. In other words, oak pollen percentages showed some connection to environment and therefore to biofacies, whereas pine in this area showed little or none.

COMPOSITAE (= ASTERACEAE) (THISTLE AND RAGWEED FAMILY) (TABLES 6.1–6.8)

Pollen of this family comprised a large proportion of all pollen in the samples, and its distribution per gram of sediment was very similar to that displayed in Figure 6.5B for total pollen and spores. Low-pollen sediments such as some of those off Galveston Island, showed a high percentage of composite pollen. That is, of the small concentration of pollen present, a high proportion was composite pollen (see Table 6.1).

GRAMINEAE (= POACEAE) (GRASS FAMILY) (TABLES 6.1–6.8; FIG. 6.7B)

Grass pollen plotted per gram (Fig. 6.7B) shows its distribution in the samples well. Large concentrations of grass pollen were found in the same areas as for total pollen, but the areas of greatest concentration were more restricted than for pine, oak, or composites. The frequency of samples with no or very little grass pollen in West Bay and upper Trinity Bay was probably some sort of response to sedimentary environment, as plants of the grass family were abundant at both locations. It is additionally significant in this connection that samples with no grass pollen were found immediately next to those with the highest readings, 10 000 or more per gram, off the Trinity River delta. The samples with high grass concentrations also contained whole grass anthers. This demonstrated the general possibility of over-representation of grass pollen in some samples obtained from the neighborhood of deltas. This is the sort of distribution that Farley and Traverse (1990) found to be potentially useful in interpreting ancient depositional environments. Pine and oak pollen, on the other hand, are so ubiquitous as not to be helpful in determining environments except, along with all palynomorphs, in terms of total concentration.

CHENOPODIACEAE-AMARANTHACEAE ('CHENO-AMS') (TABLES 6.1–6.8; FIG. 6.8A)

The periporate pollen of various representatives of these two related families are hard to separate in routine palynological analysis, and it is therefore customary to lump them as 'cheno-ams.' However, my field work in the area studied convinced me that most of the plants producing this pollen were either salt-marsh plants (*Salicornia* spp.) or weedy herbs such as *Chenopodium* spp. The distribution is most effectively displayed as the ratio to pollen sum (Fig. 6.8A). Production by *Salicornia* undoubtedly accounted for relatively abundant pollen of this type found in samples from the salty marshes of western Bolivar Peninsula. Elsewhere, relatively abundant chenopod pollen was probably the result of the very durable, thick pollen exines which the family produces. Because of this characteristic, it is a prime candidate for recycling.

TYPHA (CATTAIL) POLLEN (TABLES 6.1–6.8)

Plants of this genus were common to abundant in moist places all over the study area, even in roadside ditches. The pollen is a monoporate-reticulate form found in obligate tetrads, monads, or dyads. Despite ubiquitous occurrence of the plants, cattail pollen was relatively abundant only in the 'settling basin' of East Bay and at just a few other stations in the deltaic systems of the Trinity and San Jacinto Rivers. Many stations in marshy places were absolutely barren of *Typha* pollen. It would appear that *Typha* pollen, when found as a fossil in older sediments, could be expected to be biofacies-related. I have noticed in my studies of sediments from a few hundred to a few thousand years old in cores from the coastal plain of the eastern USA, that *Typha* pollen may

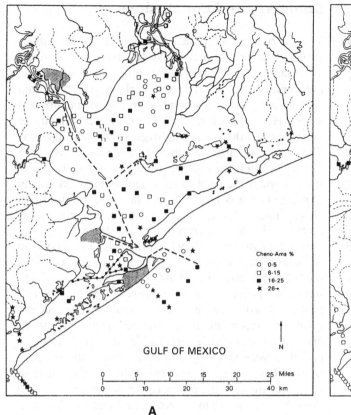

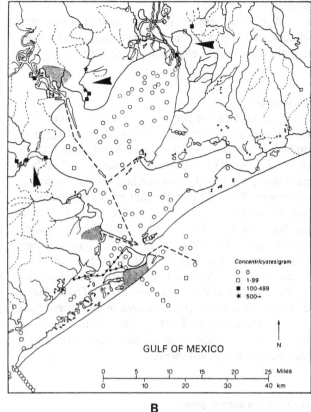

A

B

Figure 6.8.   (A) Chenopodiaceae + Amaranthaceae (cheno-ams) pollen as '%' = ratio to pollen sum; (B) *Concentricystes* (syn. *Pseudoschizaea*) sp. per gram of sediment. Arrows indicate the three regions of major occurrence. Compare with Fig. 6.4 for sample location numbers and with Tables 6.1–6.8 for basic data.

be very abundant in one sample and altogether absent in samples a few centimeters below or above. One would expect therefore that *Typha* pollen is an example of the sort of form that could be very useful for environmental interpretation.

PSEUDOSCHIZAEA (SYN. *CONCENTRICYSTES*) ALGAL ZYGOSPORES (TABLES 6.1–6.8; FIG. 6.8B)

At the time I began this investigation I did not know the identity of this microfossil form, which was moderately abundant in certain samples. Since then, it has become rather well known, and it is discussed and illustrated by Rich and Pirkle (Chap. 15, this book). It is a more or less spherical body about 40 *μ*m in diameter with a pattern of ridges and grooves reminiscent of a fingerprint. Chemically, the wall substance of this fossil type must resemble sporopollenin, since it responds to maceration in the same way and stains with basic fuchsin as do pollen grains. From study of many illustrations in the literature

at the time of this study I concluded that the form could well be a zygospore of the freshwater, green algal group, Zygnemataceae. A slide of the material was sent at that time to G. W. Prescott of Michigan State University, an authority on the group. He responded, in part, that it was '...certainly similar to a zygnemataceous zygospore, and that would be the best and about the only "guess" that I could make...' The distribution shown was very different from that of pollen in general, and corresponded to an ecological pattern that might be expected of a green algal form that grew at the edge of streams and in ponds. Cedar Bayou and Clear Lake Bayou were the two principal centers of distribution (see Fig. 6.8B and Tables 6.5, 6.8). *Pseudoschizaea* (syn. *Concentricystes*) is clearly a palynofacies indicator.

OTHER FORMS FOR WHICH DATA ARE PRESENTED IN TABLES 6.1–6.8

Carya (hickory and pecan)

The triporate pollen of wind-pollinated trees belonging to this genus was distributed very much like that of pine. It was ubiquitous in the sediments of the area. However, *Carya* pollen was absent or rare in certain groups of samples, for example, those from salt-marsh areas far from source stands of hickory, and from some sandy or shelly samples.

Ulmaceae

Elm pollen was present in most samples and was moderately abundant in some of them, especially in areas where abundant cedar elm stands were seen.

Liquidambar

The very characteristic periporate pollen of sweet gum trees had a distribution in the sediment reflecting local abundance of these trees in the Trinity River – San Jacinto River – Buffalo Bayou drainage. Relatively abundant *Liquidambar* pollen in other places reflected the sedimentary environment. For example, the factors that encouraged pollen deposition in East Bay and discouraged it in West Bay led to relatively high *Liquidambar* pollen concentrations. *Liquidambar* trees are partly wind-pollinated, partly insect-pollinated. They produce large quantities of pollen, but not as much as pine or oak trees. Very low concentrations in Galveston and Trinity Bays, and in onshore samples from the area of pine-sweet gum forests (northern half of the study area) were a result of sediment type: *Liquidambar* pollen does not occur abundantly in sandy sediments.

Taxodium (swamp cypress)

This is a tree of swamps and river bottoms. The pollen was locally concentrated in sediment associated with several of the stream courses, as well as in the 'settling basin' of East Bay, and in the offshore samples that were pollen-rich. There were many samples with absolutely no *Taxodium* pollen, even at stations with high general pollen abundance, for example, along Lone Oak Bayou on the east side of Trinity Bay. This tends to substantiate the idea previously mentioned that *Taxodium* is facies-influenced. The offshore concentrations showed, however, that this is not a simple relationship to the proximity of *Taxodium* swamps, or to rivers with swamp cypresses in the gallery forests.

Cyperaceae (sedges)

The producing plants were abundant in the wetter parts of the entire area. The large amount of sedge pollen in the Trinity River delta, corresponding parts of the San Jacinto River–Buffalo Bayou drainage and in the East Bay area was a reflection of this abundance. Sedges are not major constituents of salt marshes but are sub-dominant in freshwater marshes. The extreme paucity of sedge pollen in upper Galveston–Trinity Bay must be related to sedimentary environment; many samples were absolutely void. I have noticed in my studies of modern sediments of coastal plains in the eastern USA that sedge pollen is very erratic in occurrence.

Summary and conclusions

The distribution of pollen and spores in the surface sediment of Trinity and Galveston Bays and adjacent areas sampled in 1958–1962 was primarily a question of sedimentary factors that govern all clastic particles. Plant ecology played only a secondary role in the picture as a whole. However, some local effects related directly to autecology of the producing organisms were apparent. The probable algal zygospore, *Pseudoschizaea* (syn. *Concentricystes*), was found primarily in the vicinity of streams where the producing organisms grew. Abundant *Typha* pollen was mostly found fairly near cattail-rich areas. These 'biofacies' were, however, somewhat exceptional. The concentration of pollen and spores in the sediments studied was primarily explicable in terms of water movement and sedimentation patterns, not of wind patterns. Wind-driven water is a very significant factor in the area, however, as an agent of geological work. In that sense, wind phenomena are indirectly important to pollen sedimentation. Wind distribution is of course responsible for wide dispersal of pollen in very small quantities, as many studies have shown. Rowley and Walch (1972) showed that, at least in an exceptional case, pollen can even become airborne from the water medium as aerosol particles, and Valencia (1967) demonstrated the same thing experimentally. It should be noted that at least in the Atlantic Ocean off West Africa, Dupont et al. (1989) and Hooghiemstra (1988), among others, have indicated that wind delivery is the main source of pollen in surface sediment. Hooghiemstra found that the distribution of patterns of the pollen reflect the average atmospheric circulation, and Dupont et al. state that the northeast trade winds and African easterly jet govern pollen supply and deposition.

No evidence was seen for sedimentary sorting of pollen of various taxa. That does not mean, however, that this could not occur in some situations. In the laboratory, it is possible to concentrate different sizes of palynomorphs from each other and from debris by 'swirling' in watch glasses, using small-scale turbulence to sort the forms (see discussion in Traverse, 1988). Catto (1985) reports that such sorting occurs in silts and sands of a braided stream in northern Canada. He found that tree pollen concentrates in the silts and herbaceous pollen and spores in the sands. Holmes (1990) showed in laboratory experiments that some sorting based on size occurred as pollen moved through the flume he used.

East Bay was nearly shut off from Galveston Bay-proper at the time of this study by an oyster reef (Hanna Reef) and was filling rapidly with sediment. Because of this, it was also a 'settling basin' for the collection of pollen and other palynomorphs and palynodebris, in a manner similar to that of lee areas in Great Bahama Bank (Traverse & Ginsburg, 1966).

West Bay and lowermost Galveston Bay in the Galveston–Bolivar Peninsula area were flushed by tidal action and by high-energy wind storms, especially the winter 'northers.' For that reason, those areas were apparently not accumulating fine silty sediment in the palynomorph size range. Therefore, the surface sediment in those areas, on the whole, was poor in palynomorphs, because they were flushed out, along with other fine silt.

Some samples studied from offshore, in the Gulf of Mexico, were rich in palynomorphs and other fine sediment. Other, nearby, samples consisted of coarse, shelly sediment containing few palynomorphs. The boundary between pollen-poor and pollen-rich was quite sharp. This is in accord with sedimentation studies of alongshore currents and deposition in the area. The pollen-poor coarse sediments were from the nearshore area swept by currents. This result is interesting as one exception to the usually cited regular decline of palynomorph concentration with distance from shore.

Sediment from the bays-proper and from onshore locations such as rivers, backswamps and ponds were pollen-rich or pollen-poor, depending on the sedimentary environment. Silts from a bayou were usually pollen-rich, whereas sandy samples from the same bayou were usually pollen-poor. Sand-shell samples from all of the bays were pollen-poor, whereas silt-clay samples from nearby were often pollen-rich.

Certain pollen types were ubiquitous, for example, *Pinus* and *Quercus* (and *Carya*, to a lesser extent). That is, they were abundant in all sorts of sediment, with a few exceptions. Other types, for example, Chenopodiaceae, *Typha*, and *Taxodium* were much more abundant in certain locations than in others, a response partly to presence of source vegetation, partly to sedimentary environment. Another example of a 'facies' indicator was the probable zygnemetaceous (green algal) zygospore form, *Pseudoschizaea* (syn. *Concentricystes*), which was principally found in sediment from bayous and rivers.

In conclusion, one must compare the rather substantial seasonal fluctuations in the palynomorph load of the Trinity River and Trinity Bay I have previously reported (Traverse, 1990, 1992) with these results for the palynomorph content of the sediment produced over the whole area. Clearly the surface sediment reflected the vegetation of the area, but in a very generalized way. At the scale of the study presented, there is obviously evidence for sorting of pollen as a whole, with pollen-rich and pollen-poor environments; that is, palynolithofacies are clearly indicated. Sorting of palynomorphs, either by size characteristics for specific taxa, or according to the taxonomic groups encountered (palynobiofacies) was very limited in extent.

Acknowledgments

I am grateful to Shell Development Company for permitting the publication of the major segments of research I did as a Shell palynologist, 1955–1962. At the time of the original research so many Shell employees were helpful in so many ways, that it would be difficult to pick out individuals to acknowledge. I must be content to thank them all jointly. I also acknowledge the help of Elizabeth I. Traverse with the preparation of this manuscript. Friedemann Schaarschmidt and members of his staff at the Research Institute of the Senckenberg Museum, Frankfurt am Main, Germany, were helpful with computer, drafting and in other ways at the end of the project. Martin B. Farley and Roy J. Greenfield also assisted me with advice and information necessary for the completion of the manuscript.

References

Barbour, M. G. & Billings, W. D., eds. (1988). *North American Terrestrial Vegetation*. Cambridge: Cambridge University Press.

Bernard, H. A., Major, C. F., Jr. & Parrott, B. S. (1959). The Galveston Barrier island and environs: a model for predicting reservoir occurrence and trend. *Gulf Coast Association of Geological Societies Transactions*, **9**, 221–4.

Catto, N. R. (1985). Hydrodynamic distribution of palynomorphs in a fluvial succession, Yukon. *Canadian Journal of Earth Science*, **22**, 1552–6.

Chmura, G. L. & Liu, K. B. (1990). Pollen in the lower Mississippi River. *Review of Palaeobotany and Palynology*, **64**, 253–61.

Darrell, J. H. II (1973). Statistical evaluation of palynomorph distribution in the sedimentary environments of the modern Mississippi River delta. Ph.D. diss. Louisiana State University, Baton Rouge.

Dupont, L. M, Beug, H.-J., Stalling, H. & Tiedemann, R. (1989). 6. First palynological results from site 658 at 21° N off northwest Africa: pollen as climate indicators. *Proceedings of the Ocean Drilling Program, Scientific Results*, **108**, 93–111.

Farley, M. B. & Traverse, A. (1990). Usefulness of palynomorph concentrations in distinguishing Paleogene depositional environments in Wyoming (U.S.A.). *Review of Palaeobotany and Palynology*, **64**, 325–9.

Holmes, P. L. (1990). Differential transport of spores and pollen: a laboratory study. *Review of Palaeobotany and Palynology*, **64**, 289–96.

Hooghiemstra, H. (1988). Palynological records from northwest African marine sediments: a general outline of the interpretation of the pollen signal. *Philosophical Transactions of the Royal Society of London*, **B 318**, 431–49.

Lankford, R. R. & Rogers, J. J. W., compilers (1969). *Holocene Geology of the Galveston Bay Area*. Houston, TX: Houston Geological Society.

LeBlanc, R. J. & Hodgson, W. D. (1959). Origin and development of the Texas shoreline. *Gulf Coast Association of Geological Societies Proceedings*, **9**, 197–220.

Lohse, E. A. (1955). Dynamic geology of the modern coastal region, northwest Gulf of Mexico. In *Finding Ancient Shorelines, Society of Economic Paleontologists and Mineralogists Special Publication* **3**, 99–104.

Matty, J. M., Anderson, J. B. & Dunbar, R. B. (1987). Suspended sediment transport, sedimentation, and resuspension in Lake Houston, Texas: implications for water quality. *Environmental Geology and Water Sciences*, **10** (3), 175–86.

Mazzullo, J. (1986). Sources and provinces of late Quaternary sand on the East Texas–Louisiana continental shelf. *Geological Society of America Bulletin*, **97**, 638–47.

McEwan, M. C. (1963). Sedimentary facies of the Trinity River delta, Texas. Ph.D. diss., Rice University, Houston, TX.

Muller, J. (1959). Palynology of Recent Orinoco delta and shelf sediments. *Micropaleontology*, **5**, 1–32.

Nixon, E. S. & Willett, R. L. (1974). Vegetative analysis of the floodplain of the Trinity River, Texas. *Report for US Army Corps of Engineers, Ft. Worth District, Texas*, Contract DACW63-74-C-0030.

Penfound, W. T. & Hathaway, E. S. (1938). Plant communities in the marshlands of southeastern Louisiana. *Ecological Monographs*, **8** (1), 1–56.

Prescott, G. W. (1966). Personal communication.

Price, W. A. (1954). Shorelines and coasts of the Gulf of Mexico. In *Gulf of Mexico: Its Origin, Waters and Marine Life*, coordinater P. S. Galtstoff. US Department of the Interior: Fish and Wildlife Service Bulletin **89**, 39–62.

Reid, G. K., Jr. (1955). A summer study of the biology and ecology of East Bay, Texas, Part I. *Texas Journal of Science*, **7** (3), 316–43.

Rich, F. J. & Pirkle, F. L. (this volume).

Rowley, J. R. & Walch, K. M. (1972). Recovery of introduced pollen from a mountain glacier stream. *Grana*, **12**, 146–52.

Schuyler, A. & Traverse, A. (1990). Sedimentology of miospores in the Middle to Upper Devonian Oneonta Formation, Catskill magnafacies, New York. *Review of Palaeobotany and Palynology*, **64**, 305–13.

Shepard, F. P. (1953). Sedimentation rates in Texas estuaries and lagoons. *Bulletin of the American Association of Petroleum Geologists*, **37** (8), 1919–34.

Shepard, F. P. & Moore, D. G. (1960). Bays of central Texas coast. In *Recent Sediments, Northwest Gulf of Mexico*, ed. F. P. Shepard, F. B. Phleger & T. H. Van Andel. Tulsa, OK: American Association of Petroleum Geologists, 117–52.

Traverse, A. (1988). *Paleopalynology*. London: Unwin Hyman.

Traverse, A. (1990). Studies of pollen and spores in rivers and other bodies of water, in terms of source-vegetation and sedimentation, with special reference to Trinity River and Bay, Texas. *Review of Palaeobotany and Palynology*, **64**, 297–303.

Traverse, A. (1992). Organic fluvial sediment: palynomorphs and 'palynodebris' in the lower Trinity River, Texas. *Annals of the Missouri Botanical Garden*, **79**, 110–25.

Traverse, A. & Ginsburg, R. N. (1966). Palynology of the surface sediments of Great Bahama Bank, as related to water movement and sedimentation. *Marine Geology*, **4**, 417–59.

Valencia, M. J. (1967). Recycling of pollen from an air–water interface. *American Journal of Science*, **265**, 843–7.

Wantland, K. F. (1964). Recent foraminifers of Trinity Bay, Texas. M.A. thesis, Rice University, Houston, TX.

White, W. A., Calnan, T. R., Morton, R. A., Kimble, R. S., Littleton, T. G., McGowen, J. H., Nance, H. S. & Schmedes, K. E. (1985). *Submerged Lands of Texas, Galveston-Houston Area: Sediments, Geochemistry, Benthic Macroinvertebrates, and Associated Wetlands*. Austin, TX: Bureau of Economic Geology Special Contribution.

Tables 6.1–6.8

Basic palynological data for the sample locations displayed in Fig. 6.4. Table 6.1. Gulf of Mexico. Tables 6.2–6.4. Galveston Bay Area: Table 6.2. West Bay and associated land areas; Table 6.3. East Bay and associated land areas; Table 6.4. Central and Northern Bay. Table 6.5. Clear Lake complex. Table 6.6. Trinity Bay. Table 6.7. Lower Trinity River complex and delta. Table 6.8. Buffalo Bayou–San Jacinto River–Cedar Bayou complex. 'Sediment type' was based on gross examination in the laboratory; types are listed in order of apparent abundance. All samples are surface samples. See text for details of sample collection. 'Spores & pollen/g' means total spores and pollen per gram of sediment, calculated from air-dried samples per method described by Traverse and Ginsburg (1966). All other 'per gram' data in the tables were calculated in the same way. '%' refers to percentage of the pollen sum (=all tree and shrub pollen except *Taxodium* and *Baccharis*). In the case of cheno-ams, Compositae, Cyperaceae, Gramineae, and *Typha,* these numbers are actually ratios to the pollen sum, as these taxa are not a part of it. *=sample barren; +=no data.

Table 6.1. *Gulf of Mexico*

Sample #	66	67	68	69	70	71	72	73	74	75	76	77
Sediment type	sand shell	sand shell	sand shell	clay silt	clay silt	clay silt	clay silt	clay silt	clay silt	sand shell	sand shell	sand shell
spores & pollen/g	4	23	*	11812	11196	20940	12456	49	389	*	*	*
Concentricystes/g	0	0	*	0	0	0	0	0	0	*	*	*
fungal spores/g	1	7	*	3973	2093	3865	1987	5	46	*	*	*
leaf cuticles/g	0	1	*	30	239	819	151	1	0	*	*	*
tracheary fragments/g	4	1	*	1135	673	2501	1383	41	92	*	*	*
Carya/g	0	0	*	134	55	407	114	0	8	*	*	*
cheno-ams/g	0	0	*	1496	1128	2298	757	0	8	*	*	*
cheno-ams %	0	0	*	40	35	20	15	0	6	*	*	*
Compositae/g	2	9	*	1290	819	2665	1324	5	97	*	*	*
Compositae %	0	660!	*	35	25	24	22	11	25	*	*	*
Cyperaceae/g	0	4	*	175	218	610	246	0	0	*	*	*
Gramineae/g	0	0	*	402	546	1668	662	0	15	*	*	*
Liquidambar/g	0	0	*	31	55	264	170	0	0	*	*	*
Pinus/g	0	1	*	2167	1656	7119	3150	5	76	*	*	*
Pinus %	0	100	*	58	51	63	64	44	56	*	*	*
Quercus/g	0	0	*	1073	1056	2278	1116	7	23	*	*	*
Quercus %	0	0	*	29	32	20	23	56	17	*	*	*
Taxodium/g	0	0	*	196	546	936	170	0	0	*	*	*
Typha/g	0	0	*	61	291	540	284	0	0	*	*	*
Ulmaceae/g	0	0	*	186	109	447	132	0	15	*	*	*

Table 6.1. Continued

Sample #	78	79	80	81	82	83	84	85	86	87	88	89
Sediment type	sand, shell silt	silt, shell sand	silt sand	silt sand	silt sand	silt sand	silt sand	silt, sand shell	silt, sand shell	silt shell	sand shell	sand shell
spores & pollen/g	10	21	44	4 441	8 384	4 486	6 718	5 632	2 699	4 311	*	*
Concentricystes/g	0	0	0	8	0	15	25	0	0	0	*	*
fungal spores/g	46	3	17	203	219	107	308	88	31	35	*	*
leaf cuticles/g	5	1	6	61	0	56	234	59	4	25	*	*
tracheary fragments/g	59	1	79	384	416	214	824	343	156	219	*	*
Carya/g	0	1	0	31	66	0	74	98	31	66	*	*
cheno-ams/g	0	3	6	391	810	214	431	480	168	394	*	*
cheno-ams %	0	100	95	25	34	19	16	23	33	41	*	*
Compositae/g	0	1	7	407	744	534	1 119	392	172	258	*	*
Compositae %	0	50	105!	26	31	47	43	19	34	27	*	*
Cyperceae/g	0	1	1	23	129	122	74	59	351	53	*	*
Gramineae/g	0	1	0	222	350	366	199	372	281	259	*	*
Liquidambar/g	+	1	1	31	66	46	0	39	3	9	*	*
Pinus/g	0	0	4	1 081	131	549	1 648	1 410	289	591	*	*
Pinus %	0	0	65	68	7	48	64	69	56	62	*	*
Quercus/g	5	0	1	337	853	351	579	303	164	184	*	*
Quercus %	100	0	10	21	35	31	22	15	32	19	*	*
Taxodium/g	0	0	1	115	219	31	62	225	133	114	*	*
Typha/g	0	0	0	84	88	31	62	49	20	48	*	*
Ulmaceae/g	0	0	0	46	44	96	86	20	23	22	*	*

Table 6.2. *Galveston Bay area: West Bay and associated land areas*

Sample #	90	91	92	93	94	95	96	97	808	809	810
Sediment type	silt sand shell	silt sand shell	silt sand shell	silt shell sand	silt sand shell	silt sand shell	silt sand shell	silt shell sand	+	silt clay sand	silt clay sand
spores & pollen/g	39	3 287	2 374	809	1 675	1 369	2 120	2 163	7 542	2 199	1 416
Concentricystes/g	0	0	0	0	0	0	0	0	0	4	0
fungal spores/g	4	58	45	3	47	65	101	103	3 342	237	139
leaf cuticles/g	0	40	18	6	23	0	11	39	257	39	35
tracheary fragments/g	1	272	144	32	41	39	202	77	2 914	406	307
Carya/g	0	58	45	12	6	9	45	26	171	19	28
cheno-ams/g	0	156	153	30	152	142	135	245	643	177	118
cheno-ams %	0	11	16	6	38	34	12	23	19	17	12
Compositae/g	3	369	198	27	175	211	291	322	1 207	63	49
Compositae %	13	27	21	5	40	51	27	31	34	18	5
Cyperaceae/g	1	46	36	6	47	26	0	26	43	4	0
Gramineae/g	3	13	36	0	82	60	95	180	300	30	14
Liquidambar/g	0	6	0	8	6	13	22	13	86	15	0
Pinus/g	13	1 067	640	438	315	237	717	373	1 971	639	781
Pinus %	70	72	67	87	72	57	65	25	60	61	76
Quercus/g	5	167	144	32	47	112	258	173	857	267	101
Quercus %	27	12	25	7	11	27	23	11	26	26	20
Taxodium/g	0	17	45	3	35	17	34	26	43	8	7
Typha/g	1	150	72	9	12	21	0	26	86	45	7
Ulmaceae/g	0	35	9	6	47	30	11	26	129	19	7

Table 6.2. Continued

Sample #	814	815	820	821	822	823	824	825	826	827	828
Sediment type	silt clay sand peat	silt clay sand	silt clay	+	+	silt clay	+	silt clay sand	silt clay sand	silt clay sand	clay sand silt
spores & pollen/g	7	354	938	+	+	503	1 209	2 356	1 861	545	5 890
Concentricystes/g	0	0	0	+	+	0	0	0	0	3	0
fungal spores/g	5	97	1 132	368	25 489	439	98	146	511	24	3 337
leaf cuticles/g	0	11	78	+	+	32	0	16	0	0	36
tracheary fragments/g	0	210	1 757	278	+	331	392	421	255	15	8 485
Carya/g	0	0	78	0	359	16	33	65	0	6	36
cheno-ams/g	0	22	78	0	359	142	98	250	547	250	1 030
cheno-ams %	0	12	40	0	36	64	21	27	118!	100	65
Compositae/g	0	44	+	+		63	65	97	146	31	1 207
Compositae %	0	24	120!	+	+	36	14	14	31	23	76
Cyperaceae/g	0	0	0	0	0	0	0	16	0	18	0
Gramineae/g	5	0	0	9	359	16	33	97	36	31	249
Liquidambar/g	0	0	0	0	180	0	0	0	0	0	36
Pinus/g	0	177	195	9	+	142	359	236	319	104	1 402
Pinus %	0	91	100	51	+	64	79	35	69	77	89
Quercus/g	0	0	0	+	+	47	0	275	128	6	107
Quercus %	0	0	0	+	+	27	0	41	27	5	7
Taxodium/g	0	0	0	0	0	0	0	16	0	0	0
Typha/g	0	0	0	0	0	16	0	32	18	0	533
Ulmaceae/g	0	0	0	0	0	0	0	32	0	0	0

Table 6.3. *Galveston Bay area: East Bay and associated land areas*

Sample #	788	789	790	829	830	831	832	833	834
Sediment type	clay peat silt	silt peat clay	silt clay	silt sand clay peat	silt sand clay peat	silt sand clay shell	silt peat sand clay	silt clay sand	clay silt peat
spores & pollen/g	24 780	23 380	23 878	1 469	795	1 541	1 869	18 981	23 047
Concentricystes/g	0	0	0	0	0	0	0	0	0
fungal spores/g	1 602	819	9 196	313	210	305	426	1 086	603
leaf cuticles/g	400	46	236	0	0	0	0	384	844
tracheary fragments/g	53 383	20 967	8 073	157	0	122	125	10 624	23 037
Carya/g	234	46	78	0	0	0	0	128	241
cheno-ams/g	2 402	289	4 598	2 724	180	429	601	1 984	2 893
cheno-ams %	34	6	53	116!	72	122!	123!	34	39
Compositae/g	1 832	3 270	7 192	39	30	294	251	3 072	3 737
Compositae %	47	41	84	12	12	70	51	53	50
Cyperaceae/g	2 069	675	590	0	0	0	0	832	844
Gramineae/g	2 602	1 203	2 594	78	30	92	75	2 496	2 893
Liquidambar/g	133	0	354	0	0	31	0	192	0
Pinus/g	6 086	3 598	6 013	276	253	259	388	3 360	3 737
Pinus %	87	76	70	53	100	74	79	58	50
Quercus/g	400	241	2 004	117	0	61	75	1 800	3 014
Quercus %	6	5	23	35	0	17	15	28	41
Taxodium/g	0	0	0	0	0	0	0	448	362
Typha/g	1 335	1 301	2 594	0	0	0	0	384	844
Ulmaceae/g	+	+	+	0	0	0	0	64	121

Table 6.3. *Continued*

Sample #	835	836	837	838	839	840	841	842
Sediment type	sand silt peat clay	silt clay sand	silt shell sand clay	silt clay shell sand	silt clay	silt sand	silt clay	silt clay
spores & pollen/g	23 489	38 188	20 286	35 285	36 394	21 390	7 407	9 401
Concentricystes/g	0	64	0	0	0	0	0	0
fungal spores/g	377	512	28	1 293	4 280	126	318	275
leaf cuticles/g	283	64	59	366	486	24	53	153
tracheary fragments/g	10 334	6 340	2 221	4 324	7 101	421	1 051	2 879
Carya/g	47	320	171	73	389	126	0	122
cheno-ams/g	2 358	2 818	1 680	2 785	2 724	1 095	478	794
cheno-ams %	29	18	22	46	30	19	19	16
Compositae/g	3 254	3 586	2 620	2 272	4 864	1 852	584	855
Compositae %	37	30	33	54	54	33	22	17
Cyperaceae/g	424	512	114	733	778	84	159	92
Gramineae/g	2 095	2 626	1 082	55	5 837	942	590	763
Liquidambar/g	236	192	142	147	389	126	189	82
Pinus/g	2 672	11 559	2 286	2 623	4 119	2 942	2 203	3 662
Pinus %	55	74	83	40	50	52	61	73
Quercus/g	1 745	2 818	365	2 418	2 821	2 064	1 062	1 007
Quercus %	35	18	13	40	32	36	29	20
Taxodium/g	47	768	28	1 539	1 556	337	0	31
Typha/g	283	768	456	1 026	389	169	372	92
Ulmaceae/g	236	512	228	73	97	42	53	0

Table 6.4. *Galveston Bay area: Central and Northern Bay*

Sample #	01	02	25	26	27	28	29	30	31	32	33
Sediment type	sand silt	silt clay	silt sand	silt sand	silt sand	silt sand	silt sand	silt sand	silt sand	silt shell	silt shell
spores & pollen/g	2 901	656	4 540	6 134	9 370	5 707	867	2 024	24 158	692	2 981
Concentricystes/g	0	0	0	0	0	0	0	42	43	0	0
fungal spores/g	1 024	255	2 059	2 020	4 062	1 569	166	809	2 388	160	609
leaf cuticles/g	0	0	0	144	131	236	8	28	0	23	31
tracheary fragments/g	1 517	158	281	7 359	2 359	896	133	454	1 023	206	836
Carya/g	0	12	0	0	0	47	1	0	85	0	52
cheno-ams/g	38	49	187	144	393	330	25	128	213	100	237
cheno-ams %	14	16	9	3	14	13	6	16	11	12	25
Compositae/g	190	36	562	433	1 572	802	166	383	831	80	382
Compositae %	10	29	27	10	56	30	41	49	44	27	40
Cyperaceae/g	38	0	0	0	0	94	0	0	43	23	0
Gramineae/g	38	0	281	433	917	330	67	57	128	34	93
Liquidambar/g	38	0	94	0	262	47	17	28	43	11	31
Pinus/g	1 043	206	983	3 102	1 638	1 603	191	348	1 151	154	547
Pinus %	66	59	47	73	58	61	47	44	61	53	58
Quercus/g	493	134	842	846	393	707	158	270	490	92	258
Quercus %	31	38	40	20	14	27	39	34	26	31	27
Taxodium/g	0	0	94	0	0	0	8	0	43	0	10
Typha/g	0	12	0	0	0	47	0	0	21	0	10
Ulmaceae/g	0	0	0	144	262	141	33	128	43	23	21

Table 6.4. *Continued*

Sample #	34	35	36	37	38	39	40	41	42	43	44
Sediment type	silt shell	clay silt shell	silt shell	silt shell	silt shell	silt sand shell	silt sand shell	silt	silt	silt sand	silt sand
spores & pollen/g	3 159	3 392	854	10 191	4 747	1 698	4 336	6 660	9 776	6 323	4 105
Concentricystes/g	0	26	0	0	0	0	13	19	47	42	12
fungal spores/g	134	993	168	984	1 078	183	664	389	94	106	98
leaf cuticles/g	153	128	17	81	0	20	39	622	1 509	659	624
tracheary fragments/g	940	633	93	539	604	163	156	1 535	3 672	1 081	982
Carya/g	34	60	6	46	43	61	65	175	188	127	98
cheno-ams/g	182	205	52	297	259	122	273	292	165	403	123
cheno-ams %	17	13	23	9	15	16	15	12	3	12	5
Compositae/g	465	318	104	620	733	386	1 167	1 788	1 106	454	614
Compositae %	43	20	45	19	44	49	60	68	17	29	25
Cyperaceae/g	21	60	6	54	43	20	13	19	0	21	12
Gramineae/g	75	120	58	195	235	81	299	428	400	339	147
Liquidambar/g	16	17	0	27	43	41	52	78	118	127	123
Pinus/g	706	1 014	97	2 716	1 078	356	1 107	1 623	5 485	3 078	1 265
Pinus %	66	65	43	83	64	45	60	62	84	62	52
Quercus/g	192	342	97	317	259	264	391	505	636	1 191	823
Quercus %	18	22	43	10	15	34	21	19	10	24	34
Taxodium/g	16	34	12	88	43	0	26	16	0	0	0
Typha/g	5	1	12	40	0	0	26	39	0	0	0
Ulmaceae/g	37	26	0	61	129	41	184	136	94	85	86

Table 6.4. Continued

Sample #	45	46	47	48	49	50	51	52	53	54	55
Sediment type	silt sand	silt	silt	silt	silt sand	silt sand	silt shell sand	silt shell sand	silt shell sand	silt sand	silt shell sand
spores & pollen/g	838	10 803	2 612	2 131	4 386	9 376	12 449	18 343	11 658	8 604	3 885
Concentricystes/g	0	0	0	0	0	0	0	0	0	0	0
fungal spores/g	27	1071	506	468	614	1504	1465	2044	322	823	223
leaf cuticles/g	9	195	0	16	75	193	609	430	161	57	56
tracheary fragments/g	128	1849	111	145	509	1918	1961	1184	724	670	989
Carya/g	0	195	56	32	45	58	96	161	+	77	57
cheno-ams/g	46	681	222	194	360	733	879	861	432	670	306
cheno-ams %	11	19	22	33	16	20	15	12	10	24	18
Compositae/g	210	1752	111	387	404	1330	721	2529	2251	1430	434
Compositae %	53	50	11	67	18	37	12	36	45	52	25
Cyperaceae/g	9	97	0	16	30	154	316	377	322	52	40
Gramineae/g	46	584	0	81	210	521	1059	753	482	77	129
Liquidambar/g	18	97	0	32	15	19	90	108	241	191	88
Pinus/g	178	1752	444	178	1461	2148	3832	4492	3658	38	1013
Pinus %	45	50	44	31	61	61	66	64	73	1713	58
Quercus/g	137	823	444	210	584	945	1285	1722	724	62	418
Quercus %	34	33	44	36	26	26	22	25	14	574	24
Taxodium/g	0	0	0	16	30	77	270	215	241	21	40
Typha/g	0	97	56	0	60	135	541	430	80	230	88
Ulmaceae/g	18	195	56	48	30	154	180	108	80	115	48

Table 6.4. *Continued*

Sample #	56	57	58	59	60	61	62	63	64	65	806	807
Sediment type	silt shell sand	silt shell sand	silt shell sand	silt shell sand	silt sand shell	sand silt	shell	silt sand shell	silt sand shell	silt sand shell	clay silt sand	clay silt sand
spores & pollen/g	2 191	*	6 220	6 384	5 486	1 896	*	3 819	4 935	1 508	5 489	2 192
Concentricystes/g	0	*	0	0	0	0	*	19	11	0	83	44
fungal spores/g	189	*	1 219	1 096	1 518	566	*	662	642	57	330	385
leaf cuticles/g	89	*	182	0	51	0	*	50	146	0	1 073	105
tracheary fragments/g	246	*	1 057	575	810	149	*	666	434	325	1 156	561
Carya/g	89	*	52	82	34	37	*	85	84	0	41	35
cheno-ams/g	233	*	598	595	354	112	*	297	405	38	206	35
cheno-ams %	26	*	23	14	12	19	*	17	17	7	8	3
Compositae/g	179	*	468	740	1 046	321	*	425	647	229	908	333
Compositae %	21	*	18	25	42	35	*	25	28	40	36	28
Cyperaceae/g	45	*	78	55	152	22	*	89	68	38	124	35
Gramineae/g	112	*	236	329	102	67	*	112	118	141	413	349
Liquidambar/g	0	*	52	55	51	45	*	50	45	19	41	18
Pinus/g	290	*	1 373	2 219	1 619	520	*	1 078	1 430	257	1 321	858
Pinus %	33	*	72	74	66	58	*	63	63	60	52	93
Quercus/g	402	*	320	520	557	224	*	389	602	229	825	143
Quercus %	46	*	17	17	23	25	*	23	26	40	32	16
Taxodium/g	0	*	52	110	17	7	*	58	56	57	0	0
Typha/g	124	*	104	164	101	7	*	15	45	19	41	9
Ulmaceae/g	22	*	78	0	67	30	*	39	90	0	83	35

Table 6.5. *Clear Lake complex*

Sample #	802	803	804	805
Sediment type	silt clay sand peat	silt clay sand	silt clay sand	clay sand silt
spores & pollen/g	9 697	5 201	7 863	11 610
Concentricystes/g	128	79	150	217
fungal spores/g	2 254	1 408	1 090	1 058
leaf cuticles/g	1 063	98	101	248
tracheary fragments/g	723	860	989	1 610
Carya/g	43	860	178	93
cheno-ams/g	0	39	101	743
cheno-ams %	0	2	3	25
Compositae/g	1 361	899	1 902	3 065
Compositae %	28	39	55	73
Cyperaceae/g	43	39	76	62
Gramineae/g	340	274	203	559
Liquidambar/g	0	0	25	62
Pinus/g	866	587	1 578	2 074
Pinus %	16	25	46	50
Quercus/g	1 999	1 290	1 394	1 486
Quercus %	42	56	39	36
Taxodium/g	0	235	76	0
Typha/g	43	0	25	0
Ulmaceae/g	1 148	156	279	248

Table 6.6. *Trinity Bay*

Sample #	3	4	5	6	7	8	9	10	11	12
Sediment type	silt clay	silt shell	silt shell	silt shell	silt shell	silt peat	silt shell	silt shell	silt shell	silt shell
spores & pollen/g	6279	1527	656	6236	164	462	4418	19701	13371	+
Concentricystes/g	0	0	0	0	0	0	0	0	0	0
fungal spores/g	3540	467	341	7904	54	378	431	869	1110	10194
leaf cuticles/g	84	0	0	88	0	0	0	0	0	152
tracheary fragments/g	1686	615	478	4215	52	84	1027	1998	3228	+
Carya/g	84	37	51	176	14	32	62	87	151	152
cheno-ams/g	421	74	0	263	3	21	21	130	50	+
cheno-ams %	10	10	0	16	6	13	4	11	3	19
Compositae/g	759	49	34	966	14	11	62	130	252	1369
Compositae %	7	7	9	58	22	6	11	11	14	49
Cyperaceae/g	84	0	0	0	0	11	0	0	0	0
Gramineae/g	253	0	17	0	0	0	0	17376	10088	228
Liquidambar/g	169	0	0	176	3	0	0	0	50	304
Pinus/g	1138	474	281	703	24	42	349	673	958	1598
Pinus %	43	64	70	42	39	25	60	54	54	57
Quercus/g	1011	160	34	263	17	94	32	391	303	573
Quercus %	38	22	9	16	28	44	14	32	17	19
Taxodium/g	84	0	17	0	3	0	0	0	0	0
Typha/g	0	12	0	0	0	11	0	0	0	0
Ulmaceae/g	253	37	0	88	0	0	82	87	252	76

Table 6.6. Continued

Sample #	13	14	15	16	17	18	19	20	21	22
Sediment type	silt shell	shell silt	silt shell	silt shell	silt shell	silt shell	silt shell	silt shell	silt shell	silt shell
spores & pollen/g	2 382	15 794	2 649	1 383	2 158	2 015	1 161	1 119	10 442	5 353
Concentricystes/g	0	0	0	0	0	0	0	0	0	0
fungal spores/g	838	724	781	377	458	1 343	636	591	11 192	6 447
leaf cuticles/g	0	0	0	0	65	42	112	84	318	365
tracheary fragments/g	658	2 317	1 115	587	1 569	1 847	861	1 266	4 525	2 748
Carya/g	66	48	112	42	65	0	0	0	79	122
cheno-ams/g	66	48	56	42	65	128	0	84	397	365
cheno-ams %	6	4	6	6	6	15	0	27	13	23
Compositae/g	197	241	112	126	65	84	37	84	1 111	304
Compositae %	17	20	11	18	6	10	8	27	37	19
Cyperaceae/g	22	0	0	0	0	0	0	0	0	0
Gramineae/g	132	11 980	0	42	65	0	0	0	159	182
Liquidambar/g	66	0	0	0	65	0	0	0	79	0
Pinus/g	845	845	530	461	458	636	374	232	2 024	912
Pinus %	71	65	54	65	44	75	83	73	68	58
Quercus/g	154	97	112	210	262	42	37	42	635	365
Quercus %	13	8	11	29	25	5	8	13	21	23
Taxodium/g	0	0	0	0	0	0	0	43	159	0
Typha/g	0	0	0	0	0	42	0	0	79	0
Ulmaceae/g	0	145	167	0	65	84	0	42	159	122

Table 6.6. *Continued*

Sample #	23	24	597	598	599	600	603
Sediment type	silt shell	shell silt	sand peat silt	clay silt sand	silt clay sand	silt clay	clay silt
spores & pollen/g	8 434	8 047	2 580	43	1 974	1 600	1 679
Concentricystes/g	0	0	0	0	0	0	0
fungal spores/g	8 434	7 934	5 521	99	4 342	4 110	1 610
leaf cuticles/g	533	609	303	4	79	52	48
tracheary fragments/g	2 989	3 449	2 580	352	1 520	1 130	137
Carya/g	320	345	76	0	39	92	43
cheno-ams/g	320	460	152	0	197	37	48
cheno-ams %	10	19	15	0	11	18	5
Compositae/g	854	690	228	4	197	273	88
Compositae %	27	29	23	40	29	144!	10
Cyperaceae/g	214	115	0	0	39	0	0
Gramineae/g	747	230	0	0	0	0	20
Liquidambar/g	167	0	0	0	0	0	0
Pinus/g	1 121	1 150	530	6	178	141	639
Pinus %	36	48	54	60	26	26	75
Quercus/g	1 068	575	228	0	118	43	65
Quercus %	34	24	23	0	17	8	8
Taxodium/g	0	0	0	0	0	34	0
Typha/g	0	0	0	0	0	0	20
Ulmaceae/g	214	236	76	0	39	19	48

Table 6.7. *Lower Trinity River complex and delta*

Sample #	601	602	604	605	606	784	785	786
Sediment type	silt clay sand peat	sand	silt sand clay	silt clay sand	silt clay sand	clay sand silt	silt peat clay	silt clay peat
spores & pollen/g	1 200	1 928	643	410	104	6 354	3 056	2 420
Concentricystes/g	0	0	0	0	0	0	0	33
fungal spores/g	10 320	9 390	1 555	1 530	288	5 265	1 107	4 507
leaf cuticles/g	144	829	0	21	4	0	27	33
tracheary fragments/g	1 560	2 255	877	371	320	305	54	133
Carya/g	0	39	0	0	0	305	80	99
cheno-ams/g	0	0	0	21	4	131	80	33
cheno-ams %	0	0	0	15	14	6	9	3
Compositae/g	180	289	71	119	12	566	598	99
Compositae %	100!	30	38	92	43	26	67	9
Cyperaceae/g	0	0	0	0	0	0	27	33
Gramineae/g	0	829	0	0	0	44	54	33
Liquidambar/g	0	39	0	0	0	44	27	0
Pinus/g	36	617	116	29	4	1 175	293	431
Pinus %	20	65	62	23	14	43	44	41
Quercus/g	72	116	51	41	16	1 371	125	66
Quercus %	40	13	29	31	57	53	19	6
Taxodium/g	36	0	0	0	4	0	205	33
Typha/g	0	0	0	0	0	0	18	0
Ulmaceae/g	0	0	0	0	0	479	161	365

Table 6.7. *Continued*

Sample #	787	791	792	816	817	818	843	844
Sediment type	clay silt	clay silt	clay silt	silt clay sand	silt clay sand	silt clay sand	silt clay sand	silt clay
spores & pollen/g	1904	5111	10028	1448	6525	7664	25250	24996
Concentricystes/g	67	0	82	87	0	70	110	132
fungal spores/g	1635	6414	3324	1216	43	348	1683	5795
leaf cuticles/g	22	0	494	0	0	35	220	44
tracheary fragments/g	224	36	2225	1361	821	1114	1706	287
Carya/g	0	218	192	29	0	76	276	177
cheno-ams/g	22	0	330	145	173	330	1764	442
cheno-ams %	2	0	14	26	7	20	31	5
Compositae/g	224	725	923	87	605	826	8196	4557
Compositae %	24	39	80	16	23	38	57	47
Cyperaceae/g	22	36	191	0	0	283	418	442
Gramineae/g	45	109	907	0	432	488	2204	1194
Liquidambar/g	0	109	187	0	86	0	110	644
Pinus/g	762	1233	1159	463	2203	1836	2811	6658
Pinus %	83	64	49	89	84	84	50	68
Quercus/g	67	218	357	29	302	209	716	131
Quercus %	9	11	15	5	11	9	12	1
Taxodium/g	0	0	0	0	43	139	606	310
Typha/g	0	0	82	0	43	0	604	0
Ulmaceae/g	0	36	357	0	43	70	491	442

Table 6.8. *Buffalo Bayou – San Jacinto River – Cedar Bayou complex*

Sample #	793	794	795	796	797	798	799	800	801	819
Sediment type	silt clay sand	silt clay sand	silt clay sand	silt clay sand	silt clay	silt clay	clay silt	silt clay sand	silt clay sand	silt clay sand
spores & pollen/g	7357	6381	3356	5638	4412	11104	3818	16341	10598	316
Concentricystes/g	646	775	108	250	50	58	14	53	173	92
fungal spores/g	2062	1156	260	654	50	1038	281	4577	1990	32
leaf cuticles/g	15	25	76	0	99	298	42	0	303	0
tracheary fragments/g	369	1784	454	453	843	1483	477	1543	1231	57
Carya/g	92	58	76	252	50	0	28	160	43	8
cheno-ams/g	46	0	54	0	1154	198	548	3832	1168	8
cheno-ams %	3	0	4	0	14	32	57	30	92	7
Compositae/g	2786	1407	985	1460	644	2307	800	2714	1254	24
Compositae %	154!	50	66	62	45	65	83	65	22	21
Cyperaceae/g	369	75	43	50	99	408	84	53	173	4
Gramineae/g	262	176	195	151	392	750	239	532	476	8
Liquidambar/g	15	100	87	50	50	115	70	160	216	0
Pinus/g	816	3392	628	956	942	2769	659	2874	2465	48
Pinus %	46	49	42	40	66	77	68	69	63	41
Quercus/g	400	728	390	604	299	519	140	639	908	4
Quercus %	23	10	26	26	21	15	14	15	23	3
Taxodium/g	0	201	11	50	0	115	0	319	87	0
Typha/g	0	75	11	0	99	404	14	53	0	0
Ulmaceae/g	346	276	195	503	0	58	28	53	216	44

B Palynofacies and palynodebris sedimentation

7 The genesis and sedimentation of phytoclasts with examples from coastal environments

ROBERT A. GASTALDO

Introduction

Significant quantities of biomass are produced yearly by vegetation in terrestrial communities. The fate of the majority of these plants is to be recycled through biodeterioration and soil-forming processes. A small quantity of all biomass generated is introduced into depositional environments. There it may be subjected to a complexity of preservational mechanisms, the results of which may be the development of identifiable fossil materials. In many instances, though, the plant parts are not preserved in their entirety. The number of fossil plant assemblages (phytocoenoses) that contain exquisitely preserved elements is low when compared to those assemblages in which 'unidentifiable' detritus litters bedding surfaces. In this latter case, select plant elements, predisposed to resist biodegradation, are mechanically fragmented by physical processes operating in the specific depositional environment into which the plant parts were introduced. These meso (2 mm–200 μm) and micro-taphocoenoses ($<200\,\mu$m *sensu* Krassilov, 1975) composed of phytoclasts (*sensu* Cope, 1980) generally are overlooked by workers whose research is focused on macro-taphocoenoses. The potential data-set inherent in the nannodetrital assemblage may provide new and/or complementary information not generally available in macrofossil assemblages (see Tiffney, 1989). Palynologists, on the other hand, often analyze this data set because this particulate organic detritus is recognized as residue from palynological preparations. Based on the physical characters of this residue, recovered plant parts may be placed into either of two broadly transcribed categories, structured or amorphous organic material. Organic geochemists, on the other hand, commonly lump all these elements into the generalized term, Organic Matter (OM; see Huc, 1988), which has been shown to be comprised of various chemical structures as reflected in their pyrolysis-gas chromatography patterns (Huc *et al.*, 1985).

The elements recognized in 'nannodetrital' residues have been classified into major groups that include: structured, terrestrially sourced materials; pollen and spores; charcoal; biodegraded, terrestrially sourced materials; amorphous materials colored yellow-amber; amorphous materials colored gray; biodegraded, aqueous-sourced materials; and structured, aqueous-sourced materials (Masran & Pocock, 1981; Pocock *et al.*, 1987). Venkatachala (1981) has established categories for the differentiation of amorphous organic matter types in sediments, whereas Batten (1983) has attempted to establish an informal descriptive terminology for amorphous matter of land-plant and aquatic-plant origin, and these can be differentiated by their textural characters. It is principally the constituents of the structured, terrestrially sourced detritus category that will be addressed in this chapter. With the advent of plant taphonomic studies on the origin, assemblage composition, and destiny of terrestrial plant parts after their introduction into various depositional environments, we have begun to gain insights into those macrofossil assemblages that grace the exhibits of all major museums. As a byproduct, we have also been able to develop an understanding of those 'plant hash' assemblages once only recognized as 'comminuted detritus,' and those rocks that have been previously referred to as 'unfossiliferous' intervals.

Fate of organic detritus

Plant biomass may be subjected to a complexity of processes in terrestrial ecosystems both during and after the functional life of the plant and/or plant part (Fig. 7.1). These interactions may ultimately play a role in the final state of preservation of any particular plant part or portion thereof. It is necessary to recognize that different plant parts of different systematic affinity are composed of different chemical constituents and, as such, have a wide range of susceptibility to biodegradational, chemical, and physical deterioration. Due to the fact that different plant parts are also constructed of a variety of tissue types, specific components of any plant part may

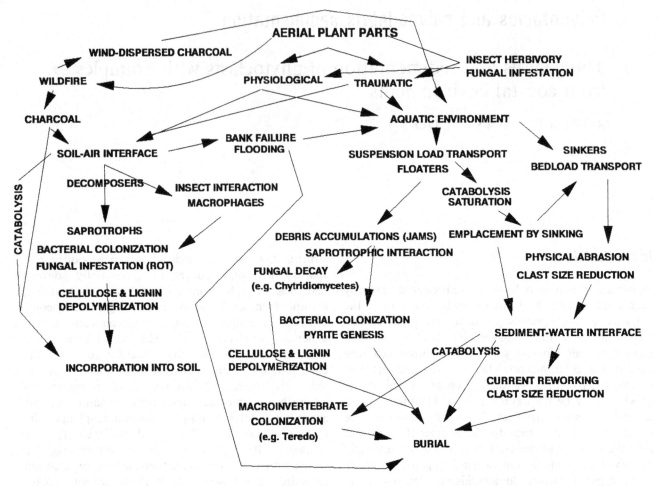

Figure 7.1. Generalized diagram outlining the fate of aerial canopy parts in terrestrial and aquatic environments.

be more resistant than others. In the case of angio-spermous leaves, for example, it will be re-emphasized that the lignin-bearing vascularization and the biopolymer (cuticle) coating the leaf epidermis are more resistant than the intervening mesophyll tissues composed of parenchyma. Physical disaggregation of such resistant parts provides much of the detritus found in processed residues.

Several recent overviews of the taphonomic processes that help shape plant fossil assemblages have provided insights into mechanisms and conditions operating on plant biomass (Gastaldo, 1988; Spicer, 1989). It is not the purpose of this contribution to repeat, in detail, the concepts introduced in these publications. Rather, a brief discussion will establish the biotic and abiotic framework that will allow an assessment of the genesis of phytoclasts.

Plant communities are heterogenous, with taxa distributed variably in space and time. Those plants living in close proximity to a depositional environment or a clastic pathway that leads into that site are more likely to contribute plant parts to that site than those plants living some distance from the area of accumulation (Spicer, 1981; Gastaldo *et al.*, 1987; Gastaldo, 1988). The generation of plant parts is a function of either

physiological or traumatic mechanisms. Physiological mechanisms include abscission of assimilating leaves and shoots, loss of lateral and annual branches, bark shedding resulting from volumetric growth, and dispersal of reproductive structures (Gastaldo, 1988). Traumatic-induced loss, in contrast, is the result of anomalous abiotic circumstances. For example, wind rarely transports canopy leaves any significant distance from the parent plant, and even the tallest temperate trees must be within approximately 50 m of an open body of water to contribute detritus to that depositional site (Ferguson, 1985). Under extreme conditions, such as gale to hurricane force winds, large areas may be defoliated, resulting in the introduction of upper story canopy biomass into adjacent depositional sites. The community is selectively sampled, either with regard to the position of the plants within the tiered community structure (understory and ground cover are rarely damaged except by branch fall from the primary canopy–Scheihing, 1980; Dittus, 1985), or to the differences in histological and morphological features between taxa in the same geographical area (Craighead & Gilbert, 1962). In either case, the plant parts arrive essentially whole in the depositional environment. Although leaves and ultimate branches may be damaged under hurricane-gale conditions by severe abrasion, transport by wind does little to fragment the plant part once disarticulated from the producing organism.

Interaction at the soil–air interface

Biomass introduced onto the soil in which the community grows is subjected to degradation. Decay of litter can be highly selective, with taxa of various histological characteristics reacting differently to deterioration processes (Ferguson, 1985; La Caro & Rudd, 1985). There is little chance of these plant parts or their structural components (e.g., more resistant tracheary elements) entering an aquatic depositional environment once resident on top of the soil. The biodegradational processes operating on these plant parts, though, provide some insight into the biochemical changes that facilitate the formation of phytoclasts. There is a wide assortment and variable proportion of the principal constituents of plant cells (cellulose, hemicellulose, and lignin) across the plant kingdom, not only with regard to systematic affinity but also to the age and maturity of the plant. In arborescent forms, an inverse relationship exits between the proportions of proteins and water soluble components, and cellulose, hemicellulose, and lignin. This is a function of age. The susceptibility of any plant part to microbial decomposition, then, is an index of its 'palatability', energy content (carbon compounds), nutritional value (such as nitrogen and phosphorus and the molecular form of the nutrient source; Ross, 1989), and the physico-chemical environment under which degradation proceeds (Swift *et al.*, 1979). In general, rates of microbial decomposition can be related to the quantity of lignin and nutrient mobilization in any particular plant part. This nutrient mobilization is partly a function of the solubility of the cellular contents, including polyphenols which act as decay-limiting compounds, and susceptibility to catabolic reactions (Swift *et al.*, 1979). In those cells where lignin plays a minor role in cellular structure, mobile elements are quickly lost (Gosz *et al.*, 1973). The bulk of the remaining plant litter, composed of lignitic cells, is decomposed via catabolic (energy-yielding enzymatic) reactions and comminutive (reduction by ingestion and digestion) processes within a few years (Swift *et al.*, 1979).

Taxonomic diversity characterizes the decomposers in terrestrial environments. These include necrotrophs, biotrophs, and saprotrophs. It is the saprotrophic microorganisms, fungi and bacteria, that are two of the principal agents that 'soften' plant tissues during decomposition. This 'softening' is the result of microbial attack, first onto the surface of the plant part, followed by substrate penetration at the cellular and molecular level (English, 1965). Fungi possess a wide spectrum of extracellular enzymes. They are responsible for mechanical penetration of cuticularized leaves, seeds, and pollen/spores by the production of 'boring' hyphae (Alexopoulos & Mims, 1979). This process has been studied extensively with respect to plant parasites, but not directly with regard to saprotrophs. The processes may be similar (Swift *et al.*, 1979). Mechanical disruption of the cuticle is sometimes preceded by enzymatic activity on cutin and cell wall polymers, but this is not requisite in all cases. Resource exploitation by fungi occurs over relatively long intervals of time. The detritus on which fungal action occurs is generally large. Spores may be produced as the resource is depleted. As will be discussed below, the action of fungi on leaf degradation in aquatic depositional environments does not appear to be the principal biotic component responsible for leaf deterioration.

Unicellular microbes interact with resources over relatively short durations, as they are adapted to surface habitats. Their exponential growth rate allows them to exploit rapidly the available resources. Bacteria do not easily penetrate resistant materials, but their small size provides an advantage in colonization of all available surface topographies. They are well adapted to occupy detritus that has a high surface to volume ratio and produce significant quantities of extracellular depolymer-izing enzymes. These substances result in the 'softening' necessary for cellulolysis. Both aerobic and anaerobic bacteria are variable in their physiological processes, and may utilize a particular enzymatic capacity in one set of conditions and not in another. Their surface-bound enzymes restrict their depolymerization to erosion of the surface in the immediate vicinity of the cell. Many bacterial species are closely associated with fungal hyphae and, as such, may occupy an adjunct role to fungi in decomposition in terrestrial environments (Swift *et al.*, 1979). Under anaerobic conditions they may be the primary agents to the almost total exclusion of other organisms. Their role may be more significant in detritus degraded in aquatic systems in that they probably initiate the process of organic maturation.

Factors affecting decomposition

Resource quality affects the rate of decomposition. Each major plant organ – leaves, stems/shoots, roots, reproductive organs – undergoes characteristically different rates of deterioration. In both temperate deciduous and tropical forests, the rate of turnover for reproductive organs is greater than that for leaves, which is greater than that for wood (Swift *et al.*, 1979). Differences in rates of degradation of specific plant parts appear to be consistent, no matter what climatic conditions prevail. Particular components of any plant part, though, may play an inhibitory role in the degradation process. The presence of intra- or extracellular organic compounds that are of a fungitoxic or fungistatic nature will prevent or retard the colonization activity of fungi. Their persistence in senescent tissues most likely affects the decomposition process. Other inhibitory components that act as protective mechanisms include coverings found on the outer surfaces of plants. These are impregnated or composed of resistant chemicals that include waxes, cutin, cutan, and suberin. These operate as barriers in two ways: first, many chemical constituents have a direct fungistatic effect; and second, the cuticle is a physical barrier because it is generally resistant to decomposition. However, recent investigations of the properties of cuticles have demonstrated that two distinctly different polymers occur. One of these polymers is easily degraded, the other not (a non-saponifiable highly aliphatic bipolymer; Nip *et al.*, 1986a,b). This fact, in itself, may bias the potential fossil record. The physical attributes of different plant parts influence decomposition. Leaf degradation may be initiated while the leaf is still functional, as in the case of fungal introduction via insect damage (Fig. 7.1; see below), but in the majority of cases degradation begins after senescence. During the functional life of a leaf, the hydrophobic character of the cuticle interferes with the development of water films. This, in turn, prevents the germination of fungal spores, their subsequent growth and development, and the potential activity of exoenzymes. In addition, epidermal hairs and scales protect stomatal openings, those potential conduits through which fungal spores could be introduced into the interior of the leaf. The original position of a leaf on a plant may also influence the leaf's potential for deterioration. Heath and Arnold (1966) have shown that in *Fagus* and *Quercus*, shade leaves (thin soft leaves from beneath the outer regions of the canopy) are more readily attacked by decomposers and degrade more rapidly than the tough, heavily cuticularized sun leaves. The thickness of the cuticle does not necessarily inhibit the mechanical penetration by fungal hyphae, as infection-pegs from germination tubes have been shown to penetrate various inert substances including gold foil and paraffin wax (Swift *et al.*, 1979). There is, though, a correlation between cuticle thickness and toughness (as measured by a penetrometer), and resistance to fungal penetration (Dickinson, 1960). Once penetration has occurred, catabolic reactions, both of resource-specific and non-resource-specific fungi and bacteria, proceed generally unabated.

The resistance of wood to decay, as compared with leaf litter, is partly due to high lignin concentrations, low concentrations of soluble carbohydrates and nutrients, thick hydrophobic bark, and the presence of a spectrum of modifiers that may be taxon specific. Some decomposers (both microbial and macrobial forms) have evolved lignases and tolerances to a wide range of aromatic modifiers that overcome some of these factors. Insects play a large role in the initial stages of wood decay, with bacteria and fungi relegated to a secondary role. It must be noted that these microorganisms are associated with insects either as a symbiotic gut flora or as a food source for larvae or nymphs (e.g., 'fungal gardens' of ant and termite colonies). Once these microorganisms are introduced into the wood by the activity of insect borers (Gastaldo *et al.*, 1989), their activities can be divided into two phases – passive colonization and active destruction (Levy & Dickinson, 1981). Extracellular enzymes are secreted, which begin to destroy the wood cell walls. Mainly gram-positive bacteria have been noted to colonize exposed woods (Clubbe, in Levy & Dickinson, 1981) but other forms have been reported (Rossell *et al.*, 1973). The main wood rotting fungi are Basidiomycetes, but wood may be infected by various other groups which Levy and Dickinson (1981) term primary molds, secondary molds, stainers, and soft rots.

The architectural framework of wood is considerably different from that of other plant organs such as leaves. Wood has a distinct anisotropy, the result of cell elongation parallel to the axis of growth and a marked localized lateral elongation of cells in a direction perpendicular to that of the primary growth. This cellular organization has a marked effect on the rate of fungal colonization. Colonization of wood by fungi occurs more rapidly along the grain (longitudinal) rather than across the grain (developing radially rather than tangentially). Cellular structure also plays a significant role in wood type and, in turn, its susceptibility to infection and degradation. For fungal decay to begin, the moisture content of the wood must be above 20% (the fiber saturation point in most woods is 30%). Where woods are immersed in waters and completely saturated, microbial activity may be inhibited, but degradatory processes can be initiated by other organisms.

Wood can be subjected to degradation processes at any time, both during life and afterwards. The effects of decay are focused on the cell walls and may include the formation of chains of cavities in the middle layer of the cell wall (soft rots), boreholes through the wall and erosion of the wall from the lumen (white rots), and chemical deterioration of the inner layers of the wall through extracellular enzymatic diffusion, resulting in a friable and disorganized appearance (brown rots; Rayner & Boddy, 1988). It has been demonstrated in the former two enzyme-retentive groups that mucilage is associated with the fungal hyphae, whereas in the latter group no mucilage is found associated with hyphae (Highley, 1976; Green, 1980). This may explain why micromorphological features of degradation can be found associated with the soft and white rots, and are absent where brown rots are the infection agent.

The chemical sequence of cellulose depolymerization is well detailed (Swift *et al.*, 1979), and the enzyme systems involved in the breakdown of other cell wall poly-saccharides is now gaining more attention. Most evidence indicates activity of multi-enzyme systems. The ability to depolymerize lignin is held by a wide variety of organisms, but it is not a common feature among decomposers. The Basidiomycetes are the only group able to metabolize the intact molecule (Swift *et al.*, 1979). Because of their ability to modify the microchemical environment in which they grow, most decomposers are not affected by changes in pH. In fact, fungi appear to be able to regulate their internal pH, maintaining it between pH 5 and 6.

Depositional environments and phytomacrodetritus

The processes and vectors of deterioration active in the terrestrial environment provide the basis from which to evaluate processes operating on plant detritus introduced to depositional regimes. Fossil plant assemblages are best preserved in depositional environments where the biological, physico-chemical, and mechanical degradation processes have been retarded or suspended for various reasons. It is the effects and efficiencies of these processes that probably constrain the type(s) and degree of maturation recognized in 'nannodetrital' residues.

Plant detritus may be introduced into an aquatic environment by various mechanisms: wind, abscission of vegetative parts that have become non-functional, dispersal of spores, pollen, fruits, seeds. For the purposes of this chapter, it will be assumed that the plant parts under consideration have been introduced by physiological processes and transported in a fluvial regime to their initial depositional site. Pollen and spore dispersal, sedimentology, and taphonomy will not be addressed here, as this topic is considered elsewhere in this volume. Many other scenarios can be visualized, but these may not be directly responsible for the development of phytoclasts (e.g., volcanogenic accumulations; see Spicer, 1989). The following descriptions and examples of plant part alteration are derived from studies in Holocene coastal and nearshore environments where degradation processes are the most active, and in the biased opinion of the author, the sites where phytoclasts are most efficiently generated.

Leaves

Upon entry into the aquatic realm, leaves undergo changes similar to those operating on the forest floor. The leaf will become saturated, with accompanying loss of soluble materials (sugars, mobile elements, and some polyphenols) from inter- and intracellular components (Fig. 7.1). The rapidity of saturation will be dictated by cuticle and epicuticular polymer thickness, stomatal and glandular hair (e.g., hydathode) density, leaf anatomy, laminar or petiolar damage (prior to or post introduction), water temperature, and water chemistry (Spicer, 1981; Ferguson & De Bock, 1983; Ferguson, 1985). Soluble chemicals will be lost to the surrounding environment until equilibrium is achieved (Nykvist, 1959a,b, 1961a,b, 1962). This progressive alteration, accompanied by microbial biodeterioration under long-term exposure, will affect the suspension load residence time of any particular leaf. Experimental data regarding flotation time of leaves in quiescent waters suggest that a complex of factors is responsible for residency time at the air-water interface (Spicer, 1981; Ferguson & De Bock, 1983; Ferguson, 1985). Flotation times vary from several hours to several weeks; thin papery leaves tend to settle before thick coriaceous leaves. Thinner leaves break down quickly and saturate faster than thicker leaves. Damage to leaves, such as injury sustained through insect damage or petiolar loss, decreases experimental flotation time due to the more rapid uptake of water. Residency time under flowing water conditions is difficult to support empirically, and most of our data are anecdotal. It is probable that residency times parallel those documented experimentally (see discussion below).

The 'softening' of leaves may be a function of the timing of leaf litter fall (Garden & Davies, 1988) and may be dramatically affected by their polyphenolic constituents. Insoluble polyphenols can delay microbial degradation due to their antifungal and antibacterial properties (Williams, 1963; Benoit, Starkey & Basaraba, 1968). This, in turn, allows catabolic reactions to dominate over the biodegradational processes. Where biodegradational processes occur in aquatic systems it has been assumed that fungal activity predominates over bacterial action (Spicer, 1989), as fungi appear to be the most important agent in leaf decay (Kaushik & Hynes, 1968, 1971). This may not necessarily be the case when the leaves have been introduced either directly into an aquatic regime with a high suspension-load, or they have been transported considerable distances. Fungal activity may play an insignificant role in degradation under these conditions.

The minimal role played by fungi in leaf degradation is exemplified in whole leaves recovered from sediments at depth from various depositional environments in the tropical Mahakam River delta, Kalimantan, Indonesia, and the temperate Mobile-Tensaw River delta, Alabama, USA (Gastaldo *et al.*, 1987; Gastaldo, 1989; Gastaldo *et al.*, in press). Figure 7.2 illustrates this phenomenon. Leaves dehisced from tropical hardwood angiosperms, recovered from a vibracore (Hoyt & Demarest, 1981) extracted in an abandoned, infilling tidal channel of the upper region of the Mahakam delta, show a distinct partitioning in the effects of fungal activity and degree

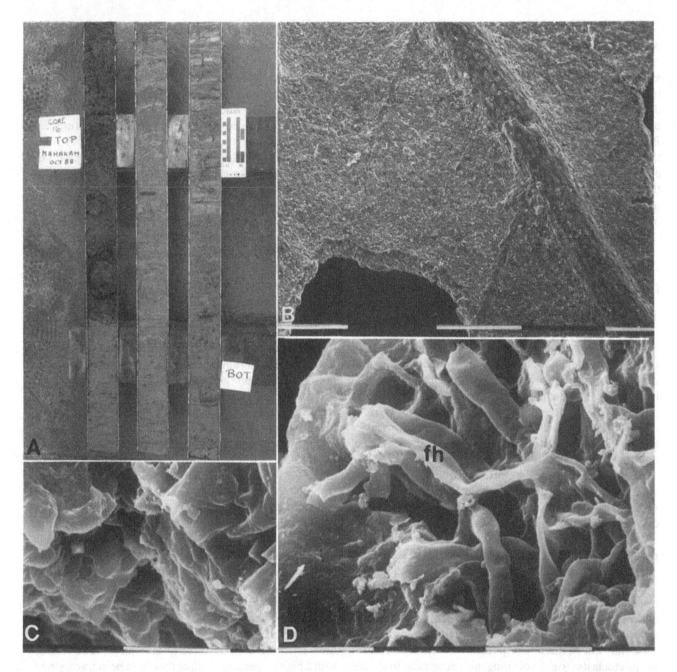

Figure 7.2. Bedded leaf-dominated litter illustrating the supressed role of fungal activity in subaqueous sites of burial. (A) Vibracore 16 recovered from an abandoned upper delta plain tidal creek showing the interbedded organic fill and tidal mud, Mahakam River delta, eastern Kalimantan, Indonesia. Vibracore lengths are one meter. Scale in cm. (B) Scanning electron micrograph (11582) of dicot leaf with insect damage recovered from vibracore 16. Insect damage in the bottom left of the lamina occurred while leaf was still attached to the tree. Bar scale equals 1 mm. (C) Scanning electron micrograph (11584) of the broken edge of the leaf a few millimeters from the site of insect damage. There is no evidence of fungal degradation in these parenchymatous cells. Bar scale 10 μm. (D) Scanning electron micrograph (11581) of the edge of insect-damaged leaf with fungal hyphae (fh) permeating the mesophyll tissue. Bar scale 10 μm.

of degradation (Fig. 7.2A). The figured leaf (Fig. 7.2B–2D) was recovered from within the upper 75 cm of core, in which aerial plant detritus was concentrated and interbedded horizontally with suspension-load mud. Total organic carbon averaged 12.4% in the sample (range 6.57–20.02%). The pH of the sample ranged from 6.48 near the surface to 6.82 at a depth of 50 cm, with accompanying Eh potentials averaging -256 mV at 24.8 °C. The leaf was partially damaged by insects prior to dehiscence from the canopy. It is common to see an entire canopy composed of these insect-damaged leaves, and it appears that they remain functional after insect attack (personal observation, 26 October 88). Their duration time in the canopy afterwards is unknown, but it must be long enough for replacement leaves to develop in growth areas. In these recovered leaves the only area affected by fungal degradation, as evidenced by the remnant fungal hyphae permeating the mesophyll tissue, is that directly adjacent to the site of injury. The fungal hyphae do not appear to have colonized or permeated the remainder of the mesophyll tissues, as other areas of the leaves examined have no evidence of fungal infection. One may suggest that fungal growth was slow and that the remainder of the leaf would eventually be degraded as infection spread. Also, if sedimentation rates in this abandoned channel parallel those in other organic-rich channel infill sequences, the 75 cm thick deposit could represent several hundred years of accumulation. This time frame would provide sufficient opportunity for fungal degradation to proceed if the chemical conditions were amenable, but moderately reducing conditions are indicated by the pH–Eh phase relationships. This, alone, may be responsible for the inhibition of fungal growth, and it appears that these environmental conditions suppressed fungal activity soon after the introduction of the leaves into the aquatic site. This does not appear to be an exception, but rather the rule.

Depositional sites into which leaves are transported have also been examined with regard to the bio-degradational processes operating within them. Gastaldo *et al.* (1987) described an allochthonous plant assemblage deposited in an interdistributary bay of the Mobile-Tensaw River delta. The plant components derived principally from levee community inhabitants were transported a minimum of 10 km in suspension load (some parts such as fascicles of *Pinus* were transported at least 60 km; the producing plants grow extrabasinally) before settling at the sediment-water interface of a crevasse splay channel. Leaves were in various stages of degradation, ranging from oxidized blackened leaves which maintained their structural integrity to leaves in which the laminae were partially decayed. Evidence of fungal activity within these leaves was not found. Leaves recovered at depth in sites that contain moderate bioturbation appear to be chemically degraded by fluctuating oxygen conditions rather than by activity of either fungi or invertebrate detritivores (Rindsberg & Gastaldo, 1989).

Plant components deposited within the major depositional environments of the Mahakam River delta are presently under study. Some of the plant parts accumulate in thick deposits (up to two meters in thickness) along the seaward margins of tidal flats (Fig. 7.3; Gastaldo *et al.*, in press). These detrital peat beaches originate in the tidally dominated interdistributary zone (Allen *et al.*, 1979). Few whole leaves are found intact in the deposit, and the mechanisms for their destruction will be discussed below. Those whole leaves that do occur in the accumulation have either been transported short distances (these are leaves tidally flushed from the mangrove swamp forests fringing the lower delta plain) or great distances (a minimum of at least 50 km originating from the upper delta hardwood community or the interior of the island of Borneo). Leaves that have remained intact during long distance transport and subsequent deposition appear similar to those recovered from aquatic environments in the Mobile-Tensaw River regime. Such oxidized, blackened leaves generally lack structural integrity, that is, they are 'softened'. The majority of whole leaves recovered from the deposit are those of the mangroves, *Avicennia* and *Rhizophora*. The parenchymatous tissues within these leaves have been degraded resulting in a cuticular envelope surrounding the xylary conducting elements (Fig. 7.3). This condition is similar to that described by Spicer (1981) for leaves resident in a limnic setting. Examination of these degraded leaves utilizing SEM reveals that very few fungal hyphae occur within the leaves, if any, and no fungal reproductive bodies are present. If degradation of these leaves was principally by fungi, some sort of spore-producing body should ensue once the food source was utilized and no longer available. Either the entire spore-producing body released spores, or remnants of the structure(s) should be present within these cuticular envelopes. Such conditions have been observed in detrital plant material collected from Florida Bay (pers. observ., Feb. 1989, Laboratoire de Sedimentologie, Université de Paris-Sud, Orsay).

Other leaf fragments display a different style of degradation, albeit without evidence of fungal infection (Fig. 7.4). These leaves/leaf fragments appear to begin degradation from areas adjacent to the venation without specificity to the level of vein architecture. Deterioration results in the separation of the intervein lamina, imparting an irregular aspect to the leaf. It is interesting to note that this not only affects the parenchymatous mesophyll tissues but also the cuticle. Nip *et al.* (1986a, 1986b, 1989) have demonstrated that two different polymers may comprise plant cuticles; one polymer that is easily degraded, the other resistant. Although geochemical data for the cuticle composition of the illustrated specimen are not available, it is probable that this style of break up may be related to chemically degradable cuticle. As discussed above, bacteria are capable of colonizing the surfaces of leaves rather than penetrating the leaf tissues. This predisposes the leaf to decay. The colonized leaf surface originally in contact with the epidermal walls may be affected, resulting in the initial degradation of the anticlinal cell walls where the cutin is first to be altered (De Vries *et al.*, 1967). Continued chemical alteration of the cell walls would lead to their separation. Because these intervenal areas are small (generally less than 1 mm^2), separation and re-entrainment in the water

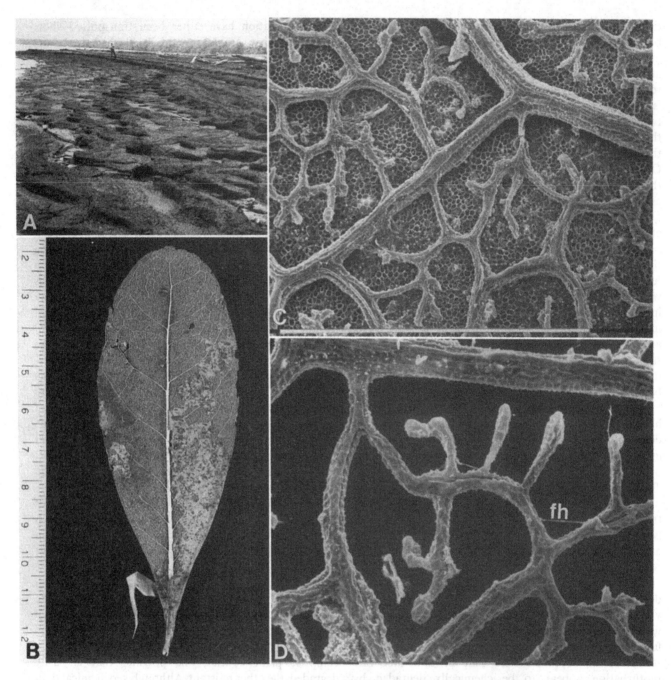

Figure 7.3. Leaf degradation processes, detrital peat beach, Tandjung Bayor, Mahakam River delta, eastern Kalimantan, Indonesia (see Gastaldo *et al.*, in press). (A) View of detrital peat beach. Photograph taken near the shoreline with the perspective looking downdrift. Note reworking of beach face by tidal activity resulting in size-sorting of detritus. Large macrodetritus is commonly found behind the beach berm. (B) Mangrove leaf recovered from the surface of the detrital peat beach. Degradation has proceeded to the point where only the tracheary elements and cuticle are preserved. Scale in cm. (C) Scanning electron micrograph (11563) of mangrove leaf with upper cuticle removed showing complete deterioration of parenchymatous tissues and accompanying absence of fungal hyphae between xylary elements. Bar scale 1 mm. (D) Scanning electron micrograph (11586) illustrating the extent of fungal infection within these leaves. Note that very few fungal hyphae (fh) occur between xylary elements and that some residual amorphous organic material is figured in the lower left. Scale 100 μm.

column would be one means to introduce structured phytoclasts into the water column.

Although it is difficult to observe the types of bacteria responsible for leaf deterioration, the products of their activity remain (Fig. 7.4B,C). The obvious byproduct of sulfur-reducing bacterial processes is the formation of pyrite framboids, signaling early biochemical (diagenetic) activity. Framboidal pyrite has been found to occur on surfaces of leaves recovered from the upper delta plain of the Mahakam. Additionally, framboids have been observed within parenchymatous tissues that border areas that have been damaged by insects. Insects, therefore, seem not only to be a vector for the introduction

Figure 7.4. Various states and byproducts of leaf degradation. Plant parts recovered from subaqueous sites in the Mahakam River delta, eastern Kalimantan, Indonesia. (A) Scanning electron micrograph (12072) showing degradation of leaf lamina and cuticle proceeding within the intervein areas of a dicotyledonous leaf recovered from a vibracore extracted from an upper delta plain tidal channel undergoing abandonment. Scale 1 mm. See Fig. 7.4B. (B) Scanning electron micrograph (12074) of the cells in the separated parenchymatous area adjacent to the resistant venation. No evidence exists such that the degradation can be attributed to fungal infection. Dicotyledonous leaf recovered from vibracore extracted from an upper delta plain tidal channel undergoing abandonment. Scale 10 μm. (C) Scanning electron micrograph (12069) of the surface of a leaf on which framboidal pyrite has resulted from bacterial activity. Pyrite also has been observed within parenchymatous tissues, especially those damaged by insects. Leaf recovered from the upper delta plain. Scale 100 μm. (D) Photograph of *Nipa* palm petiole with attached worm tubes recovered from dredge sample of upper delta plain within a hardwood swamp. Scale in cm.

of fungi into the mesophyll of the leaf, but also for the contemporaneous direct or indirect introduction of bacteria. It appears to the author that the principal biological degradation of leaves resident in aquatic regimes is the result of bacteria.

The predisposition of leaves to invertebrate detritovore activity is enhanced by microbial action (Kaushik & Hynes, 1971; Petersen & Cummins, 1974). This degradational vector, though, is operative when the leaf has settled out of suspension load into a site where invertebrates exist. Their absence, and hence inability to interact with the plant detritus, may be the result of physical and/or chemical conditions of the specific environment. These would include substrate instability, water column agitation, dysaerobic or anaerobic phases, chemical toxicity, etc. Where macroinvertebrates interact with deposited plant detritus, several mechanisms appear to be responsible for its degradation. The most obvious interaction is that of scavenging, feeding, and physical manipulation of the leaf that might introduce nanno-detrital residue either into the sediment or back into the water column. The undigested or indigestible components would be defecated from the macroinvertebrate and concentrated in the sediment either as fecal pellets or as fecal castings. A second less obvious interaction is the utilization of the detritus as a hard substrate for attachment of organisms in sediment of soupy consistency. Although this is more commonly associated with wood and stem detritus (e.g., barnacles, bivalves), leaves are also utilized (Gastaldo *et al.*, in press). For example, it is common to find worm tubes attached to leaves or leaf

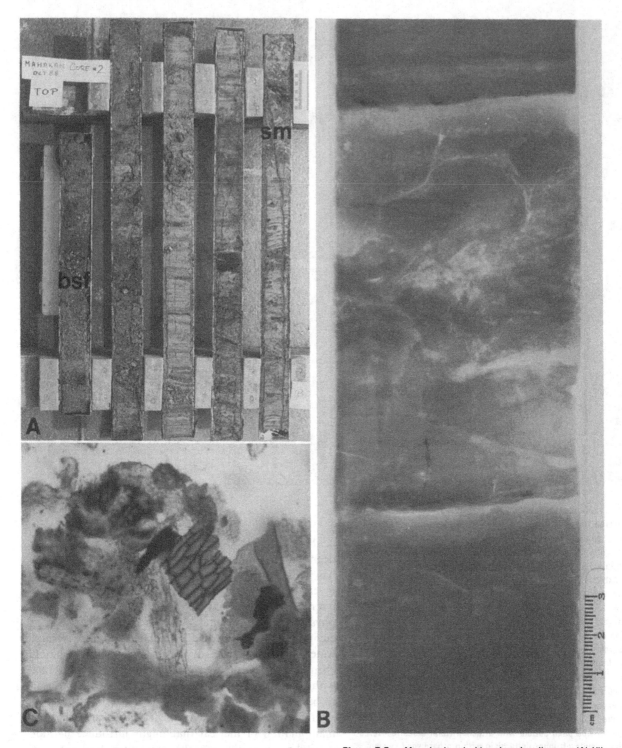

Figure 7.5. Mesodetritus in bioturbated sediments. (A) Vibracore recovered from delta front of the Mahakam River depicting various facies including bioturbated sand with shell fragments (bsf) and sandy mud with sands bioturbated into the mud (sm). Vertical burrows are a common feature in the lower part of the core. Vibracore lengths are one meter. Scale in cm. (B) X-radiograph of vibracore recovered from Conway Creek, lower delta plain, Mobile–Tensaw River delta, Alabama. Bayfill sediment is silty mud interlayered with organics and flat-laminated sand. Small burrows can be seen to penetrate the mud. Additional bioturbation in which the sediment is homogenized is also evident. Scale in cm. (C) Photomicrograph of dispersed cuticle recovered from vibracore 6 (sample 6.1/65) taken in the delta front, Mahakam River delta, eastern Kalimantan, Indonesia.

parts (both recent and fossil) recovered from estuarine and near shore environments (Fig. 7.4D).

Evidence for such nannodetrital generation comes from processed residues of moderate to heavily bioturbated sediments (Fig. 7.5). Delta front sands in the Mahakam River delta are characterized by medium-dark gray fine sand with a slight admixture of mud. Isolated or fragmented invertebrate valves are commonly found intermixed in the facies. The site is heavily bioturbated, yet the dark appearance of the sand suggests organic components dispersed throughout the sediment. The TOC (Total Organic Carbon) in these samples averages a little more than 1% (range 0.61–3.84; average 1.21%). Nannodetrital remains recovered from these sediments include cuticle (see below). Similar assemblages have been recovered from an interdistributary bay fill sequence in Chacaloochee Bay, Mobile–Tensaw delta. The bioturbated bay fill sediments vary from silty mud to sandy silt to muddy sand with TOC contents below 2% (range 0.83–1.15, average 1.25%; D. Felton, unpublished data). Little discernible plant detritus is seen in core or x-radiographs, but dispersed cuticles and few palynomorphs are recoverable (Fig. 7.5C). Homogenization of the deposits at the sediment-water interface appear to occur only where accumulations of macrodetrital plant parts are less than 2–4 cm in thickness (Rindsberg & Gastaldo, 1989).

The principal mechanism by which leaf material is broken ultimately into nannodetrital-sized fragments appears to be physical degradation. As Spicer (1989) notes, one would believe intuitively that the movement of plant parts under high flow regimes operating in fluvial systems would cause the plant parts to fracture. This might be true if a leaf were transported in the bedload and moved along the river bottom by saltation. This, though, is probably a rare occurrence for leaves (see below for wood). This might affect leaves in bedload that had been subjected to microbial attack, because under these circumstances the leaves lose their robustness (Spicer, 1981). If the plant parts were freshly abscised, their degree of fragmentation would be minimal because they retain their structural integrity (Ferguson, 1971, 1985; Spicer, 1981). Transport of leaves occurs primarily in suspension-load where the leaf is moving at the same rate and speed as the fluid surrounding it. This movement may be either through fluvial, tidal, or storm processes, and the velocities that directly affect the plant part may impart a complex history on the deposition and final burial site of that detritus. It is this complex transport and depositional history of any plant part that probably is responsible for nannodetrital generation.

Once the specific gravity of any detrital part exceeds unity, it will settle from the water column. The specific gravity of some plant parts, such as some woods (see below), is already greater than one, and they sink immediately upon introduction to the water column. In other cases, these plant fragments remain floating and undergo progressive water uptake until saturated. Spicer (1989) contends that the final settling site of such debris is determined to a large degree by the objects' submerged density and shape, the two factors affecting settling velocity and entrainment behavior. We must also consider the rate of discharge of the fluvial system and the possible total residence time in suspension-load transport before saturation of the plant part causes it to begin settling. In addition, differences in temperature in the water column and, hence, differences in water density,

will affect the site of transport within the water column. If one were to judge the quantity and quality of transported biomass by that which is observable on the water surface of either the rivers in the Mobile–Tensaw or Mahakam deltas, this biomass could not account for the estimated quantity deposited throughout these regimes. The recovery of whole leaves, intact mosses, and insect parts from depositional sites in which active riverine processes occur (lateral channel bars [Mahakam], channel bedload deposits [Mobile], distributary mouth bars [Mahakam, Mobile], tidal flats [Mahakam]), is suggestive that the plant parts were either resident in suspension load for a short time interval (days) or were never transported in bedload. It seems more likely that, as the specific gravity of these leaves changes with continued saturation, they move progressively down in the water column, being transported at different depths through time. This would be correlative with transport of woods of various densities.

It appears that settling from the water column is controlled principally by fluctuations in water velocity, particularly by velocity reduction. In the Mahakam River delta, where sedimentation is affected strongly by micro- to mesotidal processes, the lateral distributary channel bars are characterized by a cyclical depositional pattern. Although the competence of the Mahakam River is low (fine to medium sand), it has a very high capacity. The channels are extremely muddy. It is not possible to see your hand when it is held more than a few centimeters below the surface of the water. Bedload sand is deposited in migrating ripples along the inside bends of fluvial channels, where water flow is less than that on the outside of the curve. Mud and/or plant litter beds overlie these sand ripples, and are deposited in response to flood tidal energy that moves upriver (maximum spring tidal displacement is greater than two meters). The spring tidal bore causes some, but not all, of the saturated plant detritus to settle out of suspension. The litter beds are composed of leaves, mosses, and insect fragments, and they occur across the entire width of lateral channel bars. The same process may result in the settling of leaves along the shoreline on tidal flats, but grounding of this detritus at the sediment–water interface also occurs as current flow wanes over these depositional sites.

Leaves may undergo deposition and entrainment several times before final burial (Gastaldo, 1988; Spicer, 1989). Variations in discharge rates and/or migration of a channel system will re-entrain plant parts initially deposited in fluvial environments. It is this potential for continued re-entrainment of saturated leaves, whose structural integrity has already been jeopardized, that provides the biomass requisite for the genesis of vast quantities of phytoclasts.

The activity of waves and tides in coastal and nearshore shallow waters best accounts for the re-entrainment and ultimate maceration of leaves after they are transported

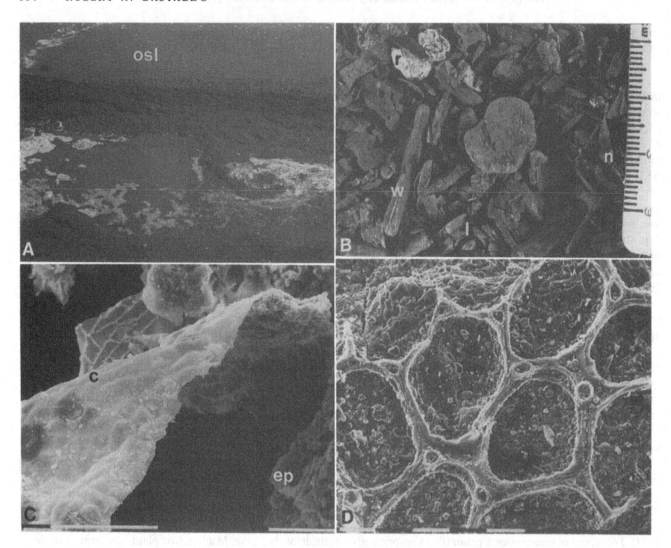

Figure 7.6. Process site and results of macrodetrital physical degradation, detrital peat beach, Tandjung Bayor, Mahakam River delta, eastern Kalimantan, Indonesia (see Gastaldo *et al.*, in press). (A) Photograph of low tide shoreline along detrital peat beach. Continued agitation of plant debris occurs in the organic slurry (osl) at the margin of the beach. (B) Photograph of select sample of detrital peat fragments from peat beach including wood (w), leaf laminae (l), fibers of a *Nipa* petiole (n), and detrital dammar (r). Scale in cm. (C) Scanning electron micrograph (12075) of a dicot leaf fragment in which the cuticle (c) is undergoing separation from the epidermis (ep). (D) Scanning electron micrograph (11606) of a leaf fragment showing epiphytic bryozoan growth on the leaf. The bryozoan colony is represented by the raised geometrical pattern. Scale 100 μm.

into the estuarine and nearshore marine environments (Rindsberg & Gastaldo, 1989; Gastaldo *et al.*, in press). It is within these sites that leaves, resident in the water column and/or resident at the sediment–water interface, are subjected to continous agitation when they are above wave base (Fig. 7.6A,B). Leaves may be brought to these sites either directly through fluvio-marine processes, by which the biomass is transported to the delta front and redistributed along the coastal zone by wave and tidal action, or after some residence time offshore (it is not uncommon to find leaf fragments on which colonies of bryozoa have developed [Fig. 7.6D], these reflecting clear quiet-water offshore conditions). In both cases, the leaves have been subjected, to some degree, to catabolic and microbial degradation subsequent to their introduction into the fluvial regime. This loss in structural integrity predisposes plant parts to mechanical fragmentation in the organic slurry that is found inside the breaker zone (Fig. 7.6A). Agitation and breakup continues while the plant part is entrained within the breaker zone, and is stopped (sometimes only temporarily) when the plant

parts are deposited on the tidal flat/peat beach during tidal flux.

The plant parts that are generated reflect the contributing biomass (Gastaldo *et al.*, in press) and include leaf petioles, intervenal leaf laminar fragments, incomplete tracheary elements (leaf venation), and isolated cuticle (Fig. 7.6B,C). The size of the leaf pieces range from rare whole leaves to fragments, the most common size being less than 1 cm in maximum dimension.

Figure 7.7. Photograph of rippled sandstone in which nannodetrital fragments are concentrated in ripple troughs. Lower Pennsylvanian 'Pottsville' Formation, Black Warrior basin, Alabama. Scale in cm.

Gastaldo *et al.* (1987) also reported that plant part and size sorting of the detritus occurs in these deposits. In the case of peat shoals in the Mobile–Tensaw delta, the 'shoreface' of the shoal (that adjacent to open water conditions and subjected to tide and wind-generated wave activity) was composed more of woody fragments than the 'back barrier' sites. These latter sites included more leaf fragments, which was interpreted to indicate that some winnowing of less dense leaf materials occurred in the swash zone. Additionally, there were more smaller plant parts (size category 0–0.5 cm) in the front of the shoal than behind its crest. Comparative data do not exist for the peat beaches in the Mahakam delta, although differences were noted to occur in particle sizes composing bedding features within the peat beach.

When examining bulk samples of these organic-rich accumulations (TOC as high as 39.4%), it is evident that plant part fragments less than or equal to 1 mm are not a large component of the deposit. In fact, they are conspicuously low quantitatively. This size category of plant fragment probably is maintained within the water column and remains in suspension-load. As tidal fluctuations rise and wane, these nannodetrital particles are transported away from their site of mechanical deterioration, and may be retained in the water column for some time before settling to the sediment–water interface. Often, this mesodetrital fraction behaves sedimentologically in a manner similar to that of mica. Once meso- and microdetrital particles sink, they may be dispersed across the bedding surface or concentrated particularly in ripple troughs (Fig. 7.7). These accumulations are often found in offshore depositional sites, at or below wave base.

The burial of plant detritus does not preclude it from degradation or further alteration. In fact, depending upon the depositional environment in which leaves accumulate, they may be more subject to continued degradation and alteration. Of the environments of deposition examined to date, sand-rich sites seem to be the environment in which these processes operate most effectively. Rindsberg and Gastaldo (1989) note that leaf litters buried in distributary mouth bar sand at a depth of less than one meter are quickly degraded where the water level fluctuates in response to variation in discharge rates, prevailing wind patterns (strong directional winds may lower water levels particularly in protected bays), and tidal cyclicity. The porosity of the sand will subject the buried litter to some flushing which may accelerate the loss of soluble compounds. The change in water level, which may subaerially expose the distributary mouth bar sand body, introduces oxygen to the litter accumulation(s) which, in turn, accelerates chemical degradation. In other instances, where litters have been buried and isolated from oxygenating waters, bacterial-induced methane production occurs. Although the distributary mouth bar is bioturbated mostly by polychaete worms, these organisms do not utilize the plant resource for food (Rindsberg & Gastaldo, 1989). Rather, when they penetrate a buried leaf, an oxidation rind occurs at the edge of the burrow in contact with the plant debris. The worm does not exploit the buried litter horizon, and this is possibly due to microenvironmental chemical conditions. Where the sand is colonized by plants (aquatics and semiaquatics), aerobic decomposition of buried leaf debris proceeds most effectively where aerenchymatous roots provide oxygen to the rhizosphere. The result of these processes is the formation of an amorphous organic-stained horizon in which the most resistant plant parts (e.g., tracheary elements and cuticles) occur. Distributary mouth bars are ephemeral coastal features, and when the site is breached by higher discharge flood waters, phytoclasts are reintroduced into the water column and transported to a new depositional site.

Other coastal sites may be subjected to tidal and wave reworking, and the leaf litters that were deposited contemporaneously with primary sedimentological structures are also reworked mechanically to nanno-detrital size. The best documented example studied, to date, occurs in Mayor Bayor within the interdistributary zone of the Mahakam River delta. The area is an inactive fluvial regime now subjected to tidally dominated processes. In the conversion from fluvial to tidal domination, the lateral channel and distributary mouth bars are exposed to higher tidal and wave energies because of the abated influence of fluvial discharge in the area. Plant litter horizons recovered from vibracores extracted from these sites are composed of nannodetrital fragments of leaves, wood, cuticle, and unidentifiable materials. Most plant parts have been physically abraded to a size equivalent to that of the fine sand comprising the clastic fraction. Again, these sites may be ephemeral, and when they are affected by higher energy, low-frequency events, phytoclasts would be entrained in suspension load and flushed from the interdistributary zone. This would result in the transport offshore of the plant fragments.

Dispersed cuticles

This structured phytoclast is generally viewed as being composed of resistant cuticles that have been separated from leaves. Most dispersed cuticles can be identified systematically, although some cannot be interpreted as parts of well definable plant organs or entities (Kerp, 1989; Upchurch, 1989). Cuticle segregation and subsequent transport theoretically results in the accumulation of isolated fragments with the same hydrodynamic properties as other parts of similar size and density. Dispersed cuticles occur within a variety of continental and marine depositional environments (open marine, deltaic and nearshore coastal, fluvial, lacustrine, paludal, and volcanogenic) and a variety of sedimentary rocks (claystone, limestone, shale, siltstone, sandstone, carbonaceous shale and coal, and tuff; e.g. Batten, 1979; Clark et al., 1986). Indeed, recovered specimen suites may represent isolated cuticles dispersed within a sedimentary environment, or they may represent residual resistant structures in an environment where degraded leaves are dispersed and not identifiable as macrofossils.

The segregation of cuticle from a leaf may occur either by biochemical activity, physical manipulation, or a combination of processes. Complete degradation of parenchymatous tissues within a plant organ can result in the isolation of the protective cuticle (Figure 7.3; Gastaldo et al., in press). This is not only applicable to aerial plant parts, but also to subterranean rootlets. For example, rootlet cuticles account for greater than 65% of the maceration residuals processed from Nipa palm and hardwood swamp soils in the Mahakam River delta. Accessory plant parts in these settings include resistant fibrous and woody components, leaf laminae, and a small proportion of dispersed leaf cuticle. This assemblage appears to be a common feature of subaerially exposed environments.

Bacterial attack on the anticlinal walls of parenchymatous tissues in leaves is probably responsible for initiation of weak areas between these cells and the overlying cuticle. Once this plane of weakness is established, separation of the cuticle can begin. In marginal sites of aquatic environments that are subjected to wetting and drying, the alternation of tissue swelling and contracting may accelerate separation (Fig. 7.8). Cuticle dispersal occurs as small fragments that are mechanically separated, either through the action of wind, moving the leaf along a subaerially exposed strandline, or in water by bedload transport by either littoral currents or tides (Fig. 7.6A,B). The density, size, and complement of accessory hairs and papillae of any cuticular fragment generated in this mode will play a role in subsequent transport and ultimate deposition.

Assessments of each major depositional environment in the Mahakam River delta have been made as to the contribution of plant part categories (unidentifiable millimeter-sized resistant detritus, mosses, woody and fibrous detritus, Nipa palm petiole, leaf laminae, dispersed cuticle, reproductive structures [fruits and seeds], roots and rootlets, and damar [resin]) to each litter-bearing accumulation. Subaerial swamps, fluvial and tidal channels, subaqueous tidal flats, interdistributary tidally influenced sites and peat beaches, and the delta front have been analyzed. Samples recovered from vibracores were hand-picked to recover isolated cuticle from dried litter accumulations. Complementary samples from each of these environments were processed for dispersed cuticle (using methods of Upchurch, 1989), even where there appeared to be no bedded or identifiable plant remains in the sediment. In general, the frequency of occurrence of isolated cuticles recovered from dried samples was low. Quantitatively, isolated cuticles comprise less than 2.3% of all plant parts recovered from bedded litters (range 0.7% in Nipa swamps and upper delta tidal channels to 5.1% in detrital peat beaches). It is difficult to assess the weight percentage of dispersed cuticle residuum from processed samples (these account for less than 0.01 gm of residue), but well preserved cuticles have been

Figure 7.8. Genesis of dispersed cuticle from leaf deterioration. Recovered leaf from tidal strandline along margin of Mobile Bay, Fairhope, Alabama, showing separation of cuticle from underlying parenchymatous and xylary tissues. This appears to be the result of alternating wetting and drying. Scale in cm.

recovered from every conceivable depositional site sampled. Some of these may not represent actual dispersed cuticles, but rather the remnants of either fragmentary or whole leaves dispersed within the sediment. The ubiquitous character of dispersed cuticle in palynological residues from various lithofacies has permitted the use of cuticles as a criterion for interpreting facies relationships (e.g., Parry *et al.*, 1981; Batten, 1982; Batten *et al.*, 1984), but their utility for facies discrimination is just now being tested.

Wood

The physical properties of any particular wood circumscribe the way in which that wood will behave during transport and subsequent interactions after initial deposition. The mechanical and nonmechanical wood properties are affected by fluctuations in the quantity of water present. Water in wood occurs both as bound water located in the cell walls and as free water within cell lumina. When the liquid phase of water exceeds a wood's fiber saturation point (generally between 25 and 30% moisture content) it accumulates in the cell lumina, and maximum moisture content is reached (Panshin & De Zeeuw, 1970). The total amount of accommodated water is limited by the void volume of the wood. Living trees will not contain more than one-half to two-thirds as much moisture as is theoretically possible. Wood that is resident in an aquatic system may become completely saturated.

The specific gravity (density index) of wood depends on the size of the cells, the thickness of cell walls, and the interrelationships between secondary xylem elements (fibers, wood parenchyma, tracheids and/or vessel elements). This characterization is based on oven dry weight, and although this is only a relative index of the cell wall materials, it provides a guide to indicate whether a particular wood is considered to be light, moderately light to moderately heavy, or heavy. Such parameters, in turn, reflect the ability of any wood to be a 'floater' or a 'sinker' upon initially entering an aquatic regime. This is important with regard to the residency time in suspension load prior to settling to the sediment–water interface. The ability of a non-degraded piece of wood to float is determined by the buoyant force that develops in response to the difference between the wood density and that of the water displaced by the fully submerged part (Panshin & De Zeeuw, 1970). Whereas woods that consist of at least two-thirds cell wall substances will sink immediately in water, most woods must first become saturated, whereupon the air spaces within are filled until the density of the wood equals or exceeds that of the displaced water. At this point, the wood moves to the bedload. It must be noted that decayed wood absorbs water more quickly than wood that has not been subjected to degradational mechanisms. Rot affects wood in a variety of ways. 'Dry rot' (attack by *Merulius lachrymans*) causes shrinkage and cracking into rectangular blocks, and the tracheid wall composition is significantly altered and grossly degraded (Harris, 1958). Wood density is changed through a reduction in dry weight. These factors will reduce residency time in suspension load because the wood will equilibrate more rapidly with the surrounding water.

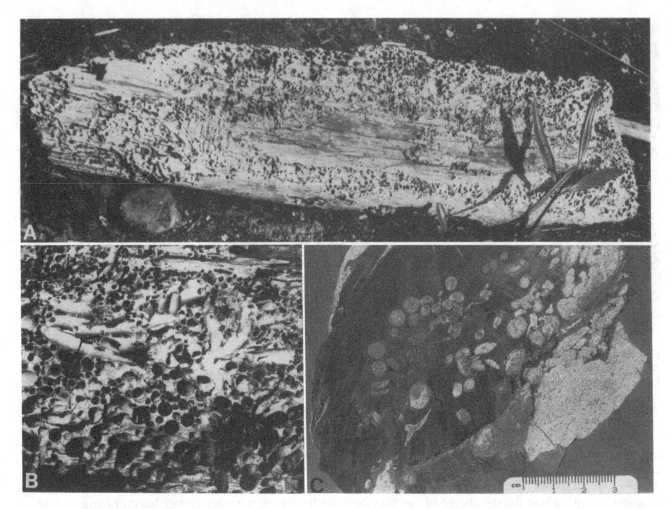

Figure 7.9. Wood as a host for endobionts. (A) Bored drift wood lying on top of detrital peat beach, Tandjung Bayor, Mahakam River delta, Indonesia. (B) Enlargement of wood borings referable to *Teredo*. (C) Transverse section of a bored Paleocene wood (boring cf. *Teredolites*), Clayton Formation, Alabama. Borings are infilled with marl, calcite, and pyrite/marcasite. Scale in cm.

The mechanisms responsible for the introduction of wood into aquatic systems are varied (physiological or traumatic loss, flood-derived, stream bank failure, etc.). Wood may be subjected to alternating periods of submergence and exposure prior to final burial. Although wood is more resistant to deterioration by microorganisms than are most other plant tissues, it may be degraded to varying degrees before introduction into and transport in an aquatic system. Insect damage can occur both to living and dead wood in emergent environments. Changes in the physical properties can be caused, for example, by beetle mining (e.g., families Scolytidae and Platypodidae), larval development under the bark or within the wood, termites (dry-wood and damp-wood types), and carpenter ants (*Camponetus*). Once wood is submerged and resident at the sediment–water interface, it can become the site of anchorage for various macroinvertebrates and also subjected to the activities of marine borers including the molluscs and crustaceans (Fig. 7.9). The molluscan genera – *Teredo*, *Bankia*, and *Martesia* – attach themselves by their caudal end to wood and begin to burrow by grinding (Turner & Johnson, 1971; Hoagland & Turner, 1981). Grinding is accomplished through the action of a pair

of toothed shell valves. It is believed that these molluscs not only use the wood for a habitat, boring into and becoming imprisoned within the wood, but also utilize it as a food source. Crustaceans use wood both as a site for inhabitation during some part of their life as well as for a food source (e.g., *Chelura*; Barnes, 1974). They bore using a pair of toothed mandibles and seldom extend their burrows very far into the wood. The role that wood plays in the dietary requirements of these animals is still speculative. Some authors contend that wood passes unaltered through their digestive tracts. Chemical analyses of wood passed through the gut of *Limnoria* (an isopod) indicates that these borers are capable of removing nearly all the hemicellulose and about half of the cellulose (Lane, 1959; Ray, 1959). In either case, nannodetrital-sized fragments of structured wood are mechanically separated during boring and can be introduced into the water

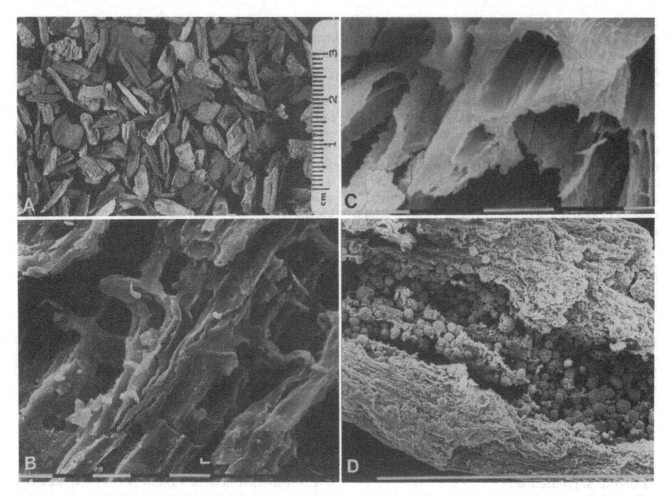

Figure 7.10. Characteristics of bedload-transported detrital woods. (A) Variously shaped wood fragments recovered from detrital peat shoal deposit, Chacaloochee Bay, Mobile–Tensaw River delta, Alabama. Wood illustrated represents the 0.5-1.0 cm size fraction. Scale in cm. (B) Scanning electron micrograph (11573) showing angularity of cell walls that have been mechanically fragmented during transport. Wood specimen recovered from detrital peat beach, Mahakam River delta, eastern Kalimantan, Indonesia. Scale 10 μm. (C) Scanning electron micrograph (12057) illustrating wood parenchyma and fibers, neither of which are infected by fungi and, hence, predisposed to microbial biochemical degradation. Fibers appear to show mechanical fracture along walls. Specimen recovered from vibracore 22.6/55. The vibracore was taken in a lateral channel bar of an active distributary channel, Mahakam River delta, eastern Kalimantan, Indonesia. Scale 10 μm. (D) Scanning electron micrograph (12061) of a wood fragment in which early diagenetic framboidal pyrite has formed in response to bacterial activity within a cavity. Specimen recovered from vibracore 8.2/15, taken in the tidally influenced lower delta plain, Mahakam River delta, eastern Kalimantan, Indonesia. Scale 1 mm.

column. Ingested wood, chemically altered through digestive processes, may also be introduced into the water column, possibly as amorphous organic matter. Consequently these phytoclasts should show some signs of organic maturation.

In most instances, though, the reduction of wood to nannodetrital-sized clasts is more a function of its long distance transport and/or interactions with bedloads of sand-sized or larger clasts. Channel lag deposits that may or may not be ephemeral are comprised of material with the highest settling velocities. Organic constituents are commonly logs and larger, robust, fruits and seeds (Spicer, 1989), and this fact has been used often to identify the bounding surfaces of fluvial architectural systems. Under unusual transport conditions, however, log accumulations may be found terminating a channel fill sequence reflecting their role in abandonment processes (Degges & Gastaldo, 1988). As discussed by Gastaldo *et al.* (1987), the residency time of wood in bedload and the distance transported in bedload affect its physical aspect. Transport of wood in suspension load and subsequent residency time at the sediment–water interface might allow for the propagation of fungal degradation, if active, thereby chemically altering the cell walls. This would predispose the wood to size and shape alteration when abraded. Woods sampled from vibracores and detrital peat beach environments have been examined. To date, these show little sign of fungal infection that might cause them to be 'softened' after which mechanical alteration of shape could proceed quickly (Fig. 7.10). Rather, it is generally believed that water retention within wood (to the point of fiber saturation) appears to be the principal mechanism responsible for softening of wood when submerged. It has been demonstrated that the addition of polar liquids to the cell wall constituents affects the microfibrillar arrangement and results in its expansion, a volumetric change in the cell wall (Panshin & De Zeeuw, 1970). The dimensional changes in any particular wood are a function of both the moisture quantity and the amount of cell wall substance. The more cell wall material present in any particular wood, the greater the dimensional changes (e.g., swelling) when moisture is added. This swelling of the cell wall and expansion of the architectural fabric predisposes the cell wall to simple fracture when subjected to mechanical stresses during bedload transport.

Bacterial colonization of wood clasts may occur, and their activity is reflected in the chemical byproducts of their respiratory activity. As demonstrated for transported leaves, the presence of early diagenetic pyrite in wood can be documented (Figure 7.10D) in certain environments in which specific chemical conditions operate. Wood containing pyrite framboids has been recovered at depth from vibracores extracted in estuarine conditions in the Mahakam River delta. Pyrite appears not to occur on exposed surface areas, but is confined to cavities within the wood where interaction with flowing waters would be minimized. These degradation products may result more in incipient permineralization near the site of pyrite formation than in large scale degradation resulting in nannodetrital production.

The physical abrasion of the cells on the exterior of any wood particle being moved in bedload mechanically breaks portions of saturated cell walls from the phytoclast. This results in the progressive rounding of wood clasts during transport and may be a relative indicator of residency time in bedload (Gastaldo et al., 1987). The alteration in clast shape from sharply angular to smooth and rounded is analogous to that of lithoclasts. The parts of cell walls and/or groups of cellular constituents broken from the phytoclast are commonly nannodetrital-sized. Their removal results in an irregularity in the form of the wood clast when viewed under SEM (Fig. 7.10B). Because this phytoclast is principally composed of residual lignin (presumably the cellulose and hemicelluloses have been degraded), the mesodetrital-sized clasts can be transported in suspension-load over significant distances and incorporated into a wide diversity of depositional environments.

Resin, copal and dammar

Resins are plant secretions that solidify when exposed to air, and approximately 10% of extant plant families living in temperate and tropical climates synthesize them (Langenheim, 1990). Resins are characterized as mixtures of either terpenoid or phenolic compounds, although the chemical composition is variable. The terpenoid resins have been examined in more detail because these commonly fossilize (Grimalt et al., 1988). Resins are common in certain gymnosperms and angiosperms, and resins from plants of different systematic affinity commonly have been given distinct appellations. All conifers produce terpenoid resins, but only the members of the Pinaceae and Araucariaceae synthesize large quantities. These secretions are generally referred to as resins or ambers. The production of resins in angiosperms is limited to a few families. Several genera of the Leguminosae with a South American–African distribution produce large quantities of diterpenoid resins, and these hardened masses are referred to as copals. Members of the tropical Burseraceae and Dipterocarpaceae produce resins that contain triterpenoids and have been termed dammars (an Indo-Malay word for resin).

These secondary substances are produced either within ducts found in various tissues/organs or in specialized surface glands. The function and ecological role of this exudate is speculative (e.g., Fraenkel, 1959; Ehrlich & Raven, 1964; Langenheim, 1990), but one thing is certain, copious quantities can be produced and accumulate on the exterior of a plant after wounding has occurred. Polymerization of these terpenoids may be initiated by light and oxidation (Cunningham et al., 1983) resulting in their hardening within weeks to months after exudation. The masses so produced are highly resistant to environmental influences (Mills et al., 1984). Resins may be dissociated from the tree in various ways, and this unstructured plant synthate may be introduced to and incorporated within various depositional regimes. These plant-derived secretions can be a common component in nannodetrital residues of fluvial and nearshore facies.

Resin ducts are most commonly found in wood. They may be a characteristic feature of some while their occurrence in other wood types is the result of injury (traumatic resin ducts). Resin ducts can be vertically oriented or oriented both vertically and horizontally within the wood, and lined with thick-walled epithelial cells (Fahn, 1969). The ducts are generally filled with synthate. Upon death of the plant, and exposure to degradation processes, the resin within the ducts undergoes polymerization and hardens in place. Pieces of wood with polymerized resin ducts may be transported in bedload and subjected to physical abrasion releasing small cylindrical masses (on the order of 300 μm in diameter) into the sediment load (Fig. 7.11). In litters recovered from vibracores extracted in fluvial, delta front, and tidal flat environments of the Mahakam River delta, small cylindrical fragments of dammar occur, in addition to angular to rounded wood fragments in which polymerized resins can be identified.

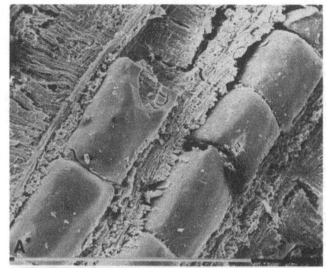

Figure 7.11. Occurrence of detrital plant resins. (A) Scanning electron micrograph of solidified dammar in resin ducts of hardwood angiosperm wood (?Dipterocarpaceae). Wood is an allochthonous fragment recovered from vibracore 22.6/55. The vibracore was taken in a lateral channel bar of an active distributary channel, Mahakam River delta, eastern Kalimantan, Indonesia. Scale 10 μm. (B) Suite of allochthonous dammar clasts taken from detrital peat beach samples, Tandjung Bayor, Mahakam River delta, eastern Kalimantan, Indonesia. Borings (b) occur in dammar of various size, and larger pieces may act as a substrate for bivalve (oy) attachment. Scale in cm.

Larger quantities of resin may be incorporated when masses of hardened exudate are introduced into the transport system. Trees damaged by trauma or insects respond by producing vast quantities of resin, deposited on the exterior of the tree around the injured site. For example, Thomas (1970) estimated that resin accumulated from *Agathis australis* in New Zealand over, presumably, the past 1000 years amounts to 25 tons per km^2 within the geographical distribution of the species. This resin polymerizes and hardens both around the wounded area and adjacent to the tree, where the viscous material has been deposited. If trees grow adjacent to waterways, resins may fall into the river system and be transported

downstream. The densities of the various terpenoid resins are all slightly greater than one. This property causes resin masses of different sizes to sink to bedload and be transported by rolling and saltation.

The dominant trees of some southeast Asian forests belong to the Dipterocarpaceae, a family of plants known to produce prodigious quantities of dammar. Although dipterocarpaceae triterpenoid resins do not polymerize as easily as the diterpenoid resins (Langenheim, 1990), angular to rounded pieces are common components of higher energy depositional sites in the Mahakam River delta (Fig. 7.11B). Where present, dammar generally accounts for less than 1% of identifiable plant parts, but in specific depositional environments (delta front, tidally influenced interdistributary area, and detrital peat beach) this component may account for as much as 4.5% of the recovered plant parts. Individual irregularly shaped pieces of dammar show features acquired during transport including rounding and fracturing. In addition, they may shown signs of macroinvertebrate interactions that include boring and utilization as a firm substrate. Coloration, translucence/opacity, and friability of the dammar do not appear to be characteristics that can be used to infer age of individual clasts. Gastaldo *et al.* (in press) report ^{14}C dates for two dammars recovered from a detrital peat beach. One piece appeared to be 'young' based on its amber coloration (dark yellowish-orange, 10 YR 6/6) and translucent appearance; another piece appeared to be 'older' based on its 'weathered' opaque appearance (bluish-white coloration 5 B 9/1). The ^{14}C age of the translucent specimen is 2645 ± 215, whereas the opaque specimen has yielded an age date of 930 ± 205.

Whitish-colored opaque dammars generally are more brittle and friable than translucent pieces. Some are easily crushed under slight pressures. The structural integrity of translucent dammars appears to be better, although these may fracture conchoidally. With regard to both conditions, nannodetrital-sized fragments can be broken from these materials during initial transport and subsequent re-entrainment in the water column. Accumulations of such detritus may remain as a residuum in palynological preparations most often identified as resinous droplets and lumps.

Charcoal

Terrestrial plant parts in various climatic regimes may be subjected to combustion during naturally occurring wildfire. Pyrolysis of unburnt plant tissues may accompany the rapid temperature increase resulting in the formation of charcoalified detritus (Chaloner, 1989), which is chemically inert (Mantell, 1968). Although charcoal is generally thought of as derived from wood, softer parenchymatous tissues of leaves and flowers may also be preserved (Scott, 1989). This physico-chemical plant byproduct can be identified by megascopic, microscopic, and chemical features (Harris, 1958, 1981; Cope, 1981; Cope & Chaloner, 1985). Megascopically, charcoal is lustrous and brittle. This latter character is best demonstrated by the development of a black 'streak' when a specimen is rubbed on paper. Charcoal fragments are commonly cuboidal, due to rigidity and the development of longitudinal and transverse shrinkage cracks during pyrolysis. Shrinkage estimates approach 45% by volume (McGinnes *et al.*, 1974). Charcoal may also form short, angular, rounded or amorphous clasts (Harris, 1958; Herring, 1985), often splintery in appearance in residues (Batten, 1973). Its density is low, the specific gravity of uncompressed fusain powder has been reported to be 0.28 (Harris, 1958). Charcoalified plant parts are relatively resistant to oxidizing chemicals (which accounts for their presence in palynological residues), have a high reflectivity, and appear porous, a reflection of the configuration of homogenized cell wall structure. Charcoal particles retain more of their original identifiable cellular constituents when found in nearshore waters, as compared to particles recovered from open ocean sites (Herring, 1985). The oldest sedimentary occurrence of charcoal in a non-carbonaceous rock has been reported from the Lower Carboniferous (Scott & Collinson, 1978). It also has been recognized in carbonaceous rock from a transitional Devonian-Carboniferous 'coal' (Warg & Traverse, 1973). It is a common component of many sedimentary regimes.

Microbial and chemical interactions with charcoal during transport, deposition and subsequent diagenesis are minimized because of its inert character, consisting of almost pure carbon (Mantell, 1968). It is easily transported both by wind and water. Large pieces of charcoal cannot be moved far by wind, and most of the charcoal generated remains near the site of pyrolysis and is incorporated into the soil. Erosion of these particulates may introduce them into fluvial and nearshore regimes, and this fraction rarely is transported beyond coastal waters (Herring, 1985).

Wind transport of charcoal is often the result of the injection of meso- and microdetrital-sized carbon particles into the air by thermal currents generated during wildfire propagation. Canopy temperatures can exceed 800 °C for short periods (one minute or less), and can be maintained at 300 °C for almost twenty minutes (Ahlgren, 1970; Rundel, 1981). Ground temperatures range between 200 and 500 °C, with temperature extremes reaching 1000 °C. Particles 1–100 μm (or larger) may remain airborne for quite some time and wood carbon may constitute up to 36% of atmospheric carbon particulate matter (Griffin & Goldberg, 1979; Herring, 1985). Eolian charcoal is transported from the continents to ocean basins via tropospheric winds operating at altitudes less than 10 km (Palmer & Northcutt, 1975). Upon introduction into the ocean waters, it is believed that sedimentation rates of the particles are accelerated by particle ingestion by zooplankton and excretion of fecal pellets (Herring, 1985).

Because of its low density, charcoalified debris may be water transported for significant distances and deposited in various aquatic environments, commonly accumulating in nearshore continental sediments. Herring (1985) believes that these coastal sites are probably the primary reservoir for charcoal in the open oceans. As Harris (1958) discusses, each fusain-rich deposit may not represent a single forest fire event, as buried charcoal may be excavated by subsequent erosional processes and redeposited in other sites. He contrasts this with other plant detritus which he believes would be destroyed by re-entrainment, but as demonstrated above, this does not necessarily occur.

Discussion

The physical and geochemical characteristics of phytoclasts have gained more attention recently because of their potential as indices for and/or sources of oil and natural gas, particularly in nonmarine and deltaic coastal settings (Combaz & de Martharel, 1978; Thomas, 1982; Nwachukwu & Barker, 1985; Bustin, 1988). Variations in abundance and type of detritus (amorphous organic material versus structured woody material; terrestrial versus open marine) with depositional environment and with stratigraphic age have been demonstrated in various geographical settings (e.g., Dow & Pearson, 1975; Hedges & Parker, 1976; Bustin, 1988). The recognition of different nannodetrital categories that can be isolated from preparation residues allows for an assessment of their provenance and resultant biofacies relationships. An understanding of the characters of the taphonomic assemblage, by assessing the individual plant part components, may provide data useful in assisting interpretations of depositional environment and studies of hydrocarbon source-rock potential.

Structured terrestrially sourced phytoclasts comprise a variety of plant part tissues and/or cells that include cuticles of various origin (leaves, aerial stems, and roots), xylary and fibrous elements (leaf venation, sclerenchymatous bundles, and secondary woods), phytoliths, resistant small fruits and seeds, and charcoalified debris. Unstructured accessory chemical exudates, such as coniferous resins, and angiospermous copals and dammars, may be important elements in the amorphous residual fraction. The characteristics of micro- and mesodetrital residue are the result of various biogeochemical and physical degradational processes acting upon plant parts throughout their functional and post-functional existence. Biodegradation may be the result of interaction by bacteria and fungi, and the role each of these decomposers may play with respect to any plant part varies. To date, plant parts recovered from various subaqueous depositional environments in temperate and tropical coastal regimes appear to be minimally affected by fungal degradation. Bacterial colonization and enzymatic degradation (catabolism possibly in response to cellular senescence) may play a more significant role in the degradation of the intracellular contents and cellulose-rich cell walls than the purported effects of fungal action.

Timing and duration of, and susceptibility to, degradation appear to play the pivotal role in the quantity and quality of plant material available for incorporation and ultimate preservation in sediments. Plant parts that are subjected to prolonged alteration processes ultimately are reduced to the nannodetrital fraction. As has been demonstrated and discussed before (Gastaldo, 1988; Spicer, 1989), resistant plant parts are more prone than less resistant parts to being incorporated into sedimentary regimes even after physically disaggregated. Less resistant tissues are those most rapidly affected after physiological or traumatic loss. Biogeochemical degradation of different plant parts occurs at different tissue and cellular levels, with parenchymatous tissues undergoing deterioration first. For leaves, cuticle chemistry may be a significant factor with regard to the systematic distribution of identifiable cuticles in processed residues. Those composed of a resistant bipolymer (Nip *et al.*, 1986a,b) do not undergo as severe deterioration as do cuticles lacking this particular bipolymer. The difference in preservation potential is aptly demonstrated in the comparison of plant leaves recovered from the Mahakam River delta cores. Within the same sample suite leaves may either be well preserved or may have undergone selective degradation, beginning at the interface between intervenal parenchymatous tissues and the xylary elements and continuing into the intervenal areas. In other instances, cuticular envelopes with venation architecture are found to coexist with leaves in various stages of deterioration (Gastaldo *et al.*, in press).

The loss of soluble substances and the oxidation (blackening) of the material seems to be a universal property after plant detritus is exposed to saturating conditions. This 'softening' of the detritus predisposes the material towards degradation to nannodetrital-sized phytoclasts through various and, at times, sustained biotic or abiotic interactions. Biotic interactions include colonization by saprotrophic bacteria and/or fungi. In addition, macroinvertebrates may utilize the plant debris for food (resulting in homogenization of the sediment through bioturbation), may use the resource as a habitat (either through epibiont growth or endobiont colonization), or a combination of these activities may exist. Macro-invertebrate behavior, including that of necrotrophs and biotrophs, may be responsible for the introduction of saprotrophs that promote tissue deterioration (e.g.,

Gastaldo *et al.*, 1989). The principal abiotic interaction apparently responsible for phytoclast generation is the physical abrasive breakup of 'softened' plant parts in moderate to high energy environments. Although this may occur during residence and transport in bedload of a fluvial or nearshore setting, it appears that in coastal sedimentary regimes the majority of mechanical size-reduction occurs within the coastal zone through reworking of previously deposited detritus. Without a doubt, macrodetritus may be subjected to a complex transport history before it is finally buried. Often plant parts may undergo deposition and re-entrainment several times, and each time the plant part is subjected to potential and continued physical breakdown. The timing of this re-entrainment may be prior to burial, shortly after initial burial (within days to decades either due to the ephemeral character of the depositional site or to subsequent high-energy storm activity), or it may occur some time afterwards (on the order of 10^3–10^4 years).

Short-term eustatic sea level fluctuations in coastal zones must not be overlooked as a mechanism in the generation and redistribution of phytoclasts. With the recognition of sequence stratigraphic processes and the concomitant development of transgressive erosional surfaces that planate former nearshore and coastal terrestrial environments (ravinement surfaces; e.g. Galloway, 1989; Larue & Martinez, 1989; Liu & Gastaldo, 1992), any plant litter that would have accumulated and buried in these sites would be reintroduced into the water column upon erosion of the place of burial. Because pieces of plant litter would have been subjected previously to various degrees of degradation by a multiplicity of processes, their erosion during shoreface transgression would place them directly in a wave zone in which they would be mechanically fragmented. Their redistribution to the subsequently developed ravinement bed or to sites offshore would be a function of coastal energy operating within any particular part of the coastal regime.

In situ degradation of buried litter may result when fluctuating water conditions introduce oxygenated waters into a site. This will result in the chemical alteration of plant parts and the development of micro- to mesodetrital-sized materials, without necessarily meaning that the physical degradational processes operated at that specific time at that specific site. The generally ephemeral nature of coastal subaqueous to subaerial sites provides a means to reintroduce the degraded (and possibly partially matured) detritus into the water column, to be deposited in other parts of the coastal regime.

Phytoclasts, such as particles of charcoal and resins, may be introduced into the sedimentary environment in a variety of ways. Inert charcoal enters the sedimentary regime most commonly through wind and/or water action, and is concentrated in nearshore environments

(Herring, 1985). Resins, including copals and dammars, are transported either within their host wood or as part of bedload transport. The concentration of these amorphous plant parts in nearshore environments probably parallels that of charcoal. Redistribution of these particles results in alteration of their form, accompanied by size reduction. This is probably a function of brittleness.

The vast majority of terrestrially derived phytoclasts recovered in coastal and nearshore facies appears to have originated from communities in which deciduous plants predominate. In the case of deltaic settings, the ecosystems contributing detritus are those of the upper delta and interior (extrabasinal) sites (Gastaldo *et al.*, 1987; Gastaldo *et al.*, in press). The herbaceous vegetation growing in 'lower delta plain' settings (marsh grasses and *Nipa* palms) have a tendency to retain their aerial parts and undergo degradation *in situ* (Gastaldo, 1989; Gastaldo, 1990). When these sites are inhabited by arborescent vegetation (e.g., Orinoco River delta; Scheihing & Pfefferkorn, 1984) the plants tend to dehisce and shed canopy parts during growth. It is more common that the deciduous elements may be transported away from their initial depositional site by tidal flushing (Dame, 1982), but this does not preclude some of the degraded herbaceous material from dispersal by this mechanism, particularly under spring tidal conditions. In fact, the various kinds of amorphous and degraded material of terrestrial origin may be generated under these circumstances. The recognized early stages of color differentiation and maturation in these particles (e.g., Caratini, Chap. 8, this volume) may reflect the various phases of the initial biochemical alteration and/or chemical manipulation of the organic matter in the depositional setting.

The loci where fluvial systems debouch into coastal environments can be acknowledged as the principal sites where large quantities of terrestrial detritus enter into the nearshore and littoral regimes. This detritus may be distributed by transport along coastlines by longshore currents and/or prevailing circulation patterns, and be deposited within various depositional settings, ranging from lagoonal to offshore barrier sands. But not all coastlines are perforated by river systems emptying into a basin. In most instances, long stretches of vegetated shoreline exist that are subjected only to the influence of tides. Differences in daily and monthly tides, and those tides amplified by localized storm conditions, generate variable conditions that affect vegetated coastal zones. When tidal velocities increase to the point where plant detritus of varied degradational states can be entrained in the water column (either moved in suspension or in bedload), litter can be flushed into nearshore sites where accumulation may result in organic-rich deposits (e.g., Risk & Rhodes, 1985; Bird, 1986). Low clastic sedimentation rates in these sites would allow for preferential degradation of the least resistant tissues, resulting in concentration of those resistant parts, such as cuticular biopolymers. What are now considered to be 'dispersed' cuticles in facies that appear organic-rich but do not 'preserve' identifiable macrofossils might represent, in part, facies in which clastic sedimentation rates are lower than rates of degradation. The fragmentary character of these sample suites may not be entirely a function of cuticle separation from leaves and differential concentration, but rather may be the result of sample preparation.

The biofacies resulting from the accumulation of phytoclasts mirror the regional terrestrial communities that flank either the distributary fluvial regimes or the marginal coastal vegetation. The temporal variations in the quality of contributed material are indicative of the changes in ecosystem structure of the provenance communities. The accessory nannodetrital components may be a signal of short-term ecosystem disruption on a localized or regional scale rather than evolutionary or long-term, climatically induced floristic turnover, as might be interpreted from the recovered palynoflora. This may be particularly applicable where palynofloras also contain quantities of charcoal (fusain). The ecological repercussions caused by wildfire have been reviewed by many authors (e.g., Rundel, 1981; Raup, 1981), and it is apparent that this disruption causes community turnover. When this is expressed on a regional scale and where k-strategy plants are eliminated, fecund r-strategists quickly become established. This alteration of community structure will be reflected quickly in changes of the palynoflora. This disruption is best manifested in contemporaneous depositional sites with the addition of wind-borne and/or water-transported charcoal fluxes. The recognition of this accessory component in higher than background concentrations might provide the clues necessary to filter out the short-term from long-term causes of vegetational change. With a developing understanding of the relationships between palynofacies generation and the taphonomic processes responsible for their nannodetrital component, a more accurate picture can be derived of their discrete utilization.

Acknowledgments

Data collected for and discussed within this chapter have been collected over the past decade with the assistance of many undergraduate students, graduate students, and colleagues too numerous to cite individually. I would like to express my appreciation to them for their help in some inhospitable localities. Phytotaphonomic research in modern environments has been supported by several agencies. Acknowledgment is made to the Donors of the Petroleum Research Fund, as administered by the American Chemical Society, for grants ACS PRF 18141 B2 and ACS PRF 20829 AC8 in partial support of these studies. The National Science Foundation is acknowledged for grant NSF EAR 8803609 that supported research in eastern Borneo. Dr. George P. Allen and Mr.

Choppin de Janvry, TOTAL, Compagnie Française des Pétroles, are gratefully acknowledged for providing logistical and field support for our vibracoring program in the Mahakam River delta, Kalimantan. Dr. Alain Yves-Huc and the Institut Français du Pétrol are thanked for their assistance in developing a cooperative project in Indonesia and financial support allowing me a residence in France to begin sample processing and preparation. Professor Bruce Purser, the Université de Paris-Sud, Orsay, is gratefully acknowledged for his cooperation and use of laboratory and electon microscopy facilities. Mr David Felton is thanked for providing data and unpublished x-radiographs of vibracores extracted in the Mobile–Tensaw River delta, Alabama, and Dr Charles E. Savrda is acknowledged for the loan of the Paleocene bored wood. The author would like to thank the editor and two anonymous reviewers for suggestions that led to the improvement of the manuscript.

References

Ahlgren, C. E. (1970). Some effects of prescribed burning on Jack Pine reproduction in Northeastern Minnesota. *Minnesota Agricultural Experimental Station Miscellaneous Report*, **94**, 1–14.

Alexopoulos, C. J. & Mims, C. W. (1979). *Introductory Mycology*. New York: John Wiley & Sons.

Allen, G. P., Laurier, D. & Thouvenin, J. P. (1979). Étude sedimentologique du delta de la Mahakam: TOTAL. *Compagnie Française des Pétroles, Paris, Notes et Mémoires*, **15**, 1–156.

Barnes, R. D. (1974). *Invertebrate Zoology*. Philadelphia: W. B. Saunders.

Batten, D. J. (1973). Use of palynologic assemblage-types in Wealden correlation. *Palaeontology*, **16**, 1–40.

Batten, D. J. (1979). Miospores and other acid-resistant microfossils from the Aptian/Albian of Holes 400A and 402A, DSDP-IPOD Leg 48, Bay of Biscay. In *Initial Reports of the Deep Sea Drilling Project, V. XLVIII*, ed. L. Montadert and D. G. Roberts. Washington, DC: US Government Printing Office, 579–87.

Batten, D. J. (1982). Palynofacies, palaeoenvironments and petroleum. *Journal of Micropalaeontology*, **1**, 107–14.

Batten, D. J. (1983). Identification of amorphous sedimentary organic matter by transmitted light microscopy. In *Petroleum Geochemistry and Exploration of Europe*, ed. J. Brooks. London: Blackwell Scientific, 275–87.

Batten, D. J., Creber, G. T. & Zhou, Zhiyan (1984). Fossil plants and other organic debris in Cretaceous sediments from Deep Sea Drilling Project: their paleoenvironmental significance and source potential: *Initial Reports of the Deep Sea Drilling Project* **80**. Washington, DC: US Government Printing Office, 629–41.

Benoit, R. E., Starkey, R. L. & Basaraba, J. (1968). Effect of purified plant tannin on decomposition of some organic compounds and plant materials. *Soil Science*, **105**, 153–8.

Bird, E. C. F. (1986). Mangroves and intertidal morphology in Western Port Bay, Victoria, Australia. *Marine Geology*, **69**, 251–72.

Bustin, R. M. (1988). Sedimentology and characteristics of dispersed organic matter in Tertiary Niger delta: Origin of source rocks in a deltaic environment. *Bulletin of the American Association of Petroleum Geologists*, **72**, 277–98.

Chaloner, W. G. (1989). Fossil charcoal as an indicator of palaeoatmospheric oxygen level. *Journal of the Geological Society of London*, **146**, 171–4.

Clark, D. L., Byers, C. W. & Pratt, L. M. (1986). Cretaceous black mud from the central Arctic Ocean. *Paleoceanography*, **1**, 265–71.

Combaz, A. & de Matharel, M. (1978). Organic sedimentation and genesis of petroleum in Mahakam delta, Borneo. *Bulletin of the American Association of Petroleum Geologists*, **62**, 1684–95.

Cope, M. J. (1980). Physical and chemical properties of coalified and charcoalified phytoclasts from some British Mesozoic sediments: an organic geochemical approach to palaeobotany. In *Advances in Organic Geochemistry, 1979, Physics and Chemistry of the Earth*, **12**, ed. A. G. Douglas and J. R. Maxwell. Oxford: Pergamon Press, 663–77.

Cope, M. J. (1981). Products of natural burning as a component of the dispersed organic matter of sedimentary rocks. In *Organic Maturation Studies and Fossil Fuel Exploration*, ed. J. Brooks. London: Academic Press, 89–109.

Cope, M. J. & Chaloner, W. G. (1980). Fossil charcoal as evidence of past atmospheric composition. *Nature*, **283**, 647–9.

Cope, M. J. & Chaloner, W. G. (1985). Wildfire, an interaction of biological and physical processes. In *Geological Factors and the Evolution of Plants*, ed. B. H. Tiffney. New Haven, CT: Yale University Press, 257–77.

Craighead, F. C. & Gilbert, V. C. (1962). The effects of Hurricane Donna on the vegetation of southern Florida. *The Quarterly Journal of the Florida Academy of Science*, **21**, 1–28.

Cunningham, A., Gay, I. D., Oehlschlager, A. C. & Langenheim, J. H. (1983). ^{13}C NMR and IR analyses of structure, aging, and botanic origin of Dominican and Mexican ambers. *Phytochemistry*, **22**, 965–8.

Dame, R. R. (1982). The flux of floating macrodetritus in the North Inlet estuarine ecosystem. *Estuarine, Coastal and Shelf Science*, **21**, 1–28.

Degges, C. W. & R. A. Gastaldo (1988). The use of fossilized logs as paleocurrent indicators in fluvial systems – A note of caution. *Geological Society of America, Abstracts with Program*, **20**(7), A264.

De Vries, H., Bredemeijer, G. & Heinen, W. (1967). The decay of cutin and cuticular components by soil microorganisms in their natural environment. *Acta Botanica Neerlandische*, **16**, 102–10.

Dickinson, S. (1960). The mechanical ability to breach the host barriers. In *Plant Pathology*, ed. J. G. Horsfall & A. E. Dimond. London & New York: Academic Press, **2**: 203–32.

Dittus, W. P. J. (1985). The influence of cyclones on the dry evergreen forest of Sri Lanka. *Biotropica*, **17**, 1–14.

Dow, W. G. & Pearson, D. B. (1975). Organic matter in Gulf Coast sediments. *Offshore Technology Conference Proceedings*, **3**, Paper 2343.

Ehrlich, P. R. & Raven, R. H. (1964). Butterflies and plants: A study of coevolution. *Evolution*, **18**, 568–608.

English, M. P. (1965). The saprophytic growth of nonkeratinophilic fungi on keratinised substrates and a comparison with keratinophilic fungi: *Transactions British Mycological Society*, **48**, 219–35.

Fahn, A. (1969). *Plant Anatomy*. Oxford: Pergamon Press.

Ferguson, D. K. (1971). *The Miocene Flora of Kreuzau. Western Germany: 1. The Leaf Remains*. Amsterdam: North Holland Publishers.

Ferguson, D. K. (1985). The origin of leaf-assemblages – new light on an old problem. *Review of Palaeobotany and Palynology*, **46**, 117–88.

Ferguson, D. K. & De Bock, P. (1983). The bryophyte and

angiosperm leaf-remains from the Holocene sediments of Deggendorf. *Documenta Naturae*, **9**, 16–33.

Fraenkel, G. (1959). The raison d'etre of secondary plant substances. *Science*, **121**, 1466–70.

Galloway, W. E. (1989). Genetic stratigraphic sequences in basin analysis I: Architecture and genesis of flooding-surface bounded depositional units. *American Association of Petroleum Geologists Bulletin*, **73**, 125–42.

Garden, A. & Davies, R. W. (1988). Decay rates of autumn and spring leaf litter in a stream and effects on growth rate of a detritovore. *Freshwater Biology*, **19**, 297–303.

Gastaldo, R. A. (1988). A conspectus of phytotaphonomy. In *Methods and Applications of Plant Paleoecology: Notes for a Short Course*, ed. W. A. DiMichele and S. L. Wing. *Paleontological Society Special Publication*, **3**, 14–28.

Gastaldo, R. A. (1989). Preliminary observations on phytotaphonomic assemblages in a subtropical/temperate Holocene bayhead delta: Mobile Delta, Gulf Coastal Plain, Alabama. *Review of Palaeobotany and Palynology*, **58**, 61–84.

Gastaldo, R. A. (1990). Phytotaphocoenoses in Late Quaternary temperate and tropical coastal deltaic regimes. Paleofloristic and paleoclimatic changes in the Cretaceous and Tertiary. In *Proceedings of the International Symposium – Paleofloristic and Paleoclimatic Changes in the Cretaceous and Tertiary, International Geological Correlation Programme Project No. 216: Global Biological Events in Earth History*, ed. F. Knobloch & Z. Kvaček. Prague: Geological Survey of Czechoslovakia, 285–90.

Gastaldo, R. A., Allen, G. P. & Huc, A. Y. (in press). Detrital peat formation in the tropical Mahakam River delta, Kalimantan, eastern Borneo: Formation, plant composition, and geochemistry. *Geological Society of America Special Paper*.

Gastaldo, R. A., Demko, T. M., Liu, Y., Keefer, W. D. & Abston, S. L. (1989). Biostratinomic processes for the development of mud-cast logs in Carboniferous and Holocene swamps. *Palaios*, **4**, 356–65.

Gastaldo, R. A., Douglass, D. P. & McCarroll, S. M. (1987). Origin, characteristics and provenance of plant macrodetritus in a Holocene crevasse splay, Mobile delta, Alabama. *Palaios*, **2**(3), 229–40.

Gosz, J. R., Likens, G. E. & Bormann, F. H. (1973). Nutrient release from decomposing leaf and branch litter in the Hubbard Brook Forest, New Hampshire. *Ecological Monographs*, **47**, 173–91.

Green, N. B. (1980). The biochemical basis of wood decay micromorphology. *Journal of the Institute of Wood Science*, **8**, 221–8.

Griffin, J. J. & Goldberg, E. D. (1979). Morphologies and origin of elemental carbon in the environment. *Science*, **206**, 563–5.

Grimalt, J. O., Simoneit, B. R. T., Hatcher, P. G. & Nissenbaum, A. (1988). The molecular composition of ambers. *Organic Geochemistry*, **13**, 677–90.

Harris, T. M. (1958). Forest fire in the Mesozoic. *Journal of Ecology*, **46**, 447–53.

Harris, T. M. (1981). Burnt ferns from the English Wealdon. *Proceedings Geological Association of London*, **92**, 47–58.

Heath, G. W. & Arnold, M. K. (1966). Studies in leaf-litter breakdown. II. Breakdown rate of 'sun' and 'shade' leaves. *Pedobiologia*, **6**, 238–43.

Hedges, J. I. & Parker, P. L. (1976). Land-derived organic matter in surface sediments from the Gulf of Mexico. *Geochimica et Cosmochimica Acta*, **40**, 1019–29.

Herring, J. R. (1985). Charcoal fluxes into sediments of the North Pacific Ocean: the Cenozoic record of burning. In *The Carbon Cycle and Atmospheric CO_2: Natural Variations Archean to Present. Geophysical Monograph*, **32**, 419–22.

Highley, T. L. (1976). Hemicelluloses of white and brown rot fungi in relation to host preference. *Material und Organismen*, **11**, 33–46.

Hoagland, K. E. & Turner, R. (1981). Evolution of woodboring bivalves. *Malacologia*, **21**, 111–48.

Hoyt, W. M. & Demarest, J. M. (1981). Vibracoring in coastal environments: The R.V. Phryne II Barge and associated coring methods. University of Delaware, Sea Grant College Program, Newark, Delaware, DEL SG 01 81.

Huc, A. Y. (1988). Sedimentology of Organic Matter. In *Humic Substances and Their Role in the Environment*, ed. F. H. Frimmel & R. F. Christman. London: John Wiley & Sons, 215–243.

Huc, A. Y., Durand, B., Roucachet, J., Vandenbroucke, M. & Pittion, J. L. (1985). Comparison of three series of organic matter of continental origin. *Advances in Organic Geochemistry, 1985. Organic Geochemistry*, **10**, 65–72.

Kaushik, N. K. & Hynes, H. B. N. (1968). Experimental study on the role of autumn leaf shed in aquatic environments. *Journal of Ecology*, **56**, 229–43.

Kaushik, N. K. & Hynes, H. B. N. (1971). The fate of the dead leaves that fall into streams. *Archives Hydrobiology*, **68**, 465–515.

Kerp, H. (1989). Cuticular analysis of gymnosperms: A short introduction. In *Phytodebris: Notes for a Workshop on the Study of Fragmentary Plant Remains*, ed. B. H. Tiffney. Paleobotanical Section of the Botanical Society of America.

Krassilov, V. A. (1975). *Palaeoecology of Terrestrial Plants: Basic Principles and Techniques*. New York: J. Wiley & Sons.

La Caro, F. & Rudd, R. L. (1985). Litter disappearance rates in a Puerto Rican montane rain forest. *Biotropica*, **17**, 269–76.

Lane, C. E. (1959). The general histology and nutrition of *Limnoria*. In *Marine Boring and Fouling Organisms*, ed. D. H. Ray. Seattle: University of Washington Press, 34–45.

Langenheim, J. H. (1990). Plant resins. *Scientific American*, **78**, 16–24.

Larue, D. K. & Martinez, P. A. (1989). Use of bed-form climb models to analyze geometry and preservation potential of clastic facies and erosional surfaces. *American Association of Petroleum Geologists Bulletin*, **73**, 40–53.

Levy, J. F. & Dickinson, D. J. (1981). Wood. In *Microbial Biodeterioration* (= *Economic Microbiology*, **6**), ed. A. H. Rose. London: Academic Press, 19–60.

Liu, Yuejin & Gastaldo, R. A. (1992). Characteristics of a Pennsylvanian ravinement surface. *Sedimentary Geology*, **77**, 197–213.

Mantell, C. L. (1968). *Carbon and Graphite Handbook*. New York: Wiley-Interscience.

Masran, T. C. & Pocock, S. A. J. (1981). The classification of plant-derived particulate organic matter in sedimentary rocks. In *Organic Maturation Studies and Fossil Fuel Exploration*, ed. J. Brooks. London: Academic Press, 145–61.

McGinnes, A. E., Szopa, P. S. & Phelps, J. E. (1974). Use of scanning electron microscopy in studies of wood charcoal formation. *Scanning Electron Microscopy/1974*, part 2, 469–476.

Mills, J. S., White, R. & Gough, L. J. (1984/85). The chemical composition of Baltic amber. *Chemical Geology*, **47**, 15–39.

Nip, M., Tegelaar, E. W., De Leeuw, J. W., Schenk, P. A. & Holloway, P. J. (1986a). A new non-saponifiable highly aliphatic and resistant biopolymer in plant cuticles: *Naturwissenschaften*, **73**, 579–85.

Nip, M., Tegelaar, E. W., Brinkhuis, H., De Leeuw, J. W., Schenk, P. A. & Holloway, P. J. (1986b). Analysis of modern and fossil plant cuticles by Curie point Py-Gc and Curie Point Py-Gc-Ms: Recognition of a new highly aliphatic and resistant biopolymer. In *Advances in Organic Geochemistry 1985*, ed. D. Leythaeuser and J. Rollkötter. *Organic Geochemistry*, **10**, 769–78.

Nip, M., De Leeuw, J. W., Schenk, P. A., Windig, W., Meuzelaar, H. L. C. & Crelling, J. C. (1989). A flash pyrolysis and petrographic study of cutinite from the Indiana Paper coal. *Geochimica et Cosmochimica Acta*, **53**, 671–83.

Nwachukwu, J. I. & Barker, C. (1985). Organic matter: size fraction relationships for recent sediments from the Orinoco delta, Venezuela. *Marine and Petroleum Geology*, **2**, 202–9.

Nykvist, N. (1959a). Leaching and decomposition of litter. I. Experiments on leaf litter of *Fraxinus excelsior*. *Oikos*, **10**, 190–211.

Nykvist, N. (1959b). Leaching and decomposition of litter. II. Experiments on leaf litter of *Pinus silvestris*. *Oikos*, **10**, 212–24.

Nykvist, N. (1961a). Leaching and decomposition of litter. III. Experiments on leaf litter of *Betula verrucosa*. *Oikos*, **12**, 249–63.

Nykvist, N. (1961b). Leaching and decomposition of litter. IV. Experiments on leaf litter of *Picea abies*. *Oikos*, **12**, 264–79.

Nykvist, N. (1962). Leaching and decomposition of litter. V. Experiments on leaf litter of *Alnus glutinosa*, *Fagus silvatica* and *Quercus robur*. *Oikos*, **13**, 232–48.

Palmer, T. Y. & Northcutt, L. I. (1975). Convection columns above large experimental fires. *Fire Technology*, **11**, 111–18.

Panshin, A. J. & De Zeeuw, C. (1970). *Textbook of Wood Technology*. Vol. 1. New York: McGraw Hill.

Parry, C. C., Whitley, P. K. J. & Simpson, R. D. H. (1981). Integration of palynological and sedimentological methods in facies analysis of the Brent Formation. In *Petroleum Geology of the Continental Shelf of North-West Europe*, ed. L. V. Illing & G. D. Hobson. London: Institute of Petroleum, 205–15.

Petersen, R. C. & Cummins, K. W. (1974). Leaf processing in a woodland stream. *Freshwater Biology*, **4**, 343–68.

Pocock, S. A. J., Vasanthy, G. & Venkatachala, B. S. (1987). Introduction to the study of particulate organic materials. *Journal of Palynology*, **23–24**, 167–88.

Raup, H. M. (1981). Physical disturbance in the life of plants. In *Biotic Crises in Ecological and Evolutionary Time*, ed. M. H. Nitecki. New York: Academic Press, 39–52.

Ray, D. L. (1959) Nutritional physiology of *Limnoria*. In *Marine Boring and Fouling Organisms*, ed. D. L. Ray. Seattle: University of Washington Press, 46–61.

Rayner, A. D. M. & Boddy, L. (1988). *Fungal Decomposition of Wood: Its Biology and Ecology*. Chichester: John Wiley & Sons.

Rindsberg, A. K. & Gastaldo, R. A. (1989). The roles of wave action, bioturbation, and diffusion of oxygen from aerenchymatous roots in degradation of leaf debris. *28th International Geological Congress Abstracts*, **2**, 701.

Risk, M. J. & Rhodes, E. G. (1985). From mangroves to petroleum precursors: an example from tropical northeast Australia. *Bulletin of the American Association of Petroleum Geologists*, **69**, 1230–40.

Ross, S. (1989). *Soil Processes: A Systematic Approach*.

London and New York: Routledge.

Rossel, S. E., Abbot, G. M. & Levy, J. F. (1973). Bacteria and wood (a review of the literature related to presence, action, and interaction of bacteria in wood). *Journal of the Institute of Wood Science*, **6**, 28–35.

Rundel, P. W. (1981). Fire as an ecological factor. In *Physiological Plant Ecology 1. Response to the Physical Environment*, ed. O. L. Lange, P. S. Nobel, C. B. Osmond & H. Ziegler. Berlin: Springer-Verlag, 501–38.

Scheihing, M. H. (1980). Reduction of wind velocity by the forest canopy and the rarity of non-arborescent plants in the Upper Carboniferous fossil record. *Argumenta Palaeobotanica*, **6**, 133–8.

Scheihing, M. H. & Pfefferkorn, H. W. (1984). The taphonomy of land plants in the Orinoco Delta: A model for the incorporation of plant parts in clastic sediments of Late Carboniferous age of Euramerica. *Review of Palaeobotany and Palynology*, **41**, 205–80.

Scott, A. C. (1989). Observations on the nature and origin of fusain. *International Journal of Coal Geology*, **12**, 443–75.

Scott, A. C. & Collinson, M. E. (1978). Organic sedimentary particles: results from Scanning Electron Microscope studies of fragmentary plant material. In *Scanning Electron Microscopy in the Studies of Sediments*, ed. W. B. Whalley. Norwich, UK: Geoabstracts, 137–67.

Spicer, R. A. (1981). The sorting and deposition of allochthonous plant material in a modern environment at Silwood Lake, Silwood Park, Berkshire, England. *United States Geological Survey Professional Paper* **1143**.

Spicer, R. A. (1989). The formation and interpretation of plant fossil assemblages. *Advances in Botanical Research*, **16**, 95–191.

Swift, M. J., Heal, O. W. & Anderson, J. M. (1979). Decomposition in Terrestrial Ecosystems. *Studies in Ecology*, **5**.

Thomas, B. M. (1982). Land-plant source rocks for oil and their significance in Australian basins. *Australian Petroleum Exploration Association Journal*, **22**, 164–70.

Thomas, B. R. (1970). Modern and fossil plant resins. In *Phytochemical Phylogeny*, ed. J. B. Harborne. London: Academic Press, 59–79.

Tiffney, B. H., ed. (1989). *Phytodebris: Notes for a Workshop on the Study of Fragmentary Plant Remains*. Paleobotanical Section of the Botanical Society of America.

Turner, R. & Johnson, A. C. (1971). Biology of marine woodboring molluscs. In *Marine Borers, Fungi and Fouling Organisms*, ed. E. B. G. Jones & S. K. Eltringham. Paris: Organization for Economic Cooperation and Development, 259–301.

Upchurch, G. R. (1989). Dispersed angiosperm cuticle. In *Phytodebris: Notes for a Workshop on the Study of Fragmentary Plant Remains*, ed. B. H. Tiffney. Paleobotanical Section of the Botanical Society of America.

Venkatachala, B. S. (1981). Differentiation of amorphous organic matter types in sediments. In *Organic Maturation Studies and Fossil Fuel Exploration*, ed. J. Brooks. London: Academic Press, 177–85.

Warg, J. B. & Traverse, A. (1973). A palynological study of shales and 'coals' of a Devonian–Mississippian transition zone, central Pennsylvania. *Geoscience and Man*, **7**, 39–46.

Williams, A. M. (1963). Enzyme inhibition by phenolic compounds. In *The Enzyme Chemistry of Phenolic Compounds*, ed. J. B. Pridham. New York: Pergamon Press, 87–95.

8 Palynofacies of some recent marine sediments: the role of transportation

CLAUDE CARATINI

Introduction

Palynological study of the organic matter content of sediments reveals information about the nature and state of preservation of constituent elements of both the sediments in general and the organic matter in particular. These observations have led to the establishment of a number of criteria to assist in reconstruction of the paleogeographical and sedimentological conditions prevailing at the time of deposition, as well as the geological history of the sediment.

Several authors have proposed relationships which may have existed between paleogeography, in the broadest sense, and the microscopic character of the sedimentary organic matter. The ultimate aim of these investigations has been to understand the origin and alteration of this organic material, and, more generally, of the sediment in which it is preserved (Haseldonckx, 1974; Hart, 1986; Pocock et al., 1987). Most often these studies have concentrated on Pre-Quaternary sedimentary formations. The paleogeography prevailing at the time of deposition can be deciphered only from reconstructions where hypotheses sometimes become important, reducing the validity of the proposed conclusions.

It was to alleviate such difficulties, and particularly to understand and consequently interpret the geological data more accurately, that the ORGON project, a series of French scientific cruise-expeditions with organic geochemical goals, was undertaken during the years 1976–1983, to study the organic content of Late Pleistocene to Holocene deposits from various ocean environments (see Fig. 8.1). Only extant phenomena which can be described or quantified are taken into consideration. This project, essentially based on observed facts, enables us to delimit, in some well-defined cases, the role and importance of the various factors affecting the nature and state of preservation of organic matter in marine sediments.

The concept of palynofacies

Palynofacies is the description, by means of transmission light microscopy, of the organic constituents of the rock and estimation of their relative proportions (paraphrase of definition by Combaz, 1964; 1980).

Organic matter

The organic matter observed does not represent the totality of organic matter present in sedimentary rock, but only the fraction which remains unhydrolysed after destruction of the mineral phase by hydrochloric and hydrofluoric acids (Durand & Nicaise, 1980). This fraction is designated as 'palynological residue' in preference to the term 'kerogen,' as the definition of the latter has become too inaccurate. 'Kerogen' is often used with different meanings in organic geochemistry and palynology. In the ORGON sediments, the palynological residues constitute an average of 60–70% of the total organic carbon fraction of the sediments and are therefore representative of the entire fraction.

Elements and criteria for characterizing a palynofacies

Combaz's (1964, section Ib) classification is adopted here. Each element is sufficiently well defined so as not to give rise to divergent interpretations by different observers. This simplified method facilitates use, without reducing possibilities of interpretation.

Structured organic matter

Structured organic matter is composed of elements which show a well-defined structure or form when viewed with a transmission light microscope. They are: (1) microfossils – pollen, spores, microplankton; (2) plant and animal

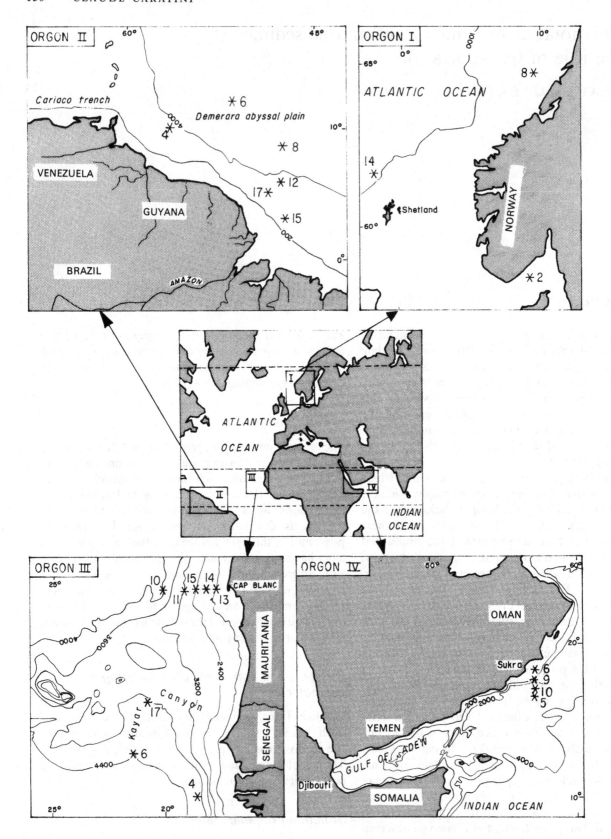

Figure 8.1. Location of the ORGON cruises and sites mentioned in the text.

debris (their dimensions and color are characters which are easy to measure or evaluate); (3) opaque particles – from reflectance examination these are found to be mostly charcoal.

Pyrite, which has not been considered in quantitative estimates, can be easily distinguished from organic matter by its crystalline, cubic or framboidal shape.

Amorphous organic matter

Amorphous organic matter is composed of particles smaller than one micron in size, making it difficult to define precisely. It is not possible to observe any structure in this fraction with the light microscope, even at the highest magnification. I have refrained from describing the different aspects in which amorphous organic matter may be present because of difficulties in agreeing to a vocabulary usable by different observers without ambiguity. Moreover, by definition, this fraction of organic matter is called 'amorphous,' indicating lack of descriptive criteria.

Numerical analysis of the palynofacies

As early as 1964, Combaz established the basis for a quantitative method to estimate the fraction of the microscopic field occupied by each category of the palynofacies, using circular charts showing different percentage classes. The results obtained by trained observers are always of the same order, which testifies to the validity of this method. For example, the palynofacies of a sample from ORGON I in the North Sea (Combaz *et al.*, 1977; see Fig. 8.2.1 – ORGON I sample KS14) is composed of 95% structured organic matter and 5% amorphous organic matter. In the Sukra transect, Gulf of Aden, ORGON IV (Caratini *et al.*, 1981; see Fig. 8.2.1 – ORGON IV sample KL5), the composition is reversed, namely, 10% and 90%, respectively.

Graphic representation of the palynofacies

The simplest way to master and compare vast quantities of data is to adopt a standardized system into which each datum is entered as it is obtained.

Choice of criteria and mode of representation

NUMERICAL CRITERIA

Proportion between amorphous and structured organic matter   The most obvious character is the relative abundance of the two main categories of organic matter, structured versus amorphous. This relationship is very significant, as Caratini *et al.* (1983) have shown that, for the majority of ORGON sediments, the structured/amorphous ratio is strongly correlated with the most useful geochemical parameters for defining sedimentary organic matter: H/C and N/C atomic ratios.

To represent this value, the increasing percentages of amorphous organic matter are plotted on the same graph axis, graduated from 0 to 100 from left to right (Fig. 8.2.I). This disposition implies that the values of structured organic matter, which are complementary, increase from right to left on this graph axis. Hence, a sample on this graph would be positioned as a function of its composition. Thus, the sample from the North Sea (cited earlier) is placed in A, while the sample from the Sukra transect appears in B.

Maximum dimension of detrital structured elements   The maximum dimension of detrital structured elements is a very useful criterion for interpreting transport and sedimentation of organic matter. This dimension, expressed in micrometers, is plotted on a logarithmic scale *y*-axis. In the example, sample A, where the largest detrital structured elements measure 60 μm, is clearly distinguishable from sample B, where they attain only 15 μm (Fig. 8.2.II).

Proportions of the different types of structured elements   The most common types of structured elements, i.e., microfossils, ligneous fragments, and opaque particles (excluding pyrite), can be represented quantitatively. For this purpose, the segment corresponding to the maximum dimension of the structured elements is divided proportionally into the different constituent percentages. This results in a stacked histogram, where the proportions of each category appear clearly (Fig. 8.2.III).

Colors   The importance of the color of organic matter in evaluating the degree of maturation and state of preservation is well known (see Corréia, 1969). Without attempting extreme precision, the major trends in the color of each type of structured organic matter may be limited to three shades: black to dark brown, yellow to light green, and an intermediate of ochre between these two extremes. Each of these shades is seen as a well-defined hatching of obvious significance, the denser the hatching, the darker the color (see Fig. 8.2). Opaque particles are distinguished by another kind of pattern.

Although, as stated above, amorphous organic matter is difficult to describe, it is nevertheless possible to estimate one character without too much controversy, namely, the color. This is especially true if we classify it into only three classes: light, dark, and intermediate. Samples A and B are examples of light and dark amorphous organic matter, respectively. To represent these colors without crowding the already dense diagram, a rectangle 5 mm high is added to the lower part of each

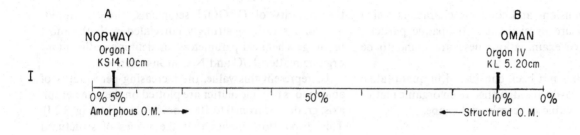

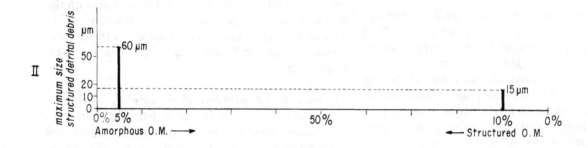

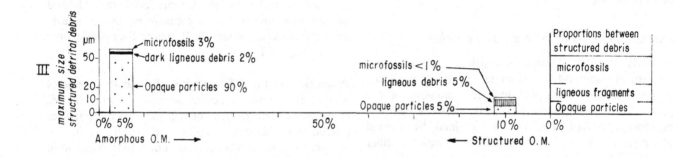

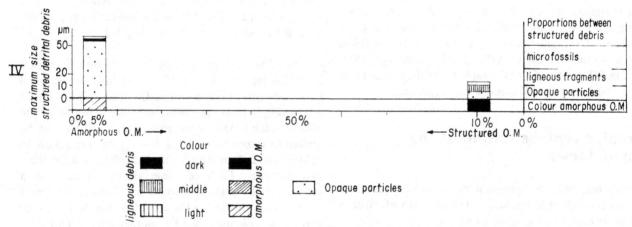

Figure 8.2. Principles of the graphic representation of palynofacies. Explanations for I, II, III, IV are in text.

bar, corresponding to the composition of the palynofacies, and explicitly hatched to demonstrate the colors (Fig. 8.2.IV).

A graphic representation of the palynofacies is obtained which is clear enough for understanding the quantitative and qualitative features, and simple enough for practical use.

Palynofacies and conditions of transportation of the organic matter

Reappraisal of some palynofacies selected from those studied following the ORGON cruises reveals the major influence of transportation conditions existing at the time

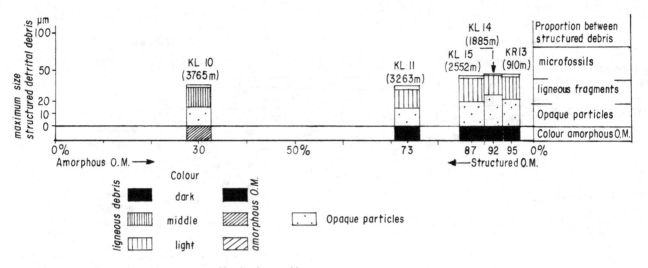

Figure 8.3. ORGON III, Cape Blanc transect, Mauritania: graphic representation of the palynofacies.

of deposition on the character of organic matter deposited in a marine environment. All the sediments are modern or very recent. In no case was the organic matter subjected to any significant diagenetic transformation. In this chapter I use only essential descriptions (for further information see Caratini *et al.*, 1975, 1979, 1981; Caratini & Coumes, 1979; Combaz *et al.*, 1977; Debyser & Gadel, 1979; Debyser *et al.*, 1978; Moyes *et al.*, 1978, 1979; Pelet *et al.*, 1981).

The role of wind

THE CAPE BLANC TRANSECT, MAURITANIA: ORGON III

The investigations carried out off Mauritania, along the Cape Blanc transect (Fig. 8.1) were aimed at studying the organic sedimentation on the continental margin of an old platform to which the vast Sahara Desert now extends. Several reasons *a priori* justified this choice (Caratini & Coumes, 1979; Pelet *et al.*, 1979):

(1)   The contribution of organic matter of continental origin is low, due to the highly scattered Saharan vegetation, as well as to the nearly total absence of a perennial hydrographic network capable of transporting sedimentary deposits to the marine environment. Thus, the organic constituents of continental origin are solely those brought by wind. These are known to be significant (Simoneit, 1977), as the trade winds sweep the Sahara continuously. The importance of these constituents was verified by aeropalynological studies during the ORGON III cruise (Caratini & Cour, 1980).

(2)   Strong and regular trade winds also play a role, although indirectly, on organic sedimentation. They induce seasonal upwellings responsible for considerable primary production, evidence of which

should be found in the form of organic matter of planktonic origin in the sediments. We could infer that this organic matter of marine origin would be predominant when compared to that transported by wind.

(3)   The Canaries current, generally oriented NE–SW, could also be the source of organic material from regions further north which it traverses. However, results obtained from the ORGON III investigations do not indicate an important role for this current as a vector for supplying organic matter.

Palynological analyses and detailed studies of sedimentary organic matter were carried out at 5 sites along the 21° N parallel (Caratini *et al.*, 1979), ranging from the slope (station 13: 910 m) to the abyssal plain (station 10: 3765 m). The corresponding palynofacies are represented in Fig. 8.3, according to the method described above. The nature of the palynological residue varies in a regular manner with increase in depth and distance from the coast.

Variation in the structured/amorphous organic matter ratio
The percentage of amorphous organic matter decreases away from the coast, from 95% in station 13, 92% in station 14, 87% in 15, 73% in 11, to only 30% in station 10. Thus, the structured/amorphous organic matter ratio increases as a function of depth and the distance from the coast. It seems as if the fragile amorphous organic matter had become altered and had disappeared in part when the movement necessary to attain marine depths increases, whereas, in contrast, the more resistant structured organic matter was relatively better preserved (Fig. 8.3). From the results of geochemical analyses (Debyser & Gadel, 1979) the amorphous organic matter is shown to be characteristically of marine origin, while the structured organic matter generally comes from the continent. Surprisingly, in a marine environment, as the depth and distance from the coast increase, the

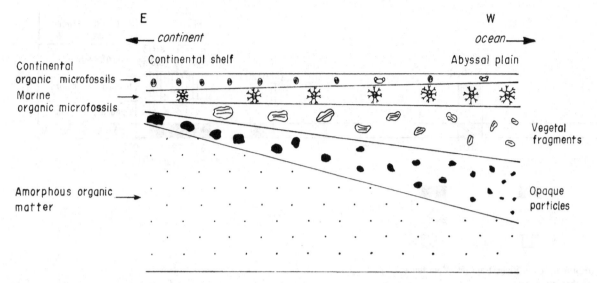

Figure 8.4. Progressive change in the composition of palynological residue from the shelf to the abyssal plain along the Cape Blanc transect, Mauritania.

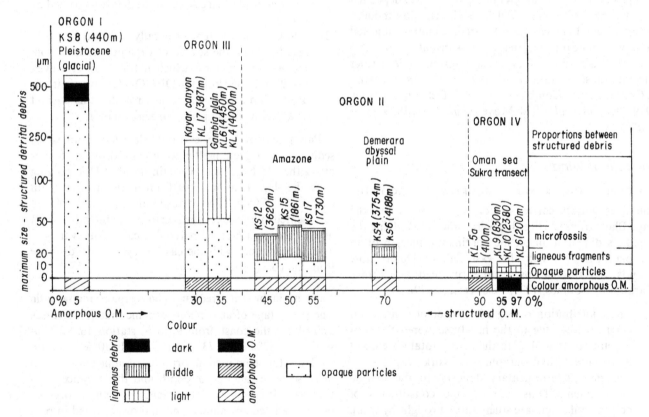

Figure 8.5. ORGON I, II, III, & IV: graphic representation of the palynofacies studies.

continental, rather than the marine markers, become more abundant. Such an apparent anomaly can be explained logically by the fact that when organic matter of continental origin reaches the marine environment it has already undergone some alteration. Hence, only the least labile fractions remain, and they will be transported further and further without any significant transformation. On the other hand, marine organic matter deteriorates

rapidly as soon as it is produced. The sedimentation pattern of pelagic to semi-pelagic materials (Moyes *et al.*, 1979) is mainly one of 'decantation' and is therefore continuous and regular, and is without noticeable reworking, which also explains the progressive changes of the organic matter seen along the transect.

Variation in the proportions of structured elements The proportions of the different types of structured elements vary depending on the site. Evolution of the composition of the palynofacies can be demonstrated as functions of depth of the sites and their distance from the coast.

Organic microfossils constitute less than 1% of the palynological residue. When this residue is concentrated for palynological analysis it is observed that organic microfossil composition varies throughout the transect. The percentage of pollen decreases seaward with a corresponding increase in dinoflagellate cysts. This quantitative progression is accompanied by qualitative changes: disaccate pollen grains, almost exclusively represented by *Pinus*, become proportionally more abundant at station 10 (the deepest), where logically the pelagic cysts of the genus *Impagidinium* replace genera found in neritic or coastal environments, such as *Tuberculodinium* or *Lingulodinium*.

The percentage of ligneous fragments and opaque particles increases with distance from shore. While in the shallower cores, KR13 and KL14, these detrital elements constitute an average of 5% of the palynological residue, their percentage progressively increases from about 10% in site KL15 to 25% in KL11, and reaches about 70% in the deeper water core, KL10. These detrital elements generally measure 5–10 μm, the maximum size decreasing progressively away from the shore, from 50 to 30 μm. The reflectance of the coal particles is generally low, less than 1.5%, which indicates low maturity. These coal particles as well as the ligneous fragments are quite comparable to the corresponding elements present in desert aerosols collected from a source region in Niger (Caratini *et al.*, 1988). This is a further proof of their Saharan origin, and of transportation by wind to the marine environment.

Variation in coloration Studies on the evolution of organic matter coloration confirm the seaward changes. Indeed, the color of the two fractions, amorphous and structured, evolves along the transect, and this evolution corresponds to an alteration towards greater depths. The color of the amorphous organic matter becomes lighter, while the ligneous fragments become darker, from light yellow in the KR13, KL14, KL15 and KL11 cores to a fairly dark ochre color in the KL10 core.

Thus, in this example, first the wind from the Sahara, then a regular decantation by pelagic to semipelagic suspensions in the sea water, explain the main patterns of transport responsible for the relationship between the sources of the organic material and the resultant composition of the sedimentary organic matter.

THE SUKRA TRANSECT, OMAN SEA: ORGON IV

The Sukra transect is located in the Oman Sea along the 57°30′ meridian (Pelet *et al.*, 1981). It is over 200 km long (Fig. 8.1), extending from the edge of the continental shelf (station 6: 200 m) to the abyssal plain (station 5: 4010 m). Like the Cape Blanc transect, this region is now bordered by a vast desert zone, and upwellings are active here also. On the other hand, the winds which give rise to the upwellings blow alternately from east and west (Currie *et al.*, 1973) and hence do not traverse the neighboring desert regions immediately north of the region studied.

In all the sites of the Sukra transect, the palynological residues are characterized by an extreme abundance of amorphous organic matter which represents 97% of the total in station 6 (the shallowest), and 90% in the abyssal plain (Figs. 8.2, 8.6.2). The structured organic matter is composed of small elements, less than 15 μm in size. In spite of such homogeneity, the ratio of amorphous/structured organic matter decreases away from the shore. This diminution is accompanied by a change in color indicating a deterioration and is comparable to that observed off Mauritania. Thus, the alteration of organic matter off Oman is the same as that described for the Cape Blanc transect.

Along the two transects investigated, Cape Blanc and Sukra, the potential sources of organic matter are similar and, consequently, the sedimentary organic matter shares common features. Their main dissimilarity is due to the difference between the eolian fluxes, heavily loaded off Sahara but not off Oman, because of the orientation of the winds in this region of the Arabian Sea.

Role of sub-marine canyons and slumping

THE KAYAR CANYON, SENEGAL, AND THE ABYSSAL PLAIN OF GAMBIA: ORGON III

The Kayar canyon is a major geomorphological feature of the continental margin off Senegal and Gambia (Caratini & Coumes, 1979). It begins near the Senegalese coast and, after a wide curve in the north, stretches over the abyssal plain of Gambia. This whole zone extends off a humid region covered by forests of the Sudanian and Guinean domains and is drained by short rivers with abundant flow. Three sites were studied in detail (Fig. 8.1): one in the middle of the canyon (station 17: 3871 m), another at its opening in the Gambia abyssal plain (station 6: 4450 m), and the third at the foot of the slope (station 4: 4000 m; see Dewaele, 1980).

Sedimentological studies of Moyes *et al.* (1979) have shown that slumping on the slope and reworking in the canyon dominate the deposition of the sediments, mainly silty clay, which is too fine for clearly revealing turbidite sequences.

The most striking feature of the palynofacies of the three cores studied is their uniformity (Fig. 8.5). Ligneous fragments, constituting 30-40% of the total organic residue, can attain dimensions up to 250 μm. They are

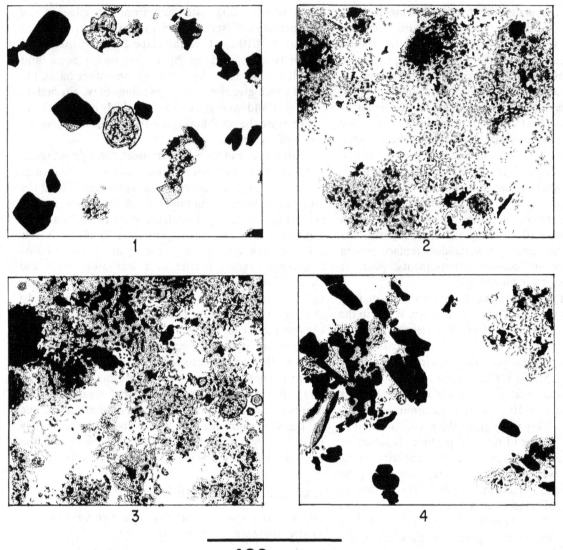

100 μm

Figure 8.6.   Magnification indicated by bar. (1) ORGON I, KS14 (0.20 m). Sea of Norway. Palynofacies with predominance of structured elements: ligneous fragments and highly colored microfossils; opaque particles. (2) ORGON IV, KL5 (0.20 m). Sea of Oman. Palynofacies dominated by amorphous organic matter. (3) ORGON III, KL14 (2.50 m). Offshore Mauritania. Cape Blanc transect. Station near the coast. Palynofacies dominated by amorphous organic matter. (4) ORGON III, KL10 (3.15 m). Offshore Mauritania. Cape Blanc transect. Station away from the coast. Palynofacies dominated by opaque particles.

light yellow in color, a good index of immaturity. The opaque particles, also comprising 30–40% of the total, belong to two granulometric classes: small-sized elements (5–15 μm), which are the most abundant, and larger elements measuring 50–100 μm. The reflectance of the large-sized particles is low, always less than 2%, indicating low maturity. The well-preserved state of structured elements is due to transportation followed by rapid deposition, the rivers being short and the continental plateau relatively narrow. The percentage of amorphous organic matter is about 30%, and the color of this fraction is quite dark, an indicator in this case of its well-preserved condition.

The heterogeneity of the opaque particles denotes a double origin, regional and probably aquatic for large-sized fragments, distant and eolian for small-sized ones. Several arguments favor an eolian origin for these small-sized fragments. They are identical to those observed in the Cape Blanc transect, where eolian origin is certain. The eolian source is also evident in the nature of the pollen associations: pollen from the Sahara predominates (about 40%); there is also a fair amount (about 5% on the average) of Mediterranean pollen.

The deposition of the sediments, dominated by the preponderant role of turbidity currents and rapid transportation by canyons, insures the preservation of the organic material, producing numerous reworkings of the sediments, and leading to their homogenization. Thus, the similarity between sites 17, 4 and 6, despite the difference in their locations in terms of both depth and distance from the coast, is remarkable and is explained by the mode of transport on the continental slope.

The role of sea level

THE AMAZON CONE AND THE ABYSSAL PLAIN OF
DEMERARA, BRAZIL: ORGON II

The Amazon River drains a huge basin now covered almost completely by tropical evergreen forests. This river then flows into the Atlantic Ocean where, in the course of geological time, it has accumulated a gigantic cone prolonged by a deep sea fan which grades laterally into a long continental rise to the abyssal plain of Demerara.

The Holocene Period   All these ocean floors are covered by a thin (a few decimeters) layer of biogenic ooze, consisting of foraminifera and pteropods. This ooze has been deposited during the Holocene by a pelagic sedimentation pattern (Moyes *et al.*, 1978), excluding the irregular fluxes of sediments by slumping. The palynofacies in all the sites studied (Fig. 8.1) are very homogeneous (Fig. 8.5). They are dominated by dark yellow amorphous organic matter (85%, on the average) and also contain small-sized (<15 μm) opaque particles (10–15%), and (rarely) light-colored (yellow to greenish-yellow) ligneous fragments. Organic microfossils are poorly represented.

Upper Pleistocene: the last glacial   The sediments deposited during the last glacial phase of the Upper Pleistocene were studied in three sites on the cone (stations 12, 15 and 17), on the lower part of the continental rise (station 4) and in the abyssal plain of Demerara (station 6). These sediments are characterized by terrigenous material in fining-up sequences (Moyes *et al.*, 1978) and are completely different from the more recent sediments.

During the last glacial, which was marked by a low sea level stand, the shoreline was located in the immediate vicinity of the platform's edge. Hence, the sediments brought by the Amazon were dumped on the cone, reaching the abyssal plain by numerous channels incised in the Amazon canyon, as well as through mass transport by turbidity currents. The palynological residues are also quite different from those of the Holocene. They share many similarities in spite of the diversity in their locations.

The structured fraction, containing ligneous and opaque particles and organic microfossils, comprises about half the total. Ligneous fragments, sometimes quite large (50 μm), dark yellow to ochre, dominate over the opaque particles, which, in contrast, are small (<20 μm) and have reflectance values ranging between 0.1 and

5.6%. Organic microfossils are almost exclusively continental in origin and are often altered. Some palynomorphs are reworked from the Paleozoic (Devonian-Carboniferous), Upper Cretaceous and Tertiary. Light-colored amorphous organic matter is indicative of a poor state of preservation.

All the characteristics described denote heterogeneity of the organic matter, which may be the result of the diversity and remoteness of the sources. The organic matter reaches the oceanic environment after considerable transport and it is, therefore, already somewhat mature before its deposition.

The detrital structured elements are smaller in the abyssal plain, the largest fragment measuring only 30 μm. Hence, the transformation of organic matter off the coast is the same as in the examples cited above, although these changes are less marked than in the Mauritania transect. On the other hand, amorphous organic matter, which quantitatively diminished away from the coast in the preceding case, becomes abundant here, comprising 70% of the organic material. This is understandable when one considers that the geochemical composition of the organic matter (Debyser *et al.*, 1978) on the whole indicates terrestrial origin. In this case, we reach the optical limit for interpreting amorphous organic matter. If it is true that in a marine environment the organic matter defined as amorphous is generally of marine origin, the amorphous organic matter could also result, at least in part, from the degradation of structured organic matter of continental origin by a process of 'destructuration.' Structured particles, especially ligneous fragments with fragmented extremities, are common. Tiny particles are liberated, which, when isolated, would be considered elements of amorphous organic matter. Thus, in this case, the optical aspect of the organic matter is modified, whereas the chemical properties do not change.

During the Holocene sea level rise and its current high stand, terrigenous sediments have been and are being trapped on the continent and continental margin. The detrital Amazonian material is no longer transported and deposited on the outer shelf, slope and abyssal plain. Hence, by modifying the conditions of transportation, the sea level fluctuations have important implications for sedimentary organic matter distribution patterns along the continental margin and even in the neighboring abyssal plain. Understanding these changes is of considerable value in deciphering past oceanographic conditions (Vail *et al.*, 1977).

The role of glaciers

THE CONTINENTAL SHELF OF NORWAY DURING A
PLEISTOCENE GLACIAL PERIOD: ORGON I (KS8)

The sediments of core KS8, 3.5 m long, taken from the upper part of the Norwegian slope off Trondheim at a depth of 440 m, were deposited during a glacial phase of the Pleistocene (Combaz *et al.*, 1977).

The palynofacies (Fig. 8.5) are dominated (70–90%) by large (up to 500 *μ*m) opaque particles. The ligneous fragments, also large, are highly colored, brown to dark. Light-colored amorphous organic matter is not well represented (<10%) (see Fig. 8.6.1 for similar palynofacies). More than 95% of the organic microfossils are reworked from Carboniferous to Neogene, with a predominance of Mesozoic fossils. Thus, the organic matter observed is heterogeneous, mainly reworked from older fossil-bearing rocks and deposited at a time when the neighboring continent was eroded by glacial processes. These high-energy processes are capable of transporting large clasts in which the enclosed organic matter remains well-preserved.

In other sediments cored in the Oslo Channel (KS2, 630 m) and at the foot of the slope (KS10, 860 m) off Norway (Fig. 8.1), the glacial phase of the Late Pleistocene is clearly distinguishable from the Holocene warming by its palynofacies, which are much more detrital during the colder phase. Moreover, the percentage of reworked palynomorphs is greatly reduced in the Holocene sediments. Thus, the palynofacies and palynological assemblages respond to the modification in the condition of transportation resulting from climatic changes.

Conclusion

The ORGON project, by describing a diversity of sedimentary environments, provides the specialist with a source of accurate references in the form of a series of well-documented models. Faced with the multiplicity and complexity of marine sedimentary environments, it is hardly justifiable to establish rules which would enable the interpretation of palynofacies with a great degree of certainty by studying only a few examples. In fact, although the relationships between the sources of organic material, the palynofacies, and the conditions of transportation sometimes seem sufficiently clear for coherent paleogeographic reconstruction, the obvious conclusion derived from our studies is the importance of distortions that could transform these relationships to the point of making them indecipherable. Such distortions are mainly caused by the numerous physical, biological and chemical transformations of the organic matter during its transport from the region of production to the site of deposition. A selective process takes place under the action of agents and the conditions of transport, which plays an important role in the final composition of the organic matter in the sediment. The example from ORGON II below provides further evidence of this process.

THE 0.80 M LEVEL OF THE KL8 C CORE (4236 M), THE ABYSSAL PLAIN OF DEMERARA: ORGON II

Core KL8 C, taken at the boundary between the abyssal plain of Demerara and the Amazonian cone (Fig. 8.1) at a depth of 4236 m, shows a distinct organic-rich layer at 0.80 m, in the Upper Pleistocene. This accumulation, about 10 cm thick, is composed of perfectly preserved ligneous fragments, including twigs and small branches up to 10 cm in length, and some entire leaves. The plant detritus is partially carbonized. Moyes *et al.* (1979) report that turbidity currents could have reached this region located nearly 600 km from the present coastline, more than 400 km from the edge of the plateau. This condition has contributed to an irregular sedimentation pattern, as indicated by numerous graded sequences, most often ending in a bed of ligneous fragments. Hence, these beds cannot be considered as exceptional deposits. This spectacular example has been emphasized, because such a feature is rarely mentioned in the literature on distribution of plant detritus in marine environments. Most often, there has only been a statement that the size of the ligneous fragments diminishes away from the coast and with water depth. In the Demerara abyssal plain, such beds of ligneous fragments obviously indicate the existence of luxurious vegetation covering the neighboring continent. The influence of the sources of organic material is therefore quite well delineated, but, here also, conditions of transportation played the primary role in the final composition and appearance of the organic material.

The most reasonable conclusion to be drawn from the palynological investigations of the ORGON sediments studied is, therefore, that the character of the palynofacies is influenced more by the agents and conditions of transportation than by the nature of the neighboring sources of the organic matter.

References

Caratini, C., Bellet, J. & Tissot, C. (1975). Étude microscopique de la matière organique: Palynologie et Palynofaciès. In *Géochimie organique des sédiments marins profonds.Orgon II, Atlantique-N.E. Brésil*. Paris: Centre National de la Recherche Scientifique, 215–47.

Caratini, C., Bellet, J. & Tissot, C. (1979). Étude microscopique de la matière organique: Palynologie et Palynofaciès. In *Géochimie organique des sédiments marins profonds. Orgon III, Mauritanie, Sénégal, îles du Cap-Vert*. Paris: Centre National de la Recherche Scientifique, 215–65.

Caratini, C., Bellet, J. & Tissot, C. (1981). Étude microscopique de la matière organique: Palynologie et Palynofaciès. In *Géochimie organique des sédiments marins profonds. Orgon IV, golfe d'Aden, mer d'Oman*. Paris: Centre National de la Recherche Scientifique, 255–307.

Caratini, C., Bellet, J. & Tissot, C. (1983). Les palynofaciès:

Représentation graphique, Intérêt de leur étude pour les reconstitutions paléogéographiques. In *Géochimie organique des sédiments marins profonds, d'Orgon à Misédor*. Paris: Centre National de la Recherche Scientifique, 327–52.

Caratini, C. & Coumes, F. (1979). La géochimie organique des sédiments marins profonds. Mission Orgon III, 1976 (Mauritanie, Sénégal, îles du Cap-Vert): le cadre géologique, structural et sédimentaire. *Revue de l'Institut Français du Pétrole*, **34**, 857–72.

Caratini, C. & Cour, P. (1980). Aéropalynologie en Atlantique oriental au large de la Mauritanie, du Sénégal et de la Gambie. *Pollen et Spores*, **22**, 245–56.

Caratini, C., Tissot, C. & Fredoux, A. (1988). Caractérisation des aérosols désertiques à Niamey (Niger) par leur contenu pollinique. *Institut Français de Pondichéry, Section Scientifique et Technique, Travaux*, **25**, 251–68.

Combaz, A. (1964). Les palynofaciès. *Revue de Micropaléontologie*, **7**, 205–18.

Combaz, A. (1980). Les Kérogènes vus au microscope. In *Kerogen. Insoluble Organic Matter from Sedimentary Rocks*, ed. B. Durand. Paris: Éditions Technip, 55–110.

Combaz, A., Bellet, J., Poulain, D., Caratini, C. & Tissot, C. (1977). Étude microscopique de la matière organique de sédiments quaternaires de la mer de Norvège. In *Géochimie organique des sédiments marins profonds. Orgon I, mer de Norvège*. Paris: Centre National de la Recherche Scientifique, 139–75.

Corréia, M. (1969). Contribution à la recherche de zones favorables à la genèse du pétrole par l'observation microscopique de la matière organique figurée. *Revue de l'Institut Français du Pétrole*, **24**, 1417–54.

Currie, R. I., Fischer, A. E. & Hargreaves, M. (1973). Arabian Sea upwelling. In *The Biology of the Indian Ocean*, ed. B. Zeitzschel. New York: Springer-Verlag, 36–51.

Debyser, Y. & Gadel, F. (1979). Géochimie des kérogénes dans les sédiments. In *Géochimie organique des sédiments marins profonds. Orgon III, Mauritanie, Sénégal, îles du Cap-Vert*. Paris: Centre National de la Recherche Scientifique, 375–403.

Debyser, Y., Gadel, F., Leblond, C. & Martinez, J. M. (1978). Étude des composés humiques, des kérogènes et de la fraction hydrolysable dans les sédiments. In *Géochimie organique des sédiments marins profonds. Orgon II, Atlantique-N.E. Brésil*. Paris: Centre National de la Recherche Scientifique, 339–54.

Dewaele, H. (1980). Contribution à l'étude des sédiments

marins profonds au large de la Mauritanie et du Sénégal (Mission Orgon III, Station 4): Dipl. E. N. S. P. M., *Rapport de l'Institut Français du Pétrole* **28 038**.

Durand, B. & Nicaise, G. (1980). Procedures for kerogen isolation. In *Kerogen. Insoluble Organic Matter from Sedimentary Rocks*, ed. B. Durand. Paris: Éditions Technip, 35–53.

Hart, G. F. (1986). Origin and classification of organic matter in the clastic systems. *Palynology*, **6**, 1–7.

Haseldonckx, P. (1974). A palynological interpretation of palaeo-environment in S.E. Asia. *Sains Malaysiana*, **3**, 119–27.

Moyes, J., Duplantier, F., Duprat, J., Faugéres, J. C., Pujol, C., Pujos-Lamy, A. & Tastet, J. P. (1979). Étude stratigraphique et sédimentologique. In *Géochimie organique des sédiments marins profonds. Orgon III, Mauritanie, Sénégal, îles du Cap-Vert*. Paris: Centre National de la Recherche Scientifique, 121–213.

Moyes, J., Gayet, J., Poutiers, J., Pujol, C. & Pujos-Lamy, A. (1978). Étude stratigraphique et sédimentologique. In *Géochimie organique des sédiments marins profonds. Orgon II, Atlantique-N.E. Brésil*. Paris: Centre National de la Recherche Scientifique, 105–56.

Pelet, R., Caratini, C. & Combaz, A. (1979). La géochimie organique des sédiments marins profonds. Mission Orgon III, 1976 (Mauritanie, Sénégal, îles du Cap-Vert): Généralités et résultats obtenus à la mer. *Revue de l'Institut Français du Pétrole*, **34**, 847–908.

Pelet, R., Caratini, C., Arnould, M., Garlenc, S., Laborde, D., Romano, J. C., Sautriot, D., DeSouza Lima, Y., Bensoussan, M., Marty, D., Bonnefonte, J. L. & Tissot, C. (1981). La géochimie organique des sédiments marins profonds. Mission Orgon IV, 1978 (golfe d'Aden, mer d'Oman). *Revue de l'Institut Français du Pétrole*, **35**, 775–83.

Pocock, S. A. J., Vasanthy, G. & Venkatachala, B. S. (1987). Introduction to the study of particulate organic materials and ecological perspectives. *Journal of Palynology*, **23–24**, 167–88.

Simoneit, B. R. T. (1977). Organic matter in eolian dusts over the Atlantic Ocean. *Marine Chemistry*, **5**, 443–64.

Vail, P. R., Mitchum, R. M., Jr., Todd, R. G., Widmeri, J. W., Thompson, S., Sangree, J. B., Bubb, J. N. & Hatelid, W. G. (1977). Seismic stratigraphy and global changes of sea level. In *Seismic Stratigraphy. Application to Hydrocarbon Exploration*. American Association of Petroleum Geologists Memoir, **26**, 49–212.

9 Maceral palynofacies of the Louisiana deltaic plain in terms of organic constituents and hydrocarbon potential

GEORGE F. HART

Introduction

Geologists have studied most aspects of sediments in order to elucidate characteristics significant in determining the environment of deposition. Traditionally the lithotype + biotype = rock approach has led to the concept of lithotopes and biotopes (environmental derivatives of lithotype and biotype), and the very successful concepts of lithostratigraphy, biostratigraphy and chronostratigraphy. It is clear that no single parameter can be used to distinguish all depositional environments, but if a number of parameters are measured, chances of error may be reduced. Unfortunately, which parameters are of most importance is unclear. Much of this uncertainty exists because sampling of the environments has been inadequate, and available data bases are poorly defined. Moreover, conclusions are rarely based upon the principles of experimental design and hypothesis testing.

The accessibility of samples strongly influences the choice of a sampling design, and, for the determination of depositional environment this is a particularly significant constraint. If outcrops are studied, a wealth of information relating to paleoenvironments becomes available; if conventional cores are studied the quantity of possible information is reduced, and even further so by sidewall cores. Unfortunately, the most abundant kind of sample available to the subsurface geologist is the collection of cuttings from the rotary bore. The information available from cuttings is indeed minimal, and even with good samples the derived information may be of reduced value due to caving and other problems connected with taking samples. Unfortunately for the subsurface geologist well cuttings with occasional sidewall cores are the rule rather than the exception.

Exploration in the subsurface and studies of modern surficial deposits show that an understanding of the facies relationships within clastic environments is fundamental to understanding the detrital models that have been proposed. The development of a new technique for determining depositional environment is important,

especially if the technique holds promise of application in the subsurface. Maceral facies analysis, utilizing the particulate organic debris characteristic of specific environments, is such a technique.

The abundance of deltaic sediments in the geological record, and the presence of their contained hydrocarbons (coal, oil and gas) are major reasons why they have been intensely studied by subsurface geologists during the past century. The early work of Gilbert (1885, 1890), Barrell (1912) and Twenhofel (1932) on ancient deltaic deposits, and of Johnston (1921, 1922) and McKee (1939) on modern deltas, stimulated a large volume of research on the Mississippi River delta, starting in the 1940s, by Fisk (1944, 1955, 1961), Welder (1959), Scruton (1960), Coleman et al. (1964), Frazier (1967), Darrell and Hart (1970), and others. In the 1960s modern deltaic work expanded to looking at different climatic settings, as is seen in the work of Wright and Coleman (1973, 1975), Galloway (1975), Coleman et al. (1974, 1980), Kuehl et al. (1982), Rhoads et al. (1985), and many others.

The principal purpose of this chapter is to show how particulate organic matter (POM) is distributed in the surficial bottom sediments of the associated environments of the Louisiana deltaic plain. It further indicates the degree to which clastic environments can be discriminated using POM. The study is based on the formerly proprietary report by Hart (1979c), and extends from the modern alluvial valley across the upper and lower deltaic plain to the active delta and near offshore environments.It builds upon the ideas developed by Hart (1971, 1975, 1976), using the classification of particulate organic matter (macerals) published by Hart (1979 a, b, 1986; see Hart et al., Chap. 17 this volume, for tabular presentation of maceral classification). Additional applications are published by Wrenn and Beckman (1981, 1982), Hart (1988), Darby and Hart (1988; Chap. 10, this volume), LeNoir and Hart (1986, 1988), and Hart et al. (1989). This region contains a great variety of deltaic subenvironments covering a broad spectrum of chemical, physical and biological variation.

Twenty-six physiographic environments were sampled at 520 stations. In all 1040 field samples were collected: two samples from each of 20 stations per physiographic environment. These physiographic areas commonly include restrictive environmental settings and to a large degree control the types of deposits that are preserved and observed in the subsurface.

Regional setting

The Louisiana deltaic plain is defined as that physiographic area comprising coastal and nearshore deposits constructed by riverborne sediments, and modified by various marine processes. The facies architecture results from a variety of interacting dynamic physical, chemical, and biological processes. These processes modify and disperse both the riverborne allochthonous sediments and the autochthonous moneran, protistan, fungal, plant and animal remains. The subareal component of this clastic system is usually divided into an upper and lower deltaic plain. The upper deltaic plain is the oldest part of the surficial clastic complex and is isolated from contemporaneous marine influences. As Coleman and Prior (1980) note, the upper deltaic plain is commonly a continuation of the alluvial valley system and is dominated by riverine processes. In contrast, the lower deltaic plain lies within the realm of rivermarine interaction.

Although the Gulf coastal plain sequences of sediment had their origin during the Jurassic Period with the initial formation of the Mississippi embayment, the modern Louisiana deltaic plain is a product of the Holocene events which began at the end of the late Wisconsin glacial episode (Fisk & McFarlan, 1955). During the Wisconsin, the Mississippi River was deeply entrenched and had prograded to the edge of the continental slope, where clastic deposition formed a thick submarine wedge. With the melting of the ice sheets sea level began to rise, and the shoreline progressively moved inland. This caused the transgression of the deltaic sediments that form modern coastal Louisiana. Stabilization of the shoreline occurred at about 6000 BC (Morgan, 1977), and the modern Louisiana deltaic plain formed from a composite of at least seven major river switches and their associated deltaic environments (Kolb & Van Lopik, 1966). This accumulation of switching depositional centers constructed a 200-mile-wide sediment platform from Vermillion Bay to the Mississippi Sound (Fisk & McFarlan, 1955), as shown in Figure 9.1.

Progradation and abandonment of the individual fluviatile-deltaic complexes comprising the Louisiana deltaic plain was a cyclic process (Coleman & Gagliano, 1964), in which two main phases are recognizable. The progradational phase caused both vertical and lateral growth of the depositional wedge. The specific characteristics of this phase were controlled by the interaction of sediment supply, rate of sedimentation, compaction, subsidence, water depth, climatic factors and coastal processes (Morgan, 1970). With excessive progradation the distributaries became over-elongated, and this caused a decline in flow gradient and sediment discharge, resulting in channel switching. Channel switching initiated depo-center abandonment. The river simply diverted along a shorter path to the Gulf of Mexico. Associated with this second phase is the subsidence of older depositional areas and the burial or destruction by reworking of previously active depositional environments.

The contemporary depositional environments of the Louisiana coastal plain show clear evidence of at least four depositional complexes in various stages of the evolutionary process described above (see Fig. 9.1).

(1) The Atchafalaya Swamp and the lakes and marshes of the Lafource Parish are the remnants of earlier depositional systems. The current major activity is burial due to compaction and subsidence, with an associated accumulation of low energy, fine grained organic sediments.

(2) The Chandeleur Islands and the sound and marshes of St. Bernard Parish represent a different kind of abandonment of an earlier depositional system. The major activity is destruction and reworking of the earlier sediments with the accumulation of arenaceous materials.

(3) The modern Mississippi (Balize) delta is an active depositional system in which development of the progradational wedge is the main theme.

(4) The modern Atchafalaya delta is a young depositional system, the birth of which mankind has witnessed (Van Heerden & Roberts, 1988).

Sampling

It is because of the synchronous existence of four deltaic complexes that it is possible to sample a large number of depositional environments representative of many facies found in the subsurface. The vastness of the area to be sampled necessitated a statistically designed sampling plan that would adequately cover the physiographic environments of the Louisiana deltaic plain. A major requirement of this design was that, irrespective of the environment, the sampling locations had to be as little influenced by man as possible. This is often difficult to accomplish in a culturally active area such as Louisiana, and approximately eight months were devoted to researching possible collecting sites. Many possible areas were rejected as overly influenced by man and not representative of the depositional processes, e.g., the

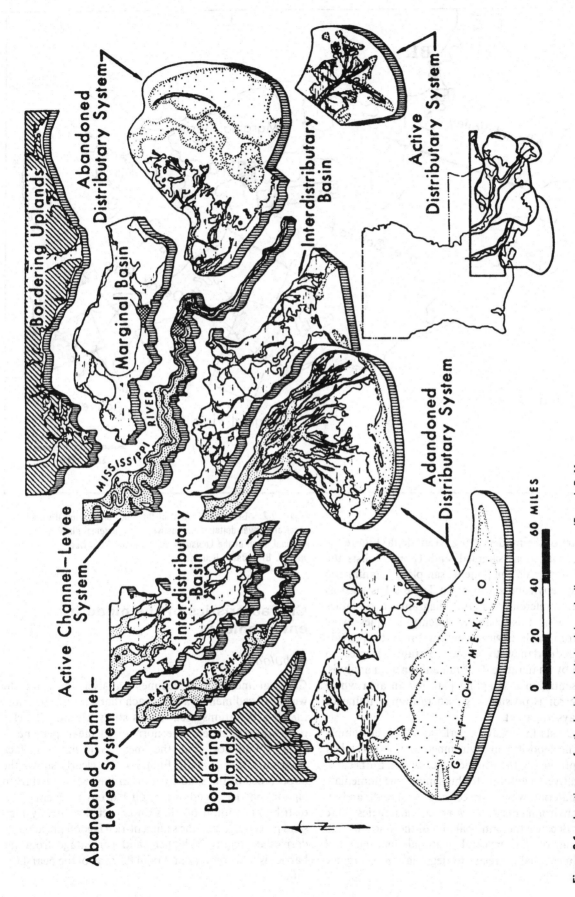

Figure 9.1. Location map for the Louisiana deltaic plain and adjacent areas. (From J. P. Morgan, Archival Files, Map Library, Louisiana State University, Baton Rouge.)

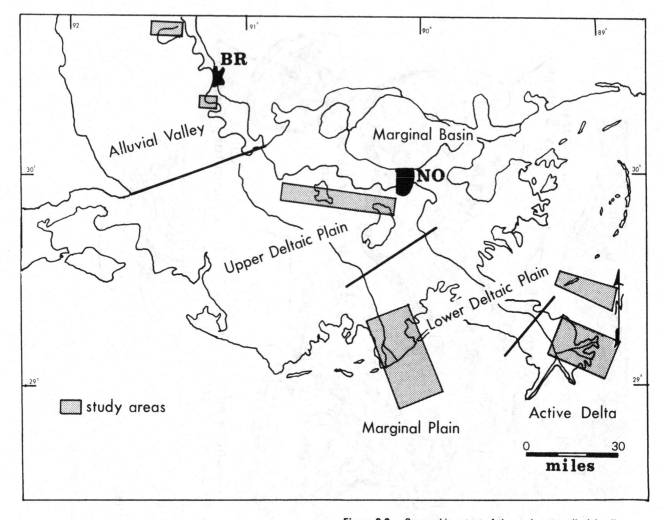

Figure 9.2. Geographic extent of the study area: alluvial valley, upper and lower deltaic plain, active delta, marginal plain, and marginal basin of the Louisiana deltaic plain. NO = New Orleans. BR = Baton Rouge.

western part of the modern Mississippi delta. Figure 9.2 shows the main areas sampled, and Table 9.1 lists the major physiographic components sampled. The numeric key for each environment given in Table 9.1 is used as a consistent reference in the diagrams and tables associated with this chapter. Twenty samples were selected from each environment in order to provide sufficient redundancy to allow aberrant samples to be rejected by an iterative discriminant function analysis. This is essential in a sampling plan for an area where sampling error is possible, e.g., where human influence is not always apparent.

Table 9.2 lists the analyses performed at each sample location. The depositional environment sample was taken from a depth below the Eh 'kick', i.e., where Eh became highly negative. In addition to this sample, the immediate surficial material was sampled and analyzed, and in aqueous environments the water characteristics were measured. Because the principal aim of the study was an investigation of the maceral material the chemical attributes measured all relate to degradation of organic matter in sediments.

Characteristics of the modern environments

Lithology

Classical methods of sediment analysis measure the weight of sediment grains in each diameter interval, and for all 520 stations used in this study a sand/silt + clay ratio was calculated. The normalized mean percentage sand abundance, and the mean silt size for each environment, is given in Figure 9.3. It clearly shows the importance of the sand fraction in the sediment-starved lagoon, barrier island and point bar environments. The distributary mouth bar has an expected relatively high sand percentage, and the sediment-rich lagoon presumably derives its relatively higher sand percentage from the barrier island complex and tidal flushing of the nearshore sands.

Table 9.1. *Major physiographic components sampled: Louisiana deltaic plain*

Lower deltaic plain		
Active delta	1.1	prodelta
	1.2	distributary mouth bar
	1.3	levee
	1.4	channel
	1.5	interdistributary bay
Shoreline complex	2.1	bay
	2.2	lake
	2.3	brackish marsh
	2.4	saline marsh
	2.5	fresh marsh
	2.6	isolated swamp
Marginal deltaic basin (sediment-starved)	3.1	back island lagoon
	3.2	barrier island
	3.3	10 foot marine
	3.4	50 foot marine
	3.5	100 foot marine
Marginal deltaic basin (sediment-rich)	4.1	back island lagoon
	4.2	barrier island
	4.3	10 foot marine
	4.4	50 foot marine
	4.5	100 foot marine

Upper deltaic plain		
Lower alluvial valley	5.1	lake
	5.2	backswamp
	5.31	levee
	5.32	dry swamp
	5.33	wet swamp
	5.34	channel
	5.35	foreswamp
	5.4	point bar
	5.5	abandoned meander loop

Organic matter in sediments has a density considerably less than that of mineral grains. Therefore, for most sediments, with only a few percent of organic matter, it can be ignored in terms of weight percent analysis. However, in sediments with considerable particulate organic matter this difference in specific gravity may be important. This is particularly so because the particle size of organic debris is often correlated with a specific depositional environment as a result of the digestive characteristics of the animals in the biocoenosis. For this reason Coulter Counter analyses were performed on all samples. The Coulter Counter measures particle volume, and because a few percent of organic matter in a sediment may have considerable volume even though it has a

negligible weight, a different aspect of the sediment is determined from that measured using sieve analysis. The measurement of the grain size volume resulted in three data sets for every sample:

(1) the volume percentage of silt and of clay, together with the silt/clay ratio;
(2) the raw (untransformed) frequency of the 1–64 μm fraction covered by standard class intervals of the Coulter Counter scale (a volume measure);
(3) the Log (10) transformed frequency of the 1–64 μm fraction (a volume measure).

The normalized mean percentage silt size values are predictable based on energy level, with the exception of the lake of the upper deltaic plain, where the higher silt size may be a result of the inclusion of a greater volume of algal organic matter (see Table 9.12).

Special lithologies of the Louisiana deltaic plain are the modern paralic organic rich environments. These peaty deposits have been extensively studied (Penfound & Hathaway, 1938; Russell, 1942; O'Neil, 1949; Palmisano, 1970; Chabreck, 1972; Chabreck & Condrey, 1979; Beckman, 1984). High concentrations of preserved organic matter can only be achieved by deposition in an environment of high productivity and predominantly reducing conditions. The organic sediments are deposited in a generally autochthonous setting of high terrestrial input and thus have a high total organic carbon content (TOC). The important factors affecting TOC are primary productivity, exportation losses, sedimentation rate, and redox potential. In coastal Louisiana these conditions are met in the extensive paralic marsh and swamp wetlands. Natural marshes alone occupy about 2.5 million acres (about 30% of Louisiana coastal region: Chabreck, 1972). An additional 1.7 million acres of ponds and lakes are active sites of high organic matter accumulation. A thick sequence of peat occurs when marsh growth and organic deposition keeps pace with subsidence. All of these environments are intimately associated with the development of the deltaic depositional complexes. There are two standard measures for the abundance of organic matter in a sediment. These are total organic carbon (TOC) and extractable organic matter (EOM). TOC was used in the present study. As seen in Figure 9.4, TOC is most abundant in the organic rich environments (with the exception of the riverine influenced foreswamp).

The organic rich deposits are either associated with active delta progradation, or occur as extensive sheets in the marginal deltaic areas. Fisk (1955) described the development of active delta interdistributary organic rich deposits. Coleman and Gagliano (1964) regard these accumulations as part of the progradational phase. The first deposits to accumulate are highly organic rich mucks overlying the silty clays of the gulf floor. As the interdistributary basin fills in, slightly brackish marsh

Table 9.2. *Analyses performed on each sample: the Louisiana deltaic plain*

Sampled population	Variables measured	Notes
Overlying water mass	pH, Eh, temperature, salinity, oxygen, turbidity	These analyses were performed because they all influence the organic activity at the surficial sediment/water-mass interface, and relate to inorganic chemical activity.
Surface sediment	pH, Eh, temperature, salinity	These analyses were performed because they all influence the organic and inorganic reactions taking place in the surficial sediment, and often are characteristic of specific environments.
Depositional sediment	macerals	Macerals were the prime objective of this study. They were analyzed to determine if the environment of deposition could be characterized by the maceral spectra, and to estimate the source rock potential of each environment.
	pH, Eh, temperature, salinity	These analyses were performed for the reasons given above.
	interstitial water sulfate and sulfides	These analyses were performed to elucidate the relationship between degradation and environment, because the bacterial activity may respond to the water sulfur chemistry.
	sand %, silt size distribution	These analyses were performed to investigate the relationship between the silt fraction of the lithotype and the environment, and also because of a possible relationship between sediment and biodegradation.
	light hydrocarbons	Methane in particular is a product of bacterial decay. The amount of methane production in an environment also is related to the source rock potential of that environment, because high methane indicates that much of the maceral material is being degraded.
	total sulfur, total organic carbon	The abundance of sulfur is related to organic processes in the environment and is important in the composition of any hydrocarbon liquids that may have their source in the deposits. Total organic carbon is an important source rock characteristic and also allows the quantitative abundance of individual macerals to be assessed.
	carbon, nitrogen	Carbon and nitrogen are used in calculating the C:N ratio of a sediment. This ratio is an estimate of the degree of biodegradation that is taking place in the sediment. C:N ratios were calculated on selected samples from each environment on the basis of average total organic carbon and range of total organic carbon values.

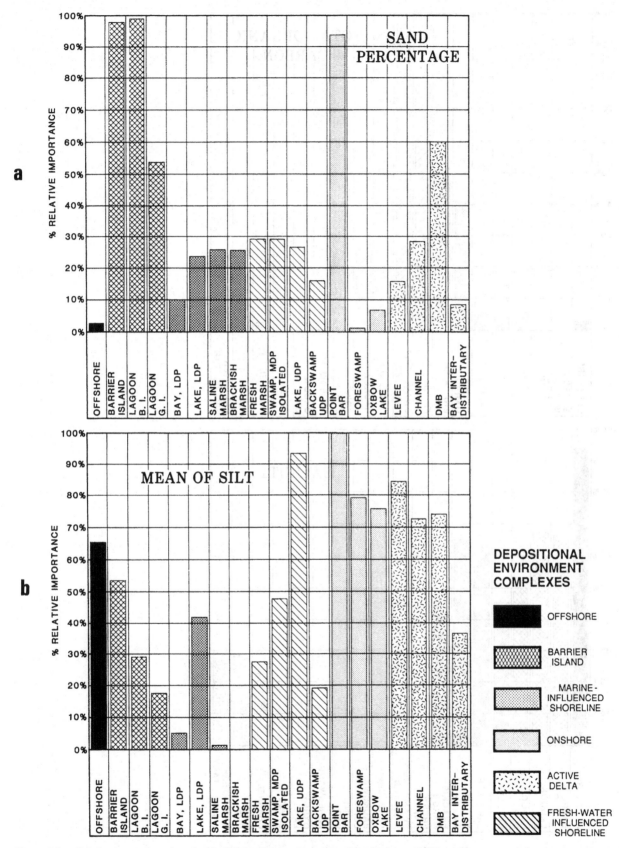

Figure 9.3. Relative importance of sand and silt in various environments of the Louisiana deltaic plain. (a) normalized mean percentage sand (by sieving); (b) normalized mean silt size (by Coulter Counter). The patterns indicate groups of related environments, as shown in the legend.

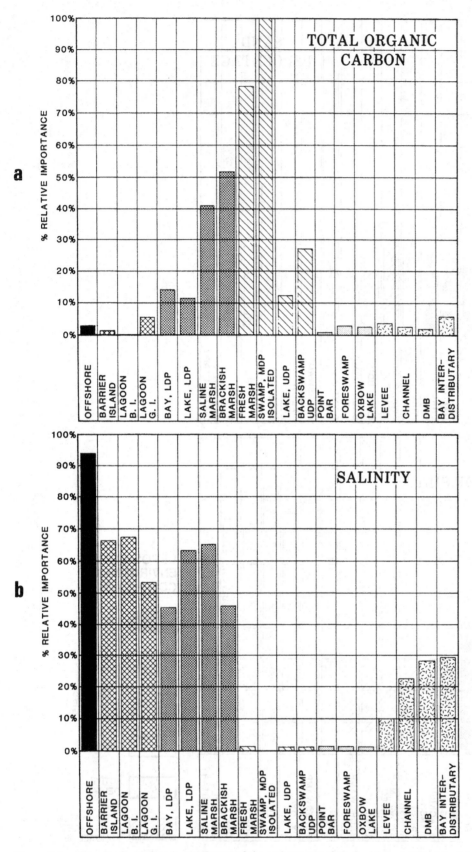

Figure 9.4. Normalized variations in constituents of the environments of the Louisiana deltaic plain: (a) TOC wt. %; (b) salinity. For explanation of patterns, see Figure 9.3.

Table 9.3. *Characteristics of the overlying water mass: Louisiana deltaic plain*

Key: A = acidic; B = basic; Br = brackish; C = clear; F = fresh; MiA = mildly acidic; MiB = mildly basic; MOx = moderately oxidized; MR = moderately reduced; MT = moderately turbid; na = not available; OL = oxygen limited; Ox = oxidized; S = saline; SBr = slightly brackish; T = turbid

Envir.	Eh	pH	Salinity	Oxygen	Turbidity
1.1	na	na	na	na	na
1.2	Ox	MiB	SBr	MOx	T
1.3	Ox	MiB	F	MOx	T
1.4	MR	MiB	F	MOx	T
1.5	MR	MiB	F	MOx	T
2.1	MR	B	Br	MOx	T
2.2	MR	B	Br	MOx	MT
2.3	MR	MiA	Br	OL	MT
2.4	MR	MiB	Br	MOx	MT
2.5	na	na	na	na	na
2.6	na	na	na	na	na
3.1	MR	B	Br	Ox	T
3.2	MR	B	Br	MOx	MT
3.3	MR	B	Br	MOx	C
3.4	MR	B	S	MOx	C
3.5	na	na	na	na	na
4.1	MR	B	Br	MOx	C
4.2	MR	B	Br	Ox	MT
4.3	MR	B	Br	Ox	C
4.4	MR	MiB	S	OL	C
4.5	na	na	na	na	na
5.1	MR	B	F	Ox	C
5.2	MR	A	F	OL	MT
5.3	na	na	na	na	na
5.4	na	na	na	na	na
5.5	MR	B	F	Ox	MT

Descriptive codes used for Tables 9.3, 9.4 and 9.5

Eh (mV)	oxidized >400	moderately reducing 100–400	reducing +100	highly reducing < −100
pH	acidic	mildly acidic 6–7	mildly basic 7–8	basic >8
salinity (%)	fresh <1	slightly brackish 1–5	brackish 5–25	saline >25
dissolved oxygen (ppm)	oxygen limiting <4	moderately oxidized 4–8	oxidized >8	
turbidity (ftu)	unturbid 0–50	moderately turbid 50–100	turbid >100	

peat accumulates in the central portion of the inter-distributary basin, with fresh marshes bordering the natural levee. Abandonment results in subsidence and the blanketing of the organic sediments with saline marsh peat and bay deposits. These interdistributary areas of peat accumulation are characterized by abrupt facies changes and rapid sedimentation rates.

The marginal deltaic, high-organic environments are laterally extensive, fine grained, and exhibit slower depositional rates but complex hydrodynamic changes. The marshes are at present depositing a blanket peat in response to the transgression of the Gulf of Mexico. Beckman (1984) showed that the thickness of the peat is controlled by the physiographic nature of the local basin.

Effect of water characteristics

Salinity is probably the most important aspect of water affecting particulate organic matter in sediments. In organic rich environments Penfound and Hathaway (1938) and Penfound (1952) divided the Louisiana marshes into four types on the basis of salinity: fresh, slightly brackish, brackish, and saline. This is a result of the interaction of tidal action and freshwater runoff. There are gradational boundaries between the marsh types on both an areal and temporal scale and only the three main types (fresh, brackish, saline) were used for the investigations of maceral distribution reported in this chapter. The characteristics of the overlying water mass are given in Table 9.3, and those of the surficial sediments in Table 9.4. There is a strong relationship between the salinity of the overlying water mass, the salinity of the surficial sediment and the salinity of the interstitial water of the depositional environment (Fig. 9.5; Tables 9.3–9.5). In the depositional environment there is a general trend for decreasing TOC content with increase in salinity (Fig. 9.6). Palmisano (1970) and Chabreck, (1972) also observed this in the surficial environment, and noted that saline marshes (average salinity 15%) contain a low diversity of plant species dominated by the halophyte, *Spartina alterniflora* Loisel. The TOC content averages 25%. The brackish marshes (average salinity 7%) are more diverse and are dominated by *Spartina patens*, in association with four to five other common species. The TOC averages 35%. The freshwater marshes are the most diverse, with 92 species dominated by *Panicum hemitomon*. TOC content reaches 50% or more.

The amount of water over an organic sediment affects the quantity of organic matter, as is seen in the scatter plot of water depth versus TOC (Fig. 9.6). Beckman (1984) pointed out that shallow water depth, or the absence of water, allow the diffusion of molecular oxygen to the surficial sediments. This encourages the development of a fungal and bacterial aerobic biocoenosis: pathways which facilitate the degradation of organic matter. This

Table 9.4. *Characteristics of surficial sediment: Louisiana deltaic plain*

Key: A = acidic; B = basic; Br = brackish; F = fresh; MA = moderately acidic; MiA = mildly acidic; MiB = mildly basic; MR = moderately reduced; N = neutral; Ox = oxidized; S = saline; SBr = slightly brackish

Environment	Eh	pH	Salinity
1.1	MR	N	S
1.2	MR	MiB	Br
1.3	MR	MiB	SBr
1.4	MR	MiB	SBr
1.5	MR	MiB	Br
2.1	MR	MiB	Br
2.2	MR	MiB	Br
2.3	MR	MiA	Br
2.4	MR	MA	Br
2.5	MR	A	F
2.6	MR	MiA	F
3.1	MR	B	Br
3.2	MR	B	S
3.3	MR	MiB	S
3.4	MR	MiB	S
3.5	MR	MiB	S
4.1	MR	MiB	Br
4.2	MR	MiB	S
4.3	MR	MiB	Br
4.4	MR	N	S
4.5	MR	MiB	S
5.1	MR	MiB	F
5.2	MR	A	F
5.3	Ox	MiA	SBr
5.4	Ox	MiA	SBr
5.5	MR	N	SBr

is seen in the fresh marsh environment (which has the lowest water level) which contain a greater abundance of infested and scleratoclastic macerals. The lower TOC of the saline marsh also may be a result of the more complete biodegradational pathway. However, there is a high exportation of organic matter from these environments (by tidal flushing), and sedimentation rates are higher.

Organic degradation

The great bulk of sedimentary organic matter produced by living systems is destroyed by the biochemical activities of monerans, protozoans, and fungi, rather than by abiologic chemical reactions. Although much of this takes place in the water column, the process of conversion of POM to DOM by bacterial action at the sediment/water interface and in the depositional environment is important in the organic chemical cycle (Degens *et al.*, 1964; Hart, 1986; Litchfield *et al.*, 1974).

Structural carbohydrates are the main energy source left in the debris that accumulates at the surface/water interface. The most abundant materials that are comparatively resistant to decay include lignin (which is the most common natural product containing aromatic rings), some proteinaceous substances, and lipids. Sporopollenin and chitin also are resistant but are not quantitatively as significant. In addition, humic substances (fulvic acid, humic acid, and humin) are quite resistant to microbial decay and may form the bulk of organic matter in both the water column and at the sediment/water interface. It is for this reason that plant debris (lignin rich), algal remains (lipid rich) and amorphous kerobitumen (often predominantly colloidal particulate humic substances) dominate POM. 'With respect to all sedimentary organic matter, the rapidity with which it passes through the surficial zone of intense degradation has a direct effect on the probability that the organic matter will be retained at depth.' (Hart, 1986)

Sieburth (1979) and Odum (1971) point out that the destruction of POM is really a bacterial-protozoan partnership, in which flagellated protozoans in the POM supplement the osmotrophic work of the monerans. Although the bulk of this activity takes place in the aerobic zone of the sediments important bacterial induced changes occur in the anaerobic zone. In continental environments fungi may be equally or more important than monera in biodegradation, but the monerans are definitely the most important degrading organisms in the marine environment. Sieburth (1979) notes that within the anaerobic zone the processes are basically the same, irrespective of habitat, and the same types of organisms are involved.

Chemically the phytoclastic macerals contain the following major polymeric biomass compounds: homo-polysaccharides (cellulose and starch), heteropolysaccharides (hemicellulose), lignins, pectins, proteins, and lipids. The homopolysaccharides are the most abundant, and cellulose comprises 45–98% of most virgin phytoclasts. In woody cell walls cellulose occurs as lignocellulose. Under aerobic conditions the fungi are extremely active in cellulose degradation.

Heteropolysaccharides (hemicellulose) may form up to 40% of phytoclastic biomass. These short, branch-chained polysaccharides are readily degraded by fungi and bacteria.

Lignin (an aromatic compound) comprises 20–30% of phytoclastic biomass. It is poorly degradable but experimentally will produce low molecular weight, aromatic compounds. This has been done at high temperatures and pH (Sleat & Mah, 1987). Under strict anaerobic conditions at least eleven degraded products of aromatic lignins can be converted to methane (Healy & Young, 1979; Healy *et al.*, 1980).

Pectin and starch are complex polysaccharides that

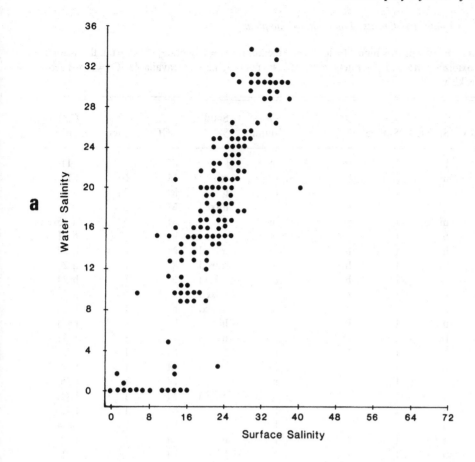

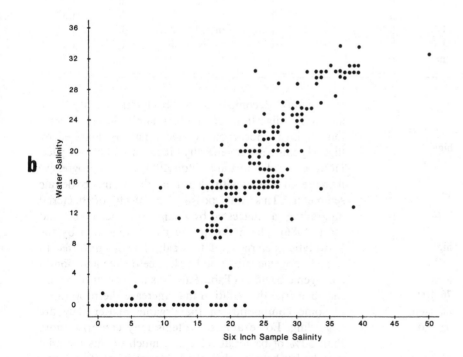

Figure 9.5. Scatter plots of constituents in environments of the Louisiana deltaic plain: (a) salinity of overlying water versus surficial bottom sediment; (b) salinity of overlying water versus depositional environment sediment.

Table 9.5. *Characteristics of depositional environment sediments: Louisiana deltaic plain*

Key: A = acidic; B = basic; Br = brackish; F = fresh; h = high; l = low; m = medium; M = moderate; MiA = mildly acidic; MiB = mildly basic; MOx = moderately oxidized; MR = moderately reduced; N = neutral; na = not available; Ox = oxidized; R = reduced; S = saline; SBr = slightly brackish

Environment	Eh	pH	Salinity	Sulfate	Sulfide	Sulfur	Org. sulfur	Sand %	TOC	Methane	C:N ratio
1.1	R	MiA	S	h	l	m	m	l	M	l	l 11
1.2	MR	MiB	Br	m	l	l	l	m	M	l	l 12
1.3	MR	MiA	Br	l	l	l	l	l	M	M	h 13
1.4	MR	N	Br	l	l	l	l	mixed	M	h	l 14
1.5	MR	MiB	Br	m	l	m	m	l	M	M	h 15
2.1	R	MiB	Br	h	l	h	h	l	m	l	h 21
2.2	R	MiB	Br	h	l	h	m	mixed	M	l	h 22
2.3	R	MiA	Br	h	l	h	h	mixed	h	l	h 23
2.4	R	MiB	S	h	l	h	h	mixed	h	l	h 24
2.5	R	A	SBr	l	l	h	h	mixed	h	h	h 25
2.6	MR	A	F	l	l	m	m	mixed	h	h	h 26
3.1	M	B	S	h	l	l	l	h	l	l	na 31
3.2	MR	MiB	Br	h	l	m	l	h	l	l	h 32
3.3	MR	MiB	S	h	l	l	l	h	l	l	l 33
3.4	R	MiB	S	h	l	m	m	l	M	l	l 34
3.5	MR	MiB	S	h	l	m	m	l	M	l	l 35
4.1	R	MiB	Br	h	l	m	l	mixed	M	l	l 41
4.2	MR	MiB	S	h	l	l	l	h	l	l	l 42
4.3	MR	MiB	Br	h	l	l	l	h	l	l	na 43
4.4	MR	MiB	S	h	l	l	l	l	M	l	l 44
4.5	MR	MiB	S	h	l	l	l	l	M	h	l 45
5.1	MR	MiA	F	l	l	l	l	mixed	M	h	h 51
5.2	MR	A	SBr	l	h	l	l	l	h	h	l 52
5.3	MR	MiA	SBr	l	l	l	l	mixed	M	h	l 53
5.4	Ox	A	SBr	l	l	l	l	mixed	l	l	l 54
5.5	MR	MiA	F	m	l	l	l	l	M	h	l 55

Descriptive codes used for Table 9.5

	low	medium	high	
Sulfate (mg/l)	0–500	500–1500	>1500	
Sulfide (wt %)	low 0–0.1	high >0.1		
Total sulfur (wt %)	low 0–0.1	medium 0.1–0.5	high >0.5	
Organic sulfur (wt %)	low 0–0.1	medium 0.1–0.5	high >0.5	
Sand %	low 0–25	medium 26–75	high 76–100	
Total organic carbon = TOC (wt %)	low 0–0.4	moderate low 0.4–0.8	moderate 0.8–5.0	high >5.0
Methane (ppm)	low 0–1000	medium 1000–2000	high >2000	
C:N ratio	low <12	high >12		

are readily decomposed by bacterial activity. The anaerobic digestion of lipids (mainly glycerol, with long chain monocarboxylic acids) may be inhibited by high H_2 content but generally takes place at a slow pace. However, Lindblom and Lupton (1961) noted a progressive increase in lipids with depth in the Orinoco deltaic sediments. This was possibly a result of bacterial 'upgrading,' as suggested by Zhang and Chen (1985) and Hart (1986). The major product after digestion by the hydrolytic bacteria are fatty acids. This is converted to a methanogenic substrate by the acetogenic and homoacetogenic bacteria (Table 9.6). It is important to realize that bacterial degradation may increase the proportion of some compounds in the organic matter they are degrading. For example, proteins may comprise more than 50% of bacterial cells, and much of this material may be left behind when the bacterial population dies.

BACTERIAL GEOCHEMISTRY AND SULFUR CONTENT

Winogradsky (1888) first noted the existence of monerans that derived energy by oxidation of sulfur compounds.

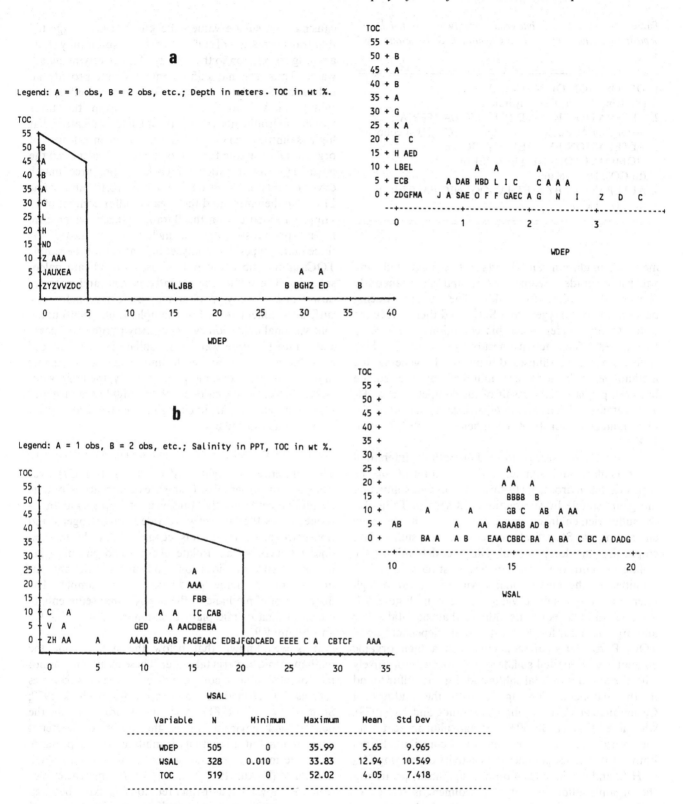

Figure 9.6.   Scatter plots of constituents in environments of the Louisiana deltaic plain: (a) water depth versus total organic carbon; (b) salinity versus total organic carbon. Plots to the right are enlargements of important areas in the main plots. WSAL = water salinity in parts per thousand; WDEP = water depth in meters; TOC = total organic carbon in wt %; obs = observation(s).

The sulfur-oxidizing bacteria involved were reviewed by Schlegel (1975). Species belonging to *Thiobacillus* are important organisms causing the oxidation of sulfide to sulfate in the oxygenic zone. Although sulfide oxidation is principally facilitated by bacterial activity, slow

Table 9.6. *Levels of bacterial activity in surficial and depositional environments (cf. Chynoweth & Isaacson, 1987, p. 3)*

1. DEPOSITION OF MACERALS
   (Proteins, Lipids, Carbohydrates)
2. ALTERATION BY HYDROLYTIC BACTERIA
   (to Organic Acids, Neutral Compounds, CO_2, H, Alcohols)
3. ALTERATION BY ACETOGENIC (and HOMOACETOGENIC) BACTERIA
   (to CO_2, H_2, Acetate)
4. ALTERATION BY METHANOGENIC BACTERIA
   (to CH_4, CO_2)

inorganic oxidation can take place. The sulfate-reducing bacteria degrade organic matter and simultaneously convert the sulfate to sulfide. The sulfate-reducing bacteria remove oxygen from SO_4^{2-} and then reduce the sulfur to S^{2-} under anaerobic conditions (Nedwell & Banat, 1981). This activity occurs to a depth at which sulfate no longer diffuses downward. In general, the methanogenic bacteria are inhibited by the sulfate bacteria, probably as a result of the competition for H_2 and acetate (Hobson *et al.* 1974; Kristjansson *et al.*, 1982; Oremland & Taylor, 1978; Schönheit *et al.*, 1982; Thauer, 1982).

Sulfur analyses were performed on both the interstitial water (sulfate and sulfide) and the sediment of the depositional environment (total sulfur, organic sulfur, and inorganic sulfide by subtraction). As seen in Table 9.5, the sulfur-rich environments are essentially the shoreline environments. Statistical analysis of the sulfur data emphasizes the dominance of salinity in determining the sulfate concentrations found in the interstitial water. High salinities in the depositional environment cause high interstitial water sulfate values as seen in Figure 9.7. Analyses of bulk sediments indicate that the total sulfur and organic sulfur for the most part are dependent upon TOC. Particularly, in sediments with a high organic content the interstitial sulfate values correlate positively with the amount of total sulfur and organic sulfur found in the sediment. This agrees with the findings of Casagrande *et al.* (1979) and Casagrande and Ng (1979), who noted that: (1) 30–50% of total sulfur in peat from the Everglades is in the organic form contributed from both plant sources and microbial activity converting SO_4 to H_2S, and (2) the formation of a significant portion of the organic sulfur content of a sediment is related to sulfate concentration. Thus, fresh waters high in organic matter and low in sulfate have low total sulfur and organic sulfur, whereas areas high in sulfate and organic matter have high total sulfur and organic sulfur (see Fig. 9.8).

Because of the relationship between sulfate concentration and salinity of the interstitial water (high salinities

causing high sulfate values), Beckman (1984) suggested that transformation of sulfate to organic sulfur may show a strong correlation to the salinity of the overlying marsh water. Thus, organic sulfur content might provide an indication of paleosalinities. TOC, organic sulfur, and salinity analysis of the 60 samples from the three marsh environments (Hart, 1979c) had indicated that high salinities relate to greater concentrations of organic sulfur. Using these data, Beckman calculated the mean TOC:sulfur ratios for each of the three marsh environments as: fresh 6.87, brackish 10.12, and saline 21.06. He then measured the organic sulfur content of 13 samples in a core from the Barataria Basin, which core he had previously shown to include a sequence of the three marsh types. His analyses indicated that the simple TOC:sulfur ratio could not be used to indicate paleosalinities. The sulfur is apparently leached and migrates downward. Moreover, sulfur may be formed below the surficial sediment zone. For example, Lord (1980) noted that seasonal oxidation events destroy pyrite and release sulfur into the pore water. The sulfate is then reduced microbially to H_2S at depth, and reacts with organic matter to produce organic sulfur. Clearly, the pore water systems of organic-rich deposits are highly reactive and play an important part in changing the chemical imprint in subsurface sediments.

BACTERIAL GEOCHEMISTRY AND METHANE CONTENT

The presence of light hydrocarbons (C1–C7) was measured on all samples. Using the occurrence of butane as an indicator of the presence of petrogenic (non-biogenic) gas, the data-set was divided into petrogenic and non-petrogenic samples. It is recognized that this method does not unequivocally isolate all non-petrogenic material, but only that it provides some indication of the amount of biogenic methane production in a sample. The distribution of methane in the depositional sediments of the Louisiana deltaic plain is summarized in Table 9.5 and Figure 9.9.

It is well known that below the sulfate zone the methanogenic bacteria take over. These monerans cannot use complex organic compounds but grow on substrates such as CO_2, H-methanol, and acetate (Gottschalk, 1979; Sorensen *et al.*, 1981). Methane production in the depositional environment is the result of bacterial degradation, but this activity is inhibited by the presence of sulfate in the interstitial water in quantities as small as 0.2 mM (Zeikus, 1977; Nikaido, 1977). The scatter plot (Fig. 9.9) shows that a distinct relationship exists between methane production and sulfate in the interstitial water. This relationship is expressed by the formula:

$$\log[CH_4] = 7.17 - 1.51 \log[SO_4^{2-}]$$

($r^2 = 0.85$), where methane and sulfate are expressed in parts per thousand.

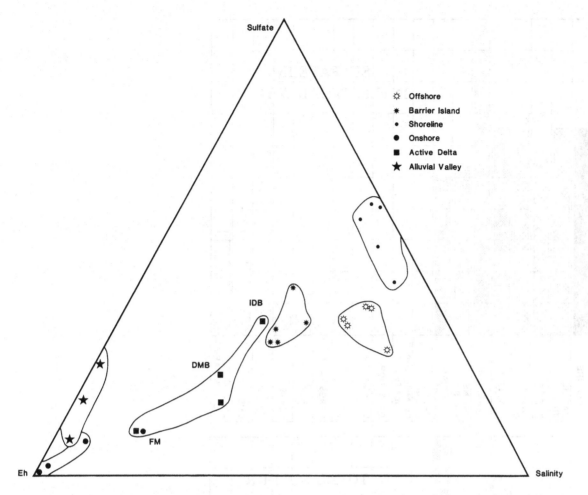

Figure 9.7. Triplot of sulfate in the interstitial water, Eh, and salinity of the environments of the Louisiana deltaic plain. DMB = distributary mouth bar; IDB = interdistributary bay; FM = fresh marsh; Eh = redox potential.

The statistical analyses performed in the present study also indicate that both sand percentage and TOC influence methane production. The highest methane production is found in freshwater environments that have high TOC and low sand content. Microbial methanogenesis occurs only under strict anaerobic conditions and involves a mixed bacterial population of obligate anaerobes acting at three levels (Chynoweth & Isaacson, 1987), as illustrated in Table 9.6. This activity occurs efficiently at a depth in the sediment at which oxygen, nitrates, and most of the sulfates have been depleted. Eh requirements in a sediment are probably less than −300 mV; pH optimally in the 6.0–7.7 range; and a temperature optimally in the 4–45 °C range (Zhang & Chen, 1985). Figure 9.10 shows the distribution of Eh and pH in the depositional environment samples studied from the Louisiana deltaic plain.

As a result of their biodegradation the hydrolytic bacteria produce organic acids, hydrogen, carbon dioxide and alcohols from the macerals (lipids, proteins, and carbohydrates). This is the rate-limiting step in the overall conversion of macerals to methane. Acetogenic and homoacetogenic bacteria then work over these compounds to produce acetate, hydrogen and carbon dioxide. This involves a complex interspecies association of bacteria primarily driven by the flow of electrons (as H_2) (Boone & Mah, 1987). The methanogenic bacteria then convert the acetate, hydrogen and carbon dioxide to methane and carbon dioxide. The final gas often includes traces of hydrogen sulfide. It should be noted that certain macerals (lignin rich organic matter) are little effected by the biodegradational processes under anaerobic conditions, even under long residence times (a factor in the formation of coals). The hydrolyzed products are peptides which are converted to amino acids and used as carbon and energy sources by the acetogenic bacteria.

MACERAL DEGRADATION

The degradation state of a sediment is an important parameter for differentiating environments of deposition. In addition, knowledge of the degree of organic degradation is useful in determining the potential of a sediment to produce hydrocarbons. Two independent

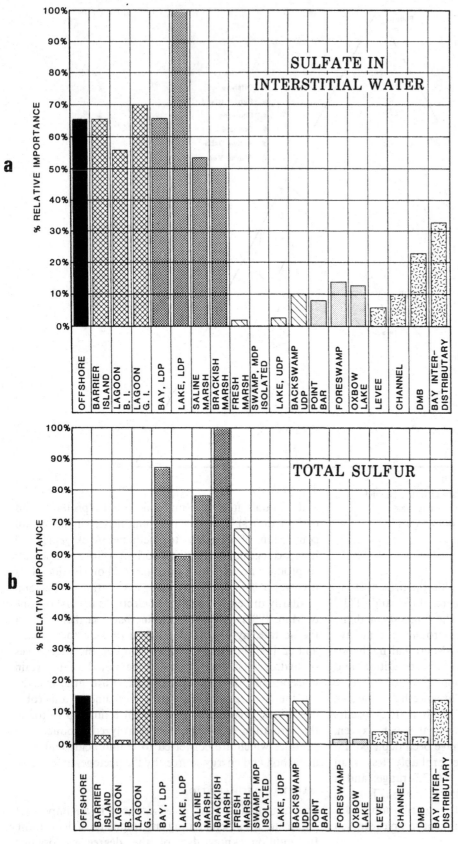

Figure 9.8. Normalized variations in constituents in environments of the Louisiana deltaic plain: (a) sulfate in the interstitial water; (b) total sulfur. For explanation of patterns, see Figure 9.3.

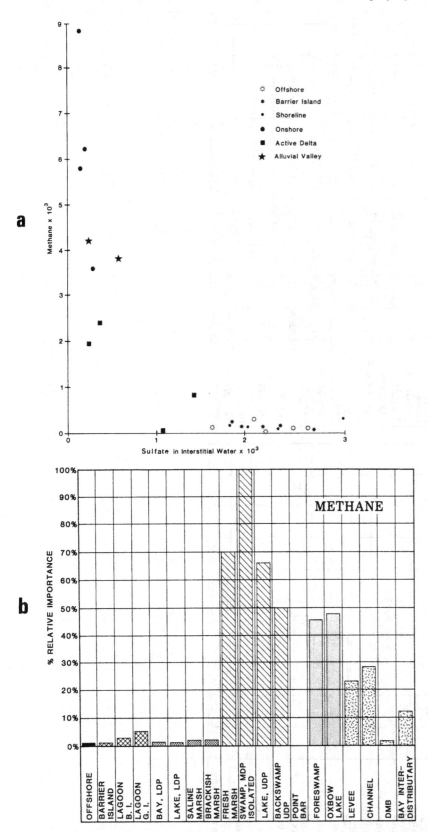

Figure 9.9. (a) Scatter diagram of methane versus sulfate in interstitial water of the environments of the Louisiana deltaic plain. Units are in parts per thousand; (b) Normalized variation in methane of the depositional environments of the Louisiana deltaic plain. For explanation of patterns, see Figure 9.3.

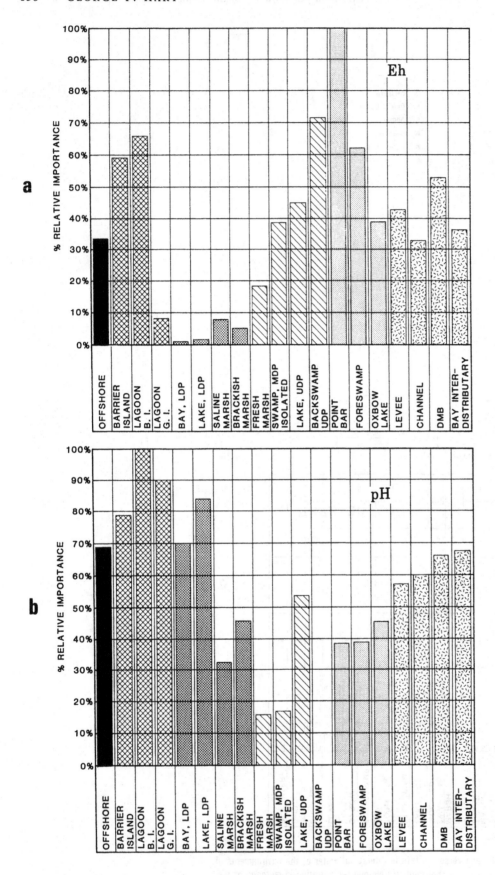

Figure 9.10. Normalized variations in constituents of the depositional environments of the Louisiana deltaic plain: (a) Eh; (b) pH. For explanation of patterns, see Figure 9.3.

Table 9.7. *Levels of degradation subenvironments of the Louisiana deltaic plain*

Key: h = high; l = low; ldp = lower deltaic plain; M = moderate; udp = upper deltaic plain.

Environment	C:N ratio level	Maceral level	Degradation level
Offshore	h	M	M–h
Barrier island	h	M–h	M–h
Lagoon (starved)	na (sand)	h	h
Lagoon (rich)	l	l	l
Bay (ldp)	l	l	l
Lake (ldp)	l	h	?
Saline marsh	l	l	l
Brackish marsh	l	l	l
Fresh marsh	l	M–l	M–l
Isolated swamp	l	M–l	M–l
Lake (udp)	l	l	l
Backswamp (udp)	h	l	?
Point bar	h	h	h
Foreswamp	h	M–h	M–h
Abandoned loop	h	h	h
Levee	l	l	l
Channel	h	M	M–h
Distributary mouth bar	h	M	M–h
Interdistributary bay	l	l	l

measures of degradation (C:N ratio and maceral index of degradation) were used to determine differences in environment of deposition with respect to degree of degradation. High C:N ratios are indicative of low biodegradation, but low C:N ratios are a result of increased bacterial degradation (Hart, 1986). In this study high C:N values are values greater than or equal to 12, and low C:N values are less than 12. The maceral index of degradation is determined by the ratio of phytoclast types in the maceral spectra (Hart, 1986; Fig. 9.3). A low degradation state is exemplified by a predominance of the well preserved phytoclasts, as compared to the infested and degraded phytoclasts. A high state of degradation of phytoclasts results in the predominance of 'degraded' compared to 'infected' and 'good' macerals in the basic degradational classification (cf. bars 1–3 in Figs. 9.12–9.15). A moderate degradational state occurs when these categories are approximately equal. Table 9.7 summarizes the results of the degradational study. Fourteen of the twenty environments exhibit agreement between the two independent measures. Only two environments (lake of the lower deltaic plain and backswamp of the upper deltaic plain) have contradictory degradational indices. Two environments cannot be positively categorized due to the moderate maceral index.

Although Table 9.7 suggests that a large overprint of the maceral degradational characteristics occurs in the depositional environment even when the material is obviously reworked (i.e., offshore environments), the operator should be prudent in using such an index without knowledge of the probable degree of reworking or recycling of the organic matter.

From the present study, the environments of high degradational activity are the abandoned meander loop (oxbow lake) and the point bar (Fig. 9.11, Table 9.7). Moderately high degradation occurs in the barrier island, foreswamp and distributary mouth bar. Possible moderately high degradation environments are the offshore marine and the channel. Low degradation is indicated for the lagoon (Grande Isle, sediment-rich), bay (lower deltaic plain), saline and brackish marsh, lake (upper deltaic plain), levee, and interdistributary bay. The fresh marsh and isolated swamp exhibit moderately low degradation. Analysis of the causes of the organic degradation are complicated by the multiplicity of variables affecting degradation. However, it can be generally stated that lithology, TOC, overlying water depth, and oxygen availability combine to determine the degree of degradation. An examination of the degradation in the marshes can be used as a general model because they show similar vegetation but differing effects of variables. According to the C:N ratio and maceral index of degradation the sequence of increasing biodegradation is from saline through brackish through fresh marsh to isolated swamp. This sequence corresponds to the salinity gradient. It is postulated that sulfate in the interstitial water (Fig. 9.8) is correlated with the salinity and acts to reduce the degree of microbial activity, by stopping biodegradation at the sulfate-reducing bacteria level. In fresh areas, biodegradation can continue beyond this level to the breakdown of organic matter by the methane-producing bacteria.

Statistical analysis of the data set

Because the prime purpose of the project was to develop a method for separating the varied depositional environments of the Louisiana deltaic plain, the iterative discriminant function method used by Hart (1979a; Hart *et al.*, 1989) was applied to the data set. The approach is multivariate, is heuristic rather than inferential (Tukey, 1962; Cooley & Lohnes, 1971), and attempts to obtain a 'purified data base' that contains populations that are statistically separable. In the present context the populations are specific depositional environments. The data were analysed using the PROC DISCRIM of the SAS (Statistical Analysis System of the SAS Institute, Raleigh, North Carolina) package. Tables 9.8–9.11 give standardized, linearized, discriminant function values for

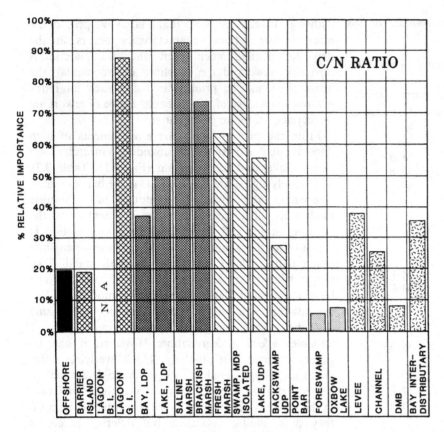

Figure 9.11. Normalized variation in degradation (C:N ratio) of the environments of the Louisiana deltaic plain. For explanation of patterns, see Figure 9.3.

Table 9.8. *Distinguishing characteristics based on discriminant functions: Louisiana deltaic plain, upper deltaic plain, lower alluvial valley. (The numbers are standardized values, as explained in text.)*

Lake

amorphous phytoclasts	2.7
algal protist	2.0
well preserved phytoclasts	1.0
amorphous indeterminate	0.9
total organic carbon	0.6

Backswamp

total organic carbon	5.6
amorphous structure phytoclasts	1.3
resin	0.9
fungi	0.1

Foreswamp

fungi	2.8
amorphous structure phytoclasts	2.3
infested phytoclasts	1.9
miospores	1.4
amorphous indeterminate	1.4
amorphous phytoclasts	0.6
dino-acritarchs	0.6
well preserved phytoclasts	0.6

Pointbar

inertite	4.0

Abandoned meander loop

amorphous structured phytoclasts	2.9
fungi	0.9
resin	0.5
well preserved phytoclasts	0.1

Table 9.9. *Distinguishing characteristics based on discriminant functions: Louisiana deltaic plain, lower deltaic plain, shoreline complex. (The numbers are standardized values, as explained in text.)*

Bay		Saline marsh	
total sulfur	15.2	total sulfur	13.0
amorphous phytoclasts	2.7	well preserved phytoclasts	3.9
infested phytoclasts	1.6	infested phytoclasts	2.0
amorphous structure phytoclasts	0.4	miospores	2.2
foraminifera	0.2	dino-acritarchs	0.6
miospores	0.02	amorphous phytoclasts	0.4
		total organic carbon	0.4
Lake		amorphous indeterminate	0.3
total sulfur	10.0	fungi	0.2
infested phytoclasts	3.6	foraminifera	0.04
miospores	0.5		
foraminifera	0.4	Fresh marsh	
amorphous phytoclasts	0.2	total organic carbon	11.8
amorphous structured phytoclasts	0.05	total sulfur	4.4
resin	0.02	infested phytoclasts	1.5
		amorphous phytoclasts	1.5
Brackish marsh		amorphous structured phytoclasts	1.1
total sulfur	14.7	resin	0.5
infested phytoclasts	5.0	miospores	0.4
total organic carbon	3.1	well preserved phytoclasts	0.2
amorphous phytoclasts	1.5	amorphous indeterminate	0.2
amorphous structured phytoclasts	1.2		
miospores	1.2	Isolated swamp	
well preserved phytoclasts	0.9	total organic carbon	17.5
amorphous indeterminate	0.7	resin	0.5
dino-acritarchs	0.3		
resin	0.3		
fungi	0.3		
foraminifera	0.2		

Table 9.10. *Distinguishing characteristics based on discriminant functions: Louisiana deltaic plain, lower deltaic plain, marginal deltaic basin. (The numbers are standardized values, as explained in text.)*

Back island lagoon (sediment-starved)		Barrier island complexes (including 10 ft marine)	
dino-acritarchs	0.8	inertite	0.5
foraminifera	0.8	foraminifera	0.3
inertite	0.4	dino-acritarchs	0.1
Back island lagoon (sediment-rich)		Offshore (including 50 ft and 100 ft marine, prodelta)	
infested phytoclasts	5.6	amorphous indeterminate	4.0
total sulfur	4.7	dino-acritarchs	1.5
miospores	1.8	amorphous phytoclasts	1.4
well preserved phytoclasts	1.7	well preserved phytoclasts	1.3
amorphous indeterminate	1.3	fungi	0.9
amorphous structured phytoclast	1.0	degraded phytoclasts	0.9
foraminifera	0.8	infested phytoclasts	0.4
dino-acritarchs	0.5	algal protista	0.3
fungi	0.1	miospores	0.2
		foraminifera	0.2

Table 9.11. *Distinguishing characteristics based on discriminant functions: Louisiana deltaic plain, lower deltaic plain, active delta. (The numbers are standardized values, as explained in text.)*

Distributary mouth bar	
inertite	0.9
resin	0.9
amorphous phytoclasts	0.8
Miospores	0.8
fungi	0.2

Levee	
resin	2.2
inertite	1.1
algal protista	0.5
amorphous phytoclasts	0.4
well preserved phytoclasts	0.5
fungi	0.1
miospores	0.1

Channel	
resin	1.2
fungi	0.8
amorphous phytoclasts	0.2
inertite	0.2

Interdistributary bay	
fungi	2.3
infested phytoclasts	1.4
amorphous phytoclasts	1.2
amorphous indeterminate	1.1
inertite	0.5
well preserved phytoclasts	0.5
resin	0.3
amorphous structured phytoclasts	0.1

each variable used in the discriminant function analysis. The program performed a test of homogeneity, and this resulted in the pooling of the covariance matrix. This means that the discriminant function is linear, and the tables can be used to estimate the individual contribution of each attribute to the overall discrimination. Therefore, the larger the number cited, the greater the contribution of that variable to delimiting the environment. The data were standardized to a mean of 0 and a variance of 1, to allow more accurate assessment of the weight each attribute has within the discriminant function.

Prior to the discriminant function analysis, 23 samples were randomly selected from the data base and removed from the set used to calculate the discriminant function. These samples were then used as a test set to assess the ability of the discriminant function to actually classify unknowns. The results show that 16 of the 23 were correctly classified (70%); four of the seven incorrect

estimates gave correct estimates as the second probable choice (87%); the remaining three of the seven incorrect estimates gave correct estimates as the third choice (100%). These results suggest that the discriminant function approach to the classification of detrital maceral facies can be used to discriminate maceral facies within an expert system (knowledge system).

Circumscription of typical samples

The maceral classification used in this study (Hart, 1979 a,b,c, 1986) divides POM into phytoclasts (plant-derived), zooclasts (animal-derived), protistoclasts (protistan-derived) and scleratoclasts (fungal-derived). Each group is subdivided further, based upon the level of decay, into well preserved (no evidence of biodegradation); poorly preserved (minor biodegradation); infested (highly disrupted cell walls); and amorphous structured (remnant cell structure only); amorphous non-structured (blocky phytoclastic mass or fluffy protistoclastic mass).

DISCRIMINANT FUNCTION CHARACTERISTICS

Tables 9.8–9.11 relate the maceral spectra to the geochemical–lithological variables in determining the depositional environment, by giving the actual weight that each variable has in discriminating the environment. For example, the importance of inertite is clearly seen to be indicative of distributary mouth bar, levee, barrier island and point bar environments. Thus, the discriminant function values are important distinguishing characteristics for each environment. From these tables some general characteristics of the main physiographic complexes can be extracted.

The offshore complex shows a diversity of macerals. An abundance of amorphous indeterminate macerals is characteristic, and this is associated with dino-acritarchs (dinoflagellates + acritarchs). Cuticular material in various states of preservation is common. Because macerals have both an allochthonous and an autochthonous source, it is not always apparent from the maceral spectra whether or not 'in situ' degradation is taking place. However, the offshore maceral spectra have a general mixture of maceral types, which suggests that most of the material is allochthonous, and little or no 'in situ' degradation is taking place. The indeterminate amorphous material is an extremely important component in the offshore environment. Unfortunately, the origin of this maceral is unknown. It may be a degradational product of cell contents, fecal material, or a precipitated complex (humic substances), or polygenetic in origin.

The barrier island complex shows the marine influence by presence of foraminiferal linings and dino-acritarchs. Inertite, suggesting oxidation, is characteristic. There are abundant well preserved and amorphous phytoclasts.

This complex is somewhat similar to the lagoon (starved) behind the island, except that in the former, indeterminate phytoclasts occur in more abundance.

The shoreline complex includes the high sulfur environments. It shows an abundance of phytoclastic material in various stages of degradation. Marine influence is seen in presence of foraminiferal linings ('microforaminifera') and dino-acritarchs in a background role, and, in the case of the back-island lagoon, the presence of amorphous indeterminate macerals. In an environment where phytoclasts are being produced and where degradation is taking place, e.g., isolated swamp, well preserved material is in low abundance and the amorphous phytoclasts are high. A similar but less pronounced situation is seen in the fresh marsh, where the amorphous indeterminate macerals are important. The brackish marsh is similar but has a lesser amount of amorphous phytoclasts and more structured infested material. The saline marsh is quite different and shows a high amount of well preserved material and low amount of structured amorphous material. The lagoon (sediment-rich) and lake (lower deltaic plain) show characteristics similar to the saline marshes surrounding them but with more infested and amorphous indeterminate material. In the bay (lower deltaic plain) the structured amorphous and amorphous protistoclasts are more abundant.

The active delta complex shows a marine influence in the interdistributary bay (amorphous indeterminate), but, in general, terrestrially derived material is dominant. Resin and inertite (both suggestive of an oxidative source for these macerals) are important, particularly in the channel-levee-distributary mouth bar environments. Well preserved, infested and amorphous phytoclasts are significant constituents in the interdistributary bay. The levee shows an abundance of well preserved phytoclasts as does the distributary mouth bar. This is expected because of the proximity of growing vegetation.

The lower alluvial valley complex shows that amorphous structured and amorphous non-structured phytoclasts are important factors in determining the environments (N.B. this is vitrinite from the point of view of organic petrology). Terrestrial derived materials are generally important throughout these environments. The swamps (isolated and back) show that resin is important, and the lake has an important contribution from protistoclasts (this does not include dino-acritarchs and amorphous indeterminate). The foreswamp contains dino-acritarchs as a minor constituent (recycled material from the river) and lacks resin. Its major constituents are fungi, amorphous structured, and infested phytoclasts, as one might expect. The abandoned meander loop is highly influenced by the surrounding swamp environment and has prominent amorphous structured phytoclasts, fungi, and resin. The point bar is distinguished by the presence of inertite (oxidation). The backswamp (upper deltaic plain), abandoned meander loop, and point bar are somewhat similar, in that they show increased abundance of the amorphous phytoclasts and amorphous structured macerals. The foreswamp has more infested phytoclasts. In general, there is some relationship between the nearness to the organic source, the degradation level and the kind of macerals that one sees in the maceral spectra.

BAR CHARTS

The discriminant function tables show those characteristics that allow the depositional environments to be separated when all environments are considered at the same time, i.e., in N-dimensional space, where N is the number of variables used for separation. The bar charts examine the individual environments one at a time, i.e., in the same way that the individual geologist would view the data. The bar charts are examined in terms of the knowledge that the geologist has regarding the processes active in the depositional environment. Table 9.12 ranks the abundance of each maceral type within each environment according to an ordinal scale of values.

The bar charts present the arithmetic means of each maceral group calculated from all samples retained after the iterative discriminant function procedure was applied to the data. The charts include three special phytoclast groups on the basis of their degradational level: good, infected, and degraded. The purpose of these is to give an overall impression of the level of degradation. Well preserved, poorly preserved and infested structural phytoclasts all could be classified as 'good' with reference to degradational level. Infested and amorphous structural phytoclasts could be classified as 'infected' as to degradational level. Amorphous structural and amorphous non-structural phytoclasts could be classified as 'degraded' with reference to degradational level. Note that the degradational categories overlap. See also the caption in Figure 9.12 for further clarification.

The Lower Deltaic Plain: marginal deltaic basins

Figure 9.12a shows the maceral spectrum for the offshore depositional environments. The phytoclast degradational level contains equal proportions of each degradational stage, and the individual phytoclastic maceral abundances are more or less equal. This is the expected phytoclast distribution with totally reworked material. The abundance of indeterminate amorphous material probably represents accumulation of humic precipitates or fecal material as suggested by Hart (1979a, 1986). The resin, miospores and fungi are a reflection of the closeness of the shoreline. The barrier island maceral spectrum shown in Figure 9.12b is more typical of a normal sequence of production and gradual degradation of phytoclasts. The high abundance of amorphous phytoclasts indicates

Table 9.12. *Rank abundance of macerals in each environment (based on average percentage abundance in each environment)*

Total count = 104 000; ordinal scale based on n^3.

Key: A = well and poorly preserved; B = infested; C = amorphous structured; D = amorphous phytoclast; E = amorphous indeterminate; F = amorphous protistoclast; G = resin; H = miospores; I = protist; J = foraminiferal linings; K = dino-acritarchs; L = fungi; ldp = lower deltaic plain; udp = upper deltaic plain

| Environment | Rank of maceral category[a] | | | | | | | | | | | | Dominant constituent |
	A	B	C	D	E	F	G	H	I	J	K	L	
Barrier island	5	4	3	5	2	4	2	3	0	2	2	3	preserved phytoclasts
Lagoon (starved)	4	3	3	5	0	5	0	3	0	0	2	2	amorphous infested
Lagoon (rich)	4	5	3	4	3	4	0	3	0	0	3	3	infested phytoclasts
Lake (ldp)	3	5	3	5	0	4	0	2	0	0	0	3	**degraded phytoclasts**
Bay (ldp)	3	4	4	4	3	4	0	2	2	0	0	2	amorphous phytoclasts
Saline marsh	5	4	2	4	2	3	0	3	2	0	0	3	preserved phytoclasts
Brackish marsh	4	5	4	4	2	4	0	2	2	0	0	4	infested phytoclasts
Fresh marsh	4	5	4	4	3	4	2	3	0	0	0	4	infested phytoclasts
Isolated swamp	4	5	3	5	0	3	2	2	0	0	2	3	**degraded phytoclasts**
Lake (udp)	4	4	3	4	3	5	2	3	3	0	2	3	amorphous (algal?)
Backswamp (udp)	4	4	4	5	2	3	2	3	0	0	0	4	amorphous phytoclasts
Point bar	4	4	4	5	0	4	2	2	0	0	0	2	amorphous phytoclasts
Foreswamp	3	5	4	4	2	4	0	3	0	0	2	4	infested phytoclasts
Abandoned loop	4	4	4	5	0	3	2	2	0	0	0	3	amorphous phytoclasts
Interdistributary bay	3	5	3	4	2	4	2	3	2	3	0	4	infested phytoclasts
Levee	4	4	3	4	2	3	3	3	2	0	0	3	phytoclasts
Distributary mouth bar	4	4	4	4	3	3	3	3	2	0	0	4	phytoclasts
Channel	4	4	3	4	2	3	2	3	0	0	0	4	phytoclasts
Offshore marine	3	4	4	4	3	5	1	2	1	1	2	3	amorphous (algal?)

(a)

Maceral %	Descriptive abundance	Ordinal scale
>81	dominant	6
27–81	abundant	5
9–26	very common	4
3–8	common	3
1–2	few	2
<1	rare	1
0	absent	0

accumulation rather than removal of that material. The associated lagoons are somewhat similar to the barrier island, with the sediment-starved lagoon containing more indeterminate amorphous material (Fig. 9.12c), and the sediment-rich lagoon showing greater influx of good phytoclasts from the nearby terrestrial source (Fig. 9.12d).

Lower Deltaic Plain: active delta

The interdistributary bay maceral spectrum (Fig. 9.13a) is similar to the sediment-rich lagoon. In both environments the relatively high amount of bio-degradational activity is suggested by the abundance of infested structured phytoclasts and fungi and the moderate accumulation of amorphous phytoclasts. The abundance of indeterminate amorphous material suggests the precipitation of humic materials. The channel and distributary mouth bar maceral spectra are similar to the interdistributary bay but show the greater influence of amorphous non-structured phytoclasts (Figs. 9.13b,c). The levee maceral spectrum is like that from the barrier island, showing a normal sequence of production and destruction of organic matter (Fig. 9.13d). The main difference is the lack of marine influence in the levee material, i.e., lack of foraminiferal linings and dinocysts, although in theory both could occur as recycled material. Also, resin is important in the levee material.

Shoreline complex

Dominant along the shoreline are the marsh environments (Figs. 9.14a–c). These are all areas of relatively high production of phytoclasts but show the differing effect of local processes. The high degradational activity is seen

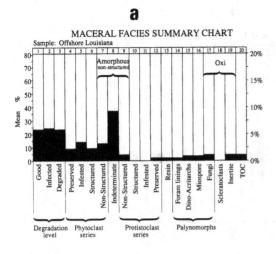

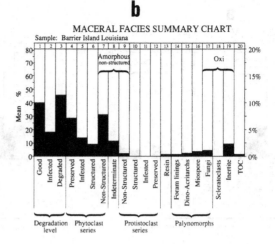

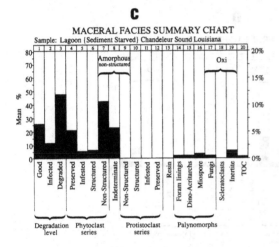

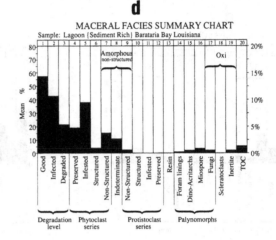

Figure 9.12. Bar charts of maceral spectra for various environments in the Louisiana deltaic plain: (a) offshore; (b) barrier island; (c) sediment-starved lagoon; (d) sediment-rich lagoon. The three bars on the left of each chart are independent summary indicators of phytoclast degradation level. Bars 4–18 show amounts of the indicated macerals. Inertite is estimated using the 'outside-the-count' method. TOC plots wt. % total organic carbon on bar 20. The method of analysis uses standard LECO equipment except where indicated otherwise. The means are percentage means. The 0–20 axis on the right applies only to TOC. 'Infested' refers to a moderately degraded maceral type. 'Infected' is a special category for level of degradation covering both infested and amorphous. See also the maceral classification of Hart *et al.* (Chap. 17, this volume).

in the abundance of infested structured phytoclasts and fungi in the fresh and brackish marsh. However, this is strongly influenced by tidal flushing in the saline marsh, which causes the good phytoclastic material, which has the shortest residence time, to be overemphasized. The isolated swamp (within the fresh marsh) shows the greater influence of biodegradation, with the accumulation of amorphous phytoclasts (Fig. 9.14d). The lake environment has high amounts of infested structured phytoclasts and

amorphous phytoclasts, because it is a region of organic decay, but not one in which the phytoclastic material is produced (Fig. 9.14e). The bay environment is probably more influenced by the mixing processes in the water, as seen in the abundance of amorphous indeterminate and general admixture of the phytoclasts (Fig. 9.14f).

The Lower Alluvial Valley

All of the maceral spectra show the importance of amorphous phytoclasts. The abundance in the point bar is particularly striking but naturally reflects the long residence time the phytoclasts have in the water prior to sedimentation (Fig. 9.15a). In the foreswamp, which is physiographically next to the point bar, the increased biodegradation is seen in accumulation of fungi and infected and degraded phytoclasts (Fig. 9.15b). The greater accumulation of degraded phytoclasts is seen in the maceral spectrum for the backswamp and oxbow (Figs. 9.15c, 9.15d). The lake is noted for the abundance of indeterminate and protistoclastic amorphous material (Fig. 9.15e). The indeterminate material is probably fecal.

a

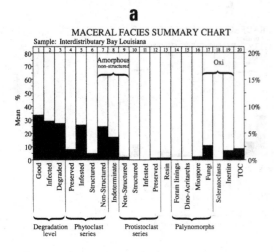

b

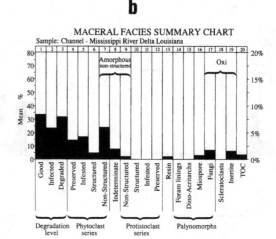

c

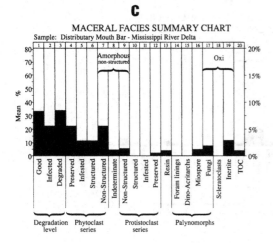

d

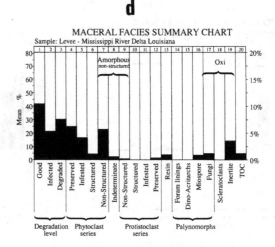

Figure 9.13. Bar charts of maceral spectra for various environments in the Louisiana deltaic plain (see also caption for Figure 9.12): (a) interdistributary bay; (b) channel; (c) distributary mouth bar; (d) levee.

SCATTER PLOTS

The scatter plots indicate that inertite and TOC are inversely related (Fig. 9.16a). This result is consonant with the view that inertite production is associated with an oxidizing environment, and TOC is most abundant in a non-oxidizing environment. The inertite is high in the point bar and levee environments, suggesting either 'in situ' oxidation or reworking–recycling. In addition, higher inertite occurs in environments where transportation is active, i.e., active delta. During the maceral counts the color of the maceral was noted following the method of Hart (1979a). Inertite and high colored macerals are most abundant in the point bar, backswamp, levee, and barrier island environments. Both 'in situ' oxidation and reworking–recycling may be the cause of high colored macerals in the sediments. Maceral color and the abundance of inertite appears to be partially a function of oxidation within the depositional environment, and, to a lesser extent, the concentration of inertite and high colored macerals may be related to recycling or reworking of sediments. Figure 9.16b is a scatter plot

relating inertite and high colored macerals. Figures 9.16a and b show the relative abundance of inertite and high colored macerals in each environment. Both the total sulfur and the TOC decrease with increased percentage of high colored macerals. The amount of high colored macerals increases as degraded phytoclasts increase.

As anticipated, high concentrations of amorphous protistoclasts tend to be associated with low abundances of amorphous phytoclasts, reflecting the predominantly marine versus terrestrial source of these macerals. The complementary relationship between the different levels of the degradation sequence in the phytoclasts simply reflects the degradation process.

Tables 9.8–9.12 show that the well preserved phytoclasts are abundant in the barrier island, levee, and saline marsh environments. These are areas of local maceral production and low degradation. Infested phytoclasts are abundant

a

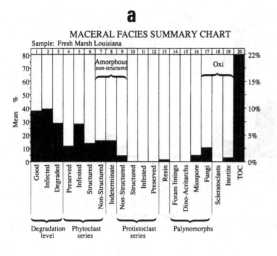

b

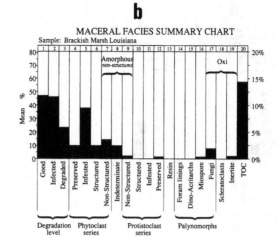

c

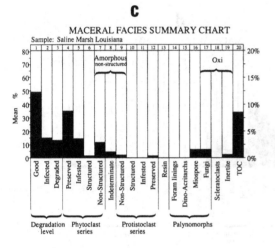

d

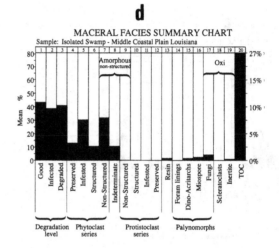

e

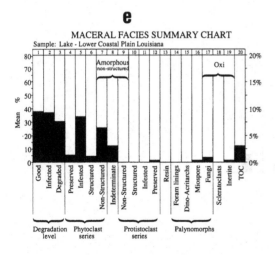

f

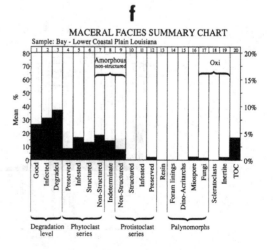

Figure 9.14. Bar charts of maceral spectra for various environments in the Louisiana deltaic plain (see also caption for Figure 9.12): (a) fresh marsh; (b) brackish marsh; (c) saline; (d) isolated swamp; (e) lower deltaic plain lake; (f) lower deltaic plain bay.

a

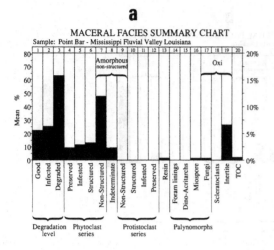

b

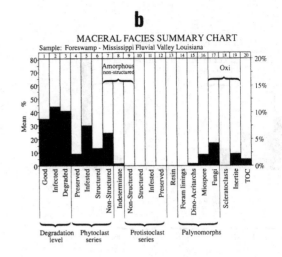

c

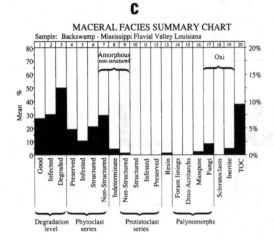

d

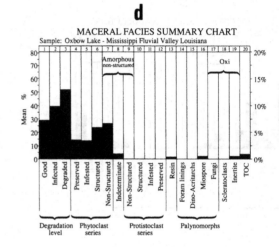

e

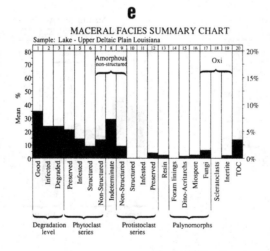

Figure 9.15. Bar charts of maceral spectra for various environments in the Louisiana deltaic plain (see also caption for Figure 9.12): (a) point bar; (b) foreswamp; (c) upper deltaic plain backswamp; (d) oxbow; (e) upper deltaic plain lake.

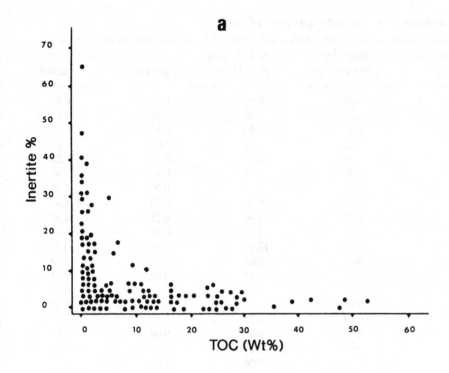

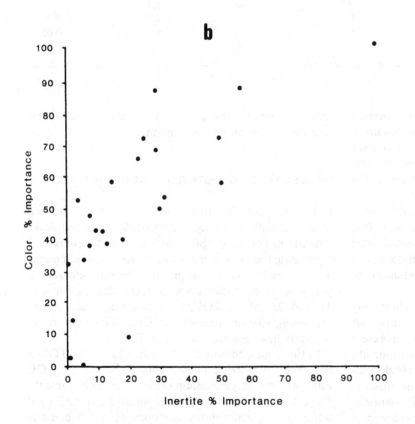

Figure 9.16. Scatter plots of constituents of the Louisiana deltaic plain: (a) inertite versus total organic carbon (TOC); (b) high colored macerals versus inertite, % importance.

Table 9.13. *Source rock parameters: Louisiana deltaic plain (maceral weight as % of TOC)*

Environment	TOC	Amorphous sapropel	Degraded phytoclasts	Non-degraded phytoclasts	Inertite	Sulfur
1.1	**1.74**	0.49	0.29	0.58	0.10	0.18
1.2	**0.65**	0.06	0.20	0.20	0.08	0.06
1.3	**1.17**	0.06	0.30	0.44	0.16	0.07
1.4	**0.99**	0.10	0.32	0.32	0.08	0.07
1.5	**1.82**	0.32	0.45	0.57	0.15	0.24
2.1	**4.01**	1.06	1.58	**1.04**	0.10	1.49
2.2	**3.15**	0.56	0.92	**1.23**	0.04	1.03
2.3	**14.23**	1.68	3.28	**6.74**	0.20	1.69
2.4	**8.0**	0.58	1.24	**4.66**	0.21	1.30
2.5	**21.8**	3.62	5.80	**7.99**	0.61	1.14
2.6	**27.4**	1.62	10.49	**12.19**	0.43	0.65
3.1	0.10	0.02	0.04	0.02	0.01	0.04
3.2	0.33	0.03	0.11	0.14	0.04	0.11
3.3	0.10	0.01	0.05	0.03	0.01	0.05
3.4	**1.03**	0.50	0.18	0.23	0.04	0.28
3.5	**1.15**	0.58	0.19	0.21	0.04	0.21
4.1	**1.53**	0.20	0.35	0.82	0.04	0.61
4.2	0.32	0.02	0.16	0.11	0.01	0.13
4.3	0.15	0.02	0.05	0.06	0.01	0.04
4.4	**0.88**	0.20	0.31	0.19	0.06	0.24
4.5	**1.19**	0.66	0.24	0.16	0.03	0.50
5.1	**3.31**	1.06	0.74	**1.11**	0.05	0.18
5.2	**7.8**	0.57	3.53	**2.08**	0.48	0.23
5.3	**0.98**	0.14	0.31	0.27	0.07	0.05
5.4	0.33	0.03	0.16	0.06	0.07	0.02
5.5	**1.03**	0.04	0.51	0.30	0.04	0.04

in the sediment-rich lagoonal environments and associated lakes, the brackish marsh, and the isolated swamps. Amorphous structured phytoclasts are never abundant but are noted because they are least important and few in number in the saline marsh. This correlates with the low degradational level of that environment.

Of the amorphous non-structured macerals, amorphous phytoclasts are abundant in the barrier island, point bar, lagoon (starved), isolated swamps, and the abandoned meander loop. Amorphous protistoclasts never dominate. Amorphous indeterminate macerals are abundant in the offshore.

Of the other maceral types, resin is important only in the levee, channel and distributary mouth bar environments, where it is common. Miospores are present either commonly or few in number in all environments. Dino-acritarchs are common only in the lagoon (sediment-rich). Foraminiferal linings are sparse in the offshore marine, common in the barrier island and essentially absent in all other environments. Other protists are most abundant in the lake (upper deltaic plain), where they are common. Fungi are very common in the high organic input environments of the brackish marsh, fresh marsh backswamp and foreswamp, and in the river-associated distributary mouth bar and channel.

Macerals and potential source rocks

There are certain limitations that appear to be environmentally controlled and are significant in petroleum generation. The total organic carbon in a sediment is used to estimate the organic matter in the rock, assuming that organic matter content is one way to estimate source potential. In most clastic source rocks there is between 0.8 and 2.0 wt % TOC, with excellent source rocks containing 10.0 or more wt % TOC. Non-source rocks generally have less than 0.4 wt % TOC.

In the deltaic environments studied the wt % TOC is not likely to increase during diagenesis and the TOC values are probably maximum values for the environments. Table 9.13 shows some important parameters (TOC, total sulfur wt %, amorphous sapropel wt %, bacterially degraded phytoclasts wt %, well preserved phytoclasts wt %, and inertite %). The wt % for maceral data is calculated as a function of the TOC as follows (inertite

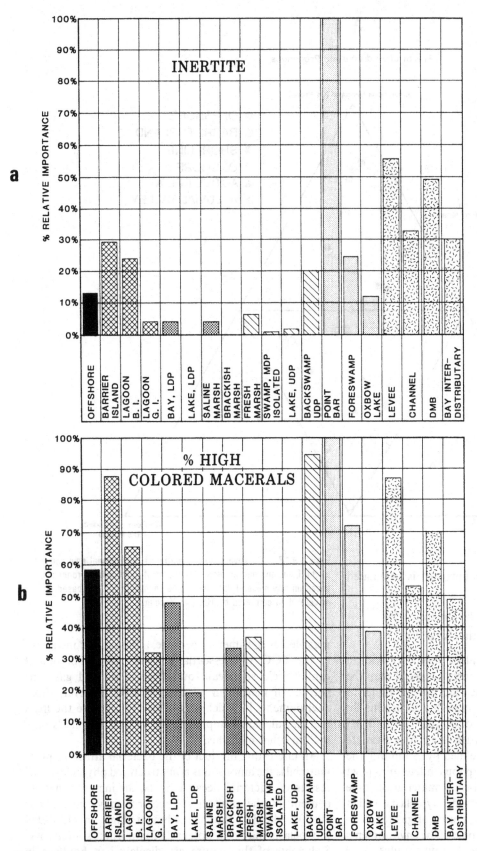

Figure 9.17.   Normalized variations of constituents of the Louisiana deltaic plain: (a) inertite; (b) high colored macerals. For explanation of patterns, see Figure 9.3.

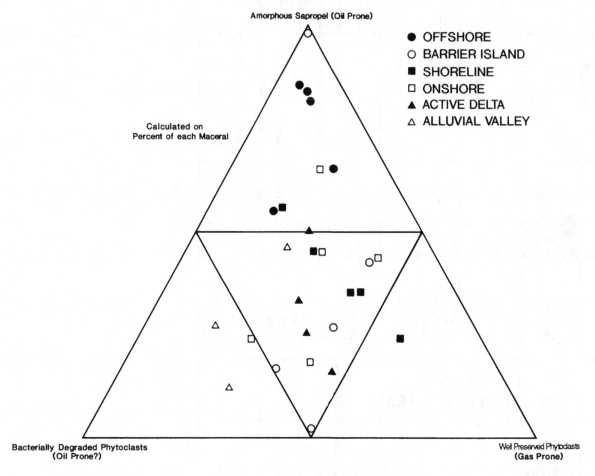

a

Triplot of Hydrocarbon Proneness

Amorphous Sapropel (Oil Prone)

● OFFSHORE
○ BARRIER ISLAND
■ SHORELINE
□ ONSHORE
▲ ACTIVE DELTA
△ ALLUVIAL VALLEY

Calculated on
Percent of each Maceral

Bacterially Degraded Phytoclasts
(Oil Prone?)

Well Preserved Phytoclasts
(Gas Prone)

Figure 9.18. Relative source-potential of depositional environments of the Louisiana deltaic plain: (a) triplot of hydrocarbon proneness, based on raw maceral data (simple % count); (b) As (a), based on maceral data corrected by TOC (wt % count).

is not involved with maceral counts but is calculated separately 'outside-the-count'):

$$\text{wt \% maceral} = \frac{\text{TOC (wt \%)} \times \text{maceral (\%)}}{\text{intertite (\%)} + 100}$$

In theory the maceral wt % values should provide a better estimate of hydrocarbon source potential than the simple TOC value, because some macerals are richer in hydrogen and more likely to generate hydrocarbons. In Table 9.13 all TOC >0.4 are displayed in the TOC column in boldface. TOC is regarded as a coarse estimate for source potential but can be refined by examining the amorphous protistoclasts + amorphous indeterminate wt % (equivalent to amorphous sapropel material of some authors). Well and poorly preserved phytoclasts, which are regarded as only a potential source of gas (Staplin, 1969), are relatively abundant in some of the environments which are potential oil source rocks (high values >1.0 wt % are displayed/in the non-degraded phytoclasts column in Table 9.13). The abundance of inertite is

regarded as detrimental because, although it contributes to the TOC, it is essentially 'dead carbon,' in that it cannot yield significant quantities of oil and gas. The percent of inertite in all cases is low; however, those environments in which it is relatively high are the fresh marsh, isolated swamp, and backswamp, which also are high in other macerals.

An interesting product of the bacterial attack of plant material is the amorphous non-structured phytoclast. The results of bacterial decay may be an increase in lipid content of the phytoclast (bacteria may have about 10–30% lipids); thus a sediment high in degraded phytoclasts may be a potentially important source rock. Sediments of this nature are displayed in boldface in Table 9.12. The tri-plots of Figures 9.17b and 9.18a show the relative hydrocarbon potential of each environment.

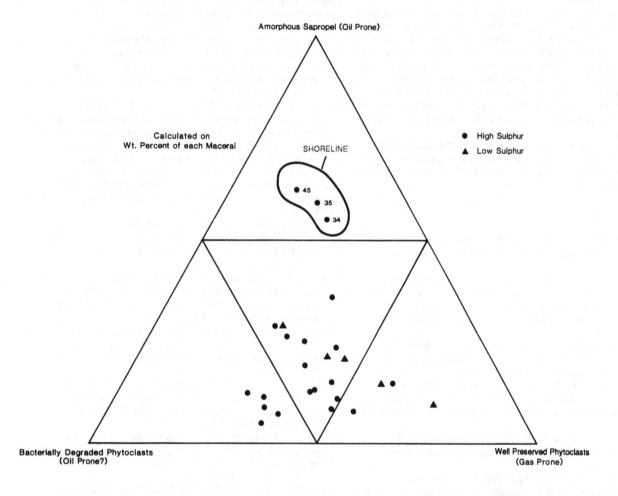

b

Triplot of Hydrocarbon Proneness

Amorphous Sapropel (Oil Prone)

Calculated on
Wt. Percent of each Maceral

SHORELINE

● High Sulphur
▲ Low Sulphur

Bacterially Degraded Phytoclasts
(Oil Prone?)

Well Preserved Phytoclasts
(Gas Prone)

In most cases the origin of the sulfur in high sulfur crude is explained by environmental conditions at the time of deposition (Tissot & Welte, 1978). Thus modern environments with high or low sulfur content partly determine the importance of sulfur in the petroleum they produce. As noted earlier, the sulfur is a product of sulfate-reducing bacteria removing oxygen from SO_4^{2-} and then reducing the sulfur to S^{2-} under anaerobic conditions. With the presence of iron the sulfur may combine to form hydrotroilite, troilite, or eventually pyrite. With less iron the sulfur may combine with residual organic matter. The present study suggests that most of the sulfur is in the form of organic sulfur, and the sulfur-rich environments are essentially the shoreline environments.

Conclusions

The maceral spectra obtained from the clastic deltaic sediments studied provide data that aid in understanding

depositional environments. Moreover, maceral spectra facilitate the identification of specific environments when used with other sedimentary parameters, e.g., total organic carbon, total sulfur.

With regard to source potential, most of the sediments studied contain high amounts of terrestrial material, and generally only offshore environments contain abundant sapropelic amorphous material. This supports the concept that, generally, nearshore marine sediments produce source rocks. However, degraded plant debris (which should be high in bacterially derived lipids) is an important component, and sediments containing such debris may be potential source rocks.

Acknowledgments

Although a number of my past students have helped me clarify my ideas on maceral analysis, I especially want to thank John Wrenn, Department of Geology and Geophysics, Louisiana State University, who has constantly encouraged me to publish this work. Through Hartax International Inc., Wrenn, John Bair, and James Hart were employed as field assistants during the collection of samples for this project; Wrenn and Scott Beckman performed much of the laboratory work needed to complete the study.

The underlying funding for this project was derived from Arco Corporation, Chevron Oil Company, Conoco Oil Company, Exxon USA, Hartax International Inc., Superior Oil Company, Tenneco Oil Company, and Texaco Inc. Without the financial support of these organizations, the field sampling and data analysis would never have been completed. Some of the geochemical analyses were contracted to Geochem Inc., Houston. Thanks go to Geoffrey Bayliss, President of Geochem Inc., for the particular attention and interest he gave to this study.

Wrenn kindly and most diligently read this manuscript and suggested improvements. I am particularly grateful to Alfred Traverse and the anonymous reviewers of this manuscript who greatly improved its readability.

References

Barrell, J. (1912). Criteria for the recognition of ancient delta deposits. *Geological Society of America, Bulletin*, **23**, 377–446.

Beckman, S. W. (1984). Paleoenvironmental reconstruction and organic matter characterizations of peats and associated sediments from cores in a portion of the LaFourche delta. Ph.D. dissertation, Louisiana State University.

Boone, D. & Mah, R. (1987). Transitional bacteria. In *Anaerobic Digestion of Biomass*, ed. D.P. Chynoweth & R. Isaacson. Amsterdam: Elsevier, 35–48.

Casagrande, D. L., Idowu, G., Friedman, A., Rickert, P., Siefert, K. & Schelnz, D. (1979). H_2S incorporation in coal precursors: origins of organic sulphur in coal. *Nature*, **282**, 599–600.

Casagrande, D. L. & Ng, L. (1979). Incorporation of elemental sulphur in coal as organic sulphur. *Nature*, **282**, 59–89.

Chabreck, R. H. (1972). Vegetation, water and soil characteristics of the Louisiana coastal region. *Louisiana Agricultural Experiment Station Bulletin*, **664**, 172.

Chabreck, R. H. & Condrey, R. E. (1979). *Common Vascular Plants of the Louisiana Marsh*. Baton Rouge: Louisiana State University Center for Wetland Resources, Sea Grant Publication LSU-T-79-03.

Chynoweth, D. P. & Isaacson, R. (1987), ed. *Anaerobic Digestion of Biomass*. Amsterdam: Elsevier.

Coleman, J. M. & Gagliano, S. M. (1964). Cyclic sedimentation in the Mississippi River Deltaic Plain. *Transactions Gulf Coast Association of Geological Societies*, **14**, 67–80.

Coleman, J. M., Gagliano, S. M. & Webb, J. E. (1964). Minor sedimentary structures in a prograding distributary. *Marine Geology*, **1**, 240–58.

Coleman, J. M. & Prior, D. B. (1980). Deltaic Sand Bodies. *American Association of Petroleum Geologists, Continuing Education Course* 15.

Coleman, J. M., Prior, D. B. & Garrison, L. E. (1980). Subaqueous sediment instabilities in the offshore Mississippi River delta. *Bureau of Land Management, Open-File report* 80-01. New Orleans, LA.

Coleman, J. M., Suhayda, J. N., Whelan, T. & Wright, L. D. (1974). Mass movement of Mississippi River Delta sediments. *Transactions Gulf Coast Association of Geological Societies*, **24**, 49–68.

Cooley, W. W. & Lohnes, P. R. (1971). *Multivariate Data Analysis*. New York: John Wiley & Sons.

Darby, J. & Hart, G. F. (1988). Recognition of depositional environments using organic petrology: The Floridian Carbonate Region. *Abstracts. 7th International Palynological Congress*, Brisbane, Australia, 34.

Darrell, J. H. & Hart, G. F. (1970). Environmental determinations using absolute miospore frequency, Mississippi River Delta. *Geological Society of America Bulletin*, **81**(8), 2513–8.

Degens, E. T., Reuter, J. H. & Shaw, K. N. F. (1964). Biochemical components in offshore California sediments and sea waters. *Geochimica et cosmochimica acta*, **28**, 45–66.

Fisk, H. N. (1944). Geological investigation of the alluvial valley of the lower Mississippi River. *Vicksburg, Mississippi River Commission, Report on Study*.

Fisk, H. N. (1955). Sand facies of recent Mississippi Delta deposits. *Proceedings 4th World Petroleum Congress*, **1-C**, 377–398.

Fisk, H. N. (1961). Bar-finger sands of the Mississippi delta. In *Geometry of Sandstone Bodies*, ed. J. A. Peterson & J. C. Osmond. Symposium, American Association of Petroleum Geologists, 29–52.

Fisk, H. N. & McFarlan, E. (1955). Late Quaternary Deltaic deposits of the Mississippi River. In *Crust of the Earth* (a symposium), ed. A. Poldervaart. *Geological Society of America Special Paper* **62**, 279–302.

Frazier, D. E. (1967). (abstract) Recent deltaic deposits of Mississippi River: their development and chronology. *American Association of Petroleum Geologists Bulletin*, **51**(10), 21–64.

Galloway, W. E. (1975). Process framework for describing the morphologic and stratigraphic evolution of deltaic depositional systems. In *Deltas*, ed. M. L. Broussard. Houston Geological Society, 87–98.

Gilbert, G. K. (1885). The topographic features of lake shores. *US Geological Survey, fifth annual report*, 69–123.

Gilbert, G. K. (1890). Lake Bonneville. US Geological Survey, Monograph Series 1.

Gottschalk, G. (1979). *Bacterial Metabolism*. New York: Springer-Verlag.

Hart, G. F. (1971). *Statistical evaluation of palynomorph distribution in the Atchafalaya Swamp*. Consultant Report. Lafayette, LA: Texaco Inc.

Hart, G. F. (1975). *Maceral analysis and Total Organic Carbon determination: Santa Maria Basin project*. Consultant Report. Baton Rouge, LA: Carbon Systems Inc.

Hart, G. F. (1976). *Maceral analysis and Total Organic Carbon determination: Kodiak Shelf Project*. Consultant Report. Baton Rouge, LA: Carbon Systems Inc.

Hart, G. F. (1979a). Maceral analysis: its use in petroleum exploration. Methods Paper 2. Baton Rouge, LA: Hartax International Inc.

Hart, G. F. (1979b). Maceral analysis and its application to petroleum exploration. (abstract) In *Petroleum Potential in Island Arcs, Small Ocean Basins, Submerged Margins and Related Areas*, Symposium UNESCO – Suva, Fiji, Abstracts, **1**, 32–3.

Hart, G. F. (1979c). *Maceral, geochemical and lithological characteristics of the Louisiana Gulf Coast*. Hartax Paper. Baton Rouge, LA: Hartax International Inc.

Hart, G. F. (1986). Origin and classification of organic matter in clastic systems. *Palynology*, **10**, 1–23.

Hart, G. F. (1988). Maceral Facies of Detrital Systems: Louisiana Deltaic Plain. Keynote Address for Symposium 26: Methods of Kerogen Analysis for Hydrocarbon Exploration. *Programme and Handbook, 7th International Palynological Congress, Brisbane*, Australia, 41.

Hart, G. F., Ferrell, R. E, Lowe, D. R. & LeNoir, E. A. (1989). Shelf sandstones of the Robulus L Zone, offshore Louisiana. Gulf Coast Section, Society of Economic Paleontologists and Mineralogists Foundation, *Seventh Annual Research Conference Proceedings*, 117–41.

Healy, J. B., Jr. & Young, L. Y. (1979). Anaerobic biodegradation of eleven aromatic compounds to methane. *Applied and Environmental Microbiology*, **38**, 84–9.

Healy, J. B., Jr., Young, L. Y. & Reinhard, M. (1980). Methanogenic decomposition of ferulic acid, a model lignin derivative. *Applied and Environmental Microbiology*, **39**, 436–44.

Hobson, P. N., Bousfield, S. & Summers, R. (1974). Anaerobic digestion of organic matter. *CRC Critical Reviews in Environmental Control*, **4**, 131–91.

Johnston, W. A. (1921). Sedimentation of the Fraser River Delta. *Geological Survey of Canada Memoirs* 125.

Johnston, W. A. (1922). The character of the stratification of the sediments in the recent delta of Fraser River, British Columbia, Canada. *Journal of Geology, British Columbia, Canada*, **30**, 115–29.

Kolb, C. R. & Van Lopik, J. R. (1966). Depositional environments of the Mississippi River Deltaic Plain, southeastern Louisiana. In *Deltas*, ed. M. L. Shirley & J. A. Ragsdale. Houston, TX: Houston Geological Society, 17–63.

Kristjansson, J. K., Schönheit, P. & Thauer, R. K. (1982). Different K_s values for hydrogen of methanogenic bacteria and sulfate reducing bacteria: an explanation for the apparent inhibition of methanogenesis by sulfate. *Archives of Microbiology*, **131**, 278–82.

Kuehl, S. A., Nittrouer, C. A. & DeMaster, D. J. (1982). Modern sediment accumulation and strata formation on the Amazon continental shelf. *Marine Geology*, **49**, 279–300.

LeNoir, E. A. & Hart, G. F. (1986). Burdigalian (Early Miocene) dinocysts from offshore Louisiana. *American Association of Stratigraphic Palynologists, Contribution Series*, **17**, 59–81.

LeNoir, E. A. & Hart, G. F. (1988). Palynofacies of some Miocene sands from the Gulf of Mexico, offshore Louisiana, U. S. A. *Palynology*, **12**, 151–65.

Lindblom, G. P. & Lupton, M. D. (1961). Microbiological aspects of organic geochemistry. In *Developments in Industrial Microbiology*. New York: Plenum Press, **2**, 9–22.

Litchfield, C. D., Munro, A. L. S., Massie, L. C. & Floodgate, C. D. (1974). Biochemistry and microbiology of some Irish Sea sediments. I. Amino-acid analyses. *Marine Biology*, **26**, 249–60.

Lord, C. J. (1980). The chemistry and cycling of iron, manganese, and sulfur in salt marsh sediments. Ph.D. dissertation, University of Delaware.

McKee, E. D. (1939). Some types of bedding in the Colorado River delta. *Journal of Geology*, **47**, 64–81.

Morgan, D. J. (1977). *The Mississippi River Delta, Legal-geomorphologic Evaluation of Historic Shoreline Changes*. Baton Rouge: Louisiana State University School of Geoscience (= *Geoscience and Man*, **16**).

Morgan, J. P. (1970). Deltas – a resume. *Journal of Geological Education*, **18**(3), 107–17.

Nedwell, D. B. & Banat, I. M. (1981). Hydrogen as an electron donor for sulfate-reducing bacteria in slurries of salt marsh sediment. *Microbial Ecology*, **7**, 305–13.

Nikaido, M. (1977). On the relation between methane production and sulfate reduction in bottom muds containing sea water sulfate. *Geochemical Journal*, **11**, 199–206.

Odum, E. P. (1971). *Fundamentals of Ecology*, 3rd edition. Philadelphia: Saunders.

O'Neil, T. (1949). *The Muskrat in Louisiana Coastal Marshes*. New Orleans: Louisiana Department of Wildlife and Fisheries.

Oremland, R. S. & Taylor, B. F. (1978). Sulfate reduction and methanogenesis in marine sediments. *Geochimica et cosmochimica acta*, **42**, 209–14.

Palmisano, A. W. (1970). Plant community soil relationships in Louisiana coastal marshes. Ph.D. dissertation, Louisiana State University, Baton Rouge, LA.

Penfound, W. T. (1952). Southern swamps and marshes. *Botanical Review*, **18**(6), 413–46.

Penfound, W. T. & Hathaway, E. S. (1938). Plant communities in the marshlands of southeastern Louisiana. *Ecological Monographs*, **8**(1), 3–56.

Rhoads, D. C., Boesch, D. F., Zhican, T., Fengshan, X., Liqiang, H. & Nilsen, K. J. (1985). Macrobenthos and sedimentary facies on the Changjing delta platform and adjacent continental shelf, East China Sea. *Continental Shelf Research*, **4**, 189–213.

Russell, R .J. (1942). Flotant. *Geographical Review*, **32**, 74–98.

Schlegel, H. G. (1975). Mechanisms of chemo-autotrophy. In *Marine Ecology*, ed. O. Kinne. London & New York: John Wiley & Sons, **2**(1): 960.

Schönheit, P., Kristjansson, J. K. & Thauer, R. K. (1982). Kinetic mechanism for the ability of sulfate reducers to out-compete methanogens for acetate. *Archives of Microbiology*, **132**, 285–8.

Scruton P. C. (1960). Delta building and the deltaic sequence. In *Ancient Sediments, Northwest Gulf of Mexico*, ed. F. P. Shepard, F. B. Phleger & T. H. van Andel. American Association of Petroleum Geologists, 82–102.

Sieburth, J. M. (1979). *Sea Microbes*. London: Oxford University Press.

Sleat, R. & Mah, R. (1987). Hydrolytic Bacteria. In *Anaerobic Digestion of Biomass*, ed. D. P. Chynoweth & R. Isaacson. Amsterdam: Elsevier, 15–33.

Sørensen, J., Christensen, D. & Jorgensen, B. B. (1981). Volatile fatty acids and hydrogen as substrate for sulfate-reducing bacteria in anaerobic marine sediment. *Applied and Environmental Microbiology*, **42**, 511.

Staplin, F. L. (1969). Sedimentary organic matter, organic metamorphism, and oil and gas occurrence. *Bulletin of Canadian Petroleum Geology*, **17**(1), 47–66.

Thauer, R. K. (1982). Dissimilatory sulphate reduction with acetate as electron donor. *Philosophical Transactions of the Royal Society of London*, **B298**, 467–71.

Tissot, B. P. & Welte, D. H. (1978). *Petroleum Formation and Occurrence*. New York: Springer-Verlag.

Tukey, J. W. (1962). The future of data analysis. *Annals of Mathematical Statistics*, **33**, 1–67.

Twenhofel, W. H., ed. (1932). *Treatise on Sedimentation*. Baltimore, MD: Williams & Wilkins.

Van Heerden, I. L & Roberts, H. H. (1988). Facies development of Atchafalaya Delta, Louisiana: a modern bayhead delta. *American Association of Petroleum Geologists Bulletin*, **72**(4), 439–53.

Welder, F. A. (1959). Processes of deltaic sedimentation in the Lower Mississippi River. *Technical Report* **12**, Coastal

Studies Institute, Louisiana State University, Baton Rouge, LA.

Winogradsky, S. (1888). *Beiträge zur Morphologie und Physiologie der Bakterien,* Vol. 1, *Die Schwefelbakterien.* Leipzig: Arthur Felix.

Wrenn, J. H. & Beckman, S. W. (1981). Maceral and total organic carbon analysis of DVDP drill core 11. In *Dry Valley Drilling Project,* ed. R. D. McGinnis. American Geophysical Union, Antarctic Research Series, **33**, 391–402.

Wrenn, J. H. & Beckman, S. W. (1982). Maceral, total organic carbon and palynological analyses of Ross Ice Shelf Project Site J9 cores. *Science,* **216**, 187–9.

Wright, L. D. & Coleman, J. M. (1973). Variations in morphology of major river deltas as functions of ocean wave and river discharge regimes. *Bulletin of the American Association of Petroleum Geologists,* **57**, 370–98.

Wright, L. D. & Coleman, J. M. (1975). Mississippi River mouth processes: effluent dynamics and morphologic development. *Journal of Geology,* **82**, 751–78.

Zeikus, J. G. (1977). The biology of methanogenic bacteria. *Bacteriological Reviews,* **41**, 514–41.

Zhang, Y.-G. & Chen, H.-J. (1985). Concepts on the generation and accumulation of biogenic gas. *Journal of Petroleum Geology* **8**(4), 405–22.

10 Organic sedimentation in a carbonate region

JASON D. DARBY and GEORGE F. HART

Introduction

The study of disseminated organic matter in sediments to determine characteristics of the depositional environment dates back to the early work in palynology. The basic reasoning is that the organic material contained in sediments derives from living organisms, and much of that material is autochthonous. Preservation of organic matter in sediments is primarily a function of the effects of the surface, depositional, and diagenetic environments upon the dead organism or parts thereof. The types of organisms and the chemical and biological alterations following death determine the types of particulate organic matter (macerals) found in the sediments. As a consequence, the maceral spectra derived from a sedimentary rock may have a strong autochthonous component which often can be used to extract information about the environment of deposition (Darrell & Hart, 1970; Hart, 1971, 1975, 1976, 1979a,b, 1986; Hart et al., 1989; LeNoir & Hart, 1986, 1988; Wrenn & Beckman, 1981, 1982). It should also be stressed that land-derived palynomorphs and other organic particles can indicate sedimentary environments due to sorting in the marine realm (Traverse & Ginsburg, 1966).

The descriptive classification system used to identify the macerals (Hart, 1979, 1986; see also Hart and Hart et al., Chaps. 9 & 17 this volume) divides them into phytoclasts (plant-derived), zooclasts (animal-derived), protistoclasts (protistan-derived) and scleratoclasts (fungal-derived), using the five-kingdom system of classification of organisms of Whittaker (1969). Based on the preservational state of cell walls, each of these biological categories consists of well-preserved (little evidence of biodegradation); poorly preserved (minor biodegradation); infested (cell walls highly disrupted by organic attack); amorphous structured (remnant cell structure only); amorphous non-structured (blocky phytoclast mass or fluffy protistoclast mass). This research extends Hart's classification of macerals from clastic depositional systems to include organic matter in carbonate environments. This involves an extension of the protistoclast classification to include the whole degradational sequence. The classification of protistoclasts into clastic systems includes well-preserved, poorly preserved, and amorphous non-structured degradation states; although other degradation stages occur, they are not distinctive enough to allow establishment of additional categories. After an initial examination of the Florida Bay samples, the classification was expanded to include infested and amorphous structured protistoclasts. Table 10.1 outlines the expanded classification.

Regional setting

The Florida carbonate region is associated with the southern tip of the Florida Plateau (Figs. 10.1, 10.2). The Straits of Florida separate it on the east and southeast from the Great Bahama Bank and Cuba (Multer, 1977; Owens, 1960). The floor of the Straits of Florida is about 1500 m below sea level between Key West, Florida, and Havana, Cuba. The warm Florida Current flows northward through this trough to become the Gulf Stream off Cape Hatteras, North Carolina. The Florida Plateau itself is only part of an even larger region of carbonate deposition which Ginsburg (1956) named the Florida–Bahamas Province. Southern Florida, the Great Bahama Bank, and the northern coast of Cuba are included in this province. The Florida carbonate region is divided into three major physiographic complexes: Florida Bay, the Florida Keys, and the reef tract (see Fig. 10.1).

Florida Bay (Fig. 10.1)

Florida Bay is a triangular, shallow water embayment of roughly 600 m² in area. It is bounded on the north by the mainland of Florida, on the south and east by the Upper Florida Keys, and on the west by the Gulf of Mexico. It connects to the Atlantic Ocean on the

Table 10.1. *Maceral classification showing the degree of degradation of maceral types. (See also Hart and Hart* et al., *Chaps 9 & 17, respectively, this volume.)*

Clast-type	Increasing degree of degradation				
Phytoclast	Well-preserved	Poorly preserved	Infested	Amorphous structured	Amorphous non-structured
Protistoclast	Well-preserved	Poorly preserved	Infested	Amorphous structured	Amorphous non-structured
Scleratoclast	Well-preserved	Poorly preserved			
Zooclast	Well-preserved	Poorly preserved			
Indeterminate					Amorphous indeterminate

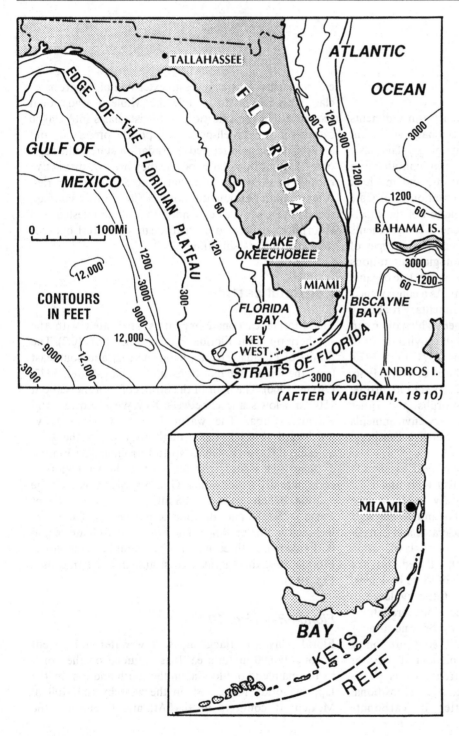

Figure 10.1. Peninsular Florida, showing margins of the Florida Plateau, and neighboring features.

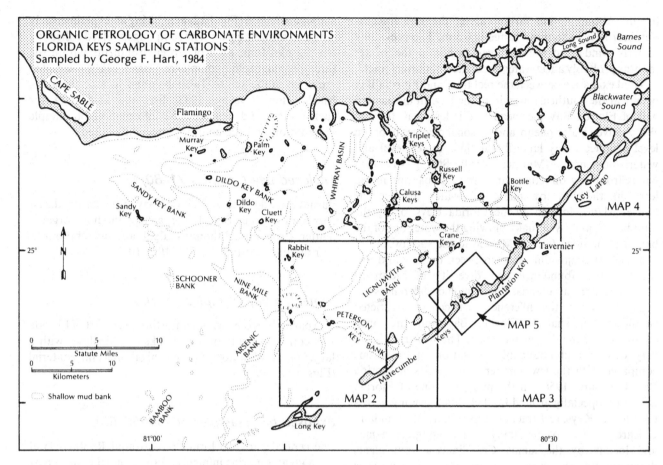

Figure 10.2. Florida carbonate region sampling areas: Florida Bay, and adjacent parts of the Florida Keys and peninsular Florida. The indicated maps (2–5) are displayed in Figures 10.3–10.6.

southeast by the channels through the Florida Keys, and to the Gulf of Mexico on the southwest by narrow channels through the almost continual mud banks. Water depths in Florida Bay range from as little as 1.2 m to 3.9 m, but most of the bay is less than 1.8 m deep. There are large areas of mud banks which are less than two feet deep and in many places they are above sea level. These mud banks produce the only 'relief' in Florida Bay. Drainage of water from the mainland into Florida Bay is an important factor in circulation, especially as this affects salinity (Ginsburg, 1956). Large seasonal and annual fluctuations of salinity occur in Florida Bay. Periods of poor circulation, coupled with seasonal freshwater runoff and evaporation, can cause either abnormally high or low salinity conditions, ranging from 10 parts per thousand to more than 50 ppt (Ginsburg, 1954; McCallum & Stockman, 1964).

The plants and protists of Florida Bay are divided into two major biocoenoses. The first is the Turtle Grass Beds of the banks and basins; the latter described as 'submarine meadows' by Thorne (1954). Several species of subaqueous angiosperms such as *Thalassia testudinum*, *Halodule wrightii*, and many algae including species of *Halimeda*, *Acetabularia*, *Penicillus*, *Udotea*, and *Sargassum*, dominate this biocoenosis. The second is the molluscan – foraminiferal biocoenosis. Lidz and Rose (1989) showed the important foraminifera in Florida Bay as *Triloculina*, *Quinqueloculina*, *Elphidium*, *Ammonia*, *Archaias* and *Miliolinella*. Ginsburg (1956) found the following genera of molluscans in Florida Bay: *Chione*, *Laevicardium*, *Tellina*, *Cerithium*, *Pinctuda*, *Bulla*, *Modulus*, *Marginella*, *Mytilus*, *Battilaria*, *Fasciolaria*, *Melampius*, *Conus*, *Arca*, *Pecten*, *Cardium*, and *Turbo*.

Reef tract (Fig. 10.1)

The reef tract is a narrow curving band 36 km wide, seaward from the Florida Keys and open to the Straits of Florida (Ginsburg, 1956). Ginsburg places the lower limit of the reef tract at 90.9 m. Within the reef tract elongate coral reefs, rocky shoals with a calc-arenite veneer, and irregularly shaped coral knolls and pinnacles create local relief up to 9.1 m above the bottom. The open circulation of the reef tract is the result of wind-driven water movements and the semidiurnal tidal exchange with the waters of the Florida Current. The tidal ranges along the edge of the reef tract decrease southward from Miami. At Miami the mean range is

0.7 m. About 52 km farther along the reef tract at Sombrero Key the mean range is 0.5 m. The prevailing easterly winds and the northeast winds in the winter months probably affect water circulation of the reef tract.

Tidal exchange between the reef tract and Florida Bay occurs at the southeast margin of the bay. In the eastern half of Florida Bay the mean tide is less than 0.15 m. Wind-driven water pileup, and seasonal changes in ocean level produce many large fluctuations in sea level and water movements (Marmer, 1954). Strong sustained winds from any direction can cause fluctuations in sea level in the bay by piling up water against the barrier islands formed by the Florida Keys, or by the mud banks, and by actually blowing the water out of the bay into the open Gulf of Mexico to the southwest.

There are major differences between the bay and reef tract in the abundance and diversity of sediment-producing flora and fauna. These distinct differences are a direct result of the differences between the restricted circulation of shallow Florida Bay and the open circulation of the deeper reef tract. The reef tract has a very diverse and abundant population of organisms compared with the few numbers and species in Florida Bay (Ginsburg, 1956). In the marginal zone of Florida Bay, and especially around tidal channels leading across the Florida Keys, reef-tract organisms may be abundant. Calcareous algae of the family Codiaceae (green algae), and locally by the family Corallinaceae (red algae) dominate the reef crest. The reef-tract biocoenosis is extremely diverse. The major calcareous elements are corals, molluscs, echinoderms, foraminifera, bryozoans, crustaceans, and a variety of worm tubes.

Florida Keys (Fig. 10.1)

This barrier island complex shows little topographic relief and is formed from carbonates of the Key Largo Formation and the Miami Formation of the Pleistocene (Sangamonian Age). An important contribution to the particulate organic matter found in the sediments of the Florida region is the mangrove swamp forest biocoenosis of the islands. This has vegetation similar to that of the saline mangrove zone of the mainland described by Craighead (1971). In this community, trees and/or shrubs such as *Rhizophora mangle* L. and *Avicennia nitida* Jacq. are common. Often associated with the arborescent vegetation are procumbent shrubs, such as *Batis maritima* L. Herbaceous dicot forms, including species of *Salicornia* and *Philoxerus* and grasses such as *Monanthochloe littoralis* Engelm. are abundant in the more open, central areas of some of the islands.

Sampling

The objective of the sampling was to obtain an unbiased collection of field samples from the modern Florida carbonate region (Fig. 10.2) area, using an experimentally designed sampling plan. The sampling involved collecting ten samples from each of ten environments within the three physiographic complexes. The localities sampled are listed in the key to Table 10.2 and include the areas listed below. (Darby, 1987, lists the field data for each sample and provides the information on sample preparation and the counting procedures used).

Enclosed bay (samples 71–80)

Barnes Sound is a large enclosed bay in the far northeast corner of Florida Bay. It has a high input of organic matter from autochthonous algal and allochthonous surrounding vegetation (Figs. 10.2, 10.5).

Protected bay (samples 91–100)

The partially protected inner northeastern part of Florida Bay consists mainly of mud banks and shallows with a large input of autochthonous algal maceral material (Figs. 10.2, 10.5).

Exposed bay (samples 50–54, 60–63)

The central part of Florida Bay around Rabbit Island has a greater marine influence than the northeastern part of the bay. It is predominantly an algal mud bank (Figs. 10.2, 10.3).

Mangrove fringed bay (samples 81–90)

Shallow bays near the Everglades having a very high input of mangrove phytoclast debris (Figs. 10.2, 10.5).

Barrier island

The varied environments of the barrier islands provide a host of variable depositional conditions. The environments sampled include: (1) intertidal zone (samples 44–46, 48, 55–57, 66–67, 69, 101); (2) key, including Key Lake (samples 41–43, 47, 58–59, 64–65, 68, 70); (3) shallow marine (samples 9–13, 36–40); (4) tidal channel (samples 26–35) (Figs. 10.2–10.4, 10.6); (5) reef and backreef (samples 1–8, 14–25; the reef and backreef probably have the least amount of maceral material preserved). (Figs. 10.2, 10.4, 10.6).

The grab samples were collected from the sediment/water interface using scuba gear. Each sample was treated with alkylbenzyldimethylammonium chloride ('Zephiran chloride') to inhibit bacterial growth, dried and sealed in a plastic container.

Table 10.2. *Discriminant function analysis of the ten environments.*

(a) Key to environmental codes

Bay complex	1.1	enclosed bay
	1.2	protected bay
	1.3	exposed bay
	1.4	mangrove fringed bay
Barrier island complex	2.1	intertidal
	2.2	key
	2.3	shallow marine
	2.4	channel
Reef tract complex	3.1	backreef
	3.2	reef crest

(b) Classification summary for calibration data. Number of observations and percents classified into environment.

From	1.1	1.2	1.3	1.4	2.1	2.2	2.3	2.4	3.1	3.2	Total
1.1	8	0	0	0	0	0	0	0	0	0	8
	100%	0%	0%	0%	0%	0%	0%	0%	0%	0%	100%
1.2	0	9	0	0	0	0	0	0	0	0	9
	0%	100%	0%	0%	0%	0%	0%	0%	0%	0%	100%
1.3	0	0	5	0	0	0	0	0	0	0	5
	0%	0%	100%	0%	0%	0%	0%	0%	0%	0%	100%
1.4	0	0	0	7	0	0	0	0	0	0	7
	0%	0%	0%	100%	0%	0%	0%	0%	0%	0%	100%
2.1	0	0	0	0	7	0	0	0	0	0	7
	0%	0%	0%	0%	100%	0%	0%	0%	0%	0%	100%
2.2	0	0	0	0	0	8	0	0	0	0	8
	0%	0%	0%	0%	0%	100%	0%	0%	0%	0%	100%
2.3	0	0	0	0	0	0	4	0	0	0	4
	0%	0%	0%	0%	0%	0%	100%	0%	0%	0%	100%
2.4	0	0	0	0	1	0	0	5	0	0	6
	0%	0%	0%	0%	16.67%	0%	0%	83.33%	0%	0%	100%
3.1	0	0	0	0	0	0	0	0	6	0	6
	0%	0%	0%	0%	0%	0%	0%	0%	100%	0%	100%
3.2	0	0	0	0	0	0	0	0	0	5	5
	0%	0%	0%	0%	0%	0%	0%	0%	0%	100%	100%
Total	8	9	5	7	8	8	4	5	6	5	65
%	12.31	13.85	7.69	10.77	12.31	12.31	6.15	7.69	9.23	7.69	100%

Statistical analysis of the data set

The prime purpose of the study was to determine the environmental effects on the distribution of macerals. Iterative discriminant function analysis (Hart, 1979a) determines the major characteristics of each environment. This same method was applied elsewhere to determine major environmental characteristics (Hart *et al.*, 1989; LeNoir & Hart, 1986, 1988). The basis of the method is to achieve a 'purified data base,' i.e., one in which each class contains typical samples. This application of data analysis, in the tradition of Tukey (1962) and Cooley and Lohnes (1971), attempts to develop an approach to scientific research based on multivariate heuristics rather than multivariate inference (see also discussion in Hart *et al.*, Chap. 17 this volume).

Initially a discriminant function procedure was applied to all 101 samples collected from the carbonate region

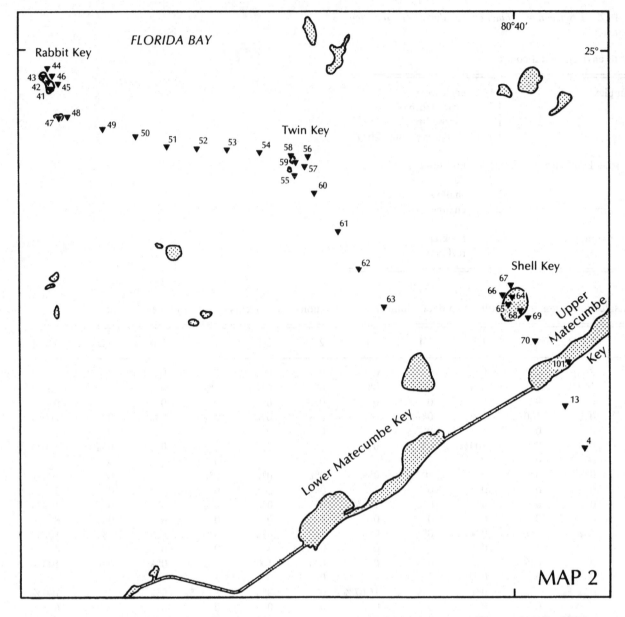

Figure 10.3.   Twin key area. (See map 2 on Fig. 10.2.)

using ten classes (the ten environments sampled). Sixty-six samples were correctly classified into their respective environments. Table 10.3 shows the results of the first iteration of the discriminant function analysis. It lists each environment along with the number of correctly and incorrectly classified samples. The aberrant samples are listed with the environment to which they were classified. After examining the results, the 35 misclassified samples were removed from the data base, and the second iteration was run. This procedure resulted in perfect discrimination in all but two classes. Table 10.4 shows results of the second iteration. It lists each environment along with the number of correctly and incorrectly classified samples. The aberrant samples are listed along with the environment to which they were classified. The

channel and backreef environments each had one sample misclassified, while all other classes had perfect discrimination. After examining the results sample 22 was removed. Sample 27 remained in the data base since the posterior probability of membership in the correct environment was high. A third iteration was run and 100% discrimination achieved in all environments, except the channel, which had one sample misclassified into the intertidal environment. This sample remained in the data base since the posterior probability of membership in the correct environment was high. The 65 remaining samples make up the purified data base for the ten environment classes (see Table 10.4). Some trends were evident in the

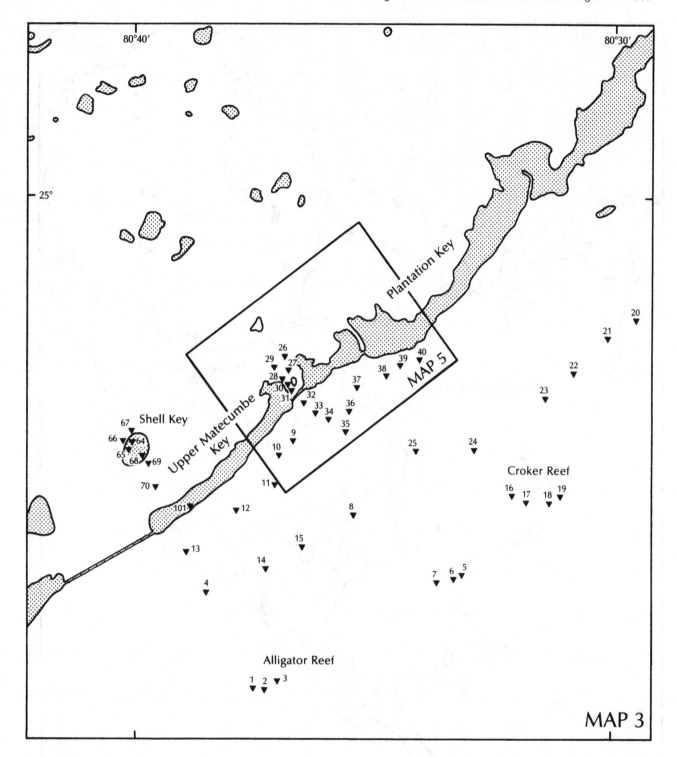

Figure 10.4.   Plantation key area. (See map 3 on Fig. 10.2.)

misclassification of the 36 samples removed from the data base. Discrimination between the reef crest and backreef environments was weak, as was the discrimination between the enclosed bay, protected bay, and mangrove fringed bay. Because of the similarity indicated by the discriminant function analysis, the channel samples were grouped with the adjacent intertidal, shallow marine, and backreef environments. The shallow marine samples were grouped with the adjacent intertidal, channel, and backreef environments. The intertidal samples were grouped with the adjacent key, shallow marine, and channel environments.

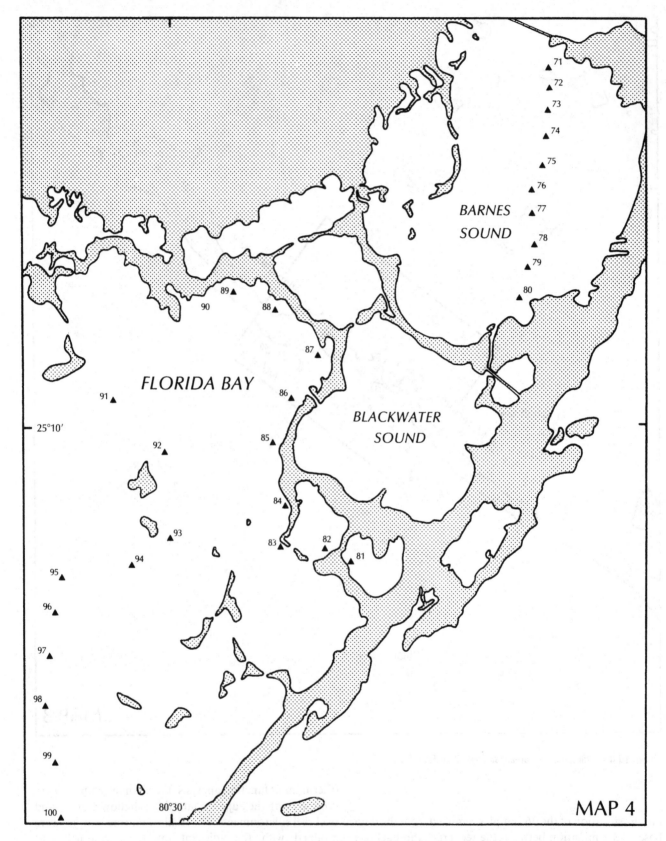

Figure 10.5.   Blackwater Sound area. (See map 4 on Fig. 10.2.)

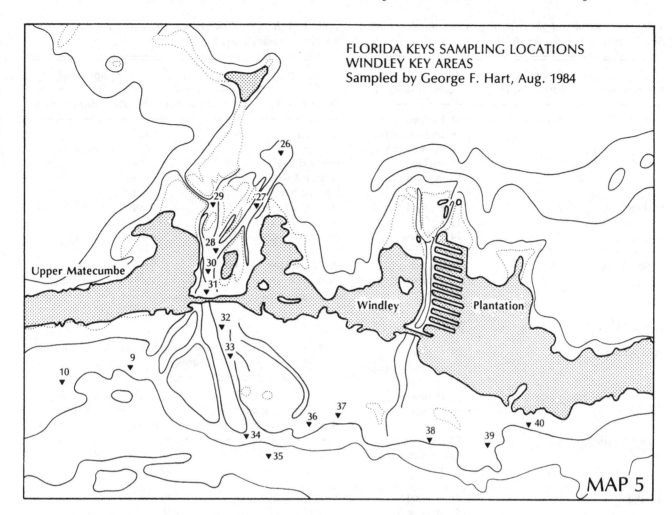

Figure 10.6. Windley key area. (See map 5 on Figs. 10.2 and 10.4.)

Based on the relationships developed by the iterative application of the discriminant function procedure, the original ten classes can be organized into five groups: (1) enclosed, protected, and mangrove fringed bays, (2) exposed bay, (3) intertidal, shallow marine, and channel, (4) key, and (5) backreef and reef crest. The discriminant function procedure was applied to the original 101 samples using these five groups as classes. Eighty-two samples were correctly classified into their respective environments. Table 10.5 shows the results of the discriminant function analysis. The 19 misclassified samples were removed from the data base and the discriminant function iterated on the 82 remaining samples. Eighty-one samples were correctly classified into their respective environments. All environments had 100% discrimination except Group 2 (exposed bay), which had one misclassified sample. Sample 54 was misclassified from Group 2 into Group 3. Table 10.6 shows the results of the second iteration. Initially sample 54 was removed from the data base and a third iteration of the discriminant function was run. This procedure

resulted in only one sample being incorrectly classified into Group 2. A fourth iteration was run resulting in no samples being incorrectly classified into Group 2. A review of the results indicated the most characteristic data base was the 82 samples used in the second iteration. These 82 samples compose the purified data base for the five environmental groups (see Table 10.7).

Finally, a discriminant function was applied to the original data base, consisting of 101 samples, using the three major physiographic complexes (bay, barrier island and reef tract) as classes. These three physiographic complexes were the major geographic divisions used in the original sampling plan and are distinguishable from each other by their unique topographic, hydrologic, and biologic conditions. The environments within each physiographic complex are:

bay = enclosed, protected, mangrove fringed and exposed bays;

barrier island = intertidal, shallow marine, channel and key;

reef tract = backreef and reef crest.

Table 10.3. *Results of the first iteration of the discriminant function procedure using ten classes*

Environment	Success ratio correct:incorrect	Aberrant samples
reef crest	5:5	1 backreef
		6 shallow marine
		7 exposed bay
		16 backreef
		18 exposed
backreef	7:3	8 exposed bay
		15 reef crest
		24 reef crest
enclosed bay	8:2	71 protected bay
		79 mangrove fringed bay
protected bay	9:1	99 intertidal
exposed bay	5:5	49 channel
		50 shallow marine
		51 reef crest
		52 channel
		53 channel
mangrove fringed bay	7:3	83 enclosed bay
		85 protected bay
		88 key
intertidal	7:4	45 channel
		55 shallow marine
		67 key
		69 mangrove fringed bay
key	8:2	64 backreef
		70 intertidal
shallow marine	4:6	12 intertidal
		13 backreef
		36 channel
		37 channel
		38 channel
		40 exposed bay
channel	6:4	29 key
		30 intertidal
		33 shallow marine
		35 backreef

Table 10.4. *Results of the second iteration of the discriminant function procedure using ten classes*

Environment	Success ratio correct:incorrect	Aberrant samples Number	Aberrant samples Environment
reef crest	5:0	none	
backreef	6:1	22	reef crest
enclosed bay	8:0	none	
protected bay	9:0	none	
exposed bay	5:0	none	
mangrove fringed bay	7:0	none	
intertidal	7:0	none	
key	8:0	none	
shallow marine	4:0	none	
channel	5:1	27	intertidal

Table 10.5. *Results of the first iteration of the discriminant function procedure using five groups*

Group	Success ratio correct:incorrect	Aberrant samples Number	Aberrant samples Environment
1	28:2	88	4
		99	5
2	4:6	49	3
		50	3
		51	5
		52	3
		53	3
		63	5
3	26:5	13	5
		35	5
		40	2
		67	4
		69	1
4	8:2	64	5
		70	3
5	16:4	6	3
		8	2
		14	3
		18	2

Eighty-one samples were correctly classified into their respective environments. Results of the discriminant function analysis are shown in Table 10.8. After removing the 20 misclassified samples from the data base the discriminant function was applied to the 81 remaining samples. This second iteration produced perfect discrimination in all three physiographic complexes. The remaining 81 samples compose the purified data base for the three physiographic complexes (see Table 10.9).

Table 10.6. *Results of the second iteration of the discriminant function procedure using five groups*

Group	Success ratio correct:incorrect	Aberrant samples Number	Aberrant samples Environment
1	28:0	none	
2	3:1	54	3
3	26:0	none	
4	8:0	none	
5	16:0	none	

Table 10.7. *Discriminant function analysis: classification summary for calibration data, number of observations and percent classification into environment*

Key to environmental group codes: 1 = enclosed + protected + mangrove fringed bays; 2 = exposed bay; 3 = intertidal, shallow marine and channel; 4 = key; 5 = backreef and reef crest

From environment	1	2	3	4	5	Total
1	28	0	0	0	0	28
	100%	0	0	0	0	100%
2	0	3	1	0	0	4
	0	75%	25%	0	0	100%
3	0	0	26	0	0	26
	0	0	100%	0	0	100%
4	0	0	0	8	0	8
	0	0	0	100%	0	100%
5	0	0	0	0	16	16
	0	0	0	0	100%	100%
Total observations	28	3	27	8	16	82
Total %	34%	4%	33%	10%	20%	100%

The general trend in the mean values and bar charts shows that the more restricted areas (protected bay, enclosed bay, and mangrove fringed bay environments) have a high amount of phytoclast input. In the enclosed bay and protected bay the degradation level is predominantly amorphous non-structured to amorphous structured clasts. The mangrove fringed bay shows a greater distribution of phytoclast material including well- and poorly preserved phytoclasts.

The more marine areas (reef crest, backreef, shallow marine and exposed bay environments) have distributions showing a high input of protistoclasts. The degradation levels are predominantly amorphous non-structured and amorphous structured. The intertidal, channel, and key environments show a fairly even distribution of protistoclast and phytoclast material. The channel and intertidal environments are slightly skewed towards the protistoclasts while the key distribution is slightly skewed towards the phytoclasts.

Circumscription of typical samples

Bar charts

Once the purified data bases were identified for the different class groupings, the maceral spectra were identified for each physiographic complex/environment. This involved calculating a mean value for each maceral type in each physiographic complex/environment. The distribution of maceral types in each physiographic complex/environment is shown using bar charts (Figs. 10.7–10.9).

Scatter plots

A scatter diagram of all samples remaining in the ten-environment purified data base allowed identification of major significant trends. (Scatter diagrams mentioned here can be consulted in Darby, 1987.) The results show a strong inverse linear relationship between the total phytoclasts and the total protistoclasts (this is the

Table 10.8. *Results of the first iteration of the discriminant function procedure using three complexes*

Physiographic complex	Success ratio correct:incorrect	Aberrant samples	
		Number	Complex
bay	31:9	49	barrier island
		50	barrier island
		51	reef crest
		52	barrier island
		53	barrier island
		54	barrier island
		62	reef crest
		63	reef crest
		88	barrier island
barrier island	34:7	13	reef crest
		35	reef crest
		40	reef crest
		48	bay
		64	reef crest
		69	bay
		70	bay
reef crest	16:4	5	barrier island
		6	barrier island
		14	barrier island
		22	barrier island

Table 10.9. *Discriminant function analysis: classification summary for calibration data, number of observations and percent classification into complexes*

Key to complex code: 1 = bay complex; 2 = barrier island complex; 3 = reef tract complex

From complex	1	2	3	Total
1	31	0	0	31
	100	0	0	100
2	0	34	0	34
	0	100	0	100
3	0	0	16	16
	0	0	100	100
Total observations	31	34	16	81
Total %	38	42	20	100

predicted result). All but three samples plot on a straight line. Similarly, plots of the three major physiographic complexes show the environments containing the samples that did not fit the general trend. All environments show the same relationship for the phytoclast versus protistoclast plot, with the barrier island physiographic complex

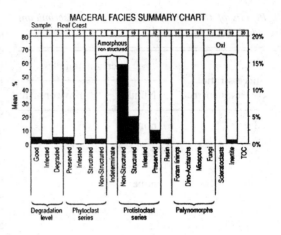

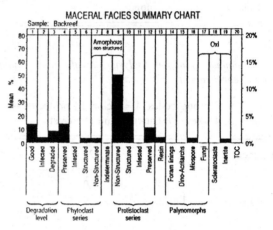

Figure 10.7. Maceral facies summary bar charts of means for reef physiographic complex.

showing three samples that did not plot on the line. Plots were then made for the ten environments. All environments show the same basic trend. Within the barrier island physiographic complex the channel environment shows three samples (31, 32, and 34; see Fig. 10.6) that do not plot on the line. The reason for the abnormality in these samples is the very high amount of amorphous indeterminate material they contain. Upon further inspection of the phytoclast/protistoclast plot it is concluded that although these samples plot under the line on which the other samples plot, they show the same trend as that line. Other samples within the channel environment (26, 27 and 28) that do not have the same high amorphous indeterminate input plot on the predicted line.

Total protistoclasts were plotted against the amorphous indeterminate material using all samples in the ten-environment purified data base. Most of the samples contain no amorphous indeterminate material and plot along the vertical axis. The samples that contain amorphous indeterminate material are all from the barrier island physiographic complex except for samples

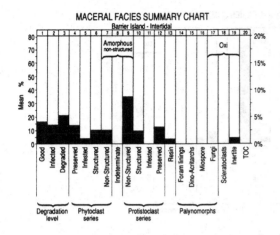

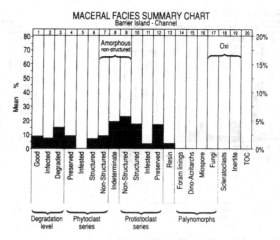

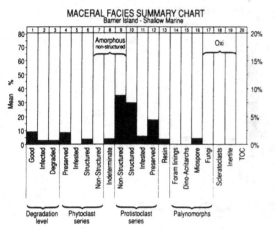

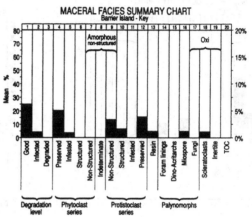

Figure 10.8. Maceral facies summary bar charts of means for barrier island physiographic complex.

2 (reef tract) and 60 (exposed bay). These have very small amounts of amorphous indeterminate material. A linear negative relationship is shown by the barrier island samples. The scatter plots of samples by environment show that the samples exhibiting this trend occur in the shallow marine, channel, intertidal, and key environments. The samples are 39 from the shallow marine, 31, 32, and 34 from the channel, 70 and 65 from the key, and 66 from the intertidal.

The final scatter plot showed the total phytoclasts versus amorphous indeterminate material using all samples in the ten-environment data base. As mentioned earlier, most samples contain no amorphous indeterminate material and plot along the vertical axis. The relationship is a slightly positive linear one.

Maceral distribution maps

Maps of the mean percent of a particular maceral in each environment sampled show the geographic distribution of the macerals. Table 10.10 summarizes the results of these distributions, and Figures 10.10–10.12 illustrate the distribution of the major kinds of macerals.

Interpretation

Reef tract physiographic complex

Sediments from the reef tract physiographic complex showed a high input (mean = 62%) of amorphous non-structured protistoclasts. Amorphous structured protistoclasts also are abundant (mean = 19%). Other maceral types which occurred with a frequency of at least 1% were: poorly preserved protistoclasts (6%), well-preserved phytoclasts (5%), well-preserved protistoclasts (2%), poorly preserved phytoclasts (1%), and amorphous non-structured phytoclasts (1%). This maceral assemblage reflects the biocoenosis of the reef tract, the spatial separation of the depositional environment from a source of terrestrial material, and the high amount of bio-degradational activity in the area. The open circulation of the reef tract maintains normal marine salinity in the environment and supplies nutrients to the area (Dole & Chambers, 1918). The varied and abundant population of organisms in this area is a direct result of these conditions. The primary contributors to the maceral assemblage are probably foraminifera, Codiaceae (green algae), and Corallinaceae (red algae), as seen by the abundance of recognizable foraminiferal linings and amorphous non-structured material of algal appearance.

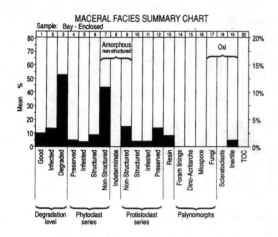

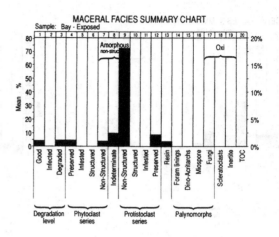

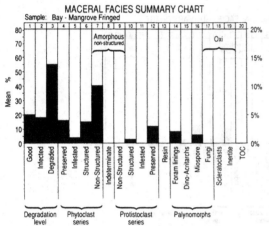

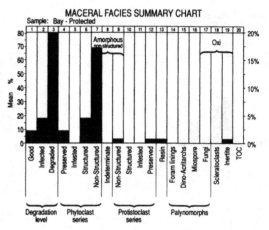

Figure 10.9. Maceral facies summary bar charts of means for bay physiographic complex.

The lack of higher marine plants, or a source of allochthonous terrestrial plant material, is reflected in the lack of phytoclast maceral material found in the sediments. A high degree of degradation activity is suggested, because 80% of the maceral material is either amorphous structured or amorphous non-structured. Such fungal and bacterial activity is expected in a well-oxygenated physiographic complex such as the reef crest.

Within the reef tract physiographic complex are the backreef and reef-crest environments. The maceral assemblages of these two environments show a high degree of similarity. Amorphous structured and amorphous non-structured protistoclasts dominate. The backreef sediments contain more total phytoclasts (15%) than the reef-crest sediments (7%). Of the phytoclasts found in the backreef, 60% are well-preserved, while 34% of the phytoclasts in the reef-crest sediments are well-preserved. This similarity in maceral assemblages between the backreef and reef-crest environments is partly a result of the similar physical conditions present in the two areas. The normal marine salinity in these environments supports similar biocoenoses, at least on the microorganism level. The dynamic community of the reef crest is significantly different from that of the backreef. The thanatocoenosis of the backreef must receive material from the reef crest, which was transported by wave action, tidal movement, and wind generated currents. The main source of nutrients for the reef is offshore of the reef crest. However, most of the waste from the eroded and decaying reef is removed from the onshore side and deposited in the backreef. Maceral analysis shows that much of this material is composed of protistoclasts. While the protistoclast components of the backreef and reef crest environment assemblages are nearly identical, the phytoclast material shows a clear difference. The backreef contains more total phytoclasts and a higher percentage of well-preserved phytoclasts. This difference can be attributed to the nearness of a source of terrestrial plant material to the backreef. Through tidal action plant material is carried from the barrier island formed by the Florida Keys to the backreef. Some of this material is carried as far as the reef crest. Having been exposed to subaqueous conditions and degradation longer than material in the backreef, it would show a higher degree of degradation than the material deposited in the backreef. This would explain the fewer well-preserved phytoclasts found in the reef crest assemblage.

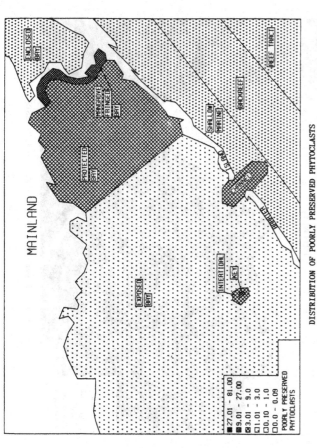

DISTRIBUTION OF POORLY PRESERVED PHYTOCLASTS

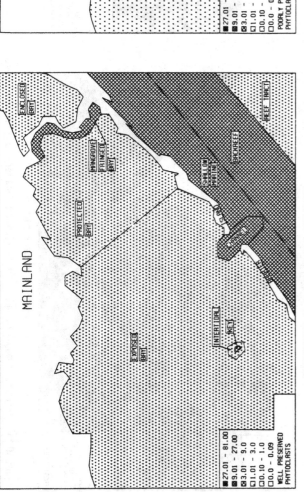

DISTRIBUTION OF WELL PRESERVED PHYTOCLASTS

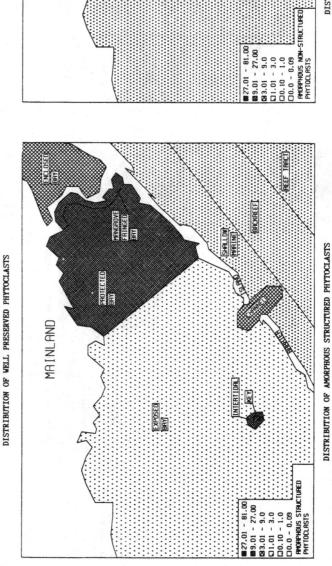

DISTRIBUTION OF AMORPHOUS NON-STRUCTURED PHYTOCLASTS

DISTRIBUTION OF AMORPHOUS STRUCTURED PHYTOCLASTS

Figure 10.10. Distribution maps of phytoclast macerals. Numbers refer to percentage counts.

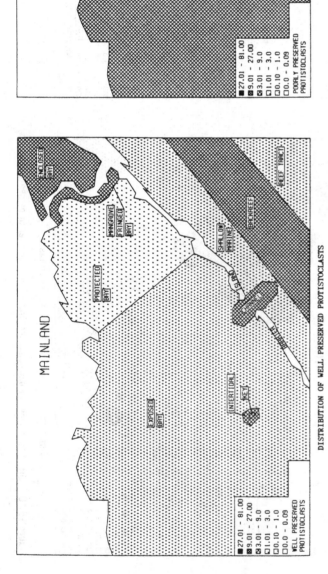

DISTRIBUTION OF WELL PRESERVED PROTISTOCLASTS

DISTRIBUTION OF POORLY PRESERVED PROTISTOCLASTS

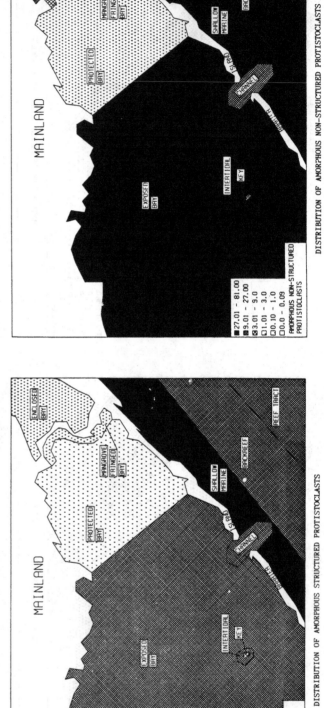

DISTRIBUTION OF AMORPHOUS STRUCTURED PROTISTOCLASTS

DISTRIBUTION OF AMORPHOUS NON-STRUCTURED PROTISTOCLASTS

Figure 10.11. Distribution maps of protistoclast macerals. Numbers refer to percentage counts.

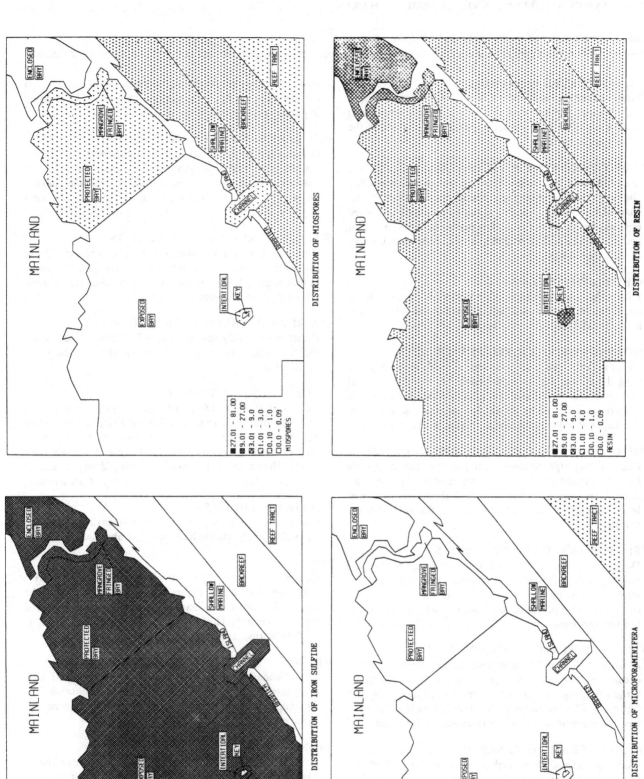

Figure 10.12. Distribution maps of iron sulfide, miospores, foraminiferal linings, and resin. Numbers refer to percentage counts.

Table 10.10. *Summary of results of the maceral distribution maps*

WELL-PRESERVED PHYTOCLASTS

Low in abundance in all environments. The mangrove fringed bay, channel, shallow marine, and reef crest environments show a slightly higher frequency of occurrence than do the other environments.

POORLY PRESERVED PHYTOCLASTS

Occur in fairly high numbers in the mangrove fringed bay and key environments. Their abundance decreases in the channel, intertidal and protected bay environments, followed by lower frequency in the shallow marine backreef, reef crest and enclosed bay. The lowest frequency of occurrence is found in the exposed bay.

INFESTED PHYTOCLASTS

Extremely low frequency of occurrence. Slightly more abundant in the enclosed bay, mangrove fringed bay, key and intertidal environments than in other areas.

AMORPHOUS STRUCTURED PHYTOCLASTS

Occur in fairly high numbers in the protected bay, mangrove fringed bay, key and intertidal environments. The channel and enclosed bay have the next highest frequency of occurrence, followed by the shallow marine, and reef crest. The exposed bay has the lowest frequency of occurrence.

AMORPHOUS NON-STRUCTURED PHYTOCLASTS

Occur in very high numbers in the protected bay, mangrove fringed bay and enclosed bay environments. The key and intertidal environments have fairly high abundances, followed by the backreef, reef crest, exposed bay and finally the shallow marine.

WELL-PRESERVED PROTISTOCLASTS

Occur in very low numbers in all environments. The enclosed bay, mangrove fringed bay, channel, backreef, key and intertidal environments have the highest frequency of occurrence. The protected bay has the lowest abundance of this maceral type.

POORLY PRESERVED PROTISTOCLASTS

A fairly even distribution among the environments. They occur in moderate numbers in the enclosed bay, mangrove fringed bay, exposed bay, key, channel, shallow marine, and reef crest. They are slightly less abundant in the protected bay and more abundant in the intertidal environment.

INFESTED PROTISTOCLASTS

Occur in extremely low numbers in the reef crest, backreef, key, intertidal, protected bay and mangrove fringed bay environments. The frequency of occurrence is slightly higher in the channel and enclosed bay environments and slightly lower in the protected bay.

AMORPHOUS PROTISTOCLASTS

Occur in very large numbers in the shallow marine environment. The reef crest, backreef, channel, exposed bay and intertidal environments contain moderately high numbers. The key has a low number, followed by the enclosed bay and mangrove fringed bay and finally the protected bay.

AMORPHOUS NON-STRUCTURED PROTISTOCLASTS

Occur in very high numbers in the reef crest, backreef, shallow marine, intertidal and exposed bay environments. The channel and enclosed bay contain a moderately high number, followed by the mangrove fringed bay and then the protected bay.

WELL-PRESERVED SCLERATOCLASTS

Occur in extremely low numbers. The enclosed bay, mangrove fringed bay, exposed bay, key and reef crest environments have a slightly higher frequency of occurrence than do the remaining environments.

POORLY PRESERVED SCLERATOCLASTS

Occur in extremely low numbers. The exposed bay has a slightly higher frequency of occurrence than do the other environments.

AMORPHOUS INDETERMINATE

Occur with the highest frequency in the channel environment, while the shallow marine environment has the next highest frequency of occurrence. This is followed by the reef crest, shallow marine, key, intertidal and exposed bay environments. The enclosed bay, mangrove fringed bay, protected bay and backreef have the lowest frequency of occurrence.

MICROFORAMINIFERA

Occur most commonly in the reef crest environment. The remaining environments have very low abundance.

INERTITE

Occurs with the highest frequency in the reef crest environment. The next highest frequencies are found in the intertidal, enclosed bay, mangrove fringed bay and protected bay environments. The lowest frequencies are seen in the key, shallow marine, and exposed bay environments.

RESIN

The abundance of resin is highest in the enclosed bay, mangrove fringed bay, key and intertidal environments. The remaining environments all have the same frequency of occurrence.

IRON SULFIDE

Present in the enclosed bay, protected bay, mangrove fringed bay, exposed bay, intertidal, key and channel environments.

MIOSPORES

Occur most commonly in the shallow marine, and key environments while the reef crest, channel, intertidal and protected bay environments have the next highest frequency of occurrence. The enclosed bay and exposed bay have the lowest frequency of occurrence.

Bay physiographic complex

The bay sediments showed a trend towards phytoclast macerals, the most frequently occurring type being amorphous non-structured phytoclasts (49%). Other maceral types with a frequency of occurrence of at least 1% were: amorphous non-structured protistoclasts (14%), amorphous structured phytoclasts (12%), poorly preserved phytoclasts (5%), poorly preserved protistoclasts (5%), well-preserved protistoclasts (4%), well-preserved phytoclasts (3%), amorphous structured protistoclasts (2%), infested phytoclasts (2%), and inertite (2%).

The bay physiographic complex includes the protected bay, enclosed bay, mangrove fringed bay, and exposed bay environments. It can be seen that phytoclasts dominate in the protected, enclosed, and mangrove fringed bays. On the other hand, protistoclasts dominate in the exposed bay. Total protistoclasts compose 91% of the maceral assemblage in the exposed bay sediments. In contrast, the protected bay assemblage contains 93% total phytoclasts. The enclosed bay assemblage contains 59% total phytoclasts and 35% total protistoclasts. The mangrove fringed bay assemblage contains 73% total phytoclasts and 20% total protistoclasts. This difference in maceral assemblages is the result of two major factors. The enclosed, mangrove fringed, and protected bays are near to sources of terrestrial plant material while the exposed bay is not. The high percentage of phytoclasts in the protected, enclosed, and mangrove fringed bays and the lack of phytoclasts in the exposed bay are the result of this presence or absence of a proximal source of higher plant material. The enclosed and mangrove fringed bays fall in the 'Runoff Zone' or area of lowest salinity. The protected bay falls in the 'Interior Zone' or area of highest salinity. The exposed bay falls in the 'Tidal Exchange Zone' or area with the most normal marine salinity (McCallum & Stockman, 1964). The normal marine salinity of the exposed bay allows for an abundant and varied biocoenosis. The abnormally low and high salinity of the enclosed, mangrove fringed, and protected bays restricts the type of organisms which can exist there. The low percentage of protistoclasts in the protected bay is attributable to the very high salinity encountered in this area. These high salinities exclude normal marine organisms. Protistoclasts in the enclosed and mangrove fringed bays may be derived from freshwater algae. These algae could exist in the low salinity of this area affected by the surface water runoff from the mainland. Note that the phytoclast macerals of the enclosed and protected bays are predominantly amorphous structured and amorphous non-structured. Well-preserved and poorly preserved phytoclasts are significant in the mangrove fringed bay. This is again attributable to the nearness of a source of terrestrial material. The mangrove fringed bay samples were collected close to the shore where the mangroves extend out into the water. It is possible to collect material in this area that is relatively 'fresh,' showing little degradation. On the other hand, the protected and enclosed bay samples are from farther out in the bays. In order for phytoclasts to reach these environments of deposition they would be exposed longer to subaqueous degradation.

Barrier island physiographic complex

The maceral assemblages found in the sediments of the barrier island physiographic complex show a high degree of variability. This variability reflects the diversity of conditions encountered in the four environments of the barrier island physiographic complex. Almost every type of maceral occurs in the sediments of the barrier island physiographic complex. Phytoclasts and protistoclasts are found at all levels of degradation in significant percentages. Total protistoclasts make up 58% of the assemblage, while total phytoclasts make up 30%.

The environments of the barrier island physiographic complex represent a variety of environmental conditions. They range from open marine (shallow marine) to terrestrial (key) to intermediate type environments (intertidal and channel). Likewise, the sediments from these environments contain a variety of assemblages. Total protistoclasts make up 85% of the maceral assemblage in the shallow marine sub-environment. Total phytoclasts make up 10% of the assemblage. This assemblage reflects normal marine conditions in this environment, as well as input from an adjacent source of terrestrial plant material. The key assemblage contains 61% total phytoclasts and 31% total protistoclasts. This environment is essentially terrestrial. Two samples were collected from small 'lakes' on the Florida Keys. Protistoclasts dominate these samples. One of these samples was removed during the discriminant function analysis. The remaining sample is probably typical of the samples containing protistoclast material in the key assemblage, although other samples showed minor percentages of protistoclasts. The intertidal environment samples were collected in that zone around several Florida Keys. The maceral assemblage in these samples represents a mixture of terrestrial input and open marine input. The assemblage from the intertidal environment contains 30% total phytoclasts and 65% total protistoclasts.

The channel environment of the barrier island physiographic complex proved to contain a very distinct assemblage of macerals. As expected it contained a mixture of protistoclasts and phytoclasts. Total phytoclasts compose 24% of the assemblage while total protistoclasts were 56% of the assemblage. Amorphous indeterminate material, rare in other environments, made up 19% of the channel assemblage. This material is so highly degraded that its origin (phytoclast or protistoclast) cannot be determined. Hart (1986) notes that much of the amorphous indeterminate material may be a humic precipitate and notes a common occurrence when waters of different salinity meet. A precipitate may occur when fresh water rich in humic substances is suddenly mixed with water of a different salinity (or pH or Eh). This is precisely what commonly happens in the channel sub-environment. It not only connects the bay with the ocean but is a major site of freshwater passage into the open water area.

Conclusions

The major objective of this study was to define the maceral characteristics of the carbonate environments of the Florida Keys. Sediments from the reef contain high percentages of amorphous structured and amorphous non-structured protistoclasts. This reflects the biocoenosis of the reef, the spatial separation of the depositional environment from any source of terrestrial material, and the high percentage of biodegradational activity in the area. The lack of higher marine plants or a source of allochthonous terrestrial plant material is reflected in the lack of phytoclasts found in the sediments. The high degradation is a function of the well-oxygenated sediments of the reef tract which promote high rates of fungal and bacterial activity. The backreef and reef crest environments show a high degree of similarity, because they both lie within the reef tract. This similarity is attributable to the backreef environment receiving material transported from the reef crest. The higher percentage of phytoclasts, especially well-preserved phytoclasts, in the backreef is the result of the proximity of the barrier island. This is a source of terrestrial plants.

Phytoclasts dominate the bay samples, and there is a significant percentage (27%) of protistoclasts. When examined separately, phytoclasts dominate the protected, enclosed, and mangrove fringed bays, but protistoclasts dominate the exposed bay. This difference in maceral assemblages is the result of two major factors. First, the enclosed, mangrove fringed, and protected bays are near to sources of terrestrial plant material, while the exposed bay is not. Second, the enclosed and mangrove fringed bays fall in the 'Runoff Zone,' or area of lowest salinity. The protected bay falls in the 'Interior Zone,' or area of highest salinity. The exposed bay falls in the 'Tidal Exchange Zone,' or area with more normal marine salinity (McCallum & Stockman, 1964). This normal marine salinity allows an abundant and varied biocoenosis to develop. On the other hand, the abnormally low and high salinities of the enclosed, mangrove fringed, and protected bays restrict the type of organisms which can exist there. The very high salinity of the protected bay exclude most organisms that would produce protistoclast macerals. The low salinities of the enclosed and mangrove fringed bays limit protistoclast producers to the freshwater algae.

The maceral assemblages found in the sediments of the barrier island physiographic complex are highly variable. Phytoclasts and protistoclasts occur at all levels of degradation and in significant percentages. This variability is a reflection of the diversity of the four environments of the barrier island. Protistoclasts dominate the shallow marine environment and the assemblage reflects the normal marine conditions. This environment also receives some terrestrial input. The key environment contains many phytoclasts as would be expected from this essentially terrestrial environment. There is also a significant percentage of protistoclast material in this assemblage which probably comes from freshwater algae in 'lakes' on the Florida Keys. The intertidal environment contains a mixture of protistoclasts and phytoclasts. This assemblage reflects the marine conditions of the environment and the nearby source of terrestrial plant material. The channel environment contains a mixture of phytoclasts and protistoclasts. There is a significant percentage of amorphous indeterminate material in this assemblage. This may be a humic precipitate resulting from the mixing of humin-rich fresh water with more marine waters.

In conclusion, we reaffirm our belief that maceral analysis is a tool that can be used in paleoenvironmental determinations. With increased use of advanced instrumental analytical techniques, such as infra-red microscopy and laser pyrolysis, more characteristics of macerals can be measured. This, coupled with progress in our knowledge of the occurrence and distribution of macerals in different depositional settings, suggests that organic petrology will play an increasingly important role in interpreting depositional architecture.

Acknowledgments

John Wrenn and Clyde Moore (Louisiana State University) kindly spent time critically reading this manuscript. I would also like to thank Alfred Traverse and the anonymous reviewers whose efforts greatly improved the readability of this chapter. We thank them for their time, effort, and critical comments. John Verlenden, editor with the Louisiana Geological Survey, kindly made corrections to the manuscript prior to submission for publication. In addition, we would like to thank SOHIO Inc. for providing financial support that allowed completion of the laboratory research at Louisiana State University. Hartax International Inc. of Baton Rouge, LA, supplied the complete set of field samples and field maps for this study. Finally, we wish to thank James R. Hart who was the field geologist and diver who collected most of the samples.

References

Cooley, W. W. & Lohnes, P. R. (1971). *Multivariate Data Analysis.* New York: John Wiley & Sons.

Craighead, F. C. (1971). *The Trees of South Florida*. Coral Gables, FL: University of Miami Press.

Darby, J. D. (1987). Organic Petrology of Carbonate Systems; Florida Bay Study. M.S. thesis. Louisiana State University, Baton Rouge.

Darrell, J. H. & Hart, G. F. (1970). Environmental determinations using absolute miospore frequency, Mississippi River delta. *Geological Society of America Bulletin*, **81** (8), 2513–8.

Dole, R. B. & Chambers, A. A. (1918). Salinity of Ocean-Water at Fowey Rocks, Florida. Washington, D.C., Carnegie Institute, Publ. 517, *Papers from Tortugas Laboratory*, **32**, 299–315.

Ginsburg, R. N. (1954). Lithification and alteration processes in South Florida carbonate deposits. Ph.D. thesis, University of Chicago.

Ginsburg, R. N. (1956). Environmental relationships of grain size and constituent particles in some South Florida carbonate sediments. *Bulletin of the American Association of Petroleum Geologists*, **40** (10), 2384–2427.

Hart, G. F. (1971). *Statistical evaluation of palynomorph distribution in the Atchafalaya Swamp*. Consultant Report. Lafayette, LA: Texaco Inc.

Hart, G. F. (1975). *Maceral analysis and Total Organic Carbon determination: Santa Maria Basin Project*. Consultant Report. Baton Rouge, LA: Carbon Systems Inc.

Hart, G. F. (1976). *Maceral analysis and Total Organic Carbon determination: Kodiak Shelf Project*. Consultant Report. Baton Rouge, LA: Carbon Systems Inc.

Hart, G. F. (1979a). *Maceral analysis: its use in petroleum exploration*. Methods Paper 2. Baton Rouge, LA: Hartax International Inc.

Hart, G. F. (1979b). Maceral analysis and its application to petroleum exploration. (abstract) In *Petroleum potential of island arcs, small ocean basins, submerged margins and related areas, Symposium UNESCO - Suva, Fiji*, Abstracts **1**, 32–33.

Hart, G. F. (1986). Origin and classification of organic matter in clastic systems. *Palynology*, **10**, 1–23.

Hart, G. F., Ferrell, R. E., Lowe, D. R. & LeNoir, E. A. (1989). Shelf sandstones of the Robulus L Zone, offshore Louisiana. *Gulf Coast Section, Society of Economic Paleontologists and Mineralogists Foundation Seventh Annual Research Conference Proceedings*, 117–41.

LeNoir, E. A. & Hart, G. F. (1986). Burdigalian (Early Miocene) dinocysts from offshore Louisiana. *American Association of Stratigraphic Palynologists, Contribution Series*, **17**, 59–81.

LeNoir, E. A. & Hart, G. F. (1988). Palynofacies of some Miocene sands from the Gulf of Mexico, offshore Louisiana, U.S.A. *Palynology*, **12**, 151–65.

Lidz, B. H. & Rose, P. R. (1989). Diagnostic foraminiferal assemblages of Florida Bay and adjacent shallow waters: a comparison. *Bulletin of Marine Science*, **44** (1), 399–418.

Marmer, H. A. (1954). Tides and Sea Level in the Gulf of Mexico. In *Gulf of Mexico: Its Origin, Waters, and Marine Life. Fishery Bulletin of the Fish and Wildlife Service*, **55**, 115–6.

McCallum, J. S. & Stockman, K. W. (1964). Water circulation: Florida Bay. In *South Florida carbonate sediments, Guidebook Field Trip No. 1*, compiler: R. N. Ginsburg. Geological Society of America, Annual Meeting 1964, University of Miami, FL, 11–13.

Multer, H. G. (1977). *Field Guide to some Carbonate Rock Environments: Florida Keys and Western Bahamas*. Kendall/Hunt, New Edition.

Owens, H. (1960). (abstract) Florida-Bahamas platform. *Bulletin of the American Association of Petroleum Geologists*, **44**, 1254.

Thorne, R. F. (1954). Flowering plants of the waters and shores of the Gulf of Mexico. In *Gulf of Mexico, its Origin, Waters and Marine Life. Fishery Bulletin of the Fish and Wildlife Service*, **55**, 193–202.

Traverse, A. & Ginsburg, R. N. (1966). Palynology of the surface sediments of Great Bahama Bank, as related to water movement and sedimentation. *Marine Geology*, **4**, 417–59.

Tukey, J. W. (1962). The future of data analysis. *Annals of Mathematical Statistics*, **33**, 1–67.

Whittaker, R. H. (1969). New concepts of kingdoms of organisms. *Science*, **163**, 150–60.

Wrenn, J. H. & Beckman, S. W. (1981). Maceral and Total Organic Carbon analysis of DVDP drill core 11. In *Dry Valley Drilling Project*, ed. L. D. McGinnis. *American Geophysical Union, Antarctic Research Series*, **33**, 391–402.

Wrenn, J. H. & Beckman, S. W. (1982). Maceral, Total Organic Carbon and palynological analyses of Ross Ice Shelf Project, Site J9 cores. *Science*, **216**, 187–9.

11 An approach to a standard terminology for palynodebris

MICHAEL C. BOULTER

Introduction

Most palynological investigations start with the maceration of sediments for the removal of their inorganic content. Not only does this separate the palynomorphs but it often leaves a residue of other organic remains which can be studied microscopically as well as chemically. There is great variation in the terms applied to these non-palynomorph remains since most of the people studying them are geologists lacking knowledge of plant anatomy and morphology. Confusing, often misleading and even wrong terms have been given to some broken plant parts. It is the purpose of this chapter to help establish a standard terminology for the microscopist.

Three different approaches are given to the study of these fragments according to the nature of the problems to be investigated and according to the background of the authors:

(1) Coal petrologists (see Brooks, 1981; Durand, 1980) still use versions of Stopes' (1935) terminology for the organic components of macerals. Teichmüller and Teichmüller (1968) related these studies to a morphological approach.

(2) Geochemical studies of kerogen (Crum-Brown, 1912) have recently been summarized by Tissot *et al.* (1974) and Burgess (1974).

(3) Paleobotanical (including paleopalynological) studies of palynodebris are the central scope of this chapter, and the visual microscopical observations so derived form the basis of these comments on their classification. It attempts to encourage a standard terminology for types of visually observed palynodebris compatible with all facies, and as far as possible to relate these types to a functional sedimentological setting for microscopical use.

'Palynodebris' is the term proposed by Manum (1976) to describe broken plant parts and other organic remains found dispersed in marine and terrestrial sediments. Other terms for the visually defined fossils have been used and often give an ambiguous or even misleading meaning. Examples of these terms, arguably less satisfactory, are 'sedimentary organic matter' (SOM), 'organic debris' (Piasecki, 1980), 'dispersed organic matter' (Cope, 1981), 'kerogen' (Brooks, 1981) and 'palynofacies' (Caratini *et al.*, 1975; Batten, 1981). Because many apparently conforming types of palynodebris are found in deposits of differing age and location there is a need for a standard terminology to facilitate comparisons and so to help paleoecological interpretation.

For strictly visual observations (by light microscopy) there are many ambiguities and uncertainties built into previous systems of naming and classifying palynodebris. For instance, Masran and Pocock (1981) nominated eight major classes including charcoal, biodegraded materials of terrestrial origin, and biodegraded materials from aqueous sources. Because of its precise meaning, charcoal is identifiable by scanning electron microscopy and geochemical testing, beyond the scope of the simple visual technique. Unfortunately some authors use the term 'charcoal' (carbonized woody plant tissue) without having obtained scientific evidence that that is what they really have. Similarly, the terms 'terrestrial' and 'aqueous' refer to the source of the palynodebris and can easily lead to tautological interpretations, unless there is controlled evidence preferably based on interdisciplinary investigations.

Other terms have been used indiscriminately, without much attention being given to how the proof of identity was made. Resin, kerogen, leaf cuticle and sapropel are four examples of terms with very precise meanings for substances which can be detected in the laboratory if suitable equipment is available. For the average visual project such testing is too time consuming, so evidence for the presence of these materials (except leaf cuticle) cannot be obtained. Without proper evidence, these precise terms should not be used. There is a need for a system of designating categories of palynodebris with terms that match the evidence and which allow other scientists to make their own interpretations.

Pocock *et al.* (1988) have offered the most recent classification of palynodebris determined by light microscopy. It is largely compatible with that proposed here. It is a modification of that of Masran and Pocock (1981) and includes nine types: structured terrestrial, spores and pollen, biodegraded terrestrial, inertinite, fungi, resins, amorphous, biodegraded aqueous and structured aqueous.

The major types of palynodebris

Recent work by Boulter and Riddick (1986) has taken these arguments into account to establish a scientifically justifiable terminology for the major types of palynodebris in a single sedimentary sequence, the Paleocene of the Forties Field in the North Sea. These are low- and high-energy sediments which may have resulted from deposition in shallow waters, nearshore conditions, submarine fan deposits, and high energy sandstone sediments. The origin of the deposits remains controversial. They may have been formed by one or by a combination of two or all three of the following processes: (1) one reminiscent of rafting from very shallow water deposits; (2) disturbance during storms in a shallow sea (McCave, 1985); (3) turbidity-like flow calamities (Carman & Young, 1981).

These deposits contain a wide range of different kinds of palynodebris which are very similar to forms found elsewhere. The Forties Field material is ideal to use as the basis for a universal palynodebris terminology, because the deposits cover a range of sediments from low-energy freshwater lake deposits to very high-energy sandstone formations. Although most kinds of broken plant parts are present, fruits and seeds are absent. Diatoms and ostracods are also not present, but the Forties Field material is nonetheless satisfactory as a basis for a universal terminology.

Boulter and Riddick (1986) began by defining as many kinds of palynodebris as possible, based on principles outlined above. This was to enable statistical analysis of the data at a later stage, which together with careful interpretation of the geological and biological factors at work, would allow split terminology to be grouped or 'lumped.' In practice, this 'lumping' occurred over two phases of the analysis. The original data comprised semi-quantitative counts of 36 kinds of palynodebris and was based on about 600 samples from four cores. Some of the original 36 kinds of palynodebris were based on dubiously valid features, but control experiments and other precautions were undertaken whenever possible to justify the initial terminology. For example, chemical treatment was kept to a minimum, using only hydrofluoric acid with no oxidizing procedures (to validate the color differences), tap water pressure during sieving was standardized (to try to make mechanical breakdown of the particles equivalent). Also, comparative studies from other deposits (Eocene Hampshire Basin, Tertiary DSDP Leg 38, etc.) were undertaken, and they showed the same kinds of debris.

Statistical analysis as a basis for terminology

The statistical analysis (using R. N. L. B. Hubbard's programs) of the Forties Field data has been described by Boulter and Riddick (1986). It led to a final recognition of only a handful of different useful forms. These are the basis of the terminology presented here and summarized in Table 11.1. At each phase the semi-quantitative counts of the different kinds of palynodebris were processed by multivariate statistical analysis. These calculations produced dendrograms summarizing the results of cluster analysis and lists of associated kinds of palynodebris resulting from principal components analysis. The results from these two different procedures show the same groupings of the original kinds.

The first phase of the statistical analysis was based on semi-quantitative counts of all 36 original kinds of palynodebris from all of the nearly 600 samples. Full details are given by Boulter and Riddick (1986). The five semi-quantitative occurrences recognized were: (1) absent, (2) <1%, (3) 1–5%, (4) 5–25%, (5) >25%.

The results show that microforaminiferal linings, reworked spores, perforated phytoclasts and fern sporangia are of little or no value. The results also suggested lumping of other types, as shown in Table 11.1. It was important to compare these objective statistical results with our more general understanding of the biological and geological processes that were involved, and happily there were no discrepancies.

The second phase of the statistical analysis was based on the same semi-quantitative data but in the now revised first phase lumped form. So, instead of the 36 variables in the first phase, here there were only 17 kinds of palynodebris. The results (see Boulter & Riddick, 1986), based on the statistical analysis of data from the same samples, show that comminuted debris, unstructured debris, degraded debris, parenchyma, brown wood, black debris and algae have very little value to help interpret the palynofacies. These types of palynodebris are shown by the statistics not to be associated in groups (principal components) and are not restricted to particular parts of the sedimentary sequences. We concluded that they have little or no value in the interpretation of the palynofacies.

More positively, other types of palynodebris show very strong association, which, furthermore, show a logical sedimentological pattern when related to each sample in the sections studied from the Forties Field Paleocene.

Table 11.1.

On the left is a list of the original 36 kinds of palynodebris used to describe palynodebris data in PHASE ONE of Boulter and Riddick's (1986) project. After objective statistical analysis of semi-quantitative data from this study, based on 600 samples with appropriate subjective assessment, a PHASE two analysis further reduced the paleoecologically significant kinds of palynodebris. These final terms are shown on the right hand side of the table. Amorphous matter and palynowafers are shown in this chapter to have the most paleoecological significance

First phase	Second phase	Final terminology	
Microforaminiferal linings . . . .<			
Comminuted debris . . . .<			
DINOFLAGELLATES			1
Jurassic miospores . . . . }JURASSIC MIOSPORES			2
?Jurassic miospores . . . . }			
other reworked spores. . . . .< tricolp(or)ates			
monocolpates . . . . }			
triporates . . . . }			
multiporates . . . . }TERTIARY POLLEN			3
Normapolles . . . . }			
inaperturates . . . . }			
bisaccates . . . . }BISACCATES & SPORES			4
spores . . . . }			
AMORPHOUS MATTER . . . . }			
associated specks }specks . . . . }	AMORPHOUS		5
free specks . . . . }	MATTER		
unstructured debris . . . .<			
degraded debris . . . .<			
perforated phytoclasts . . . .<			
yellow parenchyma . . . . }parenchyma . . . .<			
brown parenchyma . . . . }			
degraded bundles . . . . }			
yellow cuticle . . . . }leaf cuticle . . . . }			
brown cuticle . . . . }			
yellow rays . . . . }	PALYNOWAFERS		6
brown rays . . . . }well-preserved wood . . . . }			
yellow tracheids . . . . }			
brown tracheids . . . . }			
brown wood . . . .<			
black debris . . . .<			
ferns . . . .<			
fungi . . . .<			
unicellular algae . . . . }			
folded algal cells . . . . }algae . . . .<			
multicellular algae . . . . }			

< = other types which had no significance in the Forties Field material
} lumping of kinds

On this basis Boulter and Riddick chose the term 'palynowafers' to lump degraded bundles, yellow cuticle, brown cuticle, yellow rays, brown rays, yellow tracheids, brown tracheids and well-preserved wood.

The results show that amorphous matter and palynowafers (and the palynomorphs) have significant sedimentological importance and have value for interpreting marine palynofacies. These two kinds of palynodebris, and only these two kinds, can help distinguish, respectively, between low- and high-energy sedimentary conditions. The same types of palynodebris can also have relevance in other sedimentological regimes, such as nearshore and lake deposits.

Description of main types of palynodebris

The main types of palynodebris are briefly described below and illustrated in Figures 11.3–11.13. Detailed descriptions and interpretations are available in Boulter and Riddick (1986).

Amorphous matter (Figures 11.3–11.4)

This is grey to pale yellow to brown translucent structureless material varying considerably in size from a few to many hundreds of microns. The outline of the masses is usually indistinct but can range into a different type with a less granular surface and a distinct margin making the material sheet-like. Folds are often present in those forms which may have suffered less bacterial degradation during fossilization.

Forties Field amorphous matter is nearly always associated with black specks from 1 to 20 μm in diameter. Some of these are shown to be pyritic by examination under reflected light, though a lipid affinity for the others cannot be ruled out. In other localities, such as in the DSDP Leg 38 material, shallow marine sediments of the Hampshire basin and the freshwater sediments of Lough Neagh, these specks are absent. In all these localities amorphous matter is associated with low-energy sedimentation and the attendant anoxic processes of fossilization.

In the Forties Field Paleocene sediments amorphous matter is found most frequently in low-energy sediments. It is only rarely present in the sandstones.

Amorphous matter has been found and described by many authors from a wide range of deposits. It is found in archeological terrestrial deposits (R. N. L. B. Hubbard, personal communication), in low and high energy marine deposits (Staplin et al., 1973), and from sediments associated with anaerobic conditions (Masran & Pocock, 1981). Pocock et al. (1988) have given a comprehensive review of the origins and occurrence of amorphous matter.

Palynowafers (Figs 11.5–11.8)

Three main types of these are distinguished by Boulter and Riddick's analysis: degraded bundles (Fig. 11.5), leaf cuticle (Fig. 11.6) (Parry et al., 1981; Cross et al., 1966; Barnard et al., 1981; Tissot et al., 1974) and well-preserved wood (Figs 11.7, 11.8) (Cross et al., 1966). As sedimentological particles they all appear to behave in the same way as plate-like wafers. Palynowafers are comparable with some of Pocock et al.'s (1988) 'structured terrestrially sourced materials.'

The palynowafers from the Forties Field Paleocene sediments are usually up to one cell in thickness and range from about 50 μm to 500 μm in outline dimensions. Usually they show the cell outlines from leaf or wood tissue though this may not be preserved. In other localities, such as the very shallow marine water deposits

of the Hampshire Basin (southern England), they can be larger in outline dimensions and show very well preserved cellular detail. It is important to take great care with mechanical techniques when making preparations for this kind of palynodebris, as the large specimens are often brittle and can easily break.

DEGRADED BUNDLES (FIG. 11.5)

These are yellow or brown particles and show some degree of lineation suggesting the presence of fibrous structural elements such as occur in plant xylem tissue. Some cellular structure is occasionally present. They may have a transverse lineation as though they are chipped cross-sections.

LEAF CUTICLE (FIG. 11.6)

These yellow and brown sheets show epidermal cell outlines and occasionally stomata from the lower surface. Both angiosperm and gymnosperm leaf cuticles can be distinguished.

WELL-PRESERVED WOOD

The first phase analysis of the original data from the Forties Field treated this as four separate categories: brown rays (Fig. 11.7.1, 3, 5, 8, 9), brown tracheids (Fig. 11.7.2, 4, 6, 7, 10, 11), yellow rays (Fig. 11.8.1, 3, 7, 9) and yellow tracheids (Fig. 11.8.2, 4, 5, 6, 8, 10). The first statistical analysis showed no distinction between the color or the anatomy so they are lumped together as a single type, well-preserved wood.

In the Forties Field deposits palynowafers are associated with high energy sandstone deposits of great thickness. Boulter and Riddick (1986) argue that they were deposited here by the heavy sand grains falling on the flat wafer-shaped particles and pushing them down to the bottom. Such conditions are unusual, however, and cannot apply in many deposits where palynowafers are found.

Elsewhere, such as in the Hampshire Basin, palynowafers are part of a shallow marine facies, along with dinoflagellate cysts and often abundant pollen and spores (Boulter, 1987). They also occur in freshwater facies together with Azolla, Botryococcus, Scenedesmus, Pediastrum, abundant pollen and spores, foraminiferal linings, etc. Examples are in deposits from the Lough Neagh clays and the Bovey basin (Boulter 1987). The author's work on the palynodebris from DSDP Leg 38 Holes 343 and 338 (Fig. 11.1) also shows that palynowafers are abundant in terrigenous sediments and usually are absent in the pelagic oozes. In these situations there is a simpler explanation for the occurrence of the palynowafers. They have been deposited more or less autochthonously or have been transported only short distances to a non-reducing site of deposition.

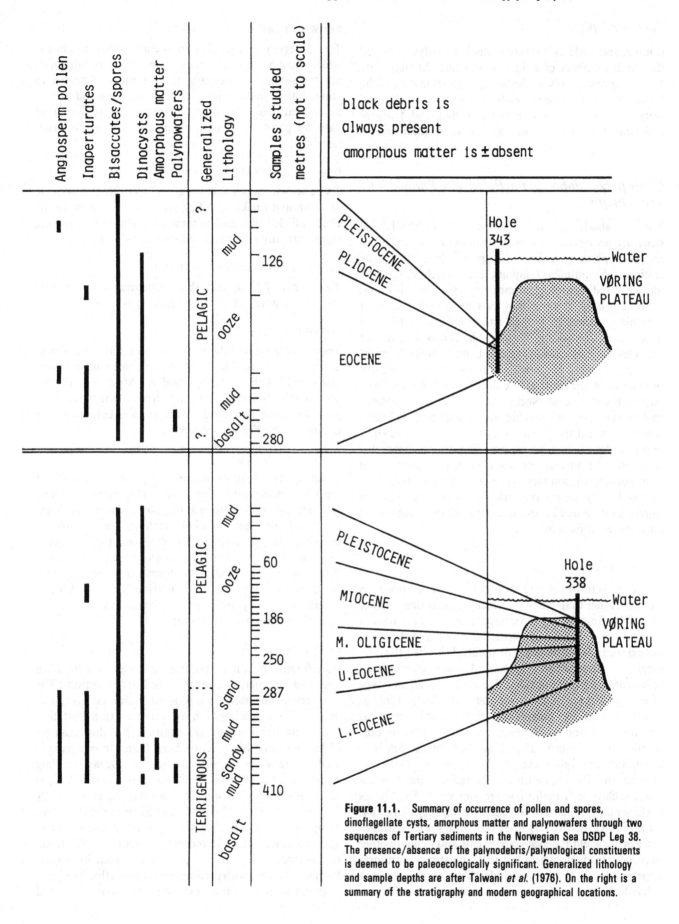

Figure 11.1. Summary of occurrence of pollen and spores, dinoflagellate cysts, amorphous matter and palynowafers through two sequences of Tertiary sediments in the Norwegian Sea DSDP Leg 38. The presence/absence of the palynodebris/palynological constituents is deemed to be paleoecologically significant. Generalized lithology and sample depths are after Talwani *et al.* (1976). On the right is a summary of the stratigraphy and modern geographical locations.

Palynomorphs

Boulter and Riddick's (1986) statistical analysis showed that high numbers of pollen, spores and dinoflagellate cysts are present only in the low-energy sediments of the Forties Field. This agrees with results from the classical work of Muller (1959), Cross *et al.* (1966), and Traverse and Ginsburg (1966), as well as with more recent work.

Other palynodebris in the Paleocene Forties Field assemblage

There are also many kinds of palynodebris with biological structure preserved, such as parts of insect exoskeletons, fungal hyphae, mycelia, and spores, fern sporangia, chitinous foraminiferal linings, etc., which have little universal sedimentological meaning as far as we know. Largely this is because they can be transported over variable distances from their site of growth and can be products of reworking. Usually, the larger and denser particles are not moved far from their source, but the rare occurrence of fern sporangia in turbidite-like sediments of the Forties Field, many kilometers from land in the Paleocene North Sea, means that exceptions to this rule may be possible. In Boulter and Riddick's (1986) statistical analysis it was confirmed that these kinds of palynodebris show no patterns that can be interpreted as a general principle. However, for review purposes, it is important to comment on some of these types, for although they may have little or no sedimentological significance in the Forties Field deposits, they might have importance elsewhere.

BLACK DEBRIS (FIG. 11.9.1, 4, 6)

This term is to be favored for purely visual palynodebris studies (rather than black wood, charcoal, fusinite, pyrite etc.). The fragments are opaque, angular, and have a sharp outline. There is no structure visible and they may have dark brown thin edges or points. It is often the only form of palynodebris associated with intrabasaltic sediments and metamorphic deposits. The former type of deposit occurs widely over the Brito-Arctic Igneous Province, and the palynology of these sediments has recently been studied (Boulter & Manum, 1989). In most of the 300 or more samples studied, black debris is abundant and often exclusively dominant. Dated at around the Paleocene/Eocene boundary, these intra-basaltic clays are roughly contemporary with the Forties Field sediments that are considered here, suggesting that all this black debris was from the same source. Both in the Forties Field Paleocene sediments and in many other sequences from elsewhere this is the most common form of palynodebris. Boulter and Riddick's work suggests that it has little other significance, at least in their study.

BROWN DEBRIS (FIG. 11.9.2, 3, 5, 7)

This category is intended to include other fragments which show no clear features of plant anatomy, but which are translucent or partially translucent and brown in color. Although the affinity with splinters of wood cannot be proven solely by light microscopy, a source from vascular plant xylem tissue is most likely, as for black debris.

DEGRADED DEBRIS (FIG. 11.10)

These brown particles are irregular in outline and variable in width and thickness. They are a common form in the Forties Field material, but there is no reason to associate them with any particular sedimentary situation.

UNSTRUCTURED DEBRIS (FIG. 11.11)

This is pale yellow, very thin, structureless material of irregular shape, but often elongate or strap-shaped.

FUNGI (FIG. 11.12.1–9, 11, 13)

These characteristically brown fungal remains consist of unicellular and multicellular hyphae, unicellular spores and multicellular fruiting bodies. Most or all are presumably chitinous. Microthyriaceous structures which grew on living and dead leaves are less common than hyphae in the Forties sediments.

ALGAE (FIG. 11.12.10)

Three types of algal remains were distinguished for the First Phase counts from the Forties material. These were unicellular algae such as *Crassosphaera*, multicellular algae colonies and synobia, such as the commonly occurring *Botryococcus* and *Pediastrum*, and algal cells (of unproven affinity to the algae) which are usually single. If they are not reworked, these forms could be important in distinguishing between marine and freshwater conditions, if they are known by other evidence not to have been transported very far.

FERNS (FIG. 11.12.12)

Rare fragments of fern sporangia annulae were found in high and low energy deposits in the Forties sections. The most complete specimen is illustrated and has no spores inside. These specimens have survived transportation with remarkably little fragmentation. Another example of fern material has recently been found in one sample studied from the marine section of Ocean Drilling Program Leg 104, Hole 643A, through the Eocene sediments of the Norwegian Sea (Manum *et al.*, 1989). Sample 643A 60/2 32 was deposited in the early stages of opening of the northernmost North Atlantic Ocean and contains very numerous specimens of *Azolla* megaspores and glochidia, presumably from freshwater basins associated with preliminary stages of the opening.

Perhaps both these examples of well preserved

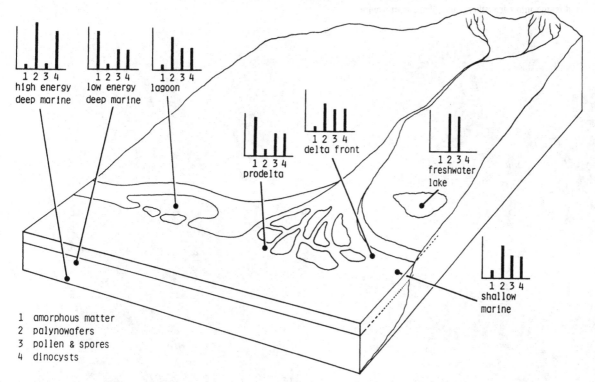

Figure 11.2. Block diagram of paleoenvironmental interpretations for palynodebris occurrences based on the evidence considered in this chapter. For each sedimentary environment a histogram shows a generalized proportional occurrence for the palynodebris/palynological constituents deemed to be paleoecologically significant: (1) amorphous matter, (2) palynowafers, (3) pollen and spores, (4) dinoflagellate cysts. The block drawing is similar to that presented by Pocock *et al.*, 1988, and the trends indicated conform with the results of those authors.

nonmarine plants in marine sediments are locally reworked from rafted blocks of terrestrial sediment floating out to sea and breaking up just before deposition. The distances travelled away from the raft could not have been very great, as in both cases the preservation is very good.

FORAMINIFERAL LININGS (FIG. 11.12.14)

These are often found in palynodebris preparations from marine sediments. They are chitinous.

PERFORATED PHYTOCLASTS (not illustrated)

These are opaque lath-shaped bodies or needles, often black in color with holes which are often structural. They may be similar to Masran and Pocock's (1981) 'vascular elements,' and hence intergrade with some elements of the categories illustrated here as palynowafers or brown and black debris.

MOSS LEAVES (FIG. 11.9.8)

Well-preserved leaf fragments attributable to *Sphagnum* were found in the Forties sediments in association with

Stereisporites spores, also thought to have been produced by *Sphagnum*. They may have accumulated in the same way as the fern fragments.

INCERTAE SEDIS (FIG. 11.9, 9, 10)

There are numerous unique specimens of unusual types which cannot fit into the kind of generalized forms described above and which may be peculiar to individual deposits.

Examples of the use of the proposed terms

Figure 11.2 is adapted from Pocock *et al.* (1988) and shows a generalization of the relationship between depositional environment (high-energy deep marine, low-energy deep marine, lagoon, prodelta, delta front, shallow marine and freshwater lake), and the occurrence of amorphous matter, palynowafers, pollen and spores, and dinoflagellate cysts.

Although this range of palynodebris is based on Paleogene material (and includes angiosperm remains lacking in, say, Jurassic deposits) the concepts drawn from it may apply equally to Mesozoic and Paleozoic sediments. This model of amorphous matter, palynowafers and palynomorphs has been tested on material from a variety of locations though few of the details have been published. These tests include preparations from deep marine sediments of the Forties Field Paleocene (Boulter

Figures 11.3–11.12. General notes: All material illustrated is from the Paleocene Forties Field, North Sea. It should be noted that the categories illustrated merge into each other: 'amorphous matter with specks,' for example, includes specimens that could also be classified as sheet-like. Each scale bar represents 25 μm.

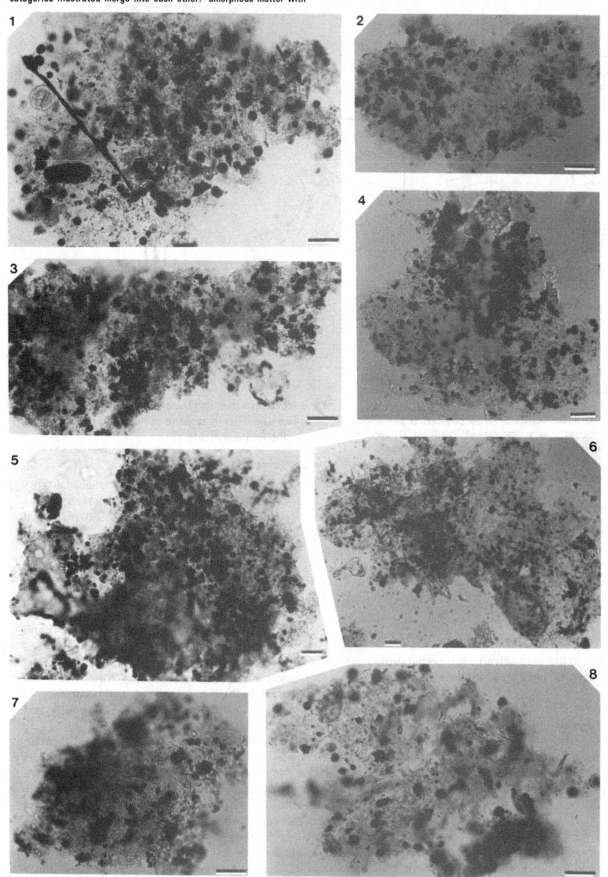

Figure 11.3.1–8. Amorphous matter with specks. There is variation in the size and frequency of the specks, which are always present. Some of the specks are pyritic.

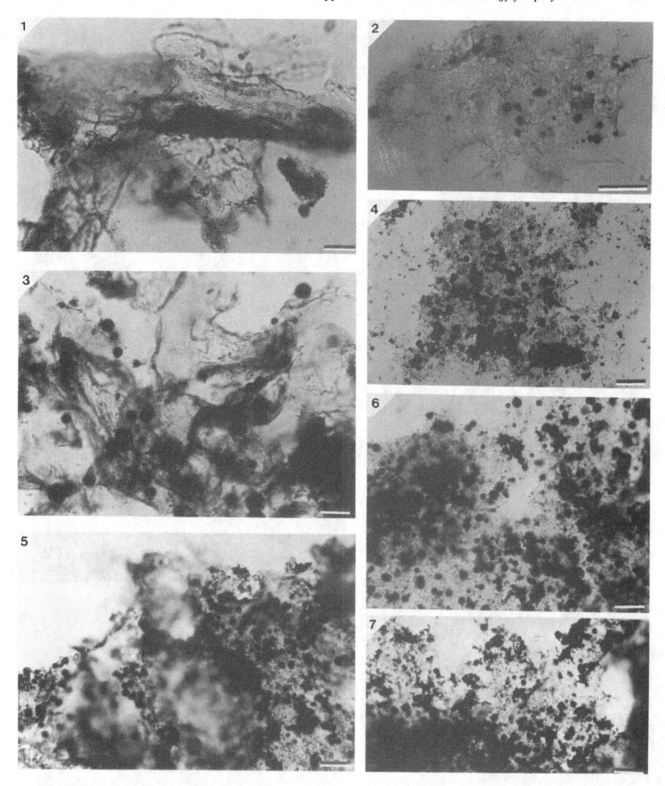

Figure 11.4.1–3. Sheet-like amorphous matter. This is thought to have a quite different origin from amorphous matter, but the two kinds invariably occur together; 4–7. Amorphous matter.

Figure 11.5.1–9. Palynowafers – degraded bundles. These always show evidence of cellular structure, usually from secondary xylem tissue.

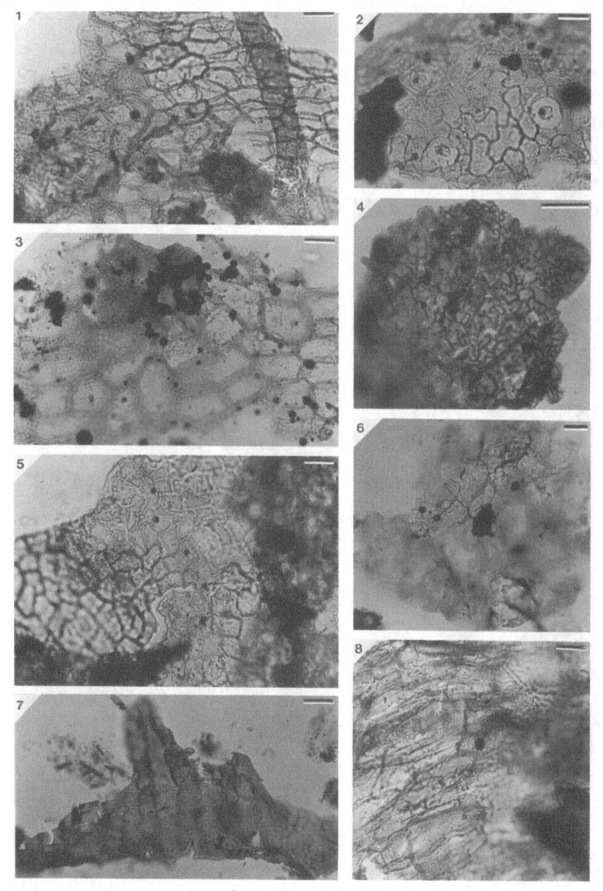

Figure 11.6.1–8.   Palynowafers – leaf cuticle. Stomata and epidermal cell remains are occasionally present.

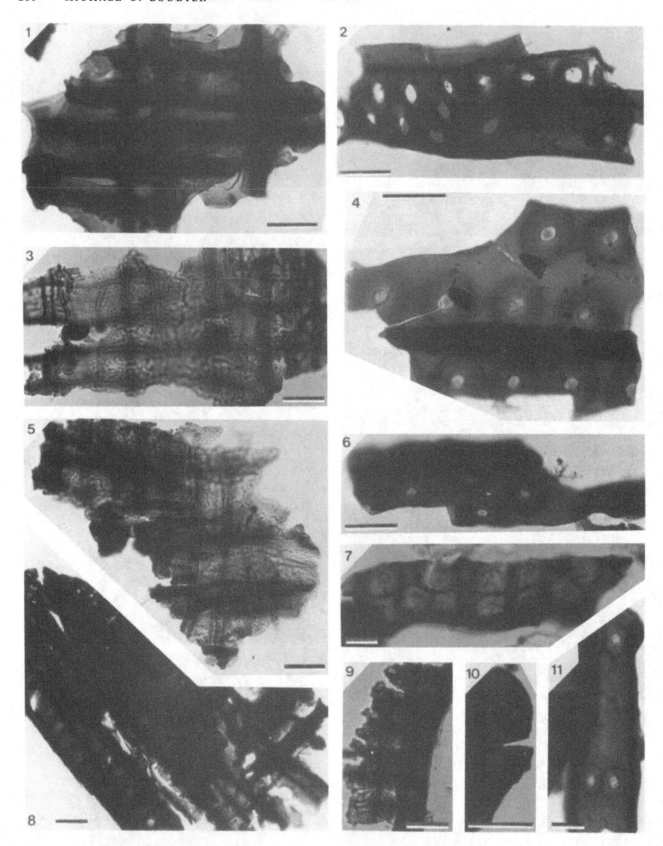

Figure 11.7.   Palynowafers — well preserved wood.
1, 3, 5, 8, 9 brown rays, usually with horizontal and vertical elements
and cross-field pitting; 2, 4, 6, 7, 10, 11 brown tracheids, usually with
bordered pitting.

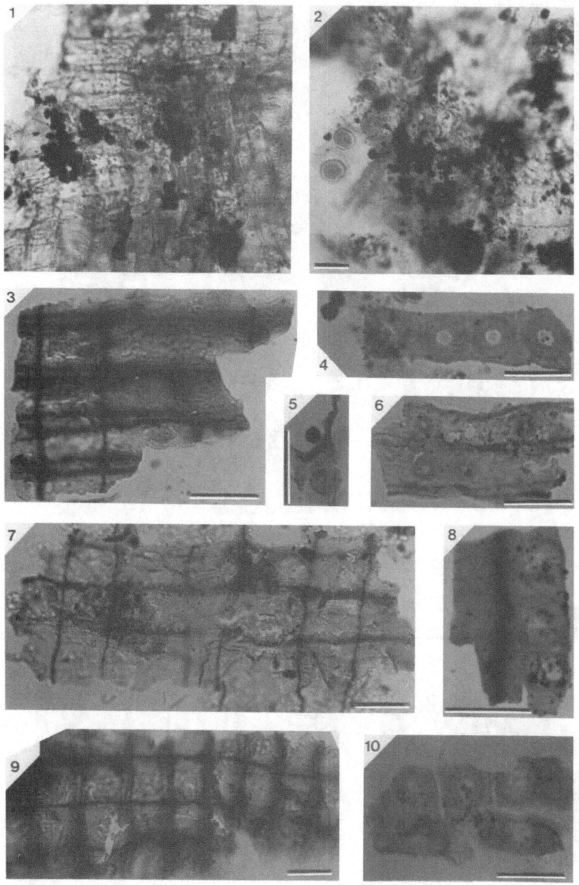

Figure 11.8. Palynowafers — well-preserved wood. 1, 3, 7, 9 yellow rays, usually with horizontal and vertical elements and cross-field pitting; 2, 4, 5, 6, 8, 10 yellow tracheids, usually with bordered pitting.

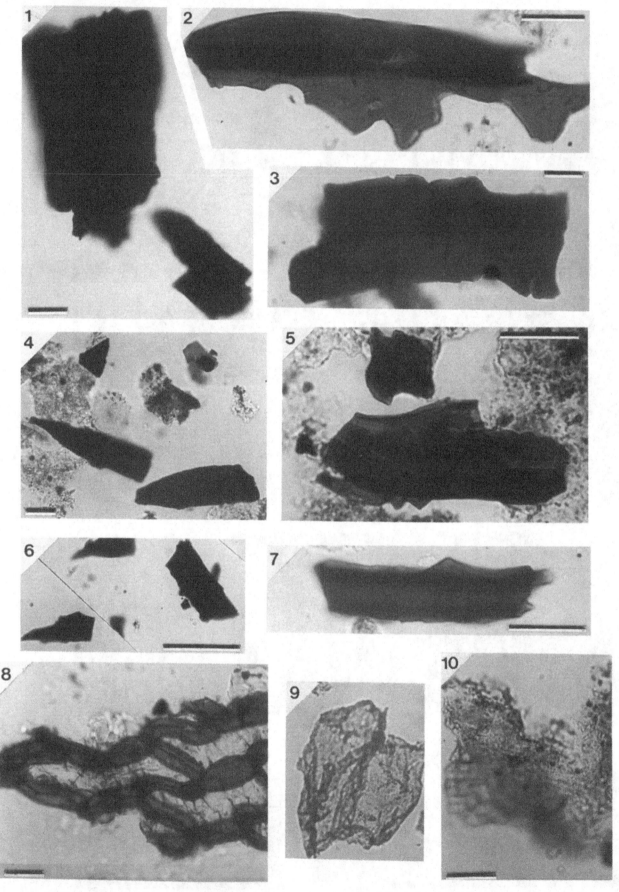

Figure 11.9.1, 4, 6. Black debris, of very variable size and shape but showing no cellular detail (6 is a collage of sections of a single print, to eliminate blank space); 2, 3, 5, 7 brown debris, of comparable size and shape, may show obscure cellular detail; 8 *Sphagnum* leaf; 9, 10 *Incertae sedis* (bar for 10 applies also to 9).

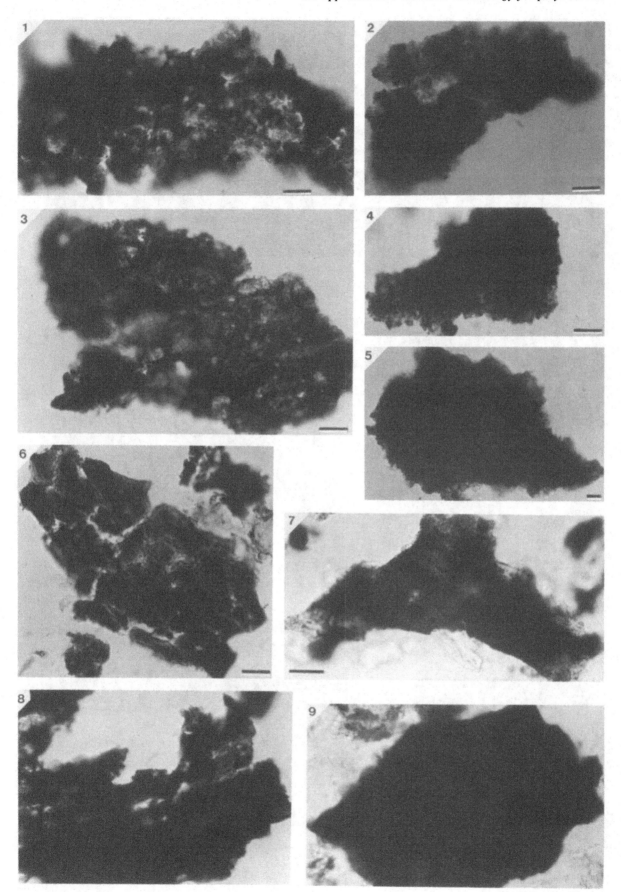

Figure 11.10.1–9   Degraded debris, usually of a very dark brown color. The surface detail, best seen in 1 and 2 merges to the opaque (6), which is difficult to distinguish from brown debris.

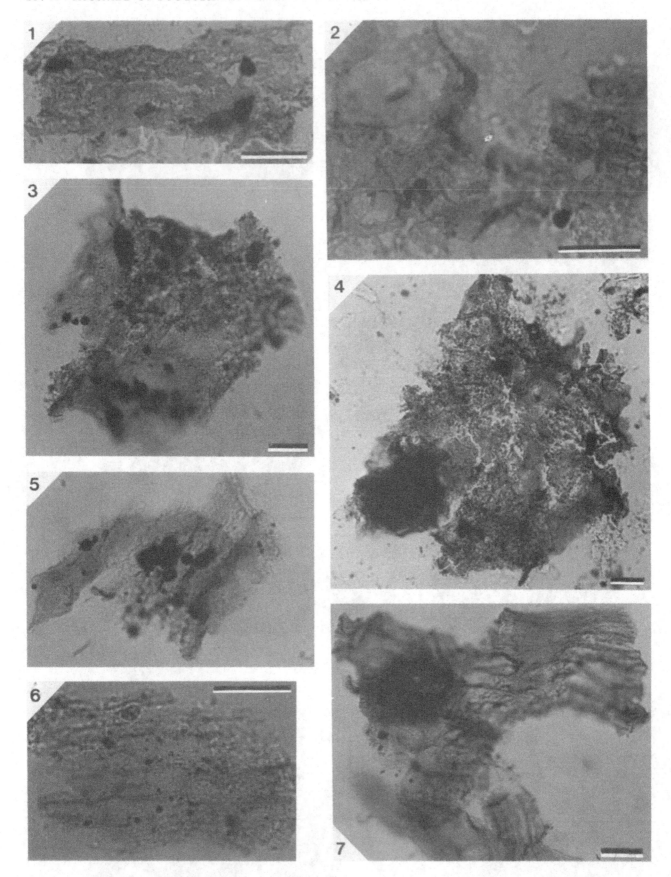

Figure 11.11.1–7   Yellow unstructured debris. Obscure cellular detail
and black specks are often apparent.

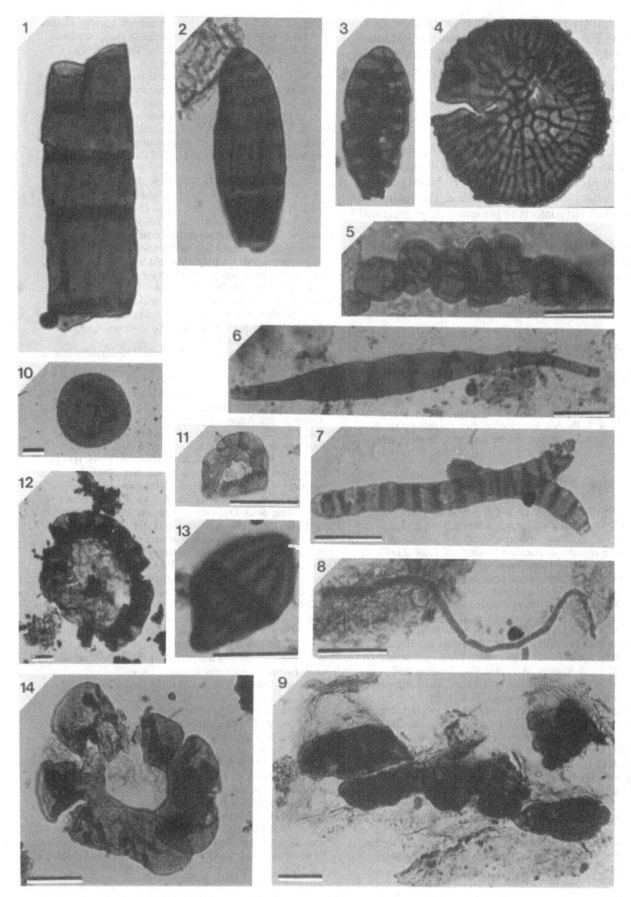

Figure 11.12.1–9, 11, 13.   Fungi; 10 Alga; 12 Fern sporangium;
14 Foraminiferal lining.

& Riddick, 1986), DSDP Leg 38 Paleogene and Neogene (Fig. 11.1) and the ODP Leg 104 Neogene (Manum *et al.*, 1989). Shallow marine sediments have been examined from the Forties Field Paleocene and the Hampshire Basin, ODP Leg 104 Holes 643A and 642D and the Paleocene/Eocene interbasaltics of the Isle of Mull. Brackish lagoonal sediments from the Hampshire Basin of southern England and freshwater and other terrestrial sediments from the Bovey Tracey Basin (Oligocene), the Lough Neagh clays (Oligocene) and Swanscombe (Pleistocene, R. N. L. B. Hubbard personal communication) have also been studied. The results are not published but conform to the terminology presented above.

Clearly, this list of applications of the model is too short to claim universal validity for the groups proposed. It is hoped that further work on visual palynodebris studies from sediments elsewhere will help take us towards a stable and reliably infomative set of terms.

Acknowledgments

The original study of palynodebris from the Paleocene of the Forties Field was financed by Shell, Esso and British Petroleum. The research assistants who obtained the data were C. Darling and A. Riddick. R. N. L. B. Hubbard undertook the statistical analysis. S. A. J. Pocock kindly offered comments on the first draft of the manuscript, thereby helping with its more general value.

References

Barnard, P. C., Collins, A. G. & Cooper, B. S. (1981). Identification and distribution of kerogen facies in a source rock horizon – examples from the North Sea Basin. In *Organic Maturation Studies and Fossil Fuel Exploration*, ed. J. Brooks. London: Academic Press, 271–82.

Batten, D. (1981). Palynofacies, palaeoenvironments and petroleum. *Journal of Micropalaeontology*, **1**, 107–14.

Boulter, M. C. (1987). Tertiary monocotyledons from aquatic environments. *Tertiary Research*, **9**, 133–46.

Boulter, M. C. & Manum, S. B. (1989). The 'Brito-Arctic Igneous Province Flora' around the Paleocene/Eocene boundary. *Proceedings Ocean Drilling Program*, **104**, 663–80.

Boulter, M. C. & Riddick, A. (1986). Classification and analysis of palynodebris from the Palaeocene sediments of the Forties Field. *Sedimentology*, **33**, 871–86.

Brooks, J. (1981). Organic maturation of sedimentary organic matter and petroleum exploration – a review. In *Organic Maturation Studies and Fossil Fuel Exploration*, ed. J. Brooks. London: Academic Press, 1–38.

Burgess, J. D. (1974). Microscopic examination of kerogen (dispersed organic matter) in petroleum exploration. *Geological Society of America Special Paper* **153**, 19–30.

Caratini, C., Bellet, J. & Tissot, C. (1975). Étude microscopique de la matière organique: Palynologie et Palynofaciès. In *Géochemie organique des sédiments marins profonds, Orgon II, Atlantique-N.E. Brésil*, ed. A. Combaz & R. Pelet. Paris: Centre Nationale de la Recherche Scientifique, 157–203.

Carman, G. J. & Young, R. (1981). Reservoir geology of the Forties oilfield. In *Petroleum Geology of the Continental Shelf of north-west Europe*, ed. L. V. Illing & G. D. Hobson. London: Heyden, 371–9.

Cope, M. J. (1981). Products of natural burning as a component of the dispersed organic matter of sedimentary rocks. In *Organic Maturation Studies and Fossil Fuel Exploration*, ed. J. Brooks. London: Academic Press, 89–110.

Cross, A. T., Thompson, G. G. & Zaitzeff, J. B. (1966). Source and distribution of palynomorphs in bottom sediments, southern part of Gulf of California. *Marine Geology*, **4**, 467–524.

Crum-Brown, A. (1912). The oil shales of the Lothians. *Memoirs Geological Survey of Scotland* **143**.

Durand, B. (1980), ed., *Kerogen – Insoluble Organic Matter from Sedimentary Rocks*. Paris: Éditions Technip.

Manum, S. B. (1976). Dinocysts in Tertiary Norwegian-Greenland Sea sediments (D. S. D. P. Leg 38) with observations on palynomorphs and palynodebris in relation to environment. *Initial Reports of the Deep Sea Drilling Project*, **38**, 897–921.

Manum, S. B., Boulter, M. C., Gunnarsdottir, H., Rangnes, K. & Scholze, A. (1989). Eocene to Miocene palynology of the Norwegian Sea (ODP Leg 104). *Proceedings Ocean Drilling Program*, **104**, 611–62.

Masran, T. C. & Pocock, S. A. J. (1981). The classification of plant-derived particulate organic matter in sedimentary rocks. In *Organic Maturation Studies and Fossil Fuel Exploration*, ed. J. Brooks. London: Academic Press, 145–76.

McCave, I. N. (1985). Hummocky sand deposits generated by storms at sea. *Nature*, **313**, 533.

Muller, J. (1959). Palynology of Recent Orinoco delta and shelf sediments. *Micropalaeontology*, **5**, 1–32.

Parry, C. C., Whitley, P. K. J. & Simpson, R. D. H. (1981). Integration of palynological and sedimentological methods in facies of the Brent Formation. In *Petroleum Geology of the Continental Shelf of north-west Europe*, ed. L. V. Illing & G. D. Hobson. London: Heyden, 205–15.

Piasecki, S. (1980). Dinoflagellate cyst stratigraphy of the Miocene Hodde and Gram Formations, Denmark. *Bulletin Geological Society of Denmark*, **29**, 53–76.

Pocock, S. A. J., Vasanthy, G. & Venkatachala, B. S. (1988). Introduction to the study of particulate organic materials and ecological perspectives. *Journal of Palynology*, **23**, 167–88.

Staplin, F. L., Bailey, N. J. L., Pocock, S. A. J. & Evans, C. R. (1973). Palynofacies. *American Association of Petroleum Geologists Symposium*, Anaheim, MS, 43–52.

Stopes, M. C. (1935). On the petrology of banded bituminous coal. *Fuel in Science and Practice*, **14**, 4–13.

Talwani. M., Udintsev, G. *et al.* (1976). *Initial Reports of the Deep Sea Drilling Project*, **38**, Washington, DC: US Government Printing Office.

Teichmüller, M. & Teichmüller, R. (1968). Cainozoic and Mesozoic coal deposits of Germany. In *Coal and Coal Bearing Strata*, ed. D. G. Murchison & T. S. Westoll. Edinburgh: Oliver & Boyd, 347–79.

Tissot, B., Durand, B., Espitalie, J. & Combaz, A. (1974). Influence of the nature and diagenesis of organic matter in the formation of petroleum. *Bulletin of the American Association of Petroleum Geologists*, **58**, 396–405.

Traverse, A. & Ginsburg, R. N. (1966). Palynology of the surface sediments of the Great Bahama Bank, as related to water movement and sedimentation. *Marine Geology*, **4**, 417–59.

12 Relationships of palynofacies to coal-depositional environments in the upper Paleocene of the Gulf Coast Basin, Texas, and the Powder River Basin, Montana and Wyoming

DOUGLAS J. NICHOLS and DAVID T. POCKNALL

Introduction

Research by Fisher and McGowen (1967) on the stratigraphy and sedimentology of the upper Paleocene part of the Wilcox Group in the Gulf Coast Basin in Texas (Fig. 12.1C) resulted in the definition of seven ancient depositional systems in the outcropping and subsurface Wilcox. Coal-forming peat had accumulated within four of them. Traverse and Nichols (1967), Nichols (1970), and Nichols and Traverse (1971) identified paleoecologically significant groups of palynomorphs within assemblages from outcropping Wilcox coal beds in an effort to distinguish palynologically the interrelated but differing depositional systems defined by Fisher and McGowen.

In the Powder River Basin of Montana and Wyoming (Fig. 12.1B), studies on lower Tertiary coal deposits of the Fort Union Formation by Flores (1981, 1983), Flores and Hanley (1984), and Hanley and Flores (1987) provided a similar stratigraphic and sedimentologic framework for palynological studies. Depositional models for coal formation in this region were proposed by Pocknall (1986a) and Pocknall and Flores (1987); see also Satchell (1984; 1985). As in the Gulf Coast study, an aspect of research on Powder River Basin coal beds was the identification of groups of palynomorphs that represent the depositional environments.

In this chapter we expand upon the interpretations of Nichols and Traverse (1971) for upper Paleocene coal beds of the Gulf Coast Basin and compare them with those obtained by Pocknall and Flores (1987) for upper Paleocene coal beds in the Powder River Basin. We use palynofacies to characterize and contrast coal-depositional environments within these regions. By comparing data and interpretations from two contemporaneous areas that differed floristically, sedimentologically, and paleoclimatically but that both produced significant deposits of coal, we explore the use of palynofacies in distinguishing the varied environments of deposition in which peat accumulates, which is relevant to the origin of major coal deposits.

Terminology

The concept of palynofacies is central to this study, and specifying the way in which we use this term is appropriate, to avoid ambiguity. We use it essentially as defined by Bates and Jackson (1987) and Traverse (1988) to refer to any group of palynomorphs that represents local environmental conditions and is not typical of the regional palynoflora. We qualify this general definition by noting that the taxa that characterize a given palynofacies are usually only a small number of the total that are present in an assemblage. Further, the characterizing taxa need not be atypical of the regional palynoflora on a presence-or-absence basis, but their relative abundances may be atypical. Characterizing taxa usually have greater relative abundance in a palynofacies than elsewhere. This definition contrasts somewhat with that of Combaz (1964) and possibly with the usage of other authors in this volume, who emphasize organic matter other than palynomorphs in describing palynofacies. In the earlier report on the palynology of Wilcox depositional environments, Nichols and Traverse (1971) redefined the term assemblage to refer to groups of palynomorphs that occur together in some preparations (samples) in addition to species common to the regional palynoflora; local sedimentary environment was the implicit cause of the co-occurrence. 'Assemblages' thus defined were said by Nichols and Traverse (1971) to reveal biofacies. Their concept (although awkwardly labeled with a well established term for all fossils found at some stratigraphic level) was the concept of palynofacies as subsequently defined, and as used here.

The precursor of the coal that constitutes most of our samples was peat. The terminology for wetland environments in which peat accumulates is used inconsistently in different parts of the world (Gore, 1983). The terms swamp, marsh, bog, fen, and mire have been used for these environments, and they are variously defined by reference to form, position of water table, kinds of plants present (species composition or stature of vegetation), sediment supply, nutrient supply, and (or)

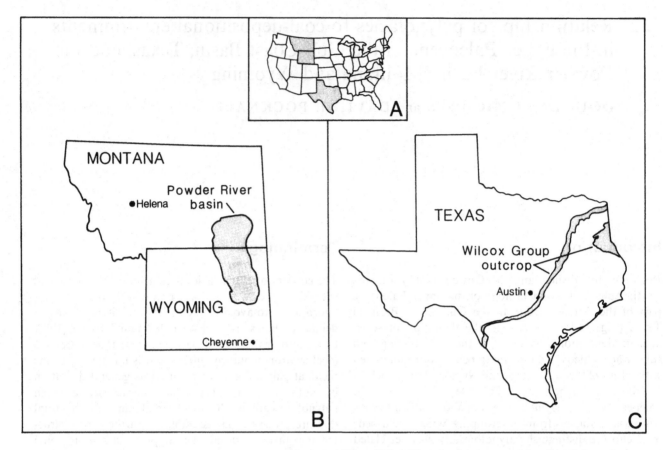

Figure 12.1. Maps showing the general locations of the regions studied. (A) Map of the USA with Montana and Wyoming (north) and Texas (south) shaded. (B) Powder River Basin (shaded) in Montana and Wyoming. (C) Gulf Coast Basin in Texas with outcrop areas of the Wilcox Group shaded.

water chemistry. For example, according to some, swamps are inhabited by trees; according to others, swamps are defined by the presence of standing water. Outside of North America the term swamp may include marsh, but peat is said not to accumulate in marshes. Bogs, which most sources agree are inhabited predominantly by mosses, may or may not have their surfaces raised above the water table (morphologies that have differing implications for rainfall and nutrient supply, but for which there may be no independent evidence in the geologic record). A major complication in applying definitions based on kinds of plants to early Tertiary environments is that reeds, grasses, and sedges (the modern plant families Gramineae and Cyperaceae) that at present dominate some wetlands evidently were absent during the Paleocene. These families have no Paleocene pollen record in North America (Muller, 1981) and presumably had not yet evolved. Botanical affinities of many Paleocene plants are uncertain, and the stature of a Paleocene plant (tree, shrub, or herb) may be unknown even when its botanical affinity can be determined. Therefore, careless usage of some of the terms, especially swamp, bog, and marsh, may have unintended implications for some readers.

The term mire is most useful because it is comprehensive and nonrestrictive. Mires include all wetland environments in which organic matter, usually peat, accumulates

(Gore, 1983). The term carries no specific implications about hydrology or ecology, although it does not prevent data on these important variables from being included in analyses and interpretations. Cowardin *et al.* (1979) classified these environments as palustrine wetlands, but we prefer the simpler term mire and use it herein as the general term for all coal-forming depositional environments. The classes of palustrine wetlands recognized by Cowardin *et al.* (e.g., scrub-shrub wetland, forested wetland, etc.) proved useful for describing some of the Paleocene environments in which coal-forming peat was deposited.

The terms coal and lignite are commonly used in discussions of the lower Tertiary sequences of the Powder River and Gulf Coast Basins, respectively. For convenience in this study, we use the more general term coal for both regions, without implying similarity in rank or other properties.

Materials and methods

Samples from the Gulf Coast Basin were collected from outcrops, coal mines, and brickyard clay pits in the upper

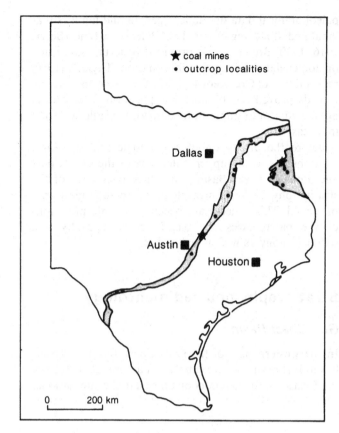

Figure 12.2. Map of part of Texas showing the study area in the Gulf Coast Basin and the sample localities. The Wilcox Group dips into the subsurface toward the south and southeast of the outcrop belt and is exposed again on the Sabine uplift in east Texas. The outcrop areas in east Texas expose deposits of a fluvial depositional system; the outcrop belt in central Texas, about 150 km northeast and southwest of Austin, exposes deposits of a delta depositional system; the outcrop in south Texas exposes deposits of a bay-lagoon depositional system. None of the sample localities was in a second bay-lagoon system mapped by Fisher and McGowen (1967) in the southern part of the Sabine uplift.

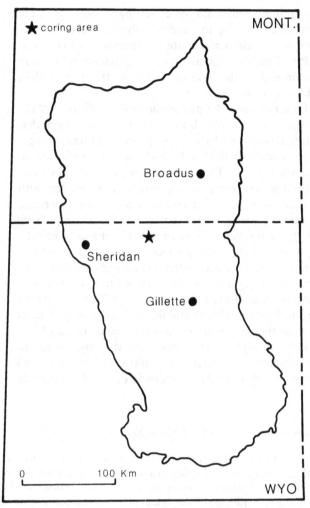

Figure 12.3. Map of the Powder River Basin, Montana and Wyoming, showing the area in which the coring sites are located. Cores penetrated the Tongue River Member of the Fort Union Formation. See Fig. 12.1B for general location of the basin.

Paleocene lower part of the Wilcox Group in the coastal plain of Texas. The general locations of sample localities are shown in Figure 12.2. Coal and claystone were sampled; the few claystone samples were partings within seams and crevasse splay deposits that had overwhelmed the mires. Influxes of clastic sediment that form partings are an integral part of coal-depositional environments, and beds overlying coal seams may contain evidence of mire history. Several of the coal samples were provided by W. L. Fisher of the Texas Bureau of Economic Geology. Samples from two strip mines were collected by Alfred Traverse of Pennsylvania State University. All samples from the Gulf Coast Basin were analyzed by Nichols. Table 12.1 summarizes data on Gulf Coast Basin samples; detailed descriptions of the localities and all data from counts were recorded by Nichols (1970).

Samples from the Powder River Basin were collected

from cores of the upper Paleocene Tongue River Member of the Fort Union Formation. The cores were provided by R. M. Flores of the U. S. Geological Survey. Figure 12.3 shows the general location of the coring sites, which were along the Powder River near Arvada, Wyoming; for detailed descriptions of the sites, refer to Pocknall (1986b) or Pocknall (1987). Partings within seams were sampled in addition to the coal. Table 12.2 summarizes data on Powder River Basin samples. All samples from the Powder River Basin were analyzed by Pocknall; count and occurrence data for all samples were recorded by Pocknall (1986b).

In both regions studied the samples were collected with the goal of obtaining representation of variations within coal beds that might reflect differences within mires. Collecting was guided by visible variations in lithology within seams, and by consideration of the independently determined sedimentologic settings. Samples were processed according to standard palynological techniques for low-rank coals and clastic sedimentary rocks (Doher, 1980). Relative abundances of palynomorphs were determined on the basis of counts of 200 to more than 1000 specimens per slide.

To understand the paleoecological significance of the palynofacies, we rely heavily on the assumption that plants represented by fossil palynomorphs had ecologic requirements similar to those of their close relatives in the modern flora. Dodd and Stanton (1981) termed this approach 'taxonomic uniformitarianism.' Consistent with this assumption, in the text we refer to the Paleocene plants by the modern names of the closest living relative. However, we recognize that our fossils actually represent early Tertiary species of modern genera or families, ancestral taxa within extant lineages; we do not mean to imply that modern species (or perhaps even modern genera) lived during the early Tertiary. Names for fossil palynomorphs and the modern plant genera or families they represent are summarized in Table 12.3.

Palynomorph morphologies are the basic data for determination of botanical affinities in both the Gulf Coast and Powder River Basins in our study. Available paleobotanical data on modern genera and families that inhabited these regions in late Paleocene time (Berry, 1916, 1930; Brown, 1962) provided little taxonomically or nomenclaturally reliable information. Taylor's (1990) reevaluation of Cretaceous – lower Cenozoic angiosperm records proved too broad for application here; it does not include, for example, floral data for the Wilcox Group in Texas.

Before the discussion of palynologic and lithologic relationships in the upper Paleocene of the Gulf Coast and Powder River Basins, summary overviews of the stratigraphy and sedimentology of these regions are presented. This is necessary because our interpretations of the palynofacies are based on stratigraphy and sedimentology as well as palynology.

Stratigraphy and sedimentology

Gulf Coast Basin

In the western part of the Gulf Coast Basin in Texas, lower Tertiary rocks are exposed in an arcuate belt from the Texas-Louisiana border on the east to the international

Table 12.1. *Summary of Gulf Coast basin samples.*

All samples from Wilcox Group. Stratigraphic position refers to position of individual sample within outcrop at given locality. Depositional environments determined by Fisher and McGowen (1967). Fluvial environments in which peat accumulated include mires on stream terraces and in abandoned tributary channels and oxbow lakes; deltaic environments were inter-distributary delta plain mires; lagoonal environments, which include both mires and marshes, were near the marine coastline.

Locality	Stratigraphic position	Lithology	Environment
AT	(S108) middle of 30-cm seam	coal	fluvial
BS	(S221) parting in coal seam 'BS'	claystone	fluvial
	(S222) upper bench of seam 'BS'	coal	
BT	(S101) middle of 15-cm seam	coal	fluvial
FC	(S103) entire 10-cm seam	coal	fluvial
GT	(S171) just above coal seam 'GT'	claystone	fluvial
	(S170) top of 75-cm seam 'GT'		
	(S169) upper part of seam 'GT'		
	(S168) middle of seam 'GT'	coal	
	(S167) lower part of seam 'GT'		
	(S166) base of 75-cm seam 'GT'		
HA	(S164) just above seam 'HA'	claystone	fluvial
	(S163) top of 30-cm seam 'HA'		
	(S162) middle of seam 'HA'	coal	
	(S161) bottom of 30-cm seam 'HA'		
HS	(S159) entire 5-cm seam	coal	fluvial

Table 12.1. *Continued*

Locality	Stratigraphic position	Lithology	Environment
HC	(S109) middle of 1.5-m seam	coal	fluvial
PS	(S225) middle of 1-m seam 'PS' (S224) base of 1-m seam 'PS'	coal	fluvial
RC	(S114) middle of 1-m seam	coal	fluvial
RCN	(S119) entire 10-cm seam	coal	fluvial
RCS	(S120) entire 10-cm seam	coal	fluvial
TCS	(S106) base of 1-m seam	coal	fluvial
DR	(S216) just above rider seam	claystone	
	(S215) top of 50-cm rider seam (S214) upper part of rider seam (S213) middle of rider seam (S212) lower part of rider seam (S211) base of 50-cm rider seam	coal	fluvial
D65	(S132) just above seam 'D65'	claystone	
	(S131) top of 3.1-m seam 'D65' (S130) above upper parting	coal	
	(S129) upper parting, seam 'D65'	claystone	
	(S128) below upper parting (S127) upper middle seam 'D65' (S126) middle of seam 'D65' (S125) lower middle seam 'D65'	coal	fluvial
	(S124) lower parting, seam 'D65'	claystone	
	(S123) below lower parting (S122) base of 3.1-m seam 'D65'	coal	
R65	(S137) just above seam 'R65'	claystone	
	(S144) top of 4-m seam 'R65' (S143) middle upper part, 'R65' (S142) lower upper part, 'R65' (S141) upper middle part, 'R65' (S140) lower middle part, 'R65' (S139) lower middle part, 'R65' (S136) middle lower part, 'R65' (S135) base of 4-m seam 'R65'	coal	deltaic
CB	(S118) top of 4-m seam 'CB' (S117) upper middle part, seam 'CB' (S116) lower middle part, seam 'CB' (S115) base of 4-m seam 'CB'	coal	deltaic
BC	(S153) middle part of 50-cm seam	coal	deltaic
NRC	(S233) top of 30-cm seam 'NRC' (S232) middle of seam 'NRC' (S231) base of 30-cm seam 'NRC'	coal	lagoonal
NRB	(S229) top of 30-cm seam 'NRB' (S228) base of 30-cm seam 'NRB'	coal	lagoonal
ZC	(S145) middle of 20-cm seam	coal	lagoonal

Table 12.2. *Summary of Powder River basin samples.*

All samples from Tongue River Member of Fort Union Formation. Stratigraphic position refers to position of individual samples within cores. Depositional environments determined by Flores (1981), Flores and Hanley (1984), and Hanley and Flores (1987). All environments were fluvial; peat accumulated in mires in flood basins of either anastomosed or meandering streams

Locality	Stratigraphic position		Lithology	Environment
D6710-E	Core 1	26.58–27.08 m		
D6710-D		27.08–27.99 m	coal	meandering
D6710-C		28.65–30.20 m		
D6710-B		32.43–32.80 m		
D6708-O	Core 2	53.87–54.60 m		
D6708-N		54.60–55.12 m		
D6708-M		57.45–57.83 m		
D6708-L		57.83–58.12 m		
D6708-K		58.12–58.55 m		
D6708-J		58.55–59.83 m	coal	
D6708-I		59.83–60.46 m		meandering
D6708-H		60.46–60.97 m		
D6708-G		60.97–61.25 m		
D6708-F		61.25–61.72 m		
D6708-E		61.84–62.18 m	carb. shale	
D6708-D		63.40–63.55 m		
D6708-C		63.26–64.46 m	coal	
D6707-Z′	Core 3	89.92–90.37 m		
D6707-Z		89.82–89.92 m		
D6707-Y		90.37–90.64 m		
D6707-X		90.64–90.91 m	coal	
D6707-W		90.91–91.89 m		
D6707-V		91.89–92.44 m		
D6707-U		93.05–93.14 m		anastomosed
D6707-T		98.45 m	carb. shale	
D6707-S		98.55–98.71 m		
D6707-R		98.71–99.33 m	coal	
D6707-Q		98.97–99.06 m		
D6707-P		99.33–99.95 m		
D6707-O		100.25–100.30 m	mudstone	
D6707-N′		149.55–149.88 m	coal	
D6707-N		149.63–149.69 m	carb. shale	
D6707-M′		149.88–150.70 m		
D6707-M		149.78–149.90 m	coal	
D6707-L		150.93–151.18 m		
D6707-K		151.70–151.79 m	shale	
D6707-J		154.63–155.52 m		
D6707-I		154.99–155.08 m		
D6707-H		155.52–156.36 m	coal	meandering
D6707-G		156.36–157.22 m		
D6707-F		157.22–157.55 m		
D6707-E		157.73–157.83 m	carb. shale	
D6707-D		158.14–158.64 m		
D6707-C		158.64–159.08 m	coal	
D6707-B		159.08–159.31 m		
D6707-A		159.20–159.29 m	carb. shale	
D6720-I	Core 4	89.62–90.29 m		
D6720-H		90.29–90.92 m		
D6720-G		90.92–91.70 m		
D6720-F		91.70–92.22 m		
D6720-E		92.43–92.62 m	coal	anastomosed
D6720-D		92.62–92.74 m		
D6720-C		92.74–92.98 m		
D6720-B		95.34–96.21 m		
D6720-A		96.21–96.61 m		

Table 12.3. *Modern generic and family names used in the text and names of corresponding fossil palynomorphs*

The palynomorphs represent early Tertiary species of the modern genera or families whose ecologic significance is assumed to have remained essentially unchanged since late Paleocene time.

Modern genus or family	Fossil palynomorph
Alnus	*Alnipollenites verus* (Potonié 1931)
Arecaceae	*Arecipites* spp.
	Calamuspollenites pertusus Elsik 1968
Betula	*Triporopollenites* sp.
Betulaceae	*Triporopollenites* spp. (see also *Alnus*, *Corylus*)
Carya	*Caryapollenites imparalis* Nichols & Ott 1978
	C. inelegans Nichols & Ott 1978
	C. veripites (Wilson & Webster 1946) Nichols & Ott 1978
	C. wodehousei Nichols & Ott 1978
Castanea	*Cupuliferoipollenites pusillus* (1934) Potonié 1960
Chenopodiaceae	*Periporopollenites* sp. and *Chenopodipollis* sp.
Hydrodictyaceae	(see *Pediastrum*)
Corylus	*Triporopollenites* spp.
Engelhardia-Alfaroa	*Momipites tenuipolus* Anderson 1960
	M. triradiatus Nichols & Ott 1978
	M. ventifluminis Nichols & Ott 1978
	M. wyomingensis Nichols & Ott 1978
Ericaceae	*Ericipites* spp.
Fagaceae	(see *Castanea*)
Glyptostrobus	*Taxodiaceaepollenites hiatus* (Potonié 1931) Kremp 1949
Juglandaceae	(see *Carya*, *Engelhardia-Alfaroa*, and *Pterocarya*)
Nyssaceae	(see *Nyssa*)
Nyssa	*Nyssapollenites thompsonianus* (Traverse 1951) Potonié 1960
Pediastrum	*Pediastrum paleogeneites* Wilson & Hoffmeister 1953
Picea	*Pityosporites* spp. (large)
Pinaceae	(see *Picea*, *Pinus*)
Pinus	*Pityosporites* spp. (small)
Platanaceae	*Tricolpites* spp. cf. *T. hians* Stanley 1965
Polypodiaceae	*Laevigatosporites* spp.
Pterocarya	*Polyatriopollenites vermontensis* (Traverse 1955) Frederiksen 1980
Sparganiaceae	*Sparganiaceaepollenites globipites* Wilson & Webster 1946
Sphagnaceae	(see *Sphagnum*)
Sphagnum	*Stereisporites* spp.
Spirogyra	*Schizophacus* spp.
Taxodiaceae	(see *Taxodium*)
Taxodium	*Taxodiaceaepollenites hiatus* (Potonié 1931) Kremp 1949

Table 12.3. *Continued*

Modern genus or family	Fossil palynomorph
Typhaceae	*Sparganiaceaepollenites globipites* Wilson & Webster 1946
Ulmus-Zelkova	*Ulmipollenites krempii* (Anderson 1960) Frederiksen 1979
	U. tricostatus (Anderson 1960) Frederiksen 1980
Ulmaceae	(see *Ulmus-Zelkova*)
Zygnemataceae	(see *Spirogyra*)

border on the south, and on the Sabine uplift in east Texas (Fig. 12.2). The Wilcox Group in Texas includes up to 1500 m of dominantly nonmarine deposits that constitute a major regressive phase of deposition; the Wilcox locally contains major deposits of coal. Stratigraphic nomenclature below the rank of group is exceedingly complex and burdened by a plethora of locally applied names. Coal beds and associated deposits discussed here that are from central and east Texas may be considered as part of the Rockdale Formation; those from south Texas may be considered as part of the Indio Formation; other local names apply, as well (Fisher, 1961). In view of the lithologic diversity and variability in definition and application of formation and member names, we apply the broader term Wilcox to all deposits discussed here. This treatment is consistent with Fisher and McGowen's (1967) emphasis on interrelation of the depositional systems within the Wilcox Group.

The term Wilcox is equated with Eocene in older literature on Gulf Coast Basin geology, but it is evident that much of the sequence is actually late Paleocene in age. This conclusion is based on biostratigraphic revisions involving foraminifers (Berggren, 1965) and nannoplankton (Hay & Mohler, 1967) from the largely marine Wilcox Group in Alabama. Fairchild and Elsik (1969) concluded on the basis of palynostratigraphy that the nonmarine, coal-bearing part of the Wilcox in Texas (=Rockdale Formation of Fisher, 1961) also is late Paleocene in age.

The depositional systems defined by Fisher and McGowen (1967), which are lithostratigraphic units, compose most of the Wilcox Group in the Gulf Coast Basin in Texas. Coal deposits are present in four of these systems: a marine-coastal delta (Rockdale delta system of Fisher and McGowen), fluvial depositional environments upstream from the delta (Mount Pleasant fluvial system of Fisher and McGowen), and lagoonal depositional environments flanking the delta to the east and west (Pendleton lagoon-bay system and Indio bay-lagoon system of Fisher and McGowen). Most of the sediments transported through the fluvial system in late Paleocene time were deposited in the delta, and the delta was the principal source of sediments deposited in the associated barrier bar, strandplain, bay-lagoon, and shelf systems.

The delta system consists of large, lobate wedges of sandstone, mudstone, and carbonaceous deposits; it accounts for 60% of the areal extent and 80% of the known volume of the Wilcox in Texas (Fisher & McGowen, 1967). It is present principally in the subsurface in the central part of the Texas coastal plain, down depositional slope and down structural dip from the thinner terrigenous clastic facies of the fluvial system, but exposures of the ancient delta accessible to surface sampling represent delta plain deposits. Delta plain deposits are characterized by narrow, thick, channel-form sand bodies within deposits of fine-grained clastics and coal. The fine-grained facies consists of mud, silt, and clay deposited as levees and crevasse splays and in flood basins on delta lobes; we observed no partings attributable to crevasse splays in the delta plain coal beds. Delta plain coal beds are tabular bodies three to five meters thick and of local extent, interpreted as having developed in interdistributary depositional environments, and more extensive tabular bodies, interpreted as having developed landward of delta destruction on foundering lobes (Fisher & McGowen, 1967). Some of these thick beds are exposed in mines. For our purposes, the most important depositional environments on the delta were the inter-distributary mires.

The fluvial system in east Texas, up paleoslope and up structural dip from the delta, includes tributary channel facies and meandering channel facies; Fisher and McGowen (1967) also distinguished a third facies that is transitional between the other two. Coal deposits characterize the tributary channel facies. The coal beds are positioned lateral to channel sand bodies that are incised into overbank mud and clay deposits. Here peat accumulated along stream terraces and in abandoned channels and oxbow lakes, and developed into elongate, narrow, sinuous deposits of coal, most of which do not exceed 2 m in thickness. An exception is an unusual 3.1 m-thick seam (locality D65, a coal mine in east Texas).

Lagoon–bay systems defined by Fisher and McGowen (1967) flanked the delta system. Deposits of these systems are present on the southern flank of the Sabine uplift in east Texas and at the southern end of the Wilcox outcrop belt in south Texas. The lagoonal deposits in both areas consist largely of claystone and mudstone but locally include thin (less than 0.5 m), discontinuous beds of impure coal. Proximity of marine sedimentary environments is indicated by occurrences of marine molluscs, foraminifers, and shark teeth in clastic rocks associated with the coal beds.

Fluvial and delta plain coals have been distinguished by Fisher and McGowen (1967) also on the basis of their composition (woody versus non-woody). Fluvial coals are variable in composition and contain high percentages of macerals of the vitrinite group, which are derived from woody tissues and indicate origin of the coal-forming peat in a forested wetland (Traverse & Nichols, 1967); these seams tend to be predominantly woody at the top. In contrast, delta coals contain relatively low percentages of vitrinitic macerals, suggesting organic production and peat accumulation in mires inhabited by herbaceous (non-woody) vegetation.

Powder River Basin

The Laramide orogeny provided source areas for more than 1500 m of fluvial-deltaic and lacustrine sediments of Paleocene age that compose the Fort Union Formation in the Powder River Basin. Major commercial coal deposits of the basin are concentrated in the Tongue River Member, the uppermost part of the Fort Union Formation. The Tongue River Member, which is up to 460 m in thickness, was deposited by both meandering and anastomosed fluvial systems (Flores & Hanley, 1984). Facies of the meandering fluvial system contain the thickest coal beds known in the Fort Union; facies of the anastomosed fluvial system contain thinner coal beds; this facies also includes some lacustrine deposits. The Fort Union is overlain by the predominantly fluvial deposits of the Paleocene and Eocene Wasatch Formation. Pocknall (1987) refined the palynostratigraphy of these rocks. His work verified the late Paleocene age of the Tongue River Member, and indicated that in some parts of the Powder River Basin, the lower part of the Wasatch Formation also is latest Paleocene in age.

Unlike the Tertiary coals of the Gulf Coast and some Upper Cretaceous coals of the Rocky Mountain region (or the Paleozoic coals of the eastern United States), the coal-forming peat deposits in the Powder River Basin accumulated entirely in inland, intermontane settings, exclusively in freshwater sedimentary environments. Two major coal-forming environments have been identified within the upper Paleocene Tongue River Member of the Fort Union Formation, as described below.

In the lower Tongue River Member, peats accumulated in extensive flood basin mires marginal to meander belts of northeast flowing fluvial channels (Flores, 1981, 1983); peat growth was sustained by a gradual and continuous rise in the water table (Pocknall & Flores, 1987). The flood basins were protected from frequent incursions of flood waters by well-developed levees that provided stable and long-lived sites for peat generation. Autocyclic shifts of channels and local overbank-crevasse splay sedimentation occasionally interrupted peat accumulation. These deposits of clastic sediment are represented by partings within the coal beds. The Anderson coal seam studied in detail by Pocknall and Flores is representative of this sedimentation regime, which is referred to as the meandering-fluvial coal-depositional environment.

Coals in the upper part of the Tongue River Member formed predominantly in mires in the flood basin of an anastomosed fluvial system that included large lakes (Flores & Hanley, 1984; Pocknall & Flores, 1987). Active detrital influxes into flood basin mires choked plant growth and diluted organic matter, producing thin, impure coals that contain numerous carbonaceous shale partings. The Smith coal interval (including coal benches and interbedded partings) studied in detail by Pocknall and Flores is representative of this sedimentation regime, which is referred to as the anastomosed-fluvial coal-depositional environment.

Palynologic and lithologic relationships

Gulf Coast Basin

MAJOR ASPECTS

In the Gulf Coast of Texas the palynoflora of the Wilcox Group includes 156 species (Nichols, 1970). Among these, five species were dominant (amounted to more than 50% of the assemblage) in one or more samples, 15 were the most abundant single species in one or more samples, and 24 had a relative abundance of at least 10% in one or more samples. Dominance of assemblages by a single species was observed in 16 assemblages from coal but in only one from claystone; correspondingly, species diversity tended to be greater in assemblages from claystone than in those from coal (Nichols & Traverse, 1971). The most common Wilcox palynomorphs represent ancestral species of the modern plant families Platanaceae (sycamore family), Betulaceae (birch family), Juglandaceae (walnut family), and Fagaceae (beech family).

Two species of *Tricolpites* (both of which are similar to *T. hians*) are the most common Wilcox palynomorphs; one or the other, or both, were present in all of the samples analyzed, and one or the other dominated the assemblages in 11 samples. These species may have botanical affinity with the modern family Platanaceae, although their simple morphology makes this uncertain; in this chapter, however, they are referred to the Platanaceae. Next in total abundance is a species of *Triporopollenites* referable to the modern family Betulaceae (= *Corylus* sp. of Nichols & Traverse, 1971). It is present in more than 90% of the samples. Other species of fossil pollen of probable betulaceous affinity also are common in Wilcox assemblages. Pollen of the genus *Momipites* (= '*Engelhardtia*' of Nichols & Traverse, 1971), which has affinity with the modern family Juglandaceae, is present in more than 90% of the samples and dominant in two of them. Two different species of *Momipites* (*M. tenuipolus* and

Table 12.4. *Summary of palynofacies in the Gulf Coast Basin coals and their characteristic palynomorphs*

The palynomorphs are illustrated in Figs. 12.4 and 12.5.

Palynofacies	Palynomorphs
Arboreal	*Cupuliferoipollenites pusillus*, *Momipites tenuipolus*, *M. triradiatus*, *Tricolpites* spp.
Betulaceae-*Sphagnum*	*Stereisporites* spp., *Triporopollenites* spp., *Ericipites* spp.
Palm-fern	*Arecipites* spp., *Calamuspollenites pertusus*, *Laevigatosporites* spp.
Bisaccate-dinocyst	*Chenopodipollis* sp., *Pityosporites* spp., Dinophyceae (gen. et sp. indet.)

M. triradiatus) were distinguished by Nichols (1973), who showed that, in addition to having different morphologies, these species have different patterns of geographic distribution and lithologic association within the Wilcox Group in Texas. Assemblages from more than 80% of the samples include *Cupuliferoipollenites pusillus* (= *Castanea* of Nichols & Traverse, 1971), and this species is dominant in about 10% of the assemblages. Its botanical affinity is with the modern family Fagaceae. Some specimens of these common Wilcox species are illustrated in Figure 12.4; see also Fairchild and Elsik (1969).

Against this background of commonly occurring species, three palynofacies (which represent plant communities in differing coal-forming environments) can be distinguished in the Wilcox Group in Texas, primarily by localized peaks in abundance of certain species. The palynomorphs that characterize these palynofacies are summarized in Table 12.4. The palynofacies represent plant communities that reflect differing local environmental conditions in mires.

ARBOREAL PALYNOFACIES

The arboreal palynofacies includes some of the most commonly occurring species in the Wilcox palynoflora. In this study, relative abundance of a single pollen species greater than 25% in a coal sample was taken to indicate that the plants that produced the pollen were common, possibly dominant, in a plant community in or adjacent to a mire. Plants represented by the arboreal palynofacies are assumed to have been trees because most living species of the families they represent are trees. The arboreal palynofacies includes pollen of fossil species of the Platanaceae, Fagaceae, and Juglandaceae. Some specimens are illustrated in Figure 12.4. Cowardin *et al.* (1979) illustrated several modern palustrine forested-wetland depositional environments that may be modern analogs in a structural sense, although the species compositions

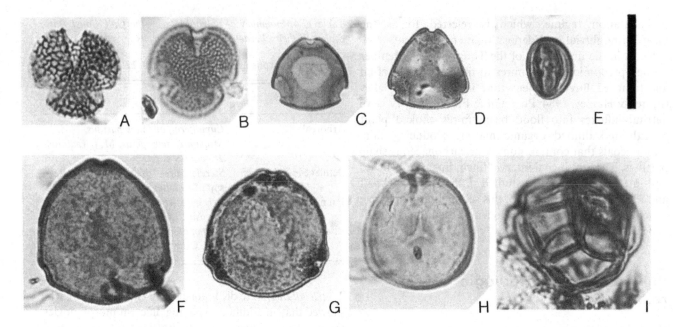

Figure 12.4. Photomicrographs (bright field) of palynomorphs characteristic of the arboreal palynofacies (A–E) and the Betulaceae–*Sphagnum* palynofacies (F–I) of the Gulf Coast Basin. (A,B) *Tricolpites* spp. (Platanaceae); (C,D) *Momipites* spp. (Juglandaceae); (E) *Cupuliferoipollenites* sp. (aff. *Castanea*, Fagaceae); (F) *Triporopollenites* sp. (cf. *Corylus*, Betulaceae); (G) *Triporopollenites* sp. (Betulaceae); (H) *Stereisporites* sp. (aff. *Sphagnum*, Sphagnaceae); (I) *Ericipites* sp. (Ericaceae). Scale bar equals 25 μm.

of the Paleocene and modern plant communities are not the same.

Pollen of the Platanaceae and Fagaceae dominate palynomorph assemblages from most of the coal beds formed in the east Texas fluvial system. The trees presumably inhabited forested wetlands upstream from the Wilcox delta system in late Paleocene time, especially along the margins of abandoned channels and oxbow lakes in which mires developed. The trees may or may not have contributed directly to the accumulation of peat, but their pollen in great abundance serves to characterize a distinctive coal-depositional environment, one associated with the tributaries of the Wilcox fluvial system. As mentioned, these coal deposits are thin, elongate, and sinuous in form; they are rich in macerals of the vitrinite group (Traverse & Nichols, 1967).

Similar depositional environments and associated plant communities are represented by the arboreal palynofacies in delta plain coal samples from central Texas, but in that area pollen of the Juglandaceae, especially *Momipites tenuipolus*, also is notably abundant. On the delta the Juglandaceae evidently was an additional major component of the forest vegetation. In some coal samples from both the fluvial and deltaic depositional environments, pollen of several species of Betulaceae is common, and although these species as well as the Platanaceae, Fagaceae, and Juglandaceae may represent arboreal vegetation, betulaceous pollen is more closely associated with other palynomorphs in a distinct palynofacies that represents a different local depositional environment, one that included shrubs and moss.

BETULACAE-*SPHAGNUM* PALYNOFACIES

Palynomorph assemblages from some coal beds that

formed in the east Texas fluvial system contain high relative abundances of pollen of Betulaceae (birch family) and common spores of *Sphagnum* (sphagnum moss). These assemblages also include low abundances of pollen tetrads referable to the Ericaceae (heath family), which are absent from other Wilcox assemblages. Some specimens are illustrated in Figure 12.4. This group of pollen and spores constitutes a distinctive palynofacies, but it may represent more than one plant community in or near coal-forming mires.

Betulaceous pollen constitutes as much as 80% of the assemblage in five samples in which it is the dominant species and would seem to represent an important species group in one of the coal-forming environments of the Wilcox Group. The betulaceous pollen can be attributed to several modern genera in the Betulaceae (including *Corylus*, as mentioned). Living genera of the Betulaceae are not known to dominate the plant communities of mires, but modern *Betula* may inhabit bottomlands along streams (Gleason & Cronquist, 1964); fossil species of the Betulaceae probably inhabited bottomlands and levees in early Tertiary time. Paleocene species of the Betulaceae inhabited the margins of Gulf Coast mires, and being anemophilous (and consequently producing vast numbers of pollen grains per plant), presumably they contributed considerable quantities of pollen to peat accumulating nearby. This pattern of distribution of trees producing betulaceous pollen is reasonable because the mires in the Wilcox fluvial system of east Texas developed in abandoned channels and oxbow lakes adjacent to major channels. There would have been an extensive network of levees of both active and abandoned channels in this depositional setting on which the trees could grow.

The relative abundance of spores of *Sphagnum* may not clearly reflect the actual importance of these plants in either plant communities or sedimentary environments. Modern species of *Sphagnum* often reproduce vegetatively, and the plants can be much more common in mires than the number of spores produced would seem to indicate (T. A. Ager, pers. commun., 1990). Similarly, the low abundance of ericaceous pollen present in samples does not truly reflect the importance of heath plants in plant communities. The Ericaceae are entomophilous, and like all such plants, they characteristically produce relatively small numbers of pollen grains. Heaths are almost always under-represented by their pollen in modern plant communities.

From the Betulaceae–*Sphagnum* palynofacies we infer that some mires in which east Texas Wilcox peat accumulated were vegetated in part by sphagnum moss and in part by a low, shrubby flora of heaths. Among modern wetland environments illustrated by Cowardin *et al.* (1979), these ancient environments may have more closely resembled modern scrub-shrub wetlands with subordinate *Sphagnum* than the moss-lichen wetlands dominated by sphagnum moss. The Paleocene mires probably were permanently water-saturated although seldom inundated. Presence of *Sphagnum* indicates acidic water in the mire (Smith, 1938). The presence of a shrubby heath flora in this depositional environment may not by itself account for the abundance of macerals of the vitrinite group (derived from wood) that are found in east Texas Wilcox coals (Traverse & Nichols, 1967). Possibly the vitrinitic macerals represent betulaceous trees that grew on the levees and that eventually fell into the mire, in addition to the woody stems of heath shrubs. Seams that are predominantly woody at the top may record terrestrialization of wetland mires, in which the mire plant communities progressed from dominantly mossy and shrubby to forested.

The Betulaceae–*Sphagnum* palynofacies is present also in some coal deposits of the central Texas Wilcox delta system. Here it is interpreted to represent depositional environments in abandoned channels similar to those that existed in the east Texas fluvial system. This suggests that the delta plain supported a greater diversity of plant communities than did the fluvial system environments upstream from the delta.

PALM–FERN PALYNOFACIES

Several assemblages from the central Texas delta system are characterized by abundances of up to 30% of pollen referable to modern palms (Arecaceae), species of *Arecipites* and *Calamuspollenites*. The assemblages also include notable abundances of fern spores, mainly species of *Laevigatosporites* (Polypodiaceae). These species define the palm–fern palynofacies. Some specimens are illustrated in Figure 12.5. The vegetation that produced this palynoflora is considered to have occupied, at least at times, the large inter-distributary areas of the delta. Fisher and McGowen (1967) envisioned these inter-distributary areas as similar to the modern 'marsh' (emergent wetland) environments of the Mississippi River delta plain, but palynofacies analysis does not support this interpretation. The palm–fern palynofacies indicates that the Wilcox delta was largely covered by communities dominated by palms; ferns and some angiosperm species were present in these communities, also. As mentioned, the plant families Cyperaceae (sedge) and Gramineae (grass) that make up emergent wetland vegetation in modern environments had not yet evolved and were absent.

We interpret the palm–fern palynofacies to represent a community of plants having mostly non-woody tissues. Although they attain tree-size, palms as monocots do not have the solid, woody stem that is produced by a bifacial cambium. This interpretation is consistent with the observation that coal beds from the delta deposits, although generally thicker, have lower relative abundances of vitrinite-group macerals than do the fluvial coal beds in east Texas (Traverse & Nichols, 1967; Nichols & Traverse, 1971). It is likely that some woody trees such as species of *Nyssa*, *Carya*, and the aforementioned Betulaceae were present in these mires in a minor role similar to that evident in the Powder River Basin (discussed later). The presence of such trees in these mires may account for some of the vitrinite found in the coals formed in interdistributary areas on the delta. As mentioned, the Betulaceae and *Sphagnum* palynofacies has a subordinate presence in the delta coal deposits; some vitrinite in delta coal may have derived from the plant community or communities represented by this palynofacies.

BISACCATE–DINOCYST PALYNOFACIES

This palynofacies is characterized primarily by the presence of rare marine dinoflagellate cysts and a greater abundance of bisaccate pollen than in other palynomorph assemblages from the Wilcox Group and secondarily by the presence of pollen (cf. Fig. 12.5E) referable to Chenopodiaceae (goosefoot family), a family whose modern species include some common coastal plants. This palynofacies represents a sedimentary environment, but not an individual plant community, and it is present mainly in samples from the bay-lagoon depositional

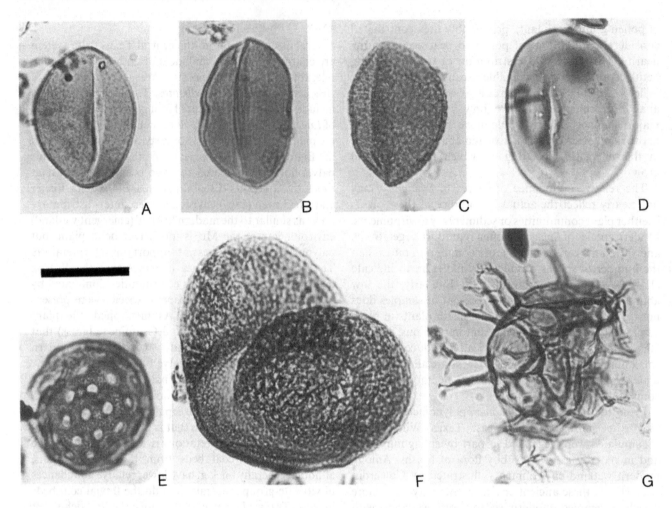

Figure 12.5. Photomicrographs (bright field) of palynomorphs characteristic of the palm-fern palynofacies (A–D) and the bisaccate-dinocyst palynofacies (E–G) of the Gulf Coast Basin. (A,B) *Arecipites* spp. (*Arecaceae*); (C) *Calamuspollenites* sp. (Arecaceae); (D) *Laevigatosporites* sp. (Polypodiaceae); (E) *Chenopodipollis* sp. (aff. Chenopodiaceae); (F) *Pityosporites* sp. (aff. *Pinus*); (G) chorate marine dinocyst (unidentified). Scale bar equals 25 μm.

system of south Texas. It is present also in a few samples from central Texas; there it is interpreted to represent deposits on distal lobes of the delta, where the delta was dissected by tidal channels or was adjacent to embayments. The dinocysts presumably were washed into lagoonal mires during high tides or storms. Bisaccate pollen referable to *Pinus* and *Picea* is often found in modern marine sediments, commonly in areas far from the location of trees producing the pollen (Traverse & Ginsburg, 1966). Bisaccate pollen presumably accumulates in marine sediments through long distance aquatic and aerial transport. The trees that produced the fossil bisaccate pollen of the Wilcox bisaccate–dinocyst palynofacies are not thought to have occupied the mires that produced the Paleocene coal. Bisaccate pollen washed into mires from the sea or perhaps was blown in. Species of Chenopodiaceae presumably would have grown in coastal dune, mudflat, and subaerial delta front depositional environments. Possibly analogous modern environments illustrated by Cowardin *et al.* (1979) are dominated by species of families that evolved in post-Paleocene time.

Thus the palynofacies in the Gulf Coast Basin to some extent represent plant communities, but to a greater extent represent depositional environments. Peat accumulated in different depositional settings, and the palynofacies within the coal reveal some of the differences in these coal-forming environments. In contrast, contemporaneous coal deposits in the Powder River Basin developed within a narrower range of sedimentological conditions, and palynofacies within those coals tend to more directly represent the Paleocene plant communities of that region.

Powder River Basin

MAJOR ASPECTS

Palynologically the coal beds of the Tongue River Member are characterized by overwhelming abundance

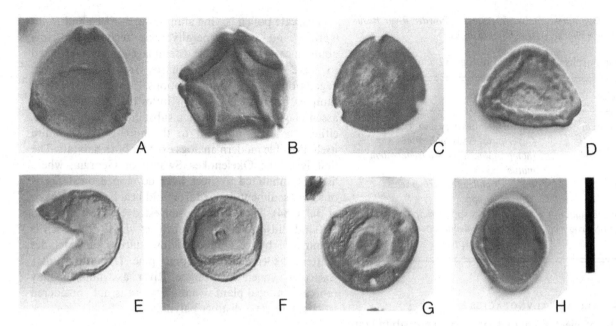

Figure 12.6. Photomicrographs (interference contrast) of palynomorphs representative of the regional forest vegetation (A–D) and palynomorphs characteristic of the mixed arboreal palynofacies (E–H) of the Powder River Basin, Montana and Wyoming. (A) *Triporopollenites* sp. (Betulaceae); (B) *Alnipollenites* sp. (aff. *Alnus*, Betulaceae); (C) *Momipites* sp. (Juglandaceae); (D) *Ulmipollenites* sp. (Ulmaceae); (E,F) *Taxodiaceaepollenites* sp. (aff. *Glyptostrobus*, Taxodiaceae); (G) *Caryapollenites* sp. (aff. *Carya*, Juglandaceae); (H) *Nyssapollenites* sp. (cf. *Nyssa*, Nyssaceae). Scale bar equals 25 μm.

of pollen of probable taxodiaceous affinity. The morphologic similarity of the pollen among genera and species within the families Taxodiaceae, Taxaceae, and Cupressaceae makes it difficult to determine the precise affinities of fossil members of these families. In this paper we follow Pocknall and Flores (1987) and use the genus name *Glyptostrobus*, although Pocknall and Flores (1987) did note that they could not rule out a close affinity between the Paleocene taxodiaceous pollen and the modern genus *Taxodium* (bald cypress). At present *Glyptostrobus* is restricted to forested wetlands in southeastern China, but megafossils of *Glyptostrobus* are present in the Paleocene and Eocene Willwood Formation in the Bighorn Basin, Wyoming (Wing, 1984), and in the Paleocene and Eocene Golden Valley Formation in the Williston Basin, North Dakota (Hickey, 1977), locations that are geographically close to the Powder River Basin. At present *Taxodium* grows almost exclusively in forested wetlands, primarily in the southeastern United States (Gleason & Cronquist, 1964). We realize that the fossils may not actually represent either of the modern genera, *Glyptostrobus* or *Taxodium*, but there is little doubt that taxodiaceous trees were the dominant plants in the late Paleocene vegetation of the Powder River Basin (Leffingwell, 1970; Tschudy

1976; Satchell, 1985; Pocknall, 1987). Pocknall and Flores concluded that these mires were not totally covered by monotypic stands of these trees, but instead had a complex pattern of plant communities. The distribution of the plant communities was governed by sedimentary environment. The characteristic palynofacies of these environments are enumerated following a discussion of the regional vegetation of the Powder River Basin in late Paleocene time, as it is represented by the palynoflora.

Palynomorphs representing species of the Betulaceae (*Betula, Alnus*), Juglandaceae (*Engelhardia–Alfaroa* group), and Ulmaceae (*Ulmus–Zelkova* group) are consistently recorded in the palynofloras throughout the Tongue River Member. They are interpreted as representing the regional vegetation. Some specimens are illustrated in Figure 12.6; see also Leffingwell (1970). Although the affinity of the fossil pollen with the genera specified is clear, it is uncertain just what role these plant taxa had in the Paleocene vegetation of the Powder River Basin. All of the genera have modern species that grow in areas that are periodically inundated, or on levees, in wetland forests, or on rolling uplands; all of these settings existed in the Powder River Basin. Fluctuations in relative abundances of pollen in coal samples suggest frequent change in the distribution of plants within the mire, supporting the concept of a mosaic pattern of distribution of plant communities within the basin, as suggested by Pocknall and Flores (1987). They considered consistent abundances of 3–5% to be indicative of a regional source for the pollen. Against this background, palynofacies representing individual plant communities in differing sedimentary environments can be distinguished. The palynomorphs that characterize these palynofacies are summarized in Table 12.5.

Table 12.5. *Summary of palynofacies in the Powder River Basin coals and their characteristic palynomorphs*

The palynomorphs are illustrated in Figs. 12.6 and 12.7.

Palynofacies	Palynomorphs
Mixed arboreal	*Caryapollenites imparalis, C. inelegans, C. veripites, Nyssapollenites thompsonianus, Taxodiaceaepollenites hiatus*
Riparian	*Polyatriopollenites vermontensis, Tricolpites* spp.
Ericaceae-*Sphagnum*	*Stereisporites* spp., *Ericipites* spp.
Pediastrum-Spirogyra	*Pediastrum paleogeneites, Schizophacus* spp.

MIXED ARBOREAL PALYNOFACIES

This palynofacies is characterized by very abundant *Glyptostrobus* and varying amounts of ancestral species of *Nyssa* (sour gum) and *Carya* (hickory, pecan). Some specimens are illustrated in Figure 12.6. The typically high abundances (greater than 80%) of *Glyptostrobus* in some assemblages from the Anderson coal seam (meandering – fluvial depositional environment) was suggested by Pocknall and Flores (1987) to indicate existence of a water table higher than the peat surface. Under such conditions there would have been areas that were permanently inundated and that would have supported almost monotypic stands of *Glyptostrobus*. The growth habit of *Glyptostrobus* trees, which have buttressed trunks and slender, conical-shaped crowns, would facilitate their flourishing in such a habitat. A closed canopy of *Glyptostrobus* trees would result in a paucity of understory plants and low influx of pollen from the regional vegetation.

At other places in the mire where drainage improved as a result of peat accumulation and the peat surface became subaerial and subjected only to infrequent inundations, perhaps during flood events, broadleaved hardwood plants evidently invaded the *Glyptostrobus* forest and a more diverse plant community developed. This community had a mosaic distribution, as described by Pocknall and Flores (1987) for the Anderson coal seam. A more open forest of *Glyptostrobus* together with *Nyssa* and *Carya* is postulated for this community. In addition, plants of lower stature may have had an ephemeral presence, probably in response to frequent minor fluctuations in the height of the water table, differential accumulation and compaction of peat, and irregular opening of the canopy in response to tree fall. Perturbations in this finely balanced system resulted in local extirpations and later recolonization of other areas where more suitable habitats prevailed.

Bisaccate pollen having affinities with *Picea* and *Pinus* (spruce and pine) generally was recorded in low abundance in Tongue River assemblages and is interpreted to have been derived primarily from the regional vegetation. There are exceptions, however. In some samples, bisaccate pollen constitutes 10–20% of the total assemblage, which indicates a substantial local presence, either on levees adjacent to the mire or in the mire itself. Possible modern analogs exist for both habitats. The first is in the Okefenokee Swamp of Georgia, where 'islands' inhabited by trees have developed on deposits of sandy sediment derived from old beach ridges (Cohen *et al.*, 1984). The second is suggested by a scrub-shrub wetland illustrated by Cowardin *et al.* (1979), which is dominated by vegetation of low stature but includes scattered specimens of pond pine (*Pinus serotina*). Bisaccate pollen in Tongue River assemblages may represent a local plant community but is not considered to characterize a distinct palynofacies.

RIPARIAN PALYNOFACIES

Assemblages recovered from partings in the Anderson coal seam (meandering–fluvial depositional environment) contain the riparian palynofacies, which is characterized by, and may be dominated by, pollen of a fossil species of *Pterocarya*, a genus of the Juglandaceae, referred to form-genus *Polyatriopollenites*. Some specimens are illustrated in Figure 12.7. This genus is extant in southeastern Asia where it occupies riparian sites along stream banks or on levees where there is good drainage (Pocknall & Flores, 1987). *Pterocarya* pollen is found in many of the coal seams in the lower part of the Tongue River Member and is well represented in the Anderson seam in particular. Where present, *Pterocarya* pollen has several stratigraphically limited peaks in relative abundance. This pattern of distribution can be used to identify periods of detrital influx into the mire and the proximity of the sample site to the margin of the mire or to fluvial channels.

Pollen of the Platanaceae tends to dominate the riparian palynofacies in the upper part of the Tongue River Member, in the interval including the Smith coal (anastomosed–fluvial depositional environment). Here the Platanaceae seem to have been more prominent than *Pterocarya*, and often grew in association with Ulmaceae (elm family) and *Alnus* (alder). Following detrital influx in both the lower and upper depositional systems of the Tongue River Member, riparian plants apparently were eliminated by recolonization and migration of major mire plant communities toward the channel (Pocknall & Flores, 1987).

ERICACEAE-*SPHAGNUM* PALYNOFACIES

The Ericaceae–*Sphagnum* palynofacies is present in the anastomosed–fluvial depositional environment of the

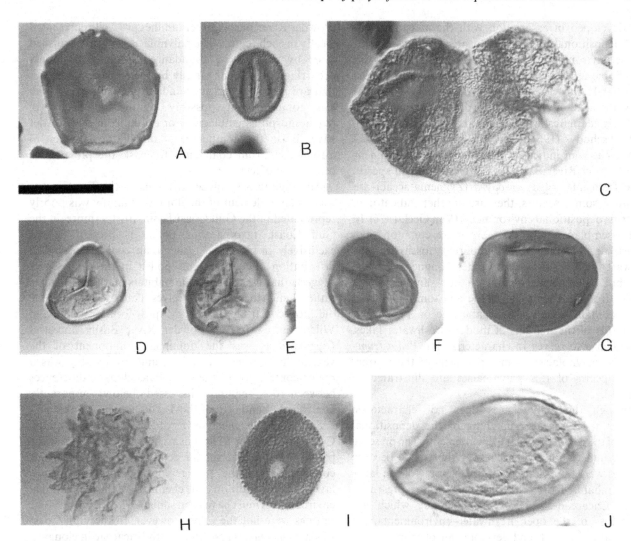

Figure 12.7. Photomicrographs (interference contrast) of palynomorphs characteristic of the riparian palynofacies (A–B), the Ericacea–*Sphagnum* palynofacies (D–G), and the *Pediastrum–Spirogyra* palynofacies (H–J) of the Powder River Basin; bisaccate pollen (C) may represent a local plant community but does not characterize a distinct palynofacies. (A) *Polyatriopollenites* sp. (aff. *Pterocarya*, Juglandaceae); (B) *Tricolpites* sp. (Platanaceae); (C) *Pityosporites* sp. (aff. *Pinus*, *Pinaceae*); (D,E) *Stereisporites* spp. (aff. *Sphagnum*, Sphagnaceae); (F) *Ericipites* sp. (Ericaceae); (G) *Laevigatosporites* sp. (Polypodiaceae); (H) *Pediastrum* sp. (Hydrodictyaceae); (I) *Sparganiaceaepollenites* sp. (aff. Sparganiaceae or Typhaceae); (J) *Schizophacus* sp. (aff. *Spirogyra*, Zygnemataceae). Scale bar equals 25 μm.

Smith coal interval of the Tongue River Member. Spores of sphagnum moss are an important component of this palynofacies and are numerically dominant in some assemblages. They are present together with spores of the fern family Polypodiaceae and pollen of Ericaceae. This palynofacies is similar to the Betulaceae–*Sphagnum* palynofacies of the Gulf Coast Basin. Some specimens are illustrated in Figure 12.7. Presence of these taxa suggests vegetation of low stature (Pocknall & Flores, 1987). This

vegetation could have existed on nutrient-poor substrates where no other plants could compete. Modern species of *Sphagnum* require acidic conditions (Smith, 1938), and it is likely that their distribution reflects the chemistry of the mire. Sphagnum moss and ferns (both terrestrial and aquatic species) influence sedimentation within the depositional environment and the development of peat. Through progressive encroachment and trapping of sediments they can cause the lake to be completely infilled. This process may ultimately result in localized doming of the surface of the mire (ombrogenous bog of Gore, 1983). Pocknall and Flores concluded, however, largely on sedimentologic evidence, that mires of the Smith coal interval were low-lying.

PEDIASTRUM–SPIROGYRA PALYNOFACIES

This palynofacies constitutes a minor component of Powder River Basin palynomorph assemblages and represents a lacustrine depositional environment rather than a mire. In late Paleocene time in the basin, ephemeral lakes developed in association with the anastomosed–fluvial depositional system. Freshwater molluscs thrived in these environments (Hanley & Flores, 1987), as did algae. Some assemblages in non-coal samples from the Smith coal interval contain *Pediastrum*, a colonial freshwater alga of the family Hydrodictyaceae that is found in modern freshwater ponds and lakes (Hutchinson, 1967). Tschudy (1961) was the first to use fossil *Pediastrum* as an indicator of sedimentary environment in the Powder River Basin region. Cysts of fossil species of the freshwater alga *Spirogyra* (Zygnemataceae) are present in some samples; they are another indicator of lacustrine depositional environments (Van Geel, 1976). In some assemblages these cysts are more common than *Pediastrum*. Fleming (1987) reported fossils of the freshwater alga *Scenedesmus* in other Tongue River lake beds, but we recognized none in our samples. Some samples from lake deposits of the anastomosed–fluvial system contain pollen of Sparganiaceae or Typhaceae, inhabitants of the margins of modern freshwater lakes. The pollen also serves to characterize the *Pediastrum–Spirogyra* palynofacies. Some specimens of the characteristic species of this palynofacies are illustrated in Figure 12.7.

Although this palynofacies does not characterize coal, *per se*, it reveals something of the depositional history of the coal-depositional environments. The presence of Sparganiaceae or Typhaceae was short-lived because the lakes were temporary. Margins of some of the lakes became inhabited by terrestrial vegetation (represented by the Ericaceae–*Sphagnum* palynofacies), which led to the loss of the open freshwater environments, the accumulation of peat, and development of mires.

Comparisons

The composition of the vegetation in the Gulf Coast and Powder River Basins was essentially similar, but there are important differences in the relative abundances of the various elements. Our study focuses on the contrasts in the plant communities that inhabited the contemporaneous depositional environments in which peat accumulated, and on the ways in which these contrasts are revealed by the palynofacies of the coal deposits. Table 12.6 summarizes the major relationships that are evident, and the following discussion enlarges on some of the more important aspects of comparison.

It must be emphasized that the palynofacies discussed here are subtle and may not be easily recognized without detailed analysis. As noted, the regional vegetation in both the Gulf Coast and Powder River Basins in late Paleocene time appears to have been composed largely of species of Betulaceae and Juglandaceae, together with Platanaceae and Fagaceae (Gulf Coast Basin) or Ulmaceae

(Powder River Basin); other families are well represented locally in each region. The palynofacies are distinguished more by the relative abundances of these commonly occurring groups than simply by their presence. Species that are present in great abundance tend to overwhelm less common but possibly significant species in palynomorph assemblages. For example, in the Powder River Basin the dominance of taxodiaceous pollen in assemblages from coal tends to mask the presence of other palynofacies.

Although in the Powder River Basin the Taxodiaceae was a major element of the flora, that family was poorly represented in the Gulf Coast Basin. Furthermore, in the Gulf Coast, taxodiaceous pollen is present almost exclusively in claystone rather than coal samples. The distribution of taxodiaceous pollen in the Gulf Coast suggests that the trees lived in habitats marginal to mires rather than in the mires themselves. Thus the Taxodiaceae did not constitute important peat-forming plants in Wilcox mires. In the Powder River Basin, ancestral *Glyptostrobus* was the dominant component of the vegetation, especially in mires, and presumably was a major contributor to the peat. Paleoclimatic differences between the regions (discussed below) apparently controlled abundance of the Taxodiaceae.

Spores of *Sphagnum* are recorded from both the Gulf Coast and Powder River Basins, apparently representing plants that fulfilled similar ecologic roles in the plant communities of the two regions. Because *Sphagnum* favors acidic conditions, the chemistry of the depositional environments must have been similar. The sedimentologic settings were not the same, however. In the Gulf Coast Basin, *Sphagnum* is associated with Ericaceae in elongate, sinuous mires that developed in abandoned channels and oxbow lakes; Betulaceae grew on levees bordering the abandoned channels and may have contributed to the accumulation of woody peat. In the Powder River Basin, *Sphagnum* is common in the early stages of development of mires of considerable lateral extent within the anastomosed–fluvial depositional environment of the upper part of the Tongue River Member. There it initially colonized the margins of freshwater lakes together with Ericaceae and Polypodiaceae. In this setting the Betulaceae were not part of the peat-forming vegetation. In both regions the vegetation represented by palynofacies characterized by sphagnum spores tended to be replaced by vegetation represented by palynofacies including arboreal pollen of forest communities, evidently as a result of terrestrialization and plant succession.

The climate in the Gulf Coast and Powder River Basins differed in late Paleocene time, but in both regions conditions were warmer and wetter than at present. The vegetation of the Powder River Basin is considered to have developed under essentially warm mesothermal climatic conditions (average temperature about 17 °C),

Table 12.6. *Summary of depositional environments, coal bed characteristics, palynofacies, and predominant plant families (or genera) in the upper Paleocene of the Gulf Coast and Powder River Basins as determined from stratigraphic, sedimentologic, and palynofacies evidence*

Environment	Coal beds	Palynofacies	Vegetation
		Gulf Coast Basin	
Fluvial	thin (<2 m), woody	Arboreal	Platanaceae, Fagaceae, Juglandaceae (*Engelhardia-Alfaroa*)
	thick (>2 m), woody	Betulaceae-*Sphagnum*	Betulaceae (*Corylus*), Sphagnaceae, Ericaceae
Deltaic	thick (3–5 m), non-woody	Palm-fern	Arecaceae, Polypodiaceae Juglandaceae (*Engelhardia-Alfaroa*)
Lagoonal	thin, high-ash	Bisaccate-dinocyst	Pinaceae, Chenopodiaceae, Dinophyceae
		Powder River Basin	
Anastomosed-fluvial	thin (<2), high-ash	Ericaceae-*Sphagnum*	Ericaceae, Sphagnaceae, Polypodiaceae
	(partings)	Riparian	Platanaceae, Ulmaceae, Betulaceae (*Alnus*)
Meandering-fluvial	thick (up to 7 m), low-ash	Mixed arboreal	Taxodiaceae, Nyssaceae, Juglandaceae (*Carya*)
	(partings)	Riparian	Juglandaceae (*Pterocarya*)

while in the Gulf Coast Basin during the late Paleocene the climate was megathermal (average temperature above 25 °C) (J. A. Wolfe, pers. commun., 1989; see also Nichols *et al.*, 1988). Some evidence of the influence of paleoclimate on the vegetation in the two regions is the greater abundance of palms in the Gulf Coast Basin than in the Powder River Basin, as indicated by relative abundance of palm pollen. Palms even contributed to the accumulation of peat on parts of the Wilcox delta plain. Today palms are widespread in subtropical and tropical regions of the world but are essentially absent in cooler climates. There is also a significant difference in species diversity between the two regions that is attributable to paleoclimate. In the coal and associated rocks of the Tongue River Member in the Powder River Basin, approximately 75 palynomorph species were recorded (Pocknall & Nichols, in press); in contrast, in the coal and associated rocks of the Wilcox Group in the Gulf Coast Basin, there were more than twice as many palynomorph species (Nichols, 1970). In modern

floras the highest diversity tends to be in the tropics and there is a progressive decrease in diversity as distance from the equator increases. Climatic control of diversity is inferred to account for this observation in the modern world and doubtless had a similar effect in North America in the late Paleocene.

Palynomorph sedimentation

In addition to the influence of paleoclimate, we interpret patterns of palynomorph species diversity as attributable to sedimentary environment in the Gulf Coast and Powder River Basins. Although fluvial depositional environments existed in both regions, they differed in detail. Two distinct fluvial environments existed in the Powder River Basin, and neither was like that of the Gulf Coast. Evidence for freshwater lake environments in the Powder River Basin and the marine coastal environment associated with the delta system and lagoons of the Gulf

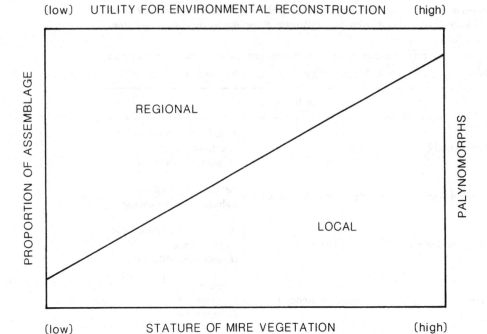

Figure 12.8.   Schematic representation of the influence of the stature of mire vegetation (especially the presence of a closed canopy in plant communities of high stature) on the utility of mire palynofloras in reconstruction of environment.

Coast Basin is clear, based on the presence of fossils of freshwater algae and marine dinoflagellates, respectively. Deltaic and lagoonal environments of the Gulf Coast had no parallel in the Powder River Basin.

Evidence from the Gulf Coast Basin shows a general trend of increasing diversity from fluvial to deltaic environments (Nichols & Traverse, 1971) that is in accord with interpretation of the fluvial depositional system as one of restricted forests and the delta depositional system as one having a variety of plant communities. Perhaps reinforcing this influence is the direct control on diversity by sedimentation discussed by Nichols and Traverse, who attributed the greater diversity of palynomorphs in Wilcox delta deposits than in fluvial deposits to be a result of greater clastic sediment input and greater numbers of transported palynomorphs in the delta depositional environment. Our samples of clastic sedimentary rocks from both the Gulf Coast and Powder River Basins exhibited higher species diversity than those obtained from coals, indicating the presence of an allochthonous palynomorph component in assemblages from allochthonous sediments. It is obvious also, of course, that peat-forming mires sometimes have a restricted flora.

The most important aspect of palynomorph sedimentation in this study is the way in which sedimentation can affect the composition of palynofacies, which in turn affects the utility of palynofacies in interpreting depositional environments. Thus it is appropriate to consider the nature of the Gulf Coast and Powder River Basin depositional environments to assess whether reasonably accurate representations of plant communities

were preserved. Because the goal of this study was to recognize palynofacies that represent different local coal-depositional environments, coal was the predominant lithology sampled. Coal may be assumed to preserve a record of local vegetation better than other kinds of sedimentary rock, but not all coals are alike. Palynomorph sedimentation (including fallout and catchment) in mires from which the coal was derived depended on the nature of the vegetation cover. An open, low vegetation would be expected to provide a catchment area in which palynomorphs from both local and regional sources would be deposited. In this setting, the regional vegetation would be well represented but could tend to mask the local vegetation, which is obviously of greater interest for interpretation of the local environment. High-standing vegetation tends to filter out palynomorphs representing the regional vegetation, especially where a forest canopy exists, and in this setting, the local vegetation would be better represented. These generalizations are depicted schematically in Figure 12.8.

In the Powder River Basin the evidence indicates that high-standing taxodiaceous trees were dominant in forested mires; they may at times have formed a canopy closed to the influx of airborne pollen from the regional vegetation. High relative abundance of taxodiaceous pollen would be predicted in these mire sediments, which is what we found in the majority of samples of coal from the Tongue River Member. The interpretation of a mixed forest community in these mires is based on observed increases in the pollen of arboreal plants such as *Nyssa* and *Carya* in some samples; both genera have modern species that inhabit wetland depositional environments and are interpreted to have been subordinate members of the forested mire community in the Powder River Basin. Thus palynomorphs deposited in these mires probably are predominantly of local origin.

Alternatively, presence of arboreal pollen suggests opening of the forest canopy, an increase in contribution from the regional vegetation. Both *Nyssa* and *Carya* are wind-pollinated, and the pollen could have been derived from trees at some distance from the site of deposition. Both of these genera are present in the Gulf Coast Basin where they are considered as part of the regional vegetation rather than as members of plant communities in coal-depositional environments.

Palynomorphs produced by riparian species or by communities inhabiting the margins of mires are common in clastic partings within coal beds in both regions. This pollen probably represents a local depositional environment, because the plants that produced it presumably colonized freshly deposited, probably fertile sediments from crevasse splay and overbank deposition. In the Powder River Basin, rivers were long-lived, deeply incised, and had well developed levees (Flores, 1983) that supported riparian plants for long periods of time (Pocknall & Flores, 1987). Species of the Betulaceae, Platanaceae and *Pterocarya* (Juglandaceae) inhabited the levees. Most of their pollen would have been deposited directly into the water, but some could have been deposited in adjacent mires. Pollen of riparian plants is rare in Powder River Basin coal samples, which suggests that most of it was filtered out by the arboreal vegetation of mires.

In both regions we interpret the abundance of *Sphagnum* spores and the presence of Ericaceae to represent similar plant communities and environments of deposition. As mentioned, *Sphagnum* may be under-represented by its spores because of its tendency to reproduce vegetatively. We interpret the common occurrence of these spores to indicate a substantial presence of sphagnum moss in both regions in late Paleocene time. Because sphagnum moss is low in stature, depositional environments in which it dwelled would be expected also to include a substantial abundance of palynomorphs contributed by the regional vegetation, perhaps especially by plants growing at the margins of the mire. Thus palynomorphs deposited in this environment were of both local and regional origin.

Finally, there are the predominantly aquatic depositional environments in each region, marine coastal lagoons in the Gulf Coast Basin and freshwater lakes in the Powder River Basin. Palynomorph sedimentation in both of these differing aquatic environments would be expected to include major contributions from non-local vegetation. Gulf Coast assemblages from the lagoonal facies contain a marked abundance of bisaccate pollen of pinaceous conifers, probably as a result of the tendency of pollen having this morphology to stay afloat and be transported in both air and water. In this instance conifers appear to be over-represented by their pollen. Rare marine dinocysts in these deposits represent a very different adjacent depositional environment. In the Powder River Basin, wind-dispersed pollen from taxodiaceous conifers accumulated in high relative abundance in lakes that existed at times during deposition of the Tongue River Member. These lacustrine deposits reflect the local dominance of taxodiaceous conifers in the Powder River Basin, in addition to preserving the record of freshwater algae in the lakes and aquatic angiosperms growing along their shores.

Conclusions

Using palynofacies, it is possible to characterize different environments within contemporaneous fluvial and deltaic systems of late Paleocene age, especially those in which coal-forming peat was deposited. In the Gulf Coast Basin the following palynofacies have been identified: arboreal, Betulaceae–*Sphagnum*, palm-fern, and bisaccate-dinocyst. In the Powder River Basin the following palynofacies have been identified: mixed arboreal, riparian, Ericaceae–*Sphagnum*, and *Pediastrum*–*Spirogyra*.

The components of the vegetation were similar in the two regions, but differed in their relative importance, as reflected by differing palynofacies. In both basins, Paleocene species of the modern family Juglandaceae were common. In the Gulf Coast Basin, fossil species of Platanaceae, Fagaceae, Betulaceae, and Sphagnaceae were dominant in the fluvial environment and were present in the deltaic environment together with Arecaceae and Polypodiaceae. Taxodiaceous conifers were present but apparently did not grow in Gulf Coast mires in any abundance. In the Powder River Basin, taxodiaceaeous conifers (ancestral *Glyptostrobus*) were a dominant constituent of a mixed forest; Betulaceae, Arecaceae, and Sphagnaceae were present as minor constituents. Riparian plants such as Platanaceae and a fossil species of *Pterocarya* (Juglandaceae) had an ephemeral presence. In both regions the over-representation in the pollen record of certain numerically dominant palynomorphs tends to obscure the significance of other plants that were important components of wetlands and marginal riparian communities, complicating recognition of the palynofacies.

The primary controls on the composition of the mire vegetation in the two regions were sedimentary environment and paleoclimate. The late Paleocene climate of the Gulf Coast Basin was megathermal; in the Powder River

Basin it was warm mesothermal. The thickest coals in the Gulf Coast developed in a deltaic depositional environment, in interdistributary mires on delta plains; the thick coals in the Powder River Basin (which in general are considerably thicker than those in the Gulf Coast) developed in a meandering–fluvial depositional environment. The thick coals of the Gulf Coast are characterized by a palm-fern palynofacies; they developed in mires inhabited largely by palms and ferns. Palms and ferns were present in the Powder River Basin but did not constitute major components of the vegetation there. The thick coals of the Powder River Basin are characterized by a mixed arboreal palynofacies; they developed in mires inhabited by an arboreal plant community dominated by taxodiaceous trees. Taxodiads were present in the Gulf Coast Basin but were not a major component of the vegetation. Aspects of the depositional environments in the two regions other than sediment-ology and paleoclimate (e.g., basin tectonics, which we have not investigated) doubtless also influenced the development of thick coal beds.

In this study of palynofacies in coal deposits from two contemporaneous sedimentary basins we have had the benefit of knowing much of the depositional history of the coal-forming peat. This allowed us to develop clearer pictures of the relationships between the vegetation and the environments of deposition than would otherwise have been possible. We have shown that determination of the distribution of palynofacies in relation to sedimentary environments can be a useful approach in geological and palynological studies and may have paleoclimatic implications. As a logical extension of this work, it should be possible in future studies to interpret similar nonmarine depositional environments in areas where detailed stratigraphic and sedimentologic information is not available.

References

Bates, R. L., & Jackson, J. A., eds. (1987). *Glossary of Geology*, 3rd edition. Alexandria, VA: American Geological Institute.

Berggren, W. A. (1965). Some problems of Paleocene-lower Eocene planktonic foraminiferal correlations. *Micropaleontology*, **11**, 278–300.

Berry, E. W. (1916). The lower Eocene floras of southeastern North America. *US Geological Survey Professional Paper* **91**.

Berry, E. W. (1930). Revision of the lower Eocene Wilcox flora of the southeastern states. *US Geological Survey Professional Paper* **156**.

Brown, R. W. (1962). Paleocene flora of the Rocky Mountains and Great Plains. *US Geological Survey Professional Paper* **375**.

Cohen, A. D., Casagrande, D. J., Andrejko, M. J., & Best, G. R. (1984). *The Okefenokee Swamp – Its Natural History, Geology, and Geochemistry*. Los Alamos, New Mexico: Wetland Surveys.

Combaz, A. (1964). Les palynofaciès. *Revue de Micropaléontologie*, **7**, 205–18.

Cowardin, L. M., Carter, V., Golet, F. C., & LaRoe, E. T. (1979). Classification of wetlands and deepwater habitats of the United States. *US Fish and Wildlife Service*, FWS/OBS-79/31.

Dodd, J. R., & Stanton, R. J., Jr. (1981). *Paleoecology, Concepts and Applications*. New York: Wiley-Interscience.

Doher, L. I. (1980). Palynomorph preparation procedures currently used in the paleontology and stratigraphy laboratories, US Geological Survey. *US Geological Survey Circular*, **830**, 1–29.

Fairchild, W. W., & Elsik, W. C. (1969). Characteristic palynomorphs of the lower Tertiary in the Gulf Coast. *Palaeontographica*, **B 128**, 81–9.

Fisher, W. L. (1961). Stratigraphic names in the Midway and Wilcox Groups of the Gulf Coastal Plain. *Transactions of the Gulf Coast Association of Geological Societies*, **11**, 263–95.

Fisher, W. L. & McGowen, J. H. (1967). Depositional systems in the Wilcox Group of Texas and their relationship to the occurrence of oil and gas. *Transactions of the Gulf Coast Association of Geological Societies*, **17**, 105–25.

Fleming, R. F. (1987). (abstract) Fossil *Scenedesmus* (Chlorophyta) and its paleoecological significance. *Palynology*, **11**, 227.

Flores, R. M. (1981). Coal deposition of fluvial paleoenvironments of the Paleocene Tongue River Member of the Fort Union Formation, Powder River area, Powder River Basin, Wyoming and Montana. In *Recent and Ancient Nonmarine Depositional Environments: Models for Exploration*, ed. F. G. Ethridge & R. M. Flores. *Society of Economic Paleontologists and Mineralogists Special Publication*, **31**, 169–90.

Flores, R. M. (1983). Basin facies analysis of coal-rich Tertiary fluvial deposits, northern Powder River Basin, Montana and Wyoming. In *Modern and Ancient Fluvial Systems*, ed. J. D. Collinson & J. Lewin. *International Association of Sedimentology Special Publication*, **6**, 501–15.

Flores, R. M. & Hanley, J. H. (1984). Anastomosed and associated coal-bearing fluvial deposits: upper Tongue River Member, Paleocene Fort Union Formation, northern Powder River Basin, Wyoming, U.S.A. In *Sedimentology of Coal and Coal-bearing Sequences*, ed. R. Rahmani & R. M. Flores. *International Association of Sedimentology Special Publication*, **7**, 85–105.

Gleason, H. A. & Cronquist, A. (1964). *The Natural Geography of Plants*. New York: Columbia University Press.

Gore, A. J. P. (1983). *Ecosystems of the World*, vol. **4A**, *Mires – Swamp, Bog, Fen, and Moor*. Amsterdam: Elsevier.

Hanley, J. H. & Flores, R. M. (1987). Taphonomy and paleoecology of nonmarine Mollusca: indicators of alluvial plain lacustrine sedimentation, upper part of the Tongue River Member, Fort Union Formation (Paleocene), northern Powder River Basin, Wyoming and Montana. *Palaios*, **2**, 479–96.

Hay, W. W. & Mohler, H. P. (1967). Calcareous nannoplankton from early Tertiary rocks at Pont Labau, France, and Paleocene-early Eocene correlations. *Journal of Paleontology*, **41**, 1505–41.

Hickey, L. J. (1977). Stratigraphy and paleobotany of the Golden Valley Formation (early Tertiary) of western North Dakota. *Geological Society of America Memoir* **150**.

Hutchinson, G. E. (1967). *A Treatise on Limnology*, vol. **2**,

Introduction to Lake Biology and the Limnoplankton. New York: John Wiley & Sons.

Leffingwell, H. A. (1970). Palynology of the Lance (Late Cretaceous) and Fort Union (Paleocene) Formations of the type Lance area, Wyoming. In *Symposium on Palynology of the Late Cretaceous and Early Tertiary*, ed. R. M. Kosanke & A. T. Cross. *Geological Society of America Special Paper* **127**.

Muller, Jan (1981). Fossil pollen records of extant angiosperms. *Botanical Review*, **47**, 1–142.

Nichols, D. J. (1970). Palynology in relation to depositional environments of lignite in the Wilcox Group (early Tertiary) in Texas. Ph.D. dissertation, Pennsylvania State University.

Nichols, D. J. (1973). North American and European species of *Momipites* ('*Engelhardtia*') and related genera. *Geoscience and Man*, **7**, 103–17.

Nichols, D. J. & Traverse, A. (1971). Palynology, petrology, and depositional environments of some early Tertiary lignites in Texas. *Geoscience and Man*, **3**, 37–48.

Nichols, D. J., Wolfe, J. A., & Pocknall, D. T. (1988). Latest Cretaceous and early Tertiary history of vegetation in the Powder River Basin, Montana and Wyoming. In *Geological Society of America Field Trip Guidebook 1988*, ed. G. S. Holden. *Colorado School of Mines Professional Contributions* **12**, 205–10, 222–6.

Pocknall, D. T. (1986a). (abstract) Composition of plant communities in and near Late Paleocene coal swamps, Fort Union Formation, Powder River Basin, Wyoming-Montana. *Palynology*, **10**, 256–7.

Pocknall, D. T. (1986b). Palynological data from the Fort Union and Wasatch Formations, Powder River Basin, Wyoming and Montana. *US Geological Survey Open-File Report* **86–117**.

Pocknall, D. T. (1987). Palynomorph biozones for the Fort Union and Wasatch Formations (upper Paleocene-lower Eocene), Powder River Basin, Wyoming and Montana. *Palynology*, **11**, 23–35.

Pocknall, D. T. & Flores, R. M. (1987). Coal palynology and sedimentology in the Tongue River Member, Fort Union Formation, Powder River Basin, Wyoming. *Palaios*, **2**, 133–45.

Pocknall, D. T., & Nichols, D. J. (in press). Palynology of coal zones of the Tongue River Member (upper Paleocene) of the Fort Union Formation, northern Powder River Basin, Montana and Wyoming. *US Geological Survey Professional Paper*.

Satchell, L. S. (1984). (abstract) Reconstruction of vegetation and environments from palynology of Paleocene coals from Wyoming, U. S. A. *Sixth International Palynological Conference Calgary 1984 Abstracts*, p. 146.

Satchell, L. S. (1985). (abstract) Climate and depositional environment of *Glyptostrobus* forest swamps that formed thick low-ash coals in the Paleocene Powder River Basin. *The Society for Organic Petrology 2nd Annual Meeting Abstracts*, 11.

Smith, G. M. (1938). *Cryptogamic Botany.* vol. 2, *Bryophytes and Pteridophytes*. New York: McGraw-Hill.

Taylor, D. W. (1990). Paleogeographic relationships of angiosperms from the Cretaceous and early Tertiary of the North American area. *Botanical Review*, **56**, 279–417.

Traverse, A. (1988). *Paleopalynology*. Boston: Unwin Hyman.

Traverse, A. & Ginsburg, R. N. (1966). Palynology of the surface sediments of Great Bahama Bank, as related to water movement and sedimentation. *Marine Geology*, **4**, 417–59.

Traverse, A. & Nichols, D. J. (1967). (abstract) Palynological and petrological characteristics of the commercial lignites of Texas. *Geological Society of America, Program 1967 Annual Meetings*, 224.

Tschudy, R. H. (1961). Palynomorphs as indicators of facies environments in Upper Cretaceous and lower Tertiary strata, Colorado and Wyoming. *Wyoming Geological Association Guidebook 16th Annual Field Conference*, 53-9.

Tschudy, R. H. (1976). Pollen changes near the Fort Union-Wasatch boundary, Powder River basin. *Wyoming Geological Association Guidebook 28th Annual Field Conference*, 73–81.

Van Geel, B. (1976). Fossil spores of Zygnemataceae in ditches of a prehistoric settlement in Hoogkarspel (The Netherlands). *Review of Palaeobotany and Palynology*, **22**, 337–44.

Wing, S. L. (1984). Relations of paleovegetation to geometry and cyclicity of some fluvial carbonaceous deposits. *Journal of Sedimentary Petrology*, **54**, 52–66.

III Reconstruction of late Cenozoic vegetation and sedimentary environments from palynological data

13 Quaternary terrestrial sediments and spatial scale: the limits to interpretation

RICHARD BRADSHAW

Introduction

The Quaternary Period can serve as a model for earlier times in demonstrating the diversity of data that can be analyzed to reveal the nature of past biota and environments. Late Quaternary palynomorphs are a particularly versatile and informative group of fossils because they are preserved in a variety of deposits, and in most cases can be referred to existing genera. Pollen can be used to infer past climates, to reconstruct former vegetation, and to provide an environmental framework for archaeological investigations, amongst other applications. The characteristics of pollen production, dispersal, deposition and preservation at each site dictate the type of application to which the individual site is suited, and determine the detail and reliability of the inferred reconstructions. In this chapter I explore some of the methodologies of environmental and community reconstruction that exploit particular dispersal and sedimentation properties of pollen grains. These same properties impose the limits for the level of detail that can be justified in the final interpretation.

The concept of a source area of pollen origin for each collecting basin is introduced by Jackson in Chapter 14 (this volume). This source area, which is controlled by the method of pollen dispersal, dictates the type of reconstruction that can be created from any given site. The size of these source areas has been investigated both theoretically and empirically. Tauber (1965) began from first principles by using the physical laws governing the dispersion of small particles. Studies of pollen dispersal from isolated trees or small, discrete stands developed this approach (e.g. Turner 1964; Janssen 1984), and Kabailiene (1969) presented a sophisticated algebraic analysis of pollen dispersal over an entire landscape. The complexity of natural vegetation makes the complete mathematical representation of pollen dispersal over the landscape extremely challenging. Prentice (1985) suggested some reasonable simplifying assumptions that complemented the more empirical studies of pollen-vegetation

relationships of Bradshaw and Webb (1985) and Jackson (Chap. 14 this volume). It is the latter, empirical approach which underlies the discussion in this chapter.

Regional climatic reconstruction is best carried out using sites with pollen source areas of at least 10^5 ha. Studies of local forest succession need a source area as small as 0.3 ha, while archaeological applications often require an intermediate area. I shall consider applications that exploit large source areas, and work down to the most spatially detailed palynological investigations.

The regional scale

Many studies have demonstrated the ability of pollen studies to resolve vegetational pattern at the regional or landscape scale. Lamb (1984) collected surface sediments from 71 lakes in Labrador, and analyzed the 28 most abundant taxa using principal components analysis (PCA) (Fig. 13.1). Five broad vegetation types were distinguished, and they served as a framework for analysis of fossil data.

Pollen data sets are multivariate, because several taxa are counted from multiple samples and locations. Analyses and comparisons of the data sets are enhanced by application of multivariate statistics (Prentice, 1986). PCA was chosen by Lamb to display the optimal two-dimensional relationships of the set of surface pollen samples, that are the palynomorphic representation of the present vegetation. Lamb (1984) distinguished zones containing similar samples from his fossil sites using a statigraphically-constrained cluster analysis (Gordon & Birks, 1972). He superimposed his zoned fossil pollen sequences onto the PCA analysis of his surface samples to assist in the reconstruction of past vegetation. Birks and Berglund (1979) used similar methods to compare zoned fossil pollen sequences from two sites in southern Sweden (Fig. 13.2). The application of multivariate statistical techniques to pollen data has been particularly valuable in revealing underlying patterns at the regional scale.

239

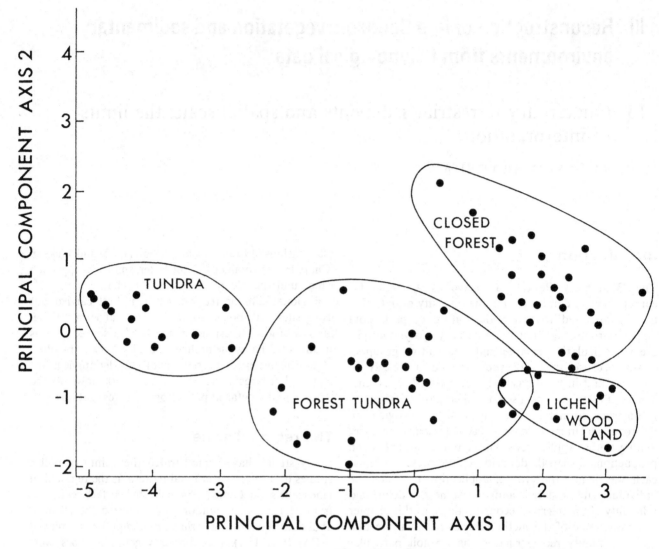

Figure 13.1. Scatter diagram showing modern pollen spectra obtained from the surface sediments of lakes in Labrador. The samples are positioned on the first two axes of a principal components analysis. The samples have been grouped into different vegetation zones. (Reproduced by permission of the author and Blackwell Scientific Publications Ltd., from Lamb, 1984).

The reconstruction of past climate from fossil pollen data is a common research goal, and a discussion of the methodologies and underlying assumptions assists in evaluation of the results. The most sophisticated methods use large sets of pollen surface samples and associated climatic data from each site as the data base for generating past climatic data from fossil pollen counts (Bartlein *et al.*, 1986; Guiot, 1987). An underlying assumption is that the spatial distribution of pollen is largely a reflection of climate at a sufficiently broad scale of observation. Bartlein *et al.* (1986) selected 830 pollen spectra from eastern and central USA and Canada, and plotted the 'response' of individual pollen types throughout this region to mean July temperature and annual precipitation (Fig. 13.3). The pollen spectra came from collecting basins ranging between 25 and 150 ha in area.

Matching a given fossil pollen spectrum to its closest modern analogue is frequently necessary in climatic and vegetational reconstruction. This is another problem involving multivariate statistics examined by Overpeck

et al. (1985). They developed and tested an elegant method of multivariate comparison using a variety of dissimilarity coefficients. They found that the squared-chord distance coefficient was best able to discriminate between samples from different regional units or formations of vegetation, and performed well when the authors attempted to 're-build' pollen diagrams using the closest modern analogue to each fossil sample. This coefficient weights the less common pollen types somewhat in the comparison, but does not give them equal importance with the dominant pollen types. This is a reasonable compromise, given that little is known about the pollen-vegetation relationships of many taxa. The squared-chord distance coefficient showed some ability to discriminate between

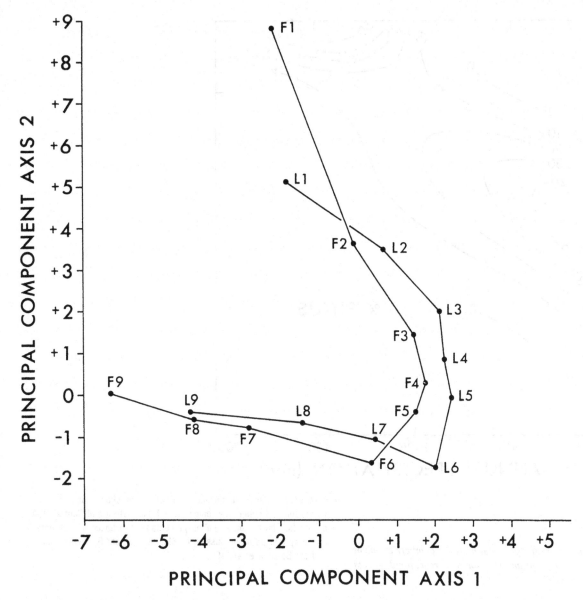

Figure 13.2. Comparison of two fossil pollen sites from southern Sweden using principal components analysis. Numerical methods were used to define the local pollen zones which are numbered F1-F9 and L1-L9 (Reproduced by permission of the author and the publisher, Universitetsforlaget, Tøyen, Norway, from Birks & Berglund, 1979, in *Boreas*, **8**, 257-79).

pollen samples from different forest types within a formation. However the majority of their 1618 modern pollen samples were collected from lakes of moderate size, imposing limits on the vegetational detail that could be resolved.

Bradshaw and Webb (1985) showed that there was a relationship between basin size and pollen source area, and that significant quantities of *Pinus* and *Quercus* pollen (among the most buoyant pollen types) travelled over 30 km to basins in the size range 30–150 ha (Table 13.1). These observations were supported by computerized

simulations that considered aerial dispersion of pollen alone, and showed that sites in this size range were well-suited for maximizing pollen representation from regional sources (Prentice 1985). Clearly basin size is an important consideration in the selection of sites for climatic reconstruction. If the pollen source area were smaller, the pollen spectra would more likely be heavily influenced by local factors such as soil type or disturbance history.

Regional pollen data are now being drawn together into large data-banks, creating the possibility of new types of research. Maps showing the changing distribution of pollen types across continents during the Holocene are proving to be valuable research tools (Huntley & Birks, 1983), and refinement of statistical techniques has led to new forms of data exploration. Jacobson *et al.* (1987) used detrended correspondence analysis (DCA) to plot

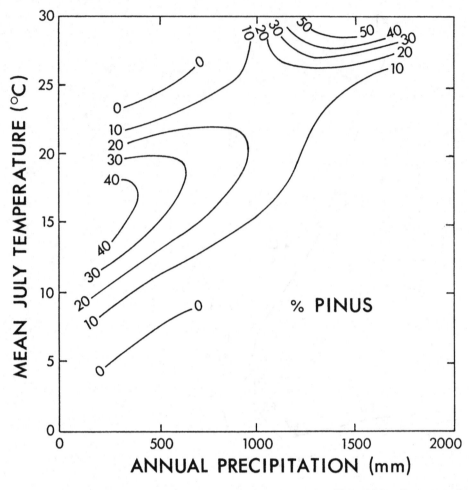

Figure 13.3. Response surface for *Pinus* pollen shown for annual precipitation and mean July temperature from Eastern and Central USA and Canada. The contoured pollen percentages are based on 830 data points (Reproduced by permission of Blackwell Scientific Publications Ltd., from Bartlein *et al.*, 1986).

Table 13.1. *Linear regression estimates (geometric mean solution) for the y-intercept and its standard error, for data sets from Wisconsin, USA*

The figures may be regarded as estimates of the percentage values of pollen that have travelled from beyond the area surveyed for vegetation (after Bradshaw & Webb, 1985).

Site size (ha)	9	9	Moss polsters
Area surveyed for vegetation (km²)	64	64	64
Number of sites	20	44	12
Pollen types			
Pinus	8.76 ± 4.48	15.71 ± 2.28	3.39 ± 8.55
Quercus	9.19 ± 2.59	7.80 ± 0.96	4.41 ± 2.30
Betula	13.52 ± 2.23	4.81 ± 2.10	1.40 ± 5.71
Tsuga	2.93 ± 0.82	1.95 ± 1.54	1.55 ± 3.56
Ulmus	3.52 ± 0.43	1.39 ± 0.44	−0.64 ± 2.06
Fagus	0.64 ± 0.14	−0.07 ± 0.31	−0.45 ± 0.88
Acer	1.09 ± 0.44	−0.40 ± 0.72	−1.69 ± 1.80

rates of palynological change at numerous sites in the USA over the last 18 000 years (Fig. 13.4). (DCA has proved to have various statistical and ecological advantages over PCA in the analysis of percentage pollen data; see Prentice, 1986). Rapid, synchronous change took place over a large region at 13 000, 12 300, and 10 000 years BP, times at which ice-volume changes were also large (Mix & Ruddiman, 1985). Shifting patterns of atmospheric circulation were the proposed driving force for both types of change (Jacobson *et al.*, 1987).

Sites that chiefly sample pollen of regional origin are best suited for the study of past climates. These regional sites also sample pollen of more local origin, and suitable site density and selection can permit analysis of more local patterns in vegetation (Fig. 13.5). Surface lake and peat samples from eastern continental USA recorded the present distribution of *Quercus* trees. Denser sampling from Wisconsin showed that state-wide patterns in

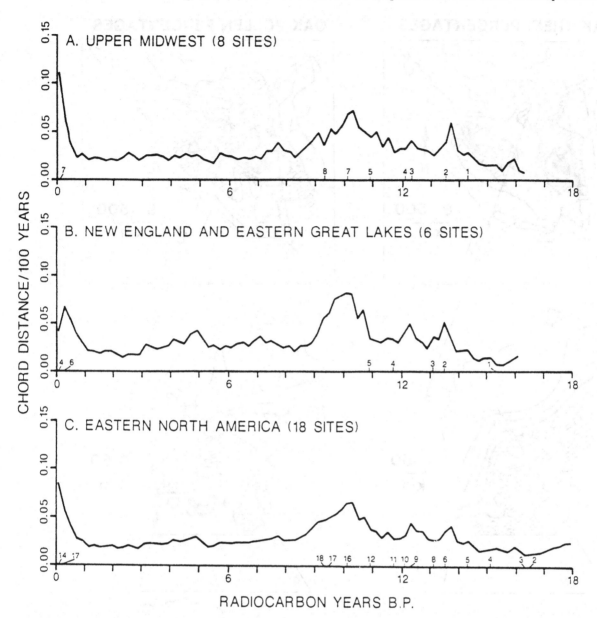

Figure 13.4.   Average rates of palynological change for differing regions within the USA during the last 18 000 years. The small number above the horizontal axis indicates the numbers of sites used to calculate the average at that point between that point and the next one to the right. (A) Upper Midwest; (B) New England and eastern Great Lakes; (C) Eastern North America (Reproduced by permission of the author from Jacobson *et al.*, 1987).

vegetation could be resolved by pollen data. Sixty-four lake and peat surface pollen samples from a single county in Wisconsin (40×30 km) all indicated the status of *Quercus* trees in the county when viewed at the regional scale (Figs. 13.5, 13.6). *Quercus* comprised 0–20% of trees within the county when surveyed with a vegetation plot size of 2800 km^2. Yet the same pollen data reflected details of the distribution of *Quercus* trees within the county when the vegetation plot size was reduced to 64 km^2

(Figs. 13.5, 13.6). At this scale *Quercus* trees were absent from some plots but comprised 60% of others. Much of the 'noise' in the pollen data at the regional spatial scale resolved into the 'signal' at the local scale (Webb *et al.*, 1978). Studies of contemporary pollen-vegetation relationships reveal the hierarchical nature of regional pollen data, which contain information about both regional and local vegetation. This characteristic becomes more apparent when peatland sites are studied in isolation.

Pollen deposition onto peatlands: the intermediate scale

Pollen grains are well preserved in peat deposits, but their modes of transport and deposition differ in several

OAK TREE PERCENTAGES

OAK POLLEN PERCENTAGES

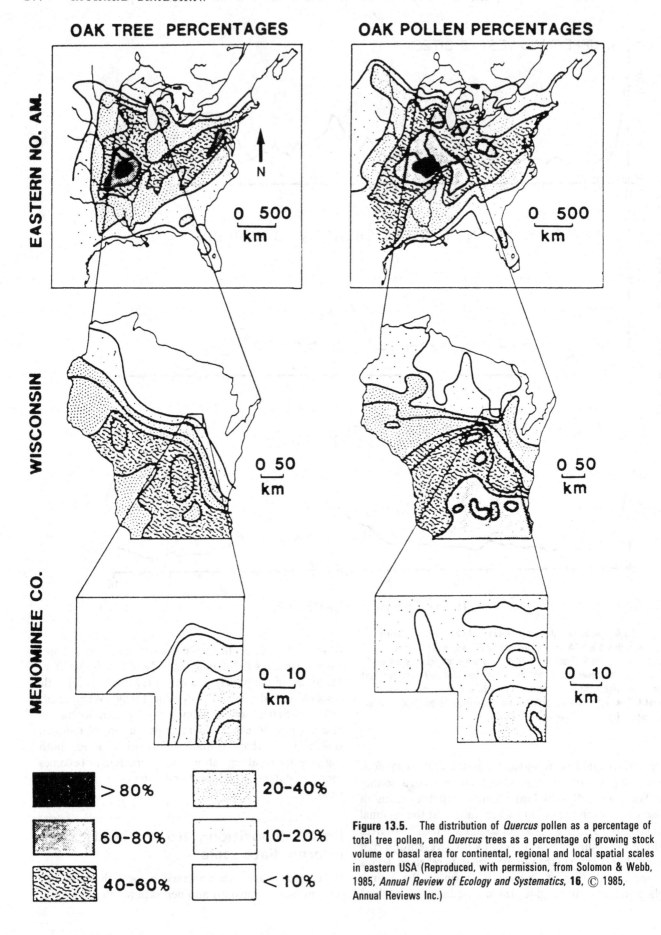

Figure 13.5. The distribution of *Quercus* pollen as a percentage of total tree pollen, and *Quercus* trees as a percentage of growing stock volume or basal area for continental, regional and local spatial scales in eastern USA (Reproduced, with permission, from Solomon & Webb, 1985, *Annual Review of Ecology and Systematics*, **16**, © 1985, Annual Reviews Inc.)

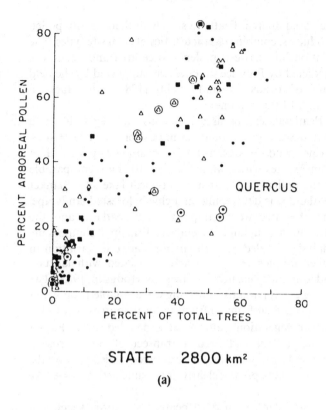

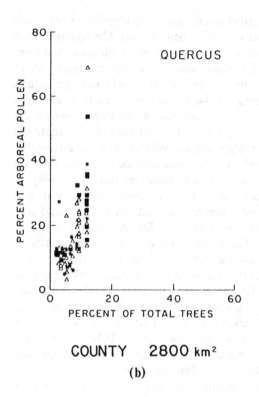

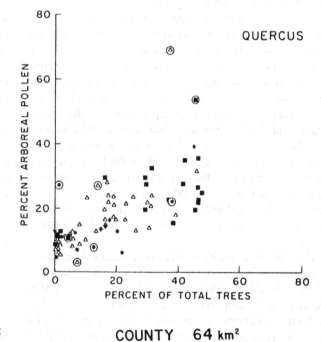

Figure 13.6. The relationship between *Quercus* pollen as a percentage of total tree pollen from lake and peatland surface samples, and *Quercus* trees as a percentage of growing stock volume or basal area at three spatial scales. (a) Data from all of Wisconsin and Michigan (upper peninsula), with 2800 km² of forest surveyed around each pollen sample; (b) Data from Menominee Indian Reservation (40 × 30 km area within Wisconsin), with the same forest area surveyed as in (a). The scatter of points can be regarded as a subset of (a); (c) The same pollen data from Menominee Indian reservation as in (b), but with only 64 km² of forest surveyed around each pollen sample. The 'noise' in the pollen data at the 2800 km² scale has resolved into a 'signal' at the local scale. Pollen data from regional sites contain information about tree distributions at both regional and local spatial scales. The different symbols refer to different sediment types.

important aspects to these processes at lake sites. All transport is aerial, and there is minimal horizontal or vertical mixing of pollen after deposition. Turner and Peglar (1988) discuss criteria by which vertical mixing might be recognised in pollen diagrams from peatlands, and present data from a peatland sampled at intervals

of only one millimeter, where mixing is undetected. By contrast, aquatic transport and thorough pre-depositional mixing are common processes in lake sites, and vertical mixing is absent only from lakes showing annual laminations in their sediments (Saarnisto, 1986).

Production of pollen of interest within the collecting

basin is a further characteristic distinguishing peatlands from lakes as sites for pollen analysis. The aquatic pollen types produced in lakes can be easily distinguished from the terrestrial pollen used to reconstruct past upland vegetation, but this distinction is less clear in peatland sites. Forested peatlands support trees such as *Picea mariana* or *Acer rubrum* that can be widespread on the landscape, and pollen of local origin may dominate the pollen assemblage. Janssen (1984) studied the relationship between modern pollen assemblages and vegetation on a Minnesotan peatland, demonstrating the ability of palynological techniques to resolve both local and regional plant communities. Non-forested peatland sites characteristic of northwest Europe recruit a larger proportion of far-travelled pollen than forested peatlands, but on-site production can still be of importance. Evans and Moore (1985) showed a linear relationship between *Calluna* pollen deposition onto peat, and the abundance of *Calluna vulgaris* plants within one meter of each site. *Calluna* is often an important pollen type in the interpretation of landscape history. Pollen from many Minnesotan peatland species are also transported laterally for only a few meters (Janssen, 1984).

Pollen accumulation rates in peat deposits do not have such temporally smoothed profiles as may be found in lakes (Fig. 13.7). Peat accumulation rates often vary in an uneven manner through time (Aaby & Tauber, 1975), and on-site pollen production can lead to occasional over-representation of one type. The simple sedimentation process avoids the problems of sediment redistribution or 'focusing' of sediment in the deepest part of the basin that can confuse interpretation of lake sites (Davis *et al.*, 1984). The peak values of pollen accumulation rate for the lake Tyotjarvi were not apparent in the adjacent peat deposit, and probably resulted from sediment focusing when most of the lake sediment and its contained pollen was deposited around the sample site, instead of being evenly distributed throughout the basin as before (Fig. 13.7; Donner *et al.*, 1978; Jacobson & Bradshaw, 1981).

The relative simplicity of the dispersal and sedimentation properties of pollen in peatland basins, in contrast to lakes, stimulated Prentice (1985) to model pollen recruitment by basins representing different sized openings in the forest canopy. He showed that the source area for light, buoyant pollen grains was far larger than that for heavy types, and that the size of the non-forested area around the site was an important influence on the effective area of vegetation sampled. His model, which assumed purely aerial pollen dispersal, distinguished two primary types of site: one within the forest canopy dominated by local production, and the other under a substantial opening in the canopy dominated by pollen of more distant origin. The pollen source area of the latter type of site increased as the opening of the tree canopy

increased in area. Peatland sites with their on-site pollen producers, combine characteristics of both site types. The relationship between pollen source area and basin size predicted by Prentice's model was supported by the field data of Bradshaw and Webb (1985), and Jackson (Chap. 14 this volume).

Peatlands can be used to reconstruct vegetation at a spatial scale that would be impossible with lake sites. Turner and Hodgson (1983) examined pollen data from 38 sites within an area of 3×10^5 ha – comparable to the pollen-source area of a typical lake site. Blanket peatland was developing on a chiefly forested landscape, and the sites were sampling both peatland and the surrounding 'upland' vegetation. Principal components analysis revealed that 'the major source of variation in pollen frequencies was clearly associated with latitude and longitude "location"' (Turner & Hodgson, 1983). The sites were apparently collecting a reasonable proportion of their pollen from sufficiently local sources to resolve former vegetation patterns at a detailed scale. Earlier Turner (1970) had used a transect of sites across a peatland to pinpoint a clearance episode on the surrounding uplands: clearly impossible with a single lake site.

Pollen deposition onto peatlands records vegetation with sufficient spatial resolution to complement archaeological investigations. Ancient field systems have been buried by blanket peat in County Mayo on the Atlantic coast of Ireland (Caulfield, 1978). Pollen analysis of the peat, and the underlying soil profiles, revealed a complex interaction between climate and human activity in the spread of blanket peat (O'Connell, 1986). Peat formation began in the basin at 4000 years BP, but sufficiently dry areas remained to permit farming phases at 2750, 1950 and 1000 BP. These phases are recorded in the form of cereal and weed pollen in a deep peat section, and in soil profiles from within the cultivated fields. Cultivation was abandoned at each site as peat spread to engulf the fields, suggesting a relationship between peat initiation and human disturbance. The recruitment of pollen into the peat in this example was from a sufficiently local source area to provide a reference for the very localized point samples represented by the soil profiles.

Under the forest canopy: the local scale

Andersen (1970), working in Denmark, was the first to demonstrate the correspondence between pollen samples taken at single points on the forest floor, and the forest composition within the surrounding 20–30 m. Bradshaw (1981), and Heide and Bradshaw (1982) confirmed his observations in England and the USA, and further showed that the pollen-vegetation relationships at this micro-scale differed from those established at the regional

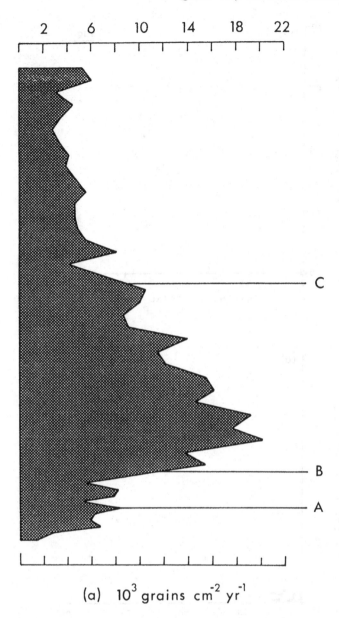

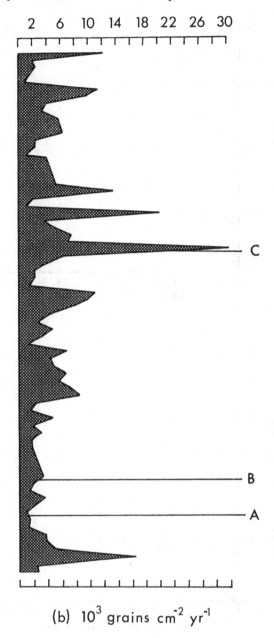

Figure 13.7. Pollen accumulation profiles for (a) sediments of a lake, and (b) an adjoining peat deposit in Finland. Note the relatively high internal variability in the accumulation rate for the peat section. Points A, B and C are synchronous at the two sites. (Reproduced by permission of the Quaternary Research Center, University of Washington and Academic Press, Inc., after Donner *et al.*, 1978, from Jacobson & Bradshaw, 1981.)

scale for the USA (Webb *et al.* 1981; see Fig. 13.8). Far-travelled pollen occurs in sites beneath the forest canopy, but is numerically overwhelmed by the on-site, local production. Suitable fossil sites such as small, wet hollows, mor humus profiles ('...dry, acidic accumulation of partially decomposed leaf litter of low biological activity': Bradshaw, 1988, p. 729), and buried soils document vegetational processes that are undetected in

more conventional sites. Sites under the forest canopy can present problems in the form of hiatuses or poor pollen preservation (Table 13.2), but the excellent spatial resolution arising from their small source areas makes them useful partners to regional pollen studies (Bradshaw, 1988). These types of sites are sensitive recorders of poorly dispersed pollen which is rarely recovered from lakes or peatlands.

Local sites act as a conceptual bridge between the regional scale covering thousands of hectares, and the scale of the 0.01 ha woodland quadrat of the contemporary plant ecologist. For example, ecologists had speculated for many years about the origin of relatively species-rich woodlands of native species on small, isolated islands in lakes of western Ireland. A number of authors suggested

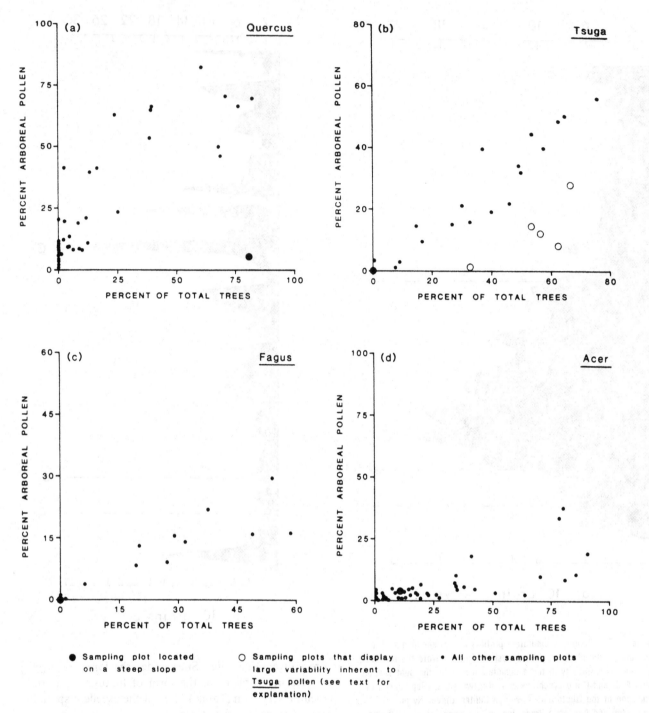

● Sampling plot located
on a steep slope

○ Sampling plots that display
large variability inherent to
Tsuga pollen (see text for
explanation)

• All other sampling plots

Figure 13.8. Scatter diagrams of pollen percentages against tree basal area percentages for trees growing within 20 m of the pollen samples, from Wisconsin and Upper Michigan, USA (Reproduced by permission of Elsevier Science Publishers from Heide & Bradshaw, 1982).

that they were relict examples of a formerly more widespread vegetation type. Regional pollen diagrams from the area lacked the spatial sensitivity to resolve the debate, but pollen profiles from deep, mor humus layers clearly revealed the recent vegetation history (Hannon & Bradshaw, 1989). An earlier species-rich woodland type had been destroyed by fire, and an open, disturbed community of *Calluna vulgaris* and *Pteridium aquilinum* had flourished for a while. About 300 years ago, a secondary succession was initiated involving the trees

Taxus baccata, *Sorbus aucuparia*, *Quercus petraea*, *Betula pubescens* and *Ilex aquifolium*. The present woodland was of recent origin arising after a period of human disturbance. The pollen in the humus was predominantly of local origin. Tree pollen types predominate in a landscape devoid of trees except on the

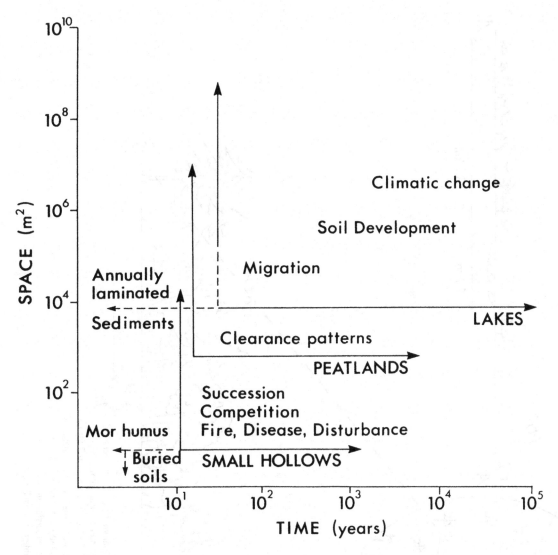

Figure 13.9. Lake and peat sites record vegetational change over large areas. Sites under the forest canopy record some of the same changes, but additionally resolve processes that occur at the scale of the individual woodland stand.

scattered islands. The enormous concentrations of pollen found, confirm that the pollen is of local origin. Low levels of far-travelled *Pinus* pollen, enabled correlation to be made with regional sites. *Ilex aquifolium* is the second most abundant pollen type at the present time with values exceeding 40% of total pollen. *Ilex* is primarily insect pollinated, and in regional diagrams is only sporadically recorded at trace levels.

Local sequences such as this one have opened a new field of application for palynology based on the exploitation of the dispersal properties of pollen. Pollen deposited very close to its point of origin can reveal details of forest dynamic processes of great interest to plant ecologists, but unanticipated by palynologists working at the regional scale.

Table 13.2. *Properties of importance for the successful pollen analysis of material that has accumulated under closed forest canopies*

	Small hollows	Mor humus	Soils
Length of record (years)	10^3	$10^2–10^3$	10^1
Continuity of record	may be patchy	good	poor
Temporal resolution (years/sample)	10–100	1–10	–
Mixing or downwash of pollen	minimal	minimal	yes
Pollen preservation	usually good	good	poor
Site availability	restricted	restricted	widespread

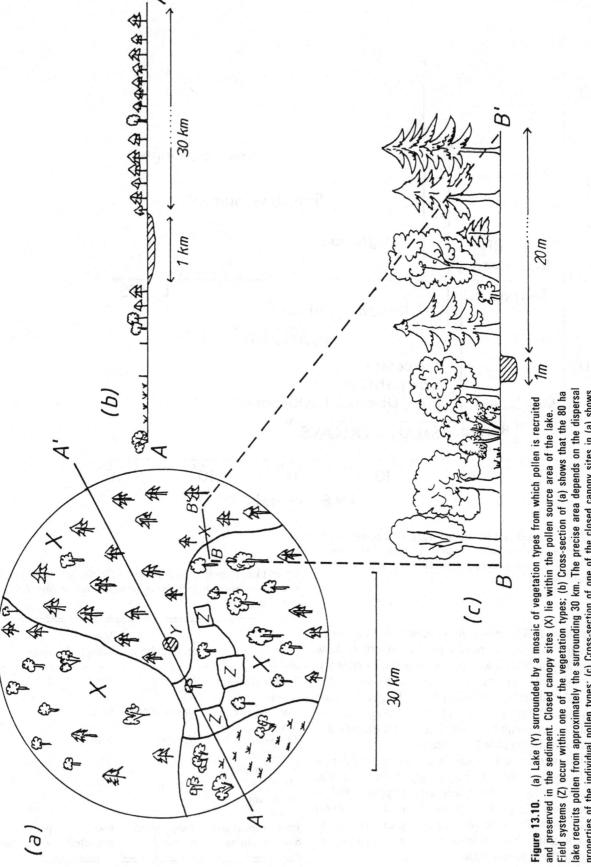

Figure 13.10. (a) Lake (Y) surrounded by a mosaic of vegetation types from which pollen is recruited and preserved in the sediment. Closed canopy sites (X) lie within the pollen source area of the lake. Field systems (Z) occur within one of the vegetation types; (b) Cross-section of (a) shows that the 80 ha lake recruits pollen from approximately the surrounding 30 km. The precise area depends on the dispersal properties of the individual pollen types; (c) Cross-section of one of the closed canopy sites in (a) shows that pollen is chiefly recruited from the surrounding 20 m, permitting spatial precision in the reconstruction of vegetation (Reproduced by permission of Kluwer Academic Publishers from Bradshaw, 1988).

The synthesis of scales

Pollen arriving at any site where it will be preserved has originated from sources at varying distance from that site. Pollen is transported different distances in different ways (Tauber, 1965; Jacobson & Bradshaw, 1981). Pollen travels up to hundreds of kilometers carried by winds in the atmosphere; aquatic transport into lakes is over shorter distances; most pollen is blown or washed just tens of meters to its final resting place, while a significant proportion simply falls to the ground from its parent plant. Judicious site selection can emphasize the sources desired to illuminate the problem under investigation. Processes operating at a number of spatial and temporal scales can be studied (Fig. 13.9). Lake and peatland sites show a certain degree of overlap in their modes of pollen recruitment and the processes that they can be used to investigate. Small hollows, mor humus and soils operate at a very distinctive local scale (Fig. 13.10), primarily recruiting pollen less than 30 m from its source. This fine spatial scale opens the door to a whole new area of investigation that is still in its infancy.

References

Aaby, B. & Tauber, H. (1975). Rates of peat formation in relation to degree of humification and local environment, as shown by studies of a raised bog in Denmark. *Boreas*, **4**, 1–17.

Andersen, S. T. (1970). The relative pollen productivity of North European trees, and correction factors for tree pollen spectra. *Danmarks Geologiske Undersøgelse* II, **96**, 1–99.

Bartlein, P. J., Prentice, I. C. & Webb, T., III (1986). Climatic response surfaces from pollen data for some eastern North American taxa. *Journal of Biogeography*, **13**, 35–57.

Birks, H. J. B. & Berglund, B. (1979). Holocene pollen stratigraphy of southern Sweden: a reappraisal using numerical methods. *Boreas*, **8**, 257–79.

Bradshaw, R. H. W. (1981). Modern pollen representation factors for woods in south-east England. *Journal of Ecology*, **69**, 45–70.

Bradshaw, R. H. W. (1988). Spatially-precise studies of forest dynamics. In *Handbook of Vegetation Science*, Vol.7. *Vegetation History*, ed. B. Huntley & T. Webb, III. Dordrecht: Kluwer Academic Publishers, 725–51.

Bradshaw, R. H. W. & Webb, T. III (1985). Relationships between contemporary pollen and vegetation data from Wisconsin and Michigan, USA. *Ecology*, **66**, 721–37.

Caulfield, S. (1978). Neolithic fields: the Irish evidence. In *Early Land Allotment in the British Isles. A Survey of Recent Work*, ed. H. C. Bowen & P. J. Fowler. B. A. R. British series, **48**, 137–43.

Davis, M. B., Moeller, R. E. & Ford, J. (1984). Sediment focusing and pollen influx. In *Lake Sediments and Environmental History*, ed. E. Y. Haworth & J. W. G. Lund. Leicester: University of Leicester Press. 261–93.

Donner, J. J., Alhonen, P., Eronen, J., Jungner, H. & Vuorela, L. (1978). Biostratigraphy and radiocarbon dating of the Holocene lake sediments of Tyotjarvi and the peats in the adjoining bog Varrassuo west of Lahti in southern Finland. *Annales Botanici Fennici*, **15**, 258–80.

Evans, A. T. & Moore, P. D. (1985). Surface pollen studies of *Calluna vulgaris* (L.) Hull and their relevance to the interpretation of bog and moorland pollen diagrams. *Circaea*, **3**, 173–8.

Gordon, A. D. & Birks, H. J. B. (1972). Numerical methods in Quaternary palaeoecology. I: Zonation of pollen diagrams. *New Phytologist*, **71**, 961–79.

Guiot, J. (1987). Late Quaternary climatic change in France estimated from multivariate pollen time series. *Quaternary Research*, **28**, 100–18.

Hannon, G. E. & Bradshaw, R. H. W. (1989). Recent vegetation dynamics on two Connemara Lake islands, western Ireland. *Journal of Biogeography*, **16**, 75–81.

Heide, K. M. & Bradshaw, R. H. W. (1982). The pollen-tree relationship within forests of Wisconsin and Upper Michigan, U.S.A. *Review of Palaeobotany and Palynology*, **36**, 1–23.

Huntley, B. & Birks, H. J. B. (1983). An atlas of past and present pollen maps for Europe: 0–13,000 years ago. Cambridge: Cambridge University Press.

Jacobson, G. L. & Bradshaw, R. H. W. (1981). The selection of sites for paleovegetational studies. *Quaternary Research*, **16**, 80–96.

Jacobson, G. L., Webb, T., III & Grimm, E. C. (1987). Patterns and rates of vegetation change during the deglaciation of eastern North America. In *North America and Adjacent Oceans During the Last Deglaciation*, vol. K-3, ed. W. F. Ruddiman & H. E. Wright. Boulder, Colorado: Geological Society of America, 277–88.

Janssen, C. R. (1984). Modern pollen assemblages and vegetation in the Myrtle Lake peatland, Minnesota. *Ecological Monographs*, **54**, 213–52.

Kabailiene, M. V. (1969). On formation of pollen spectra and restoration of vegetation. *Transactions of the Institute of Geology (Vilnius)*, **11**, 1–147. (In Russian with English and Lithuanian summaries.)

Lamb, H. F. (1984). Modern pollen spectra from Labrador and their use in reconstructing Holocene vegetational history. *Journal of Ecology*, **72**, 37–59.

Mix, A. C. & Ruddiman, W. F. (1985). Structure and timing of the last deglaciation; oxygen-isotope evidence. *Quaternary Science Reviews*, **4**, 59–108.

O'Connell, M. (1986). Reconstruction of local landscape development in the post-Atlantic based on palaeoecological investigations at Carrownaglogh prehistoric field system, County Mayo, Ireland. *Review of Palaeobotany and Palynology*, **49**, 117–76.

Overpeck, O., Webb, T., III & Prentice, I. C. (1985). Quantitative interpretation of fossil pollen spectra: dissimilarity coefficients and the method of modern analogs. *Quaternary Research*, **23**, 87–108.

Prentice, I. C. (1985). Pollen representation, source area, and basin size: towards a unified theory of pollen analysis. *Quaternary Research*, **23**, 76–86.

Prentice, I. C. (1986). Multivariate methods for data analysis. In *Handbook of Holocene Palaeoecology and Palaeohydrology*, ed. B. E. Berglund. New York: John Wiley & Sons, 775–97.

Saarnisto, M. (1986). Annually laminated lake sediments. In *Handbook of Holocene Palaeoecology and Palaeohydrology*, ed. B. E. Berglund. New York: John Wiley & Sons, 343–70.

Solomon, A. M. & Webb, T., III (1985). Computer aided

reconstruction of late-Quaternary landscape dynamics. *Annual Review of Ecology and Systematics*, **16**, 63–84.

Tauber, H. (1965). Differential pollen dispersion and the interpretation of pollen diagrams. *Danmarks Geologiske Undersøgelse* II, **89**, 1–69.

Turner, J. (1964). Surface samples from Ayrshire, Scotland. *Pollen et Spores*, **6**, 583–92.

Turner, J. (1970). Post-Neolithic disturbance of British vegetation. In *Studies in the Vegetational History of the British Isles*, ed. D. Walker & R. G. West. Cambridge: Cambridge University Press. 97–116.

Turner, J. & Hodgson, J. (1983). Studies in the vegetational history of the northern Pennines. III. Variations in the composition of the mid-Flandrian forests. *Journal of Ecology*, **71**, 95–118.

Turner, J. & Peglar, S. M. (1988). Temporally-precise studies of vegetation history. In *Handbook of Vegetation Science*, Vol.7, *Vegetation History*, ed. B. Huntley & T. Webb, III. Dordrecht: Kluwer Academic Publishers, 753–77.

Webb, T., III, Laseski, R. A. & Bernabo, J. C. (1978). Sensing vegetational patterns with pollen data: choosing the data. *Ecology*, **59**, 1151–63.

Webb, T., III, Howe, S., Bradshaw, R. H. W. & Heide, K. M. (1981). Estimating plant abundances from pollen percentages: the use of regression analysis. *Review of Palaeobotany and Palynology*, **34**, 269–300.

14 Pollen and spores in Quaternary lake sediments as sensors of vegetation composition: theoretical models and empirical evidence

STEPHEN T. JACKSON

Introduction

Virtually all applications of Quaternary pollen analysis, whether ecological, climatological, archeological, or stratigraphic, are ultimately concerned with inferring temporal changes in vegetation composition from pollen assemblages. Stratigraphic changes in pollen assemblages are assumed to record changes in vegetation composition, which in turn provide information about changes in species distribution, prevailing climate, human activities, and cultural resources. Within certain spatial and temporal realms, spatially separated sites contain similar stratigraphic changes in pollen assemblages. These similarities form the basis for classical pollen-stratigraphic zonation, and for chronostratigraphic correlation when independent evidence (e.g., radiocarbon dates) indicates that the zones are contemporaneous among nearby sites. The synchroneity derives from the pollen sequences at the individual sites having recorded the same or similar changes in vegetation composition.

The interest in studying pollen sequences that record past vegetational changes has led Quaternary palynologists to seek sedimentary deposits with good pollen preservation, and stable, continuous, datable deposition. Preservation must be sufficiently good that most pollen grains can be reliably assigned to morphotypes corresponding to extant plant taxa at or below the family level. A stable depositional environment is required to ensure that changes in pollen assemblages can be attributed to vegetational changes rather than depositional episodes. Continuous, independently datable records are desirable for an uninterrupted record of vegetational changes at a site, for comparison of records among sites, and for linking the pollen record to other kinds of data (paleontologic, paleoclimatic, geologic, archeologic, biogeographic).

Quaternary palynologists bring additional criteria to bear in evaluating pollen records and potential sampling sites. These criteria concern not only the type of deposit studied, but the way in which the pollen assemblages are measured (i.e., pollen percentages versus pollen accumulation rates; choice of pollen sum) and how the data are presented (e.g., stratigraphic diagrams, isopoll maps, space–time contour plots). First, the pollen assemblages must be related to vegetation composition at the spatial scale of interest. A second, corollary criterion is that spatial variation in pollen assemblages must be related to spatial variation in vegetation composition. A third requirement is that within-site variation in pollen assemblages be substantially less than the variation among sites attributable to vegetational variation. All of these considerations for pollen data and site selection can be identified in Von Post's seminal 1916 lecture on the use of pollen data for stratigraphic correlation (Von Post, 1967).

These considerations have inevitably led Quaternary palynologists to studies of the depositional properties of various sedimentary basins, the relationships between pollen assemblages and source vegetation, and the spatial scales at which pollen assemblages record vegetation composition. Research has been motivated by the uniformitarian precept that we must understand what aspects of the modern vegetation are recorded by pollen assemblages in order to make appropriate and accurate inferences about past vegetation pattern and composition from the fossil assemblages. The need for effective ways to extract vegetational information from pollen assemblages has led to concerns about how best to measure and present pollen data (e.g., percentages versus accumulation rates, calibrated versus uncalibrated data, stratigraphic profiles versus isopoll maps).

Most early studies of Quaternary pollen sequences focused on bog sediments (Erdtman 1931, 1943; Cain 1939), but within the past 30 years lake sediments have emerged as a primary source of Quaternary pollen data (e.g., Webb, 1985; Webb & Webb, 1988). Lake sediments are especially well suited for providing detailed, accurate records of vegetation change. Lake stratification ensures that the lake bottom is oxygen-depleted during much of the year, inhibiting biological decomposition of recently

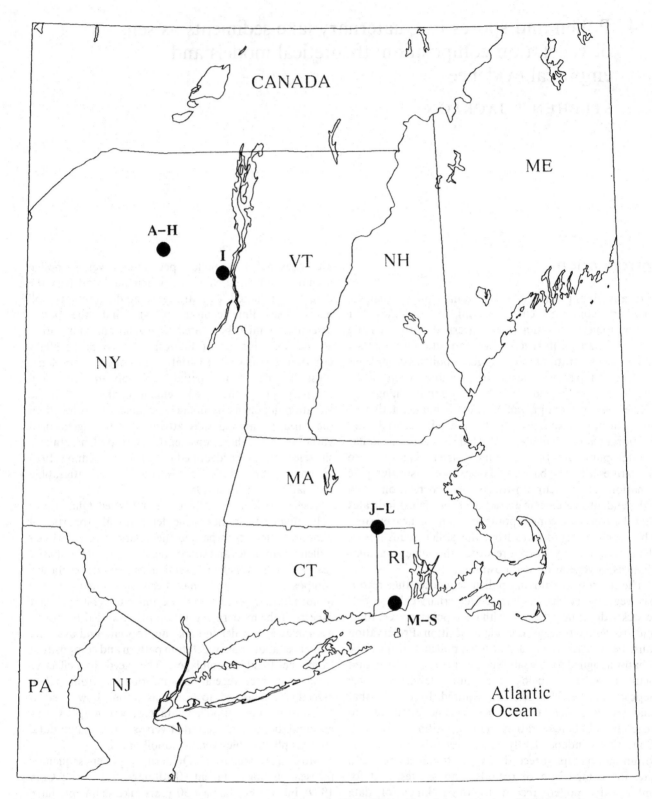

Figure 14.1. Map of the northeastern United States showing locations of the lake-study sites used in Table 14.1 and in Figures 14.5, 14.12–14.14. Reproduced from Jackson (1990) by permission of Elsevier Science Publishers.

deposited pollen grains. Relatively rapid and continuous accumulation of sediments entombs the pollen grains in anoxic sediments where they can be preserved indefinitely. These processes result in long, continuous pollen records with high temporal resolution (10–100 years). Because most late Quaternary lake sediments contain much organic matter and no intrusive roots or burrows, radiocarbon dating can provide a reliable absolute sediment chronology.

The bottom of a lake provides a relatively stable depositional environment, both spatially and temporally. For this reason, changes in pollen assemblages can usually be assumed to indicate changes in the composition of the pollen intercepted by the lake surface, rather than changes in depositional regime. Although important exceptions to this assumption occur, especially when pollen accumulation rates are used, the exceptions can be recognized or at least anticipated in practice. Investigators can make choices about how they 'measure' pollen assemblages (e.g., percent versus accumulation rates) and thus maximize the yield of vegetational information (signal) relative to depositional information (one source of noise).

In this chapter I discuss pollen assemblages, particularly those from lake sediments, from the standpoint of how they record vegetational composition and pattern. First, I take an historical perspective and review the empirical basis for the assumption that pollen assemblages are related to vegetational composition. Then I review conceptual models of pollen/vegetation relationships used to calibrate (qualitatively or quantitatively) pollen assemblages in terms of vegetation composition, and note the assumptions and approximations underlying the models. Those assumptions are discussed in more detail in subsequent sections. I adopt the perspective that all interpretations of pollen assemblages, whether qualitative or quantitative, are inevitably based on critical assumptions concerning the way in which vegetation composition is recorded by pollen data. Throughout, I try to show how theoretical and empirical studies of pollen/vegetation relationships can guide appropriate choice of assumptions in interpreting pollen data.

Many of the issues discussed in this chapter are illustrated, using data from my recent studies comparing modern pollen assemblages from sediments of 19 small lakes (0.1–0.5 hectares) in the northeastern United States with qualitative and quantitative data on the composition of the forest vegetation surrounding the lakes (Fig. 14.1). Jackson (1990) describes the study sites, methods, and results in more detail.

Empirical justification of pollen analysis

Historical perspective

The fundamental assumption underlying most applications of Quaternary pollen analysis can be expressed as a simple relationship,

$$P = f(V), \qquad (14.1)$$

where P represents a given sedimentary pollen assemblage and V is the composition of vegetation surrounding the depositional site at the time of deposition. This relationship is implicit in Von Post's 1916 lecture (Von Post, 1967) and in most subsequent discussions of the theory of pollen analysis (e.g., Erdtman 1931, 1943, Cain 1939, 1944, Fagerlind 1952, Davis 1963). The assumption has intuitive appeal because the pollen grains that comprise an assemblage are produced by the vegetation, so they should provide information about that vegetation. In application, of course, we need a more precise understanding of how information about vegetation is recorded by pollen assemblages. A key aim of this chapter is to define more precisely the vegetation term in the equation above.

Several processes influence the relationship between vegetation and pollen assemblages, including pollen production, dispersal, deposition, preservation, and identification (Fagerlind, 1952; Webb & McAndrews, 1976). These processes are in turn governed by a wide array of factors ranging from physical (meteorological, hydrological, topographic) and chemical (pH, redox conditions) to biological (physiological, morphological, ecological) and even behavioral. Therefore, temporal and spatial changes in pollen assemblages need not correspond directly to vegetational changes. These potential confounding factors have long been recognized by palynologists (Cain, 1939; Erdtman, 1943; Fagerlind, 1952; Davis, 1963; Von Post, 1967; Faegri & Iversen, 1975). What is the basis for the confidence with which Quaternary palynologists have inferred vegetational changes from pollen data?

Most of the empirical justification for Quaternary pollen analysis has been based on a derivative form of the axiom,

$$\Delta P = f(\Delta V), \qquad (14.2)$$

where ΔP = spatial or temporal variation among pollen assemblages, and ΔV = spatial or temporal variation in vegetation composition. This expression is advantageous because it leads directly to observable consequences: spatial and temporal variations in pollen assemblages should correspond meaningfully to spatial and temporal variations in vegetation composition. Von Post clearly recognized this and used it as a central part of his empirical argument justifying pollen analysis in 1916 (Von Post, 1967). His 1916 transect of pollen diagrams across southern Sweden revealed high *Fagus* pollen percentages at southern sites and high *Picea* percentages to the north during the late Holocene (Fries, 1967; Von Post, 1967), a pattern consistent with the modern distributions of these tree types. Szafer's (1935) pioneering use of isopoll maps showed a clear correspondence between late Holocene pollen abundance patterns and modern ranges of *Picea* and *Fagus* in Poland. The first

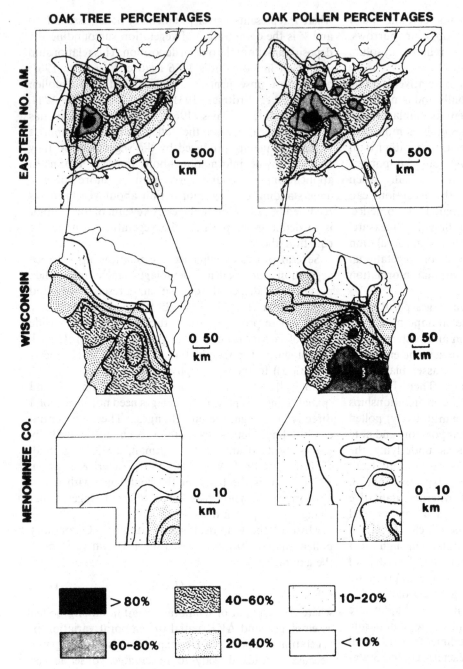

OAK TREE PERCENTAGES   **OAK POLLEN PERCENTAGES**

Figure 14.2. Maps comparing spatial patterns of *Quercus* tree abundance (% basal area) with modern *Quercus* pollen abundance (% of arboreal pollen) at three different spatial scales. Data are from Delcourt *et al.* (1984) and Bradshaw and Webb (1985). Reproduced with the authors' permission from Solomon and Webb (*Annual Review of Ecology and Systematics*, **16** © 1985, by permission of Annual Reviews, Inc.).

fossil pollen records published from eastern North America (Sears, 1935) also showed geographic patterns clearly interpretable in terms of modern vegetation patterns. Further support came from studies of modern pollen assemblages that showed geographic patterns consistent with regional and subcontinental vegetation patterns (e.g., Aario, 1940; Leopold, 1957; Davis, 1967a; Wright, 1967; Lichti-Federovich & Ritchie, 1968).

More recently, isopoll maps and numerical analyses have illustrated spatial patterns in modern and fossil pollen assemblages that correspond respectively to observed modern or plausible past vegetation patterns

(Fig. 14.2) (Webb, 1974b; Davis & Webb, 1975; Webb & McAndrews, 1976; Prentice, 1978; Webb *et al.*, 1978a,b, 1983a,b; Huntley & Birks, 1983; Delcourt *et al.*, 1984; Overpeck *et al.*, 1985). Many studies indicate that variation among pollen assemblages can in some cases be related to vegetation patterns at spatial scales as fine as 10^2–10^3 m (Janssen, 1966, 1984; Andersen, 1970; Calcote & Davis, 1989; Gaudreau *et al.*, 1989; Jackson, 1990).

Temporal variations in pollen assemblages have also been important in providing empirical support for pollen analysis. Palynologists have been reassured by general correspondence between stratigraphic patterns in post-glacial pollen assemblages and postglacial vegetational changes that were inferred or expected from other lines of evidence (e.g., phytogeography, paleoclimatology, geomorphology). For example, disjunct distributions of arctic/alpine, boreal, and prairie plant species and communities in glaciated portions of eastern North America led nineteenth and early twentieth century biogeographers and ecologists to conclude that: (1) arctic and boreal vegetation had colonized glaciated terrain following ice retreat; (2) a warm, dry 'xerothermic' interval led to the eastward expansion of prairie into Indiana, Ohio, and Michigan during the postglacial period; (3) subsequent cooling and moistening resulted in westward contraction of prairie to its modern limits (Gray, 1858, 1878; Adams, 1905; Gleason, 1909, 1923; Braun, 1928). These events were widely accepted well before the first pollen studies in the region (Gleason, 1923; Clements, 1934; Raup, 1937; Sears, 1942; Cain, 1944), and led biogeographers to expect to find certain key features in pollen diagrams of the region if the method worked.

Pollen diagrams produced during the 1930s from sites in the midwestern United States did in fact show patterns that were interpreted as indicating both early dominance of boreal conifers and at least one postglacial 'xerothermic' interval (Sears, 1935, 1938). These studies are primitive by today's standards in terms of both methodology (Shane, 1989) and interpretation (Wright, 1976; Webb *et al.*, 1987). Furthermore, some of the premises by which early phytogeographers and ecologists inferred vegetational history are questionable (Cain, 1944; Deevey, 1949; Cushing, 1965; Davis, 1965); in many cases the vegetational changes observed in the pollen sequences were anticipated for the wrong reasons. Many disjunct populations have been shown to be adventives responding to recent climatic or edaphic changes rather than relicts of past climates (Turesson, 1927; Wright, 1968; Watts, 1979). Similarly, paleoecological studies indicate that some disjunct plant communities such as the subalpine spruce-fir forests of the Adirondack and White Mountains have existed in their present configuration for fewer than 3000 years (Davis *et al.*, 1980; Jackson, 1989; Jackson & Whitehead, 1991). Nevertheless, the apparent consistency between the pollen and phytogeographic evidence was crucial to North American acceptance of pollen analysis as a valid method during the 1930s and 1940s (Transeau, 1935; Cain, 1944; Deevey, 1949). Although some ecologists took issue with specific details of interpretation, they did not question the general validity of pollen analysis in view of its general correspondence with biogeographic antecedents (Braun, 1950, 1955).

Within the past few decades, additional support has come from observed correspondence between pollen and plant macrofossil sequences (Fig. 14.3) (Watts & Winter, 1966; Birks & Mathewes, 1978; Watts, 1979; Davis, *et al.*, 1980; Jackson, 1989; Jackson & Whitehead, 1991), detection of historically documented events such as forest diseases and land clearance in pollen sequences (Anderson, T. W., 1974; Brugam, 1978; Van Zant *et al.*, 1979), and pollen evidence of prehistoric land clearance and agriculture in association with archeological evidence (Iversen 1941, 1973; Burden *et al.*, 1986; Leyden, 1987; Behre, 1988; McAndrews, 1988; McAndrews & Boyko-Diakonow, 1989). Furthermore, many of the major temporal changes observed in pollen diagrams are consistent with climatic changes inferred from independent evidence (lake levels, oxygen isotope records, glacial geology, marine paleontology, paleoclimate modeling) (Bradbury *et al.*, 1981; Dean *et al.*, 1984; Ruddiman & Wright, 1987; COHMAP, 1988). While none of these temporal correspondences between pollen and other data is by itself crucial to current acceptance of the fundamental assumption that we can reliably infer vegetational change from changes in pollen assemblages, they collectively increase our confidence in that assumption.

Measuring pollen abundance: percentages versus accumulation rates

The two most widely used ways of measuring abundance of a pollen morphotype from sediments are: (1) as a percentage of a designated pollen sum (e.g., tree pollen, terrestrial pollen); (2) as a pollen accumulation rate in the sediment (number of grains accumulated/cm²/year). Pollen accumulation rates are calculated by dividing pollen concentration (number of grains/cm³) by the sediment deposition time (yr/cm), which is the reciprocal of the sediment accumulation rate (cm/yr).

At first appearance, pollen accumulation rates would seem to be the method of choice for measuring pollen assemblages, because the respective abundances of the individual pollen taxa should be independent. In fact, when pollen accumulation rates were first determined following the advent of ^{14}C dating (Davis & Deevey, 1964; Davis, 1967b), the method was widely viewed as liberating palynology from the percentage constraints that had vexed palynologists (Fagerlind, 1952; Davis, 1963) since Von Post's 1916 lecture. However, pollen accumulation rates in sediments are really a proxy variable for pollen flux density at the lakewater surface (i.e., the number of grains intercepted per unit area of lake surface per year), which is the variable of direct

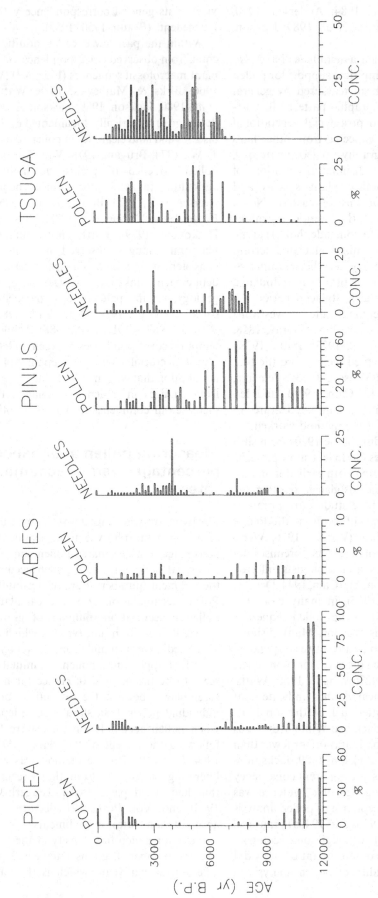

Figure 14.3. Composite stratigraphic profile of pollen percentages and needle concentrations (#/100 cm³) in a sediment core from Heart Lake in the Adirondack Mountains of New York. Pollen percentages are based on a sum of all arboreal pollen types. Data from Jackson (1989) and Jackson and Whitehead (1991).

interest for vegetational inference (Davis *et al.*, 1984). The use of a proxy variable requires empirical demonstrations that the variable actually measures what it is intended to measure. The utility of pollen accumulation rates should be judged not by an abstract notion of intrinsic superiority to percentage data, but by how effectively and unambiguously vegetational information can be extracted relative to pollen percentages.

Detailed studies of pollen accumulation rates and lake sedimentation (Davis *et al.*, 1973, 1984; Pennington, 1973, 1979; Likens & Davis, 1975; Davis & Ford, 1982) have indicated that pollen accumulation rates are highly sensitive to within-lake depositional processes, and hence provide a highly and often irretrievably distorted record of pollen flux density at the lake surface. Sediment focusing, the tendency for deeper areas of lakes to accumulate sediments at faster rates than shallower areas, results in wide variations in pollen accumulation rates at sampling sites within and among lakes (Davis *et al.*, 1984). The magnitude and spatial pattern of focusing can also vary within a lake through time, causing changes in pollen accumulation rates at a coring site that are unrelated to vegetational changes.

Within forested regions, variations in modern pollen percentages among lakes are much more clearly related to spatial vegetational patterns than are variations in pollen accumulation rates. Davis *et al.* (1973) showed that modern pollen accumulation rates for tree taxa from 29 lakes in Michigan corresponded poorly with composition of forests within 22 km of the lakes. In contrast, pollen percentage data from the same region (including the same lakes) showed clear spatial variation closely related to variation in forest composition (Webb, 1974b; see also Webb *et al.*, 1981; Bradshaw & Webb, 1985).

Although pollen accumulation rates have performed poorly as measures of vegetational variation within forested regions, they have been successful in discriminating major vegetation types that differ substantially in pollen source strength (number of dispersible grains produced per unit area of landscape per year). In regions where pollen source strength is low (owing to low net primary productivity, predominance of animal-pollinated taxa, or both), pollen percentage data can be ambiguous because locally produced pollen types are outnumbered by pollen transported from distant, more productive vegetation. For example, pollen spectra in subarctic tundra lakes have 20–60% tree pollen (*Pinus, Picea*), and the assemblages are similar to those from subarctic forest-tundra and open boreal forest (Lichti-Federovich & Ritchie, 1968). However, pollen accumulation rates are one to two orders of magnitude lower in tundra than in forested regions (Davis *et al.*, 1973). In this case, differences in pollen flux density to lake surfaces among the regions are sufficiently large to outweigh the variation

in pollen accumulation rates induced by local depositional processes.

Pollen accumulation rates will certainly continue to be useful in coarse discriminations of major vegetation types in the fossil record (e.g., Davis, 1967b; Ritchie, 1984). However, pollen percentages are more clearly sensitive to spatial vegetational variations within regions (Fig. 14.2) (Webb, 1974b), and have the most promise for providing detailed, quantitative reconstructions of forest composition. Pollen percentages are of course subject to distortion by depositional processes, e.g., differential resuspension and redeposition (Davis, 1968, 1973; Davis & Brubaker, 1973), and by differential preservation within and among lakes (Cushing, 1967), and can vary substantially within lakes (R. B. Davis *et al.*, 1969; M. B. Davis *et al.*, 1971; Dodson, 1977). However, the variation attributable to depositional processes is small relative to that attributable to variations in source vegetation (Fig. 14.2) (Webb, 1974a; Webb & McAndrews, 1976; Webb *et al.*, 1978b). In the rest of this chapter, I concentrate on how pollen percentage data record variations in vegetation composition.

Composition of pollen assemblages and the pollen sum

Pollination biology and pollen representation

Utilization of pollen percentage data requires that one or more pollen sums be specified for calculation of percentages. A pollen sum usually comprises some subset of the entire array of pollen and spore morphotypes found in sediments. In this section, I briefly review the kinds of plants that are (and are not) typically represented by pollen in lake sediments, and then discuss the considerations involved in designating an appropriate pollen sum.

With few exceptions, most of the palynomorphs accumulated in Quaternary lake sediments consist of pollen from terrestrial, wind-pollinated (anemophilous) plants. Pollen from anemophilous plants dominate pollen assemblages even in wet tropical regions where the forest canopy is dominated by insect-pollinated (zoophilous) trees (e.g., Kershaw & Hyland 1975; Bush & Colinvaux, 1988; Colinvaux *et al.*, 1988). Although some zoophilous and amphiphilous (both wind- and insect-pollinated) taxa frequently occur in temperate zone pollen assemblages (e.g., *Tilia, Acer, Salix*), they are consistently underrepresented compared to their abundance in vegetation (e.g., Webb *et al.*, 1981; Delcourt *et al.*, 1984; Bradshaw & Webb, 1985).

The primacy of anemophilous pollen in lake sediments is directly related to three aspects of the pollination biology of the plants: pollen production, liberation, and

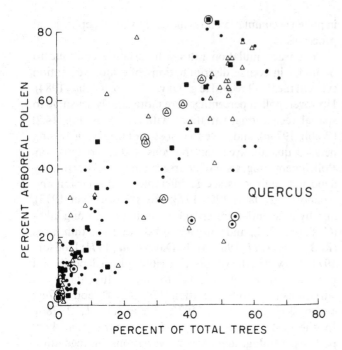

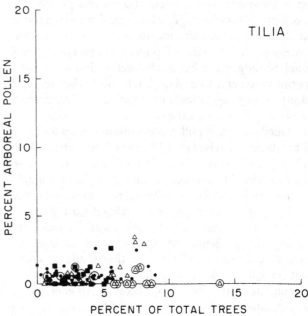

Figure 14.4. Scatter diagrams comparing modern pollen percentages and percent tree basal area for *Quercus*, an anemophilous tree genus, and *Tilia*, an entomophilous tree genus. Pollen percentages are based on an arboreal pollen sum. Pollen data are from surface sediments of 181 lakes in Wisconsin and Upper Michigan. Tree abundances are based on forest inventories within a 30 km radius of each lake. Different symbols denote different lake sizes and morphometries. Reproduced with permission of the authors and Elsevier Science Publishers from Webb *et al.* (1981).

dispersal. First, pollen productivity (number of grains produced per unit plant abundance per unit time) is typically much higher in anemophilous plants than zoophilous plants (Whitehead, 1983). Second, the floral or microsporangial architecture of anemophilous plants is adapted to facilitate liberation of pollen by wind (Niklas, 1985; Crane, 1986), whereas zoophilous flowers have various adaptations to ensure that the pollen stays put in the flower until the appropriate vector (insect, bird, or bat) arrives. Thus, anemophilous pollen is released into the environment at large where it is potentially susceptible to wind, gravity, and water transport to lakes, while zoophilous pollen enters the outside environment aboard animals that generally avoid landing on or in lakes. Finally, anemophilous pollen is aerodynamically better suited for transport by wind than zoophilous pollen (Whitehead, 1969, 1983). Pollen grains of zoophilous plants are frequently large and dense, and often tend to clump together, so even if they do escape from flowers into the air they are not likely to travel far before they settle.

Scatter plots comparing pollen percentages and tree abundance percentages for *Quercus* and *Tilia*, which are respectively anemophilous and entomophilous, illustrate the effects of pollination biology on pollen representation (Fig. 14.4). Pollen of *Tilia* was unrecorded at many sites where *Tilia* trees were abundant nearby. Another example is shown by the different representation of pollen from local entomophilous and distant anemophilous herbs and shrubs in seven small lakes (M-S) in southern Rhode Island (Table 14.1). The margins of all seven ponds are lined with dense, overhanging *Clethra alnifolia* and Ericaceae (*Vaccinium corymbosum, Kalmia latifolia, K.*

angustifolia, Rhododendron maximum). These entomophilous species also occur in the understory of the surrounding *Quercus* forests. Yet these taxa comprise a minute portion of the pollen assemblages (Table 14.1). In contrast, anemophilous ruderal herbs (*Ambrosia, Rumex, Plantago*), which are absent from the pond margins and surrounding forests but locally abundant on open land farther away (beyond 300–1500 m), are abundantly represented (Table 14.1).

Another aspect of pollination biology that is especially relevant in forested regions is where the pollen is liberated. Flowers of angiospermous trees originate from buds on distal shoots (i.e., near the crown surface), and so pollen grains are easily entrained by turbulent, high-velocity winds in and above the canopy. In contrast, understory shrubs and forest-floor herbs release pollen in an environment where wind speeds are low and pollen is not likely to travel far (Tauber, 1965). Handel (1976) showed that for two anemophilous forest-floor herbs, *Carex platyphylla* and *C. plantaginea*, little pollen is dispersed more than a few meters from the male spikelets.

Data on *Lycopodium* spores provide an illustration of the poor dispersal of ground-released pollen and spores to lake sediments in forested regions (Table 14.1).

Table 14.1. *Percentage of selected non-arboreal pollen and spores in surface sediments of 19 small lakes in southern New England and northern New York*

For locations of the sites (A–S), see Fig. 14.1. Ericaceae, *Clethra*, and ruderal pollen percentages are based on a sum including all terrestrial pollen types. *Lycopodium* percentages are based on a sum of all terrestrial pollen types plus all pteridophyte spores except *Isoetes*. Terrestrial pollen sums ranged from 638 to 1011 (mean = 727). Ruderal taxa include *Ambrosia*, *Rumex*, and *Plantago*. *Lycopodium* includes *L. clavatum*, *L. complanatum/tristachyum*, *L. obscurum*, and *L. lucidulum*.

Site	Ericaceae	*Clethra*	Ruderals	*Lycopodium*
A	0.3	0.0	4.3	0.7
B	0.7	0.0	0.5	0.1
C	0.0	0.0	4.1	0.9
D	0.4	0.0	2.8	0.4
E	0.2	0.0	3.5	0.6
F	0.2	0.0	4.0	0.4
G	0.5	0.0	2.6	0.3
H	0.3	0.0	2.7	0.7
I	0.2	0.0	1.6	0.5
J	0.0	0.2	5.0	0.5
K	1.2	0.0	5.1	0.6
L	1.6	0.0	5.2	0.6
M	1.5	0.0	5.5	0.3
N	1.2	0.3	7.4	0.5
O	1.9	0.1	9.6	0.9
P	1.1	0.1	8.9	0.5
Q	1.9	0.0	8.6	0.3
R	2.3	0.1	9.2	0.1
S	1.2	0.0	6.6	0.4

Lycopodium clavatum, *L. lucidulum*, and *L. obscurum* plants are locally abundant on the forest floor in hemlock and northern hardwoods forests within 50–100 m of eight small lakes in northern New York (Fig. 14.1.A–H). *Lycopodium complanatum* and *L. obscurum* plants occur within 50–100 m of the seven southern Rhode Island sites (Fig. 14.1.M–S). *Lycopodium* produces huge quantities of wind-dispersed spores, in sufficient numbers to have once supported a commercial spore industry for photographic flash powders, fireworks, and pill coatings (Clute, 1905). Despite this copious production, few spores are dispersed to sediments of the lakes (Table 14.1).

Most aquatic plant pollen can be assumed to derive from plants growing within the lake. Many submersed and floating-leaved aquatic plants are insect-pollinated (Nymphaeaceae, *Utricularia*, *Eriocaulon*) or have water-dispersed pollen (*Ruppia*). Although some important taxa are anemophilous (Haloragaceae, *Potamogeton*, *Brasenia*) (Cook, 1988), pollen production is typically low (many are facultatively cleistogamous) and dispersal is often limited (e.g., Osborn & Schneider, 1988). Many emergent

aquatics are anemophilous and highly productive of pollen (*Typha*, *Sparganium*, Gramineae, Cyperaceae); trace amounts of these pollen types are often found in lakes where they do not grow. However, they are rarely abundant except in lakes where the plants occur locally (e.g., Janssen, 1967a; McAndrews, 1969). A more serious problem is determining whether pollen of such taxa as Gramineae and Cyperaceae represent local aquatic plants or more distant terrestrial plants (e.g., from disturbances, prairies, tundra). Plant macrofossils can help in this discrimination (Baker *et al.*, 1987; Jackson *et al.*, 1988).

Representation of aquatic plants in pollen assemblages is highly variable, depending mainly on the abundance and extent of aquatic plants in the lake and on whether the pollen sample is from benthic or littoral sediments (Janssen, 1966) (Fig. 14.5). Pollen analysis of littoral sediments can provide information on changes in aquatic vegetation resulting from succession and watershed disturbance (Janssen, 1967a; Birks *et al.*, 1976; Jackson *et al.*, 1988).

Designating the pollen sum

The total pollen and spore assemblage of a lake sediment sample consists of a wide variety of taxa deriving from different kinds of vegetation and different distances from the lake. Selection of the specific subset of taxa to include in a particular pollen sum should be guided by two considerations: (1) the paleoecological question(s) of interest, and (2) the filtering out of noise unrelated to that question. Faegri (1966, p. 136) expressed this idea well: 'Pollen sums must be adapted to the problem they are supposed to elucidate, and then the basic rule is extremely simple: they should contain the pollen grains from those plants that are of interest in elucidating the actual question. Any grains may be brought in and any excluded....'

The examples cited in the previous discussion illustrate these considerations. Four different pollen sums were used: tree pollen (Figs. 14.2, 14.3, 14.4), terrestrial pollen (trees + shrubs + herbs) (Table 14.1), pteridophyte spores plus terrestrial pollen (Table 14.1), and aquatic plus terrestrial pollen (Fig. 14.5). When aquatic plants were of primary interest, aquatic pollen percentages were expressed as a percentage of aquatic and terrestrial pollen types, based on the assumption that terrestrial pollen flux density is approximately constant among the lakes. However, aquatic taxa were excluded from the sum when terrestrial vegetation was of interest (Table 14.1), because the variation in aquatic pollen abundances (Fig. 14.5) would cause variation in percentages of the terrestrial types wholly unrelated to vegetational differences. Similarly, herbs and shrubs were excluded from the sum when tree abundances in predominantly forested regions were being considered (Figs. 14.2, 14.3, 14.4). The

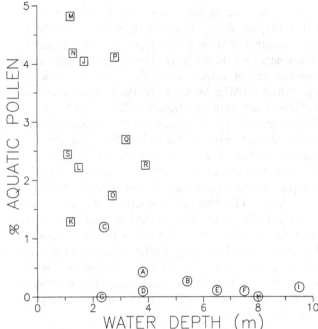

Figure 14.5. Scatter diagram comparing percent aquatic pollen with maximum water depth for 19 small lakes (0.1–0.5 hectares) in southern New England (square symbols) and northern New York (circles). Site letters correspond to those in Figure 14.1 and to Jackson (1990). Percent aquatic pollen is based on a sum of all terrestrial pollen types (trees, shrubs, herbs) plus all obligately aquatic pollen types (*Potamogeton, Typha/Sparganium, Typha latifolia, Sagittaria, Brasenia, Nuphar, Nymphaea, Menyanthes, Nymphoides, Myriophyllum, Utricularia*). Note the negative relationship, which corresponds to the generally negative relationship between water depth and aquatic plant abundance and productivity (Hutchinson 1975). Note also the relatively wide variation in aquatic pollen percentages at low water depths, which corresponds in part to between-lake variation in composition, pollen production, and especially abundance of aquatic vegetation. Sites A, C, D, and G are acidic kettle ponds with highly colored water and steep sides, and hence are poorly suited for aquatic vegetation (Jackson & Charles, 1988). Aquatic vegetation is absent or restricted to scattered populations of *Nuphar* and *Utricularia*. In contrast, the southern New England sites (J-S) are chemically and morphometrically better suited for aquatic plants and have extensive, diverse aquatic vegetation. However, autochthonous aquatic pollen is uncommon compared to allochthonous terrestrial pollen, despite the small lake size and abundant aquatic growth.

substantial differences in ruderal pollen from distant sources between northern New York (Table 14.1.A–I) and southern New England (Table 14.1.J–S) sites would have substantial effects on tree pollen percentages unrelated to tree abundance near the ponds.

Specific pollen taxa can be included or excluded from pollen sums based on the plant growth form, ecology, pollination mechanism, pollen preservability in sediments, and assumed or observed proximity to depositional sites (e.g., Janssen, 1966; Andersen, 1970). Certain pollen types display high-frequency variation in space and time owing to differential preservation, local overrepresentation, low production and/or dispersal, and other factors. Such

'low-fidelity' pollen types provide little information on abundance of the taxon in vegetation, and can introduce spurious noise confounding the interpretation of other taxa. Rare or consistently underrepresented taxa may best be excluded from the pollen sum in both calibrations (Parsons *et al.*, 1983) and in quantitative or graphical comparisons among sites (e.g., difference diagrams, isopoll maps). The overall aim in choosing a pollen sum should be to maximize the signal-to-noise ratio of the data for the specific problem of interest.

Pollen–vegetation calibration: general considerations

Von Post and other early palynologists observed that modern pollen assemblages varied systematically with vegetation composition, but that particular taxa were consistently overrepresented or underrepresented in pollen assemblages (Erdtman, 1931, 1943; Cain, 1939; Fagerlind, 1952; Von Post, 1967). Studies comparing modern pollen assemblages from individual bogs, lakes, or moss polsters with quantitative data on composition of nearby vegetation (Carroll, 1943; Hansen, 1949; Cain, 1953; Potzger *et al.*, 1956; Davis & Goodlett, 1960) were followed by attempts at developing calibration factors (R-values) capable of expressing pollen percentages in terms of relative abundance of taxa in the source vegetation (Curtis, 1959; Davis, 1963; Janssen, 1967b; Comanor, 1968; Livingstone, 1968; Whitehead & Tan, 1969). The calibration factors consisted of taxon-specific coefficients (R_{ij}) that could be derived and applied according to a simple model,

$$p_{ij} = R_{ij} v_{ij}, \qquad (14.3)$$

in which p_{ij} = the abundance of taxon j in the pollen assemblages at site i; v_{ij} = the abundance of taxon j in the vegetation within some specified area (A_n) surrounding site i; and R_{ij} = the ratio of v_{ij} to p_{ij}, which is the slope of the straight line between point (v_{ij}, p_{ij}) and the origin (0,0).

When applied to fossil pollen data, an implicit assumption was made that $R_{ij} = R_j$, i.e., R_j holds for all sites. Because these early calibration studies were restricted to single lakes, bogs, or forest stands, the calibration factors (R_j) contained no estimates for the inherent variability in pollen/vegetation relationships and were statistically unreliable (Livingstone, 1968, 1969; Parsons & Prentice, 1981; Parsons *et al.*, 1983). These early calibration attempts were also limited by failure of the calibration model to account adequately for pollen originating from outside the area over which vegetation composition was measured (Faegri, 1966; Parsons & Prentice, 1981), and by lack of clear criteria to guide the choice of appropriate size of A_n (Livingstone, 1969).

Andersen (1970) first proposed that pollen/vegetation

relationships could be described by a linear regression model,

$$p_{ij} = r_j v_{ij} + p_{oj}, \qquad (14.4)$$

where r_j = the slope coefficient for taxon j, and p_{oj} = y-intercept for taxon j (representing pollen originating from plants growing outside of area A_n).

The regression model includes an explicit background component (p_o), and its coefficients are estimated empirically from multiple paired pollen-vegetation measurements, allowing determination of standard errors for the calibration parameters. Andersen's (1970) original application was concerned with moss polsters in closed forest, and used semi-absolute pollen data and absolute tree abundance data rather than percentages. The model has since been widely applied to pollen percentage data from lake sediments (Webb *et al.*, 1981; Delcourt *et al.*, 1984; Bradshaw & Webb, 1985; Delcourt & Delcourt, 1987; Schwartz, 1989; Jackson, 1990). However, the model as applied to percentage data does not correct for the 'Fagerlind effect' whereby linear relationships between variables measured on an absolute scale become nonlinear when the data are converted to percentages (Fagerlind, 1952; Webb *et al.*, 1981; Prentice & Webb, 1986). Webb *et al.* (1981) demonstrated that the Fagerlind effect should be unimportant when tree percentages for individual taxa are all low to moderate (<20–30%).

An alternative calibration model, the 'extended R-value' model, has been developed to correct for the Fagerlind effect (Parsons & Prentice, 1981; Prentice & Parsons, 1983; Prentice, 1986a; Prentice & Webb, 1986). This model is similar to equation (14.4) above in that it has slope and y-intercept terms, but it also includes for each taxon a site-dependent factor that varies according to the abundance-weighted representation coefficients of all the other taxa represented in the vegetation at each site. Thus, for a site with abundant trees of a poorly represented taxon, the slope terms for the other taxa at the site will be scaled up; conversely if the site is dominated by trees of a well-represented type, the slope terms for the other taxa will be scaled down. The parameters for this model can be estimated simultaneously using a maximum likelihood technique (Prentice & Parsons, 1983).

The extended R-value model has been applied to several data sets (Parsons *et al.*, 1980; Prentice & Webb, 1986; Prentice *et al.*, 1987; Jackson, unpublished). Prentice and Webb (1986) applied the extended R-value model to the data sets of Bradshaw and Webb (1985) and showed that the magnitude of the Fagerlind effect was modest but significant. The study also demonstrated that relationships between pollen and tree abundance percentages could be made more nearly linear by application of the extended R-value model.

Accurate estimation and application of calibration equations to pollen data requires maximizing the signal-to-noise ratio (Webb *et al.*, 1978b) of the pollen-vegetation relationship. Distortions owing to the Fagerlind effect comprise one source of noise, which can potentially be eliminated by application of the extended R-value model to calibration. Noise may also derive from how we choose to measure v_{ij} and p_{ij}. Two decisions are particularly critical with respect to v_{ij}: the aspect of plant abundance used to estimate v_{ij}, and the area (A_n) surrounding the study sites within which to estimate v_{ij}. Site selection (e.g., lake versus bog, large versus small) and choice of pollen sum (i.e., which taxa to include) are important to estimation of p_{ij}. These decisions are interrelated. For example, choice of lake size may have a major effect on choice of A_n, as will be discussed below.

The repeated attempts at pollen/vegetation calibration over the past three decades have played crucial roles in the recognition of these choices and their consequences. Comparative studies of modern pollen and vegetation (Webb *et al.*, 1978b, 1981; Parsons *et al.*, 1980; Bradshaw & Webb, 1985; Prentice *et al.*, 1987; Jackson, 1990) and development of theory on pollen source area and dispersal (Tauber, 1965; Kabailiene, 1969; Jacobson & Bradshaw, 1981; Prentice, 1985) have also increased awareness of the importance of these choices to pollen/vegetation calibration and to interpretation of the pollen record in general. Inappropriate choices concerning v_{ij}, p_{ij}, and A_n can result in poorly estimated calibration coefficients, with consequent uncertainties and misguided inferences. In the absence of explicit calibration efforts, interpretation of pollen data nonetheless involves assumptions (explicit or implicit) concerning the aspect of vegetation recorded by the pollen assemblage, and the spatial scale represented by the pollen assemblage (i.e., the pollen source area). Inappropriate choices of assumptions, whether qualitative or quantitative, and whether explicit or implicit, can yield vague or inaccurate interpretations. In the following sections, I discuss theoretical and empirical considerations guiding the appropriate choice of assumptions concerning v_{ij} and A_n. I will not discuss pollen/vegetation calibrations *per se*; the interested reader is referred to recent reviews by Birks and Gordon (1985) and Prentice (1986a).

Pollen source strength as a measure of plant abundance

The abundance of a plant taxon on the landscape can be perceived and described in a variety of ways. Ecologists have expressed plant abundance per unit area in terms of stem density, bud density, total biomass, leaf biomass, vertically projected areal cover, total leaf area, stem area, net primary production, wood volume, and frequency (likelihood of occurrence in a plot of given size). Pollen source strength, the number of pollen grains of a taxon produced per unit area of land per unit time, is just another way to describe plant abundance. Pollen source strength (integrated over some area A_n and inversely weighted by distance from the site) is how plant abundance is perceived by pollen assemblages. All else held constant (e.g., dispersal, deposition), the temporal changes seen in pollen diagrams should correspond to changes in the respective pollen source strengths of the taxa involved.

Pollen source strength would seem to be the ideal way of expressing v_{ij}, both in calibrations and in more qualitative interpretations. However, because pollen source strength is difficult or impossible to measure directly, and anyway is not how ecologists and palynologists are used to thinking about plant abundance, it becomes necessary to select a variable to stand in for pollen source strength. A suitable proxy variable would be some easily obtainable measure of plant abundance that is both intuitively appealing (i.e., similar or identical to more conventional measures) and, most importantly, is demonstrably closely related to pollen production.

Von Post originally proposed that in forested regions, pollen assemblages measure the relative areal crown coverage of tree species (Andersen, 1970). Crown coverage is probably a reasonable general approximation to pollen source strength for many species. Pollen source strength for a given species should be directly related to the number of flowers or inflorescences per unit area. Because of the modular architecture of flowering plants (Harper 1977, 1981), flower density should be related to the density of distal buds, which should in turn increase as a function of crown area. The generalization may also apply to many non-forest vegetation types, especially those dominated by shrubs and/or graminoids. Flower production should be closely related to net primary production and leaf biomass, both of which should be positively related to vertically projected areal coverage for shrubs, graminoids, and many herbs.

Because crown coverage is inconvenient to measure in forest vegetation, basal area (cross-sectional area of the trunk 1.3 m above the ground) and its allometric derivative, wood volume, have been used in most quantitative comparisons of pollen and forest composition data. Andersen (1970) showed a close correspondence between basal area and crown cover for six deciduous tree genera in Denmark, which has supported widespread application of basal area as a stand-in for crown area. Basal area has the dual advantages of being easily measured in the field and of being widely used in commercial and governmental forest inventories. How well does basal area work in theory and in practice as a measure of pollen source strength?

In using basal area, we are implicitly assuming a linear model of the form,

$$s_{jk} = m_j(b_{jk} - t_j), \qquad (14.5)$$

in which s_{jk} = the pollen produced by an individual tree

k of taxon j (grains produced/year); b_{jk} = the basal area of an individual tree k of taxon j (cm^2); m_j = a slope coefficient relating pollen production to basal area for individual trees of species j; t_j = a taxon-specific threshold coefficient (cm^2) (most trees must attain a certain size before they produce pollen or seeds; Fig. 14.6a); and $b_{jk} \geq t_j$.

Pollen source strength of a uniform forest stand is assumed to follow the equation,

$$S_{ij} = \sum_{k=1}^{n_{ij}} \frac{s_{jk}}{a_i} = \sum_{k=1}^{n_{ij}} \frac{m_j(b_{jk} - t_j)}{a_i} \qquad (14.6)$$

where S_{ij} = the pollen source strength for taxon j in a forest stand i (grains produced/hectare/year); a_i = the total area of stand i (hectares); n_{ij} = the total number of trees of taxon j in stand i. Thus, pollen source strength for a tree taxon in a stand should increase as a function of reproductively 'mature' basal area, which can be increased either by increasing the density of reproductively mature trees (n_{ij}/a_i), or by increasing the size of individual trees (b_{jk}).

Addition of mature trees of any taxon to a forest stand should increase the pollen source strength of that taxon, up to a limit at which competition suppresses primary production and growth (Harper, 1977). This maximum density sets an upper limit to the pollen source strength attainable by a species in a stand. The effect of increasing basal area of individual trees on pollen source strength will depend upon the value of the slope coefficient m_j and upon the underlying assumption of equation (14.5) that pollen production of individual trees increases linearly as a function of basal area. This assumption is reasonable insofar as we are justified in assuming that (1) pollen production is linearly or at least monotonically related to such variables as crown area, leaf biomass, and net primary production, and (2) the latter variables are in turn linearly or monotonically related to basal area. These assumptions probably hold for many tree taxa (Fig. 14.6) (Sarvas, 1962, 1968; Whittaker & Woodwell, 1968; Kinerson & Fritschen, 1971; Snell & Brown, 1978; Cooper, 1981; Halldin, 1985; Waring & Schlesinger, 1985; Koop, 1989). However, there may be important departures for some taxa.

Cooper (1981) presented data indicating that leaf biomass of *Populus tremuloides* and *P. grandidentata* increased relatively little with tree age beyond ~30 years, and even decreased after 50 years. In contrast, trunk biomass increased rapidly and steadily up to 70 years of age. This effective decoupling of leaf and trunk biomass suggests that pollen production may not be related to basal area in this genus. Similarly, seed and cone production increase rapidly with age and tree height in young trees of *Abies veitchii*, but remain constant in trees more than 60 years old, despite continued tree growth (Kohyama, 1982). In these cases, pollen source strength

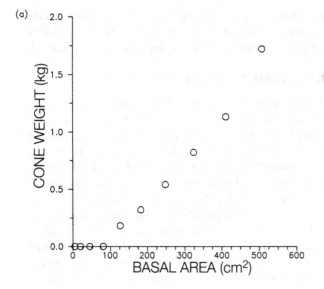

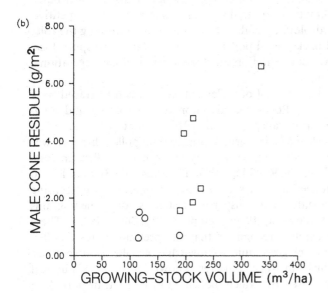

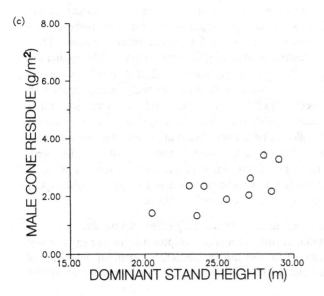

may depend more on density of mature trees than on total basal area.

Dioecious trees present a special problem, because only male individuals will fit model 14.5, and sexes of individual trees are not usually distinguished in forest surveys. Many important tree taxa of eastern North America are obligately or facultatively dioecious, including *Acer rubrum*, *A. negundo*, *A. saccharinum*, *Fraxinus americana*, *F. pennsylvanica*, *Fraxinus nigra*, *Juniperus virginiana*, *Nyssa sylvatica*, *Populus grandidentata*, *P. tremuloides*, and *Salix nigra* (Wright, 1953; Fowells, 1965). This is probably a minor problem when vegetation sampling areas and pollen source areas are sufficiently large to smooth over fine-scale heterogeneity in sex ratios. In such cases, model 14.6 provides a reasonable approximation, with S_{ij} downscaled to some degree according to the sex ratio. The problem is potentially more serious for taxa and pollen assemblages with small pollen source areas, where a small number of trees are involved. High-frequency spatial variations in sex ratios may contribute to the scatter in pollen/vegetation relationships from moss polsters and small hollows.

The slope and threshold coefficients (m_j and t_j) undoubtedly vary among taxa. A consequence of

Figure 14.6. (a) Relationship between total dry weight of female cones and basal area for *Abies balsamea* trees in even-aged stands in New Brunswick, based on data presented in Baskerville (1965, table I). Cone weights represent averages from between three and 22 trees in each size category (101 trees total). Note the threshold ($\sim$100 cm^2) below which female cones are not produced, and the linear pattern at sizes above the threshold. If allocation of biomass to male cone production follows a similar pattern, then equations (14.5) and (14.6) will provide a good description. (b) Relationship between growing-stock volume and male cone production for ten monospecific stands of *Pinus sylvestris* in Finland (Sarvas 1962). Circles represent stands in northern Finland (>66° N latitude); squares represent stands in southern Finland (<62° N). Although the method for estimating growing-stock volume is not described by Sarvas (1962), it is conventionally defined as the wood volume in the tree bole, and is closely correlated with basal area (Wenger 1984). Male cone production is expressed as mean dry weight of male cone residues collected over 3–8 years in 4–12 seed traps (0.5 m^2) randomly located in each stand. According to Sarvas (1962), much of the variation in growing-stock volume among these stands is related to site fertility and climate. Data are from Tables 1 and 4 in Sarvas (1962). (c) Relationship between dominant stand height and male cone production for ten monospecific stands of *Picea abies* in southern Finland (Sarvas 1968). Dominant stand height is defined as the mean height of the mature trees in a stand. Male cone production is expressed as mean dry weight of male cone residues collected over 8–13 years in seed traps (0.5 m^2) randomly located in each stand. Although the number of traps per stand is not specified by Sarvas (1968), it appears to be $\sim$15. Most of the variation among these stands in dominant height is related to stand age (Sarvas 1968). Individual tree height is closely related to basal area in many tree taxa (Whittaker & Woodwell 1968; Kinerson & Fritschen, 1971; Whittaker *et al.*, 1974), including *Picea abies* (Koop 1989). Data are from Table 2 in Sarvas (1968).

variation in the slope coefficient is that responses of pollen production to changes in basal area will vary among species. A low slope for a taxon would imply that pollen source strength is more closely related to adult tree density than to total basal area. Differences in the threshold coefficient introduce noise in that pollen-producing individuals of low-threshold taxa may be excluded from vegetation surveys, while non-producing individuals of high-threshold taxa may be included. The slope and threshold coefficients may also be stand-specific for many taxa. Variations in site conditions and stand structure can affect both reproductive allocation (Kozlowski, 1971; Harper, 1977) and the allometric relationship between basal area and canopy leaf area (Waring et al., 1982; Waring & Schlesinger, 1985).

Despite these uncertainties and potential sources of error, basal area performs reasonably well in practice as a proxy for pollen source strength, as can be seen in Figures 14.2, 14.4, and 14.12–14.14, and in the numerous pollen-vegetation studies where it has been applied (Andersen, 1970; Webb et al., 1981; Heide & Bradshaw, 1982; Bradshaw, 1981; Delcourt et al., 1984; Bradshaw & Webb, 1985; Prentice et al., 1987; Jackson, 1990). Overestimation or underestimation of tree abundance owing to inappropriate threshold coefficients (which are implicit in the choice of the minimum size of tree to measure) are probably compensated in part by the relatively greater impact of large trees on basal area estimates and in part by correlation between density of large trees and small trees of a species when estimated over a large area. Within-species variations in slope and threshold coefficients are also undoubtedly smoothed to some extent over large areas. In the case of species with low slope coefficients, basal area may serve as a reasonable proxy variable for density; density and basal area are usually correlated to some degree.

Pollen source strength is an aspect of vegetation that differs in important ways from those aspects commonly studied and quantified by ecologists. Basal area provides a useful first approximation to pollen source strength, and will undoubtedly continue to be used because of its convenience, wide application, and demonstrated success. However, the assumptions and potential sources of error in using basal area need to be recognized and studied further. Potential errors may be especially serious at fine spatial scales (e.g., gap-phase patches, forest stands, topo-edaphic complexes), where high-frequency variation in tree age, stand structure, disease, herbivory, and site quality may result in high-frequency variation in slope and threshold coefficients both within and among species. Such variation is probably increasingly smoothed and stabilized at coarser spatial scales. However, for pollen morphotypes consisting of more than one species (i.e., most trees), additional variation may be introduced as scales increase to include ranges or abundance peaks of more than one species, owing to differences among species in pollen production. Pollen production within a species may also decrease as range boundaries are approached.

The pollen source area

All vegetational inferences from fossil pollen data, whether qualitative or quantitative, are based on assumptions concerning the pollen source area, the area from which most of the pollen of each taxon derives. These assumptions have a major impact on the calculation and application of pollen-vegetation calibrations; choice of A_n, the assumed pollen source area, directly affects both the calibration parameters and the goodness of fit of the calibration model (Bradshaw & Webb, 1985; Prentice et al., 1987; Jackson, 1990). Assumptions concerning A_n may not be explicit in more qualitative interpretations, but they are manifest in the choice of mechanisms to which vegetational changes are attributed (climatic, edaphic, disturbance, successional, competitive, hydrological) and in interpretations concerning the roles of background pollen (p_o), long-distance transport, local overrepresentation, and local versus regional vegetational change.

The concept of pollen source area can be traced back to Von Post's consideration of site spacing and long-distance transport in his 1916 transect in Sweden (Von Post, 1967). The need to understand pollen dispersal was repeatedly emphasized by some early pollen analysts (Cain, 1939, 1944, 1953; Erdtman, 1943) and led to pioneering efforts at measuring or calculating settling velocities and dispersal distances of pollen grains (Dyakowska, 1936; Rempe, 1937; Buell, 1947). These researchers recognized that interpretation of fossil pollen assemblages requires knowledge of pollen source areas, which in turn requires knowledge of the physical processes by which pollen grains are transported to depositional basins.

The following section briefly reviews aspects of the physics of pollen dispersal in the atmosphere and discusses implications for pollen representation. Then contrasting models of pollen transport to lake basins are discussed. Each emphasizes different mechanisms for pollen dispersal and displays different consequences with respect to pollen source areas and representation. Each of these models is evaluated, using empirical information on pollen dispersal and source areas, and the circumstances under which each model is most useful are discussed. Finally noted are some of the implications of what is currently known about pollen source areas with respect to interpretation of fossil pollen data.

Physics of atmospheric pollen dispersal

Pollen grains of most anemophilous plants range in size from 10 to 100 μm in diameter. Terminal velocities of particles within this size range can be predicted by the

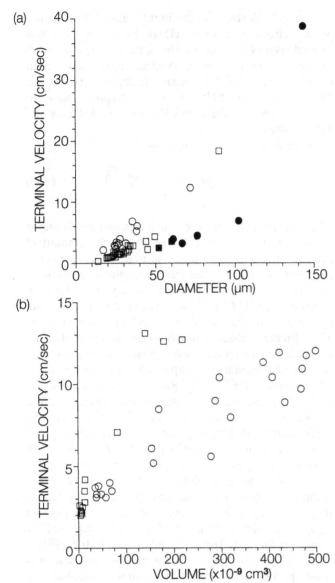

Figure 14.7. (a) Scatter plot of measured terminal velocity of pollen grains versus average grain diameter. Circles represent data for 18 tree species from Dyakowska (1936, Table III); squares represent data for 39 tree, shrub, and herb taxa from Durham (1946, Table V). Solid symbols represent conifer taxa with bisaccate pollen grains (*Pinus, Abies, Picea*). Despite the obvious differences between the two data sets, and despite the likelihood of substantial measurement errors for both variables in both data sets, both show positive relationships between the two variables in accord with predictions of the Navier-Stokes equation (14.7) (see also Figure 3 in Gregory, 1973). Relative terminal velocities can be accurately predicted from relative pollen grain size. (b) Scatter plot of measured terminal velocity of pollen grains versus average grain volume. The data consist of 36 tree species from Eisenhut (1962, tables 2 and 9). Circles represent conifers with bisaccate grains; squares represent angiosperms and other conifers.

Navier–Stokes equation, which for a spherical particle takes the form,

$$v_s = \frac{2r^2 g(p_o - p)}{9\mu} \qquad (14.7)$$

where v_s = terminal velocity (cm/sec); r = particle radius (cm); g = gravitational acceleration constant (981 cm/sec^2); p_o = particle density (g/cm^3); p = density of medium (1.27×10^{-3} g/cm^3 for air); and μ = viscosity of medium (1.8×10^{-4} g/cm/sec for air at 18 °C) (Gregory, 1973). This equation has two major consequences. First, v_s will increase linearly as particle density (p_o) increases. Second, v_s will increase with the square of the particle radius (r). Because the range of pollen and spore densities is narrow (0.4–1.0 g/cm^3) (Eisenhut, 1961; Gregory, 1973), variations in particle density will have modest effects on terminal velocities, while even modest variations in pollen size will lead to large differences in terminal velocity.

Most pollen grains, of course, are not perfect spheres, and terminal velocities are difficult to measure accurately (Durham, 1946; Gregory, 1973; Niklas, 1984). Those measurements that exist, however, indicate that terminal velocity varies widely among pollen types, and increases with pollen size (Fig. 14.7) (Knoll, 1932; Dyakowska, 1936; Durham, 1946; Eisenhut, 1962; Niklas & Paw U, 1983). Correction factors for non-spherical particles are available (Chamberlain, 1975), and an alternative formulation of equation (14.7) predicts v_s based on particle mass and volume rather than density and radius. This formulation can be applied to non-spherical particles when p_o is much larger than p (Vogel, 1981), a condition that holds for pollen grains. Although these possibilities have not yet been thoroughly explored for pollen grains, they should be valuable in checking the accuracy of experimental determinations and in modeling pollen dispersal (e.g., Crane, 1986).

A consequence of variation in terminal velocity among pollen types is that in moving air, pollen grains with low v_s will tend to travel farther before settling than pollen grains with high v_s. In moving air with laminar flow and constant speed, dispersal and deposition of pollen from a point source can be described by a simple model,

$$d_h = \frac{d_v u}{v_s}, \qquad (14.8)$$

where d_h = horizontal distance from the source at which pollen grains are deposited at the ground surface (cm); d_v = elevation of the source above the ground surface (cm); and u = wind velocity (cm/sec). This model is heuristically useful in showing that as terminal velocity increases, horizontal travel distance will decrease. Dispersal distances of pollen grains from point sources have in fact been observed to vary predictably among pollen types according to their size (Wright, 1952; Raynor et al., 1972a, 1974b, 1975). However, equation (14.8) provides an inadequate description of pollen dispersal for most situations. First, most air flow is turbulent, so trajectories of pollen dispersal will be more complex than implied by the model. Also, the model predicts that relatively little pollen will be deposited on the ground at intermediate distances between the source and d_h (i.e.,

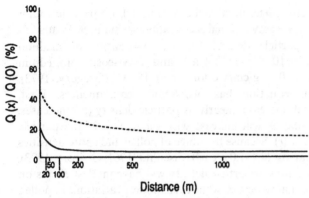

Figure 14.8. Graph plotting $Q(x)/Q(O)$, the proportion of pollen remaining airborne at distance x from a point source, based on equation 14.9. Assuming that wind velocity $= 3$ m/sec, the dashed line shows the pattern expected for a pollen morphotype with $v_g = 3$ cm/sec, while the solid line shows the pattern for $v_g = 6$ cm/sec. Note the steep decline in the function at small distances from the source, and the marked differences in the patterns produced by small differences in v_g. Modified from Prentice (1985), and used with permission of the author and Quaternary Research Center, University of Washington.

there will be a substantial 'skip distance' between the source and the point of deposition). Such a pattern, however, is seldom observed. Instead, a leptokurtic pattern, in which pollen deposition decreases rapidly near the source and then remains steady but low at greater distances, has been repeatedly observed in studies by plant geneticists (Bateman, 1947; Griffiths, 1950; Colwell, 1951; Wright, 1952; Wang *et al.*, 1960; Silen, 1962; Koski, 1970; Kozumplik & Christie, 1972; Levin & Kerster, 1974), aerobiologists (Raynor *et al.*, 1968, 1972a,b), and paleoecologists (Buell, 1947; Turner, 1964; Janssen, 1966, 1984; Tinsley & Smith, 1974). Also, the airborne concentration of pollen and spores typically decreases rapidly with distance from a point source (Raynor *et al.*, 1969, 1972a,b, 1975, 1976).

High concentrations of pollen deposited near the source might be partly explained by clumping of pollen grains at the time of release (Andersen, 1970). Tonsor (1985) has demonstrated that a steep gradient in pollen deposition within one meter of *Plantago* plants represents in part a gradient in size of pollen clumps, with large clumps falling closest to the source plant. Pollen clumping has been observed in many anemophilous plants (Andersen, 1967, 1970, 1974) and may be common. However, the majority of anemophilous pollen grains probably leave the anthers or microsporangia as individuals.

The rapid decrease in airborne pollen concentration with distance from the source can best be explained by three-dimensional diffusion of the pollen cloud in turbulent air. Diffusion, accompanied by a constant rate of particle sedimentation, should yield a leptokurtic deposition pattern similar to that observed in field studies.

Sutton (1947) developed equations to predict the diffusion of an airborne gaseous cloud from elevated and ground-level point sources. The Sutton equations, based on eddy diffusion theory (Sutton, 1932), have been modified for particulate dispersal and deposition (Gregory, 1973; Chamberlain, 1975), and applied to pollen dispersal (Eisenhut, 1961; Tauber, 1965; Kabailiene, 1969; Prentice, 1985, 1988b).

In the ground-level formulation,

$$Q(x) = Q(O) \exp\left(\frac{-4v_g x^{n/2}}{nu\sqrt{\pi C_z}}\right) \tag{14.9}$$

where $Q(x) =$ the amount of pollen remaining airborne at distance x from a point source; $Q(O) =$ the amount of pollen emitted by the point source; $n =$ a turbulence parameter (~ 0.25 in neutral atmospheric conditions, i.e., stable air with no inversion); $C_z =$ a vertical diffusion coefficient (~ 0.12 m$^{1/8}$ in neutral conditions); and $v_g =$ velocity of deposition of the pollen grains (Tauber, 1965; Prentice, 1985). Velocity of deposition is the rate at which particles are removed from a moving parcel of air by sedimentation, impaction, and interception (Chamberlain, 1967, 1975). Removal by interception is probably unimportant in most natural situations, and the limited empirical evidence available indicates that the role of impaction in pollen deposition is minor compared to that of sedimentation (Chamberlain & Little, 1981). Thus, velocity of deposition (v_g) can as a first approximation be regarded as equivalent to terminal velocity (v_s) (Prentice, 1988). In situations where impaction is important, v_g will still increase as a function of v_s; large particles are more likely to be captured by impaction than small ones (Chamberlain & Little, 1981).

The Sutton formulation for a ground-level source is most appropriate for pollen dispersal in forested as well as unforested regions, because in forests most pollen is released in the canopy, which is effectively the 'ground level' for the landscape surface (Prentice, 1985, 1988b). This formulation correctly predicts a strongly leptokurtic relationship between $Q(x)$ and distance (Fig. 14.8), whereas the elevated-source version predicts a substantial skip distance between the source and the area of maximum deposition (Tauber, 1965; Prentice, 1985).

Equation (14.9) is based on a substantial body of theoretical and empirical studies of atmospheric dispersal (Tauber, 1965; Gregory, 1973; Chamberlain, 1975; Prentice, 1985; Okubo & Levin, 1989). The model has several important consequences relevant to pollen dispersal to and deposition in lakes. First, deposition of pollen from a point source (e.g., an individual tree) will be highest close to the source and fall off rapidly thereafter. Therefore, any given point on the landscape sufficiently close to individual pollen-producing plants will have substantial pollen deposition from those plants.

Similarly, deposition from an areal source (e.g., a forest stand or shrub patch) will be highest near the source. However, pollen deposition at a given distance from an areal source will be substantially greater than deposition at an equivalent distance from a point source. Because of the leptokurtic pattern of pollen dispersal, $Q(x)$ will have an extended, near-asymptotic tail (Fig. 14.8). Deposition outside an areal source will include pollen from the distribution tails of all the individual plants in the stand, rather than one in the case of a single, point-source plant.

On a vegetated landscape, a point may fall well within the 'pollen shadow' (i.e., the steeply descending portion of the $Q(x)$ function) of one or a few individual plants and thus receive substantial pollen deposition from those plants. However, regardless of its proximity to individual plants, any point on the landscape will be under the pollen distribution-tails of an extremely large number of plants on the surrounding landscape. Thus, pollen deposition can be substantial even at considerable distances from vegetated areas (e.g., in lake centers) (Bonny, 1976, 1978, 1980; Tauber, 1977; Bonny & Allen, 1984).

Another important implication of equation 14.9 is that the shape of the function $Q(x)$ will depend on the deposition velocity of the pollen grains (Fig. 14.8). Deposition velocities of pollen grains will in fact influence dispersal distances and the length and shape of the 'pollen shadow' around source plants and source vegetation.

Atmospheric dispersal of pollen grains is a complex process, involving liberation, diffusion, advection, and deposition in a turbulent, three-dimensional atmosphere over a heterogeneous three-dimensional landscape. However, as pointed out by Prentice (1985) and others (Tauber, 1965; Gregory, 1973; Levin & Kerster, 1974), the theory described above goes far toward explaining observed features of pollen dispersal and deposition. Accordingly, the theory has much to tell us about how pollen assemblages record vegetation composition. Prentice (1985, 1988b) has used equation 14.9 as the basis for a predictive model of pollen dispersal to depositional basins that will be discussed in detail below.

Models of pollen transport to lakes

THE 'POLLEN RAIN' MODEL

The classical model of pollen transport to depositional basins (Faegri & Iversen, 1975), envisions a diurnal rhythm in which daytime convective lofting of pollen into the troposphere is followed by nighttime still-air deposition of pollen onto the landscape. While suspended in the turbulent air masses, the pollen is mixed with pollen originating from other areas, resulting in a more or less homogeneous, integrated atmospheric pollen assemblage.

The pollen remains suspended throughout the day, during which it is transported over distances of 50–100 km. At night, the pollen settles onto the landscape in a uniform 'pollen rain', although some differential settling may occur (i.e., pollen grains with sufficiently low terminal velocities may not reach the ground before being lofted and transported further the following day). Also, pollen in the troposphere will be periodically 'scavenged' and brought to the ground by falling rain (McDonald, 1962; Tauber, 1965). The model assumes that relatively little pollen is transported horizontally near the landscape surface (Faegri & Iversen, 1975), and so the Sutton equations are not relevant.

According to this model, the pollen assemblage of a lake or peatland should represent an integrated sample of the pollen emitted by plants over an area 10–100 km in diameter. All points on the landscape within a region receive the same pollen rain, and hence variations among pollen assemblages from sites within a region (i.e., within 10–100 km of each other) should result primarily from local depositional anomalies or from local over-representation by overhanging trees or shrubs (Faegri & Iversen, 1975). Therefore, a key testable prediction of this model is that variations in vegetation composition at subregional spatial scales (i.e., < 10–100 km) should not be detectable in pollen assemblages, except for the high-frequency, fine-scale patterns attributable to local overrepresentation.

The 'pollen rain' model as outlined above is no longer tenable on either theoretical or empirical grounds. Tauber's (1965, 1967) cogent critique of the model's physical and meteorological assumptions is borne out by studies indicating that much pollen is neither deposited directly below the source plants nor lofted into high-altitude air masses, but is dispersed horizontally within a few meters of the landscape surface (Raynor *et al.*, 1969, 1970, 1973, 1974b, 1975; Koski, 1970). Furthermore, the model is inconsistent with repeated recent demonstrations that vegetation patterns at spatial scales intermediate between regional and local (i.e., between 100 m and 50 km) can be detected with pollen data from lake and peatland sediments (Fig. 14.2) (Brubaker, 1975; Jacobson, 1979; Webb *et al.*, 1978a, 1983b; Bradshaw & Webb, 1985; Gaudreau *et al.*, 1989; Jackson, 1990).

Although the pollen rain model can no longer be retained as the primary model for pollen transfer, it has some limited applications, and can be used as a component of a more comprehensive model (e.g., Tauber, 1965; see below). Convective lofting of pollen and horizontal transport at high altitudes in the troposphere are well documented (Rempe, 1937; Lanner, 1966; Raynor *et al.*, 1974a; Faegri & Iversen, 1975; Mandrioli *et al.*, 1984), and a more or less uniform background 'rain' of pollen transported at high altitudes from distant areas may well occur. However, long-distance pollen transport

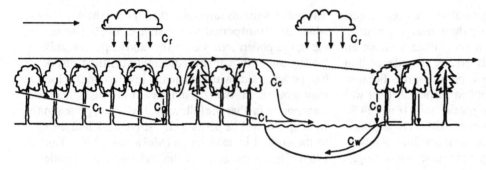

Figure 14.9. Composite model of pollen transfer in a forested region to the forest floor (left) and to a lake (right), after Tauber (1965, 1977) and Prentice (1985). Arrows looping through the tree crowns represent pollen exchange between the trunk-space and canopy components. Letter symbols refer to equation (14.10) in the text. Modified from Prentice (1985), and used with permission of the author and Quaternary Research Center, University of Washington.

can take place in air moving near the land surface rather than at high altitudes (e.g., Christie & Ritchie, 1969).

If the concept of a spatially uniform pollen rain is decoupled from the mechanisms of lofting and high-altitude transport, it can continue to be useful in some specific contexts. For example, the deposition of *Ephedra* and *Sarcobatus* pollen grains over the northeastern United States may be approximately uniform, at least when averaged over a few years. In regions characterized by relatively high pollen productivity and dispersibility (e.g., temperate and boreal forests and grasslands), there is likely to be a spatially uniform background deposition of pollen from both regional and distant sources. However, superimposed on this smooth surface are trends and patterns owing to pollen dispersal over shorter distances and representing finer-scale vegetation patterns. Janssen's exemplary studies of Minnesota lakes and peatlands show how the pollen sum can be manipulated to reveal both the uniform regional pollen rain and fine-scale anomalies owing to more-local pollen dispersal and vegetation patterns (Janssen, 1966, 1984). The uniform pollen rain concept may be most useful in regions with low pollen productivity and substantial deposition from distant sources (e.g., arctic tundra), although fine-scale patterns in pollen assemblages related to more-local vegetation patterns may still occur (Ritchie *et al.*, 1987).

THE COMPOSITE POLLEN TRANSFER MODEL

Tauber (1965, 1977) proposed a schematic model of pollen transfer to depositional basins that has been widely adopted, with some modification (Fig. 14.9) (Jacobson & Bradshaw, 1981; Prentice, 1985). The total pollen assemblage (*P*) accumulated in sediments of a basin is viewed as the sum of several components differing in the physical processes of pollen transport and in the proximity of the source plants to the collecting basin:

$$P = C_g + C_t + C_c + C_r + C_w \qquad (14.10)$$

(Tauber, 1965, 1977; Jacobson & Bradshaw, 1981).

The gravity component (C_g) consists of pollen dropped vertically onto a basin surface from overhanging plants (Andersen, 1967, 1970, 1974; Krzywinski, 1977, Jacobson & Bradshaw, 1981). The pollen in the gravity component usually falls while entrapped in anthers, microsporangia, raindrops, or other large objects, or in clusters large enough that under typical wind conditions the downward component of motion is much greater than the horizontal. Much of the pollen in the gravity component originates from local plants (within a few meters of lakeshores), although the component undoubtedly includes many pollen grains from more distant sources adhering to flowers, leaves, and other expendable plant organs (Tauber, 1965; Prentice, 1985). This component is roughly equivalent to the locally over-represented pollen of Faegri and Iversen (1975).

Pollen in the trunk-space component (C_t) is carried below the forest canopy by low-velocity winds (0.5–1.5 m/sec), and is obviously restricted to forested regions. Because of the low wind speeds, pollen dispersal distances are assumed to be short; Tauber (1965) used equation (14.8) to suggest distances on the order of 100–1000 meters depending on terminal velocity. Therefore, pollen grains dispersed onto a lake surface by trunk-space winds from adjacent forest should derive mainly from trees within tens to hundreds of meters of the lake. However, the trunk-space component also contains pollen grains from greater distances that have penetrated the canopy via turbulence or settling (Figs. 14.9, 14.11) (Raynor *et al.*, 1974b, 1975; Prentice, 1985).

The canopy component (C_c) consists of pollen carried by turbulent winds immediately above the forest canopy. The higher wind speeds (2–6 m/sec) provide a strong horizontal component to pollen motion, and hence pollen grains may travel far from source plants before settling or impacting. Pollen grains could originate from plants growing from within a few meters to tens of kilometers from the lake. The Sutton equation (14.9) applies directly to the canopy component (Tauber, 1965; Prentice, 1985).

As discussed above, some pollen is entrained into higher-altitude air by convection, where it can be carried long distances owing to high wind velocities. Tauber (1965) suggested that most pollen deposition from high altitudes would result from scavenging by raindrops. Rainfall will also cause deposition of pollen from air nearer the ground (i.e., the canopy component) (McDonald, 1962). The 'rainout' component (C_r) can thus include pollen from plants ranging from hundreds of meters to hundreds of kilometers away.

Finally, the waterborne component (C_w), consisting of pollen brought into the lake by influent streams and/or slopewash, contributes pollen originating at least proximally from within the lake watershed. Much of this pollen must originate from trees growing within the watershed, but some undoubtedly derives from more-distant sources, reaching the watershed via rainout and winds above the canopy and in the trunk space.

The composite model of pollen transfer emphasizes the multiplicity of mechanisms by which pollen grains are transported to depositional basins (Fig. 14.9). It is especially useful in showing that spatial and temporal variations in the relative importance of the various modes of pollen transfer may result in substantial variations in pollen source area and, accordingly, variations in the spatial scale at which the pollen assemblages record vegetation composition (Tauber, 1965; Jacobson & Bradshaw, 1981). Tauber's (1965) original conception has given rise to more-detailed theoretical and empirical considerations of pollen source area. Currently, models of pollen source area fall into three categories, each emphasizing different components of pollen transfer. One model, originally conceived by Tauber (1965) and modified somewhat by Jacobson and Bradshaw (1981), places particular emphasis on variations in the relative contributions of C_c, C_t, and C_g with changes in basin surface area. A second model assumes that C_c is the primary component regardless of basin size and uses equation (14.9) to predict pollen source areas (Prentice, 1985, 1988b). The third model assumes that C_w is the predominant component and that most of the pollen in C_w originates from plants in the basin catchment (Bennett, 1983, 1986).

THE TAUBER/JACOBSON-BRADSHAW MODEL

This model is based on two key assumptions: (1) the trajectory of air masses moving over a forest canopy will deflect downward only gradually after passing over a break in the canopy (e.g., a clearing, lake, or peatland), and (2) most of the pollen suspended at any given point in the trunk space of forests (C_t) derives from nearby trees (within a few hundred meters). Because above-canopy air descends at a gentle angle over a canopy opening, it will not reach the ground or lake surface for some distance, estimated by Tauber (1965, 1967a) to be on the order of several hundred meters. Therefore, for lakes smaller than a few hundred meters in diameter, air from above the canopy will not reach the water surface before reaching the opposite shore, and accordingly, pollen entrained in above-canopy air masses (C_c) will not be deposited on the lake surface. Tauber (1965) suggested that most pollen deposited in small basins would be derived from trunk-space winds blowing over the basin surface (i.e., C_t). Jacobson and Bradshaw (1981) pointed out that the gravity component from trees overhanging the basin edge (C_g) would also be magnified.

Jacobson and Bradshaw (1981) summarized predictions of the model with respect to the relationship between basin size and pollen source area (Fig. 14.10). Two key predictions are: (1) pollen source area will increase as basin size increases, and (2) for basins less than ~ 1 hectare, pollen assemblages will be dominated by the trunk-space and gravity components, and hence most pollen will derive from within a few hundred meters of the basin (Fig. 14.10) (Jacobson & Bradshaw, 1981). Jacobson and Bradshaw (1981) and Tauber (1965, 1977) emphasized that for a given basin, pollen source areas might differ among taxa according to their dispersal properties and to distance of populations from the basin.

The Tauber/Jacobson-Bradshaw model can be evaluated by how well its predictions match empirical estimates of pollen source areas, and how well its assumptions are supported by studies of pollen dispersal and forest meteorology. Pollen source areas can be empirically estimated when modern pollen assemblages are graphically and statistically compared with forest composition from within different radii around the lakes (Bradshaw & Webb, 1985; Prentice et al., 1987; Jackson, 1990). Scatter plots of pollen percentages versus tree percentages can be compared to determine the vegetation-sampling radius of best linear fit, which should correspond approximately to the radius from which most of the pollen originates (e.g., Fig. 14.12). Goodness of fit can also be determined by linear correlation, and the y-intercept term (p_o) from linear regression (in equation 14.4) can serve as an estimate of the amount of pollen deriving from outside the vegetation-sampling radius (Webb et al., 1981; Bradshaw & Webb, 1985).

Recent studies of this type indicate that pollen source areas generally increase with increasing basin size (Bradshaw & Webb, 1985; Jackson, 1990). However, this observation is also consistent with an alternative model (Prentice, 1985; see below). I recently tested the second prediction by estimating pollen source areas for tree taxa at the 19 lakes (0.1–0.5 hectares) shown in Figure 14.1 (Jackson, 1990). Although pollen source radii of some taxa appeared to be on the order of 100 m (e.g., Fig. 14.13), these taxa comprised a relatively small portion of the assemblages (10–40%). The pollen assemblages were

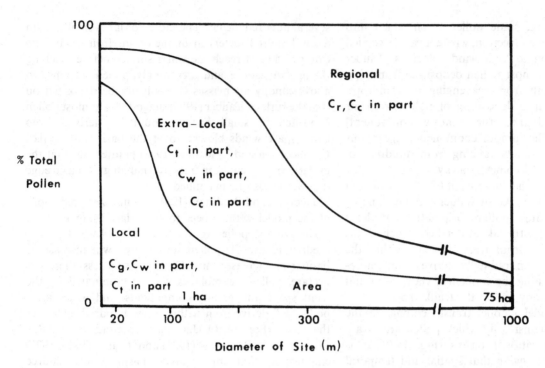

Figure 14.10. Graphical model of Jacobson and Bradshaw (1981) showing hypothesized relationship between basin size and proportional representation of pollen from different components and from different distances from the basin. Reproduced with permission of the authors and Quaternary Research Center, University of Washington, from Jacobson and Bradshaw (1981).

dominated by *Pinus*, *Betula*, and *Quercus*, and those taxa had pollen source radii of at least 1000 m (Fig. 14.14). Overall, the pollen assemblages were not nearly as sensitive to forest composition within a few hundred meters of the lakes as would be expected if the trunk-space and gravity components were dominant.

Failure of the model to predict correctly the pollen source areas observed in my study indicates that one or both of the critical assumptions underlying the model do not hold. Most pollen deposited on the lake surfaces must have been carried by canopy winds, or pollen carried in the trunk-space must have derived from much broader areas than assumed by the model.

Tauber (1965) originally justified the first assumption on Sutton's equation (14.9) for an elevated source, assuming that wind velocities below the source are equivalent to those at or above the source. However, the greater the differential between wind velocities above the forest canopy and in the lee of the forest, the more rapidly the above-canopy air mass will descend toward the ground or lake surface. Tauber (1965) argued that winds blowing through the trunk-space of forests would be sufficiently strong to extend horizontally over lake surfaces, inhibiting downward diffusion of canopy air. Tauber (1977) later noted that trunk-space wind velocities would differ among forests depending on stem and foliage density, and would frequently be insufficient to prevent rapid descent of above-canopy air in openings. Figure 14.11a shows that in a *Pinus*-dominated forest, wind velocities are reduced in the trunk space, and above-canopy air reaches the ground surface within 20 m of the forest edge (Raynor, 1971). An elegant, low-cost

pollen-trapping experiment by Currier and Kapp (1974) showed that pollen carried from above the canopy reached the surface of a small pond (~0.1 hectares) within a few meters of its windward shore. The trunk-space component of pollen transfer may not be dominant at many or even most small basins.

The assumption that the trunk-space component is dominated by local pollen is also questionable. Experimental studies by Raynor *et al.* (1974b, 1975) indicate that there is a continuous interchange of pollen across the forest canopy (Fig. 14.11b,c). Thus, much of the pollen in the trunk-space component at any given point has settled or been blown from above the canopy and hence originates from plants growing at distances greater than a few hundred meters. This is manifest in the substantial amounts of pollen originating from distant sources that are observed in moss polsters and hollows under closed canopy (e.g., Andersen, 1970; Bradshaw, 1981; Heide & Bradshaw, 1982; Chen, 1988). The trunk-space and canopy components are fundamentally similar in that they both represent distance-weighted integrations of the pollen liberated from over a broad area. The trunk-space component may be somewhat more heavily weighted toward local trees than the canopy component, but this weighting is evidently not as strong as assumed by the Tauber/Jacobson-Bradshaw model.

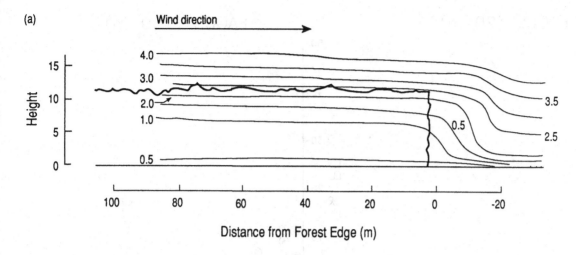

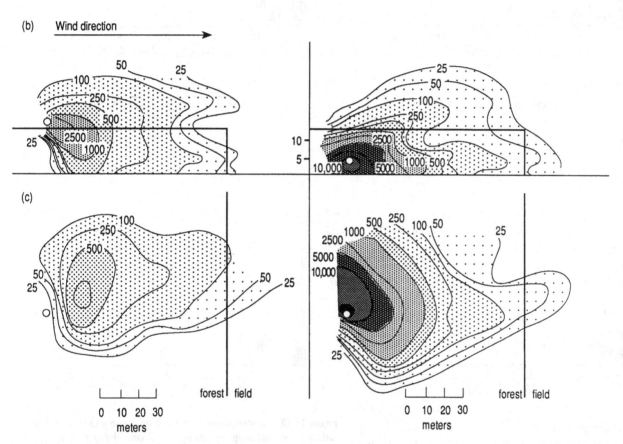

Figure 14.11. (a) Vertical profile of a pine forest and adjacent field, showing isopleths of wind velocity (m/sec) for winds blowing perpendicular to the forest edge toward the field. Crown surface and edge of the pine stand are denoted by the dark, roughened outline. Note the uniformly low wind velocities in the trunk-space, and the rapid increase within and above the canopy. Also note the rapid increase in wind velocities near the ground in the field within a few meters of the forest edge, indicating rapid descent of above-canopy air. Based on data from Raynor (1971). (b) Vertical profile of a pine forest and adjacent field showing isopleths of marked airborne pollen concentration (grains/m³) during release from point sources within the trunk-space and above the canopy. Location at which marked pollen was released is indicated by open circle. Left diagram shows pattern during release of marked pollen within the trunk-space (3.5 m above the ground); right diagram shows pattern during release above the canopy (14.0 m). Note rapid vertical diffusion from the trunk-space source, with considerable pollen entrained in winds above the canopy. Note also rapid vertical diffusion of pollen from the elevated source, with large amounts of pollen from above the canopy being moved laterally in the trunk-space. Based on data from Raynor *et al.* (1975). (c) Horizontal (plan) view of a pine forest and adjacent field 1.75 m above the ground during release of marked pollen (see above). Note rapid diffusion of pollen both downwind and laterally. Together, (b) and (c) show that airborne pollen is continually exchanged across the canopy owing to mechanical and convective turbulence (Raynor *et al.*, 1974b, 1975). Based on data from Raynor *et al.* (1975).

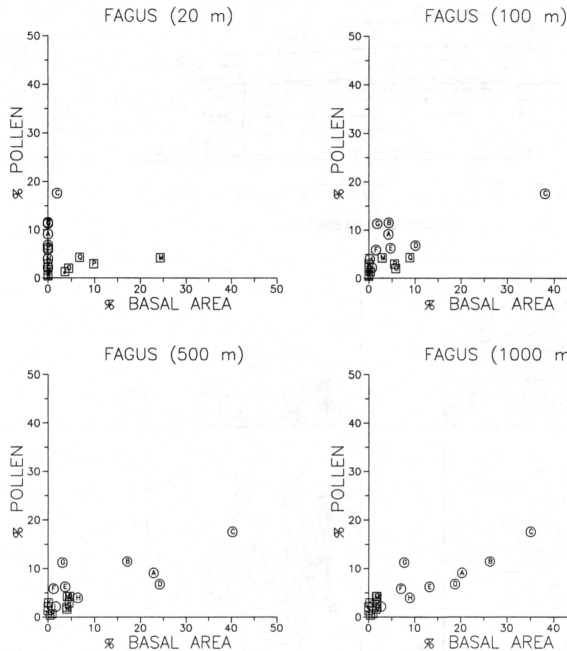

Figure 14.12. Scatter diagrams showing relationship between *Fagus* surface pollen percentage and percent basal area of *Fagus* trees within 20, 100, 500, and 1000 m of 19 small lakes (Fig. 14.1) in southern New England (square symbols) and northern New York (circles). Note the increasingly good linear fit as the vegetation-sampling radius increases to 1000 m. The optimal fit and low *y*-intercept (p_o) at 1000 m indicate that most *Fagus* pollen derives from within 1000 m of the lakes. Reproduced from Jackson (1990), with permission of Elsevier Science Publishers.

PRENTICE'S MODEL

This model assumes that the dominant mode of pollen transfer to lakes is from winds above the canopy, and hence is based on the ground-level formulation of Sutton's equation (14.9) (Prentice, 1985, 1988b). Prentice's model predicts the proportion $F(x)$ of pollen of a taxon deposited in a given basin originating from within a given distance x of the basin as a function of deposition velocity (v_g), above-canopy wind velocity (u), and basin radius (r):

$$F(x) = 1 - e^{-75(v_g/u)(x^{1/8} - r^{1/8})} \qquad (14.11)$$

If windspeed is assumed to be constant at some average value (Prentice (1985) suggests 3 m/sec), the model

predicts that (1) for a lake of given size, pollen source area will increase with decreasing settling velocity of pollen grains and therefore with decreasing grain size, and (2) for a given pollen morphotype, pollen source area

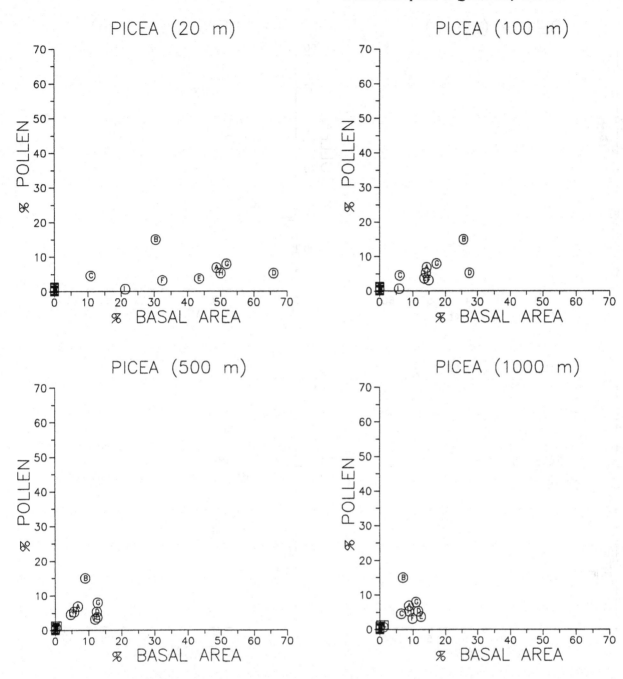

Figure 14.13. Scatter diagrams showing relationship between *Picea* surface pollen percentage and percent basal area of *Picea* trees within 20, 100, 500, and 1000 m of 19 small lakes (Fig. 14.1) in southern New England (square symbols) and northern New York (circles). The best linear fit is at 100 m. Note also the low *y*-intercept (i.e., low p_o). Reproduced from Jackson (1990), with permission of Elsevier Science Publishers.

will increase with increasing lake size. The effect of basin size is strongest at small sizes (<1 hectare). The first prediction follows directly from the Sutton equation (14.9) and the Navier-Stokes equation (14.7). The second prediction is based on the leptokurtic pattern of pollen from a point source. As lake size increases, increasingly

large areas of the lake are located outside the 'pollen shadows' of nearby plants, and hence fall entirely under the summed distribution tails of plants growing over an extremely large area.

Prentice's model is consistent with available empirical data on pollen source areas of lakes (Prentice, 1985, 1988b; Bradshaw & Webb, 1985; Prentice *et al.*, 1987; Gaudreau *et al.*, 1989; Jackson, 1990, 1991). It correctly predicts the observed effect of basin size on pollen source area. It also predicts the systematic differences in pollen source area observed among many pollen morphotypes. For example, inferred pollen source areas for small lakes are lowest for *Picea* (Fig. 14.13), intermediate for *Fagus*

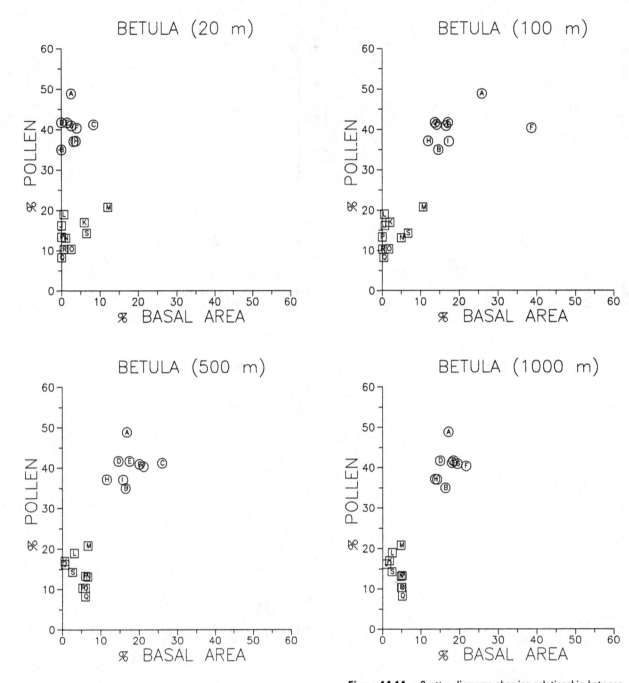

Figure 14.14. Scatter diagrams showing relationship between *Betula* surface pollen percentage and percent basal area of *Betula* trees within 20, 100, 500, and 1000 m of 19 small lakes in southern New England (square symbols) and northern New York (circles). The optimal linear fit is at 1000 m, but note the high *y*-intercept, indicating substantial amounts of *Betula* pollen from beyond 1000 m. Reproduced from Jackson (1990), with permission of Elsevier Science Publishers).

(Fig. 14.12), and highest for *Betula* (Fig. 14.14), which is consistent with differences in size and terminal velocities (Jackson, 1990). Quantitative model predictions of pollen source areas fit empirical estimates reasonably well (Prentice, 1985; Prentice *et al.*, 1987; Jackson, 1990, 1991).

Prentice's model summarizes much of what we know about atmospheric dispersal of pollen, and provides a good fit to available data. Although the model needs more comprehensive tests and better empirical estimates of its parameters, it is currently the best available model for estimating pollen source areas for basins where atmospheric transport is dominant. The model undoubtedly simplifies a highly complex set of processes that vary

spatially and temporally. Prentice (1985) has noted that the model can be modified if necessary to incorporate vegetational heterogeneity, variation in wind direction and velocity, and depositional processes. As more is learned about the dispersal and deposition of pollen, it may become necessary to further modify the model, or to replace or complement it with alternative models.

THE BASIN WATERSHED MODEL

Bennett (1983, 1986) has recently suggested that most pollen arriving in lakes is transported by streams and surface runoff rather than wind, and therefore derives from plants growing within the lake watersheds. Several recent studies have shown that significant quantities of pollen can be transported to lakes by permanent streams (Peck, 1973; Tauber, 1977; Pennington, 1979; Bonny, 1980). Significant waterborne pollen transport may occur at lakes where permanent inflowing streams drain large watersheds, as is the case with these studies. However, results of these studies do not necessarily apply to all or even most lakes (e.g., Clark, 1988).

If most of the pollen transported to a lake derives from plants growing within the watershed, then pollen source areas should not differ appreciably among taxa, and should be equivalent to the lake watershed. The demonstrations that pollen source areas differ predictably among taxa according to atmospheric dispersal properties, and in some cases are tens of kilometers in radius (Bradshaw & Webb, 1985; Prentice *et al.*, 1987; Jackson, 1990, 1991), indicate that the predominant mode of pollen transport to most of the lakes studied is atmospheric.

A component of Bennett's (1983, 1986) argument for the primacy of waterborne transport is that he observed substantial differences in pollen chronologies at two sites less than three kilometers apart in southeastern England. However, these sites lack inflowing streams, and transport by surface runoff may be negligible in view of the low topographic relief of the watersheds (< 15 m relief in watersheds of 1–4 km^2; Bennett, 1986). Interpretation of differences in pollen chronologies among closely spaced sites in terms of non- or partially overlapping source areas (Bennett, 1983, 1986; Delcourt *et al.*, 1986) depends critically on accurate age models (Jacobson, 1979). If chronological control is secure, size differences among basins can lead to substantial differences in contemporaneous pollen assemblages (e.g., Heide, 1984). Finally, differences in composition of forests immediately surrounding the lakes can lead to substantial differences in pollen assemblages (Jackson, 1990). There is no necessary need to invoke waterborne transport to explain differences among adjacent pollen assemblages.

Waterborne transport is probably the dominant mode only in certain restricted situations (permanent inflow streams, high watershed area:lake area ratio, steep watershed slopes, high hydraulic retention time). Even in these situations, the watershed may not serve as the best estimator of the pollen source area in terms of the vegetation that produced the pollen. As discussed previously, much of the pollen deposited on the land surface has been transported in the atmosphere from distant sources, so much of the pollen carried in streams must derive from plants outside the watershed.

Pollen source area and pollen-vegetation calibration

As discussed above, both theoretical and empirical studies indicate that pollen morphotypes vary substantially but predictably in terms of pollen source areas, and that source areas for individual pollen morphotypes vary among basins differing in surface area. This has obvious implications for application of pollen-vegetation calibrations; the calibrations will be useful only insofar as the vegetation sampling radius A_n approximates the effective pollen source area. Inappropriate choice of A_n will yield excessive scatter in the pollen-vegetation relationships and, consequently, poorly estimated calibration coefficients. This is inevitable if a single vegetation-sampling area is used for all taxa and all basin sizes (e.g., Delcourt & Delcourt, 1985, 1987); the vegetation-sampling radius may be appropriate for some taxa but not for others.

The problem posed by variable pollen source areas might not be too serious in situations where vegetation composition is homogeneous at spatial scales encompassing the range of variation in pollen source areas. In such cases, estimates of v_{ij} would not differ appreciably whether they were measured within 100 m, 1 km, 10 km, or 100 km of lake margins, and hence the calibration coefficients would be independent of A_n. Spatial autocorrelations between v_{ij} at different A_n probably do occur in certain regions at certain spatial scales. Knowledge of the circumstances under which this might occur, along with better understanding of pollen source areas, might allow more robust application of some or all of the calibrations of Delcourt *et al.* (1984) and Webb *et al.* (1981) to specific situations.

Vegetation on the landscape is often patterned in an irregular fashion, owing to environmental mosaics and gradients and to disturbances. Depositional basins are not uniformly distributed across the landscape, nor are they located randomly with respect to vegetation patterns. Lake locations are often biased toward certain topographic, geomorphic, and hydrological settings, and lakeshores frequently differ from more distant sites in soil moisture, insolation, wind exposure, and other factors. Therefore, local abundance of a species near lakes is not necessarily correlated with abundance at broader scales (Fig. 14.15). Even when abundances at different scales are positively correlated, the relationship may or may not approximate a line with slope of ~ 1 and y-intercept of ~ 0, which is necessary if samples representing local abundance are to be used as direct estimates at broader scales (or vice versa). Trees of a taxon may be locally overrepresented or underrepresented at paleoecological

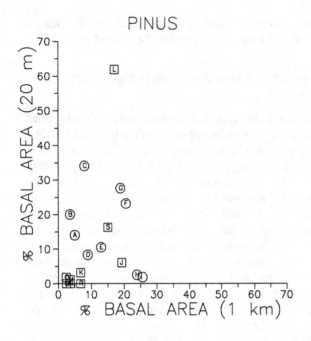

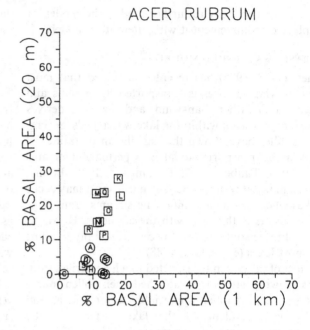

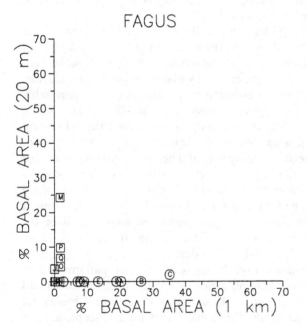

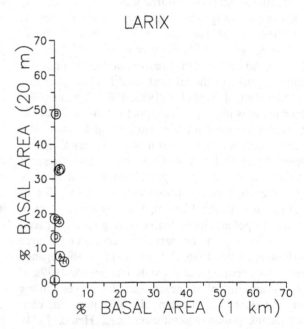

sampling sites relative to their abundance in regional vegetation (Fig. 14.15).

The examples in Figure 14.15 may be extreme because they compare lakeshore vegetation to more-regional vegetation, but they illustrate the general point that lakes are predisposed to have certain taxa growing nearby and others farther away. The phenomenon also occurs at broader scales because of coarser vegetational mosaics and gradients. For example, most of the data points in Figures 14.12–14.14 shift positions along the x-axis as the vegetation-sampling radius increases from 100 to 500 m and from 500 to 1000 m. The scatter plots of Bradshaw and Webb (1985) and Prentice et al. (1987) illustrate this

Figure 14.15. Scatter plots showing percent basal area of trees measured within 20 meters of 19 small lakes (Fig. 14.1) versus percent basal area of trees estimated within 1 km of the same lakes. Southern New England sites are indicated by squares, northern New York sites by circles. Note the overall lack of correlation for *Pinus*. Trees of this genus are locally over-represented at some sites, under-represented at others, and equally represented at still others. In the case of *Acer rubrum*, New England sites show modest local over-representation but are well correlated. New York sites show local under-representation. *Fagus* trees are locally over-represented at some New England sites, and are locally under-represented at all New York sites. Note strong local over-representation for *Larix*. Only New York sites are shown.

at even greater radii (~ 5 km to 100 km). The changes in v_{ij} as A_n varies inevitably affect the calibration parameters. The discrepancies between the calibration parameters obtained by Schwartz (1989) and those of Bradshaw and Webb (1985) from the same region may result primarily from Schwartz' use of a 5-km vegetation sampling radius in contrast to the 30-km radius of Bradshaw and Webb (1985). Many of the sites used by Schwartz (1989) are within 10–15 km of major ecotones.

A further consequence of vegetational heterogeneity is that the composition of vegetation near a lake will inevitably affect representation of more distant taxa when pollen percentages are being considered (Jackson, 1990). If a lake falls within the pollen shadows of highly productive trees, the pollen contribution of more distant trees will be masked. Conversely, local concentrations of trees with low pollen production and/or poor dispersal may amplify percentages of more productive taxa farther away.

The concept of pollen source area as an area in the literal sense may no longer be adequate; it may be more appropriately thought of as a volume or topological surface, because of the distance-weighting imposed by the leptokurtic nature of pollen dispersal. Nearby plants will have an inordinately large influence on a pollen assemblage, regardless of whether they produce large or small amounts of well- or poorly dispersed pollen. Prentice (1986a, 1988b) has suggested that because pollen abundance of a taxon represents a distance-weighted average of a taxon's abundance in the surrounding vegetation, the appropriate value of v_{ij} to use in calibrations is a distance-weighted average, with the particular distance weighting determined by the taxon's dispersal properties and the basin size via equation (14.11). This approach has not yet been tested, although efforts are currently being made to do so using the data of Jackson (1990).

Ecological interpretation of fossil pollen assemblages

Pollen assemblages sense vegetation composition and pattern in a unique manner. Accordingly, our interpretations of the fossil pollen record should be guided by our knowledge of what aspects of vegetation are (and are not) recorded by the data. In many cases, such familiar ecological concepts as population growth, competition, stand dynamics, and succession may be misapplied to interpretation of pollen sequences from lake sediments. Just as we now recognize that we cannot reliably infer vegetational history from disjunct plant distributions, we need to acknowledge that pollen sequences do not provide records of population dynamics *sensu stricto*.

Pollen assemblages do not sample any single population or forest stand, but rather integrate tree abundance and forest composition over an area transcending the limits of what ecologists conventionally think of in their population and stand models. The integration is shaped by the individual pollen dispersal functions and the spatial patterns of vegetation. In a few cases, a fine-scale vegetational change may occur over a broad area in a sufficiently short time to allow meaningful application of ecological concepts designed for fine-scale phenomena. For example, the *Tsuga* decline recorded ~ 4800 yr BP over much of eastern North America (Davis, 1981; Allison *et al.*, 1986) can be partly interpreted in terms of stand dynamics, succession, and competitive release. However, most of the vegetational changes recorded by pollen sequences represent much broader temporal and spatial scales.

The fossil pollen record is a potentially rich source of information on vegetation dynamics. However, the concepts inherited from conventional ecology are inadequate to fully extract and understand that information. Ecologists and paleoecologists need to develop and apply interpretational models and metaphors appropriate to the vegetation-sensing properties of fossil pollen data. Paleoecologists have made some steps toward this goal (Prentice, 1986b, 1988a; Davis, 1986; Dexter *et al.*, 1987; Jacobson *et al.*, 1987; Huntley, 1988; Webb, 1988; Jackson & Whitehead, 1991). The rapidly developing theory of landscape ecology offers additional possibilities for appropriate linkages between ecological theory and paleoecological data.

Conclusions

The relationship between pollen assemblages and their source vegetation is complex but comprehensible. The complexity derives from the numerous physical and biological processes intervening between the vegetation and the pollen assemblages. The comprehensibility derives from the fact that the effects of these processes can be predicted from theory, and is supported by empirical correspondences between pollen assemblages and vegetation composition (e.g., Fig. 14.2).

Each individual plant taxon has unique properties of pollination biology, pollen productivity, and pollen dispersibility. Also, each plant taxon has unique habitat requirements and tolerances, which influence the spatial patterns of its populations with respect to depositional basins, and which also determine the kinds of environmental forcing factors to which it responds. The aspect of plant abundance (e.g., density versus basal area) recorded by pollen abundance may differ among taxa. The abundance of each pollen morphotype in a pollen assemblage, therefore, may carry a unique message in terms of size, location, and areal coverage of plant populations and in terms of the contemporary environment.

For example, an unproductive, poorly dispersed taxon tending to grow in wet soils (e.g., *Larix*) provides information on tree abundance within a few tens of meters of the lakeshore and on local hydrological conditions. In contrast, a productive, well-dispersed taxon growing mainly on uplands (e.g., *Pinus*) provides a distance-weighted average of tree abundance within tens of kilometers of a site, and tells us more about regional temperature, soil moisture, and disturbance.

Individual basins also have unique pollen sampling properties (Jacobson & Bradshaw, 1981), and hence differ in the vegetational information recorded in the pollen assemblages. Differences in basin size affect the pollen source areas of individual pollen types. Basin morphometry influences the extent of sediment mixing and redeposition, which in turn affects the degree of temporal integration of the pollen assemblage (Davis, 1974). Basin morphometry also dictates the rate and magnitude of sediment focusing, and morphometry and sediment accumulation rates influence the preservation state of the pollen. The topographic and hydrological setting of individual basins determines the nearby vegetation patterns and hence the kinds of local distortions and biases present in the pollen assemblage. The physical structure of vegetation surrounding the basin and the topographic setting can affect local wind patterns and pollen dispersal to the lake surface. Inflowing streams can distort the pollen assemblage in poorly understood ways.

Despite the variability among taxa and among sites, and despite the complex nature of the various processes intervening between source vegetation and pollen assemblages, at appropriate spatial and temporal scales we can extract clear vegetational signals that are robust with respect to the various potential confounding factors (e.g., Fig. 14.2). The pollen data are clearly telling us much about the distribution and abundance of plant taxa on the landscape. The systematic biases and distortions in the pollen data are comprehensible in terms of the reproductive biology of plants, the physics of pollen dispersal and deposition, the ecology of vegetation, and the depositional properties of lakes and wetlands. Distortions and biases, once understood, can be corrected for qualitatively or quantitatively.

Paleoecologists have recognized the need for a fundamental theoretical and empirical basis for interpreting pollen assemblages since Von Post's 1916 lecture. We have made considerable progress since then both in understanding the mechanisms responsible for distortions and biases in the pollen record, and in developing models that help us to filter out the effects of those distortions and to extract the wealth of information on vegetation composition and pattern contained in pollen assemblages. Further development, testing, and application of those models should be a high priority for Quaternary paleoecologists. Further development and validation of the theory underlying Quaternary pollen analysis will benefit not only paleoecology but also the many other disciplines (ecology, conservation biology, biogeography, climatology, archeology) that rely on information about Quaternary vegetation dynamics.

Acknowledgments

This research was supported by a National Science Foundation Post-Doctoral Fellowship in Environmental Biology (Grant No. BSR-8600182) to S. T. Jackson and by a grant from the NSF Ecology Program (BSR-8906486) to S. T. Jackson and T. Webb, III. My thinking on the theory of pollen analysis has profited from discussions and/or correspondence with Richard H. W. Bradshaw, James S. Clark, Margaret B. Davis, Denise C. Gaudreau, I. Colin Prentice, Thompson Webb, III, Peter S. White and Donald R. Whitehead. I thank Thompson Webb, III and Martin B. Farley for critical comments on the manuscript, and Robyn O'Reilly for assistance with illustrations.

I dedicate this chapter to Stanley A. Cain, in recognition of his pioneering efforts in North American paleoecology and his early advocacy of detailed comparative studies of modern pollen assemblages and vegetation. The first draft of this review was written in December 1989, exactly 50 years after publication of Cain's articulate summary of the potentialities and pitfalls of pollen analysis as applied to ecology and biogeography. A revised version of that paper appeared as a chapter in his 1944 text, *Foundations of Plant Geography*, and in that form served as my introduction to paleoecology.

References

Aario, L. (1940). Waldgrenzen und subrezenten Pollenspektren in Petsamo Lappland. *Annales Academie Scientiarium Fennicae*, A **54**(8), 1–120.

Adams, C. C. (1905). The postglacial dispersal of the North American biota. *Biological Bulletin*, **9**, 53–71.

Allison, T. D., Moeller, R. E. & Davis, M. B. (1986). Pollen in laminated sediments provides evidence for a mid-Holocene forest pathogen outbreak. *Ecology*, **67**, 1101–5.

Andersen, S. T. (1967). Tree-pollen rain in a mixed deciduous forest in South Jutland (Denmark). *Review of Palaeobotany and Palynology*, **3**, 267–75.

Andersen, S. T. (1970). The relative pollen productivity and pollen representation of north European trees, and correction factors for tree pollen spectra determined by surface pollen analyses from forests. *Danmarks Geologiske Undersøgelse*, Series II, **96**, 1–199.

Andersen, S. T. (1974). Wind conditions and pollen deposition in a mixed deciduous forest. I. Wind conditions and pollen dispersal. *Grana*, **14**, 57–63.

Anderson, T. W. (1974). The chestnut pollen decline as a time horizon in lake sediments in eastern North America. *Canadian Journal of Earth Sciences*, **11**, 678–85.

Baker, R. G., Horton, D. G., Kim, H. K., Sullivan, A. E., Roosa, D. M., Witinok, P. M. & Pusateri, W. P. (1987). Late Holocene paleoecology of southeastern Iowa: development of riparian vegetation at Nichols Marsh. *Proceedings of the Iowa Academy of Science*, **94**, 51–70.

Baskerville, G. L. (1965). Estimation of dry weight of tree components and total standing crop in conifer stands. *Ecology*, **46**, 867–9.

Bateman, A. J. (1947). Contamination of seed crops. II. Wind pollination. *Heredity*, **1**, 235–46.

Behre, K.E. (1988). The role of man in European vegetation history. In *Vegetation History*, ed. B. Huntley and T. Webb, III. Dordrecht: Kluwer Academic Publishers, 633–72.

Bennett, K. D. (1983). Devensian late-glacial and Flandrian vegetational history at Hockham Mere, Norfolk, England. I. Pollen percentages and concentrations. *New Phytologist*, **95**, 457–87.

Bennett, K. D. (1986). Competitive interactions among forest tree populations in Norfolk, England, during the last 10,000 years. *New Phytologist*, **103**, 603–20.

Birks, H. H. & Mathewes, R. W. (1978). Studies in the vegetational history of Scotland. V. Late Devensian and early Flandrian pollen and macrofossil stratigraphy at Abernethy Forest, Inverness-shire. *New Phytologist*, **80**, 455–84.

Birks, H. H., Whiteside, M. C., Stark, D. M. & Bright, R. C. (1976). Recent paleolimnology of three lakes in northwestern Minnesota. *Quaternary Research*, **6**, 249–72.

Birks, H. J. B. & Gordon, A. D. (1985). *Numerical Methods in Quaternary Pollen Analysis*. London: Academic Press.

Bonny, A. P. (1976). Recruitment of pollen to the seston and sediment of some Lake District lakes. *Journal of Ecology*, **64**, 859–87.

Bonny, A. P. (1978). The effect of pollen recruitment processes on pollen distribution over the sediment surface of a small lake in Cumbria. *Journal of Ecology*, **66**, 385–416.

Bonny, A. P. (1980). Seasonal and annual variation over 5 years in contemporary airborne pollen trapped at a Cumbrian lake. *Journal of Ecology*, **68**, 421–41.

Bonny, A. P. & Allen, P. V. (1984). Pollen recruitment to the sediments of an enclosed lake in Shropshire, England. In *Lake Sediments and Environmental History*, ed. E. Y. Haworth & J. W. G. Lund. Leicester, UK: University of Leicester Press, 231–59.

Bradbury, J. P., Leyden, B., Salgado-Labouriau, M., Lewis, Jr., W. M., Schubert, C., Binford, M. W., Frey, D. G., Whitehead, D. R. & Weibezahn, F. H. (1981). Late Quaternary environmental history of Lake Valencia, Venezuela. *Science*, **214**, 1299–1305.

Bradshaw, R. H. W. (1981). Modern pollen-representation factors for woods in south-east England. *Journal of Ecology*, **69**, 45–70.

Bradshaw, R. H. W. & Webb, T., III (1985). Relationships between contemporary pollen and vegetation data from Wisconsin and Michigan, USA. *Ecology*, **66**, 721–37.

Braun, E. L. (1928). Glacial and post-glacial plant migrations indicated by relic colonies of southern Ohio. *Ecology*, **9**, 284–302.

Braun, E. L. (1950). *Deciduous Forests of Eastern North America*. New York: MacMillan.

Braun, E. L. (1955). The phytogeography of unglaciated eastern United States and its interpretation. *Botanical Review*, **21**, 297–375.

Brubaker, L. B. (1975). Postglacial forest patterns associated with till and outwash in northcentral Upper Michigan. *Quaternary Research*, **5**, 499–527.

Brugam, R. B. (1978). Pollen indicators of land-use change in southern Connecticut. *Quaternary Research*, **9**, 349–62.

Buell, M. F. (1947). Mass dissemination of pine pollen. *Journal of the Elisha Mitchell Scientific Society*, **63**, 163–7.

Burden, E. T., McAndrews, J. H. & Norris, G. (1986). Palynology of Indian and European forest clearance and farming in lake sediment cores from Awenda Provincial Park, Ontario. *Canadian Journal of Earth Sciences*, **23**, 43–54.

Bush, M. B. & Colinvaux, P. A. (1988). A 7000-year pollen record from the Amazon lowlands, Ecuador. *Vegetatio*, **76**, 141–54.

Cain, S. A. (1939). Pollen analysis as a paleo-ecological research method. *Botanical Review*, **5**, 627–54.

Cain, S. A. (1944). *Foundations of Plant Geography*. New York: Harper & Row.

Cain, S. A. (1953). Field work in 1953: the problem of pollen representation. *Asa Gray Bulletin*, **2**, 299–303.

Calcote, R. R. & Davis, M. B. (1989). Comparison of pollen in surface samples of forest hollows with surrounding forests. *Bulletin of the Ecological Society of America*, **70** (Supplement), 75.

Carroll, G. (1943). The use of bryophytic polsters and mats in the study of recent pollen deposition. *American Journal of Botany*, **30**, 361–66.

Chamberlain, A. C. (1967). Deposition of particles to natural surfaces. In *Airborne Microbes*, ed. P. H. Gregory & J. L. Monteith. Cambridge: Cambridge University Press, 138–64.

Chamberlain, A. C. (1975). The movement of particles in plant communities. In *Vegetation and the Atmosphere*, ed. J. L. Monteith. New York: Academic Press, **I**, 155–203.

Chamberlain, A. C. & Little, P. (1981). Transport and capture of particles by vegetation. In *Plants and their Atmospheric Environments*, ed. J. Grace, E. D. Ford, & P. G. Jarvis. Oxford: Blackwell, 147–73.

Chen, Y. (1988). Pollen source and distribution in a forest hollow in Sylvania, Michigan, U.S.A. *Pollen et Spores*, **30**, 95–110.

Christie, A. D. & Ritchie, J. C. (1969). On the use of isentropic trajectories in the study of pollen transports. *Naturaliste Canadien*, **96**, 531–49.

Clark, J. S. (1988). Particle motion and the theory of charcoal analysis: source area, transport, deposition, and sampling. *Quaternary Research*, **30**, 67–80.

Clements, F. C. (1934). The relict method in dynamic ecology. *Journal of Ecology*, **22**, 39–68.

Clute, W. N. (1905). *The Fern Allies*. New York: Frederick A. Stokes.

COHMAP Members. (1988). Climatic changes of the last 18,000 years: observations and model simulations. *Science*, **241**, 1043–52.

Colinvaux, P. A., Frost, M., Frost, I., Liu, K.-B. & Steinitz-Kannan, M. (1988). Three pollen diagrams of forest disturbance in the western Amazon basin. *Review of Palaeobotany and Palynology*, **55**, 73–81.

Colwell, R. N. (1951). The use of radioactive isotopes in determining spore distribution patterns. *American Journal of Botany*, **38**, 511–23.

Comanor, P. L. (1968). Forest vegetation and the pollen spectrum: an examination of the usefulness of the R value. *New Jersey Academy of Science Bulletin*, **13**, 7–19.

Cook, C. D. K. (1988). Wind pollination in aquatic angiosperms. *Annals of the Missouri Botanical Garden*, **75**, 768–77.

Cooper, A. W. (1981). Above-ground biomass accumulation and net primary production during the first 70 years of succession in *Populus grandidentata* stands on poor sites in northern Lower Michigan. In *Forest Succession: Concepts and Application*, ed. D. C. West, H. H. Shugart & D. B. Botkin. New York: Springer-Verlag, 339–60.

Crane, P. R. (1986). Form and function in wind dispersed

pollen. In *Pollen and Spores: Form and Function*, ed. S. Blackmore & I. K. Ferguson. Linnean Society Symposium Series Number 12. London: Academic Press, 179–202.

Currier, P. J. & Kapp, R. O. (1974). Local and regional pollen rain components at Davis Lake, Montcalm County, Michigan. *Michigan Academician*, **7**, 211–25.

Curtis, J. T. (1959). *The Vegetation of Wisconsin: an Ordination of Plant Communities*. Madison: University of Wisconsin Press.

Cushing, E. J. (1965). Problems in the Quaternary phytogeography of the Great Lakes region. In *The Quaternary of the United States*, ed. H. E. Wright, Jr. & D. G. Frey. Princeton, NJ: Princeton University Press, 403–16.

Cushing, E. J. (1967). Evidence for differential pollen preservation in late Quaternary sediments in Minnesota. *Review of Palaeobotany and Palynology*, **4**, 87–101.

Davis, M. B. (1963). On the theory of pollen analysis. *American Journal of Science*, **261**, 897–912.

Davis, M. B. (1965). Phytogeography and palynology of northeastern United States. In *The Quaternary of the United States*, ed. H. E. Wright, Jr. & D. G. Frey. Princeton, NJ: Princeton University Press, 377–401.

Davis, M. B. (1967a). Late-glacial climate in northern United States: a comparison of New England and the Great Lakes region. In *Quaternary Paleoecology*, ed. E. J. Cushing and H. E. Wright, Jr. New Haven, CT: Yale University Press, 11–43.

Davis, M. B. (1967b). Pollen accumulation rates at Rogers Lake, Connecticut, during late- and postglacial time. *Review of Palaeobotany and Palynology*, **2**, 219–30.

Davis, M. B. (1968). Pollen grains in lake sediments: redeposition caused by seasonal water circulation. *Science*, **162**, 796–9.

Davis, M. B. (1973). Redeposition of pollen grains in lake sediment. *Limnology and Oceanography*, **18**, 44–52.

Davis, M. B. (1981). Outbreaks of forest pathogens in Quaternary history. *Proceedings IVth International Palynological Conference, Lucknow (1976–1977)*, **3**, 216–27.

Davis, M. B. (1986). Climatic instability, time lags, and community disequilibrium. In *Community Ecology*, ed. J. Diamond and T. J. Case. New York: Harper and Row, 269–84.

Davis, M. B. & Brubaker, L. B. (1973). Differential sedimentation of pollen grains in lakes. *Limnology and Oceanography*, **18**, 635–46.

Davis, M. B., Brubaker, L. B. & Beiswenger, J. M. (1971). Pollen grains in lake sediments: pollen percentages in surface sediments from southern Michigan. *Quaternary Research*, **1**, 450–67.

Davis, M. B., Brubaker, L. B. & Webb, T., III (1973). Calibration of absolute pollen influx. In *Quaternary Plant Ecology*, ed. H. J. B. Birks & R. G. West. Oxford: Blackwell, 9–25.

Davis, M. B. & Deevey, E. S. (1964). Pollen accumulation rates: estimates from late-glacial sediment of Rogers Lake. *Science*, **145**, 1293–95.

Davis, M. B. & Ford, M. S.(J.) (1982). Sediment focusing in Mirror Lake, New Hampshire. *Limnology and Oceanography*, **27**, 137–50.

Davis, M. B. & Goodlett, J. C. (1960). Comparison of the present vegetation with pollen-spectra in surface samples from Brownington Pond, Vermont. *Ecology*, **41**, 346–57.

Davis, M. B., Moeller, R. E. & Ford, J. (1984). Sediment focusing and pollen influx. In *Lake Sediments and Environmental History*, ed. E. Y. Haworth & J. W. G. Lund. Leicester, UK: University of Leicester Press, 261–93.

Davis, M. B., Spear, R. W. & Shane, L. C. K. (1980). Holocene climate of New England. *Quaternary Research*, **14**, 240–50.

Davis, R. B. (1974). Stratigraphical effects of tubificids in profundal lake sediments. *Limnology and Oceanography*, **19**, 466–88.

Davis, R. B., Brewster, L. A. & Sutherland, J. (1969). Variation in pollen spectra within lakes. *Pollen et Spores*, **11**, 557–72.

Davis, R. B. & Webb, T., III. (1975). The contemporary distribution of pollen from eastern North America: a comparison with the vegetation. *Quaternary Research*, **5**, 395–434.

Dean, W. E., Bradbury, J. P., Anderson, R. Y. & Barnosky, C. W. (1984). The variability of Holocene climate change: evidence from varved lake sediments. *Science*, **226**, 1191–4.

Deevey, E. S. (1949). Biogeography of the Pleistocene. Part I: Europe and North America. *Bulletin of the Geological Society of America*, **60**, 1315–1416.

Delcourt, H. R. & Delcourt, P. A. (1985). Comparison of taxon calibrations, modern analogue techniques, and forest-stand simulation models for the quantitative reconstruction of past vegetation. *Earth Surface Processes and Landforms*, **10**, 293–304.

Delcourt, P. A. & Delcourt, H. R. (1987). *Long-term Forest Dynamics of the Temperate Zone: A Case Study of Late-Quaternary Forests in Eastern North America*. New York: Springer-Verlag.

Delcourt, P. A., Delcourt, H. R., Cridlebaugh, P. A. & Chapman, J. (1986). Holocene ethnobotanical and paleoecological record of human impact on vegetation in the Little Tennessee River Valley, Tennessee. *Quaternary Research*, **25**, 330–49.

Delcourt, P. A., Delcourt, H. R. & Webb, T., III (1984). Atlas of paired isophyte and isopoll maps for important eastern North American tree taxa. *Contributions Series*, **14**, Dallas, Texas: American Association of Stratigraphic Palynologists.

Dexter, F., Banks, H. T. & Webb, T., III (1987). Modeling Holocene changes in the location and abundance of beech populations in eastern North America. *Review of Palaeobotany and Palynology*, **50**, 273–92.

Dodson, J. R. (1977). Pollen deposition in a small closed drainage basin lake. *Review of Palaeobotany and Palynology*, **24**, 179–93.

Durham, O. C. (1946). The volumetric incidence of atmospheric allergens. III. Rate of fall of pollen grains in still air. *Journal of Allergy*, **17**, 70–8.

Dyakowska, J. (1936). Researches on the rapidity of the falling down of pollen of some trees. *Bulletin de l'Académie Polonaise des Sciences et des Lettres*, **B 1**, 155–68.

Eisenhut, G. (1961). Untersuchungen über die Morphologie und Ökologie der Pollenkörner heimischer und fremdländischer Waldbäume. *Forstwissenschaftliche Forschungen*, **15**, 1–68. (Translation by P. Jaumann).

Erdtman, G. (1931). Pollen-statistics: a new research method in paleo-ecology. *Science*, **73**, 399–401.

Erdtman, G. (1943). *An Introduction to Pollen Analysis*. Waltham, MA: Chronica Botanica.

Faegri, K. (1966). Some problems of representativity in pollen analysis. *Palaeobotanist*, **15**, 135–40.

Faegri, K. & Iversen, J. (1975). *Textbook of Pollen Analysis*. 3rd edition. Copenhagen: Munksgaard.

Fagerlind, F. (1952). The real significance of pollen diagrams. *Botaniska Notiser*, **105**, 185–224.

Fowells, H. A., ed. (1965). Silvics of forest trees of the United States. *United States Department of Agriculture Handbook*, **271**. Washington, DC.

Fries, M. (1967). Lennart Von Post's pollen diagram series of 1916. *Review of Palaeobotany and Palynology*, **4**, 9–13.

Gaudreau, D. C., Jackson, S. T. & Webb T., III (1989). Spatial scale and sampling strategy in paleoecological studies of vegetation patterns in mountainous regions. *Acta Botanica Neerlandica*, **38**, 369–90.

Gleason, H. A. (1909). Some unsolved problems of the prairies. *Bulletin of the Torrey Botanical Club*, **36**, 265–71.

Gleason, H. A. (1923). The vegetational history of the Middle West. *Annals of the Association of American Geographers*, **12**, 39–85.

Gray, A. (1858). Diagnostic characters of new species of phenogamous plants, collected in Japan by Charles Wright, Botanist of the U.S. North Pacific Exploring Expedition (Published by request of Captain John Rodgers, Commander of the Expedition.) With observations upon the relations of the Japanese flora to that of North America, and of other parts of the Northern Temperate Zone. *Memoirs of the American Academy of Arts and Sciences*, **6**, 377–452.

Gray, A. (1878). Forest geography and archeology. *American Journal of Science*, **114**, 85–94; 183–96.

Gregory, P. H. (1973). *The Microbiology of the Atmosphere*. 2nd edition. New York: John Wiley & Sons.

Griffiths, D. J. (1950). The liability of seed crops of perennial ryegrass (*Lolium perenne*) to contamination by windborne pollen. *Journal of Agricultural Science*, **40**, 19–38.

Halldin, S. (1985). Leaf and bark area distribution in a pine forest. In *The Forest-Atmosphere Interaction*, ed. B. A. Hutchinson & B. B. Hicks. Dordrecht: D. Reidel, 39–58.

Handel, S. N. (1976). Restricted pollen flow of two woodland herbs determined by neutron-activation analysis. *Nature*, **260**, 422–3.

Hansen, H. P. (1949). Pollen content of moss polsters in relation to forest composition. *American Midland Naturalist*, **42**, 473–9.

Harper, J. L. (1977). *Population Biology of Plants*. London: Academic Press.

Harper, J. L. (1981). The concept of population in modular organisms. In *Theoretical Ecology*, 2nd edition, ed. R. M. May. Sunderland, MA: Sinauer Associates, 53–77.

Heide, K. M. (1984). Holocene pollen stratigraphy from a lake and small hollow in north-central Wisconsin, USA. *Palynology*, **8**, 3–20.

Heide, K. M. & Bradshaw, R. H. W. (1982). The pollen-tree relationship within forests of Wisconsin and Upper Michigan, U.S.A. *Review of Palaeobotany and Palynology*, **36**, 1–23.

Huntley, B. (1988). Europe. In *Vegetation History*, ed. B. Huntley & T. Webb, III. Dordrecht: Kluwer Academic Publishers, 341–83.

Huntley, B. & Birks, H. J. B. (1983). *An Atlas of Past and Present Pollen Maps for Europe: 0–13,000 Years Ago*. Cambridge: Cambridge University Press.

Hutchinson, G. E. (1975). *A Treatise on Limnology*. Volume III. *Limnological Botany*. New York: John Wiley & Sons.

Iversen, J. (1941). Landnam i Danmarks stenalder. *Danmarks Geologiske Undersøgelse*. II. Raekke, 66.

Iversen, J. (1973). The development of Denmark's nature since the last glacial. *Danmarks Geologiske Undersøgelse*. V. Raekke, 7–C.

Jackson, S. T. (1989). Postglacial vegetational changes along an elevational gradient in the Adirondack Mountains (New York): a study of plant macrofossils. *New York State Museum Bulletin*, **465**.

Jackson, S. T. (1990). Pollen source area and representation in small lakes of the northeastern United States. *Review of Palaeobotany and Palynology*, **63**, 53–76.

Jackson, S. T. (1991). Pollen representation of vegetational patterns along an elevational gradient. *Journal of Vegetation Science*, **2**, 613–24.

Jackson, S. T. & Charles, D. F. (1988). Aquatic macrophytes in Adirondack (New York) lakes: patterns of species composition in relation to environment. *Canadian Journal of Botany*, **66**, 1449–60.

Jackson, S. T., Futyma, R. P. & Wilcox, D. A. (1988). A paleoecological test of a classical hydrosere in the Lake Michigan Dunes. *Ecology*, **69**, 928–36.

Jackson, S. T. & Whitehead, D. R. (1991). Holocene vegetation patterns in the Adirondack Mountains. *Ecology*, **72**, 641–53.

Jacobson, G. L., Jr. (1979). The palaeoecology of white pine (*Pinus strobus*) in Minnesota. *Journal of Ecology*, **67**, 697–726.

Jacobson, G. L., Jr. & Bradshaw, R. H. W. (1981). The selection of sites for paleovegetational studies. *Quaternary Research*, **16**, 80–96.

Jacobson, G. L., Jr., Webb, T., III & Grimm, E. C. (1987). Patterns and rates of vegetation change during the deglaciation of eastern North America. In *North America and Adjacent Oceans During the Last Deglaciation*, ed. W. F. Ruddiman & H. E. Wright, Jr. Boulder, CO: Geological Society of America, 277–88.

Janssen, C. R. (1966). Recent pollen spectra from the deciduous and coniferous deciduous forests of Northeastern Minnesota: a study in pollen dispersal. *Ecology*, **47**, 804–25.

Janssen, C. R. (1967a). A postglacial pollen diagram from a small *Typha* swamp in northwestern Minnesota, interpreted from pollen indicators and surface samples. *Ecological Monographs*, **37**, 145–72.

Janssen, C. R. (1967b). A comparison between the recent regional pollen rain and the sub-recent vegetation in four major vegetation types in Minnesota (U.S.A.). *Review of Palaeobotany and Palynology*, **2**, 331–42.

Janssen, C. R. (1984). Modern pollen assemblages and vegetation in the Myrtle Lake Peatland, Minnesota. *Ecological Monographs*, **54**, 213–52.

Kabailiene, M. V. (1969). On formation of pollen spectra and restoration of vegetation. *Transactions of the Institute of Geology (Vilnius)*, **11**, 1–148. (In Russian with English and Lithuanian summaries.)

Kershaw, A. P. & Hyland, B. P. M. (1975). Pollen transfer and periodicity in a rain-forest situation. *Review of Palaeobotany and Palynology*, **19**, 129–38.

Kinerson, R. & Fritschen, L. J. (1971). Modeling a coniferous forest canopy. *Agricultural Meteorology*, **8**, 439–45.

Knoll, F. (1932). Über die Fernverbreitung des Blütenstaubes durch den Wind. *Forschungen und Fortschritte; Nachrichtenblatt der deutschen Wissenschaft und Technik*, **8**, 301–2.

Kohyama, T. (1982). Studies on the *Abies* population of Mt. Shimagare. II. Reproductive and life history traits. *Botanical Magazine (Tokyo)*, **95**, 167–81.

Koop, H. (1989). *Forest Dynamics*. Berlin: Springer-Verlag.

Koski, V. (1970). A study of pollen dispersal as a mechanism of gene flow in conifers. *Communicationes Instituti Forestalis Fenniae*, **70** (4), 1–78.

Kozlowski, T. T. (1971). *Growth and Development of Trees*. Volume II. *Cambial Growth, Root Growth, and Reproductive Growth*. New York: Academic Press.

Kozumplik, V. & Christie, B. R. (1972). Dissemination of

orchard-grass pollen. *Canadian Journal of Plant Sciences*, **52**, 997–1002.

Krzywinski, K. (1977). Different pollen deposition mechanism in forest: a simple model. *Grana*, **16**, 199–202.

Lanner, R. M. (1966). Needed: a new approach to the study of pollen dispersion. *Silvae Genetica*, **15**, 50–2.

Leopold, E. B. (1957). Some aspects of late-glacial climate in eastern United States. *Veröffentlichungen des Geobotanischen Institutes Rubel in Zürich*, **34**, 80–5.

Levin, D. A. & Kerster, H. W. (1974). Gene flow in seed plants. *Evolutionary Biology*, **7**, 139–220.

Leyden, B. W. (1987). Man and climate in the Maya Lowlands. *Quaternary Research*, **28**, 407–14.

Lichti-Federovich, S. & Ritchie, J. C. (1968). Recent pollen assemblages from the western interior of Canada. *Review of Palaeobotany and Palynology*, **7**, 297–344.

Likens, G. E. & Davis, M. B. (1975). Post-glacial history of Mirror Lake and its watershed in New Hampshire, U.S.A.: an initial report. *Verhandlungen Internationalen Vereiningung Limnologie*, **19**, 982–93.

Livingstone, D. A. (1968). Some interstadial and postglacial pollen diagrams from eastern Canada. *Ecological Monographs*, **38**, 87–125.

Livingstone, D. A. (1969). Communities of the past. In *Essays in Plant Geography and Ecology*, ed. K. N. H. Greenidge. Halifax: Nova Scotia Museum, 83–104.

Mandrioli, P., Negrini, M. G., Cesari, G. & Morgan, G. (1984). Evidence for long range transport of biological and anthropogenic aerosol particles in the atmosphere. *Grana*, **23**, 43–53.

McAndrews, J. H. (1969). Paleobotany of a wild rice lake in Minnesota. *Canadian Journal of Botany*, **47**, 1671–9.

McAndrews, J. H. (1988). Human disturbance of North American forests and grasslands: the fossil pollen record. In *Vegetation History*, ed. B. Huntley and T. Webb, III. Dordrecht: Kluwer Academic Publishers, 673–97.

McAndrews, J. H. & Boyko-Diakonow, M. (1989). Pollen analysis of varved sediment at Crawford Lake, Ontario: evidence of Indian and European farming. In *Quaternary Geology of Canada and Greenland. The Geology of North America*, Volume K-1, ed. R. J. Fulton. Boulder, CO: Geological Society of America, 528–30.

McDonald, J. E. (1962). Collection and washout of airborne pollen and spores by raindrops. *Science*, **135**, 435–7.

Niklas, K. J. (1984). The motion of windborne pollen grains around conifer ovulate cones: implications on wind pollination. *American Journal of Botany*, **71**, 356–74.

Niklas, K. J. (1985). The aerodynamics of wind pollination. *Botanical Review*, **51**, 328–86.

Niklas, K. J. & Paw U, K. T. (1983). Conifer ovulate cone morphology: implications on pollen impaction patterns. *American Journal of Botany*, **70**, 568–77.

Okubo, A. & Levin, S. A. (1989). A theoretical framework for data analysis of wind dispersal of seeds and pollen. *Ecology*, **70**, 329–38.

Osborn, J. M. & Schneider, E. L. (1988). Morphological studies of the Nymphaeaceae *sensu lato*. XVI. The floral biology of *Brasenia schreberi*. *Annals of the Missouri Botanical Garden*, **75**, 778–94.

Overpeck, J. T., Prentice, I. C. & Webb, T., III (1985). Quantitative interpretation of fossil pollen spectra: dissimilarity coefficients and the method of modern analogs. *Quaternary Research*, **23**, 87–108.

Parsons, R. W., Gordon, A. G. & Prentice, I. C. (1983). Statistical uncertainty in forest composition estimates obtained from fossil pollen spectra via the R-value model. *Review of Palaeobotany and Palynology*, **40**, 177–89.

Parsons, R. W. & Prentice, I. C. (1981). Statistical approaches to R-values and the pollen-vegetation relationship. *Review of Palaeobotany and Palynology*, **32**, 127–52.

Parsons, R. W., Prentice, I. C. & Saarnisto, M. (1980). Statistical studies on pollen representation in Finnish lake sediments in relation to forest inventory data. *Annales Botanici Fennici*, **17**, 379–93.

Peck, R. M. (1973). Pollen budget studies in a small Yorkshire catchment. In *Quaternary Plant Ecology*, ed. H. J. B. Birks & R. G. West. Oxford: Blackwell, 43–60.

Pennington, W. (1973). Absolute pollen frequencies in the sediments of lakes of different morphometry. In *Quaternary Plant Ecology*, ed. H. J. B. Birks & R. G. West. Oxford: Blackwell, 79–104.

Pennington, W. (1979). The origin of pollen in lake sediments: an enclosed lake compared with one receiving inflow streams. *New Phytologist*, **83**, 189–213.

Potzger, J. E., Courtemanche, A., Sylvio, Br. M. & Hueber, F. M. (1956). Pollen from moss polsters on the mat of Lac Shaw Bog, Quebec, correlated with a forest survey. *Butler University Botanical Studies*, **13**, 24–35.

Prentice, I. C. (1978). Modern pollen spectra from lake sediments in Finland and Finnmark, north Norway. *Boreas*, **7**, 131–53.

Prentice, I. C. (1985). Pollen representation, source area, and basin size: toward a unified theory of pollen analysis. *Quaternary Research*, **23**, 76–86.

Prentice, I. C. (1986a). Forest-composition calibration of pollen data. In *Handbook of Holocene Palaeoecology and Palaeohydrology*, ed. B. E. Berglund. Chichester: John Wiley & Sons, 799–816.

Prentice, I. C. (1986b). Some concepts and objectives of forest dynamics research. In *Forest Dynamics Research in Western and Central Europe*, ed. J. Fanta. Wageningen: Pudoc, 32–41.

Prentice, I. C. (1988a). Palaeoecology and plant population dynamics. *Trends in Ecology and Evolution*, **3**, 343–5.

Prentice, I. C. (1988b). Records of vegetation in time and space: the principles of pollen analysis. In *Vegetation History*, ed. B. Huntley and T. Webb, III. Dordrecht: Kluwer Academic Publishers, 17–42.

Prentice, I. C., Berglund, B. E. & Olsson, T. (1987). Quantitative forest-composition sensing characteristics of pollen samples from Swedish lakes. *Boreas*, **16**, 43–54.

Prentice, I. C. & Parsons, R. W. (1983). Maximum likelihood linear calibration of pollen spectra in terms of forest composition. *Biometrics*, **39**, 1051–7.

Prentice, I. C. & Webb, T., III (1986). Pollen percentages, tree abundances and the Fagerlind effect. *Journal of Quaternary Science*, **1**, 35–43.

Raup, H. M. (1937). Recent changes of climate and vegetation in southern New England and adjacent New York. *Journal of the Arnold Arboretum*, **18**, 79–117.

Raynor, G. S. (1971). Wind and temperature structure in a coniferous forest and a contiguous field. *Forest Science*, **17**, 351–63.

Raynor, G. S., Hayes, J. V. & Ogden, E. C. (1969). Areas within isopleths of ragweed pollen concentrations from local sources. *Archives of Environmental Health*, **19**, 92–8.

Raynor, G. S., Hayes, J. V. & Ogden, E. C. (1974a). Mesoscale transport and dispersion of airborne pollens. *Journal of Applied Meteorology*, **13**, 87–95.

Raynor, G. S., Hayes, J. V. & Ogden, E. C. (1974b). Particulate dispersion into and within a forest. *Boundary-Layer Meteorology*, **7**, 429–56.

Raynor, G. S., Hayes, J. V. & Ogden, E. C. (1975).

Particulate dispersion from sources within a forest. *Boundary-Layer Meteorology*, **9**, 257–77.

Raynor, G. S., Ogden, E. C. & Hayes, J. V. (1968). Effect of a local source on ragweed pollen concentrations from background sources. *Journal of Allergy*, **41**, 217–25.

Raynor, G. S., Ogden, E. C. & Hayes, J. V. (1970). Dispersion and deposition of ragweed pollen from experimental sources. *Journal of Applied Meteorology*, **9**, 885–95.

Raynor, G. S., Ogden, E. C. & Hayes, J. V. (1972a). Dispersion and deposition of corn pollen from experimental sources. *Agronomy Journal*, **64**, 420–7.

Raynor, G. S., Ogden, E. C. & Hayes, J. V. (1972b). Dispersion and deposition of timothy pollen from experimental sources. *Agricultural Meteorology*, **9**, 347–66.

Raynor, G. S., Ogden, E. C. & Hayes, J. V. (1973). Dispersion of pollens from low-level, crosswind line sources. *Agricultural Meteorology*, **11**, 177–95.

Raynor, G. S., Ogden, E. C. & Hayes, J. V. (1976). Dispersion of fern spores into and within a forest. *Rhodora*, **78**, 473–87.

Rempe, H. (1937). Untersuchungen über die Verbreitung des Blütenstaubes durch die Luftströmungen. *Planta*, **27**, 93–147.

Ritchie, J. C. (1984). *Past and Present Vegetation of the Far Northwest of Canada*. Toronto, Ontario: University of Toronto Press.

Ritchie, J. C., Hadden, K. A. & Gajewski, K. (1987). Modern pollen spectra from lakes in arctic western Canada. *Canadian Journal of Botany*, **65**, 1605–13.

Ruddiman, W. F. & Wright, H. E., Jr., eds. (1987). *North America and Adjacent Oceans During the Last Deglaciation*. Volume K-3, *The Geology of North America*. Boulder, CO: Geological Society of America.

Sarvas, R. (1962). Investigations on the flowering and seed crop of *Pinus sylvestris*. *Communicationes Instituti Forestalis Fenniae*, **53**(4), 1–198.

Sarvas, R. (1968). Investigations on the flowering and seed crop of *Picea abies*. *Communicationes Instituti Forestalis Fenniae*, **67**(5), 1–84.

Schwartz, M. W. (1989). Predicting tree frequencies from pollen frequency: an attempt to validate the R value method. *New Phytologist*, **112**, 129–43.

Sears, P. B. (1935). Types of North American pollen profiles. *Ecology*, **16**, 488–99.

Sears, P. B. (1938). Climatic interpretation of postglacial pollen deposits in North America. *Bulletin of the American Meteorological Society*, **19**, 177–85.

Sears, P. B. (1942). Xerothermic theory. *Botanical Review*, **8**, 708–36.

Shane, L. C. K. (1989). Changing palynological methods and their role in three successive interpretations of the late-glacial environments at Bucyrus Bog, Ohio, U.S.A. *Boreas*, **18**, 297–309.

Silen, R. R. (1962). Pollen dispersal considerations for douglas-fir. *Journal of Forestry*, **60**, 790–5.

Snell, J. A. K. & Brown, J. K. (1978). Comparison of tree biomass estimators – DBH and sapwood area. *Forest Science*, **24**, 455–7.

Solomon, A. M. & Webb, T., III (1985). Computer-aided reconstruction of Late-Quaternary landscape dynamics. *Annual Review of Ecology and Systematics*, **16**, 63–84.

Sutton, O. G. (1932). A theory of eddy diffusion in the atmosphere. *Proceedings of the Royal Society of London*, **A 135**, 143–65.

Sutton, O. G. (1947). The problem of diffusion in the lower atmosphere. *Quarterly Journal of the Royal Meteorological Society*, **73**, 257–76.

Szafer, W. (1935). The significance of isopollen lines for the investigation of the geographical distribution of trees in the post-glacial period. *Bulletin de l'Académie Polonaise des Sciences et des Lettres, Serie B: Sciences Naturelles*, **1**, 235–9.

Tauber, H. (1965). Differential pollen dispersion and the interpretation of pollen diagrams. *Danmarks Geologiske Undersøgelse*, Raekke 2, Number 89.

Tauber, H. (1967). Differential pollen dispersion and filtration. In *Quaternary Paleoecology*, ed. E. J. Cushing & H. E. Wright, Jr. New Haven, Connecticut: Yale University Press, 131–41.

Tauber, H. (1977). Investigations of aerial pollen transport in a forested region. *Dansk Botanisk Arkiv*, **32**, 1–121.

Tinsley, H. M. & Smith, R. T. (1974). Surface pollen studies across a woodland/heath transition and their application to the interpretation of pollen diagrams. *New Phytologist*, **73**, 547–65.

Tonsor, S. J. (1985). Leptokurtic pollen-flow, non-leptokurtic gene-flow in a wind-pollinated herb, *Plantago lanceolata* L. *Oecologia (Berlin)*, **67**, 442–6.

Transeau, E. N. (1935). The prairie peninsula. *Ecology*, **16**, 423–37.

Turesson, G. (1927). Contributions to the genecology of glacial relics. *Hereditas*, **9**, 81–101.

Turner, J. (1964). Surface sample analyses from Ayrshire, Scotland. *Pollen et Spores*, **6**, 583–92.

Van Zant, K. L., Webb, T., III, Peterson, G. M. & Baker, R. G. (1979). Increased *Cannabis/Humulus* pollen, an indicator of European agriculture in Iowa. *Palynology*, **3**, 227–33.

Vogel, S. (1981). *Life in Moving Fluids: the Physical Biology of Flow*. Boston, MA: Willard Grant.

Von Post, L. (1967). Forest tree pollen in south Swedish peat bog deposits. *Pollen et Spores*, **9**, 375–401. (Translation by M. B. Davis and K. Faegri, Introduction by K. Faegri and J. Iversen).

Wang, C.-W., Perry, T. O. & Johnson, A. G. (1960). Pollen dispersion of slash pine (*Pinus elliottii* Engelm.) with special reference to seed orchard management. *Silvae Genetica*, **9**, 78–86.

Waring, R. H. & Schlesinger, W. H. (1985). *Forest Ecosystems: Concepts and Management*. Orlando, FL: Academic Press.

Waring, R. H., Schroeder, P. E. & Oren, R. (1982). Application of the pipe model theory to predict canopy leaf area. *Canadian Journal of Forest Research*, **12**, 556–60.

Watts, W. A. (1979). Late Quaternary vegetation of central Appalachia and the New Jersey coastal plain. *Ecological Monographs*, **49**, 427–69.

Watts, W. A. & Winter, T. C. (1966). Plant macrofossils from Kirchner Marsh, Minnesota – a paleoecological study. *Geological Society of America Bulletin*, **77**, 1339–60.

Webb, R. S. & Webb, T., III (1988). Rates of sediment accumulation in pollen cores from small lakes and mires of eastern North America. *Quaternary Research*, **30**, 284–97.

Webb, T., III (1974a). A vegetational history from northern Wisconsin: evidence from modern and fossil pollen. *American Midland Naturalist*, **92**, 12–34.

Webb, T., III (1974b). Corresponding distributions of modern pollen and vegetation in lower Michigan. *Ecology*, **55**, 17–28.

Webb, T., III (1985). A global paleoclimatic data base for 6000 yr B.P. *United States Department of Energy Report* TR-018. Washington, DC.

Webb, T., III (1988). Eastern North America. In *Vegetation*

History, ed. B. Huntley & T. Webb, III. Dordrecht: Kluwer Academic Publishers, 385–414.

Webb, T., III, Bartlein, P. J. & Kutzbach, J. E. (1987). Climatic changes in eastern North America during the past 18,000 years; comparisons of pollen data with model results. In *North America and Adjacent Oceans During the Last Deglaciation*, ed. W. F. Ruddiman & H. E. Wright, Jr. Boulder, CO: Geological Society of America, 447–62.

Webb, T., III, Cushing, E. J. & Wright, H. E., Jr. (1983a). Holocene changes in the vegetation of the Midwest. In *Late Quaternary Environments of the United States*. Volume 2, *The Holocene*, ed. H. E. Wright, Jr. Minneapolis: University of Minnesota Press, 142–65.

Webb, T., III, Laseski, R. A. & Bernabo, J. C. (1978b). Sensing vegetational patterns with pollen data: choosing the data. *Ecology*, **59**, 1151–63.

Webb, T., III, Howe, S. E., Bradshaw, R. H. W. & Heide, K. M. (1981). Estimating plant abundances from pollen percentages: the use of regression analysis. *Review of Palaeobotany and Palynology*, **34**, 269–300.

Webb, T., III & McAndrews, J. H. (1976). Corresponding patterns of contemporary pollen and vegetation in central North America. *Geological Society of America Memoir*, **145**, 267–99.

Webb, T., III, Richard, P. J. H. & Mott, R. J. (1983b). A mapped history of Holocene vegetation in southern Quebec. *Syllogeus*, **49**, 273–336.

Webb, T., III, Yeracaris, G. Y. & Richard, P. (1978a). Mapped patterns in sediment samples of modern pollen from southeastern Canada and northeastern United States. *Géographie Physique et Quaternaire*, **32**, 163–76.

Wenger, K. F., ed. (1984). *Forestry Handbook*, 2nd edition. New York: John Wiley & Sons.

Whitehead, D. R. (1969). Wind pollination in the angiosperms: evolutionary and environmental considerations. *Evolution*, **23**, 28–35.

Whitehead, D. R. (1983). Wind pollination: some ecological and evolutionary perspectives. In *Pollination Biology*, ed. L. Real. New York: Academic Press, 97–108.

Whitehead, D. R. & Tan, K. W. (1969). Modern vegetation and pollen rain in Bladen County, North Carolina. *Ecology*, **50**, 235–48.

Whittaker, R. H., Bormann, F. H., Likens, G. E. & Siccama, T. G. (1974). The Hubbard Brook ecosystem study: forest biomass and production. *Ecological Monographs*, **44**, 233–52.

Whittaker, R. H. & Woodwell, G. W. (1968). Dimension and production relations of trees and shrubs in the Brookhaven forest, New York. *Journal of Ecology*, **56**, 1–25.

Wright, H. E., Jr. (1967). The use of surface samples in Quaternary pollen analysis. *Review of Palaeobotany and Palynology*, **2**, 321–30.

Wright, H. E., Jr. (1968). The roles of pine and spruce in the forest history of Minnesota and adjacent areas. *Ecology*, **49**, 937–55.

Wright, H. E., Jr. (1976). The dynamic nature of Holocene vegetation: a problem in paleoclimatology, biogeography, and stratigraphic nomenclature. *Quaternary Research*, **6**, 581–96.

Wright, J. W. (1952). Pollen dispersal of some forest trees. *United States Forest Service, Northeastern Forest Experiment Station Paper* **46**.

Wright, J. W. (1953). Notes on flowering and fruiting of northeastern trees. *United States Forest Service, Northeastern Forest Experiment Station Paper* **60**.

15 Paleoecological interpretation of the Trail Ridge sequence, and related deposits in Georgia and Florida, based on pollen sedimentation and clastic sedimentology

FREDRICK J. RICH and FREDRIC L. PIRKLE

Introduction

Late Cenozoic deposits of the Atlantic coastal plain of the southeastern United States include a large volume of very heterogeneous sediments. Heavy mineral-bearing quartz sands are a constituent of these sediments, as are peats, peaty sands and clays, marine and brackish water clays and silts, and shell beds. Herrick (1965) describes the lithology of Pleistocene deposits of coastal Georgia:

> The Pleistocene deposits in updip areas consist predominantly of fine- to medium-grained, subangular, arkosic, cherty, sparsely phosphatic sand interbedded with minor amounts of very thin-bedded white micaceous, sandy kaolin. In downdip areas much of the sand is apparently replaced by tongues of dark-brownish-gray to black, blocky, rather tough, sandy, coarsely micaceous, lignitic, locally fossiliferous clay. Beneath the lignitic clay is a basal unit consisting of subangular to subrounded, sparsely phosphatic sand and gravel ....

Where they have been sampled, many of the sediments have proven to be very productive palynologically. The abundant vegetation of the coastal plain, high water tables, and extensive peat deposits would seem to make such a place an ideal setting for palynological investigations. However, most of the late Cenozoic strata of the Atlantic coastal plain in the Southeast have not been studied. Even surface deposits have received little attention, as was attested to by Davis and Webb (1975). Their attempt to correlate the distribution of modern pollen with extant vegetation failed in the Southeast, because the sampling grid there was 'extremely sparse.' Even the distribution of pollen from such common plants as the hollies (*Ilex*: Aquifoliaceae) could not be mapped, because too few data points existed in the Southeast.

This chapter helps explain the paleoecological and sedimentological evolution of Cenozoic deposits from the coastal plain of the southeastern US, and we also analyze the origins and composition of several sand bodies, and evaluate the palynological composition of the associated organic sediments. The sedimentological and palynological studies were mostly undertaken independently, to test whether the results and interpretations would corroborate one another. We believe they do. The pollen and other organic particles are part of a rather orderly, though dynamic, depositional system, dominated by inorganic clastic sediments. It is reasonable to expect that, after careful analysis of all the sediment components, their depositional histories should be consistent with one another.

General geology and physiography of the region

The area of study in this chapter lies between Starke, Florida, and the mouth of the Cooper River near Charleston, South Carolina. It extends westward from the present-day shoreline to the foot of the Orangeburg Escarpment in South Carolina/Georgia and to the western slope of Trail Ridge in Georgia/Florida (Fig. 15.1). The area includes the region studied by Hoyt and Hails (1969) in their paper on Pleistocene shorelines, as well as the region mapped by Clark and Zisa (1976) as the Barrier Island Sequence District of Georgia. Raisz (1957) refers to the area as the Lower Pine Belt in South Carolina and the Low Terraces in Georgia. The significant geomorphological features in this region include a series of sand ridges and intervening terraces which developed along a succession of late Cenozoic coastlines. Trail Ridge (Fig. 15.2) is the most prominent of these ancient ridges and is one of the primary focal points of this chapter. This ridge and some of the less obvious ridges which lie between it and the Atlantic Ocean are of considerable economic importance because of their heavy mineral content (Pirkle *et al.*, 1991).

Studies of ridges provide clues to the late Cenozoic development of the coastal plain. Much speculation in the geologic literature purports to explain the origin of the sand ridges and other late Cenozoic features of the

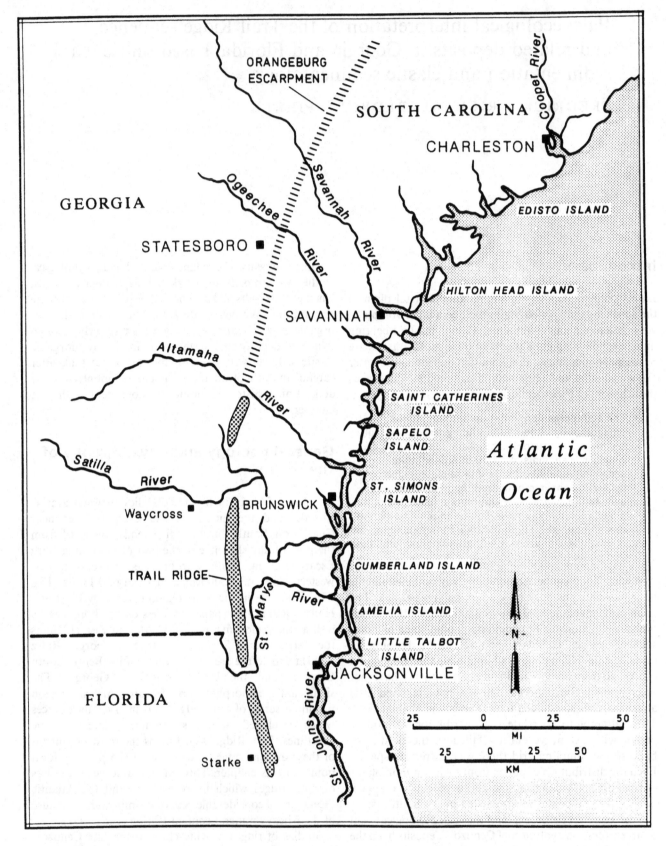

Figure 15.1. General location of the study area.

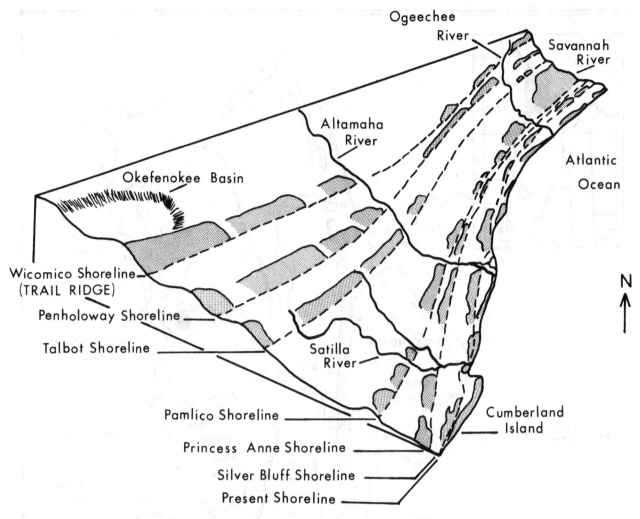

Figure 15.2. Coastal ridge complex of Georgia (from Georgia Dept. of Natural Resources, 1976).

coastal plain. Important contributions include those of Lyell (1846), Veatch and Stephenson (1911), Cooke (1945), MacNeil (1950), Hoyt and Hails (1969) and Huddlestun (1988). Nevertheless, the area remains essentially unknown. This chapter represents an attempt to use palynology to gain some insight into the geologic history of the region.

Geology of Trail Ridge

Trail Ridge is one of the most prominent geomorphological features of the southern Atlantic coastal plain. The ridge extends from the Altamaha River in Georgia southward for about 160 km to northern Putnam County in Florida (Fig. 15.1) and is between one and two kilometers wide. The crest of the ridge stands 46 m above sea level in Georgia and rises to elevations of 77 m in Bradford and Clay counties in Florida. The elevation above the adjacent coastal plain may not be immediately apparent, but the fact that the ridge is covered with pine-oak-palmetto forest rather than cypress-tupelo swamp is

obvious. This is especially true in Georgia, where Trail Ridge lies as a relatively high, dry, sandy surface immediately east of the Okefenokee Swamp (Figs. 15.2, 15.3).

Trail Ridge is composed of diverse sediments, but is dominated by a lower unit of clayey sands and sandy clays that locally contains important occurrences of lignitic peat and peaty sands, and an upper unit of quartz sand that in some localities contains commercial quantities of heavy minerals. This two-fold sedimentary sequence has been referred to by Pirkle and Yoho (1970) and Pirkle, Yoho and Hendry (1970) as the Trail Ridge Sequence. Considerations of the genetic relationship between the peat and the sand constitute an important part of this chapter.

The upper sandy portion of the Trail Ridge Sequence has been the focus of numerous papers (e.g., Spencer, 1948; Pirkle *et al.*, 1970; Pirkle, 1975; Pirkle *et al.*, 1977). In many areas the sands contain as much as 3–4% heavy minerals with a variety of compositions. The heavy mineral suite is dominated, however, by titanium-bearing minerals which include ilmenite and leucoxene. The

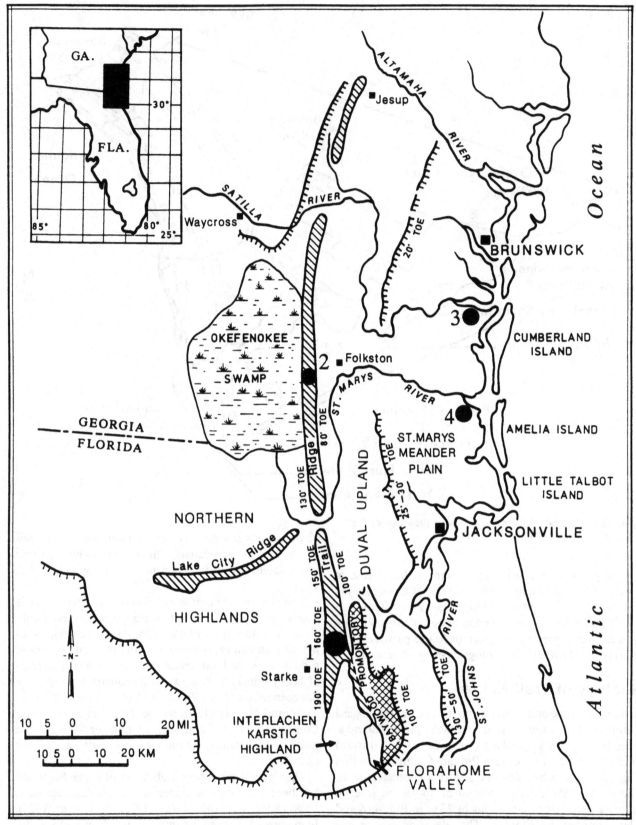

Figure 15.3.   General locations from which samples discussed in this investigation were collected. Samples from sites 1 and 2 are discussed in the text as Trail Ridge samples while samples from sites 3 and 4 are classified as samples from beneath the Pamlico and Princess Anne terraces.

Table 15.1. *Relative abundances (%) of the dominant huminite macerals from cores TR1 and TR2.*

Note abundance of detrital macerals.

TR1 (depth in meters)	Ulminite	Humodetrinite	Gelinite	Textinite
15.6	30.4	54.3	1.6	3.3
15.7	63.3	16.3	2.1	1.4
15.8	22.0	64.2	0.8	3.3
16.0	13.0	74.5	1.3	1.6
16.2	19.7	58.8	2.0	1.9
16.3	15.5	71.1	2.5	4.1
16.5	8.7	83.7	1.0	2.8
16.7–17.4	13.8	77.8	0.7	3.6
17.4–17.5	44.9	44.0	0.5	5.1

TR2 (depth in meters)				
16.7–16.9	38.6	52.1	1.9	2.6
16.9	19.8	69.8	1.4	1.2
17.1	27.1	62.6	0.9	3.3
17.3	14.5	66.0	3.0	3.3
17.4–17.6	11.0	75.1	2.6	4.8
17.6–17.8	17.3	69.9	1.7	5.4
17.8–18.0	4.6	79.8	2.7	1.6
18.0–18.3	10.0	77.6	4.2	1.4
18.4	12.4	76.0	3.0	1.2
18.7	16.6	74.6	1.8	1.9
18.8	27.4	65.9	1.0	2.3

concentration of the titanium minerals is such that the Trail Ridge sands constitute one of the most important domestic sources of titanium dioxide ore in the US.

Geologists associated with various mining companies (chiefly E. I. DuPont de Nemours & Co.), the University of Florida, the Florida Geological Survey, and the US Geological Survey have written extensively on the geology of Trail Ridge and the characteristics of its sands. The prevailing theory now suggests that the sands were deposited in a coastal environment, and that eolian processes were dominant at least in the southern area of the ridge where sands accumulated in a dune field (Pirkle, 1984; Force & Rich, 1989).

The lower lignitic peat portion of the Trail Ridge Sequence has not been studied as intensively as the overlying sands. Recent work by Rich and various co-workers, however, has done much to improve our understanding of the origin of the peat. An initial investigation of the palynological and petrographic properties of the Trail Ridge peat was conducted by Rich *et al.* (1978). They determined the huminite reflectance of one sample of the peat to have a mean maximum value of 0.274%. Their sample was also shown to be composed of a mixture of detrital humic constituents and highly

gelified woody tissue. The reflectivity of the huminites and the generally highly altered aspect of the sediment led Rich and his colleagues to consider the peat intermediate in diagenetic development between peat and lignite or brown coal.

Since 1978, more samples of the Trail Ridge peat have been analyzed. Two cores (TR1 and TR2) of the sediment obtained jointly by the Florida Geological Survey, the University of Florida, and DuPont were taken from property adjacent to DuPont's Trail Ridge Plant site in northern Florida (Figs. 15.3, 15.4). Samples from these cores were analyzed petrographically at the Coal Research Section (now Energy and Fuels Research Center) of The Pennsylvania State University. Highly altered, detrital humodetrinite dominates (Table 15.1). Gelified cell walls (ulminite) are of secondary importance.

Analysis of additional samples randomly recovered from various depths and geographic locations on Trail Ridge (Figs. 15.3–15.7) further substantiate the generally detrital nature of the peat deposit. Solid pieces of wood sometimes are found in the peat beds, or periodically occur, surrounded by sand and clay within Trail Ridge sediments. These pieces usually display the well-structured fabric of coalified secondary wood, otherwise known as texto-ulminite or textinite. Such woody samples are apparently uncommon, however, in comparison to the large volume of detrital organic remains.

Huminite reflectance values for the random suite of samples are in agreement with the value presented by Rich *et al.* (1978). Reflectivities for five of the samples range from 0.2% to 0.278% (Table 15.2). These values are high in comparison with peat, and overlap with the low end of the range of values for lignites and brown coals.

The age of Trail Ridge continues to be a matter of debate. Its sedimentological, physiographical, and geomorphological relationships with other deposits and land forms in the southeastern US have lead various authors to propose ages ranging from Miocene to Pleistocene. Rich (1985) attempted to date the lignitic peat by palynological means, but was able to conclude only that it does not appear to be Miocene, and it has only a few characteristics in common with Pleistocene deposits. Carbon isotope dates recently obtained from a number of samples, including a portion of one of the two cores collected by DuPont, the University of Florida and the Florida Geological Survey, show ages ranging from 33 000 to greater than 45 000 years. These values show that Trail Ridge and its associated peat are not of latest Pleistocene age.

As previously mentioned, the sands of Trail Ridge accumulated at least in part within a coastal dune field. Data collected by Force and Rich (1989) show that the sedimentological interface between the sands and the peats of the Trail Ridge Sequence has characteristics which support the eolian origin of the sands. Isolated quartz grains have been found in thin sections to be

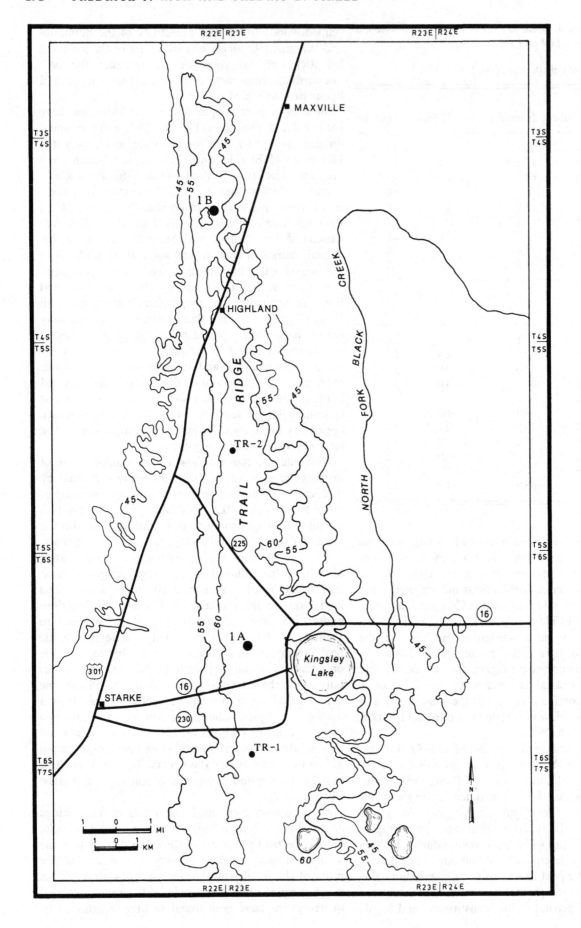

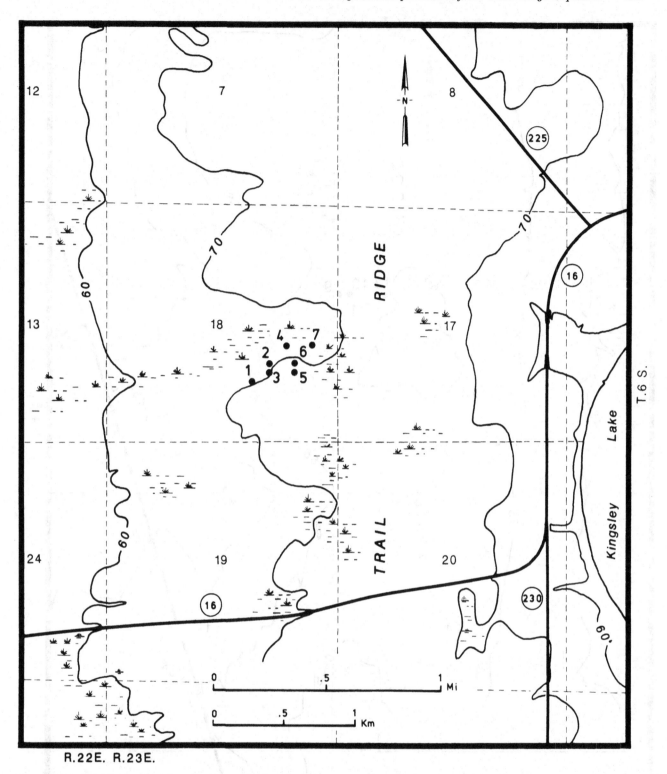

Figure 15.5. Detailed location of samples from site 1A (Fig. 15.4). A few contour lines (in meters) are shown.

Figure 15.4. General location of sample sites associated with site 1 (Fig. 15.3). The locations of TR1 and TR2 are shown on the figure. Site 1 has been divided into site 1A and site 1B. Figures 15.5 and 15.6 show the locations of individual samples found in sites 1A and 1B. A few contour lines (in meters) are shown.

disseminated and embedded within the peat. The organic matter has been draped around the grains as a result of post-depositional compaction. The occurrences demonstrate that the grains of sand and the peat fragments accumulated at the same time. Both sand grain shapes and surface characteristics are similar to those of the sands overlying the peat. The grains are believed to

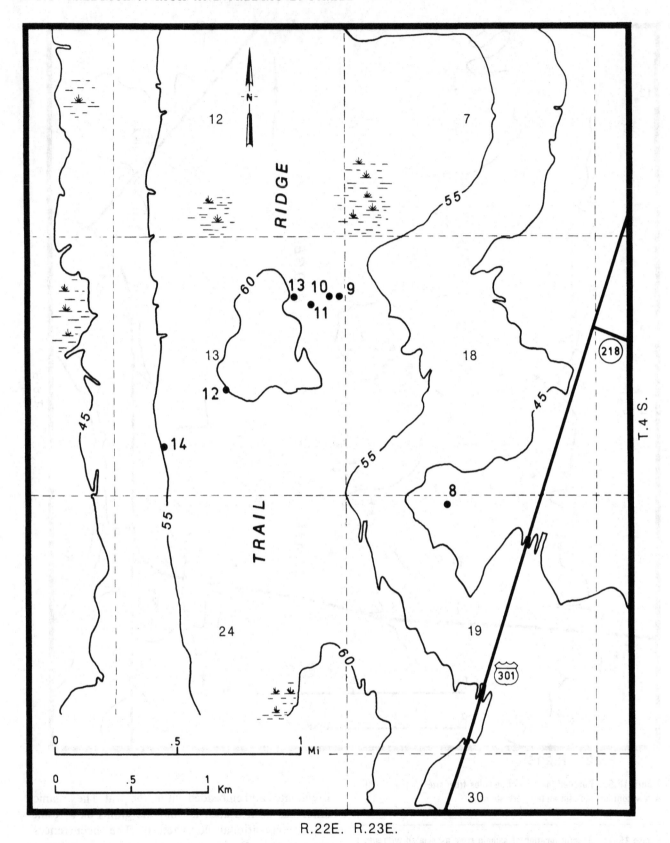

Figure 15.6. Detailed location of samples from site 1B (Fig. 15.4). A few contour lines (in meters) are shown.

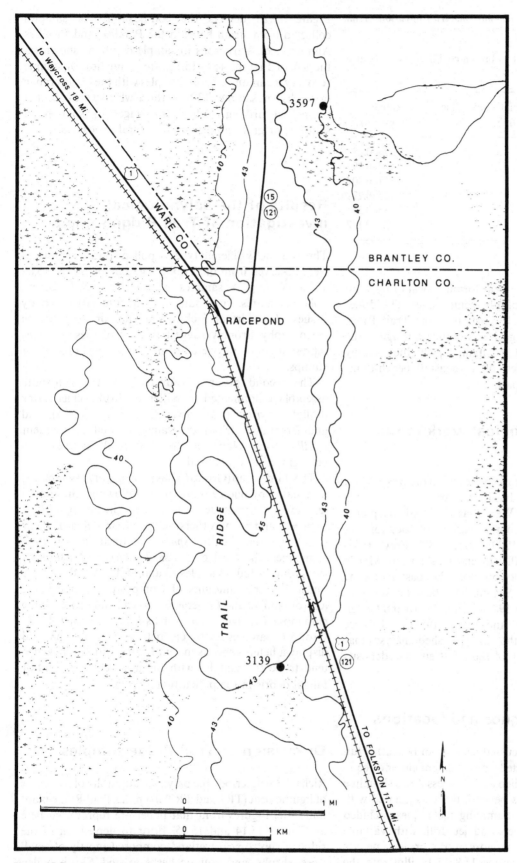

Figure 15.7.   Detailed location of samples from site 2 (Fig. 15.3). A few contour lines (in meters) are shown.

Table 15.2. *Reflectivities of huminite macerals in various Trail Ridge samples.*

Note that change from peat to brown coal/lignite occurs at a value of about 0.266%.

Sample No.	Mean maximum value (percent)	Range (percent)
12	0.2	0.174–0.233
11	0.207	0.134–0.293
10	0.231	0.151–0.307
9	0.236	0.175–0.271
5068A	0.278	0.235–0.34
3597	0.161	0.12 –0.192

have been blown into the peat-forming swamp by wind passing over the crests of advancing dunes. The dunes which produced the sandy portion of the Trail Ridge Sequence were therefore generally of the same age as the swamp which produced the lignitic peat; they simply represent different facies of a coastal beach-dune, field-swamp/marsh complex.

Previous palynological work in the region

Previous palynological work in the region of investigation described in this chapter has been sparse, but there have been important studies. Watts has analyzed the pollen content of sediments from a variety of locations in Georgia (Watts, 1969, 1970, 1971, 1980). Frey (1952) conducted a study of Pleistocene strata from Myrtle Beach, SC, north of Charleston, along the coast. Holocene peats in the Okefenokee Swamp have been analyzed by Cohen (1975), Rich (1979, 1984a,b), and Fearn and Cohen (1984). Rich (1985), Rich and Pirkle (1988), and Force and Rich (1989) present the only published analyses that deal with the palynology of the late Cenozoic deposits treated in this report.

Sampling techniques and locations

Most of the samples discussed came from beneath Trail Ridge; others were collected from sediment either directly associated with the Pamlico and Princess Anne shoreline ridges or from sediments beneath those associated with these ridges (Fig. 15.3). Sampling techniques included coring, split spoon, auger, and jet drill, with variable sample recovery, and therefore more or less precise stratigraphic control. Figures 15.8–15.13 illustrate the palynological compositions of Trail Ridge samples.

Figure 15.20 depicts the compositions of samples collected from strata beneath the Pamlico and Princess Ann surfaces. Variations in sample depth, as shown on the left of Figures 15.8–15.13, are a function of sample recovery techniques. Those samples with the largest range of depths (i.e., sample 3139) are those which were collected during jet drilling. Others are auger or split-spoon samples, or grab samples from natural exposures.

Results of the palynological investigation of Trail Ridge samples

Three distinctly different types of pollen/spore assemblages were observed from the Trail Ridge samples (Figs. 15.8–15.13). Group A samples contain very little or no pollen or spores. These samples consist mostly of woody tissues. The virtual lack of palynomorphs and the petrography strongly suggest that all of these samples represent debris recovered from single logs, branches or stumps.

The second group (Group B) has a pollen/spore assemblage dominated by pollen of shrubs, chiefly *Ilex* (holly), *Myrica* (wax myrtle), *Corylus* (hazel) and undifferentiated triporate grains, as well as frequent *Cyrilla* (ti-ti). None of the Group B samples was petrographically analyzed.

The third group (Group C) has pollen/spore assemblages dominated by one or more of the following: Gramineae (grasses), Compositae (composites, including asters, ragweed, and many others), Cyperaceae (sedges), *Pinus* (pine), *Quercus* (oak), *Sphagnum* (peat moss), and, sometimes, the algal cyst, *Pseudoschizaea*. A subset of this group includes samples 13 and 14. Both of these have unusually large amounts of Compositae, Umbelliferae (carrot family), and Cyperaceae pollen, and spores of the club moss, *Lycopodium*, as well as *Pseudoschizaea*. Those Group C samples that were petrographically analyzed display a heterogeneous maceral composition. They also tend to be the samples with the most charcoal and/or fungal debris in the pollen preparations.

Discussion of Trail Ridge samples

Rich (1985) described the palynostratigraphy of two cores of lignitic peat (TR1 and TR2) from the Trail Ridge plant mine site. Figures from that paper are reproduced here as Figures 15.14 and 15.15. Prior to deposition of the ridge, a subtropical wetland composed mostly of small trees, shrubs, and aquatic herbs existed. Shrub pollen clearly was dominant, although sedges, *Sphagnum* and

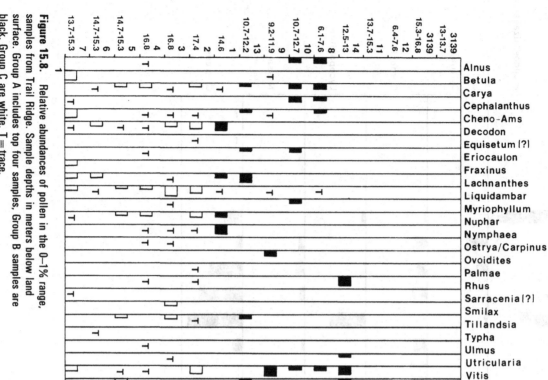

Figure 15.8. Relative abundances of pollen in the 0–1% range, samples from Trail Ridge. Sample depths in meters below land surface. Group A includes top four samples, Group B samples are black, Group C are white. T = trace.

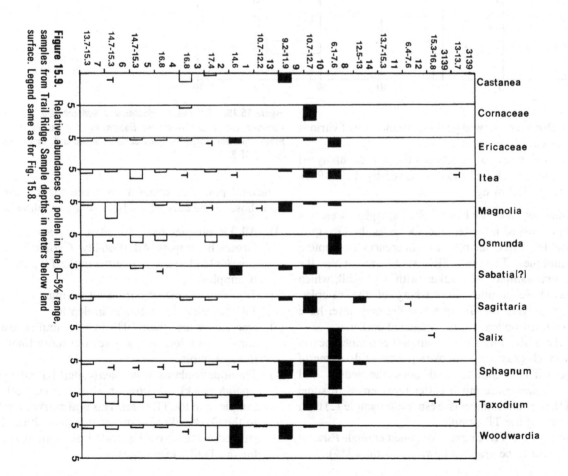

Figure 15.9. Relative abundances of pollen in the 0–5% range, samples from Trail Ridge. Sample depths in meters below land surface. Legend same as for Fig. 15.8.

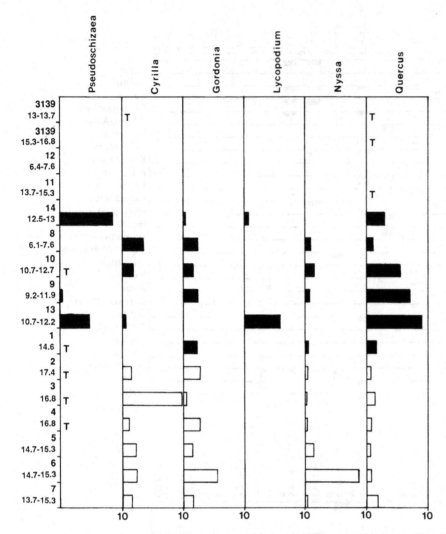

Figure 15.10. Relative abundances of *Pseudoschizaea*, *Cyrilla*, *Gordonia*, *Lycopodium*, *Nyssa* and *Quercus* found in samples from Trail Ridge. Note abundance of *Pseudoschizaea* in Group B. Legend same as for Fig. 15.8.

the fern, *Osmunda*, showed peaks of abundance at various levels in each of the cores.

Several differences exist between the samples analyzed in the current work and those observed by Rich (1985), including the following:

(1) None of the TR1 or TR2 samples were as impoverished in pollen and spores as the Group A samples. This could be due to differences in sampling techniques: TR1 and TR2 were cores, while the current samples were taken with a jet drill, which may have resulted in washing of the samples. Alternatively, the Group A samples may never have contained pollen, as was suggested earlier.

(2) None of the TR1 or TR2 samples contained nearly as much grass or composite pollen as do some of the current samples, and both the sedges and Umbelliferae are much better represented in some of the current samples (see especially sample 13) than in any of the TR samples.

(3) None of the TR samples contained enough *Pseudoschizaea* to be graphed (always less than 1%).

Several points of similarity between the TR samples and those of the current study include the following:

(1) All TR samples have abundant *Myrica*, as do the Group B samples. Additionally, *Cyrilla*, *Ilex*, and *Corylus* tend to occur together in the TR and Group B samples.

(2) *Taxodium* is uncommon in all samples and in the TR1 samples above 16.3 m in depth.

(3) *Quercus* is less than 10% in all samples, usually considerably less. *Pinus* is seldom more than 15% in any sample.

(4) The aquatic plants *Nymphaea* (waterlily) and *Nuphar* (spadderdock) are rare or absent from both our samples and the TR cores. This is in marked contrast to the Okefenokee Swamp marsh peats Rich (1979) studied, which have been used as modern analogs for the Trail Ridge deposits.

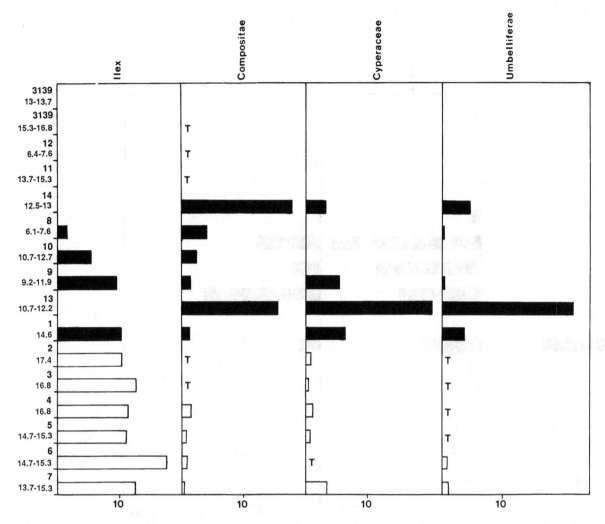

Figure 15.11. Relative abundances of *Ilex*, Compositae, Cyperaceae and Umbelliferae found in samples from Trail Ridge. Note particular abundance of the latter three in Group B. Legend same as for Fig. 15.8.

Depositional interpretations of Trail Ridge samples

Group A

Depositional environments for the lower part of the Trail Ridge Sequence apparently varied widely. Those samples which contain little or no pollen and are composed mostly of woody tissue fragments (Group A) were evidently derived from pieces of wood incorporated within the peat bed underlying the Trail Ridge sand body, or were deposited as individual pieces of wood within the sand. Peats from the Okefenokee Swamp commonly contain such pieces of wood, especially if the peats are derived from swamp forest vegetation. Rich (1979) and Cohen (1975) have both discussed the locations and characteristics of Okefenokee cypress peat deposits which are in some ways our most suitable analog for the woody portions of the Trail Ridge deposit.

Group B

The vegetation in Group B samples of this report is virtually the same as the shrub-swamp peat vegetation described by Rich (1985). These seem to be the most characteristic vegetation types of the Trail Ridge lignitic peats. The predominance of pollen from insect-pollinated (entomophilous) shrubs is most important in reconstructing the vegetation reflected by Group B samples. An especially useful taxon in this respect is *Ilex*. Several species are common in the Southeast. Among those that might be expected to have contributed to the Trail Ridge peats are *I. cassine* and *I. coriacea*. All species of holly are insect-pollinated, chiefly by bees. The close relationship which holly has with bees as pollinating agents has resulted in holly producing pollen not discharged freely from the flowers to be carried by air currents. Rather, the pollen which is not carried by bees or other insects

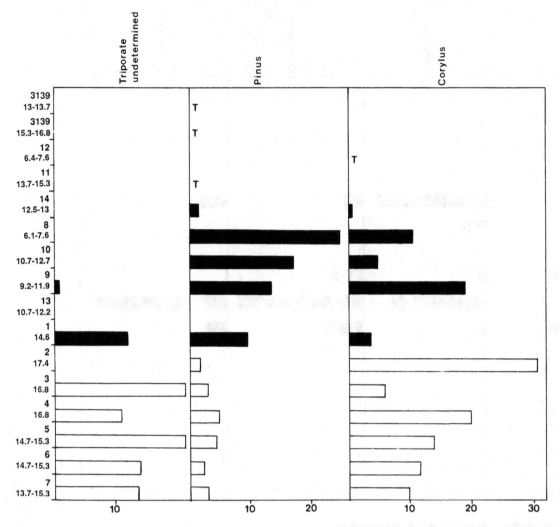

Figure 15.12.   Relative abundances of undetermined triporates, *Pinus* and *Corylus* found in samples from Trail Ridge. Legend same as for Fig. 15.8.

accumulates in large concentrations right where plants grow (Rich, 1979). An abundance of *Ilex* pollen is, therefore, good evidence that holly grew at the site of deposition. The abundance of *Ilex* pollen in Group B sediments is very strong evidence that those peats were produced by a holly-bearing shrub-swamp community.

Additional entomophilous shrub-swamp taxa include *Cyrilla* and *Gordonia* (red bay). Both have been used by Rich (1979, 1985) and Rich and Spackman (1979) as indicators of the shrub-swamp habitat (Fig. 15.16). The small quantities of *Magnolia* and *Nyssa* (tupelo) pollen present in the Trail Ridge samples were probably also locally produced and preserved as elements of the shrub community.

Myrica and *Corylus* are wind-pollinated (anemophilous) and the pollen would be expected in a variety of sediments, but their abundances in the Trail Ridge samples (as high as 20–30%) suggest local derivation of the pollen. The presence of *Myrica* is, in any case, consonant with the appearance of *Ilex* and *Cyrilla*, because the three usually occur together in modern swamps.

The low percentages of pine and oak pollen in Group B sediments may be the result of sheltering of the sites by the shrubby vegetation. That, as well as local over-production of shrub pollen such as *Ilex* and *Myrica*, could have obscured oak and pine. It is also possible that the presence of small amounts of pine and oak pollen means that neither taxon grew near the sites of deposition.

Group C

Group C samples could be classified within deposits broadly identified as marsh peats. The abundant representation of sun-loving herbaceous plant types (especially grasses, sedges, and composites), the presence of pollen and spores derived from plants which live only on wet soil, e.g., *Sphagnum*, *Sagittaria* (arrowhead or duck potato), and *Lycopodium*, and relatively high quantities of wind-blown pollen from *Quercus* and *Pinus*, all support

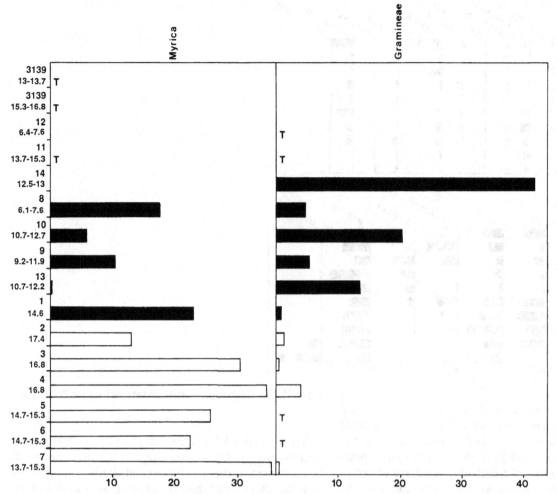

Figure 15.13. Relative abundances of *Myrica* and Gramineae in Trail Ridge samples. Note abundance of the former in Group C, while grasses are most abundant in Group B. Legend same as for Fig. 15.8.

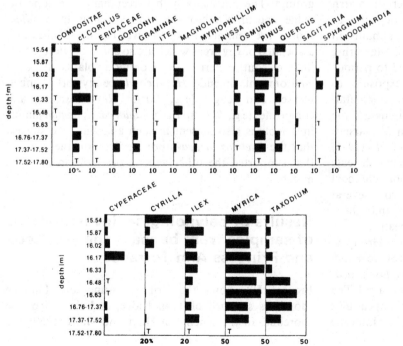

Figure 15.14. Palynological composition of core TR1 (from Rich, 1985).

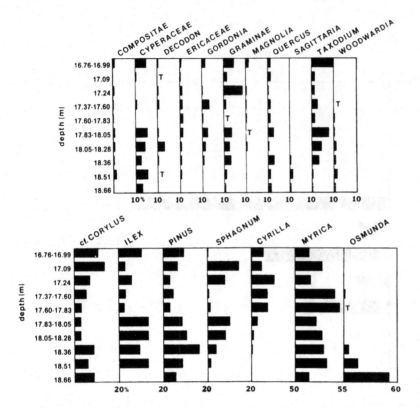

Figure 15.15. Palynological composition of core TR2 (from Rich, 1985).

such an interpretation. Rich (1979) and Rich and Spackman (1979) have described palynological contrasts that exist between shrub and marsh peats in the Okefenokee Swamp and have shown that marsh peats there typically contain the highest percentages of pollen derived from aquatic plants, as well as pine and oak. The latter produce copious wind-blown pollen deposited in open areas of Okefenokee Swamp (Fig. 15.16). The abundance of those pollen types in Group C samples is, therefore, a good indication of open-water or marsh conditions existing during Trail Ridge peat deposition.

The frequency of charcoal and fungal remains in Group C preparations shows that the plants living during deposition of those sediments were subjected to periods of dryness during which burning and exposure to extensive fungal decay were important environmental/ diagenetic factors. Cohen (1974) has discussed the relationship between fires and perpetuation of marshes in Okefenokee Swamp, while Spackman et al., (1976) illustrate the abundance of charcoal bands in marsh peats from the same area. Rich (1979) quantified both charcoal and fungal remains in peat samples taken from several cores in Okefenokee Swamp, and showed that those constituents are most abundant in marsh peats.

The *Pseudoschizaea*-bearing samples from Group C stand out as representing a peculiar variation of the marsh environment. Little is known about the *Pseudoschizaea*-producing organism, except that it evidently is algal. The cysts have a characteristic fingerprint-like appearance (Fig. 15.17) and are considered abundant in Holocene

sediments of the Atlantic coastal plain (T. Ager, personal communication, 1987). Christopher (1976) describes fossil forms related to *Pseudoschizaea* and clarifies its relationship to other algal fossils, as well as showing that *Concentricystes* is a synonym of *Pseudoschizaea*.

The palynological remains with which *Pseudoschizaea* is associated suggest deposition on moist, but not sodden, ground. The unusually high percentages of composite, sedge, and umbelliferous pollen, as well as *Lycopodium* spores, are reminiscent of the wet barrow ditches and associated boggy areas which are so common along roads in the southeastern US. Common plants in such environments include *Hydrocotyl* (pennywort; Umbelliferae), and sedges (e.g., *Carex*, *Rhynchospora*), and composites (e.g., *Bidens*). *Lycopodium carolinianum* almost invariably is also present in such a setting. We suggest that the *Pseudoschizaea*-producer lived on the wet soil that supported the other genera, probably in an open marsh.

Results of palynological investigations of samples from beneath the Pamlico and Princess Ann Terraces

Figure 15.2 shows the Princess Anne and Pamlico shorelines seaward of Trail Ridge, and the Wicomico shoreline. In the scheme of Hoyt and Hails (1969), the

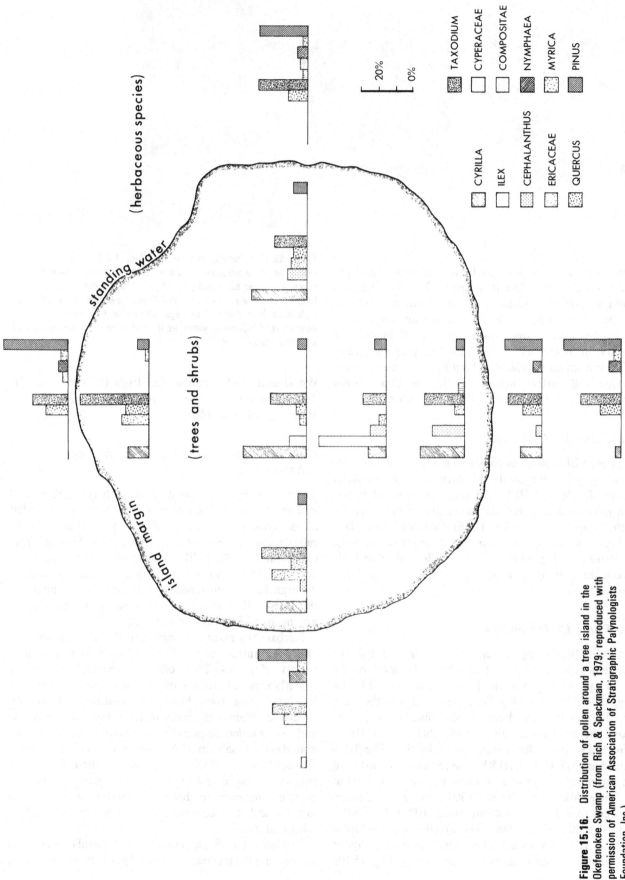

Figure 15.16. Distribution of pollen around a tree island in the Okefenokee Swamp (from Rich & Spackman, 1979; reproduced with permission of American Association of Stratigraphic Palynologists Foundation, Inc.).

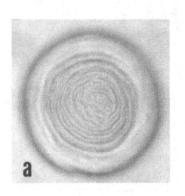
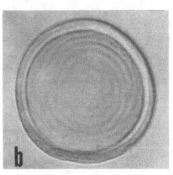
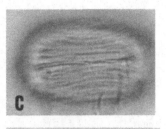

Figure 15.17. *Pseudoschizaea* morphology. (a) Polar view, high-focus. Note anastomosing ridges at the pole and apparently concentric ridges surrounding it; (b) As (a), mid-focus. Note circular form and relatively thick wall; (c) Different specimen, equatorial view, high-focus. Note that surface ridges lie parallel to a median line of dehiscence; (d) Same specimen as (c), mid-focus, to demonstrate wall structure and thickness.

Pamlico shoreline is younger than Trail Ridge, and the Princess Anne shoreline is younger than the Pamlico shoreline. Samples collected recently from beneath the Pamlico Terrace (Cabin Bluff area in Camden County in southeastern Georgia and Reids Bluff in Nassau County in northeastern Florida) are palynologically different from those typical of Trail Ridge. These samples provide still further insights as to the depositional behavior of some pollen near the marine coast.

The Cabin Bluff samples

The Cabin Bluff samples consisted of (1) sandy clay with black, angular fragments of carbonaceous material (sample D-14), and (2) bits of angiosperm wood mixed with gray sand and mollusc fragments (sample 5068A). Both samples (Figs. 15.3, 15.18) contained abundant pyrite, either as minute dispersed euhedra or as large framboids. The presence of the shells and framboids makes these samples very unlike typical Trail Ridge samples.

The Reids Bluff samples

The Reids Bluff samples consisted of a suite of different sediment types collected at or above the level of the St. Marys River upstream from Fernandina, Florida (Fig. 15.3). The river there lies at sea level, and the bluff, actually the bisected Pamlico shoreline, rises steeply above the river on its south bank. Pirkle *et al.* (1991) provide a stratigraphic description of the bluff. The Reids Bluff samples included: (1) RB-1: sandy clay, washed from the surface of a piece of *Taxodium* wood collected about one meter above river level; (2) RB-2: clayey sand taken adjacent to a *Taxodium* stump about 100 m E of RB-1 and two meters above river level; (3) RB-3: sediment from within a *Mercenaria* shell, taken from a prominent oyster and clam lens that overlies the tree stumps (Fig. 15.19).

Wood and a *Mercenaria* shell from the site were ^{14}C dated. The wood is $> 38\,130$ years old, while the shell is $36\,000 \pm 610$ years old.

Role of Chenopodiaceae/Amaranthaceae (cheno-ams)

The pollen/spore composition of both the Cabin Bluff and the Reids Bluff samples is very different from that of the Trail Ridge samples. Figure 15.20 illustrates the relative abundances of several of the taxa identified from Cabin Bluff and Reids Bluff. Comparison of Figures 15.20 and 15.8–15.13 shows that pollen of Chenopodiaceae/Amaranthaceae (cheno-ams) is present or abundant in the Cabin Bluff and Reids Bluff samples and rare or absent in the Trail Ridge samples.

Herbaceous plants belonging to the Chenopodiaceae and Amaranthaceae families are found frequently in a variety of places. They often are weedy invaders of recently exposed ground, or inhabit transitory, marginal habitats along river banks or eroding and shifting shorelines. Common genera include *Atriplex*, *Salicornia* and *Amaranthus* (especially *A. cannabinus*), which are abundant in salt marshes along the Atlantic coast (Radford *et al.*, 1968). The pollen of these families is periporate and easily identified as a group. It is very difficult, however, to distinguish pollen of the various genera, and it is conventional to refer to them all as 'cheno-ams.'

Frederiksen (1985) makes some very useful observations concerning the paleoecological value of cheno-am pollen.

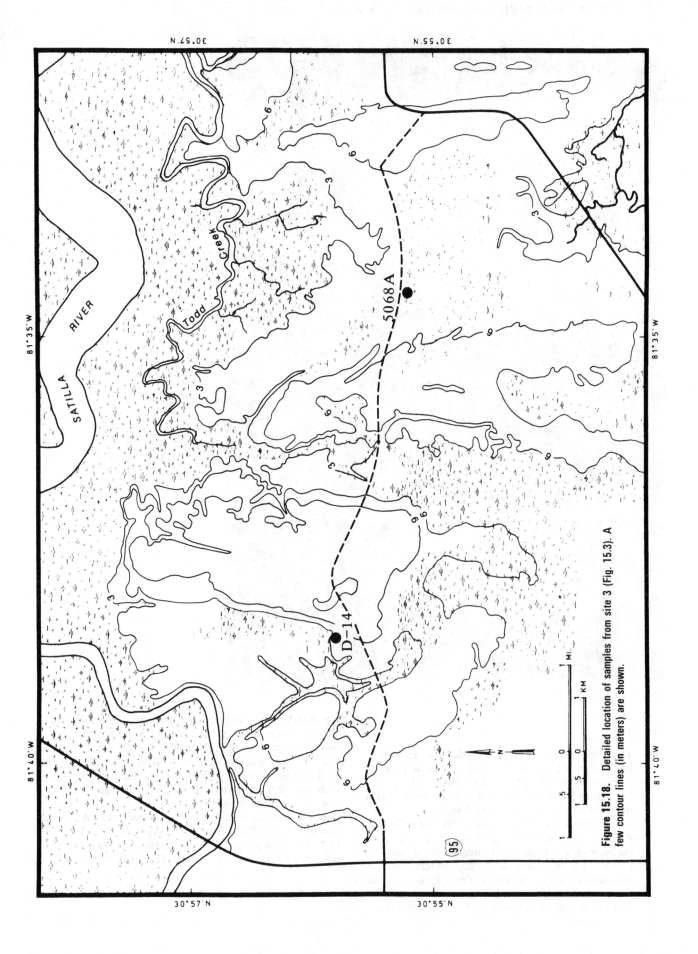

Figure 15.18. Detailed location of samples from site 3 (Fig. 15.3). A few contour lines (in meters) are shown.

Figure 15.19. View of Reids Bluff. Sediments from the river level upward are loose quartz sand (A), oyster/clam lens (B), blue-gray clayey sediments (C), and an upper oyster lens (D). A thin zone of blue-gray clayey sediments underlies the lower oyster lens, and about 1.5–2 m of blue-gray clayey sediments overlie the upper oyster lens. These upper blue-gray clayey sediments are overlain by loose quartz sand that extends to the land surface. Sample RB-1 was collected about 1 m above the river level within the loose quartz sand labeled A. Sample RB-2 was collected approximately 100 m east of RB-1, also within unit A. Sample RB-3 came from a *Mercenaria* found within unit B. From river level to ridge top is approximately 18 m.

He states:

> The paleoecology of cheno-am pollen is complicated by the fact that although Chenopodiaceae is mainly a family of xerophytes [dry land dwellers] and halophytes [salty soil dwellers], Amaranthaceae has a more variable ecology. Some authors have demonstrated that cheno-am pollen in Lower Tertiary deposits is associated with salty, usually coastal environments... High relative frequencies of cheno-am pollen are as characteristic of many modern tidal marshes as massive amounts of *Rhizophora*-type pollen are in many mangrove sediments. Several authors have reported abundant cheno-am pollen in samples that

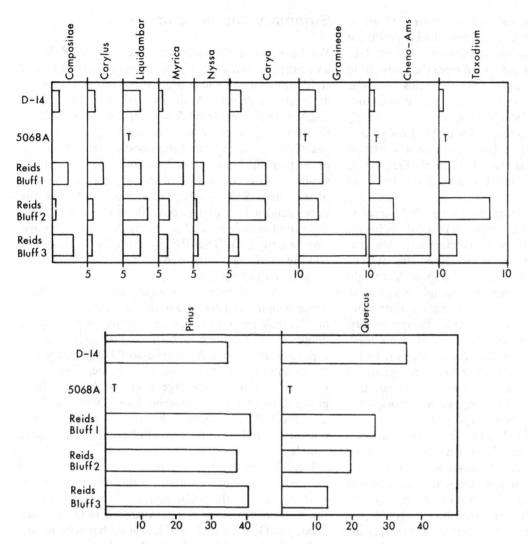

Figure 15.20. Relative abundances of several important taxa found in samples from Reids Bluff and Cabin Bluff.

also contain dinoflagellates, acritarchs, and other marine fossils as well as having other characteristics of coastal sediments....

Pollen sample 5068A was produced by dissolving shells and shell fragments washed from a larger bulk sample. Too few pollen/spores were present to count, but among the ones which were present were cheno-ams. The shells demonstrate marine influence in this sample, and the cheno-am pollen supports this conclusion. Sample D-14 had 1.6% cheno-ams, more than most of the other samples discussed here. Coupled with the presence of microforams and pyrite framboids, this leads to the conclusion that D-14 sediment accumulated at a coastal site under marine influence.

The suite of samples from Reids Bluff is also interesting from this point of view. The sediment in-filling of the *Mercenaria* shell from Reids Bluff (RB-3) contained 17.9% cheno-am pollen; the shoreline vegetation near

where the clam was buried obviously included a large number of cheno-am plants, probably *Salicornia*. Even the sediment washed from *Taxodium* wood (RB-1) and the sediment collected from around the tree stump (RB-2) contain measurable amounts of cheno-am pollen (2.8% and 6.6%, respectively). They show that cheno-ams grew in the vicinity of the cypress trees as the marine encroachment which buried them was in progress. Perhaps the pollen might have been produced by cheno-ams of an inland type, not associated with a marine environment, but the sedimentological and stratigraphic context of the site makes this seem unlikely.

Additional observations on the pollen composition of the Cabin Bluff and Reids Bluff samples

During the course of analyzing the Cabin Bluff and Reids Bluff samples for their pollen content and relating these results to framboid and microforam abundance, it was discovered that four otherwise unremarkable pollen types appeared to be most abundant: *Pinus*, *Quercus*, *Liquidambar* (sweet gum), and *Carya* (hickory). Figure 15.20 illustrates the relative abundances of these four genera in samples D–14 and 5068A from the Cabin Bluff area and in the three samples from the Reids Bluff site. The distribution of these four genera in Trail Ridge sands was reported by Rich (1985; see also Figs. 15.14, 15.15): (1) *Pinus* is typically less than 20% in Trail Ridge core samples (usually <20% in TR2, and always <10% in TR1); (2) *Quercus* is less than 10% in all Trail Ridge samples; (3) *Liquidambar* and *Carya* are less than 1% in all Trail Ridge samples.

Therefore, as these types are rare in strictly nonmarine sediments, their presence in marine or brackish sediments suggests they act as marine markers. While their true value as coastal/maritime forest indicators remains to be evaluated, there is strong reason to suspect they might be useful as such. Brush (1988), for example, presents the results of her work in establishing pollen abundance gradients in the Potomac estuary and Chesapeake Bay. She states that the gradients of plants of *Pinus taeda* (loblolly pine), *Liquidambar* and *Carya* adjacent to the estuarine tributaries '...are similar to their pollen gradients in surface sediments. Pollen distributions suggest that pollen, and hence, fine sediment, are not transported far in the estuary before being deposited.'

While the Cabin Bluff and Reids Bluff samples probably did not accumulate in an estuarine environment, we contend that the pollen dispersal potential is much the same for these taxa whether the coast is estuarine or barrier strand-dominated. Observation of palynological residues from salt marsh sediments collected between Sea Island, Georgia, and the mainland substantiate this claim; pyrite framboids, microforams, *Carya*, *Liquidambar*, *Pinus*, *Quercus* and cheno-ams are all found together.

Still further evidence that *Carya*, *Liquidambar* and *Quercus* are compatible with a marine coastal environment has been produced by Darrell (1989). Pratt *et al.* (1989) and Darrell jointly undertook a study of pollen and vertebrate fossil remains from the Early Miocene Marks Head Formation in southeastern Georgia. Pratt *et al.* identified from that formation a small but diverse assemblage of vertebrates which they found to be consistent with a subtropical nearshore marine depositional environment. Darrell observed pollen/spores from the same sediments and found that *Quercus*, *Ulmus*, Chenopodiaceae, Gramineae and Cyperaceae dominate the assemblage, with lesser quantities of *Carya*, *Liquidambar* and fern spores. The palynological composition of the sample led Darrell to postulate a very nearshore marine environment for the Marks Head samples. Darrell's data, coupled with those of Brush, and our independent observations all point to the same conclusion: in southeastern North America high relative abundances of *Quercus*, *Pinus*, *Carya* and *Liquidambar* may characterize deposits of shallow marine or brackish water sediments.

Summary and conclusions

We have noted that sediments beneath Trail Ridge and other coastal plain ridges in northern Florida and southern Georgia include peats with varying degrees of diagenetic alteration. A number of these peats have reached lignitic maturity ($R_0 = 0.23$) and have ^{14}C ages of 34 000 years or more. The sediments containing peat/lignite beneath Trail Ridge constitute the lower unit of a twofold sedimentary sequence of Trail Ridge sediments. The upper unit of the sequence consists of quartz sand. Sedimentological studies show that the upper sand unit is partly eolian in origin, representing deposition in dune fields. Apparently the dunes migrated over swampy areas. Thus the underlying peaty sediments and the overlying quartz sands of Trail Ridge would be of the same general age.

Three pollen/spore assemblages occur in the Trail Ridge samples, and are used to define plant communities in this paleoecological synthesis. One assemblage is dominated by *Ilex*, *Myrica*, *Corylus* and undifferentiated triporates, and *Cyrilla*. A second assemblage is dominated by Gramineae, Compositae, Cyperaceae, *Pinus* and *Quercus*. A third assemblage contains the taxa from group 2, as well as *Lycopodium*, Umbelliferae, and the algal fossil, *Pseudoschizaea*. The latter assemblages were not discovered in previous studies of Trail Ridge samples.

Identification of the ancient plant communities is based on taxonomic data as well as the sedimentological characteristics of the palynomorphs. Previous work on the dispersal characteristics of these taxa, in Okefenokee Swamp in Georgia, and comparison of that information with data from other authors, allows for the definition of three plant communities from Trail Ridge: (1) shrubby swamp community, with *Ilex*, Myrica, *Corylus* and *Cyrilla*; (2) marsh community with shallow standing water and Gramineae, Cyperaceae, Compositae and allochthonous *Pinus* or *Quercus*; (3) marsh community on damp soil, with Compositae, Umbelliferae, Cyperaceae, *Lycopodium* and the *Pseudoschizaea*-producing organism. The petrographic and palynological heterogeneity of the peat/lignite illustrate the diversity of wetland communities that were covered by advancing Trail Ridge dunes.

Sand ridges lying east of Trail Ridge (i.e., the Pamlico and Princess Anne shoreline ridges) contain or overlie organic-rich sediments which are generally similar palynologically to those of Trail Ridge. Notable exceptions are that *Pinus*, *Quercus*, *Liquidambar*, *Carya* and Chenopodiaceae/Amaranthaceae are unusually abundant at the Pamlico and Princess Anne localities. The presence of pyrite framboids and/or shell fragments in these same sediments indicates strong marine influence. Other data support the suggestion that these pollen taxa may be useful indicators of deposits accumulating along strand lines in shallow marine waters near a maritime forest.

Acknowledgments

The authors wish to acknowledge the generous support offered by E. I. DuPont de Nemours and Co. In particular, Richard Doerr provided encouragement for this study. We are also indebted to the Department of Geology and Geological Engineering, South Dakota School of Mines and Technology, where much of the research was done. The Department of Geology and Geography of Georgia Southern University extended additional support toward completion of this chapter. Typists Donna Cain and Shawn Hutchison of Georgia Southern and Patti Crawford of DuPont deserve particular thanks. Appreciation is also extended to Danusia Goodman for aid in design and drafting of the figures.

References

Brush, G. S. (1988). (abstract) Transport and deposition of pollen in an estuary. *Abstracts, 7th International Palynological Congress, Brisbane*, 20.

Christopher, R. A. (1976). Morphology and taxonomic status of *Pseudoschizaea* Thiergart and Frantz ex R. Potonié emend. *Micropaleontology*, **22**(2), 143–50.

Clark, W. Z. & Zisa, A. C. (1976). Physiographic map of Georgia. Georgia Geological Survey.

Cohen, A. D. (1974). Possible influences of subpeat topography and sediment type upon the development of the Okefenokee Swamp-marsh complex of Georgia. *Southeastern Geology*, **15**(3), 141–51.

Cohen, A. D. (1975). Peats from the Okefenokee Swamp-marsh complex. *Geoscience and Man*, **11**, 123–31.

Cooke, C. W. (1945). Geology of Florida. *Florida Geological Survey Bulletin* **29**.

Darrell, J. H. (1989). A preliminary palynological investigation of the Marks Head Formation (Miocene), Effingham County, Georgia. *Georgia Journal of Science*, **47**(1), 20.

Davis, R. B. & Webb, T., III (1975). The contemporary distribution of pollen in eastern North America: a comparison with the vegetation. *Quaternary Research*, **5**, 395–434.

Fearn, L. B. & Cohen, A. D. (1984). Palynologic investigations of six sites in the Okefenokee Swamp. In *The Okefenokee Swamp, its Natural History, Geology, and Geochemistry*, ed. A. D. Cohen, D. J. Casagrande, M. J. Andrejko & G. R. Best. Los Alamos, NM: Wetland Surveys, 423–43.

Force, E. R. & Rich, F. J. (1989). Geologic evolution of Trail Ridge eolian heavy mineral sand and underlying peat, northern Florida. *US Geological Survey Professional Paper* **1499**.

Frederiksen, N. O. (1985). Review of early Tertiary sporomorph paleoecology. *American Association of Stratigraphic Palynologists Contribution Series* **15**.

Frey, D. G. (1952). Pollen analysis of the Horry Clay and a seaside peat deposit near Myrtle Beach, South Carolina. *American Journal of Science*, **250**, 212–25.

Georgia Department of Natural Resources (1976). *Wetlands and Geologic Resources*. Atlanta: Department of Natural Resources, Office of Planning and Research.

Herrick, S. M. (1965). A subsurface study of Pleistocene deposits in coastal Georgia. *Georgia Geological Survey, Information Circular* **31**.

Hoyt, J. H. & Hails, J. R. (1969). Pleistocene shorelines in a relatively stable area, southeastern Georgia, U.S.A. *Giornale di Geologia* (2) **XXXV**, fasc. IV, 105–17.

Huddlestun, P. F. (1988). A revision of the lithostratigraphic units of the coastal plain of Georgia – the Miocene through Holocene. *Georgia Geological Survey Bulletin* **104**.

Lyell, C. (1846). *Travels in North America*, vol. 1. New York: Wiley & Putnam.

MacNeil, F. S. (1950). Pleistocene shorelines in Florida and Georgia. *US Geological Survey Professional Paper* **221-F**, 95–112.

Pirkle, E. C., Pirkle, W. A. & Yoho, W. H. (1977). The Highland heavy-mineral sand deposit on Trail Ridge in northern peninsular Florida. *Florida Bureau of Geology Report of Investigation* **84**.

Pirkle, E. C. & Yoho, W. H. (1970). The heavy-mineral ore body of Trail Ridge, Florida. *Economic Geology*, **65**, 17–30.

Pirkle, E. C., Yoho, W. H. & Hendry, C. W., Jr. (1970). Ancient sea-level stands in Florida. *Florida Bureau of Geology Bulletin* **52**.

Pirkle, F. L. (1975). Evaluation of possible source regions of Trail Ridge sands. *Southeastern Geology*, **17**, 93–114.

Pirkle, F. L. (1984). Environment of deposition of Trail Ridge sediments as determined from factor analysis. In *The Okefenokee Swamp, Its Natural History, Geology, and Geochemistry*, ed. A. D. Cohen, D. J. Casagrande, M. J. Andrejko & G. R. Best. Los Alamos, NM: Wetland Surverys, 629–50.

Pirkle, F. L, Pirkle, E. C. & Reynolds, J. G. (1991). Heavy mineral deposits of the southeastern Atlantic Coastal Plain. In *Proceedings of the Symposium on the Economic Geology of the Southeastern Industrial Minerals*, ed. S. J. Pickering, Jr. *Georgia Geological Survey Bulletin* **120**, 15–41.

Pratt, A. E., Petkewich, R. M. & Morgan, G. (1989). Fossil vertebrates from the Marks Head Formation (Early Miocene) of Southeastern Georgia. *Georgia Journal of Science*, **47**(1), 20.

Radford, A. E., Ahles, H. E. & Bell, C. R. (1968). *Manual of the Vascular Flora of the Carolinas*. Chapel Hill: University of North Carolina Press.

Raisz, E. (1957). *Landforms of the United States*, 6th rev. ed., Scale ca. 1:4,435,200. Cambridge, MA: originally published by the author.

Rich, F. J. (1979). The origin and development of tree islands in the Okefenokee Swamp, as determined by peat petrography and pollen stratigraphy. Ph.D. dissertation, Pennsylvania State University, University Park.

Rich, F. J. (1984a). Ancient flora of the eastern Okefenokee Swamp as determined by palynology. In *The Okefenokee Swamp, its Natural History, Geology, and Geochemistry*, ed. A. D. Cohen, D. J. Casagrande, M. J. Andrejko & G. R. Best. Los Alamos, NM: Wetland Surveys, 410–22.

Rich, F. J. (1984b). Development of three tree islands in the Okefenokee Swamp, as determined by palynostratigraphy and peat petrography. In *The Okefenokee Swamp, its Natural History, Geology, and Geochemistry*, ed. A. D. Cohen, D. J. Casagrande, M. J. Andrejko & G. R. Best. Los Alamos, NM: Wetland Surveys, 444–55.

Rich, F. J. (1985). Palynology and paleoecology of a lignitic peat from Trail Ridge, Florida. *Florida Bureau of Geology Information Circular* **100**.

Rich, F. J. & Pirkle, F. L. (1988). (abstract) Paleoecological interpretation of the Trail Ridge sequence, Florida, based on pollen sedimentation and clastic sedimentology. *Abstracts, 7th International Palynological Congress, Brisbane*, 139.

Rich, F. J. & Spackman, W. (1979). Modern and ancient

pollen sedimentation around tree islands in the Okefenokee Swamp. *Palynology*, **3**, 219–26.

Rich, F. J., Yeakel, J. D. & Gutzler, R. Q. (1978). (abstract) Trail Ridge buried peat: its significance as a coalification intermediate. *Geological Society of America Abstracts with Program*, **10**(7), 478.

Spackman, W., Cohen, A. D., Given, P. H. & Casagrande, D. J. (1976). *The Comparative Study of the Okefenokee Swamp and the Everglades – Mangrove Swamp – Marsh Complex of Southern Florida*. University Park: Pennsylvania State University, Coal Research Section.

Spencer, R. V. (1948). Titanium minerals in Trail Ridge, Florida. *US Bureau of Mines Report of Investigations* **4208**.

Veatch, J. O. & Stephenson, L. W. (1911). Preliminary report on the geology of the coastal plain of Georgia. *Georgia Geological Survey Bulletin* **26**.

Watts, W. A. (1969). A pollen diagram from Mud Lake, Marion County, North-central Florida. *Geological Society of America Bulletin*, **80**, 631–42.

Watts, W. A. (1970). The full glacial vegetation of northwestern Georgia. *Ecology*, **51**, 17–33.

Watts, W. A. (1971). Postglacial and interglacial vegetation history of southern Georgia and central Florida. *Ecology*, **52**:4, 676–90.

Watts, W. A. (1980). Late Wisconsin climate of northern Florida and the origin of species-rich deciduous forest. *Science*, **210**, 325–7.

IV Application of data on palynosedimentation to solution of geological problems

A Sedimentary cycles

16 Palynology of sedimentary cycles

DANIEL HABIB, YORAM ESHET and ROBERT VAN PELT

Introduction

The palynology of sedimentary cycles is based on the study of dispersed organic matter in rocks of various depositional lithofacies. This chapter describes how different kinds of organic particles can be used to detect depositional environments and their migration in response to change in sea level relative to the land surface. Episodes of marine transgression and regression affect the distribution and concentration of organic particles such as pollen grains and spores, algal fossils, the conductive tissue and cuticle of land plants, recycled palynomorphs and inertinite, and fecal amorphous debris. Several examples illustrate the evidence that palynology provides for cycles of transgression and regression. These were selected from contrasting lithological types, for example, siliciclastic sand to lime mud; different depositional environments, for example, shallow shelf versus carbonate platform; and different geologic ages, for example, Middle Jurassic, Late Cretaceous-Danian. An example of Permian-Triassic age illustrates how reworked palynomorphs can be used to detect cyclicity in the sedimentary record.

Organic particle sedimentology

The vast majority of particulate organic matter deposited in marine sediments consists of admixtures of material of both terrestrial and marine origin. The mixing of material from both sources occurs on the deep sea floor as well as on the shallow shelf. Commonly, one or two components predominate over others, and these can be used to trace the principal origin of the organic matter. Other parameters, such as the diversity and morphological sorting of organic particles, can be used to interpret hydrodynamic and ecological processes operating in the area of deposition.

Land plant particles

One quantitatively important kind of organic material is land plant detritus and other particles of nonmarine origin, such as fungal spores and freshwater algae. Land plant materials include vascular and other 'woody' tissue, cuticle, resins, pollen grains and pteridophyte (mostly fern) spores. In a series of articles published in the late 1960s, the widespread distribution of terrigenous organic matter was reported in sediments deposited in modern shallow-water environments (Traverse & Ginsburg, 1966; Cross et al., 1966) and in Cretaceous sediments deposited on the deep ocean bed (Habib, 1968). These studies showed the relative importance of the influx of terrigenous particles transported by fluvial discharge systems, their morphological sorting in marine dynamic systems, and their widespread distribution, reflecting episodes of deposition of terrigenous materials well beyond their points of origin. For example, Hay and Honjo (1989) commented on the widespread occurrence of terrigenous organic matter in black laminae in the Black Sea basin, at distances of approximately 1000 km from points of river discharge.

Certain pollen grains are more widely distributed than others in sediments. These are the 'windblown' grains dispersed from anemophilous conifers and flowering plants. Their buoyancy is increased by the possession of air bladders, such as in the bisaccate pollen of pine (Pinus), or by their smooth spheroidal form, as in the pollen of the grasses (Family Gramineae). Pollen grains of these plants are produced in large numbers for widespread dispersal, which favors wind pollination. They become even more buoyant when they settle in marine waters. On the Great Bahama Bank, a carbonate platform impoverished of terrigenous mineral sediment, the distribution and sedimentation of pine pollen are controlled by the path and intensity of marine currents (Traverse & Ginsburg, 1966). These pollen grains remain suspended in the marine currents, and become abundant in surface sediment where the current velocity diminishes,

311

and they are deposited. Other pollen grains, and especially the spores produced by ferns and other lower vascular plants, are deposited closer to the producing plants and are concentrated sedimentologically with terrigenous mineral and other organic detritus by a fluvial or fluviodeltaic discharge system. In the siliciclastic depositional environments of the Gulf of California, terrigenous organic matter is concentrated and morphologically diversified in the fine silt to clay size fraction of sediments in delta mouths of the Colorado River and smaller intermittent streams (Cross *et al.*, 1966). Tracheal vascular tissue, cuticle, inertinite and modern and recycled pollen grains and spores are most abundant in the fine clastics around the delta mouths. They are redistributed and morphologically sorted by currents in the Gulf of California that pass through submarine canyons to offshore deeps. Grains of pine pollen are abundant in the deltaic sediments and, because they are morphologically adapted for widespread dispersal, they remain relatively abundant farther offshore as the total amount of terrigenous organic matter diminishes. Cross *et al.* (1966) stressed the importance of the influx of terrigenous organic matter carried by streams into the Gulf of California and, especially, the role of marine currents in sorting pollen morphotypes as they are carried into less turbulent offshore environments. Terrigenous organic matter may be so widespread and distantly removed from its source that it deposits in relative abundance on the deep ocean bed adjacent to continental margins (Habib, 1968).

In the shallow marine environment, the predominance of land plant detritus is used to define the vascular tissue facies (Habib & Miller, 1989). Vascular tissue is abundant, and occurs with pollen grains, spores, resins, cuticular tissue, reworked palynomorphs and inertinite. In fluviodeltaic and prodeltaic environments, stream transported particles are relatively abundant, along with 'windblown' pollen morphotypes, in this organic facies.

Inertinite

The second quantitatively important kind of organic matter is inertinite. In the general sense, inertinite is that microscopic refractory material of black color which ranges in form from structureless shards to particles which can still be recognized as the altered conductive tissue of land plants. It includes fusinite and micrinite, since they behave collectively with inertinite as sedimentary particles. These organic particles are carbonized and resemble wood charcoal. They have been incompletely oxidized, either by fire in swamps and forests or through burial and recycling. Because it is more or less chemically inert, inertinite is commonly recycled and thus is ubiquitous in marine sediments which contain organic matter. It can become abundant in the vascular tissue facies where it has been transported as detritus along with other land plant particles. It is sparse in the amorphous debris facies. It defines the *inertinite facies* (Habib & Miller, 1989) when it is predominant. Generally, this facies is recovered from small residues, which suggests its occurrence is due not to absolute increase in the inertinite but mostly to the absence of other organic particles. The inertinite facies generally contains few palynomorphs. Pollen grains and spores are usually represented, and there may be relatively few specimens of as many as 10 to 15 dinoflagellate species. Occasionally, there may be many specimens of just one or two species. Reworked palynomorphs are more frequent in the inertinite facies.

Amorphous debris

A third kind of organic matter predominates in marine sediments. It is derived directly and indirectly from the productivity of marine organic-walled plankton. Organic matter of marine origin consists principally of dinoflagellate cysts (dinocysts), other algal cysts of uncertain taxonomic affinity (acritarchs), and the amorphous debris associated with fecal pellets consisting of organic matter of the same appearance. The abundance of this debris in discrete stratigraphic intervals defines the *amorphous debris facies*. It is in this facies that dinoflagellate species tend to be most numerous.

Amorphous debris is fine, granular organic matter which is optically translucent and pale to bright yellow in color when it has not been altered by recycling or by thermal diagenesis. It is abundant in the amorphous debris facies, comprising more than 90% of the total organic matter. In unfiltered residues, it commonly masks the presence of other particles. It occurs in marine sediments containing high percentages (5–20%) of organic carbon, and is hydrogen-rich. Its association with fecal pellets composed of the same material in both the shallow-marine and deep-sea record (Habib, 1982a, 1983; Rullkötter *et al.*, 1987; Habib & Miller, 1989) strongly suggests that it is produced as zooplankton feces in areas of high phytoplankton productivity (Tissot & Pelet, 1981). In the deep-sea record of the Cretaceous Period, abundant, well-preserved amorphous debris and infrequent fecal pellets occur in organic-rich, noncalcareous, hemipelagic clays (Habib, 1979, 1982b). Pellet microfossils in these sediments are close in shape, size and organic content to the modern pellets produced by feeding motile phytoplankton cells to the copepod *Cyclops sentifer* (Porter & Robbins, 1981). The paucity of structured organic matter in the pellets is attributed to the digestion of cellulosic cells, which would leave little optical evidence of the digested organism. Nevertheless, high-resolution transmission electron microscopy at magnifications of × 20 000– × 50 000 reveals that thermally unaltered amorphous debris of various geologic ages consists of lamellar membranes, structures attributed directly to phytoplankton cell walls (Raynaud *et al.*, 1988, 1989).

Fecal pellets consisting of organic matter settle in sediment traps at bathypelagic depths of the North Atlantic (Honjo, 1980). They may remain intact or, commonly, the encasing membrane ruptures to disperse amorphous debris. Deep diving zooplankton may excrete fecal pellets at bathypelagic depths (Vinogradov & Tseitlin, 1983). Zooplankton grazing activity is greatest in areas of high phytoplankton productivity. It follows that if the phytoplankton is organic-walled, there would also be a greater production of fecal amorphous debris in corresponding areas of the shallow sea and of the deep ocean. For example, siliceous-walled phytoplankton would not produce fecal amorphous debris. Thus, fecal debris should be abundant in sediments underlying zones of upwelling adjacent to continental margins. Caratini *et al.*, (1975) investigated the Cap Blanc transect in the upwelling zone opposite Mauritania during the French oceanographic expedition ORGON III. Abundant and well-preserved amorphous debris was recovered from surface sediments closest to the area of high primary productivity (see also Caratini, Chap. 8 this volume).

Palynological evidence indicates that abundant and well-preserved amorphous debris in marine sediments is the indirect product of the activities of marine organic-walled phytoplankton. Amorphous debris is also produced in lakes and other nonmarine aqueous environments where there is plankton productivity and feeding. However, this debris would not be transported in abundance and in a good state of preservation to the marine environment. Also, even in those areas where lakes empty through a relatively short distance into the coastal ocean, such as in the Gulf of Tehuantepec (Mexico), the debris is admixed with abundant land plant detritus (Habib *et al.*, 1971). Amorphous debris may also form from the degradation of land plant materials, but this debris can easily be distinguished by its poor preservation, darker color and association with poorly preserved pollen grains and with inertinite. Recycled amorphous debris of this kind has been identified in the deep-sea record in a Cretaceous interval referred to as 'poorly preserved xenomorphic facies' (Habib & Drugg, 1987).

Dinoflagellates

Dinoflagellate species are an important component of the amorphous debris facies. Modern marine dinoflagellates occupy a wide variety of shelf and oceanic habitats. Together with other phytoplankton they produce the primary biomass in the sunlit waters of the world ocean. However, it is only the dinoflagellate cyst wall which is capable of being fossilized. Perhaps 20% of known living marine species produce dinocysts (Dale, 1976) and these

are the meroplanktonic forms which occupy shallow environments adjacent to continental and island margins (Wall *et al.*, 1977). Quantitative analyses of the distribution of dinoflagellate species in modern marine sediments show biogeographic patterns which relate to hydro-dynamic and climatological parameters (Wall *et al.*, 1977), making dinoflagellates valuable for interpreting paleo-ecology based on fossil content. According to Wall *et al.*, (1977), different dinoflagellate species become relatively abundant in a seaward succession. Certain species were found to be most abundant in estuarine environments, whereas others attained maximum abundance in open shelf sediments. Dale (1983) tabulated the latitudinal distribution of viable dinocysts and showed that they could be used to distinguish biogeographic zones originally defined from molluscan assemblages. Harland (1988) used dinoflagellate species to interpret paleo-climatic change in a core of late Quaternary age from the North Sea.

The change in number of dinoflagellate species can be related to independent evidence of change in depositional environments. Within a region of relatively uniform climate there exists an orderly relationship between the size of an area and the number of species present (MacArthur & Wilson, 1967). An expanded continental shelf formed as a result of marine transgression provides a larger number of habitats and an increase in the number of meroplanktonic species; regression limits the amount of shelf area and reduces the number of species (Habib & Miller, 1989). The regularity of the pattern that smaller areas have fewer species than larger areas is true for many taxonomic groups, which suggests that such a pattern conforms to basic principles of ecology (Krebs, 1985). The number of dinoflagellate species may also increase in an inshore to offshore direction, due mainly to the addition of offshore species not found in inshore environments. The seaward increase may also be due to the hydrodynamic displacement of estuarine species and, possibly, of recycled dinocyst specimens (Wall *et al.*, 1977). The expanded continental shelf would be the environment in which there is an increased number of dinoflagellates and other organic-walled phytoplankton and, thus, increased zooplanktonic grazing activity and increased generation of fecal amorphous debris. The reduced shelf environment would limit the production of the debris, but would also place the depositional locus closer to the shore line, closer to the prograding delta (vascular tissue facies) or closer to the lagoon or bay (inertinite facies). Other factors which control productivity, such as the influx of nutrients by rivers and by upwelling, must also be considered.

Dinocysts occur in abundance in the deep ocean bed adjacent to continental margins. In these sediments they were evidently displaced by currents which mixed shelf and upper slope assemblages in an oceanward direction,

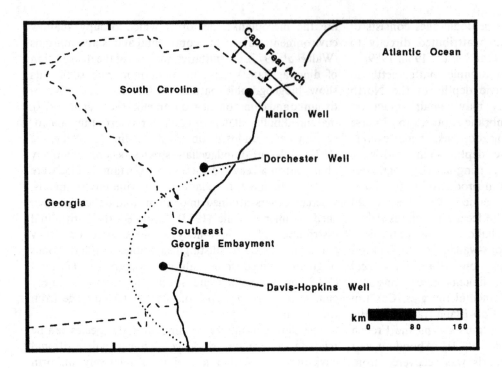

Figure 16.1. Location of the Marion, Dorchester and Davis-Hopkins wells. Marion well (Britton Neck Test Hole, Marion-78) drilled at 35° 51′ 43″ N, 79° 19′ 50″ W, Marion County, SC. Dorchester well (St. George Test Hole, Dorchester-211) drilled at 33° 09′ 25″ N, 80° 31′ 18″ W, Dorchester County, SC. Davis-Hopkins well (C. D. Hopkins No. 1 Test Hole) drilled at 31° 32′ 30″ N, 81° 43′ 45″ W, Wayne County, GA. The three wells are aligned from the southwestern flank of the Cape Fear arch to the central part of the Southeast Georgia embayment.

and were then displaced farther downslope by gravity currents. Thus, dinocysts are also concentrated as sedimentary particles in clays and silts in the deep-sea record, where they may be admixed with terrigenous organic matter. Despite this displacement from their biocoenosis, empirical study of the dinoflagellate record in the deep-sea depositional setting indicates that it is valuable in interpreting Cretaceous stratigraphy (Williams, 1975). Study of Deep Sea Drilling Project sections reveals that the stratigraphy is homotaxial through a wide area of the western North Atlantic (Habib & Drugg, 1987), and the fluctuation in number of species through the Berriasian-Cenomanian interval agrees with seismic stratigraphic evidence of sea level change at continental margins (Vail *et al.*, 1977). Further, the dinoflagellate stratigraphy of the Berriasian-Aptian interval in the western North Atlantic correlates well with the magnetic polarity-reversal stratigraphy and with relative sedimentation rates of this interval (Ogg, 1987). Adjacent to the Iberian Peninsula in the eastern Atlantic, dinoflagellate evidence of Aptian age (Drugg & Habib, 1988) is supported by the well-defined polarity Chron MΦ (Ogg, 1988). Thus, the magnetic stratigraphy suggests that the dinoflagellate homotaxy is time correlative.

Transgression, progradation and regression in shallow marine siliciclastic sediments

The cycle of transgression and regression in the shallow shelf siliciclastic facies is recorded in a subsurface

sequence of Campanian to early Danian age in the southeastern United States. Three wells were cored in the southern sector of the Atlantic coastal plain; the Marion, Dorchester and Davis-Hopkins wells (Fig. 16.1). Habib and Miller (1989) related depositional organic facies to paleoenvironments distinguished on the basis of lithofacies and interpreted the sequence of events in the context of migrating environments. The sediments were deposited in a marginal basin formed in the tectonic setting of the Cape Fear arch and Southeast Georgia embayment. Figure 16.2 schematically illustrates the sequence of transgression and regression from the palynological and lithofacies evidence in the three wells, which were correlated on the basis of their biostratigraphic age. Environments ranged from the largely subaerial, fluvial, sand-dominated, lower deltaic plain, to coastal environments such as delta front, tidal flat, lagoon and inter-distributary bay/estuary, and then to the marine proximal prodelta and inner neritic shallow shelf. Figure 16.3 graphically illustrates the migration of the environments distinguished in the wells of Figure 16.2.

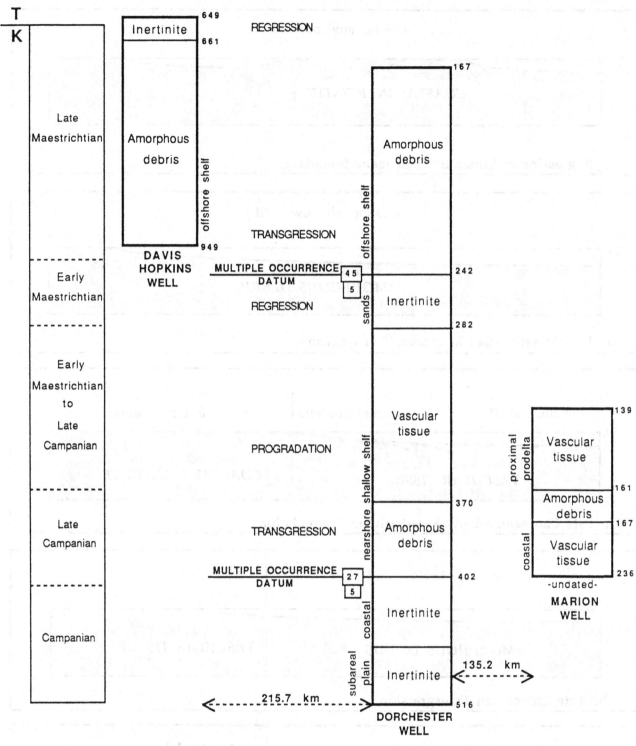

SE Georgia Embayment   Cape Fear Arch

Figure 16.2. Biostratigraphic correlation of depositional organic facies and lithofacies in Davis-Hopkins, Dorchester and Marion wells. Depths (in meters) are listed to the right of each column, but they are not necessarily to scale. Sequence of transgression and regression indicated between Davis-Hopkins and Dorchester wells. Numbers to the left of the Dorchester column represent dinoflagellate species abundance. Thus, there are 45 species in the multiple occurrence datum at 242 m compared to five species immediately below.

Marion well

The Marion well is situated on the flank of the Cape Fear arch (Fig. 16.1), closest to the basin margin. Both the lithofacies and palynological evidence point to a more landward locus of deposition. The lower deltaic plain facies in the lower part of the well is undated; samples

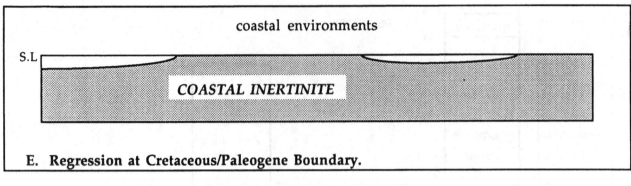

coastal environments

S.L

COASTAL INERTINITE

E. Regression at Cretaceous/Paleogene Boundary.

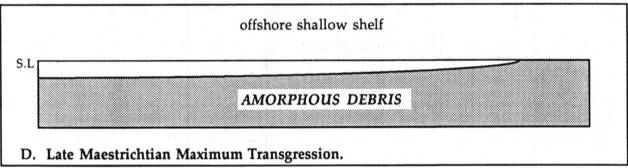

offshore shallow shelf

S.L

AMORPHOUS DEBRIS

D. Late Maestrichtian Maximum Transgression.

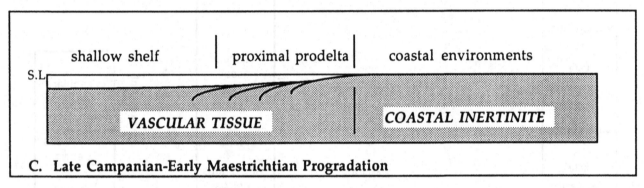

shallow shelf | proximal prodelta | coastal environments

S.L

VASCULAR TISSUE   **COASTAL INERTINITE**

C. Late Campanian-Early Maestrichtian Progradation

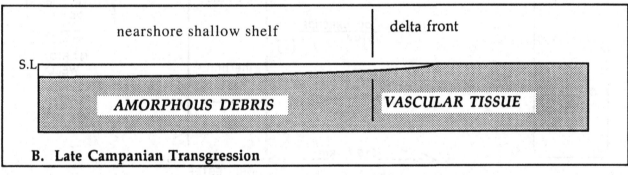

nearshore shallow shelf | delta front

S.L

AMORPHOUS DEBRIS   **VASCULAR TISSUE**

B. Late Campanian Transgression

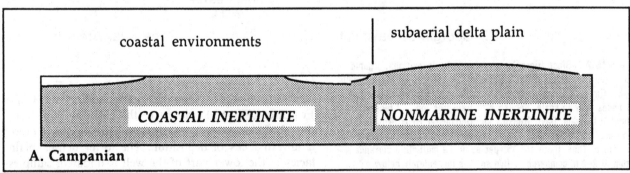

coastal environments | subaerial delta plain

COASTAL INERTINITE   **NONMARINE INERTINITE**

A. Campanian

yield meager residues of nonmarine inertinite barren of palynomorphs. The first fossiliferous samples occur in the vascular tissue facies in delta-front environments of late Campanian age (Fig. 16.2). The assemblages are dominated by fern spores and Normapolles angiosperm pollen grains. The number of dinoflagellate species fluctuates dramatically (from two to 26) in this interval, but even where there are only a few species in a sample there is an abundance of specimens of *Palaeohystrichophora infusorioides*, *Chatangiella* spp., *Gillinia hymenophora*, and *Palaeoperidinium pyrophorum*, an assemblage which suggests coastal or nearshore habitats. The coastal amorphous debris facies occurs above in a six meter interval of late Campanian age. It possesses abundant fecal debris, some admixed land plant detritus and abundant *P. pyrophorum*. The section directly above, of late Campanian-early Maestrichtian age, contains the vascular tissue facies developed in a proximal prodelta environment.

Dorchester well

The Dorchester well is basinward of the Marion well.

Figure 16.3. Schematic sequence of the transgressive and regressive environments illustrated in the columns in Figure 16.2. (A) Coastal inertinite and nonmarine inertinite represent coastal environments and subaerial delta plain, respectively, in the Dorchester well. Coastal environments include lagoon, tidal flat and interdistributary bay, which were not distinguished palynologically. Coastal inertinite is distinguished from nonmarine inertinite by the presence of dinoflagellate species. (B) The Late Campanian transgression is represented by the amorphous debris facies in the Dorchester well and by the vascular tissue facies in the Marion well. The delta front organic matter of the Marion well consists of diverse and well-preserved land plant particles. The shallow shelf lithofacies consists of abundant and well-preserved fecal debris, dinoflagellate species, and some Normapolles pollen in a nearshore depositional locus. (C) Progradation of prodeltaic vascular tissue facies in the regression of Late Campanian to Early Maestrichtian age. Land plant detritus is characteristic of the proximal prodelta but also extends into the shallow shelf lithofacies represented in the the Dorchester well. The regression terminates in the migration of coastal environments (and coastal inertinite) to the depositional locus of the Dorchester well at the end of the Early Maestrichtian. (D) Maximum transgression of the investigated interval, as depicted in sediments of Late Maestrichtian age in the Dorchester and Davis-Hopkins wells. The amorphous debris facies is rich in dinoflagellate species and fecal amorphous debris. The farther offshore setting of the Davis-Hopkins well explains why it contains very few pollen grains or spores. (E) Regression, resulting from sea level fall, in sediments at the Cretaceous/Paleogene boundary in the Davis-Hopkins well. Evidence of regression in the latest Maestrichtian is evident in the uppermost 20 m of both the Davis-Hopkins and Dorchester wells (Habib & Miller, 1989), in which the number of dinoflagellate species decreases systematically although still within the shallow shelf amorphous debris facies. The absence of the vascular tissue facies at the Cretaceous/Paleogene boundary suggests that the regression was not due to deltaic progradation.

The increase in number of dinoflagellate species upsection and the succession of organic and lithologic facies indicate an increasingly marine environment as the sequence becomes younger. However, the trends are not continuous but fluctuate stratigraphically. Two intervals defined by the amorphous debris facies correspond to episodes of marine transgression in the Dorchester well (Fig. 16.4). The amorphous debris facies is further distinguished stratigraphically by the occurrence of a multiple occurrence datum at its base. A multiple occurrence datum is a horizon containing a sudden and substantial increase in the number of dinoflagellate species from the immediately subjacent facies. The late Campanian episode of transgression occurred in a nearshore, shallow shelf environment. The multiple occurrence datum at the base of the interval (402.1 m) is characterized by an increase from five species to 27. Amorphous debris is well-preserved and abundant, with fecal pellets and abundant dinoflagellate specimens dominated by *P. infusorioides* and *Hystrichosphaerina varians*. There is some admixture of land plant detritus, particularly several Normapolles pollen species. The transgressive amorphous debris facies in the Dorchester well correlates in age with the thin coastal amorphous debris facies and vascular tissue facies of the Marion well (Fig. 16.2).

The second interval defined by the amorphous debris facies occurs in the episode of maximum transgression developed in an expanding offshore (but still inner neritic) shelf of late-early to early-late (hereafter called middle) Maestrichtian age (Fig. 16.4). Palynological evidence supports the farther offshore setting of the shallow shelf. There is very little terrigenous organic matter; some scattered inertinite and Normapolles grains occur, as well as 'windblown' bisaccate taxa, for example, *Rugubivesiculites*, *Pinuspollenites*, which are sparse. Amorphous debris is abundant and the interval contains the largest number of dinoflagellate species recorded in the investigated section. The datum defining the base of the amorphous debris facies interval is characterized by an increase to 45 species from five in the immediately subjacent facies. The number of species in the interval ranges from 39 to 45, and is dominated by a plexus of chorate dinocysts with discrete morphotypes, e.g., *Glaphyrocysta retiintexta* and *Areoligera medusettiformis*. In the Dorchester well *G. retiintexta* first appears as a minor constituent in the late Campanian transgressive episode. It becomes a dominant component in the maximum transgression. *P. infusorioides* last appears in the multiple occurrence datum at the base of this interval (313.3 m), which suggests that its extinction may have been related in part to processes connected with the maximum transgression.

The lower part of the section in the Dorchester well is characterized by nonmarine inertinite in the subaerial deltaic plain facies. This interval is Campanian and is

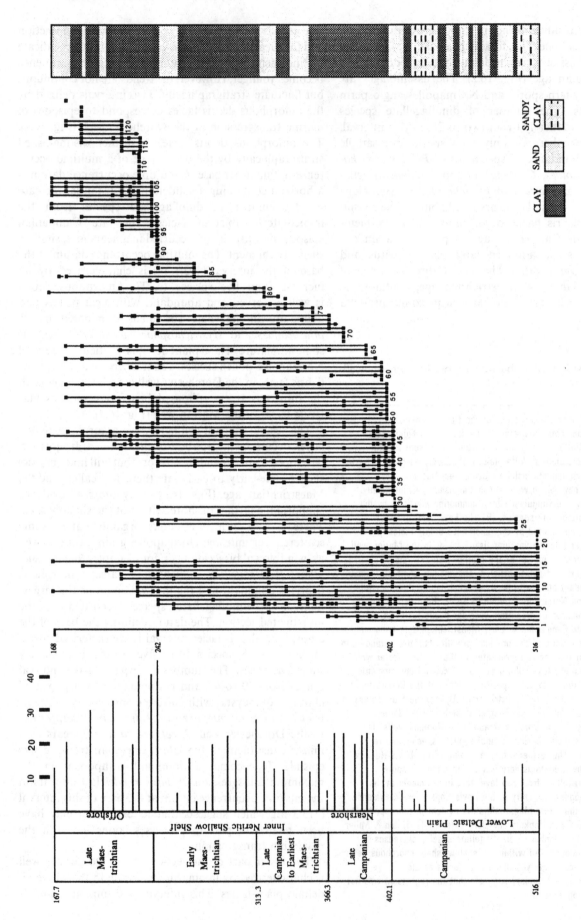

Figure 16.4. Stratigraphic occurrences and ranges of dinoflagellate and other algal species in the Dorchester well. Multiple occurrence datums are evident at 402 m and at 242 m. Another multiple occurrence datum is indicated at 313 m. The total number of species in each sample is illustrated in the column to the left of the stratigraphic occurrences chart. Species are listed in Table 16.1.

Table 16.1. *List of species presented (A) in the same numerical order as in Figure 16.4, and (B) alphabetically. See also Habib and Miller (1989)*

(A) [numerical]

1 Dinoflagellate sp. A
2 *Chatangiella tripartita* (Cookson and Eisenack) Lentin and Williams
3 *Palaeoperidinium pyrophorum* (Ehrenberg) Sarjeant
4 *Cyclonephelium distinctum* Deflandre and Cookson
5 *Palaeoperidinium* sp. A
6 *Palaeohystrichophora infusorioides* Deflandre
7 *Xenascus ceratioides* (Deflandre) Lentin and Williams
8 *Millioudodinium* sp. A
9 *Dinogymnium euclaense* Cookson and Eisenack
10 *Pediastrum* sp. A
11 *Paralecaniella* sp. A
12 *Areoligera* sp., cf. *A. coronata* (Wetzel) Lejeune-Carpentier
13 *Cerodinium boloniense* (Riegel) Riegel and Sarjeant
14 *Exochosphaeridium bifidum* (Clarke and Verdier) Clarke, Casey, Sarjeant and Verdier
15 *Gillinia hymenophora* Cookson and Eisenack + cf. *Gillinia* sp.
16 *Hystrichosphaeridium palmatum* (White) Downie and Sarjeant
17 *Spinidinium ornatum* (May) Lentin and Williams
18 *Subtilisphaera* sp. A
19 *Spinidinium echinoideum* (Cookson and Eisenack) Lentin and Williams
20 *Canningia colliveri* Cookson and Eisenack
21 *Manumiella* sp. A
22 *Palaeoperidinium* sp. B
23 *Chlamydophorella nyei* Cookson and Eisenack
24 *Microdinium* spp.
25 *Chatangiella victoriensis* (Cookson and Manum) Lentin and Williams
26 *Dinogymnium albertii* Clarke and Verdier
27 Dinoflagellate sp. B
28 *Odontochitina operculata* (Wetzel) Deflandre and Cookson
29 *Cordosphaeridium* sp. A
30 *Dinogymnium nelsonense* (Cookson) Evitt, Clarke and Verdier
31 *Senoniasphaera* sp. A
32 *Tanyosphaeridium* spp.
33 *Fromea amphora* Cookson and Eisenack
34 *Senegalinium dubium* Jain and Millepied
35 *Hystrichokolpoma rigaudiae* Deflandre and Cookson
36 *Hystrichodinium pulchrum* Deflandre
37 Palynomorph sp. A
38 *Odontochitina costata* Alberti
39 *Cordosphaeridium varians* May
40 *Hystrichosphaerina varians* (May) Lentin and Williams
41 *Alterbidinium minus* (Alberti) Lentin and Williams
42 *Exochosphaeridium phragmites* Davey, Downie, Sarjeant and Williams
43 *Dinogymnium* sp. A (Habib and Drugg, 1987, pl. 7, fig. 5)
44 *Spiniferites multibrevis* (Davey and Williams) Below
45 *Palambages* sp. A (McIntyre, 1974, pl. 13, fig. 11)
46 *Cordosphaeridium fibrospinosum* Davey and Williams
47 *Cerodinium striatum* (Drugg) Lentin and Williams

48 *Dinogymnium westralium* (Cookson and Eisenack) Evitt, Clarke and Verdier
49 *Spinidinium microceratum* Stanley + *S. densispinatum* Stanley
50 *Riculacysta* sp. A
51 *Dinogymnium digitus* (Deflandre) Evitt, Clarke and Verdier
52 *Deflandrea microgranulata* Stanley
53 *Palambages* sp. B (McIntyre, 1974, pl. 13, fig. 12)
54 *Fromea fragilis* (Cookson and Eisenack) Stover and Evitt
55 *Spiniferites ramosus ramosus* (Ehrenberg) Loeblich and Loeblich
56 *Senegalinium bicavatum* Jain and Millipied
57 *Andalusiella* sp. A
58 *Glaphyrocysta retiintexta* (Cookson) Stover and Evitt + *G. reticulosa* (Gerlach) Stover and Evitt
59 *Callaiosphaeridium asymmetricum* (Deflandre and Courteville) Davey and Williams
60 *Palaeoperidinium* sp. C
61 *Alterbidinium acutulum* (Wilson) Lentin and Williams
62 *Cribroperidinium wetzelii* (Lejeune-Carpentier) Helenes
63 *Trichodinium* spp.
64 *Andalusiella polymorpha* (Malloy) Lentin and Williams
65 *Isabelidinium* sp. A
66 *Senegalinium laevigatum* (Malloy) Bujak and Davies
67 *Hystrichosphaeridium tubiferum* (Ehrenberg) Deflandre
68 *"Chytroeisphaeridia solida"* (Wilson, 1974)
69 *Microdinium setosum* Sarjeant
70 *Dinogymnium denticulatum* (Alberti) Evitt, Clarke and Verdier
71 Palynomorph sp. B
72 *Diconodinium wilsonii* Aurisano
73 *Areoligera* sp. A
74 *Paralecaniella indentata* (Deflandre and Cookson) Cookson and Eisenack
75 *Coronifera oceanica* Cookson and Eisenack
76 *Cerodinium pannuceum* (Stanley) Lentin and Williams
77 *Palaeocystodinium reductum* May
78 *Ellipsodinium rugulosum* Clarke and Verdier
79 *Spongodinium delitiense* (Ehrenberg) Deflandre
80 *Pierceites pentagona* (May) Habib and Drugg
81 *Dinogymnium acuminatum* Evitt, Clarke and Verdier
82 *Samlandia angustivela* (Deflandre and Cookson) Eisenack
83 *Oligosphaeridium complex* (White) Davey and Williams
84 *Chytroeisphaeridia baetica* Riegel
85 *Deflandrea? macrocysta* Cookson and Eisenack
86 *Spiniferites ramosus granosus* (Davey and Williams) Lentin and Williams
87 *Kleithriasphaeridium truncatum* (Benson) Stover and Evitt
88 *Deflandrea* sp. B
89 *Impagidinium* sp. A
90 *Deflandrea* sp. C
91 *Senoniasphaera* sp. B
92 *Isabelidinium balmei* (Cookson and Eisenack) Stover and Evitt
93 *Cerodinium diebelii* (Alberti) Lentin and Williams

Table 16.1. *Continued*

(A) [numerical] *Continued*

94 *Andalusiella* sp. B
95 *Chatangiella robusta* (Benson) Stover and Evitt
96 *Cerodinium speciosum* (Alberti) Lentin and Williams
97 *Fromea laevigata* (Drugg) Stover and Evitt
98 *Areoligera senonensis* Lejeune-Carpentier
99 *Deflandrea* sp. D
100 *Systematophora placacantha* (Deflandre and Cookson) Davey, Downie, Sarjeant and Williams
101 *Catillopsis* sp. A
102 *Deflandrea galeata* (Lejeune-Carpentier) Lentin and Williams
103 *Areoligera medusettiformis* (Wetzel) Lejeune-Carpentier
104 *Palaeocystodinium gabonense* Stover and Evitt
105 *Phelodinium magnificum* (Stanley) Stover and Evitt
106 *Deflandrea* sp. E
107 *Spiniferites fluens* (Hansen) Stover and Williams
108 *Deflandrea dilwynensis* Cookson and Eisenack
109 *Ascodinium* sp. A

110 cf. *Spiniferites ramosus* (Ehrenberg) Loeblich and Loeblich
111 *Tanyosphaeridium xanthiopyxides* (Wetzel) Stover and Evitt
112 *Cyclapophysis monmouthensis* Benson
113 *Manumiella seelandica* (Lange) Bujak and Davies em. Firth
114 *Palaeoperidinium basilium* (Drugg) Drugg
115 *Cerodinium cordiferum* (May) Lentin and Williams
116 *Palynomorph* sp. C
117 *Cordosphaeridium* sp. B
118 *Palaeocystodinium benjaminii* Drugg
119 Palynomorph sp. D
120 *Cleistosphaeridium* sp. A
121 *Cassidium?* sp. A
122 *Gerdiocysta cassiculus* (Drugg) Liengjarern, Costa and Downie
123 *Palynodinium grallator* Gocht
124 *Palaeocystodinium* sp. A
125 *Cordosphaeridium* sp. C

(B) [alphabetical]

61 *Alterbidinium acutulum* (Wilson) Lentin and Williams
41 *Alterbidinium minus* (Alberti) Lentin and Williams
64 *Andalusiella polymorpha* (Malloy) Lentin and Williams
57 *Andalusiella* sp. A
94 *Andalusiella* sp. B
103 *Areoligera medusettiformis* (Wetzel) Lejeune-Carpentier
98 *Areoligera senonensis* Lejeune-Carpentier
73 *Areoligera* sp. A (May, 1980, pl. 7, fig. 18)
12 *Areoligera* sp. cf. *A. coronata* (Wetzel) Lejeune-Carpentier
109 *Ascodinium* sp. A
59 *Callaiosphaeridium asymmetricum* (Deflandre and Courteville) Davey and Williams
20 *Canningia colliveri* Cookson and Eisenack
121 *Cassidium?* sp. A
101 *Catillopsis* sp. A
13 *Cerodinium boloniense* (Riegel) Riegel and Sarjeant
115 *Cerodinium cordiferum* (May) Lentin & Williams
93 *Cerodinium diebelii* (Alberti) Lentin & Williams
76 *Cerodinium pannuceum* (Stanley) Lentin & Williams
47 *Cerodinium striatum* (Drugg) Lentin & Williams
95 *Chatangiella robusta* (Benson) Stover and Evitt
2 *Chatangiella tripartita* (Cookson and Eisenack) Lentin and Williams
25 *Chatangiella victoriensis* (Cookson and Manum) Lentin and Williams
23 *Chlamydophorella nyei* Cookson and Eisenack
84 *Chytroeisphaeridia baetica* Riegel
68 "*Chytroeisphaeridia solida*" (Wilson, 1974)
120 *Cleistosphaeridium* sp. A
46 *Cordosphaeridium fibrospinosum* Davey and Williams
39 *Cordosphaeridium varians* May
29 *Cordosphaeridium* sp. A
117 *Cordosphaeridium* sp. B

125 *Cordosphaeridium* sp. C
75 *Coronifera oceanica* Cookson and Eisenack
62 *Cribroperidinium wetzelii* (Lejeune-Carpentier) Helenes
112 *Cyclapophysis monmouthensis* Benson
4 *Cyclonephelium distinctum* Deflandre and Cookson
108 *Deflandrea dilwynensis* Cookson and Eisenack
102 *Deflandrea galeata* (Lejeune-Carpentier) Lentin and Williams
85 *Deflandrea? macrocysta* Cookson and Eisenack
52 *Deflandrea microgranulata* Stanley
88 *Deflandrea* sp. B
90 *Deflandrea* sp. C
99 *Deflandrea* sp. D
106 *Deflandrea* sp. E
72 *Diconodinium wilsonii* Aurisano
1 Dinoflagellate sp. A
27 Dinoflagellate sp. B
81 *Dinogymnium acuminatum* Evitt, Clarke and Verdier
26 *Dinogymnium albertii* Clarke and Verdier
70 *Dinogymnium denticulatum* (Alberti) Evitt, Clarke and Verdier
51 *Dinogymnium digitus* (Deflandre) Evitt, Clarke and Verdier
9 *Dinogymnium euclaense* Cookson and Eisenack
30 *Dinogymnium nelsonense* (Cookson) Evitt, Clarke and Verdier
48 *Dinogymnium westralium* (Cookson and Eisenack) Evitt, Clarke and Verdier
43 *Dinogymnium* sp. A (Habib and Drugg, 1987, pl. 7, fig. 5)
78 *Ellipsodinium rugulosum* Clarke and Verdier
14 *Exochosphaeridium bifidum* (Clarke and Verdier) Clarke, Casey, Sarjeant and Verdier
42 *Exochosphaeridium phragmites* Davey, Downie, Sarjeant and Williams

Table 16.1. Continued

(B) [alphabetical] *Continued*

33 *Fromea amphora* Cookson and Eisenack	123 *Palynodinium grallator* Gocht
54 *Fromea fragilis* (Cookson and Eisenack) Stover and Evitt	37 Palynomorph sp. A
97 *Fromea laevigata* (Drugg) Stover and Evitt	71 Palynomorph sp. B
122 *Gerdiocysta cassiculus* (Drugg) Liengjarern, Costa and Downie	116 Palynomorph sp. C
15 *Gillinia hymenophora* Cookson and Eisenack + cf. *Gillinia* sp.	119 Palynomorph sp. D
	74 *Paralecaniella indentata* (Deflandre and Cookson) Cookson and Eisenack
58 *Glaphyrocysta retiintexta* (Cookson) Stover and Evitt + *G. reticulosa* (Gerlach) Stover and Evitt	11 *Paralecaniella* sp. A
36 *Hystrichodinium pulchrum* Deflandre	10 *Pediastrum* sp. A
35 *Hystrichokolpoma rigaudae* Deflandre and Cookson	105 *Phelodinium magnificum* (Stanley) Stover and Evitt
16 *Hystrichosphaeridium palmatum* Deflandre and Courteville	80 *Pierceites pentagona* (May) Habib and Drugg
67 *Hystrichosphaeridium tubiferum* (Ehrenberg) Deflandre	50 *Riculacysta* sp. A
40 *Hystrichosphaerina varians* (May) Lentin and Williams	82 *Samlandia angustivela* (Deflandre and Cookson) Eisenack
89 *Impagidinium* sp. A	56 *Senegalinium bicavatum* Jain and Millepied
92 *Isabelidinium balmei* (Cookson and Eisenack) Stover and Evitt	34 *Senegalinium dubium* Jain and Millepied
65 *Isabelidinium* sp. A	66 *Senegalinium laevigatum* (Malloy) Bujak and Davies
87 *Kleithriasphaeridium truncatum* (Benson) Stover and Evitt	31 *Senoniasphaera* sp. A
113 *Manumiella seelandica* (Lange) Bujak and Davies, em. Firth	91 *Senoniasphaera* sp. B
21 *Manumiella* sp. A	19 *Spinidinium echinoideum* (Cookson and Eisenack) Lentin and Williams
69 *Microdinium setosum* Sarjeant	49 *Spinidinium microceratum* Stanley + *S. densispinatum* Stanley
24 *Microdinium* spp.	17 *Spinidinium ornatum* (May) Lentin and Williams
8 *Millioudodinium* sp. A	107 *Spiniferites fluens* (Hansen) Stover and Williams
38 *Odontochitina costata* Alberti	110 cf. *Spiniferites ramosus* (Ehrenberg) Loeblich and Loeblich
28 *Odontochitina operculata* (Wetzel) Deflandre and Cookson	86 *Spiniferites ramosus granosus* (Davey and Williams) Lentin and Williams
83 *Oligosphaeridium complex* (White) Davey and Williams	55 *Spiniferites ramosus ramosus* (Ehrenberg) Loeblich and Loeblich
118 *Palaeocystodinium benjaminii* Drugg	79 *Spongodinium delitiense* (Ehrenberg) Deflandre
104 *Palaeocystodinium gabonense* Stover and Evitt	18 *Subtilisphaera* sp. A
77 *Palaeocystodinium reductum* May	100 *Systematophora placacantha* (Deflandre and Cookson) Davey, Downie, Sarjeant and Williams
124 *Palaeocystodinium* sp. A	111 *Tanyosphaeridium xanthiopyxides* (Wetzel) Stover and Evitt
6 *Palaeohystrichophora infusorioides* Deflandre	32 *Tanyosphaeridium* spp.
114 *Palaeoperidinium basilium* (Drugg) Drugg	63 *Trichodinium* spp.
3 *Palaeoperidinium pyrophorum* (Ehrenberg) Sarjeant	7 *Xenascus ceratioides* (Deflandre) Lentin and Williams
5 *Palaeoperidinium* sp. A	
22 *Palaeoperidinium* sp. B	
60 *Palaeoperidinium* sp. C	
45 *Palambages* sp. A (McIntyre, 1974, pl. 13, fig. 11)	
53 *Palambages* sp. B (McIntyre, 1974, pl. 13, fig. 12)	

correlated lithologically with the stratigraphically corresponding, undated, interval in the lower part of the Marion well (Fig. 16.2). It differs in the occurrence of several layers containing coastal inertinite of an interdistributary bay or estuarine environment. This variation is probably due to the more basinward location of the Dorchester well. The coastal inertinite facies is developed in the Campanian episode which predates the Late Campanian, nearshore, transgressive, amorphous debris facies, and in the early Maestrichtian episode, which predates the middle to late Maestrichtian maximum transgressive facies. In the Dorchester well, the vascular tissue facies

extends through most of an interval dated late Campanian to early Maestrichtian. It is well-developed in the nearshore, shallow shelf lithofacies. The biostratigraphic correlation of this lithofacies with that of the proximal prodelta lithofacies of the Marion well indicates progradation of the vascular tissue facies beyond its initial depositional locus into the shallow shelf.

Davis-Hopkins well

The Davis-Hopkins well is the most basinward of the three wells. The investigated section in this well ranges in age from late Maestrichtian to early Danian. The Cretaceous/Tertiary (K/T) boundary is identified by the first appearance of the dinoflagellate *Danea californica*. The organic facies in the late Maestrichtian is equivalent to that of the late Maestrichtian in the Dorchester well (Fig. 16.2). It represents an offshore neritic shelf facies characterized by abundant and well-preserved amorphous debris, numerous dinoflagellate species and good examples of fecal pellets. Pollen grains are rare, consisting of a few bisaccates and pollen of proteaceous angiosperms. Normapolles pollen and fern spores are virtually absent. Dinoflagellate species fluctuate somewhat through the section but remain numerous. In several intervals there are fewer species associated with poorly-preserved amorphous debris and admixed inertinite, which suggests some cyclicity in the late Maestrichtian. In the latest Maestrichtian, beginning at 20 m below the K/T boundary, there is a decline from 40 to 18 species within the amorphous debris facies. There is also a dramatic decrease in the number of specimens, especially of *G. retiintexta* and *A. medusettiformis*. These species are apparently absent in the inertinite facies, which developed in the approximate position of the K/T boundary. For example, *D. californica* first appears in the inertinite facies. The reduction in the number of dinoflagellate species, dramatic decrease in abundance of *G. retiintexta* and *A. medusettiformis*, and presence of the inertinite facies is considered palynological evidence of regression in the approximate position of the K/T boundary. Brinkhuis and Zachariasse (1988) used specific dinoflagellate associations and increased amounts of land plant detritus to reflect the short episode of rapidly falling sea level at the K/T boundary. The boundary was identified at the level of extinction of Cretaceous foraminiferal species, and was coincident with the first appearance of *Danea californica*. Hultberg and Malmgren (1986) recorded a decrease in dinoflagellate diversity (with a concomitant increase in *Spiniferites*) accompanying the latest Maestrichtian lowering of sea level. The inshore-offshore dinoflagellate diversity gradients established by Wall *et al.*, (1977) were cited by Hultberg and Malmgren as evidence that the decrease in diversity indicated a trend to more inshore environments at the K/T boundary.

Discussion

The Dorchester well exemplifies best the palynological evidence of marine transgression (Fig. 16.4). The investigated section in this well represents the longest span of geologic time, and correlates well with the facies of the Marion and Davis-Hopkins wells. The correspondence of palynological and lithofacies evidence of transgression in both the nearshore and offshore neritic shelf is direct. In both episodes of transgression, the first palynological evidence of transgression is signalled by an interval of the amorphous debris facies distinguished at its base by the substantial increase in number of dinoflagellate species. In the nearshore transgression of late Campanian age, this multiple occurrence datum is comprised in part by the first appearances of species which are known to have an older Cretaceous history. The nearshore facies contains some admixture of terrigenous organic matter, including the displaced freshwater alga, *Pediastrum*. The middle Maestrichtian multiple occurrence datum contains the last appearances of many species, including *P. infusorioides* which became extinct at this time. In addition, both transgressive events are characterized by species which predominate in the assemblages. The late Campanian transgression is characterized by abundant *P. infusorioides* in the nearshore shelf, and by abundant *P. pyrophorum* in the coastal facies (Marion well). In the late Maestrichtian transgression, *G. retiintexta* and *A. medusettiformis* predominate in both the Dorchester and Davis-Hopkins wells. In the investigated sections, the increase in number of dinoflagellate species signals transgression. The larger maximum number of species in the late Maestrichtian transgression (45) may indicate that it was larger in geographic extent than the late Campanian transgression (27 species). The associated abundance of fecal amorphous debris is a logical consequence of increased zooplankton grazing of motile organic-walled plankton.

The first palynological signal for regression is more difficult to identify precisely. In the Dorchester well, regression is evident in the progradation of the vascular tissue facies beyond the prodelta. Regression is evident also in the occurrence in sequence of the inertinite facies. However, the stratigraphic decrease in number of dinoflagellate species in the upper parts of the intervals distinguished by the amorphous debris facies may indicate the first evidence of regression. In the uppermost part of the Dorchester well (Fig. 16.4), in sediments of latest Maestrichtian age, dinoflagellate species decline in number towards the top of the section. The stratigraphic decrease in number of species in the Davis-Hopkins well, and reduction in the number of specimens of the species which characterize the offshore amorphous debris facies, precedes the inertinite facies at the approximate position of the K/T boundary. The numerical decrease of species in the latest Maestrichtian of both wells, still within the amorphous debris facies, may reflect diminishing habitats in a contracting shelf environment which continued to accumulate fecal amorphous debris.

Transgression and regression in shallow marine carbonate sediments

The palynology of sedimentary cycles in lime accumulating environments was studied in outcrop and in subsurface

STUDY AREA OF THE TWIN CREEK LIMESTONE

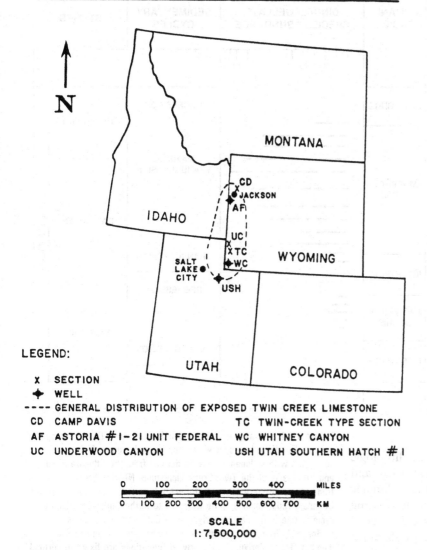

LEGEND:

X   SECTION
✦   WELL
----   GENERAL DISTRIBUTION OF EXPOSED TWIN CREEK LIMESTONE
CD   CAMP DAVIS   TC   TWIN-CREEK TYPE SECTION
AF   ASTORIA #1-21 UNIT FEDERAL   WC   WHITNEY CANYON
UC   UNDERWOOD CANYON   USH UTAH SOUTHERN HATCH #1

SCALE
1:7,500,000

Figure 16.5. Geographic distribution of Twin Creek Limestone indicated by dashed line in westernmost WY, easternmost ID and northeastern UT. Outcrop sections and cored wells indicated by graphic and letter symbols. Camp Davis outcrop section located at northern end of study area.

sections of the Middle Jurassic Twin Creek Limestone (Van Pelt, 1990). It is one of several formations which are situated in the Idaho-Wyoming-Utah Overthrust Belt, formed during the Laramide Orogeny. The formation extends in a north-south trending belt located in westernmost Wyoming, easternmost Idaho, and northeastern Utah (Fig. 16.5). Six sections were studied, three each from outcrop and from cored intervals in wells.

The Twin Creek Limestone represents carbonate platform environments ranging from hypersaline lagoon at its base, grading upward to littoral and then to offshore shelf. This sequence is repeated in two cycles of transgression. The lower part of the formation (and lower

sedimentary cycle) correlates with the hypersaline to nearshore facies of the Gypsum Spring Formation in northern and south-central Wyoming. The upper sedimentary cycle correlates with the marine shelf facies of the 'lower Sundance' Formation of northern Wyoming (Wright, 1973). Evidence of transgression is shown by the migration in sequence of environments identified from bivalve associations and from lithofacies, and from the geographic extent of the various members (and lithofacies) which comprise the Twin Creek Limestone (Imlay, 1967, 1980; Wright, 1973). Figure 16.6 illustrates the stratigraphic occurrence of the seven members and the two cycles of transgression and regression, as revealed in the outcrop section at Camp Davis, Wyoming (Van Pelt, 1990). It also shows the numerical distribution of dinoflagellate species and the stratigraphic occurrence of depositional organic facies.

The first episode of marine transgression and regression

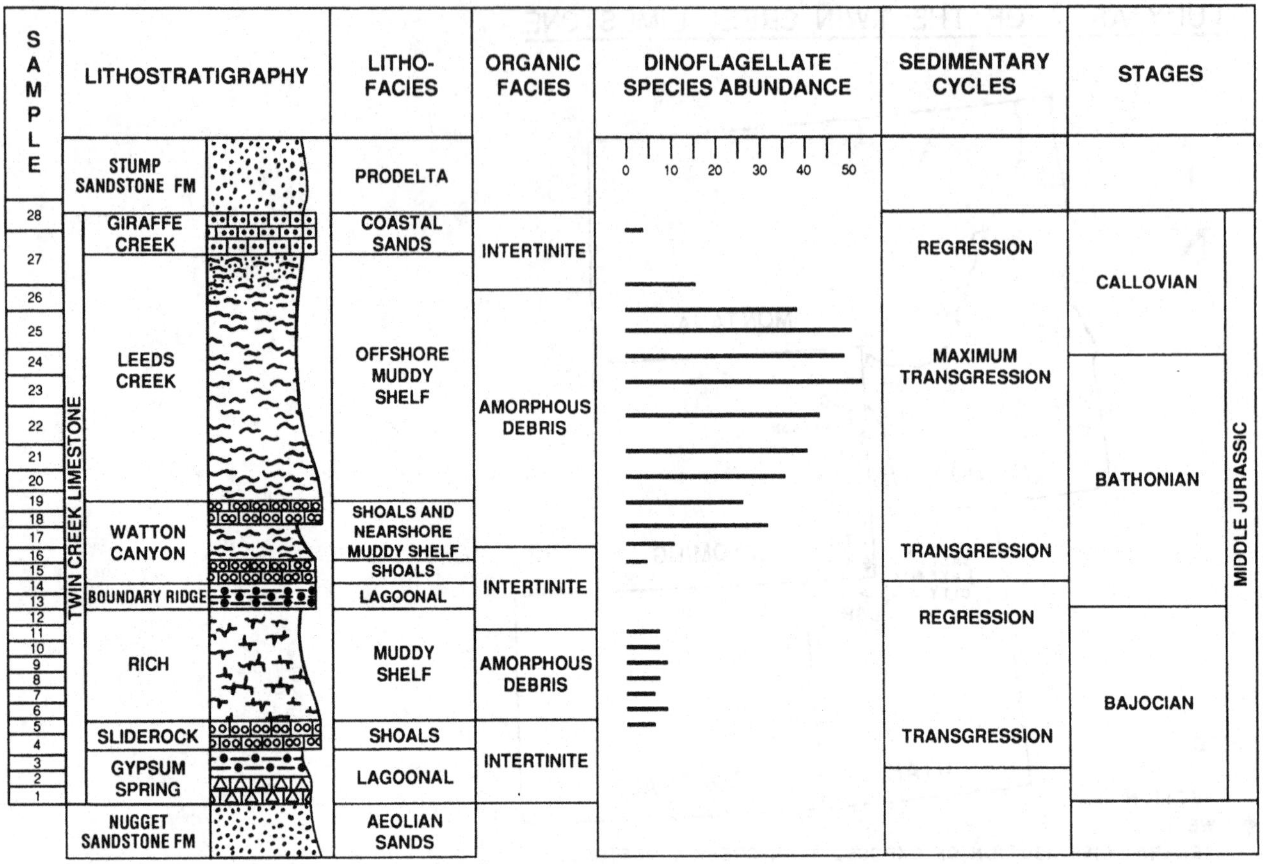

Figure 16.6. Stratigraphy and palynology of Camp Davis outcrop section. Dinoflagellate species abundance and organic facies are correlated with sedimentary cycles derived from the lithostratigraphy and lithofacies of the Twin Creek Limestone. Nearer shore transgressive cycle extends conformably from Sliderock member to the Boundary Ridge member (late Bajocian to Bathonian). The offshore cycle is late Bathonian to early/middle Callovian and extends conformably from the Watton Canyon member to the Giraffe Creek member. The underlying and overlying formations are listed in vertical sequence.

is late Bajocian to early Bathonian. The lithologic sequence is gradational in the Sliderock, Rich and Boundary Ridge members. However, the lithologic boundary between the Sliderock and the underlying Gypsum Spring is sharp. The Sliderock is a dark gray, oosparitic, limestone formed in a high energy shoal (or bank) environment. It grades upward into the shaly limestone of the Rich member. The Rich represents the subtidal shelf, on which lime mud was accumulating during transgression. The upper part of the Rich member indicates the return to the nearshore environment in regression (Imlay, 1967). The amount of silt and sand increases as the shaly limestone grades upward to the reddish calcareous siltstones of the Boundary Ridge member. Oscillation ripples and horizontal burrows in the Boundary Ridge represent lagoonal coastal environments. The second, and major cycle of transgression and regression ranges from late Bathonian to late Early Callovian (Imlay, 1967, 1980). The second sequence of deposition is similar to the first. There is a sharp boundary between the regressive Boundary Ridge member and the overlying Watton Canyon member. The basal part of the Watton Canyon member represents a high-energy carbonate environment as shown by crossbedded oosparites and oobiosparites, and is succeeded by an episode of rapid deepening during transgression, resulting

in an expanding shelf environment. The record of transgression continues into the overlying Leeds Creek member. This thick sequence of calcareous shale was deposited in a farther offshore, outer shelf setting. The maximum marine transgression occurred during the early Callovian and is evident in the middle to upper part of the Leeds Creek. The evidence of maximum transgression is based on the association of shelly macrofossils and on the geographic extent of this lithofacies. A change in lithology in the upper part of this member indicates regression. There is an increase in the amount of sand and silt in the lime mud with a corresponding decrease in the diversity of macrofossils. There is a gradational change to the crossbedded and ripple-marked sandy limestone of the overlying Giraffe Creek member. The Giraffe Creek is the uppermost member of the Twin Creek

Limestone, and represents the late Early Callovian regression in a very shallow coastal (possibly intertidal) environment. The palynology indicates a similar history (Fig. 16.6). The inertinite facies occurs in lagoonal and shoal environments of the Gypsum Spring and Sliderock members, respectively. Palynomorphs are very few and poorly preserved. The palynomorphs consist mostly of small acritarchs, bisaccate pollen and fungal spores. They are fragmented in the Sliderock, which suggests microscopic-scale abrasion in this high-energy depositional environment. The amorphous debris facies is developed in the transgressive shelf environment recorded in the lower part of the Rich member. The dinoflagellate *Sentusidinium* spp. appears, along with *Diacanthum filapicatum* and *Mendicodinium* sp. ('*M. scrobiculatum*' of Van Pelt, 1990) which become abundant. Land plant detritus is sparse, except for local abundance of the windblown *Classopollis*. Although there is an interval of well-developed amorphous debris facies, there is no multiple occurrence datum at its base; only seven species of plankton (including acritarchs and the Tasmanaceae) occur. This small increase (from one species in the subjacent inertinite facies) may reflect a nearshore locus of deposition. The upper part of the Rich member consists of the inertinite facies developed during regression. The inertinite facies extends upwards through the terminal, regressive Boundary Ridge member. Inertinite is abundant, and there is an absence of dinocysts except for rare occurrences of *Sentusidinium*.

The sequence of organic facies and dinoflagellate species abundance is best developed in the cycle of maximum transgression of late Bathonian-early Callovian age to the regression in the late Early Callovian. The inertinite facies continues in the lower part of the Watton Canyon member. This oolitic shoal facies contains very few dinoflagellates (*Sentusidinium* sp., *D. filapicatum*) and few bisaccate pollen grains and *Classopollis*. The first palynological evidence of marine transgression is the well-developed amorphous debris facies, found in the rapidly accumulating lime mud deposited on the expanding shelf. The base of the interval of abundant amorphous debris is marked by the datum which indicates increase in the number of dinoflagellate species. The number of species increases from 11 to 30 in an interval 6 m thick. The number of species continues to increase in the farther offshore depositional environment represented by the middle parts of the Leeds Creek member, to a maximum of 50 in the maximum transgression. The number of dinoflagellate species then declines within the amorphous debris facies and is reduced to a minimum in the inertinite facies of the Giraffe Creek member, which represents the return to coastal environments.

In the outcrop at Camp Davis, certain dinoflagellate associations are evident in the record of both episodes of transgression and regression (Fig. 16.7). *Sentusidinium* occurs in the oolitic shoal facies, despite the paucity of other species. The genus becomes abundant in the Bathonian transgression, but remains abundant in the Callovian regression (upper part of the Giraffe Creek) in the apparent absence of other taxa. *Diacanthum filapicatum* becomes most abundant in both episodes of transgression. *Rhynchodiniopsis cladophora*, *Glomodinium evittii*, *Endoscrinium* sp. A, *Valensiella ovula* and *Ellispsoidictyum gochtii* become most abundant in the maximum transgression, and decline dramatically in the subsequent regression. On the other hand, *Atopodinium prostatum* becomes most abundant in the amorphous debris facies of the uppermost part of the Leeds Creek member, which is coincident with regression within the marine shelf.

Comparison of siliciclastic and carbonate environments

The stratigraphic fluctuation in the number of dinoflagellate species, and the sequence of depositional organic facies, display patterns which correspond to cycles of transgression and regression. These patterns are evident in both siliciclastic and carbonate shallow shelf environments, despite their different geologic ages. The amorphous debris facies occurs in the marine shallow shelf. The inertinite facies occurs in coastal environments which range from tidal flat and lagoon to oolitic shoals or banks. In the deltaic plain sequence, it occurs also in interdistributary bays and as nonmarine inertinite in the subaerial delta plain.

In both carbonate and siliciclastic depositional settings, dinoflagellate species vary in numerical abundance in relation to the sequence of transgression or regression. Dinoflagellate species increase in number during transgression and tend to be most numerous in the amorphous debris facies. Their numbers increase dramatically in a transgressive episode at the base of the interval defined by the amorphous debris facies. The multiple occurrence datum which appears at this boundary provides the first palynological evidence of transgression. Dinoflagellate species are most numerous in a multiple occurrence datum when the datum indicates the maximum transgression of the investigated interval (Fig. 16.4). They may otherwise increase in numerical abundance when the maximum transgression is recorded stratigraphically above the multiple occurrence datum (Fig. 16.6).

In either case, dinoflagellate species are most numerous in the best developed transgression within a given sequence and, thus, are potentially valuable for detecting it. Dinoflagellates may also provide the first palynological evidence of regression. In both depositional settings, the sequence of regression is indicated by the change from the older marine shelf accumulating amorphous debris to the younger coastal environments collecting inertinite.

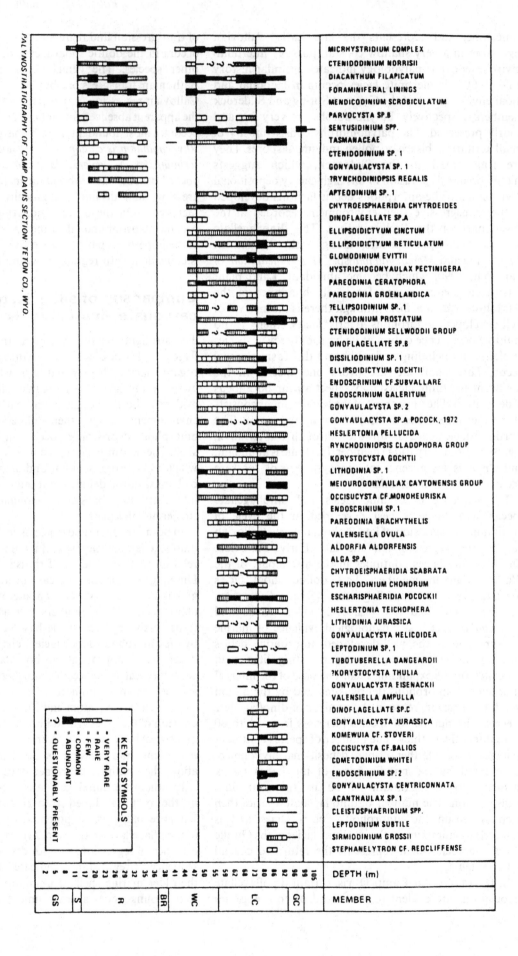

Figure 16.7. Dinoflagellate stratigraphy of the Camp Davis outcrop section in Teton County, WY. Multiple occurrence datum is well-defined in the transgression recorded in the Watton Canyon member of the Twin Creek limestone. The relative abundance of species is depicted. Note the dramatic decrease in the number of species in the Giraffe Creek member. Member designations at bottom of figure: GC = Giraffe Creek; LC = Leeds Creek; WC = Watton Canyon; BR = Boundary Ridge; R = Rich; S = Sliderock; GS = Gypsum Spring.

Dinoflagellate species begin to decrease in number systematically within the amorphous debris facies, which may be due to the decreasing area of the shelf during the beginning of regression. Species numbers then become fewest in the coastal inertinite facies, which is usually accompanied by an increase in land plant detritus. In general there are the fewest number of dinocyst specimens in the coastal inertinite facies, although there are good examples in both the siliciclastic and carbonate depositional environments where very few species are represented by abundant dinocyst specimens, e.g., *Gillinia hymenophora* and *Sentusidinium* spp. Smelror and Leereveld (1989) have suggested that genera such as *Sentusidinium* are generally characteristic of coastal to inner shelf environments. There is also an apparent relationship between depositional environment and the relative abundance of dinocysts of certain species. Examples are *Palaeohystrichophora infusorioides* and *Palaeoperidinium pyrophorum* in the nearshore and coastal facies of Late Cretaceous age; *Glaphyrocysta retiintexta* and *Areoligera medusettiformis* in the offshore neritic facies of the Late Cretaceous age; *Diacanthum filapicatum* in the transgressive shelf facies of late Bajocian and early Callovian ages; *Rhynchodiniopsis cladophora* in the maximum transgressive outer shelf environment of early Callovian age; and *Atopodinium prostatum* in the regressive shelf environment of early to middle Callovian age.

There are differences in the palynology of the siliciclastic and carbonate settings. The vascular tissue facies is not evident in the environments recorded in the Twin Creek Limestone. The lack of this facies is due to the lack of deltaic sedimentation on the carbonate shelf. There is very little terrigenous mineral sediment in any facies of the Twin Creek and very little vascular tissue, and few fern or other trilete spores occur (Van Pelt, 1990). Pollen grains do occur in the Twin Creek Limestone. These are almost exclusively of 'windblown' *Classopollis* and bisaccates. The vascular tissue facies occurs in siliciclastic deposits representing coastal delta front and marine prodelta environments of early Maestrichtian age, and in coeval adjacent shelf sediments palynologically prograded beyond the prodelta (Habib & Miller, 1989). Fern spores morphologically attributed to the modern family Schizaeaceae are known to occur in the Middle Jurassic of North Africa (Dörhöfer, 1979) and in the Middle Jurassic of the North Atlantic sequence (Habib, 1983). One does not expect to find corresponding stream-delivered components in the modern Bahama Bank, but such components are present in the fluviodeltaic discharge system in the Gulf of California (Cross *et al.*, 1966). The multiple occurrence datum is not well developed in the carbonate shelf environment. It is best expressed in the siliciclastic sedimentary setting. The datums in siliciclastic sediments

may have been accentuated by the seaward displacement of specimens of coastal species and by recycling of penecontemporaneous deposits by marine currents (Wall *et al.*, 1977).

Implications of the deep-sea record

Organic particles are widespread in specific intervals of Cretaceous age recovered from the deep-sea record. In the paleobathymetrically deep North Atlantic Cretaceous, virtually all palynomorphs were derived from continental margins. These include pollen grains, spores and other fossils that were produced in the land vegetation and dinocysts of the marine meroplanktonic dinoflagellates which occupied the continental shelf and slope. Recycled inertinite was admixed with land plant detritus and carried to bathypelagic depths. Inertinite may also have formed in situ as diagenetic residue in hematitic red clay deposits (Habib, 1983). The distribution of land plant particles shows morphological sorting in the North Atlantic Cretaceous, in trends away from continental margins. Similar trends are evident in the distribution of dinoflagellate species of Neocomian age (Habib & Drugg, 1987, p. 755).

Amorphous debris is being produced as fecal pellets in the modern North Atlantic Ocean, and therefore its geographic occurrence in the ocean basin should not be as limited as other organic facies. The middle Cretaceous marine-carbon event, which culminated in the Cenomanian-Turonian global transgression, indicates that sediments deposited during these intervals, which are especially rich in amorphous debris, formed in a wide range of environments in both marginal seas and open oceans (Schlanger & Jenkyns, 1976). Dinocysts and other palynomorphs are rare or absent in the deposits that accumulated on the deep ocean floor far removed from the continent. On the other hand, dinocyst-rich amorphous debris was most likely derived from processes operating close to the continental margin. Perhaps the debris was displaced by currents on a continental shelf, the area of which was expanding during transgression, or perhaps it was deposited directly into zones where currents are upwelling against the continental shelf. Abundant and well-preserved amorphous debris was deposited in the zone of upwelling opposite Cap Blanc, Mauritania (Caratini *et al.*, 1975). Dinocysts are abundant in this sediment and are represented by species of *Impagidinium*. Wall *et al.* (1977) reported that dinocysts in this genus are limited to (and occur in abundance in) modern pelagic sediments occurring near the outer continental shelf and beyond. They did not specifically relate the abundance of *Impagidinium* (as *Leptodinium*) to upwelling, so it is probable that these dinocysts were displaced as detrital particles into the zone of upwelling

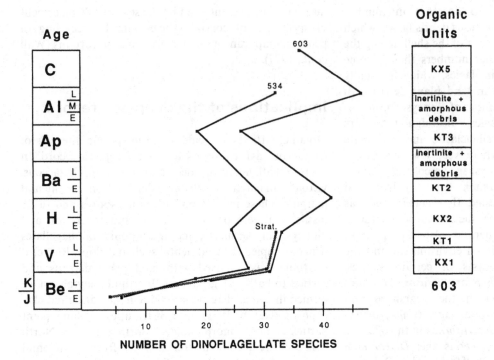

Figure 16.8.   Dinoflagellate species abundance curves for Deep Sea Drilling Project sites 534 and 603 along the western margin of the North Atlantic. Site 534 drilled at 28° 20.6′ N, 75° 22.9′ W in 4971 m water depth. Site 603 drilled at 35° 29.71′ N, 70° 1.71′ W in 4633.5 m water depth. The stratotype (strat.) curve is derived from outcrop Stage stratotype and ammonite dated sections of shallow-water origin in the Berriasian, Valanginian and Hauterivian of Tethyan Southern Europe (Habib & Drugg, 1983). Organic facies stratigraphy of site 603 modified from Habib and Drugg (1987). Interval from unit KT1 to just below KX5 is largely terrigenous, even in the presence of amorphous debris. Units KX1 (late Berriasian) and KX5 (late Albian–late Cenomanian) contain amorphous debris facies almost identical to that of shallow water environments. Jurassic/Cretaceous boundary is situated at the boundary between early and late Berriasian, after Haq et al., (1987).

from outer shelf or upper slope environments, and that the amorphous debris was produced from the digestion of motile, organic-walled plankton that would otherwise not be fossilized.

Drilling by the Deep Sea Drilling Project (DSDP) along the deep western margin of the North Atlantic revealed a sequence of palynological events which appears to coincide with episodes of sea level change. Figure 16.8 shows the numerical distribution of early Cretaceous dinoflagellate species at two DSDP sites, as well as their distribution in the shallow-water facies recovered from outcrops of the stratotype of the Neocomian stage and other ammonite-dated sections in southern Europe (Habib & Drugg, 1987). The two DSDP sites represent deep water environments of different depositional lithofacies. Site 534 was drilled near the center of the Blake-Bahama Basin, and recovered an early Cretaceous section of largely continuous hemipelagic carbonate and carbonaceous clay with episodic deposition of limestone and sandstone turbidites (Sheridan et al., 1983). Site 603 was drilled on the lower continental rise opposite Cape Hatteras, North Carolina (Wise et al., 1986). In contrast to site 534, drilling at site 603 intersected an extensive deep-sea fan complex of Hauterivian to Aptian age.

The species-abundance curves for sites 534 and 603 compare closely, and there is an almost precise match in the curves between the Berriasian-Hauterivian shallow-water facies and the deep-water facies in the Berriasian-Hauterivian interval at Site 603 (Fig. 16.8).

There are very few dinoflagellate species in the early Berriasian. A major influx of species, represented almost

exclusively by first appearance datums, occurs in the late Berriasian interval of the three curves. This is followed by a net increase in the number of dinoflagellate species in the early Valanginian and a minor increase, or slight decrease, in the late Valanginian and early Hauterivian. At sites 534 and 603, there is a further substantial increase in the early Barremian. This trend is interrupted within the interval of late Barremian to Aptian by an episode of apparent extinctions (last appearance datums). The maximum number of dinoflagellate species recorded at both sites occurs in the late Albian. The species-abundance curves compare favorably with the eustatic sea level curve proposed for this time period by Haq et al. (1987). Sea level was low in the early Berriasian and rose in the late Berriasian. It fell sharply in two short-term episodes in the Valanginian and then rose to a high in the early Barremian. It fell again in the Aptian and then rose to the maximum sea level of the Cretaceous Period in the late Cenomanian. However, the rise to this maximum level was already in evidence in the late Albian.

Figure 16.8 illustrates the stratigraphy of the depositional organic facies at site 603 in the interval from late Berriasian to the Cenomanian-Turonian boundary. Habib and Drugg (1987, Appendix B, p. 765–6) described the organic facies and plotted organic stratigraphic units against the dinoflagellate stratigraphy. The influx of dinoflagellate species in the multiple first occurrence datum of late Berriasian age occurs at the base of an interval defined by abundant and well-preserved amorphous debris and associated fecal pellets (KX1 of Fig. 16.8 and of back-cover range chart in Habib & Drugg, 1987). An interval of well-developed and abundant land plant detritus (KT1) follows in the middle part of the Valanginian (early to late Valanginian). This is followed by an interval (KX2) of poorly preserved amorphous debris and palynomorphs, which includes samples rich in terrigenous particles.

The interval from the late Hauterivian to early Albian is characterized by the occurrence of terrigenous organic matter of various proportions and amounts. In the early Barremian, a large amount of land plant detritus (exinitic facies of Habib & Drugg, 1987) was delivered to the submarine fan environment in clayey sediments situated in poorly consolidated to unconsolidated sand and silt turbidites. Palynomorphs are abundant, and are poorly sorted and diversified. They include a variety of well-preserved fern spores, sporangial clusters of fern spores, and even megaspores. This is followed in the late Barremian by intervals of poorly preserved amorphous debris and inertinite. The second major episode of terrigenous influx occurs in the Aptian (KT3 of Habib & Drugg, 1987), within massive Bouma-sequence sand turbidites (Wise *et al.*, 1986). Fern spores are abundant in this interval as well, although the assemblages are not as diversified as those in the early Barremian. Terrigenous matter continues to be predominant in residues of early to middle Albian age, in the form of tracheal inertinite.

The second, and major, interval of abundant and well-preserved amorphous debris occurs within unit KX5 (Fig. 16.8), which is equivalent in age with the maximum sea level rise of the Cretaceous Period. Numerous dinoflagellate species occur in small residues of amorphous debris in the lower (late Albian) part of KX5; the Cenomanian part of this unit is characterized by large residues of abundant amorphous debris. Wise *et al.* (1986) report organic carbon values up to 13.6% in the Cenomanian interval. Dinoflagellate species are most numerous at the base of this unit in two closely spaced multiple occurrence datums of late Albian age, but decline in number in the amorphous debris-rich (and organic carbon-rich) part adjacent to the Cenomanian-Turonian boundary.

The similarity of the dinoflagellate species curves for the shallow water and deep ocean facies of equivalent Early Cretaceous age suggests that the evidence used herein for transgression and regression is applicable in paleobathymetrically deep environments close to the continental margin. The similarity persists despite the different lithological character of sites 534 and 603. Dinocyst specimens were apparently contemporaneously displaced from the upper continental margin into various deep-water depositional settings. Dinoflagellate species appear to have varied in number according to the record of sea level change. They are most numerous in multiple occurrence datums at the base of intervals defined by fecal amorphous debris in both the late Berriasian and late Albian, as they are in the shallow-water datums of the late Campanian and middle Maestrichtian (Habib & Miller, 1989). There is also an internally consistent correspondence between the maximum number of dinoflagellate species in a single section and independent evidence of maximum transgression (or maximum sea level rise). The late Albian multiple occurrence datum signals the beginning of the maximum sea level rise, which culminated in the late Cenomanian. The reduction in number of species in the Cenomanian, although still numerous, may be the result of migration of the depositional locus away from continental margin processes during the maximum transgression. The kind and amount of terrigenous organic matter in clayey lithological types provide background evidence of terrigenous sedimentation. The large amount of poorly sorted land plant materials in the submarine fan environments of site 603 suggests rapid sedimentation. The occurrence of megaspores and sporangia in deep ocean facies is sedimentologically unusual; these spores are usually present only in terrigenous rocks which indicate rapid sedimentation. In the early Cretaceous, clayey sediments in terrigenous turbidites also contain many fern spores (Habib & Drugg, 1987).

A similar history of early to middle Cretaceous events is evident at site 398, which was drilled by the Deep Sea Drilling Project at the opposite margin of the North Atlantic, adjacent the Iberian Peninsula southwest of Vigo, Spain (Sibuet *et al.*, 1979). The depositional history is similar to that of site 603. The interval of Barremian to Albian age is characterized by terrigenous turbidites in a submarine fan setting, and is succeeded in the late Albian-Cenomanian by hemipelagic clays deposited during the maximum transgression and sea level rise. The stratigraphic occurrence of organic particles of both marine and nonmarine origins compares favorably with the geochemical assessment of the nature and amount of organic carbon in the section (Fig. 16.9). According to Deroo *et al.* (1978), terrestrial organic carbon values in excess of 2.5% are associated with numerous fern spores and other land plant particles in terrigenous turbidite

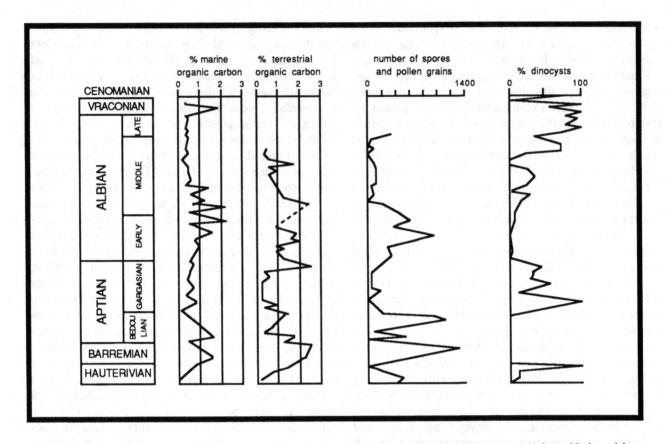

Figure 16.9. Correlation of palynological data with the weight percentage occurrence of organic carbon in the Barremian-Cenomanian interval at Deep Sea Drilling project Site 398. Total number of palynomorphs consists almost entirely of pollen grains and spores. The amplitude of terrigenous organic particles corresponds closely to the number of pollen grains and spores, and to the amount of organic carbon derived from terrigenous mineral sediment. Dinocysts are infrequent throughout the section, but dinoflagellate species increase in percentage (in the virtual absence of pollen) in the late Albian-Cenomanian episode of rich marine amorphous debris. High organic carbon percentages in this hemipelagic sediment result from the abundance of amorphous debris. Site 398 was drilled at 40° 57.6′ N, 10° 43.1′ W in a water depth of 3910 m.

episodes of Barremian, early Aptian and early to middle Albian age. The largest amount of land plant detritus occurs in exinitic samples of Barremian and early Aptian age (Habib, 1979). Pteridophyte spores (mainly ferns and lycopods) are especially numerous. They are represented by as many as 25 species and virtually mask the presence of dinocysts. Figure 16.9 illustrates the abrupt termination of terrestrial carbon in the middle Albian. Organic carbon values are low but are now entirely of marine origin. Marine carbon values increase in the late Albian and Vraconian. The late Albian marine carbon event is matched by the almost complete cessation of the land plant materials which were abundant in the earlier Cretaceous. Palynomorphs are few in abundant amorphous debris in the Cenomanian, but are represented in the late Albian almost entirely by marine dinocysts, except for a few windblown bisaccates and *Classopollis*.

The records at site 398 and 603 reflect an early Cretaceous history of deltaic progradation which extended seaward virtually to the present shelf/slope break at the margins of continents which ringed the North Atlantic (Ryan & Cita, 1977). The late Albian-Cenomanian episodes of marine carbon-rich intervals of amorphous debris coincide with evidence of sea level rise and transgression of the continental margins around the North Atlantic. The organic geochemistry at site 398 complements the palynology. Both the early and middle

Cretaceous events are carbonaceous, but the carbon is of different kinds and origins. The highly carbonaceous episodes of Barremian, Aptian and Albian age are the result of the rapid supply of large amounts of land plant particles. The middle Cretaceous marine carbon event is the result of the significant increase in organic-walled plankton productivity associated with the maximum transgression which drowned continental margins, and the consequent abundance of zooplankton fecal matter. The high organic carbon values in the Cenomanian are matched directly with the abundance of fresh amorphous debris. Dinocysts may be plentiful or sparse, depending in part on the proximity and pathway of marine currents.

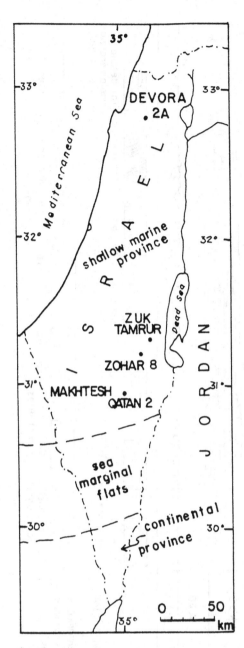

Figure 16.10. Location of the Makhtesh-Qatan 2, Zohar 8, Zuk Tamrur and Devora 2A wells. Generalized inner neritic environments in the Permian and Triassic of Israel illustrated. Arabian Craton to the south; Tethyan seaway to the northwest. Locus of deposition of the Devora 2A well is farthest offshore. (Reproduced from Eshet *et al.,* 1988.)

Reworked palynomorphs in nearshore environments

Reworked palynomorphs may be quantitatively important in sedimentary cycles. In the Permo-Triassic sequence in Israel, clastic, carbonate and evaporite sediments were deposited in a variety of environments which ranged from terrestrial to nearshore marine. Eshet *et al.* (1988)

discovered that the proportion of reworked specimens in palynomorph assemblages varied according to cycles of transgression and regression. Reworked palynomorphs were distinguished by their opacity or blackened color in otherwise relatively well-preserved assemblages (Eshet *et al.*, 1988, plate 1). Where there was sufficient morphology preserved so that they could be identified taxonomically, the palynomorphs were shown to be of older age. For example, chitinozoan species of Silurian and Devonian age and acritarchs of Silurian age were found in the Triassic sequence.

Four wells were investigated (Fig. 16.10). Triassic seas transgressed southwestward in three cycles of sedimentation, from the Tethys Ocean to the north towards the landmass of the Arabian Craton in the south. Figure 16.11 illustrates the lithostratigraphy and lithofacies in the north-south trend, and shows the occurrence of reworked palynomorphs in the Makhtesh-Qatan 2 well. The relative percentage of reworked palynomorphs corresponds closely with the lithofacies evidence of transgression and regression. Percentages are highest in intervals of regression and are lowest in the transgressive intervals. Although these organic particles were transported with terrigenous sediments, they are abundant also when the regression is represented by sediments other than clastic rocks. They are most abundant in the clastic-dominated regressions of the Zafir and Gevanim Formations, but are also abundant in the anhydrite-dominated Mohilla Formation (Fig. 16.11).

According to Eshet *et al.* (1988), the amplitude of curves showing the abundance of reworked organic particles is reduced in the direction away from their source. Figure 16.12 illustrates the curves for the four investigated wells. The same trends are evident, although the amplitude is most reduced in the northernmost (Devora 2A) well. Nevertheless, the same trends are evident in this well, even though it lacks the clastic lithologies of the southern wells. Reworked palynomorphs still record the regression of the Mohilla Formation in the Devora 2A well. They occur in higher relative percentages in the evaporite facies characterized by anhydrite lithology and euryhaline species of ostracodes (Gerry *et al.*, 1991).

Discussion and conclusions

Palynological evidence is useful for detecting cycles of transgression and regression in a variety of sediments deposited in shallow-marine environments. This evidence is also useful in selected deep-sea depositional environments situated relatively close to continental margins. It is important to study the total organic matter recovered from thermally immature residues in order to determine depositional, not diagenetic processes. Recent studies of the geographic distribution of modern dinocysts in

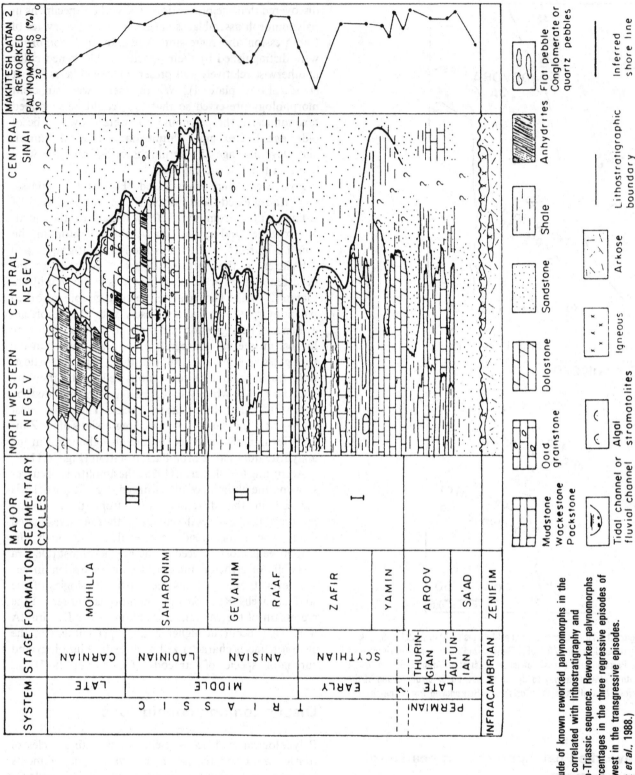

Figure 16.11. Amplitude of known reworked palynomorphs in the Makhtesh-Qatan 2 well correlated with lithostratigraphy and lithofacies of the Permo-Triassic sequence. Reworked palynomorphs occur in the highest percentages in the three regressive episodes of Triassic age and are lowest in the transgressive episodes. (Reproduced from Eshet *et al.*, 1988.)

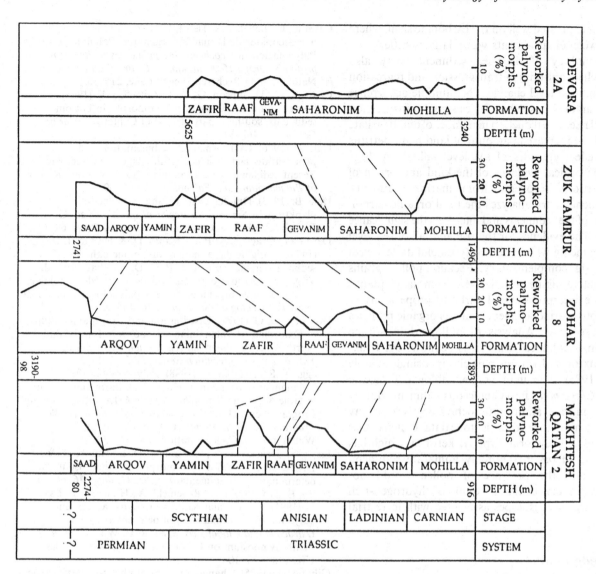

Figure 16.12. Correlation of reworked palynomorph curves in the four wells. The same trends and correlations with transgression and regression are evident, even though the amplitude is diminished in the Devora 2A well, farthest offshore. (Reproduced from Eshet *et al.*, 1988.)

surface sediments indicates that dinoflagellates are useful for distinguishing depositional environments. The examples from the Jurassic and Cretaceous are proposed as evidence that the number of dinoflagellate species increases in continental shelf environments that expand in area during transgression and that the number is reduced as the shelf area becomes smaller during regression, similar to the principle of ecology discussed by Krebs (1985). Amorphous debris of zooplankton fecal origin is abundant where dinoflagellate species are most numerous. However, the vast majority of fecal debris probably formed from a variety of motile organic-walled plankton that would not otherwise have contributed to the fossil record. In both the shallow-marine and deep continental-rise environments, the abundance of poorly sorted land plant detritus is directly associated with deltaic, prodeltaic and submarine fan depositional processes. Because terrigenous organic matter may extend well beyond a particular depositional lithofacies,

it records transgression and regression in offshore sections, even where lithological evidence is lacking. This is evident in the Black Sea sediments as well as in the vascular tissue facies in the Dorchester well and in the regressive intervals in the Devora 2A well. Inertinite is ubiquitous in sediments containing organic matter. It predominates in nonmarine and coastal environments otherwise impoverished of organic matter.

The multiple occurrence datum provides the first palynological evidence of transgression in the marine environment. It is a datum occurring at the base of an interval rich in amorphous debris. It is expressed best in siliciclastic rocks, which suggests that it represents mixing of the palynology of different sedimentary facies. This datum is not expressed well on the carbonate shelf, where mixing between environments was relatively little.

The palynology of deep-sea sediments may also indirectly reflect episodes of transgression and regression resulting from sea level change. The multiple occurrence datum and amorphous debris facies signal transgression and sea level rise. The reduction in number of dinoflagellate species and the increase in amount of land plant detritus indirectly reflect episodes of sea level fall. Palynology provides the optical evidence of the kind and origin of organic particles recovered from marine sediments. Organic geochemistry analyzes the total organic carbon composition of the sediments and interprets the significance of isotopes. The evidence for the rise and fall of sea level is most convincing where the two types of data test each other and are complementary. Because pollen grains and conductive tissue are derived from land plants, they test the interpretation of carbon-isotope analysis. Carbonaceous marine sediment contains organic particles of different sources. Analysis of bulk carbon-isotope composition is enhanced when it is acknowledged that there are mixtures of various particles in a single sample (Gilmour, 1985). The influx of 'windblown' bisaccate pollen and *Classopollis* into Cretaceous sediments during episodes of marine amorphous debris deposition may have compromised the carbon-isotope data. Pollen exines are not lignitic. They form Type III kerogen, which has an H/C ratio of 0.8 in thermally immature sediment. Thus, although the presence of pollen would not significantly influence the amount of hydrogen-rich matter, it introduces isotopes associated with terrestrial carbon.

Acknowledgments

We are indebted to James A. Miller of the US Geological Survey, who first suggested the palynological study of core samples from South Carolina and Georgia. We are grateful also to Stephen R. Jacobson (Chevron Oil Field Research Company) and Robert P. Wright (Chevron USA, Inc.), for their support of the study of the Twin Creek Limestone. Alfred Traverse (Pennsylvania State University) kindly invited us to submit this chapter for inclusion in this volume. We thank him, Jim Miller, Jan Jansonius (Esso Resources Canada, Ltd.) and the anonymous reviewers who criticized the manuscript for us. Much of the research contained within this chapter has been supported both directly and indirectly by the US National Science Foundation, particularly grants OCE-8303695 and EAR-8903522. Elizabeth A. Parascandolo (Queens College of the City University of New York) prepared the samples for palynological study and drafted the computer graphics.

References

Brinkhuis, H. & Zachariasse, W. J. (1988). Dinoflagellate cysts, sea level changes and planktonic foraminifers across the Cretaceous-Tertiary boundary at El Haria, northwest Tunisia. *Marine Micropaleontology*, **13**, 153–91.

Caratini, C., Bellet, J. & Tissot, C. (1975). Étude microscopique de la matière organique: Palynologie et Palynofaciès. In *Géochemie organique des sédiments marins profonds. Orgon II, Atlantique-N.E. Brésil*. Paris: Centre National de la Recherche Scientifique, 215–65.

Cross, A. T., Thompson, G. G. & Zaitzeff, J. P. (1966). Source and distribution of palynomorphs in bottom sediments, southern part of Gulf of California. *Marine Geology*, **4**, 467–524.

Dale, B. (1976). Cyst formation, sedimentation and preservation: factors affecting dinoflagellate assemblages in Recent sediments from Trondheimsfjord, Norway. *Review of Palaeobotany and Palynology*, **22**, 39–60.

Dale, B. (1983). Dinoflagellate resting cysts: 'benthic plankton'. In *Survival Strategies of the Algae*, ed. G. A. Fryxell. Cambridge: Cambridge University Press, 69–136.

Deroo, G., Graciansky, P. C., Habib, D. & Herbin, J. P. (1978). L'origine da la matière organique dans les sediments cretaces du site I. P. O. D. 398 (haut-fond de Virgo): correlations entre les donnes de la sédimentologie, de la géochimie organique et de la palynologie. *Bulletin de la Société Géologique de France*, **20**, 465–9.

Dörhöfer, G. G. (1979). Distribution and stratigraphic utility of Oxfordian to Valanginian miospores in Europe and North America. *American Association of Stratigraphic Palynologists Contribution Series*, **5B**(2), 101–32.

Drugg, W. S. & Habib, D. (1988). Palynology of the Valanginian-Barremian in Hole 638B, Barremian-Albian in Hole 641C and Turonian in Hole 641A, Ocean Drilling Program Leg 103, In *Proceedings of the Ocean Drilling Program, Scientific Results 103*, ed. G. Biollot, E. L. Winterer, *et al*. College Station, TX, 429–32.

Eshet, Y., Druckman, Y., Cousminer, H. L., Habib, D. & Drugg, W. S. (1988). Palynomorphs and their use in the determination of sedimentary cycles. *Geology*, **16**, 662–5.

Gerry, E., Honigstein, A., Rosenfeld, A., Hirsch, F. & Eshet, Y. (1991). The Carnian salinity crisis: ostracodes and palynomorphs as indicators of paleoenvironment. In *Ostracoda and Global Events*, ed. R. C. Whatley *et al*. Tenth Symposium on Ostracoda (Aberystwyth, Wales, UK, 1988), 87–100.

Gilmour, I. (1985). Changing weight of photosynthesis in the ocean. *Nature*, **315**, 184.

Habib, D. (1968). Spores, pollen, and microplankton in deep-sea cores from the Horizon Beta outcrop. *Science*, **162**, 1480–1.

Habib, D. (1979). Sedimentology of palynomorphs and palynodebris in Cretaceous carbonaceous facies south of Vigo Seamount. In *Initial Reports of the Deep Sea Drilling Project 47*, ed. J. C. Sibuet & W. B. F. Ryan. Washington, DC: US Government Printing Office, 451–65.

Habib, D. (1982a). Sedimentation of black clay organic facies in a Mesozoic oxic North Atlantic. *Third North American Paleontological Convention, Proceedings*, **1**, 217–20.

Habib, D. (1982b). Sedimentary supply origin of Cretaceous black shales. In *Nature and Origin of Cretaceous Carbon-Rich Facies*, ed. S. O. Schlanger & M. B. Cita. London: Academic Press, 113–27.

Habib, D. (1983). Sedimentation rate-dependent distribution of organic matter in the North Atlantic Jurassic-Cretaceous, In *Initial Reports of the Deep Sea Drilling Project 76*, ed. R. E. Sheridan, F. M. Gradstein, *et al*. Washington, DC: US Government Printing Office, 781–94.

Habib, D. & Drugg, W. S. (1983). Dinoflagellate age of Middle Jurassic-Early Cretaceous sediments in the

Blake-Bahama Basin. In *Initial Reports of the Deep Sea Drilling Project* **76**, ed. R. E. Sheridan, F.M. Gradstein, *et al.* Washington, DC: US Government Printing Office, 623–38.

Habib, D. & Drugg, W. S. (1987). Palynology of sites 603 and 605, Leg 93, Deep Sea Drilling Project, In *Initial Reports of the Deep Sea Drilling Project* 93, ed. J. E. van Hinte, S. W. Wise, Jr., *et al.* Washington, DC: US Government Printing Office, 751–75.

Habib, D. & Miller, J. A. (1989). Dinoflagellate species and organic facies evidence of marine transgression and regression in the Atlantic Coastal Plain. *Palaeogeography, Palaeoclimatology, Palaeoecology,* **74**, 23–47.

Habib, D., Thurber, D., Ross, D. & Donahue, J. (1971). Holocene palynology of the Middle America Trench near Tehuantepec, Mexico. *Geological Society of America, Memoir,* **126**, 233–61.

Haq, B. U., Hardenbol, J. & Vail, P. R. (1987). Chronology of fluctuating sea levels since the Triassic. *Science,* **235**, 1156–66.

Harland, R. (1988). Dinoflagellates, their cysts and Quaternary stratigraphy. *The New Phytologist,* **108**, 111–20.

Hay, B. & Honjo, S. (1989). Particle deposition in the present and Holocene Black Sea. *Oceanography,* **2**, 26–31.

Honjo, S. (1980). Material fluxes and modes of sedimentation in the mesopelagic and bathypelagic zones. *Journal of Marine Research,* **38**, 53–97.

Hultberg, S. U. & Malmgren, B. A. (1986). Dinoflagellate and planktonic foraminiferal paleobathymetrical indices in the boreal uppermost Cretaceous. *Micropaleontology,* **32**, 316–23.

Imlay, R. W. (1967). Twin Creek Limestone (Jurassic) in the western interior of the United States. *US Geological Survey, Professional Paper* **540**.

Imlay, R. W. (1980). Jurassic paleobiogeography of the conterminous United States in its continental setting. *US Geological Survey, Professional Paper* **1062**.

Krebs, C. J. (1985). *Ecology. The Experimental Analysis of Distribution and Abundance.* 3rd ed. New York: Harper and Row.

MacArthur, R. H. & Wilson, E. O. (1967). *The Theory of Island Biogeography.* Princeton, NJ: Princeton University Press.

Ogg, J. G., 1987. Early Cretaceous magnetic polarity time scale and the magnetostratigraphy of D.S.D.P. sites 603 and 534, western central Atlantic. In *Initial Reports of the Deep Sea Drilling Project* 93, ed. J. E. van Hinte, S. W. Wise, Jr., *et al.* Washington, DC: US Government Printing Office, 849–79.

Ogg, J. G. (1988). Early Cretaceous and Tithonian magnetostratigraphy of the Galicia Margin (Ocean Drilling Program Leg 103). In *Proceedings of the Ocean Drilling Program, Scientific Results* **103**, ed. G. Boillot, E. L. Winterer, *et al.* College Station, TX, 659–82.

Porter, K. G. & Robbins, E. I. (1981). Zooplankton fecal pellets link fossil fuel and phosphate deposits. *Science,* **212**, 931–3.

Raynaud, J. F., Lugardon, B. & Lacrampe-Couloume, G. (1988). Observation de membranes fossiles dans la matière organique 'amorphe' de roches mères de pétrole. *Centre Recherches Academie Science Paris,* **307**(II), 1703–9.

Raynaud, J. F., Lugardon, B. & Lacrampe-Couloume, G. (1989). Structures lamellaires et bacteries, composants

essentiels de la matière organique amorphe des roches mères. *Bulletin Centres Recherches Exploration-Production Elf-Aquitaine,* **13**(I), 1–21.

Rullkötter, J., Mukhopadhyay, P. K. & Welte, D. H. (1987). Geochemistry and petrography of organic matter from D.S.D.P. Site 603, lower continental rise off Cape Hatteras, In *Initial Reports of the Deep Sea Drilling Project* 93, ed. J. E. van Hinte, S. W. Wise, Jr., *et al.* Washington, DC: US Government Printing Office, 1163–76.

Ryan, W. B. F. & Cita, M. B. (1977). Ignorance concerning episodes of ocean-wide stagnation. *Marine Geology,* **23**, 197–215.

Schlanger, S. O. & Jenkyns, H. C. (1976). Cretaceous oceanic anoxic sediments: causes and consequences. *Geologie en Mijnbouw,* **55**, 179–84.

Sheridan, R., Gradstein, F., *et al.*, eds. (1983). *Initial Reports of the Deep Sea Drilling Project* **76**. Washington, DC: US Government Printing Office.

Sibuet, J. C., Ryan, W. B. F., *et al.* (1979). *Initial Reports of the Deep Sea Drilling Project* **47**. Washington, DC: US Government Printing Office.

Smelror, M. & Leereveld, H. (1989). Dinoflagellate and acritarch assemblages from the late Bathonian to early Oxfordian of Montague Crussol, Rhone Valley, southern France. *Palynology,* **13**, 121–42.

Tissot, B. & Pelet, R. (1981). Sources and fate of organic matter in ocean sediments. In *Publications du 26ème congrès géologique international: Colloque c4, Géologie des océans, Oceanologica Acta* **4**, supplement, 97–103 (French summary).

Traverse, A. & Ginsburg, R. N. (1966). Palynology of the surface sediments of Great Bahama Bank, as related to water movement and sedimentation. *Marine Geology,* **4**, 417–59.

Vail, P. R., Mitchum, R. M. & Thompson, S. (1977). Seismic stratigraphy and global changes of sea level, part 4: global cycles of relative changes of sea level. *American Association of Petroleum Geologists Memoir,* **26**, 83–97.

Van Pelt, R. (1990). Palynology of the Twin Creek Limestone. Ph.D. dissertation, Graduate School of the City University of New York.

Vinogradov, M. E. & Tseitlin, V. B. (1983). Deep-sea pelagic domain (aspects of bioenergetics) In *Deep Sea Biology,* ed. G. T. Rowe. *The Sea,* **8**. New York: John Wiley & Sons: 123–66.

Wall, D., Dale, B., Lohmann, G. P. & Smith, W. K. (1977). The environmental and climatic distribution of dinoflagellate cysts in modern marine sediments from regions in the North and South Atlantic Oceans and adjacent seas. *Marine Micropaleontology,* **2**, 121–200.

Williams, G. L. (1975). Dinoflagellate and spore stratigraphy of the Mesozoic-Cenozoic, offshore eastern Canada. *Geological Survey of Canada, Paper* **74-30**(2), 107–61.

Wise, S. W., Jr., *et al.* (1986). Mesozoic-Cenozoic clastic depositional environments revealed by D.S.D.P. Leg 93 drilling on the continental rise off the eastern United States. In *North Atlantic Palaeoceanography,* ed. C. P. Summerhayes & N. J. Shackleton. *Geological Society Special Publication,* **21**. Oxford: Blackwell Scientific.

Wright, R. P. (1973). Marine Jurassic of Wyoming and South Dakota, Its Paleoenvironments and Paleobiogeography. *Museum of Paleontology, Papers in Paleontology,* **2**. Ann Arbor: University of Michigan.

B Sequence stratigraphy and sedimentation of organic particles

17 Particulate organic matter, maceral facies models, and applications to sequence stratigraphy

GEORGE F. HART, MARK A. PASLEY and WILLIAM A. GREGORY

Introduction

Organic matter found dispersed in aqueous systems and in sedimentary rocks consists of organic compounds derived from both autochthonous and allochthonous sources. The form of this material ranges from well-preserved particulate organic matter (POM or macerals) to dissolved organic matter (DOM). The total mass of this organic matter is very large, and McIver (1967) notes that it is 500–1000 times the amount found in coal, and 10 000 times the amount in all organisms living at the present time. Both Weeks (1958) and Hunt (1962) estimated the total mass to be 3.2×10^{15} metric tons.

The microscopic analysis of POM has gained widespread acceptance as a reliable means of assessing the petroleum source potential and thermal maturity of rocks in sedimentary basins. Numerous studies, using various classifications of organic matter, indicate that a relationship exists between POM and petroleum source potential (e.g., Burgess, 1974; Dow, 1977; Teichmüller & Wolf, 1977; Robert, 1981; Mukhopadhyay et al., 1985; Senftle et al., 1987). Thermal maturity measurements in the form of vitrinite reflectance and transmitted color are routine in petroleum source rock evaluation (Staplin, 1969; Bostick, 1971, 1974; Castaño & Sparks, 1974; Dow, 1977; Peters et al., 1977; Hunt, 1979; Dow & O'Connor, 1982; Saxby, 1982; Stach et al., 1982; Pearson, 1984; Tissot & Welte, 1984; Waples, 1985).

Contemporary depositional environments contain many clues to the origin of the sediments accumulating on the depositional surface. Unfortunately, rocks rarely preserve sufficient information to reconstruct accurately the original depositional environment. Information is lost principally at the sediment–water interface and during the diagenetic stage of lithogenesis. It is not always possible to determine the detailed environment from a rock, but only to offer a number of possible alternatives because of this information loss. This factor led stratigraphers to substitute the word 'facies' for the word 'environment,' when discussing depositional settings.

Organic petrology is useful in sedimentologic interpretations, and the concept of organic facies and how they relate to lithofacies is important in this regard. In the past stratigraphic palynologists have concentrated their studies on that part of the dispersed POM that consisted of whole fossils such as acritarchs, dinocysts, miospores, and megaspores. 'Palynofacies' are facies based on the distribution of palynomorphs (whole fossils). More recently the organic debris derived principally from algae, fungi, and plants have been classified and proven useful in many areas of stratigraphic analysis (Hart, 1979a,b,c, 1982, 1986). 'Maceral facies' are facies defined by the distribution of POM in a rock unit. Both maceral facies and palynofacies can be regarded as subsets of 'organic facies' (Jones, 1987; Tyson, 1987). Jones' definition of organic facies is a very satisfactory definition from the point of view of the stratigraphic palynologist, because it emphasizes similarity of organic facies to lithofacies and biofacies as mappable stratigraphic units that are distinguished by the composition of their constituent organic matter. Several workers have shown that both maceral facies and palynofacies provide valuable information concerning paleoenvironmental conditions (Summerhayes, 1981; Wrenn & Beckman, 1981; Habib, 1982; Rullkötter & Mukhopadhyay, 1986; Tyson, 1987; LeNoir & Hart, 1988; Habib & Miller, 1989; Hart et al., 1989; Crumière et al., 1990; Ganz et al., 1990; Stein & Littke, 1990).

Because environments can be distinguished by their chemical, physical, and biological attributes, the underlying premise in maceral analysis is that each clast type, which has a specific genetic origin and chemical composition, is affected differently in different depositional environments. Moreover, each clast is affected by changing conditions during burial, again principally according to chemical changes. These chemical changes may be caused by physical (especially thermal), biological (especially biodegradational), or chemical (especially pore-water interaction) activity. In addition to the chemical changes that imprint the effects of the environment and burial

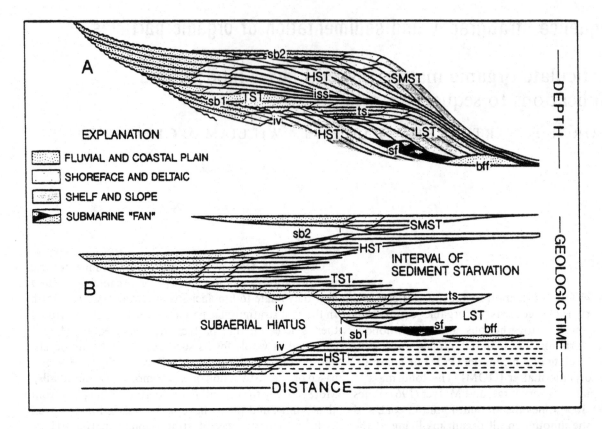

Figure 17.1. Schematic sequence stratigraphic diagrams as presented by Christie-Blick (1991); used by permission of Elsevier Science Publishers. Figure was modified from the original by Vail (1987). Vertical dimensions are depth (A) and time (B), with basinward direction to the right in both diagrams. Note internal arrangement of systems tracts (SMST = shelf margin, HST = highstand, TST = transgressive, LST = lowstand) and their bounding surfaces (sb = sequence boundary, ts = transgressive surface, iss = interval of sediment starvation) within the depositional sequence. Interval of sediment starvation contains the downlap surface, which forms the boundary between the transgressive and the highstand systems tracts. Other abbreviations are: iv = incised valley, sf = slope fan, bff = basin floor fan.

history on the individual macerals, each particle is affected by the dynamic forces within the depositional system (transportation and deposition of both allochthonous and autochthonous components, reworking, and recycling). The end result is that a specific environment shows a maceral spectrum that reflects the origin and history of the organic matter in the sediment. Interpreting this maceral spectum can provide insights into the original environment of deposition and is a critical part of modern facies analysis. The maceral facies can provide information on the age, depositional environment, thermal history and burial diagenesis of a sediment. Many workers have noted the coincidence between the occurrence of organic-rich intervals and deposition during transgression of marginal and epicontinental seas (Schlanger & Jenkyns, 1976; Hallam & Bradshaw, 1979; Demaison & Moore, 1980; Jenkyns, 1980; Leggett, 1980; Arthur & Silva, 1982; Rice & Gautier, 1983; Dow, 1984; Pratt, 1984; Arthur *et al.* 1985, 1987, 1990; de Boer, 1986; Flexer *et al.*, 1986; Jones, 1987; Summerhayes, 1987; Dean & Arthur, 1989; Ganz *et al.*, 1990; Lipson-Benitah *et al.*, 1990; Stefani & Burchell, 1990; Crumièr *et al.*, 1990; Wignall, 1991). Recent work has examined the connection between stratigraphic principles and the deposition of organic matter and petroleum source rocks (LeNoir & Hart, 1987, 1988; Hart *et al.*, 1989; Habib & Miller, 1989, Leckie *et al.*, 1990; Pasley & Hazel, 1990; Stefani & Burchell, 1990; Bohacs & Isaksen, 1991; Creaney *et al.*, 1991; Pasley,

1991; Pasley *et al.*, 1991; Wignall, 1991; Posamentier & Chamberlain, in press).

Sequence stratigraphy provides solutions to some of the problems associated with classical approaches to lithostratigraphy, especially those pertaining to time-rock relationships. The fundamental unit in sequence stratigraphy is the depositional sequence. An ideal depositional sequence consists of three depositional systems tracts that are, in ascending order: (1) lowstand (or shelf margin), (2) transgressive, (3) highstand (see more formal treatments by Haq *et al.*, 1987; Posamentier & Vail, 1988; Van Wagoner *et al.*, 1988; Vail & Wornardt, 1990). This is illustrated in Figure 17.1. A unique stacking pattern of depositional systems characterizes each of the systems tracts. These stacking patterns develop in response to changing volumes of accommodation space

(Jervey, 1988). These packages are separated by physical surfaces (unconformities and their correlative conformities, as well as flooding surfaces) that are perceived to have chronostratigraphic significance. The recognition of lithofacies as integral parts of larger depositional systems (Fisher & McGowen, 1967) was of fundamental importance. Stacking of depositional systems is initially progradational and becomes more aggradational in the lowstand systems tract. The transgressive systems tract is characterized by an overall retrogradational (or backstepping) stacking pattern and the subsequent highstand systems tract is characterized by a stacking of depositional systems that is initially aggradational but becomes dominantly progradational. Recognition of these stacking patterns on well logs, especially gamma ray logs, provides the foundation for regional sequence stratigraphic interpretations at scales below the limit of seismic resolution (Vail, 1990; Vail & Wornardt, 1990; Van Wagoner et al., 1990).

The relationship between stacking patterns of depositional systems (or depositional systems tracks) and organic matter type and preservation may be useful as a tool to enhance sequence stratigraphic interpretations. This approach was employed in the Paleocene-Eocene of Louisiana (Gregory, 1991; Gregory & Hart, MS), the Eocene-Oligocene of Alabama (Pasley & Hazel, 1990; Pasley, 1991), and the Cretaceous of the San Juan Basin in New Mexico (Pasley, 1991). Integration of data from biostratigraphy, sedimentology, and regional lithostratigraphy with information derived from study of maceral facies can provide greater resolution in the identification of critical surfaces that bound systems tracts within a depositional sequence.

The studies of organic matter deposition within a sequence stratigraphic framework indicate that the type and preservation of organic matter deposited in the marine realm is related to stacking of depositional systems, and consequently, the depositional systems tract in which it is deposited (Gregory & Hart, 1990; Pasley & Hazel, 1990; Bohacs & Isaksen, 1991; Creaney et al., 1991; Pasley, 1991). It is apparent that this relationship is primarily the result of changing amounts of terrigenous influx into the neritic environments. During progradation in the lowstand or highstand systems tracts, organic matter is typically of terrestrial origin, whereas mostly autochthonous (marine) organic matter is deposited on the shelf during an overall backstepping of depositional systems (transgressive systems tract). In general, maceral facies deposited in the lowstand and highstand systems tracts are characterized by abundant and well-preserved phytoclasts. Shelf sediments deposited in the transgressive systems tract contain abundant amorphous non-structured protistoclasts and sparse, highly degraded phytoclasts. This increased degradation of terrestrial organic matter in the transgressive systems tract is apparently related to

decreased rates of terrigenous sedimentation during transgression (Pasley et al., 1991).

This chapter reviews much of the work pertaining to maceral facies and stratigraphic analysis that has been completed by the principal author and his students during the past 20 years. Since the initial work on maceral facies (Hart, 1971) the method has proven useful for various scenarios (Hart, 1979a,b,c, 1982, 1984, 1986, 1991; Wrenn & Beckman, 1981, 1982; Pavlik, 1981; Smith, 1981; Wrenn, 1982; Beckman, 1985; LeNoir & Hart, 1987, 1988; LeNoir, 1987; Darby, 1987; Gregory, 1987; Pasley et al., 1991; Hart et al., 1989; Gregory & Hart, 1990; Pasley & Hazel, 1990; Darby & Hart, Chap. 10 this volume; Gregory, et al., 1991.) These studies have shown that, within the context of classical stratigraphy, maceral assemblages can be applied to problems involving the general application of Walther's Law, including sequence stratigraphic concepts.

The classification of organic matter

Hart (1979a) defined a maceral as an organic fragment of plant, protist, or other organism found dispersed in sediments. Particulate, solid, organic or mineraloid substances and amorphous precipitates are macerals, whereas mineralized shell materials are not. This definition of maceral is within the original scope of Stopes (1935).

Macerals found in dispersed sediments are classified according to the system proposed by Hart (1979a, 1982, 1986) and expanded by Darby (1987); Darby & Hart (Chap. 10 this volume), and Pasley et al. (1991). This classification is outlined in Table 17.1 The anatomy of a maceral spectrum is provided in Figure 17.2. The spectra are presented as a bar chart with four major components:

(1) Degradation level. This is shown in columns 1–3 of the bar chart. These columns allow a visual estimate of the level of degradation that compares very well to estimates made using the conventional C/N (carbon to nitrogen) ratios of limnologists (Hart 1979b, 1986). The units are percentages based on the mean value per sample (for a single sample they are simply the percentage value). The visual degradational level is derived from the continuous sequence of phytoclasts shown in Table 17.1. The 'good' phytoclasts are the sum of the well-preserved, poorly preserved, and infested phytoclasts. The 'infected' phytoclasts are the sum of the infested and amorphous structured phytoclasts. The 'degraded' phytoclasts are the sum of the amorphous structured and amorphous non-structured phytoclasts. The reason for overlap is that the base maceral categories actually come from a continuous sequence, and

Table 17.1. *Classification of macerals*

BIOLOGIC ORIGIN

		PHYTOCLAST	PROTISTOCLAST	SCLERATOCLAST	INDETERMINATE
PRESERVATIONAL STATE	WELL PRESERVED	Angular in outline with good structural framework and internal cell structure. Evidence for bacterial or fungal attack is absent or sparse.	Cell outline distinct and all characteristics of algal or protozoan cell are present. No fungal or bacterial attack. Cell contents may be preserved.	Cell outline and structural framework distinct. Characteristics of fungal remains (hyphae, spores) are obvious.	
	POORLY PRESERVED	Angular outline with good structural framework and internal cell structure. Initial stages of fungal attack on cell walls and bacterial scarring.	Cell outline distinct and characteristics of algal or protozoan cell are present. Bacterial attack evident but fungal degradation rare.	Cell outline and structural framework distinct but bacterial scarring and pitting present. Despite bacterial attack, fungal origin obvious.	
	INFESTED	Angular in outline with structural framework obvious. Cell walls highly disrupted by fungal attack and infestation by fungal hyphae present.	Cell outline distinct and characteristics of algal or protozoan cell are present. Bacterial attack advanced and fungal infestation present. *		
	AMORPHOUS STRUCTURED	Angular in outline and general structural framework present. Cell walls almost completely destroyed. Bacterial and fungal attack in advanced stages.	Cell outline and evidence of protist origin only faintly preserved. Bacterial pitting and scarring total. Evidence of fungal infestation may be present. *		Cell outline and structural framework absent. Contains small masses of fungal hyphae, fungal spores, and/or other maceral debris.
	AMORPHOUS NON-STRUCTURED	Blocky outline with no structural framework. No cell walls present. Fungal and bacterial attack not obvious because whole mass is amorphous.	Cell outline and structural framework absent. Maceral recognized by "fluffy" or "cloudlike" gross outline.		
		<u>Also:</u> Inertinite Well Preserved Sporinite Poorly Preserved Sporinite	<u>Also:</u> Dinocysts Microforams		

* Not originally described by Hart (1986) but added by Darby and Hart (this volume) after study of modern carbonate environments.

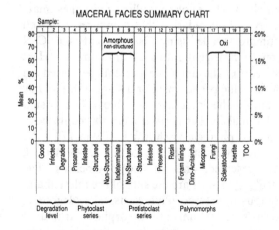

Figure 17.2. Maceral spectrum chart used to display maceral facies data. Here and in all the other figures using this model, TOC = Total Organic Carbon; Oxi = Oxygenic environments.

overlapping groups seem to provide more sensible results for continuous sequences when displayed as a histogram.

(2) Maceral components. This is shown in columns 4–19 of the bar chart (Fig. 17.2). The units are percentages based on the mean value per sample (for a single sample they are simply the percentage value). Columns 4–7 represent the continuous phytoclast sequence of Table 17.1. The preserved phytoclasts are a combination of the well-preserved and the poorly preserved phytoclasts. The layout of the maceral spectrum allows the total amount of structured to amorphous material to be seen which allows a rapid assessment of the terrestrial versus marine component. Columns 7–9 (Fig. 17.2) show the three groups of amorphous macerals represented in

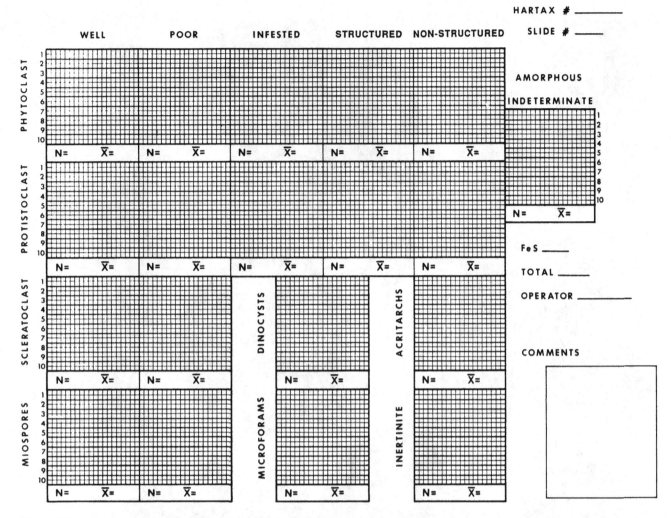

Figure 17.3.   Counting sheet for maceral facies analysis (from Hart, 1979a).

Table 17.1 by the phytoclast, protistoclast, and indeterminate columns. This allows a rapid assessment of the source rock potential of the sediment. Columns 9–12 (Fig. 17.2) are the protistoclast degradational series and columns 13–18 contain resin and the more commonly occurring organic microfossil groups.

(3) Inert component. This is shown in column 19 (Fig. 17.2). The overall maceral spectra are based on a maceral count of 100+ macerals excluding inertite. The inertite is thus an 'outside-the-sum' measure, i.e., the total number of inertite grains observed at the point all others totalled the selected maceral count.

(4) Total organic carbon. This is seen in column 20 of the bar chart. By convention this is measured using a Leco System. If a Rock-Eval pyrolysis method is used this fact is noted by an asterisk.

Figure 17.3 shows a maceral counting sheet (Hart,

1979a). A place is available for all of the maceral groups of Table 17.1. The dinocysts and acritarchs are separated during counting, and two degradational levels are available for miospores. A place is available for a Boolean value of presence/absence of iron sulfide. This chart allows the macerals to be counted according to their color using a color scheme of 1–10: three levels of yellow, three levels of orange, three levels of brown, and black.

Although some photographs illustrating the maceral types have been published (Hart, 1986) the original publication was not widely distributed (Hart, 1979a). Figures 17.4–17.10 are the original plates from that publication.

Development of an understanding of maceral facies

The approach to understanding the distribution of organic matter in sediments and especially the determination of biofacies (including maceral facies) has undergone continual evolution during the past thirty

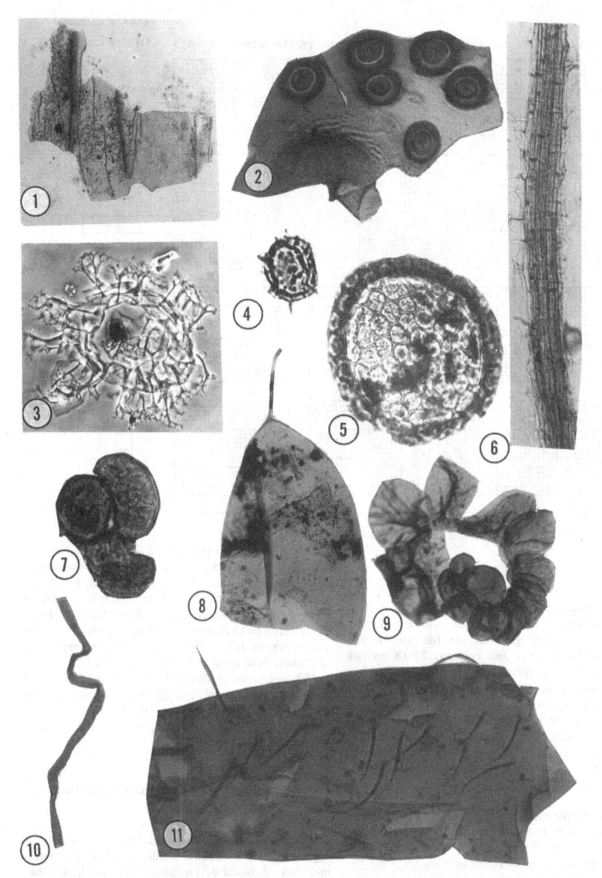

Figure 17.4.   Well-preserved macerals. (1) well-preserved phytoclast specimen shows initial stages of bacterial degradation; (2) well-preserved phytoclast; (3) well-preserved dinocyst; (4) well-preserved acritarch; (5) well-preserved miospore; (6) well-preserved phytoclast; (7) microforaminiferal lining; (8) well-preserved scleratoclast (fungal sporangia); (9) microforaminiferal lining; (10) well-preserved scleratoclast (fungal hypha); (11) well-preserved zooclast (insect part).

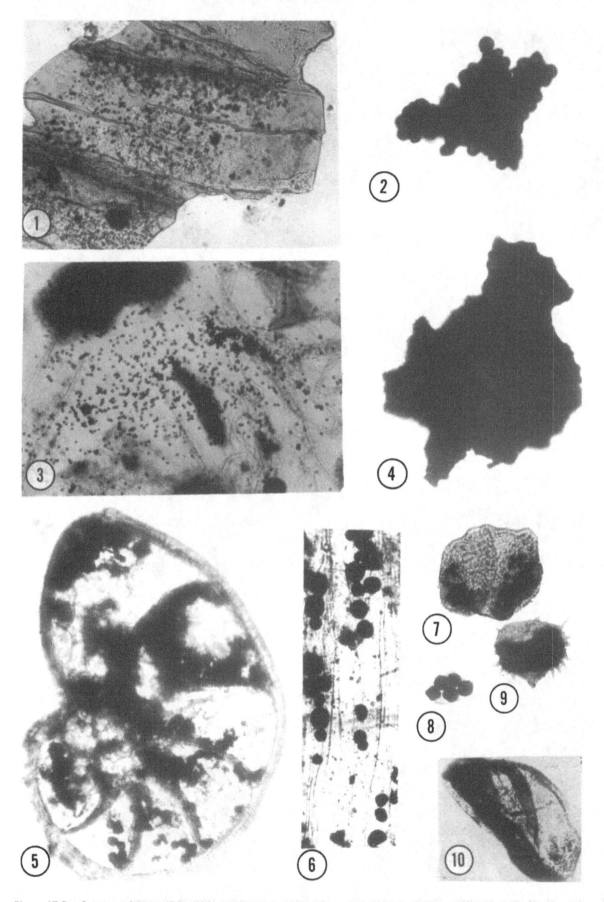

Figure 17.5. Presence of iron sulfide. (1) bacterially produced iron sulfide on surface of phytoclast; (2) iron sulfide nodule; (3) bacterially produced iron sulfide within phytoclast; (4) phytoclast totally coated and impregnated with iron sulfide; (5) bacterially produced iron sulfide within microforaminifera; (6) enlargment showing shape of iron sulfide granules; (7) iron sulfide within disaccate miospore; (8) iron sulfide within algal (?) cell; (9) iron sulfide within acritach; (10) iron sulfide totally impregnating surface of protist. This gives the impression of a highly carbonized maceral when observed under light microscopy.

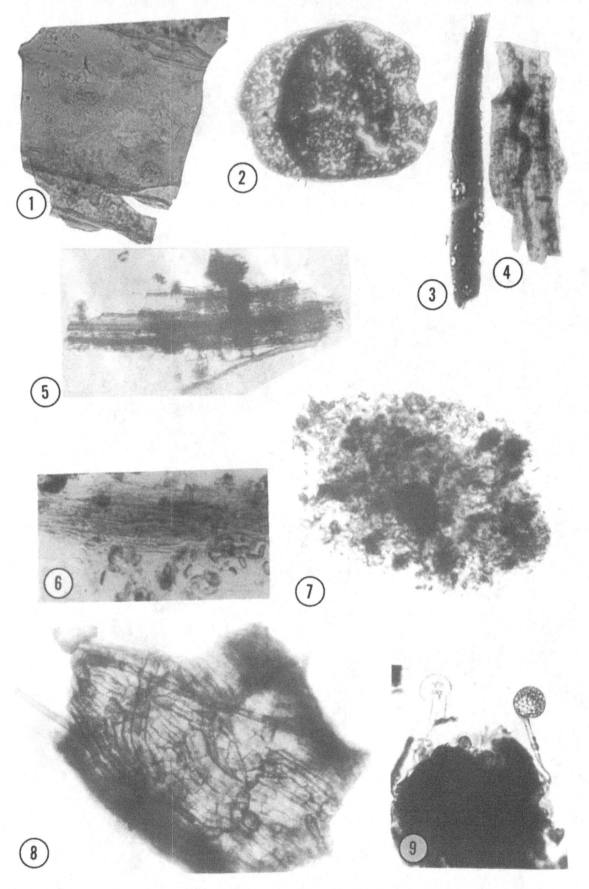

Figure 17.6. Bacterial and fungal infestation. (1) phytoclast showing scarring after attack by iron sulfide bacteria; (2) miospore showing typical bacterial degradational surface patterns; (3) resinous phytoclast showing surface bacterial attack; (4) phytoclast showing penetration by fungal hyphae; (5) phytoclast infested by fungi showing how fungal growth within the cells fractures the phytoclast; (6) phytoclast infested with fungi that have recently released fungal spores; (7) amorphous indeterminate material infested by fungi; (8) phytoclast (?) totally infested by fungi; (9) amorphous phytoclast infested by fungi.

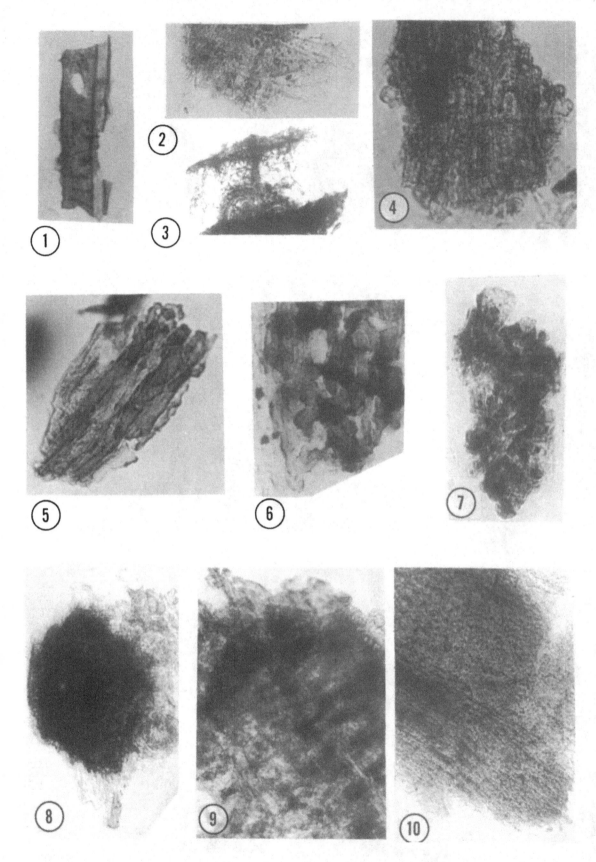

Figure 17.7. Degraded phytoclasts; (1) poorly preserved phytoclast; (2) infested phytoclast; (3) poorly preserved phytoclast; (4) infested phytoclast; (5) amorphous structured phytoclast; (6) amorphous structured phytoclast; (7) amorphous structured phytoclast – almost completely amorphorized; (8) amorphous non-structured mass (dark body) attached to amorphous structured phytoclast (light background to the right); (9) amorphous structured phytoclast showing bacterially derived (?) amorphous masses within structure; (10) amorphous structured phytoclast showing granulation produced by bacterial attack.

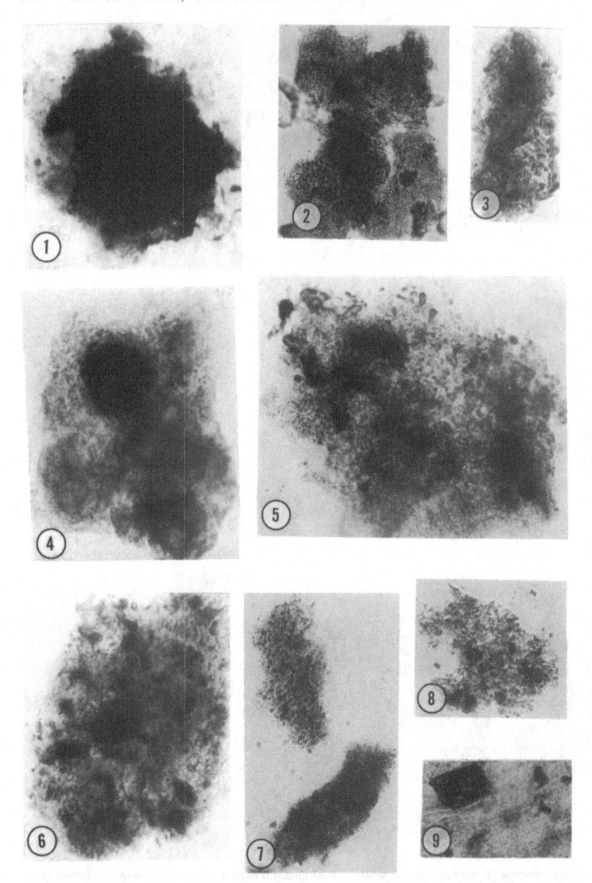

Figure 17.8. Amorphous macerals. (1) amorphous non-structured phytoclast; (2) amorphous non-structured phytoclast; (3) amorphous non-structured phytoclast; (4) amorphous non-structured phytoclast showing bacterially derived (?) masses; (5) amorphous non-structured phytoclast; (6) amorphous indeterminate – infested with fungal and bacterial material; (7) amorphous indeterminate (lower photograph may be a fecal pellet); (8) amorphous indeterminate; (9) amorphous indeterminate (enlargement showing incorporation of small phytoclasts).

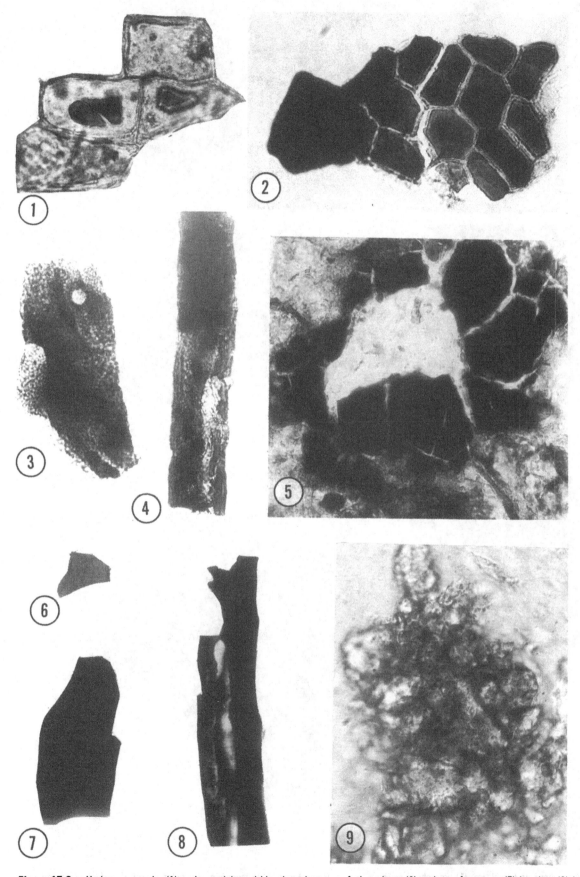

Figure 17.9. Various macerals. (1) resin particles within phytoclast; (2) resin particles within phytoclast; (3) resinous phytoclast; (4) resinous phytoclast; (5) organic precipitate within cell (stomata?) of phytoclast; (6) resinous fragment; (7) inertite; (8) inertite; (9) receptorclast fine-grained organic debris attached to clay particle (by precipitation?).

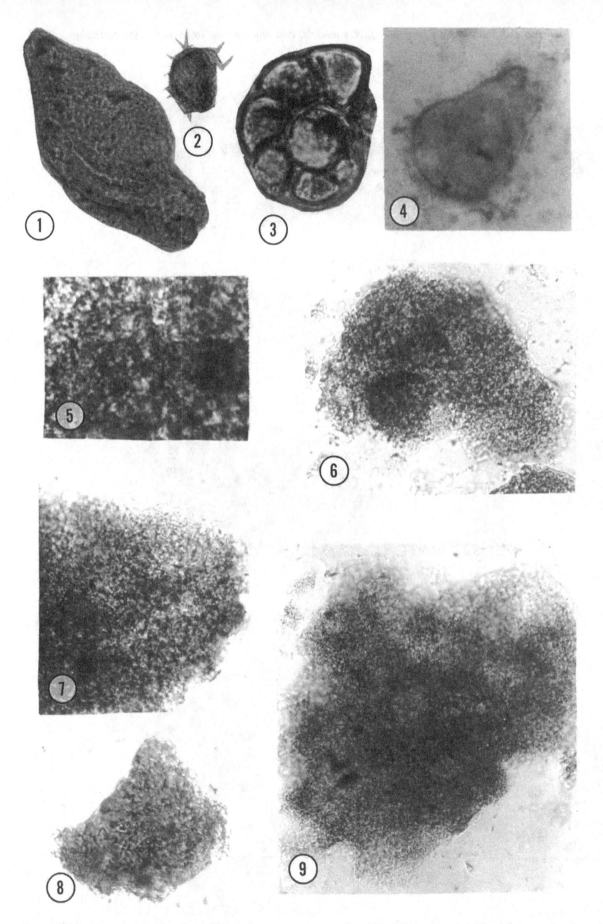

Figure 17.10. Amorphous protistoclasts. (1) amorphous mass showing generally smooth outline and bacterial pitting typical of an amorphous protistoclast; (2) amorphous material within an acritarch; (3) amorphous material within a microforam; (4) amorphous material within and extruding from an algal protist. (Illustrations 17.10.5–9 are Green River Shale macerals.) (5) enlargement of surface of amorphous protistoclast; (6) amorphous protistoclast (compare Fig. 17.8.4); (7) amorphous protistoclast; (8) amorphous protistoclast; (9) amorphous protistoclast.

years. Initially, a simplistic graphical and cartoon approach was used to characterize individual rock types and sequences. Such methods were often supplemented by simple statistical procedures such as similarity indices (Hart, 1964, 1965, 1969) and chi-squared tests (Hart, 1969). In the late sixties and early seventies most statistical methods were based on univariate techniques and especially ANOVA (Darrell, 1973; Darrell & Hart, 1970), covariance analysis (Christopher & Hart, 1971), and regression analysis (Schilling & Hart, 1973; Hart, 1969). Today, multivariate methods are prominent (Hart *et al.*, 1989; LeNoir & Hart, 1987, 1988), and these are finding graphical expression using the new tools of scientific visualization. Statistical procedures help in ascertaining whether or not certain differences observed among rock types are important, and allow a meaningful understanding of the role of varying transport and depositional factors on the dispersion and ultimate distribution of organic matter in sediments. However, perhaps the most important facet of the earlier work was the recognition that there are factors which will alter both the real and apparent relative abundance of particulate organic matter in sediments.

In the ideal situation a sediment sample reflects the exact assemblage which was present at the surface at the time of deposition. However, this never occurs, and in order to evaluate the relationship the sample maceral spectra has to the population maceral spectra, the sources of variation which affect organic matter sedimentation must be determined and understood. It is the very act of recognizing the sources of variation that leads to an understanding of why differences occur among biofacies.

This kind of analytical approach to facies analysis can be described as follows. If one assumes that the ideal situation exists, then a deterministic model can be established for the distribution of each kind of organic matter, as:

$$Y = f[X],$$

where Y is the relative abundance of a maceral in sample; X is the relative abundance of the maceral in the original population.

However a more realistic approach would be to use the linear additive model:

$$Y_i = u + e_i,$$

where Y_i is the relative abundance of the maceral observed in the ith sample taken from the original population; u is the true average abundance of the maceral in the environment at the time the sample was being deposited; e_i is a random error effect, NID $(0, \sigma_2)$ (NID = normally and independently distributed; cf. Christopher & Hart, 1971).

To this basic model additional sources of variation can be added to attain an even more realistic model to describe the relationship between the sample and the original population. This is done by partitioning the error term (e_i) to include other effects on the model. For example, day to day, month to month, season to season, or cyclical changes in the relative abundance of macerals can be included as definite terms in such a model. The linear additive model approach represents a way of thinking about a problem in a stochastic world rather than a statistical approach to data analysis. To reiterate: the important aspect of the linear additive model approach to stratigraphy in general is that it forces the observer to consider all of the sources of variation effecting a depositional system prior to sampling and this improves understanding of why differences occur.

The sources of variation affecting particulate organic matter are of two major types: (1) analytical factors, such as sample collecting procedures, laboratory contamination, and slide preparation procedures; (2) natural factors, such as the chemical, physical and biological forces acting within the depositional environment.

It is the latter group of factors that is of interest from the viewpoint of facies analysis. Facies interpretation of maceral spectra must take into account these sources of variation and attempt to determine the importance of each effect (e.g., isolate the specific 'signal' that can aid in determining the degree of degradation in the original environment from the 'noise' of all of the other effects).

Maceral facies models

The present view of the relationship between maceral spectra and modern depositional environments emphasizes the interplay between terrestrial and aqueous derived organic matter and highlights the importance of continental input into marine environments as an important source of understanding local palaeogeographic variation. The approach used in maceral facies is not unlike that used in more conventional facies models, because maceral facies models are based on observations from modern depositional environments and comparisons to ancient examples. It is anticipated that, with further refinement, maceral facies models will meet many of the criteria for facies models discussed by Walker (1984), Blatt *et al.* (1980), and Miall (1984).

Modern maceral facies have been studied from a variety of environments in the Gulf Coastal Plain, and some of these results are discussed below.

Deep basin maceral facies

Jendrzejewski (1976) and Jendrzejewski and Hart (1978) noted that the deep basin of the Gulf of Mexico contained very little particulate organic matter. This is partially because the Gulf is a low productivity basin, and partially

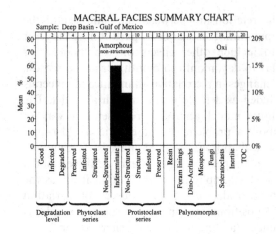

Figure 17.11. Maceral facies of the Deep Gulf Basin.

a result of the slow sedimentation rate and great depth. Much of the POM probably is removed by bacterial decay as it sinks through the water column, and as a result of the long residence period in the surficial zone where biodegradation is highest. The macerals are essentially amorphous indeterminate and amorphous protistoclasts, with varying amounts of protistoclastic material in general (Fig. 17.11). Most of this material is probably fecal in origin, as suggested by Hart (1986), and is essentially autochthonous, being derived from the phytoplankton and nekton of the overlying water column. There is a distinct allochthonous component consisting of recycled material from the major rivers that drain the hinterland and reworked material from the shoreline. Study of siliceous and acid resistant material shows that the western and eastern parts of the Gulf of Mexico are quite different in their thanatocoenosis. The western Gulf is essentially a closed system influenced largely by localized effects and containing many components from the Mississippi River. The eastern Gulf is an open system influenced by water masses of the Atlantic Ocean. Reworked miospores were noted exclusively in the western Gulf, and recycled material is of greatest abundance offshore of the Mississippi River system. Using coccoliths, Pierce (1975) and Pierce and Hart (1978, 1979) showed that the important areas of influx of allochthonous organic matter into the deep basin are the main river systems. In these areas the proportion of allochthonous continental component increases, being highest in the deltaic deposits and of varying but major importance in the neritic environments.

Neritic maceral facies

Studies of maceral facies from modern middle and outer neritic environments have not been published, although many have been studied. The work of Hart (Chap. 9 this volume) covers some of the inner neritic region.

Additional, unpublished work awaiting further compilation includes studies of the Mississippi River Plume between the mouth of South-West Pass and Atchafalaya Bay, and bottom samples extending from California to Alaska along the west coast. There is considerable variation in the neritic region. Some typical (?) maceral spectra are given in Hart (Chap. 9 this volume).

Prodelta maceral facies

The macerals of prodeltaic sediments were examined by Hart (1979c) and are discussed in Hart (Chap. 9 this volume). In his study of the palynomorphs of the modern Balize delta of the Mississippi River, Darrell (1973) found only one form statistically significant in the prodelta environment. This implies that conventional palynology provides no useful information for this particular facies.

Deltaic maceral facies

A minor report on POM from the Balize delta was published by Darrell and Hart (1970) on the absolute abundance of miospores in the environments of the Modern delta, and by Chmura and Liu (1990) on miospores within the waters of the Mississippi River. Darrell's (1973) dissertation on the modern delta of the Mississippi River is the main study on palynomorphs from that region. Smith (1981) added dimensionality to Darrell's work by completing an analysis of palynomorph distribution through the last deltaic lobe of the delta. All of these studies dealt with conventional palynology. Hart (1979c; Chap. 9 this volume) completed an extensive study of the distribution of POM in the Louisiana deltaic plain which included work on the active delta and illustrates characteristic maceral spectra.

Swamp and marsh maceral facies

Studies on the macerals of the swamps and marshes were completed by Hart (1979) and Beckman (1985). Piel (1965) presented a short study on selected marsh and bay environments, and Hart (1971) investigated POM in the Atchafalaya Swamp from both surficial sediments and cores.

Beckman (1985) concentrated on the marshes and studied a variety of organic types in an effort to understand the diagnostic characteristics of marsh depositional environments. Using Q-weighted pair grouped cluster analysis of a distance function, the important effect of salinity and water body type on maceral facies similarity was studied. This aspect of his work is summarized in Figure 17.12, which shows that seven clusters were formed from an analysis of samples through three wells. The samples from each cluster correlated with the depositional environment. The marsh

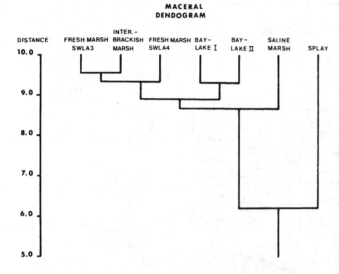

Figure 17.12. Cluster analysis of maceral data from the La Fourche Delta (from Beckman, 1985).

samples formed four clusters and the non-marsh samples three clusters. The saline marsh is unique, with a high abundance of poorly preserved phytoclasts, infested phytoclasts and amorphous indeterminate phytoclasts. The fresh and intermediate-brackish marshes are characterized by high abundance of amorphous non-structured macerals, and are notably devoid of scleratoclasts. The splay sediments are very distinctive in their high abundance of amorphous structured phytoclasts and inertite, with lower abundance of scleratoclasts and amorphous protistoclasts. The two lacustrine clusters are closely related and have a relatively high abundance of scleratoclasts and amorphous protistoclasts. Characteristic swamp and marsh maceral spectra are given by Hart (Chap. 9 this volume).

Alluvial valley maceral facies

Maceral spectra from the alluvial valley are recorded by Hart (1979c; Chap. 9 this volume).

Facies maps

An important step in building three-dimensional models of the subsurface is the development of an understanding of the areal variation of maceral types throughout an adjacent set of environments. Figure 17.13 shows the distribution of maceral facies as a theoretical map of a deltaic plain based on the work of Hart (1979c). Visualization tools are currently being applied to these maps to create a three-dimensional artificial reality model of the distribution of organic matter in deltaic systems. Their interpretation within a framework that includes an understanding of the active depositional processes within

each environment provides a strong insight into how maceral facies develop.

Ancient examples of maceral facies

Holocene

In an attempt to understand the significance of Darrell's work on the Balize delta, Smith (1981) studied the palynomorphs through the last deltaic lobe (Boothville borehole). The subsurface environments were defined by sedimentological and x-ray radiography analysis by Coleman and Roberts (unpub.). A very significant conclusion of Smith's work was that the abundance of recycled palynomorphs was often equal to or greater than the abundance of extant taxa. Smith did not study all palynomorphs present in the samples but selected a subset of taxa that were known to range into the Holocene-Modern epochs. An analysis of covariance procedure was performed to investigate the possibility that lithology was influencing the distribution of the taxa. The model partitioned the effect of lithology and depth, and for those taxa that showed a significant distribution with depth independent of lithology, regression curves were plotted to investigate the degree of depth response (linear, quadratic, cubic), using the ideas of Christopher and Hart (1971). In addition, a one-way analysis of variance (completely randomized design) was used to test for the significance in distribution of all palynomorphs among subsurface environments.

Miocene

LeNoir (1987), LeNoir and Hart (1987, 1988), and Hart *et al.* (1989) investigated the maceral spectra from two neritic (Robulus #2 and #5) sands deposited during the Miocene Epoch, which provide insight into the nature of organic matter deposition in the outer neritic environment. These publications discuss the variation occurring in a sandstone sequence and show how maceral analysis can be used to understand the distribution of permeability and porosity in an interval dominated by sandy lithofacies. Study of the macerals and the associated dinocysts showed that areas of high permeability and porosity (and therefore areas of hydrocarbon occurrence) corresponding to areas where high-energy influxes of sediment were indicated in the sequence. These high-energy pulses and their associated facies probably derive from the localized changes in the ancient Mississippi River system corresponding to the influx of terrestrially derived material.

The organic matter is ranked as thermally immature to marginally mature, based on vitrinite reflectance, T_{max}, and TAI (thermal alteration index) values. Incidently, because much of the maceral material is oxidized and

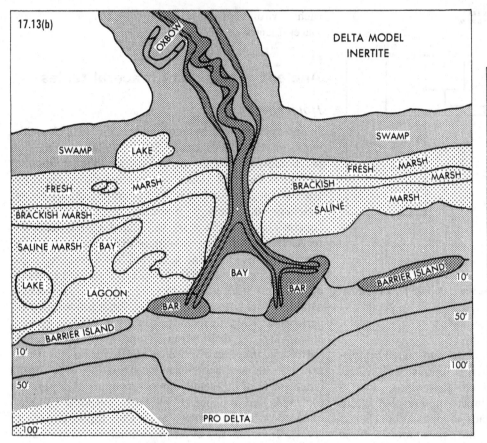

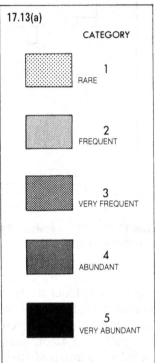

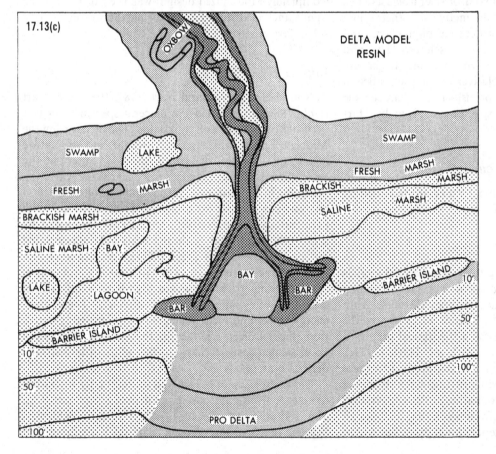

Figure 17.13. Model for spatial distribution of macerals. (a) key to ordinal scale of values; (b) inertite; (c) resin; (d) miospores; (e) well-preserved phytoclasts; (f) infested phytoclasts; (g) amorphous structured phytoclasts; (h) amorphous non-structured phytoclasts; (i) scleratoclasts; (j) amorphous indeterminate; (k) amorphous protistoclasts; (l) foraminiferal linings.

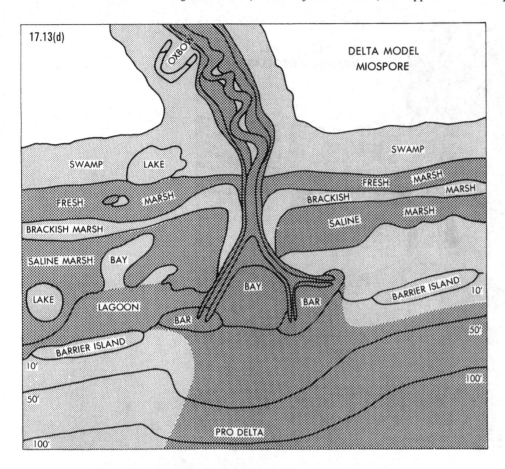

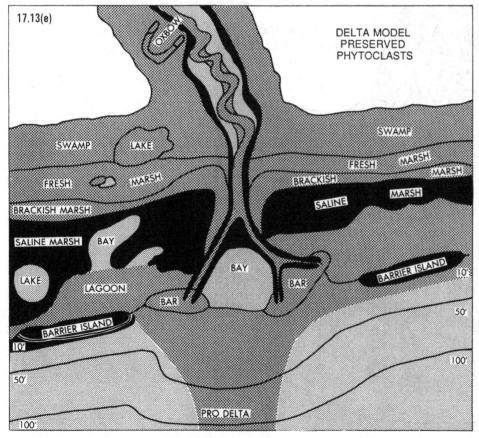

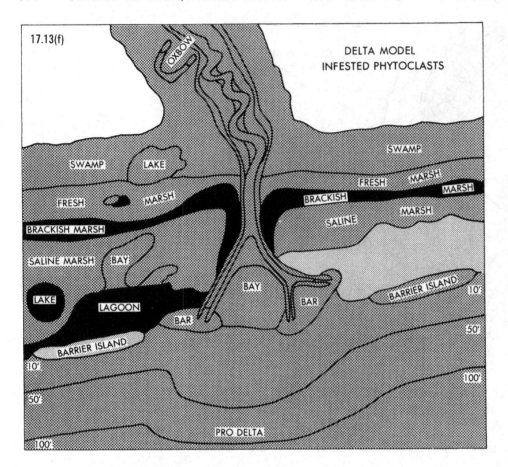

17.13(f)

DELTA MODEL
INFESTED PHYTOCLASTS

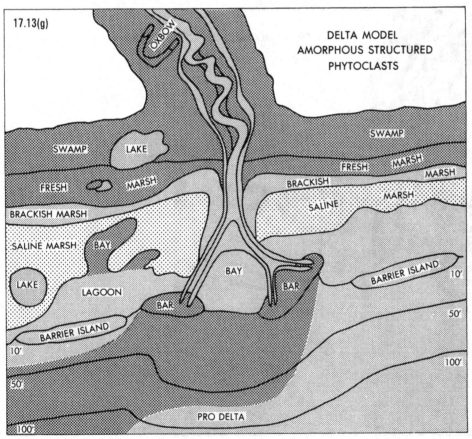

17.13(g)

DELTA MODEL
AMORPHOUS STRUCTURED
PHYTOCLASTS

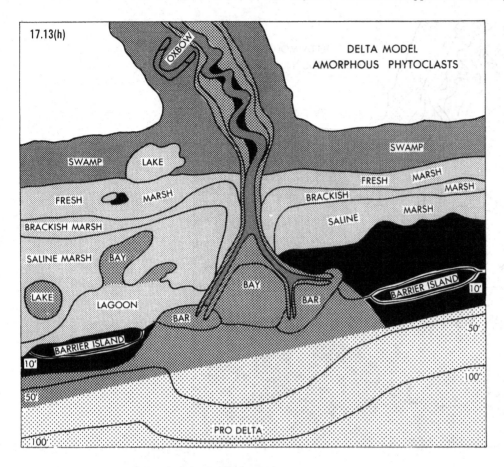

17.13(h)

DELTA MODEL
AMORPHOUS PHYTOCLASTS

OXBOW

SWAMP

SWAMP   LAKE

FRESH   MARSH

FRESH   MARSH
BRACKISH   MARSH
SALINE   MARSH

BRACKISH MARSH

SALINE MARSH   BAY

LAKE   BAY

LAKE   LAGOON   BAY   BAR

BAR   BARRIER ISLAND   10'

BARRIER ISLAND   50'

10'

50'   100'

100'   PRO DELTA

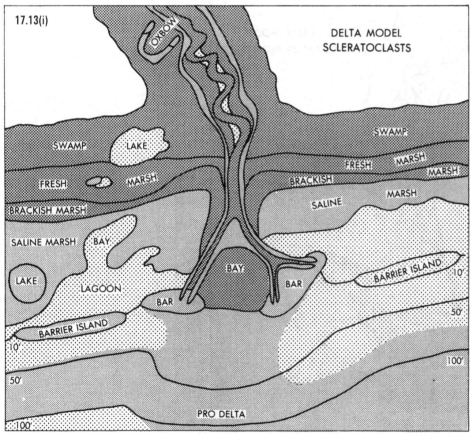

17.13(i)

DELTA MODEL
SCLERATOCLASTS

OXBOW

SWAMP

SWAMP   LAKE

FRESH   MARSH

FRESH   MARSH
BRACKISH   MARSH
SALINE   MARSH

BRACKISH MARSH

SALINE MARSH   BAY

LAKE   BAY

LAKE   LAGOON   BAY   BAR

BAR   BARRIER ISLAND   10'

BARRIER ISLAND   50'

10'

50'   100'

100'   PRO DELTA

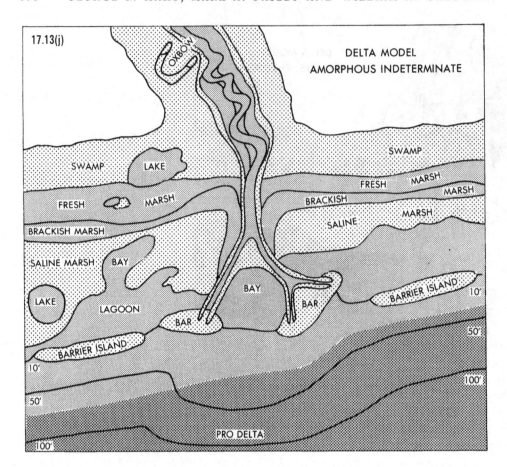

17.13(j)

DELTA MODEL
AMORPHOUS INDETERMINATE

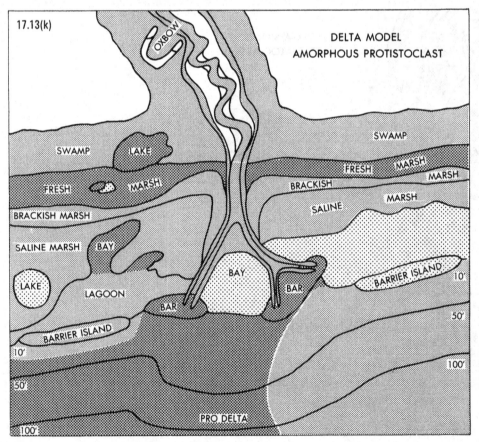

17.13(k)

DELTA MODEL
AMORPHOUS PROTISTOCLAST

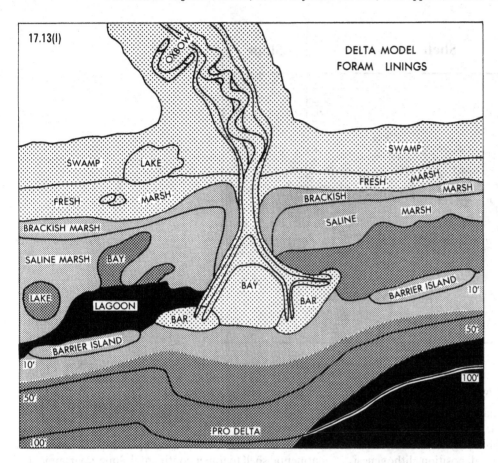

reworked, and the maceral spectra are dominated by terrestrial phytoclasts, the hydrocarbon generative potential of the sediments is low. Low hydrogen indices from programmed pyrolysis confirm that this maceral facies has no meaningful source potential (Pasley *et al.*, 1991).

Paleocene-Eocene

Gregory (1987, 1991), Gregory and Hart (1990, 1992) and Gregory *et al.* (1991) examined macerals from the Wilcox Group. Variations seen in the vertical distribution of the maceral spectra were largely controlled by the influence of a fluvial plume associated with the eastern lobe of the Holly Springs delta complex and correlative Upper Wilcox delta complexes, and again illustrate the significance of the terrestrial component. The important macerals were scleratoclasts, protists, miospores, and amorphous indeterminate. The least important were microforaminiferal linings and amorphous non-structured protistoclasts. Figure 17.14 shows the interpretation of maceral distribution in a fluvial plume-deltaic setting. Comparison with previous studies suggests that Gregory and Hart's (1990) maceral facies 1, 4, 5, & 7 are comparable with a deltaic fluvial plume in which terrestrial derived macerals

(phytoclasts + scleratoclasts + miospores) dominate and protistoclasts are less than 3%. Similarly, maceral facies 2 and 6 represent a greater marine influence, as shown by the relatively higher percentage of protistoclasts (5–11%). Maceral facies 3 and 8 are similar to the offshore mixing zone (water depths of 50–100 feet) of Hart (1979c; Chap. 9 this volume).

Gregory *et al.* (1991) studied Wilcox (Late Paleocene–Early Eocene) and Sparta (Middle Eocene) middle neritic samples using fluorescent light microscopy on the macerals, and programmed pyrolysis of the associated whole rock samples. Transmitted light maceral analysis indicated amorphous macerals of mixed origin dominate the samples investigated. There was a poor correlation between the quantity of amorphous macerals and the hydrogen index. Fluorescent microscopy of the macerals indicated that much of the amorphous matter fluoresced to some degree. A significant change occurs as the oil window is reached. The miospore and the dinoflagellate-acritarch macerals indicate that the moderately fluorescent macerals increase and highly fluorescent macerals decrease in quantity as thermal maturity is achieved. These changes are associated with changes in chemical composition of sporopollenin as it is heated above 435 °C.

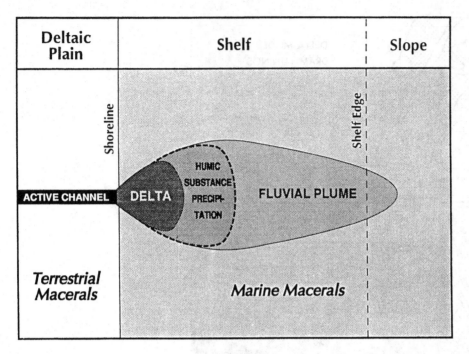

Figure 17.14. Plume interpretation from Wilcox Delta (from Gregory & Hart, 1991; used by permission of American Association of Stratigraphic Palynologists Foundation).

Reworked and recycled maceral material

Reworking is the moving of material along a single time plane involving a series of depositional and erosional events, while recycling is the moving of material across time planes which involve the deposition, lithogenesis and consolidation of a sediment prior to erosion and involvement of material in a second cycle. Within the deltaic framework, recycled material as seen in modern samples is found principally in the channel, levee, distributary mouth bar, prodelta and neritic sediments. The river plume deposits contain abundant recycled palynomorphs and are dominated by terrestrial organic matter.

Gregory and Hart (1992) used both the recycled palynomorphs and the percentage of terrestrial palynomorphs to develop an initial predictive model for recognizing sea level changes within a sequence. They studied the three-component system beginning with a fall in relative sea level: (1) the basal lowstand (LST) or shelf margin systems tract (SMST), (2) the transgressive systems tract (TST), and (3) the highstand systems tract. The units are genetically related in that the sediments were deposited during a single cycle of sea level fall and rise when viewed within a sequence stratigraphic framework as given in Figure 17.1.

Palynological studies interpreted within a stratigraphic context involve the construction of sea level curves based on the ratios of marine to terrestrial palynomorphs (Van Pelt & Habib, 1988), or are concerned with quantitative palynofacies analyses interpreted in a depositional sequence context (Ford & Goodman, 1988; Goodman, 1990). The initial model developed by Gregory & Hart

(MS) concentrated on marine deposition in the outer continental shelf to upper continental slope environments and is especially applicable to areas with extreme amounts of clastic sediment input in a deltaic setting such as the Wilcox Group of Louisiana. Figure 17.15 shows the theoretical distribution developed for the model.

During a relative sea level fall associated with a low stand systems tract, the position of the shoreline moves basinward, and the fluvial systems are forced to adjust their equilibrium profile downward (Haq *et al.*, 1987; Vail, 1987; Van Wagoner *et al.*, 1987, 1988). Two types of deposits are possible, depending on the amount of the relative sea level fall. A large drop in relative sea level results in the formation of a Type 1 sequence boundary (see Vail, 1987, 1990), with the associated incisement of fluvial valleys in sediments of the continental shelf. A location seaward of the shoreline position at the maximum fall in sea level and in close proximity to a fluvial source should experience marine sedimentation including turbidites. The typical sequence of lowstand deposits begins with the initial deposition of clastic sediments on the basin floor in the form of a basin floor fan. Following deposition of the basin floor fan, clastic sediments are laid down on top of the fan with proximal portions deposited on the continental slope. These sediments are the lowstand prograding wedge or complex.

Assuming that sea level does not fall below the shelf break, the expected organic matter on the outer continental shelf and upper slope should contain a marine

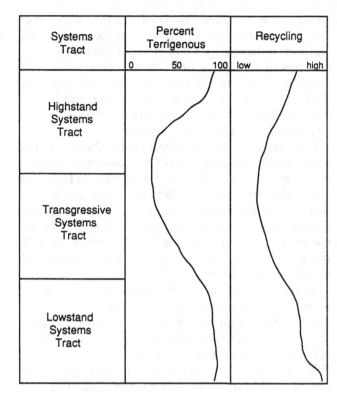

Systems Tract	Percent Terrigenous			Recycling	
	0	50	100	low	high
Highstand Systems Tract					
Transgressive Systems Tract					
Lowstand Systems Tract					

Figure 17.15.   Initial model for transgressive–regressive cycle using recycled palynomorphs.

component, but dominated by terrigenous forms. The marine organic matter should display low diversity due to the large amount of fine-grained suspended material in the water column resulting in stressed environmental conditions and generally is poorly preserved due to mechanical and biological degradation. The terrigenous component of the organic matter will contain a spectrum of environmental conditions ranging from shoreline (swamps and lowlands near the delta) to hinterland (uplands, possibly mountainous regions near the headwaters). The influence of climate on the terrigenous organic matter depends on the width of the climatic zones. A more equable climate causes the broadening of floral provinces and results in lower species diversity due to the palynoflora being derived from a single or few floral provinces. Conversely, a colder climate results in a contraction of floral provinces and higher species diversity due to the derivation of the palynoflora from several floral provinces. Because of the complicating factors of the number of environments contributing palynomorphs into the terrigenous palynoflora and the influence of climatic changes, it is usually not practical to use the makeup of the terrigenous palynoflora to perform direct, detailed paleoenvironmental interpretations.

Degradational conditions affecting the palynoflora during deposition of the LST can vary substantially. High sedimentation rates with rapid burial of the palynomorphs

can produce remarkably well-preserved floras even under aerobic conditions by removing the palynomorphs from the surficial zone of bioturbation and degradation (Hart, 1982, 1986). Low sedimentation rates and slow burial often accompanied by bioturbation can result in a severely degraded palynoflora with some of the more delicate forms becoming unrecognizable.

The percentage of recycled forms is expected to be high in the sediments deposited in the LST. The obvious source of the recycled forms would be the sediments formerly in the incised valleys, and undoubtedly some of the recycled forms come from these sediments. The age of the recycled forms derived from these sediments depends on the severity of incisement. A small drop in base level or a larger drop over a short time with associated low levels of incisement would produce a recycled flora that probably could not be distinguished from the 'normal' flora. A large drop with deep incisement may or may not produce recognizable recycled forms. In areas which previously had high sedimentation rates, even extreme incisement may not remove enough sediments to downcut into significantly older sediments. Conversely, if previous sedimentation rates were not as high, formations may be removed and a clearly recognizable recycled flora produced. The second source of recycled palynomorphs that could produce a greater diversity of recycled forms is from sediments in the middle and upper reaches of the fluvial system. Increased erosion from the lowering of base level and fluvial rejuvenation in the hinterland should result in increased erosion of older sediments throughout the fluvial system. In order for these forms to be included in sediments deposited during the sea level lowstand they must be transported directly into the depositional basin (Gregory, 1991).

In areas of the outer continental shelf and upper continental slope laterally removed from the incised valleys and their associated deposits, a more 'normal' marine palynoflora is expected. However, in an area with high terrigenous input considerable amounts of silt and clay sized material in suspension due to either longshore drift or direct input from an adjacent landmass are moved around by currents, resulting in a substantial terrigenous component to the palynoflora.

If the drop in relative sea level was not large, i.e., was not close to or below the shelf break, a Type 2 sequence boundary is formed and a shelf margin systems tract (SMST) will be the expected deposit (Van Wagoner *et al.*, 1987). Similar results would occur if sedimentation and subsidence rates were high enough to keep pace with the sea level fall. In this type of deposit, significant incisement does not occur, and most of the sediment is deposited on the continental shelf in the form of aggradational to progradational shelf edge deltas.

The expected response of the organic matter to this type of situation would be the same as for the lowstand prograding wedge, except that the deposits are shifted to the continental shelf. One would expect fewer marine components, except in localized areas that are not influenced by the progradational wedge. The response of the recycled palynomorphs would be similar, with one exception. There would not be any forms resulting from the formation of incised valleys. The source of recycled palynomorphs would be from erosion in the hinterland. In this case, the recycled palynomorphs can provide an indication of the source of the fluvial sediments (provenance).

Once the rate of sea level rise is greater than the subsidence rate, the relative sea level begins to rise. This causes a landward shift in the shoreline, which marks the end of the lowstand systems tract and beginning of the transgressive systems tract. Associated with the landward shift in the shoreline is the creation of a ravinement or transgressive surface. This surface results from the erosion of the shoreface and beach deposits as the ideal beach-shoreface profile is shifted landward and upwards in response to the rise in sea level (Kraft et al., 1987; Nummedal et al., 1988, 1989a,b; Nummedal, 1990). The sediments are transported onto the shelf and eventually deposited in the form of shelf sand ridges or sheets in accommodation space made available by the upward shift in the beach-shoreface profile.

In areas influenced by fluvial deposition, specifically the incised valleys cut during the lowstand, the transport of sediment through the valleys and out onto the upper continental slope decreases. Sediment deposition begins in the incised valleys in order to fill the available accommodation space. The transgressive systems tract is composed of a series of backstepping parasequences capped by a surface of maximum starvation or downlap surface. This surface represents the lowest sedimentation rates in the sequence and results from the trapping of most terrigenous material in the incised valleys and other coastal environments as sea level continues to rise. During this interval deposition on the outer continental shelf and upper continental slope consists mostly of pelagic sedimentation (Swift, 1976; Jervey, 1988; Posamentier & Vail, 1988; Roberts & Coleman, 1988; Baum & Vail, 1988). In the outer shelf and upper slope environments the return to pelagic sedimentation signals the beginning of the formation of a condensed section (Loutit et al., 1988).

In response to marine transgression, a definite change in the organic matter is found in offshore environments. There is a dominance of marine organic matter in the transgressive systems tract (TST). Much of the terrigenous component of the palynoflora is expected to be trapped in the estuarine and nearshore environments along with the majority of the sediments. In the offshore environments, the marine component of the palynoflora is dominant both in frequency and diversity. Terrigenous organic matter present in offshore sediments of the TST could be derived from erosion associated with the formation of the transgressive or ravinement surface, or could represent suspended load that escaped from the estuarine and nearshore environments during periods of high fluvial discharge. In sediments directly associated with the formation of the transgressive or ravinement surface (shelf sand ridges), the organic matter may reflect the terrigenous nature of these sediments.

Estimates of 5–10 m of sediment redistributed into the lower shoreface and back barrier environments have been postulated by Kraft et al. (1987) for the modern Delaware coast. This reworking of sediments should not have a significant effect on the recycling of recognizably older palynomorphs, and the amount of recycled organic matter is expected to decrease in sediments of the TST due to the trapping of sediments in the estuarine and nearshore environments.

As the rate of sea level rise decelerates, sedimentation is once again able to overcome sea level rise, and the fluvial systems begin to prograde onto the shelf. The early part of the highstand systems tract is dominantly aggradational with minor progradation, while the late HST is dominantly progradational with minor aggradation (Posamentier et al., 1988; Posamentier & Vail, 1988). This results in an aggradational to progradational parasequence stacking pattern. In the outer continental shelf and upper continental slope environments the early part of the HST will be marked by continued condensed section deposition until the fluvial systems can prograde out to these more distal environments. The HST is topped by either a Type 1 or Type 2 sequence boundary (cf. Vail, 1987, 1990) associated with the next sea level fall.

The response of the organic matter at the onset of progradation will vary depending on proximity to the deltas. Locations more distal to the delta will experience continued marine dominance in the palynoflora, while those in a more proximal position will see a change from marine-dominated to a more mixed maceral spectra. This change will continue, with eventual dominance of terrigenous organic matter as progradation continues.

The response of the recycled element of the palynoflora is difficult to predict. The frequency and diversity of recycled forms should increase with the influx of clastic sediments into the offshore depositional environments. The ability to recognize the recycled flora in HST sediments depends to a large degree on the source of the clastic sediments and the richness and diversity of the contained organic matter. A comparison of the percent terrigenous palynomorphs, marine diversity, total number recycled, recycled diversity and systems tract interpretation was given by Gregory and Hart (1991), who indicated that, although there are recycled palynomorphs in most samples, those samples with high levels of recycled palynomorphs are associated with dominant terrigenous macerals.

Sequence stratigraphic applications

Gregory (1989, 1991) interpreted the sequence within a sequence stratigraphic framework of the Ragley Lumber #D1 well from the Paleocene-Eocene of Louisiana. In southwestern Louisiana during the Paleocene, a large amount of terrigenous material was introduced into the Gulf of Mexico Basin. The paleogeographic and paleoenvironmental setting suggests the deposits should be dominated by LST and HST sediments with TST sediments, consisting of relatively thin units representing the toes of backstepping parasequences. In some samples dominated by terrigenous material, as much as 98% of the palynoflora was terrigenous in origin, while in marine-dominated samples 40–46% of the palynoflora was terrigenous in origin.

A direct comparison of the sequence stratigraphic interpretations for the Ragley well and the percent terrigenous palynomorphs and marine diversity indicates that a direct relationship between systems tract and organic matter is unclear. However, when electric log patterns are interpreted to represent parasequence sets and parasequences using the concepts of Van Wagoner *et al.* (1987, 1988) and are compared to the organic matter attributes, a definite relationship is evident. In general, samples from progradational electric log stacking patterns are terrigenous-dominated and have low to high marine diversity, while samples from retrogradational electric log stacking patterns exhibit marine dominance to strong marine influence and have high marine diversity. In general, the retrogradational electric log stacking patterns are interpreted to correspond to fining upwards, lithologic units deposited during a relative sea level rise. Samples from above the top of the Wilcox Group within the Ragley Lumber Well represent a relatively condensed interval corresponding to a regional transgression prior to the onset of deltaic progradation of the Sparta Formation. The samples from this interval exhibit marine dominance to strong marine influence, have high marine diversity and moderate amounts of recycled palynomorphs.

Most recently Pasley (1991) has undertaken a more definitive study of the organic matter and sequence stratigraphic concepts. This work centered on two regions: (1) the San Juan Basin, and (2) St. Stephens Quarry section of Alabama.

THE SAN JUAN BASIN

The San Juan Basin of northwestern New Mexico and southwestern Colorado (Figure 17.16) was the site of thick (> 1500 m) accumulation of predominantly clastic sediments in the Western Interior Seaway during the Late Cretaceous Period. Sediments were deposited in multiple transgressive and regressive cycles, which resulted in the intertonguing of nonmarine and marine rocks (Pike, 1947;

Molenaar, 1983; Kauffman, 1977, 1985, 1988). Sequence stratigraphic principles have been recently applied to these deposits with particular emphasis on the Tocito Sandstone–Mancos Shale interval (Nummedal *et al.*, 1988, 1989a,b; Nummedal, 1990; Riley & Nummedal, 1989, 1991; Valasek, 1991). Samples were studied from the Bisti oil field, and two cored wells, Newsom A3E (A20) and Angel Peak B#37, which are located on the cross-section given in Figure 17.17.

The Bistri oil field section is summarized in Figure 17.18. The Upper Cretaceous, lower Mancos Shale of the San Juan Basin represents deposition during a sea level lowstand. The overlying Tocito Sandstone, is a transgressive deposit in the form of shelf sand ridges, and the upper Mancos Shale represents transgressive deposits in an offshore environment. The sequence stratigraphic framework developed for the Upper Cretaceous strata in the San Juan Basin of New Mexico is summarized for the Angel Peak and Newsom wells in Figures 17.19 and 17.20, respectively. Although five depositional sequences are depicted, only the middle three are discussed here. For the purposes of discussion, these are labeled Sequence A (Bridge Creek Limestone Member–Lower Mancos Shale), Sequence B (Lower Mancos Shale–Juana Lopez Member), and Sequence C (Tocito Sandstone–Upper Mancos Shale).

Maceral data are plotted against stratigraphic position for the Bisti, Angel Peak and Newsom cores in Figures 17.21, 17.22, and 17.23, respectively. Several important differences are evident upon comparison of the organic matter from the regressive lower Tocito interval to that found in the transgressive upper Tocito and overlying Mancos Shale. Petrographically, most of the organic matter in the regressive lower Tocito interval at Bisti oil field is of terrestrial origin (Figure 17.24). Conversely, the most common macerals found in the transgressive upper Mancos at Bisti are amorphous non-structured protistoclasts. A similar relationship between stratigraphic position and maceral facies exists in the Tocito–Mancos interval in the Newsom and Angel Peak cores (Figs. 17.25, 17.26). A distinct decrease in the amount of terrestrial organic matter is observed as one moves up the section across the transgressive surface and into deposits associated with the transgression.

The Mancos Shale below the Tocito Sandstone is dominated by terrigenous palynomorphs (mean = 66.4%), as are the samples from the mudstone interbeds within the Tocito Sandstone (mean = 67.2%). However, the composition of the palynofloras is not the same. The samples from the Mancos Shale below the Tocito Sandstone contain a more diverse and well-preserved terrigenous palynoflora than do the samples from the mudstone interbeds within the Tocito Sandstone. The samples from the mudstone interbeds within the Tocito Sandstone have higher percentages of fungal spores and

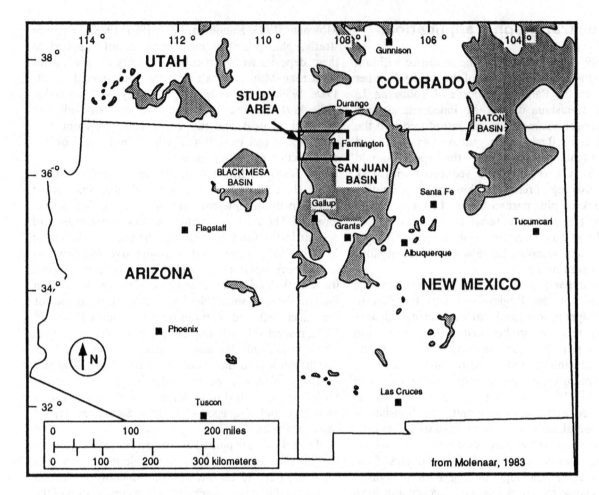

Figure 17.16. Location map for the San Juan Basin in northwestern NM and southwestern CO, USA. Shaded areas denote outcrops of Upper Cretaceous rocks. Rectangle corresponds to study area. (adapted from Molenaar, 1983).

small angiosperm pollen grains. The larger angiosperm pollen grains are poorly preserved, as is the marine component of the palynoflora from the mudstone interbeds within the Tocito Sandstone. The Mancos Shale above the Tocito Sandstone is clearly dominated by marine palynomorphs, with a mean percent terrigenous palynomorphs equal to 38.8%. In addition, the samples from above the Tocito Sandstone contain abundant microforaminiferal linings. Because of generally poor preservation, recycled palynomorphs could not be clearly identified in the samples from the Bisti oil field. Not only the amount of terrestrial organic matter, but also the preservational state of the phytoclasts found in the Tocito–Mancos interval, varies with stratigraphic position (Figs. 17.25, 17.26). The phytoclasts in the regressive portion are better preserved (higher degradational index values = D-values) than those found within and above the transgressive upper Tocito. In the Bisti oil field samples, most of the amorphous macerals (both phytoclast and protistoclast) in the regressive lower Tocito are weakly or non-fluorescent. In contrast, amorphous macerals in the transgressive Mancos Shale above the Tocito Sandstone commonly exhibit strong fluorescence. Samples from the mudstone interbeds within the upper Tocito

Sandstone possess intermediate fluorescence levels. This comparison is displayed graphically in Figure 17.27.

The available information derived from the Bisti oil field provides additional insight into sequence stratigraphic characteristics of palynofacies. In an area like the Cretaceous western interior seaway, where the amount of terrigenous input was low in comparison with Gulf of Mexico standards, the expected marine to terrigenous palynomorph ratios are expected to be skewed towards the marine end. This expected result is, in fact, observed. The maximum terrigenous dominated percentage is 73%, in comparison with 98% for the Ragley Lumber #D1 samples and 95% for the St. Stephens Quarry samples. The results for the marine to terrigenous palynomorph ratios indicate that the distribution of marine- and terrigenous-dominated intervals is not uniform. There is a definite correspondence between marine-dominant intervals and the Mancos Shale samples above the Tocito Sandstone and the terrigenous-dominated intervals and the Mancos Shale below the

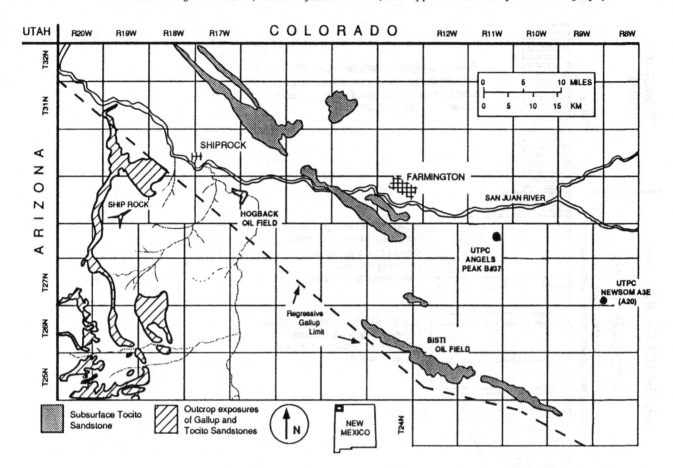

Figure 17.17. Location map for samples from the San Juan Basin. Outcrop samples from Hogback oil field and cores from Bisti oil field are located close to the most basinward extent of the regressive Gallup Sandstone. Cores from UTPC Angel Peak B #37 and Newsom A3E (A20) are located in a more basinward position.

Tocito Sandstone. This relationship of marine dominance associated with the transgressive systems tract is predicted by the model, as is the terrigenous dominance associated with the lowstand systems tract in a shoreline proximal environmental setting. The association of the mudstone interbeds within the Tocito Sandstone with terrigenous dominance can be interpreted in the following way. The formation of the transgressive or ravinement surface and subsequent deposition of the Tocito Sandstone resulted in the reworking of lowstand deposits with its associated terrigenous dominant palynoflora. This palynoflora was then incorporated into the sediments deposited during the early stages of the transgression associated with the formation of the ravinement surface and deposition of the Tocito Sandstone and accompanying mudstone interbeds. The poor preservation and restricted diversity of the palynoflora from the mudstone interbeds is interpreted to result from the mechanical and biological degradation of the organic material, of which the palynomorphs are a component, during reworking, transport and deposition of the sediments.

The organic matter variations deposited in the regressive lower Tocito interval compared with those found in the overlying transgressive sediments reveals two important differences. First, as a percentage of the total POM, the transgressive sediments (upper Tocito

interval and Mancos Shale directly above) contain less terrestrial organic matter (TOM). The decrease in TOM occurs at the transgressive surface that separates sediments deposited during regression from those associated with transgression and is related to lowering of terrigenous influx to the shelf during transgression of the shoreline. Second, the phytoclasts are more degraded in the transgressive deposits. Increased degradation probably results from slower sedimentation rates and an input of reworked material from the underlying regressive sediments. As a consequence, the phytoclasts deposited on the transgressive shelf are exposed to biodegradation for a longer period of time. These two differences can be seen in phytoclast abundances as they change from dominantly well-preserved phytoclasts in the samples below the upper Tocito interval to mostly amorphous structured and non-structured phytoclasts in the samples above. The same relationship between stratigraphic position and the relative amount and preservation of terrestrial organic matter can be seen in the Newsom and

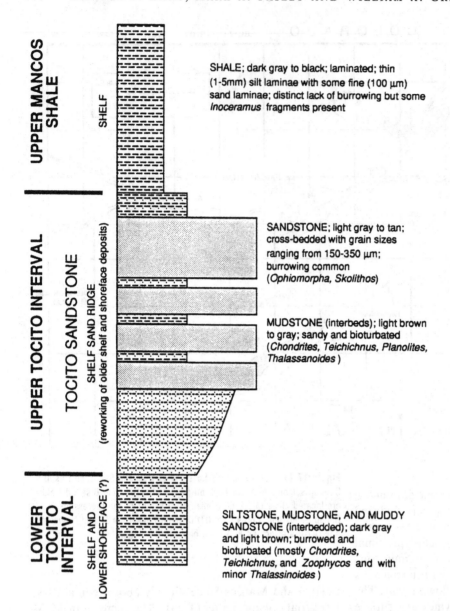

Figure 17.18. Summary of lithostratigraphy and sedimentology of Tocito–Mancos interval at Bisti oil field, NM. Figure represents information compiled from Pasley *et al.* (1991) and Bergsohn (1988).

Angel Peak cores. The change in organic matter across the transgressive surface at the base of the upper Tocito is consistent across locations in the San Juan Basin. This suggests that the change from well-preserved, terrestrially dominated organic matter in the regressive sediments to that which is mostly amorphous and marine (autochthonous) in origin in the transgressive sediments is not a local phenomenon.

Programmed pyrolysis and total organic carbon (TOC) measurements were made on the core samples from Bisti oil field, UTPC Newsom A3E (A20), and UTPC Angel Peak B#37. The pyrolysis data from the Bisti oil field are plotted on a modified van Krevelen diagram (Fig. 17.28) and against stratigraphic position (Fig. 17.29). Because no thermal maturity difference across the Bisti oil field is apparent, the modified van Krevelen diagram (for explanation of diagram see Tissot & Welte, 1984,

p. 151; cf. McIver, 1967) in Figure 17.28 shows a definite compositional difference in the organic matter, based on stratigraphic position. Using the nomenclature of Jones (1987), this transition is equivalent to a change from organic facies C (regressive lower Tocito) to BC (mudstone interbeds within the upper Tocito) to B (transgressive Upper Mancos). Descriptions of these organic facies by Jones (1987) closely match the observations made in this study at Bisti oil field (Table 17.2).

The change in Hydrogen Index (HI) with stratigraphic position for the Tocito–Mancos interval in the Newsom and Angel Peak cores is shown in Figures 17.30 and 17.31, respectively. In all three locations, an increase in

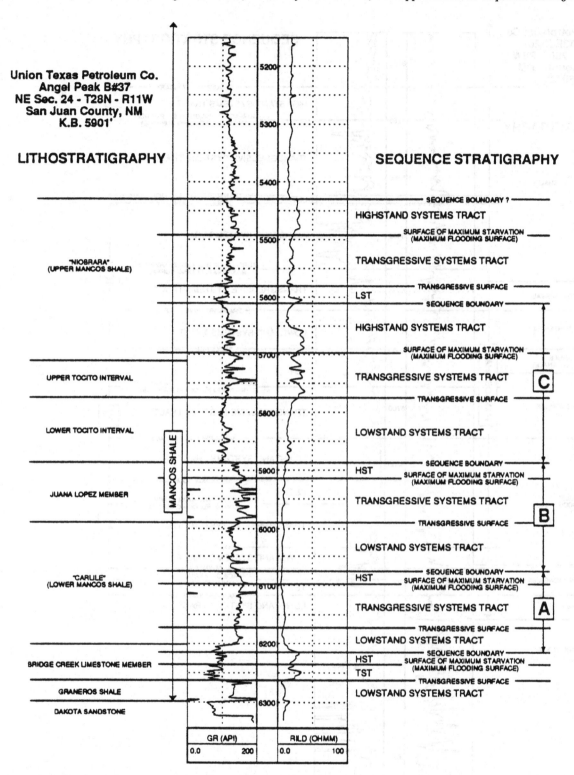

Figure 17.19. Lithostratigraphy and sequence stratigraphy for the UTPC Angel Peak B #37 well. Sequence stratigraphic interpretation based on well log cross-sections, core descriptions, and published stratigraphic results.

HI (or S2) coincides with the change from regressive sediments below to transgressive sediments above. This change in composition is accompanied by a similar change in the amount of organic matter preserved in these sediments as measured by TOC (Figs. 17.32–17.34). The regressive lower Tocito is relatively lean in organic carbon content, whereas the transgressive upper Mancos

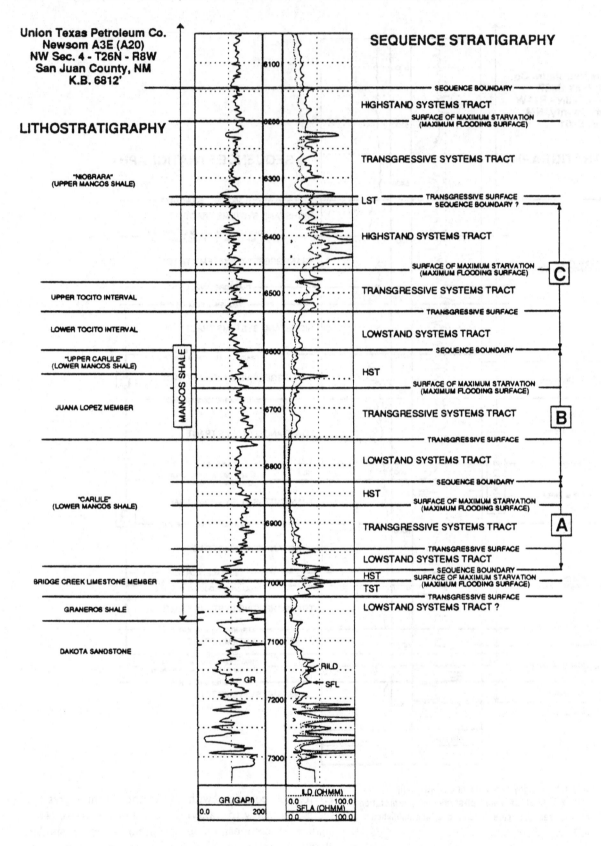

Figure 17.20. Lithostratigraphy and sequence stratigraphy for the UTPC Newsom A3E (A20) well. Sequence stratigraphic interpretation based on well log cross-sections, core descriptions, and published stratigraphic results.

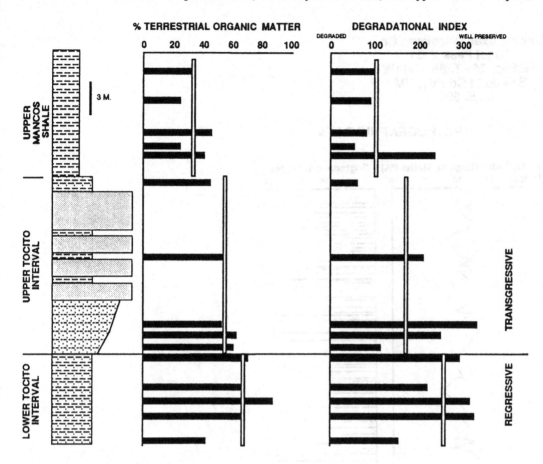

Figure 17.21. Maceral data plotted against stratigraphic position for Bisti samples with vertical bars representing group means.

could be considered organic-rich (commonly greater than 2% by wt).

The pyrolysis results and total organic carbon measurements reveal meaningful differences in the composition and amount of organic matter deposited in transgressive and regressive shelf sediments. Plots of HI versus TOC can be used to infer the amount of terrestrial input and bottom water oxygen conditions in the various depositional environments (Dean *et al.*, 1986; Dean & Arthur, 1989). The plot of these parameters presented by Dean *et al.* (1986) for Cretaceous Western Interior sediments is reproduced here in Figure 17.32A. A similar trend can be recognized in the data from the Bisti oil field (Fig. 17.32B), the Newsom core (Fig. 17.33), and the Angel Peak core (Fig. 17.34). The Tocito–Mancos samples plot as separate fields on the HI versus TOC diagrams according to stratigraphic position. The change from regressive lower Tocito to the overlying transgressive sediments is especially distinct. Inferences concerning terrestrial organic carbon input using this diagram as proposed by Dean *et al.* (1986) and Dean and Arthur (1989) are corroborated by the amount of terrestrial organic matter from petrographic analysis (% TOM). The

transgressive upper Mancos plots as the uppermost field (Figs. 17.32–17.34) and possesses the lowest TOM values. The reverse is true for the regressive lower Tocito, since these samples plot as the lowest field and contain the most TOM. Bottom water oxygen conditions can also be inferred from these diagrams, and the results appear to agree with the ichnofossils present in the Tocito–Mancos interval. Recent work has shown that trace fossil assemblages are closely related to bottom water oxygen conditions (Bromley & Ekdale, 1984; Jordan, 1985; Savrda & Bottjer, 1986; Ekdale, 1985, 1988; Ekdale & Mason, 1988). The regressive lower Tocito is commonly burrowed and contains abundant *Chondrites*, common *Teichichnus*, *Planolites* and *Zoophycos*, and minor *Thalassinoides* (Bergsohn, 1988). This sessile deposit feeding-dominated assemblage represents aerobic to dysaerobic bottom water conditions (Ekdale & Mason, 1988). In contrast, the lack of noticeable burrowing in the transgressive upper Mancos Shale above the Tocito is indicative of dysaerobic to anaerobic conditions. This change in bottom water oxygen conditions with stratigraphic position can be seen in Figures 17.32–17.34.

Several distinct differences are seen in the organic matter preserved in the Tocito–Mancos interval. The transgressive Mancos Shale above the Tocito Sandstone contains the least TOM and the most amorphous

Union Texas Petroleum Co.
Angel Peak B#37
NE Sec. 24 - T28N - R11W
San Juan County, NM
K.B. 5901'

PETROGRAPHIC DATA

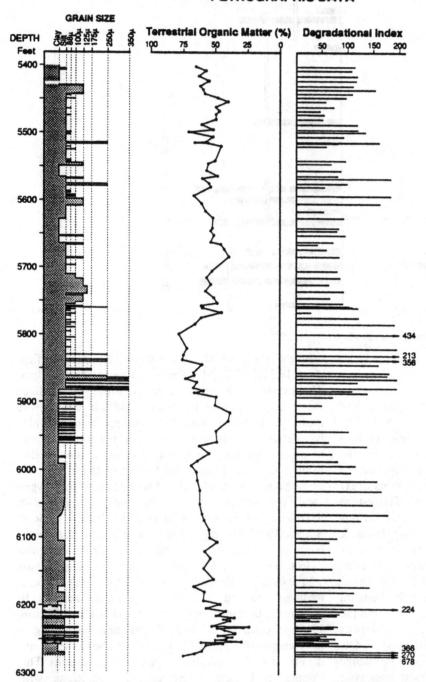

Figure 17.22. Data from maceral analysis plotted against core depth and grain size for UTPC Angel Peak well. Degradational Index increases (numerically) as degradation of phytoclasts decreases (i.e., when terrestrial organic matter is well preserved, Degradational Index is high).

Union Texas Petroleum Co.
Newsom A3E (A20)
NW Sec. 4 - T26N - R8W
San Juan County, NM
K.B. 6812'

PETROGRAPHIC DATA

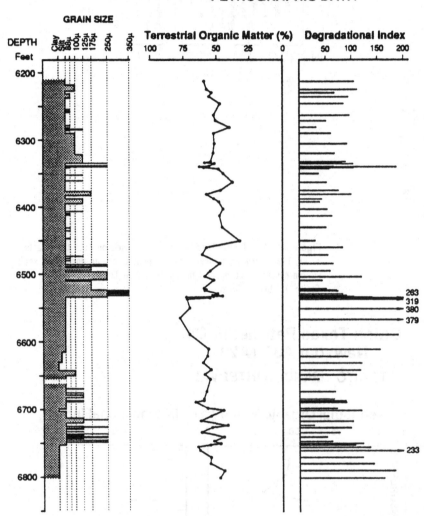

Figure 17.23. Data from maceral analysis plotted against core depth and grain size for UTPC Newsom well. Degradational Index increases (numerically) as degradation of phytoclasts decreases (i.e., when terrestrial organic matter is well-preserved, Degradational Index is high).

non-structured protistoclasts. Terrestrial organic matter found in the transgressive shale is highly degraded. Where thermal maturity is sufficiently low (e.g., Bisti oil field), fluorescence intensities are noticeably higher in the amorphous macerals from the transgressive shales. The organic matter in the transgressive Mancos is more hydrogen-rich, as evidenced by programmed pyrolysis

measurements (HI and S2). Results from pyrolysis-gas chromatography confirm that organic matter in the transgressive Mancos Shale above the Tocito Sandstone is hydrogen-rich (Pasley *et al.*, 1991). The transgressive shales have consistently higher TOC values. Finally, based on trace fossil assemblages and organic matter present, bottom water conditions during deposition of the transgressive Mancos Shale above the Tocito were the least oxygenated and may have been anoxic. This confirms the ideas presented by Rice and Gautier (1983) that marine shales in the Cretaceous Western Interior are of two types: (1) thin, transgressive shales, which

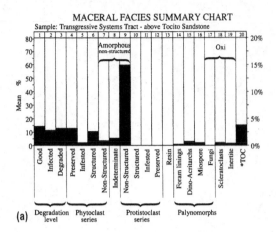

(a)

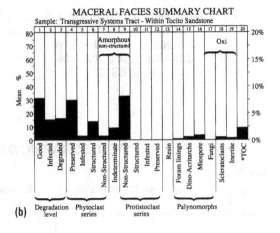

(b)

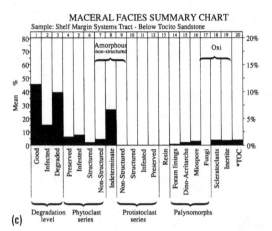

(c)

Figure 17.24. Tocito Sandstone section based on data from Pasley *et al.*, 1991, showing transgressive systems tract and underlying shelf margin system tract maceral facies. (a) above Tocito; (b) within Tocito; (c) below Tocito.

Union Texas Petroleum Co.
Newsom A3E (A20)

TOCITO - MANCOS INTERVAL

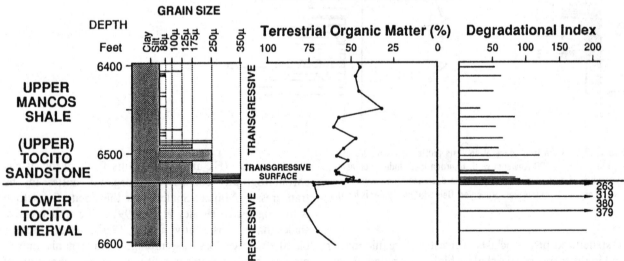

Figure 17.25. Data from maceral analysis plotted against core depth and grain size for the Tocito–Mancos interval in the UTPC Newsom well. Note lower TOM and *D*-values above transgressive surface at base of upper Tocito. Regressive sediments below this surface contain abundant, well-preserved phytoclasts. Compare to same interval in Angel Peak Core.

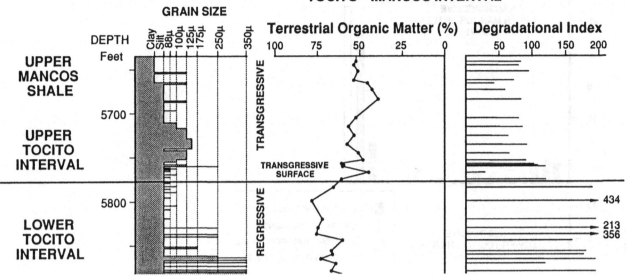

Figure 17.26. Data from maceral analysis plotted against core depth and grain size for the Tocito–Mancos interval in the UTPC Angel Peak well. Compare to same interval in Newsom core.

Table 17.2. *Comparison of stratigraphic position and organic matter characteristics at Bisti oil field with observations on organic facies by Jones (1987)*

Table modified from Pasley *et al.* (1991); used by permission of Pergamon Press.

STRATIGRAPHIC POSITION	ORGANIC MATTER CHARACTERISTICS (BISTI OIL FIELD)	ORGANIC FACIES (JONES, 1987)
ABOVE (Transgressive Upper Mancos directly above Tocito Sandstone)	Mostly strongly fluorescent amorphous non-structured protistoclasts. Minor terrestrial organic matter (TOM = 25-46%) present and is highly degraded. Hydrogen Indices are the highest in the Mancos/Tocito interval (HI = 433-623).	**ORGANIC FACIES B** (HI = 400-650) "...volumetrically the most important of the oil-prone sourcerocks and is the source of the majority of the world's oilfields." "Organic Facies B is laminated - well bedded.." "... is often found in transgressive marine shales deposited in shallow water, and is likely to be interbedded with less oil-prone facies due to fluctuations in water bottom anoxia"
WITHIN (mudstone interbeds in transgressive upper Tocito Sandstone)	Sub-equal terrestrial and marine organic matter (TOM = 45-63%). Intermediate fluorescence, degradation and HI (311-413). Intermediate between shales below and above primarily because of transgressive reworking of older (regressive) shelf and shoreface deposits.	**ORGANIC FACIES BC** (HI = 250-400) "The facies is often deposited under an oxic water column in a fine-grained system of siliceous clastics....". "The OM is usually a mixture of partially biodegraded terrestrial and algal material."
BELOW (regressive lower Tocito interval)	Terrestial organic matter dominant (TOM = 41-88%) and well preserved. Amorphous material fluoresces weakly or not at all. Hydrogen Indices low (HI = 135-263).	**ORGANIC FACIES C** (H I = 125-250) "...is the "gas-prone" facies of the literature. The OM is dominated by terrestrial debris in various stages of oxidation. Large volumes of Organic Facies C were deposited on the Tertiary and Mesozoic shelves and slopes of continental margins."

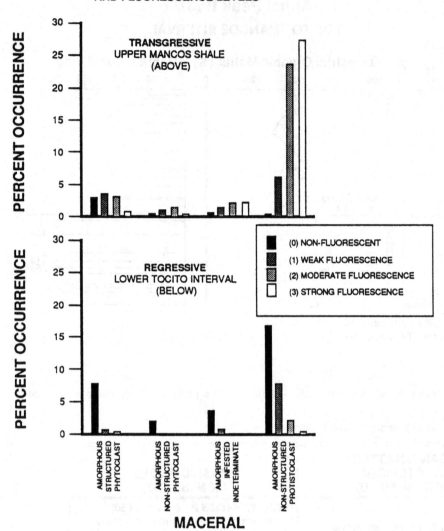

AVERAGE AMORPHOUS MACERAL COMPOSITION AND FLUORESCENCE LEVELS

Figure 17.27. Bar chart of percent occurrence for the amorphous macerals in the Tocito–Mancos interval at Bisti oil field. Each maceral is divided into four fluorescence levels. Amorphous material is typically non-fluorescent or weakly fluorescent in regressive lower Tocito, but moderate and strong fluorescence is most common for the amorphous material in the transgressive Upper Mancos above the Tocito Sandstone. (modified from Pasley *et al.*, 1991; reprinted with permission of Pergamon Press PLC.)

contain abundant marine organic matter and form major oil source beds; (2) thick, progradational shales, which are coarser grained and contain smaller amounts of organic matter that is mostly terrestrial in origin.

THE ST. STEPHENS QUARRY

The St. Stephens Quarry material is from Washington County, Alabama (Fig. 17.35), and is Eocene-Oligocene in age (Jackson and Vicksburg Groups) (Fig. 17.36). The sediments are typically fine-grained and calcareous, and were deposited in shelf environments that received varying amounts of terrigenous input. Paleobathymetric studies have indicated that depositional environments ranged from inner to outer neritic (MacNeil, 1944; Murray, 1961; Hazel *et al.*, 1980; Loutit *et al.*, 1983, 1988; Baum & Vail, 1988). Because the main depocenter for Jackson and Vicksburg sediments was in Louisiana and Texas, the uppermost Eocene and lowermost Oligocene rocks of southwestern Alabama are located in a region

that was transitional between thick, predominantly clastic accumulations in the west, and carbonate deposition to the east (MacNeil, 1944; Murray, 1961; Huddlestun & Toulmin, 1965; Galloway, 1989). Paleobathymetric studies indicate that paleoenvironments varied from inner to outer neritic in response to changes in sea level (Hazel *et al.*, 1980; Loutit *et al.*, 1983; Baum & Vail, 1988; Loutit *et al.*, 1988).

Based on regional stratigraphic studies, two important surfaces associated with transgression of the shoreline are found in the section at St. Stephens Quarry. First,

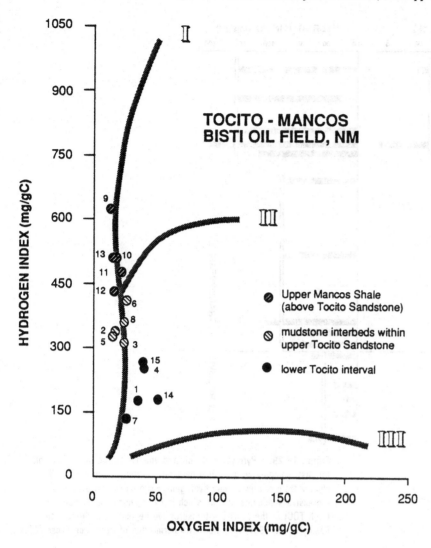

Figure 17.28. Programmed pyrolysis data for the Bisti oil field samples plotted on a modified van Krevelen diagram. Legend indicates stratigraphic position with respect to upper Tocito Sandstone for the various samples. (From Pasley *et al.*, 1991; reprinted with permission of Pergamon Press PLC.)

the upper contact of the Shubuta Clay is marked by a laterally continuous but thin (<2 cm) shell hash that contains considerable amounts of glauconite, pyrite, and phosphate. Second, the contact between the Red Bluff Clay and the overlying Mint Springs Marl is a transgressive surface that represents southeast to northwest transgression over the Forest Hill–Red Bluff deltaic system (MacNeil, 1944; Murray, 1961; Hazel *et al.*, 1980; Mancini & Tew, 1988; Dockery, 1982, 1990; Pasley & Hazel, 1990). Therefore, this surface separates regressive, fine-grained, prodeltaic sediments (Red Bluff Clay) from transgressive, calcareous, shelf deposits (Mint Springs Marl, Marianna and Glendon Limestones).

Much has been published on both the regional stratigraphic relationships (e.g., MacNeil, 1944; Murray, 1961; Deboo, 1965; Hazel *et al.*, 1980), and the sequence stratigraphy of the Eocene–Oligocene strata exposed at this important location (Mancini *et al.*, 1987; Baum & Vail, 1988; Loutit *et al.*, 1988; Mancini & Tew, 1988, 1990; Pasley & Hazel, 1990). When the results of these studies are combined with those from other parts of the

Gulf Coast (Galloway, 1989; Coleman, 1990; Coleman & Galloway, 1990; Dockery, 1990), a better sequence stratigraphic interpretation of this interval emerges.

The Pachuta Marl and Shubuta Clay were deposited in the transgressive systems tract. The upper boundary of this systems tract is the starvation surface (shell hash) found at the top of the Shubuta Clay, where, using graphic correlation techniques, a definite marine hiatus can be resolved (Pasley & Hazel, 1990). This hiatus represents approximately 300 000 years and is equivalent to the upper 11.3 m of the Shubuta Clay in its type area along the Chickasawhay River in eastern Mississippi. This equivalence is important, because all of the earlier work on the sequence stratigraphy of the St. Stephens section placed the Red Bluff Clay/Bumpnose Limestone and Red

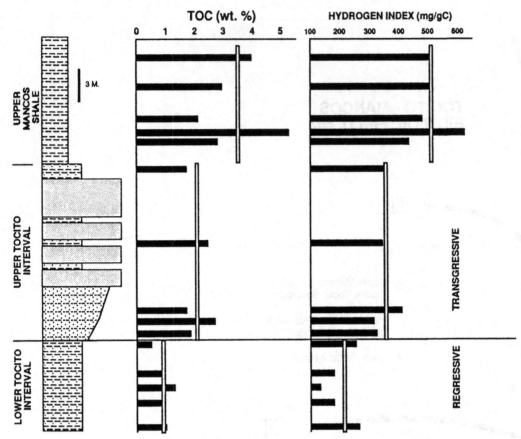

Figure 17.29. Pyrolysis data plotted against stratigraphic position for Bisti samples with vertical bars representing group means. Note change from relatively lean hydrogen-poor organic matter in regressive sediments to that which is hydrogen-rich and abundant (high TOC) in the overlying transgressive deposits. (Modified from Pasley *et al.*, 1991; reprinted with permission of Pergamon Press PLC.)

Union Texas Petroleum Co.
Newsom A3E (A20)
TOCITO - MANCOS INTERVAL

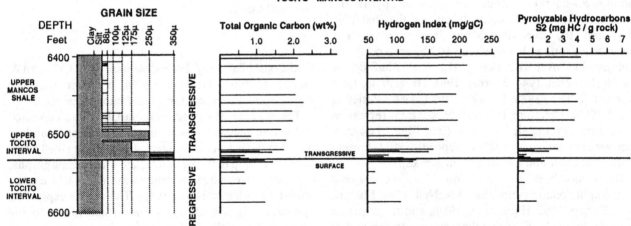

Figure 17.30. Data from programmed pyrolysis (TOC, HI, and S2) plotted against core depth and grain size for Tocito–Mancos shale interval in UTPC Newsom well. Note lower TOC and HI (S2) in regressive sediments. Compare to same interval at Bisti oil field and in Angel Peak core.

Union Texas Petroleum Co.
Angel Peak B#37

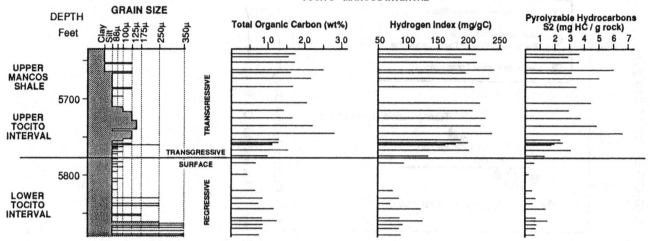

TOCITO - MANCOS INTERVAL

Figure 17.31. Data from programmed pyrolysis plotted against core depth and grain size for the Tocito–Mancos interval in the UTPC Angel Peak well. Note lower TOC and HI (S2) in regressive sediments and distinct increase at transgressive surface. Compare to same interval at Bisti oil field and in Newsom core. Hydrogen indices from samples with less than 0.5% TOC are not reliable and are not included in this figure (see Peters, 1986).

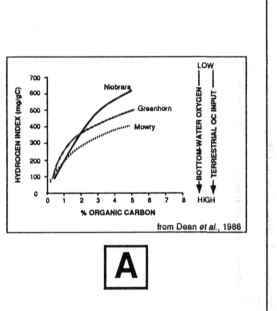

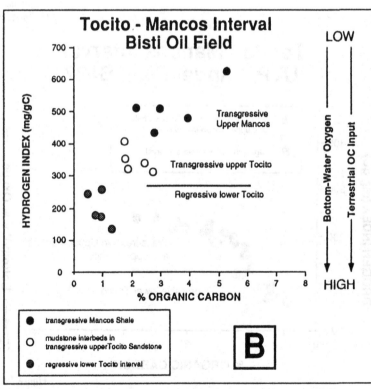

Figure 17.32. Relationship between hydrogen index (HI) and total organic carbon (TOC) for Cretaceous Western Interior sediments. (A) Diagram modified from Dean *et al.* (1986), showing generalized curves for Niobrara, Greenhorn, and Mowry; (B) Data from Bisti oil field plotted on similar diagram. Each stratigraphic position plots as a separate field with boundary between regressive and transgressive sediments especially distinct. Inferences proposed by Dean *et al.* (1986) concerning bottom water oxygen and terrestrial organic carbon appear correct based on maceral analysis and ichnofossil assemblages. (From Pasley *et al.*, 1991; reprinted with permission of Pergamon Press PLC.) Compare to similar plots for Tocito–Mancos interval in Newsom and Angel Peak cores.

Tocito - Mancos Interval
UTPC Newsom A3E (A20)

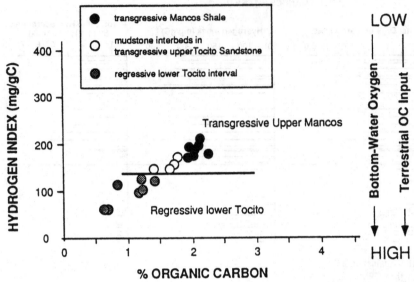

Figure 17.33. Relationship between hydrogen index (HI) and total organic carbon (TOC) for samples from the Tocito–Mancos interval in Newsom core. Regressive and transgressive sediments plot as separate fields. Maceral analysis and ichnofossil assemblages confirm inferences proposed by Dean *et al.* (1986) concerning bottom water oxygen and terrestrial organic carbon input. Compare to similar plots for Tocito–Mancos interval at Bisti oil field and in Angel Peak core.

Tocito - Mancos Interval
UTPC Angel Peak B#37

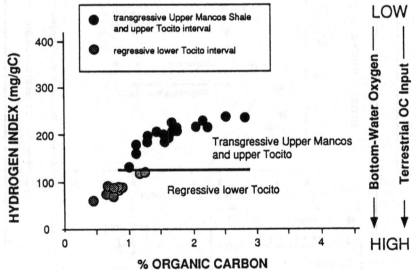

Figure 17.34. Relationship between hydrogen index (HI) and total organic carbon (TOC) for samples from Tocito–Mancos interval in Angel Peak core. Regressive and transgressive sediments plot as separate fields. Maceral analysis and ichnofossil assemblages confirm inferences proposed by Dean *et al.* (1986) concerning bottom water oxygen and terrestrial organic carbon input. Compare to similar plots for Tocito–Mancos interval at Bisti oil field and in Angel Peak core.

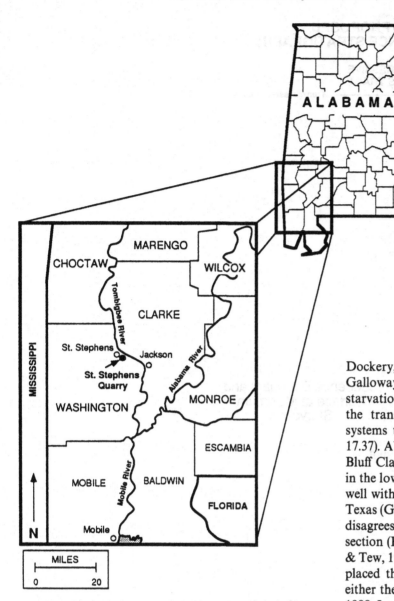

Figure 17.35. Location map of St. Stephens Quarry section, Washington County, AL, USA (from Pasley & Hazel, 1990; used by permission of the Gulf Coast Association of Geological Societies).

Bluff Clay interval above the shell hash in the highstand systems tract. Correlation to well log cross-sections presented by Dockery (1990) reveals that the hiatus at the top of the Shubuta at St. Stephens contains not only the Eocene–Oligocene epochal boundary, but also the entire highstand systems tract. Deposition in the highstand systems tract of this particular sequence occurred further updip (or closer to the supply of sediment) and is represented by the upper Shubuta that is missing in St. Stephens Quarry. Deposition resumed at St. Stephens only after relative sea level was lowered, and the Forest Hill–Red Bluff interval prograded into the basin. These sediments rest on a recognizable subaerial unconformity in Texas, Louisiana, and Mississippi (Murray, 1961;

Dockery, 1986, 1990; Galloway, 1989; Coleman & Galloway, 1990). Therefore, the surface of maximum starvation at St. Stephens Quarry represents the top of the transgressive systems tract, the entire highstand systems tract, and the sequence boundary (Figs. 17.36, 17.37). Above this surface, the lowermost Oligocene Red Bluff Clay/Bumpnose Limestone interval was deposited in the lowstand systems tract, an observation that agrees well with findings from Mississippi (Dockery, 1990) and Texas (Galloway, 1989; Coleman & Galloway, 1990), but disagrees with earlier work on the St. Stephens Quarry section (Baum & Vail, 1988; Loutit *et al.*, 1988; Mancini & Tew, 1988, 1990; Pasley & Hazel, 1990). These workers placed the sequence boundary higher in the section at either the base of the Mint Springs Marl (Baum & Vail, 1988; Loutit *et al.*, 1988; Mancini & Tew, 1988, 1990), or in the Red Bluff Clay (Pasley & Hazel, 1990).

Regional stratigraphic relationships discussed by MacNeil (1944), Hazel *et al.* (1980) and Pasley and Hazel (1990) show that the base of the Mint Springs cannot be the sequence boundary, because it is time-transgressive. This surface climbs stratigraphically landward, and, as a result, not all rocks below it are older than all rocks above it (see discussion by Nummedal & Swift, 1987, on the diachronous nature of transgressive surfaces). The base of the Mint Springs Marl at St. Stephens is a transgressive surface, and is interpreted to represent the boundary between the lowstand systems tract (Red Bluff Clay) and the overlying transgressive systems tract (Mint Springs Marl). The correlative transgressive surface in Mississippi is found in the upper portion of the Forest Hill Sand. According to MacNeil (1944) and Dockery (1982), the upper Forest Hill is estuarine and contains a fauna similar to the Mint Springs Marl. These sediments

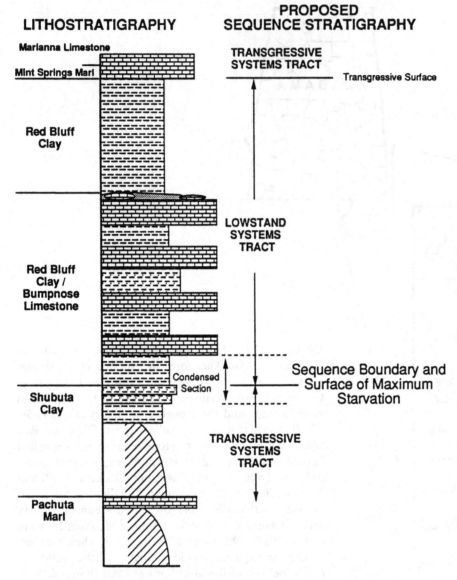

Figure 17.36. Sequence stratigraphic interpretation for the Eocene–Oligocene section at St. Stephens Quarry. This interpretation integrates findings from earlier work on this section (see Pasley & Hazel, 1990), with those from Mississippi (Dockery, 1990). Location of condensed section faces based on gamma ray data presented by Loutit *et al.* (1988). Actual condensation of time resolved as a marine hiatus at surface of maximum starvation (top of Shubuta Clay) by Pasley & Hazel (1990).

are interpreted to represent the lower part of the transgressive systems tract. The lithostratigraphic contact between the Mint Springs Marl and the Forest Hill Sand in Mississippi is a transgressive surface (probably a ravinement surface) but, because it has little chronostratigraphic significance, does not form the lower boundary of the transgressive systems tract.

Study of the particulate organic matter at St. Stephens Quarry reveals a direct relationship between organic matter type, preservation, and transgression of the shoreline (Fig. 17.38). Much like the transgressive surface at the base of the upper Tocito interval, the transgressive surface at the base of the Mint Springs Marl is marked by a drastic reduction in the amount of terrestrial organic matter. This reduction is related to the transgression of the Forest Hill–Red Bluff deltaic deposits that are present in Mississippi. The Red Bluff Clay in St. Stephens Quarry was deposited as the fine-grained equivalent to this delta

system and contains abundant well-preserved phytoclasts. The calcareous Mint Springs Marl was deposited on the shelf as the delta was transgressed and terrigenous sediment and total organic matter supply greatly reduced.

Organic matter in the shell hash associated with the starvation surface at the top of the Shubuta Clay in St. Stephens Quarry records the effects of sediment starvation. The particulate organic matter in this thin deposit is mostly autochthonous (amorphous nonstructured protistoclasts), and any TOM that is present

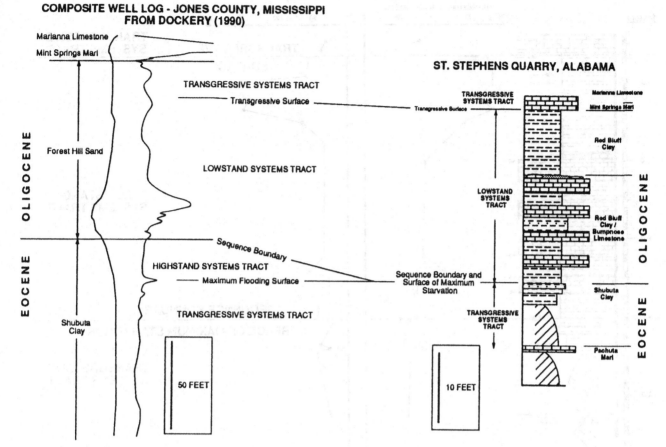

Figure 17.37. Correlation of sequence stratigraphy at St. Stephens Quarry to equivalent rocks in Mississippi. Composite well log section based on material from Dockery (1990) represents a section a short distance downdip from the outcrop on the Chickasawhay River. Red Bluff Clay is a distal facies of the deltaic Forest Hill (see MacNeil, 1944). Highstand systems tract in Mississippi is represented by upper Shubuta Clay, and in St. Stephens by the marine hiatus. Transgressive surface at the base of the Mint Springs at St. Stephens separates lowstand from transgressive systems tracts and is correlated to the upper part of the Forest Hill that contains a 'Mint Springs fauna' (MacNeil, 1944; Dockery, 1982). Note difference in scales as section in Mississippi is much thicker.

in the shell hash is highly degraded (Fig. 17.39). The increase in TOM above this surface (Fig. 17.40) is indicative of renewed terrigenous influx to the shelf following this transgression. This increase is slight in the Red Bluff Clay/Bumpnose Limestone (see Fig. 17.40.b,c), because terrigenous sediment influx remained relatively low until sediments directly associated with the local Forest Hill delta reached the St. Stephens Quarry area.

The maceral data (% TOM, and D) from the section at St. Stephens Quarry are plotted against the stratigraphic position for each sample (Fig. 17.38). Distinct changes in organic matter are observed at two surfaces associated with transgressions of the shoreline. The shell hash associated with the starvation surface found at the top

of the Shubuta Clay is low in terrestrial organic matter (TOM), and a general increase in TOM is observed as one moves up through the interfingering Red Bluff Clay/Bumpnose Limestone and overlying Red Bluff Clay. The transgressive surface at the contact between the Red Bluff Clay and the thin tongue of the Mint Springs Marl is marked by a sharp decrease in TOM.

Changes in the degradational index (D) also occur at surfaces associated with transgressions. Terrestrial organic matter is highly degraded (low D-value) in samples from the Pachuta Marl and Shubuta Clay. The phytoclasts are the most degraded and D attains a minimum in the shell hash (starvation surface). Samples above the shell hash contain increasing amounts of well-preserved phytoclasts, and, consequently, display increasing D-values. High D-values for the samples from the lower part of the regressive Red Bluff Clay indicate that terrestrial macerals in this part of the section have undergone little biodegradation. The two samples in the uppermost portion of the Red Bluff Clay, however, contain more degraded phytoclasts and lower D-values. As a result, the transgressive surface at the base of the Mint Springs Marl is marked by only a small decrease in the degradational index.

The samples from the Pachuta Marl, Shubuta Clay, Red Bluff Clay/Bumpnose Limestone and Mint

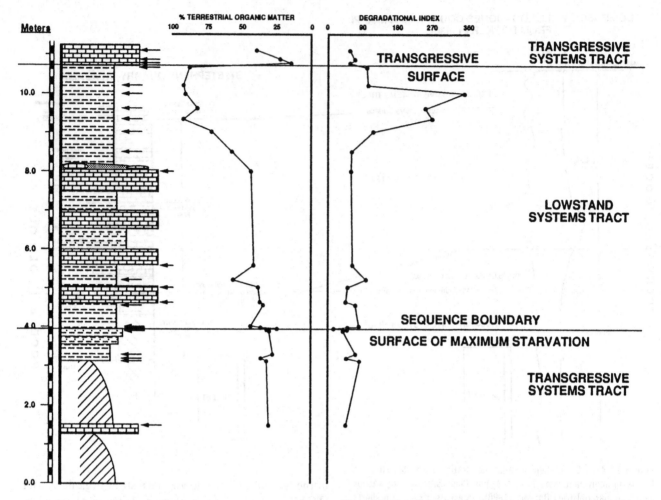

Figure 17.38. Maceral data plotted against stratigraphic position and sequence stratigraphy for the section at St. Stephens Quarry.

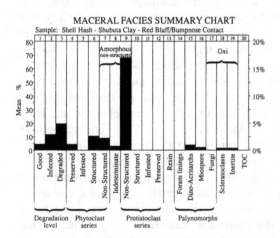

Figure 17.39. Average maceral composition for Shell Hash–Shubuta Clay/Red Bluff–Bumpnose contact at St. Stephens Quarry.

Springs Marl intervals are characterized by a relatively diverse, well-preserved marine-dominated palynomorph assemblage (mean % terrigenous palynomorphs = 34.1%). The Red Bluff Clay above the Red Bluff Clay/Bumpnose Limestone interval is characterized by a diverse, well-

preserved terrigenous-dominated palynoflora with a mean equal to 83.7%.

The section exposed at St. Stephens Quarry represents an area laterally removed from the areas of major deltaic progradation to the west and at the boundary with the carbonate province to the east (Glawe, 1969). The lateral proximity of deltaic areas is expected to result in higher terrigenous percentages during lowstand and highstand deposition. This is observed for the Red Bluff Clay above the Red Bluff Clay/Bumpnose Limestone interbeds. The low mean percent terrigenous palynomorphs for the Red Bluff Clay/Bumpnose Limestone interbeds is interpreted to result from a time delay for sediments resulting from deltaic progradation in the west to reach the St. Stephens Quarry area. The low mean percent terrigenous palynomorphs characterizing the Pachuta Marl, Shubuta Clay, Red Bluff Clay/Bumpnose Limestone and Mint Springs Marl is interpreted to result from the proximity of the study area to the carbonate province to the east and the alternating influence of this province on the study area associated with high sea level intervals.

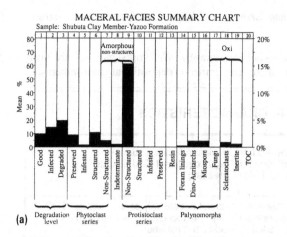

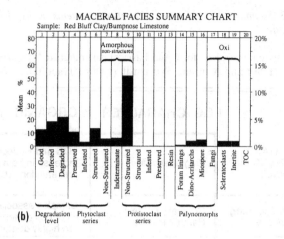

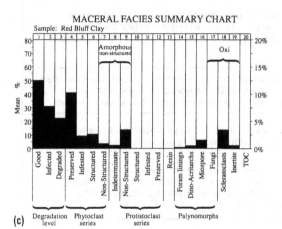

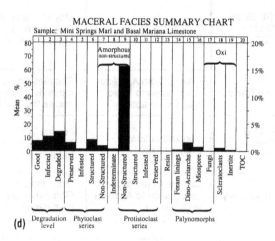

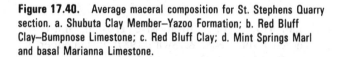

Figure 17.40. Average maceral composition for St. Stephens Quarry section. a. Shubuta Clay Member–Yazoo Formation; b. Red Bluff Clay–Bumpnose Limestone; c. Red Bluff Clay; d. Mint Springs Marl and basal Marianna Limestone.

Because of sparse abundance and distribution of the recycled palynoflora, interpretations on its significance would be unwise.

Maceral spectra from the St. Stephens Quarry indicate a relationship between the type and preservation of POM and depositional systems tracts (Fig. 17.40; see also Table 17.3). Maceral assemblages in the transgressive systems tract are dominated by amorphous non-protistoclasts, which are thought to be marine algal in origin. Phytoclasts that are present in this systems tract are highly degraded. An example of this maceral facies can be observed in the transgressive systems tracts from two different sequences, Shubuta Clay and Mint Springs Marl. Sediments in the lowstand systems tract (Red Bluff Clay) contain abundant phytoclasts that are commonly well-preserved. This relationship between organic matter type and preservation and depositional systems tract is apparently a result of the relative importance of progradational depositional systems in each systems tract. Shelf sediments in the lowstand and

highstand systems tracts are influenced heavily by progradation (hence their aggradational or progradational stacking patterns) and, as a consequence, contain considerable amounts of well-preserved phytoclasts. Autochthonous organic matter is much more common in the transgressive systems tract, because progradation of depositional systems onto the shelf is of lesser extent (which results in an overall retrogradational stacking pattern).

Comparison of the sequence stratigraphic interpretation presented in Figure 17.36 with that presented by Pasley and Hazel (1990) reveals a distinct difference in the placement of the sequence boundary. The fact that Pasley and Hazel's integration of organic petrologic data into sequence stratigraphy failed to recognize that the Red Bluff Clay/Bumpnose Limestone and Red Bluff Clay were deposited in the lowstand systems tract rather than in the highstand systems tract illustrates two important points. First, organic matter from the highstand and lowstand systems tracts can appear similar, a result not entirely surprising when one considers that these systems tracts share similar stacking patterns. This implies that, although the use of data from the characterization of organic matter in sequence stratigraphy is very helpful

Table 17.3. *Summary of the relationship between systems tracts, their bounding surfaces, well log character, and organic matter characteristics*

This table is envisaged to form the basis of an integrated approach to sequence stratigraphy. Well log signatures are taken from Vail (1990) and Vail and Wornardt (1990). ('Tract' refers to a group of contemporaneous depositional environments.)

	LOG CHARACTER	ORGANIC MATTER CHARACTERISTICS
HIGHSTAND SYSTEMS TRACT	Progradational	TOC and HI decreases upward from maximum flooding surface. Increase in both amount of terrestrial organic matter (TOM) and its preservation. (Degradational Index increases)
Maximum Flooding Surface	Generally Highest Gamma Ray Count	Samples associated with this surface do not necessarily contain highest TOC or HI but do generally exhibit the least TOM. Phytoclasts associated with MFS are highly degraded (Dedgradational Index at a minimum).
TRANSGRESSIVE SYSTEMS TRACT	Retrogradational (backstepping)	May contain highest TOC and HI and, consequently, the highest source rock potential (if sufficiently thick). Terrestrial organic matter decreases upward toward maximum flooding surface. Phytoclasts are highly degraded.
Transgressive Surface	Change from Aggradational below to Retrogradational above	Samples above this surface contain considerably lower amounts of TOM than those directly below. Also, change in Degradational Index is pronounced across this surface as phytoclasts are more degraded in overlying samples.
LOWSTAND SYSTEMS TRACT	Aggradational to Progradational	Lowest TOC and HI values are recorded in samples from this systems tract. Terrestrial organic matter is abundant and well preserved (Degradational Index high).
Sequence Boundary	Change from Progradational below to Aggradational above	Most difficult surface to recognize using organic matter characteristics. General increase in Degradational Index as phytoclasts become better preserved in LST above the boundary than in the HST below.

in the identification of transgressive surfaces and surfaces of maximum starvation, identification of sequence boundaries requires considerable amounts of regional stratigraphic information. It is important to remember that organic matter changes that occur on the shelf are related to transgressions and regressions of the shoreline and not necessarily to the downward shifts in coastal onlap that are associated with sequence boundaries. Second, just as the ideal cyclothems described by Wanless and Weller (1932) are seldom observed at one location, not all vertical sections (either boreholes or outcrops) contain all of the components of a particular depositional sequence. The starvation surface at the top of the Shubuta that is readily identified with organic petrology and biostratigraphy (Pasley & Hazel, 1990) represents the entire highstand systems tract and sequence boundary. Rigid application of the sequence stratigraphic model led previous workers to assign the Red Bluff Clay–Bumpnose Limestone and Red Bluff Clay interval to the highstand systems tract, based on identification of the surface of maximum starvation and condensed section. Hence, the section at St. Stephens Quarry demonstrates that modification of the stratigraphy by sediment starvation, such as that documented in southern Alabama by

Coleman (1990), or transgressive erosion, must be considered (Nummedal & Swift, 1987; Posamentier & James, 1991; Riley & Nummedal, 1991).

Conclusions

From the work of Habib and Miller (1989), Leckie *et al.* (1990), Pasley and Hazel (1990), Pasley *et al.* (1991), Posamentier and Chamberlain (in press), and the studies presented here, it has become apparent that the nature of organic matter deposition has stratigraphic significance. For example, Habib and Miller (1989) concluded that changes in POM can be useful in delineating transgressive and regressive intervals. They demonstrated that transgressive sediments contained a larger number of dinoflagellate species and a maceral facies dominated by amorphous debris. Maceral facies in regressive sediments contained more terrestrially derived POM (vascular tissue and inertite). Their findings are in agreement with those reported here. Important differences clearly can be observed petrographically in the POM that is deposited in transgressive intervals when compared to the underlying regressive sediments. These differences suggest that less terrestrial organic matter is delivered to the shelf during transgression of the shoreline, and that the terrestrial organic matter deposited during transgression is reworked from below, spends more time at the sediment–water interface, and/or is exposed to more biodegradation during transport. These observations are not unexpected when one considers the nature of transgressive shelf sedimentation. During transgression of the shoreline, terrigenous sediment is deposited in estuarine and coastal environments and, consequently, little is introduced to the shelf (Swift, 1976; Dominguez *et al.*, 1987; Nummedal & Swift, 1987; Jervey, 1988; Loutit *et al.*, 1988; Posamentier & Vail, 1988; Roberts & Coleman, 1988; Leithold & Bourgeois, 1989). Studies of modern estuaries have proven them to be efficient sediment traps (McCave, 1973; Nichols & Biggs, 1985). Of special interest to this discussion is the work of LeBlanc *et al.* (1989), who found that the amount of terrestrial organic matter decreases steadily seaward in the estuarine environments of Negro Harbor, Nova Scotia. Using fatty acid and ^{13}C measurements, they reported that the organic matter in sediments 3 km from where the river entered the estuary was over 70% terrestrial in origin. Sediments 12 km from the river mouth, however, contained only 10% terrestrial organic matter. Similar results have been reported for the estuarine environs in the Bay of Biscay (Fontugne & Jouanneau, 1987), Gulf of St. Lawrence (Lucotte *et al.*, 1991), and the Amazon (Showers & Angle, 1986). This work on modern depositional environments suggests that, during transgressions, terrestrial organic matter is trapped in estuarine environments and is probably deposited on the shelf only after spending considerable time exposed to bacterial and fungal degradation.

Examination of pyrolysis data in the context of sequence stratigraphy (Figs. 17.29, 17.30) reveals that changes in the amount and composition of organic matter are related to the depositional systems tract in which the organic matter is found. Organic matter deposited on the shelf in the lowstand and highstand systems tracts is less hydrogen-rich than that deposited in the transgressive systems tract. In addition, shelf sediments in the transgressive systems tract tend to be more organic-rich (higher TOC). This suggests that both the organic richness and organic facies, as defined by Jones (1987), change predictably in a sequence stratigraphic framework. These changes in the amount and composition of organic matter are identified when samples below transgressive surfaces are compared to those above. For example, the transgressive sediments in the Tocito–Mancos interval of the San Juan Basin contain more organic carbon and yield relatively high amounts of pyrolyzable hydrocarbons ('facies B & BC' of Jones, 1987). Similar results have been reported for transgressive sediments in the Canadian portion of the Cretaceous Western Interior (Leckie *et al.*, 1990;

Posamentier & Chamberlain, in press). The lower Tocito in the San Juan Basin and examples from the Gulf of Mexico (Dow, 1984; Pasley *et al.*, 1991) indicate that sediments deposited on the shelf during regression contain organic matter that is relatively hydrogen-poor ('facies C or CD' of Jones). These variations in organic matter between transgressive and regressive sediments can be recognized when pyrolysis data (HI and TOC) are plotted against stratigraphic position and imply that transgressive shelves are more prone to experience anoxic bottom water conditions. This agrees with the model discussed by Wignall (1991), in which transgression introduces low-oxygen water from deeper parts of the basin. According to this model, the combination of subsidence, sea level rise, and decreased sediment supply during transgression results in expansion of the 'puddle' of deep water into shallower water settings.

The implication of maceral facies analysis to petroleum source rock prediction can be seen when one examines the Mancos Shale. It has been suggested that the Mancos Shale interval contains the major petroleum source rocks for the San Juan Basin (Ross, 1980; Rice, 1983). Sufficient data were generated during the course of this study to evaluate properly the petroleum source potential of a large portion of the Mancos Shale. The lower Tocito interval in both the Newsom and Angel Peak cores and at Bisti oil field possesses no appreciable source potential (organic facies 'C' of Jones). The same is true for the Carlile interval of the Lower Mancos Shale directly below the Juana Lopez Member (Semilla equivalent). These shales are low in TOC and HI, and the most common macerals are of terrestrial origin. In the Bisti samples, fluorescent intensities for the amorphous material in the regressive lower Tocito are, at best, low. However, the Carlile interval above the Bridge Creek Member (between 6100 ft and 6150 ft in the Angel Peak core), the lower part of the Juana Lopez Member, and the Upper Mancos Shale directly above the Tocito Sandstone (Mulatto Tongue equivalent) possess excellent source potential. These transgressive, marine shales contain sufficient organic matter that is hydrogen-rich. Petrographically, the organic matter in the transgressive Upper Mancos (Mulatto Tongue) fluoresces strongly in the Bisti samples and is mostly marine algal in origin (amorphous non-structured protistoclast). These findings agree with the assertion that most of the petroleum source rocks in the Cretaceous Western Interior are transgressive in nature (Rice & Gautier, 1983).

One of the appealing aspects of sequence stratigraphy is that it provides a framework for predictions concerning the nature of sedimentary basin fill (e.g., Vail, 1987; Jervey, 1988; Posamentier & Vail, 1988; Vail & Wornardt, 1990). While these predictive capabilities have been applied to reservoir facies, information concerning detailed source rock predictions in a sequence stratigraphic framework is lacking. Previous workers have suggested that, in a given depositional sequence, the condensed section facies should contain the best marine source rocks (Vail, 1987). However, in the San Juan Basin, some intervals (the Carlile above the Bridge Creek, the lower Juana Lopez, and the Mulatto Tongue equivalent of the Upper Mancos) possess considerable source potential and are not exclusively condensed sections. These units were deposited as offshore marine sediments in the transgressive systems tracts. The surfaces of maximum sediment starvation in each of the sequences are generally found at the top of the intervals containing the optimum source potential. In some cases, potential source sediments directly overlie the transgressive surface that forms the lower boundary of the transgressive systems tract. Values for TOC and HI (and, as a consequence, source potential) generally decrease as one moves upward from the surface of maximum starvation into the highstand systems tract. Studies on the Cretaceous in Canada also found source potential in transgressive shales below the condensed section associated with the downlap surface and only a short distance above the transgressive surface (Leckie *et al.*, 1990; Posamentier & Chamberlain, in press). The models described by Wignall (1991) similarly indicate that organic-rich facies are most common in the transgressive systems tract and are not symmetrically distributed about the surface of maximum starvation. These results, in conjunction with the findings from the San Juan Basin and the St. Stephens Quarry section, suggest that the best marine petroleum source rocks are found in the offshore, fine-grained intervals of the transgressive systems tract below the condensed section facies that contains the downlap surface.

Integration of maceral facies data with sedimentologic observations indicates that organic facies deposited in shelf environments change predictably across depositional surfaces. Both the Eocene–Oligocene boundary section at St. Stephens Quarry in southwestern Alabama and the Upper Cretaceous of the San Juan Basin of New Mexico demonstrate these changes. The amount of terrestrial organic matter decreases markedly as a percentage of the total organic matter, as one moves upward from sediments deposited during regression below a transgressive surface into those that are associated with regional transgression above. In deposits associated with transgressions, phytoclasts are highly degraded and, where thermal maturity is sufficiently low, amorphous macerals exhibit high fluorescent intensity. Shelf sediments deposited during regression typically contain abundant well-preserved phytoclasts and weakly fluorescent amorphous macerals. Surfaces formed on the shelf by the starvation of sediment also can be recognized more clearly when data from organic matter characterization are used. Maceral analysis reveals that sediments associated with these surfaces contain less terrestrial

organic matter than sediments above and below, and that phytoclasts associated with starvation surfaces are the most degraded. Sediments above the starvation surface contain increasing amounts of well-preserved terrestrial organic matter.

In addition to changes in maceral facies, both the amount of organic matter (% TOC) and its composition (as measured by the Hydrogen Index = HI) change across surfaces. Transgressive shelf sediments generally contain more organic carbon and yield more pyrolyzable hydrocarbons than do their regressive counterparts. When TOC and HI are plotted against one another, transgressive sediments plot separately from the underlying regressive deposits. As suggested by Dean *et al.* (1986) and Dean and Arthur (1989), this procedure reveals information concerning terrestrial organic carbon input and bottom water oxygen conditions. This information, along with observations on trace fossil assemblages, suggests that transgressive shelves more commonly experience dys-aerobic (= between aerobic and anaerobic) or anaerobic bottom water conditions than those that are influenced by prograding depositional systems.

Finally, it is clear from the comparison of the organic matter found in regressive and transgressive sediments that maceral facies deposited on the shelf are related to the amount of terrigenous sediment that reaches the shelf. This finding has special relevance to the lithogenetic approach employed in sequence stratigraphy. Because each systems tract in a sequence exhibits a characteristic stacking of depositional systems, the amount of terrigenous sediment delivered to the shelf varies with changing systems tracts. Therefore, a direct relationship between type and preservation of organic matter and position within the depositional sequence can be discerned. Lowstand systems tracts generally contain organic matter that is both terrestrial in origin and well-preserved. Both TOC and HI values in these sediments are low. In contrast, shelf deposits in the transgressive systems tract contain sparse, highly degraded phytoclasts and exhibit high TOC and HI values. Sediments from the highstand systems tract typically display a gradual decrease in TOC and HI and an increase in the amount of well-preserved phytoclasts as one moves upward from the surface of maximum starvation (also known as the maximum flooding surface or downlap surface). This relationship between maceral facies and position within the depositional sequence makes an integrated approach to sequence stratigraphy possible. The combination of organic matter characterization with data more commonly used in sequence stratigraphy (well log correlations, core and outcrop descriptions, biostratigraphy, and regional stratigraphic relationships) provides greater resolution in the location of critical surfaces (transgressive surfaces and surfaces of maximum starvation) that bound depositional systems tracts with the depositional sequence. Location of these surfaces is important to sequence stratigraphic studies concerned with intervals that are below the limit of seismic resolution. This integrated approach to sequence stratigraphy is summarized in Table 17.3.

References

Arthur, M. A., Dean, W. E., Pollastro, R. M., Claypool, G. E., & Scholle, P. A. (1985). Comparative geochemical and mineralogical studies of two cyclic transgressive pelagic limestone units, Cretaceous Western Interior Basin, U.S. In *Fine-grained Deposits and Biofacies of the Cretaceous Western Interior Seaway: Evidence of Cyclic Sedimentary Processes*, ed. L. M. Pratt, E. G. Kauffman & F. B. Zelt. SEPM Field Trip Guidebook No. 4, 1985 Midyear Meeting, 16–27.

Arthur, M. A., Jenkyns, H. C., Brumsack, H.-J., & Schlanger, S. O. (1990). Stratigraphy, geochemistry, and paleoceanography of organic carbon-rich Cretaceous sequences. In *Cretaceous Resources, Events, and Rhythms: Background and Plans for Research*, ed. R. N. Ginsburg & B. Beaudoin. Dordrecht, The Netherlands: Kluwer Academic Publishers, 75–119.

Arthur, M. A., Schlanger, S. O. & Jenkyns, H. C. (1987). The Cenomanian-Turonian Oceanic Anoxic Event, II. palaeoceanographic controls on organic-matter production and preservation. In *Marine Petroleum Source Rocks*, ed. J. Brooks & A. J. Fleet. Oxford: Blackwell Scientific, *Geological Society Special Publication* **26**, 401–20.

Arthur, M. A. & Silva, I. P. (1982). Development of widespread organic carbon-rich strata in the Mediterranean Tethys. In *Nature and Origin of Cretaceous Carbon-rich Facies*, ed. S. O. Schlanger & M. B. Cita. London: Academic Press, 7–54.

Baum G. R. & Vail P. R. (1988). Sequence stratigraphic concepts applied to Paleogene outcrops, Gulf and Atlantic basin. In *Sea-level Changes: an Integrated Approach*, ed. C. K. Wilgus, B. H. Hastings, C. G. Kendall, H. W. Posamentier, C. A. Ross & J. C. Van Wagoner. *Society of Economic Paleontologists and Mineralogists, Special Publication* **42**, 309–27.

Beckman S. W. (1985). Paleoenvironmental reconstructions and organic matter characterizations of peats and associated sediments from cores in a portion of the LaFourche Delta. Ph.D. dissertation, Louisiana State University.

Bergsohn, I. (1988). Lithofacies architecture of the Tocito Sandstone, northwest New Mexico. M.S. thesis, Department of Geology and Geophysics, Louisiana State University.

Blatt, H., Middleton, G. V. & Murray, R. (1980). *Origin of sedimentary rocks.* 2nd ed. Englewood Cliffs, NJ: Prentice-Hall.

de Boer, P. L. (1986). Changes in the organic carbon burial during the Early Cretaceous. In *North Atlantic Palaeoceanography*, ed C. P. Summerhayes & N. J. Shackleton. Oxford: Blackwell Scientific, *Geological Society Special Publication* **21**, 321–31.

Bohacs, K. M. & Isaksen, G. H. (1991). (abstract) Source quality variations tied to sequence development: integration of physical and chemical aspects, Lower to Middle Triassic, western Barents Sea. *Bulletin of the American Association of Petroleum Geologists*, **75**, 544.

Bostick, N. H. (1971). Thermal alteration of clastic particles as an indicator of contact and burial metamorphism in sedimentary rocks. *Geoscience and Man*, **3**, 83–92.

Bostick, N. H. (1974). Phytoclasts as indicators of thermal metamorphism, Franciscan assemblage and Great Valley sequence (upper Mesozoic), California. In *Carbonaceous Materials as Indicators of Metamorphism*, ed R. R. Dutcher, P. A. Hacquebard, J. M. Schopf & J. A. Simon. *Geological Society of America Special Paper* **153**, 1–17.

Bromley, R. G. & Ekdale, A. A. (1984). Chondrites: a trace fossil indicator of anoxia in sediments. *Science*, **224**, 872–4.

Burgess, J. D. (1974). Microscopic examination of kerogen (dispersed organic matter) in petroleum exploration. In *Carbonaceous Materials as Indicators of Metamorphism*, ed. R. R. Dutcher, P. A. Hacquebard, J. M. Schopf & J. A. Simon. *Geological Society of America Special Paper* **153**, 19–30.

Castaño, J. R. & Sparks, D. M. (1974). Interpretation of vitrinite reflectance measurements in sedimentary rocks and determination of burial history using vitrinite reflectance and authigenic minerals. In *Carbonaceous Materials as Indicators of Metamorphism*, ed. R. R. Dutcher, P. A. Hacquebard, J. M. Schopf & J. A. Simon. *Geological Society of America Special Paper* **153**, 31–52.

Chmura G. L. & Liu, K-B (1990). Pollen in the lower Mississippi River. *Review of Palaeobotany and Palynology*, **64**, 253–61.

Christie-Blick, N. (1991). Onlap, offlap, and the origin of unconformity-bounded depositional sequences. *Marine Geology*, **97**, 35–56.

Christopher R. A. & Hart G. F. (1971). A statistical model in palynology. *Geoscience and Man*, **3**, 49–56.

Coleman, J. & Galloway, W. E. (1990). (abstract) Sequence stratigraphic analyses of the lower Oligocene Vicksburg Formation of Texas. In *Sequence Stratigraphy as an Exploration Tool – Concepts and Practices in the Gulf Coast*, ed. J. M. Armentrout, J. M. Coleman, W. E. Galloway & P. R. Vail. Eleventh Annual Research Conference, Gulf Coast Section Society of Economic Paleontologists and Mineralogists Foundation, Program and Extended Abstracts, 99–112.

Coleman, J. L., Jr. (1990). (abstract) Sequence stratigraphy of the Tertiary of the north central Gulf coastal plain: what goes on between the seismic reflectors. In *Sequence Stratigraphy as an Exploration Tool – Concepts and Practices in the Gulf Coast*, ed. J. M. Armentrout, J. M. Coleman, W. E. Galloway & P. R. Vail. Eleventh Annual Research Conference, Gulf Coast Section Society of Economic Paleontologists and Mineralogists Foundation, Program and Extended Abstracts, 87–98.

Coleman J. & Roberts H. (unpublished). Sedimentary analysis of the Boothville Borehole based on X-radiographic analysis.

Creaney, S., Passey, Q. R. & Allan, J. (1991). (abstract) Use of well logs and core data to assess the sequence stratigraphic distribution of organic-rich rocks. *Bulletin of the American Association of Petroleum Geologists*, **75**, 557.

Crumière, J. P., Crumière-Airaud, C., Espitalié, J. & Cotillon, P. (1990). Global and regional controls on potential source-rock deposition and preservation: the Cenomanian-Turonian Oceanic Anoxic Event (CTOAE) on the European Tethyan margin (southeastern France). In *Deposition of Organic Facies*, ed A. Y. Huc. Tulsa, OK: *American Association of Petroleum Geologists Studies in Geology* **30**, 107–18.

Darby, J. D. (1987). Organic petrology of carbonate systems: Florida Bay study. M.S. thesis, Louisiana State University.

Darrell, J. H. (1973). Statistical evaluation of palynomorph distribution in the sedimentary environments of the modern Mississippi River Delta. Ph.D. dissertation, Louisiana State University.

Darrell, J. H., & Hart, G. F. (1970). Environmental determinations using absolute miospore frequency, Mississippi River Delta. *Geological Society of America Bulletin*, 81(8), 2513–8.

Dean, W. E. & Arthur, M. A. (1989). Iron-sulfur-carbon relationships in organic-carbon-rich sequences I: Cretaceous Western Interior Seaway. *American Journal of Science*, 289, 708–43.

Dean, W. E., Arthur, M. A. & Claypool, G. E. (1986). Depletion of ^{13}C in Cretaceous marine organic matter: source, diagenetic, or environmental signal? *Marine Geology*, 70, 119–57.

Deboo, P. B. (1965). Biostratigraphic correlation of the type Shubuta Member of the Yazoo Clay and Red Bluff Clay with their equivalents in southwestern Alabama. *Geological Survey of Alabama, Bulletin* 80.

Demaison, G. J. & Moore, G. T. (1980). Anoxic environments and oil source bed genesis. *Bulletin of the American Association of Petroleum Geologists*, 64.8, 1179–1209.

Dockery, D. T., III (1982). Lower Oligocene Bivalvia of the Vicksburg Group in Mississippi. *Mississippi Bureau of Geology, Bulletin* 123.

Dockery, D. T., III (1986). Punctuated succession of Paleogene mollusks in the northern Gulf Coastal Plain. *Palaios*, 1, 582–9.

Dockery, D. T., III (1990). (abstract) The Eocene-Oligocene boundary in the northern Gulf – a sequence boundary. In *Sequence Stratigraphy as an Exploration Tool – Concepts and Practices in the Gulf Coast*, ed. J. M. Armentrout, J. M. Coleman, W. E. Galloway & P. R. Vail. Eleventh Annual Research Conference, Gulf Coast Section Society of Economic Paleontologists and Mineralogists Foundation, Program and Extended Abstracts, 141–50.

Dominguez, J. M. L., Martin, L. & Bittencourt, A. C. S. P. (1987). Sea-level history and Quaternary evolution of river mouth-associated beach-ridge plains along the east-southeast Brazilian coast: a summary. In *Sea-Level Fluctuation and Coastal Evolution*, ed. D. Nummedal, O. H. Pilkey & J. D. Howard. Tulsa, OK: *Society of Economic Paleontologists and Mineralogists Special Publication* 41, 115–27.

Dow, W. G. (1977). Kerogen studies and geological interpretations. *Journal of Geochemical Exploration*, 7, 79–99.

Dow, W. G. (1984). Oil source beds and oil prospect definition in the Upper Tertiary of the Gulf Coast. *Gulf Coast Association of Geological Societies Transactions*, 34, 329–39.

Dow, W. G. & O'Connor D. I. (1982). Kerogen maturity and type by reflected light microscopy applied to petroleum exploration. In *How to Assess Maturation and Paleotemperatures. Society of Economic Paleontologists and Mineralogists Short Course* 7, 133–57.

Ekdale, A. A. (1985). Trace fossils and mid-Cretaceous anoxic events in the Atlantic Ocean. In *Biogenic Structures: Their Use in Interpreting Depositional Environments*, ed. H. A. Curran. Tulsa, OK: *Society of Economic Paleontologists and Mineralogists Special Publication* 35, 333–42.

Ekdale, A. A. (1988). Pitfalls of paleobathymetric interpretations based on trace fossil assemblages. *Palaios*, 3, 464–72.

Ekdale, A. A. & Mason, T. R. (1988). Characteristic trace-fossil associations in oxygen-poor environments. *Geology*, 16, 720–3.

Fisher, W. L. & McGowen, J. H. (1967). Depositional systems in the Wilcox Group of Texas and their relationship to occurrence of oil and gas. *Gulf Coast Association of Geological Societies Transactions*, 17, 105–25.

Flexer, A., Rosenfeld, A., Lipson-Benitah, S. & Honigstein, A. (1986). Relative sea level changes during the Cretaceous in Israel. *Bulletin of the American Association of Petroleum Geologists* , 70, 1685–99.

Fontugne, M. R. & Jouanneau, J.-M. (1987). Modulation of the particulate organic carbon flux to the ocean by a macrotidal estuary: evidence from measurements of carbon isotopes in organic matter from the Gironde system. *Estuarine, Coastal and Shelf Science*, 24, 377–87.

Ford, L. N., Jr. & Goodman, D. K. (1988). (abstract) Ecostratigraphic analysis of dinoflagellate data from the Nanjemoy Formation (Ypresian, Maryland, U.S.A.). *Palynology*, 12, 238.

Galloway, W. E. (1989). Genetic stratigraphic sequences in basin analysis II: application to northwest Gulf of Mexico Cenozoic Basin. *Bulletin of the American Association of Petroleum Geologists*, 73, 143–54.

Ganz, H. H., Luger, P., Schrank, E., Brooks, P. W. & Fowler, M. G. (1990). Facies evolution of Late Cretaceous black shales from southeast Egypt. In *Deposition of Organic Facies*, ed. A. Y. Huc. Tulsa, OK: *American Association of Petroleum Geologists Studies in Geology* 30, 217–29.

Glawe L. N. (1969). *Pecten perplanus* stock (Oligocene) of the southeastern United States. *Geological Survey of Alabama Bulletin* 91.

Goodman, D. K. (1990). (abstract) Depositional sequences and the interpretation of biological data in stratigraphy. *Palynology*, 14, 212.

Gregory, W. A., Jr. (1987). Organic petrology of some argillaceous sediments of the Wilcox Group, Allen Parish, Louisiana. M.S. thesis, Louisiana State University.

Gregory, W. A., Jr. (1991). Taxonomy and biostratigraphy of Sabinian palynomorphs from the Wilcox Group (Paleocene-Eocene Epochs) of Southwestern Louisiana. Ph.D. dissertation, Louisiana State University.

Gregory, W. A., Jr. & Hart, G. F. (1990). Subdivision of Wilcox Group (Sabinian) argillaceous sediments using particulate organic matter. *Palynology*, 14, 105–21.

Gregory, W. A., Jr. & Hart, G. F. (1992). Towards a predictive model for the palynologic response to sea level changes. *Palaios*, 7, 3–33.

Gregory, W. A., Jr., Chinn, E. W., Sassen, R. & Hart, G. F. (1991). Fluorescent microscopy of particulate organic matter: Sparta Formation and Wilcox Group, South Central Louisiana. *Organic Geochemistry*, 17(1), 1–9.

Habib, D. (1982). Sedimentary supply origin of Cretaceous black shales. In *Nature and Origin of Cretaceous Carbon-rich Facies*, ed. S. O. Schlanger & M. B. Cita. London: Academic Press, 113–27.

Habib, D. & Miller, J. A. (1989). Dinoflagellate species and organic facies evidence of marine transgression and regression in the Atlantic Coastal Plain. *Palaeogeography, Palaeoclimatology, Palaeoecology*, 74, 23–47.

Hallam, A. & Bradshaw, M. J. (1979). Bituminous shales and oolitic ironstones as indicators of transgressions and regressions. *Journal of the Geological Society of London,* **136**, 157–64.

Haq, B. U., Hardenbol, J. & Vail, P. R. (1987). Chronology of fluctuating sea levels since the Triassic. *Science,* **235**, 1156–67.

Hart, G. F. (1964). A review of the classification and distribution of the Permian miospore taxon: *Disaccate striatiti,* 5th International Congress of Stratigraphy and Geology of the Carboniferous, *Comptes Rendus, Paris,* **3**, 1171–99.

Hart, G. F. (1965). *The Systematics and Distribution of Permian Miospores.* Johannesburg, South Africa: Witwatersrand University Press.

Hart, G. F. (1969). (abstract) Palynoflora and changing paleogeography in a South African Permian coal basin. *Abstracts with Programs for 1969,* Geological Society of America, South Central Section, **2**:13.

Hart, G. F. (1971). *Statistical evaluation of palynomorph distribution in the Atchafalaya Swamp.* Consultant Report. Lafayette, LA: Texaco Inc.

Hart, G. F. (1979a). *Maceral analysis: its use in petroleum exploration.* Methods Paper 2. Baton Rouge: Hartax International.

Hart, G. F. (1979b). (abstract) Maceral analysis and its application to petroleum exploration. In *Petroleum Potential in Island Arcs, Small Ocean Basins, Submerged Margins, and Related Areas.* Symposium UNESCO-Suva, Fiji, Abstracts, **1**, 32–3.

Hart, G. F. (1979c). *Maceral, Geochemical and Lithological Characteristics of the Louisiana Gulf Coast.* Baton Rouge: Hartax International.

Hart, G. F. (1982). Classification and origin of macerals in clastic systems. (abstract) *Palynology,* **6**, 283.

Hart, G. F. (1984). Organic petrology in petroleum exploration. *Third Latin American Congress of Paleontology, Proceedings,* 631–41.

Hart, G. F. (1986). Origin and classification of organic matter in clastic systems. *Palynology,* **10**, 1–23.

Hart, G. F., Ferrell, R. E., Lowe, D. R. & LeNoir, E. A. (1989). Shelf sandstones of the Robulus L zone, offshore Louisiana. Gulf Coast Section, Society of Economic Paleontologists and Mineralogists Foundation Seventh Annual Research Conference, *Proceedings,* 117–41.

Hazel, J. E., Mumma, M. D. & Huff, W. J. (1980). Ostracode biostratigraphy of the lower Oligocene (Vicksburgian) of Mississippi and Alabama. *Gulf Coast Association of Geological Societies Transactions,* **30**, 361–401.

Huddlestun, P. F. & Toulmin, L. D. (1965). Upper Eocene–lower Oligocene stratigraphy and paleontology in Alabama. *Gulf Coast Association of Geological Societies Transactions,* **15**, 155–9.

Hunt, J. M. (1962). *Proceedings of the International Scientific Oil Conference,* Budapest.

Hunt, J. M. (1979). *Petroleum Geochemistry and Geology.* San Francisco: W. H. Freeman.

Jendrzejewski, J. E. (1976). Diatoms and other siliceous biogenic remains from surficial bottom sediments of the deep Gulf of Mexico. Ph.D. dissertation, Louisiana State University.

Jendrzejewski, J. E. & Hart, G. F. (1978). Distribution of siliceous microfossils in the surficial bottom sediments of the Gulf of Mexico. *Palynology,* **2**, 159–66.

Jenkyns, H. C. (1980). Cretaceous anoxic events: from continents to oceans: *Journal of the Geological Society of London,* **37**, 171–88.

Jervey, M. T. (1988). Quantitative geological modeling of siliciclastic rock sequences and their seismic expression. In *Sea-level Changes: an Integrated Approach,* ed. C. K. Wilgus, B. H. Hastings, C. G. Kendall, H. W. Posamentier, C. A. Ross & J. C. Van Wagoner. *Society of Economic Paleontologists and Mineralogists, Special Publication* **42**, 47–69.

Jones, R. W. (1987). Organic facies. In *Advances in Petroleum Geochemistry,* ed. J. Brooks & D. Welte. London: Academic Press, **2**, 1–90.

Jordan, D. W. (1985). Trace fossils and depositional environments of Upper Devonian black shales, east–central Kentucky, U.S.A.. In *Biogenic Structures: Their Use in Interpreting Depositional Environments,* ed. H. A. Curran. Tulsa, Oklahoma: *Society of Economic Paleontologists and Mineralogists Special Publication* **35**, 279–98.

Kauffman, E. G. (1977). Geological and biological overview: Western Interior basin. *Mountain Geologist,* **14**, 75–100.

Kauffman, E. G. (1985). Cretaceous evolution of the Western Interior basin of the United States. In *Fine-grained Deposits and Biofacies of the Cretaceous Western Interior Seaway: Evidence of Cyclic Sedimentary Processes,* ed. L. M. Pratt, E. G. K. & F. B. Zelt. *Society of Economic Paleontologists and Mineralogists Field Trip Guidebook* **4**, 1985 Midyear Meeting, iv–xiii.

Kauffman, E. G. (1988). Concepts and methods of high-resolution event stratigraphy. In *Annual Review of Earth and Planetary Sciences,* ed G. W. Wetherill, A. L. Albee & F. G. Stehli. Palo Alto, CA: *Annual Reviews, Inc.,* **16**, 605–54.

Kraft, J. C., Chrzastowski, M. J., Belknap, D. F., Toscano, M. A. & Fletcher, C. H. (1987). The transgressive barrier-lagoon coast of Delaware: morphostratigraphy, sedimentary sequences and response to relative rise in sea level. In *Sea Level Fluctuation and Coastal Evolution,* ed. D. Nummedal, O. H. Pilkey & J. D. Howard. *Society of Economic Paleontologists and Mineralogists Special Publication* **41**, 129–43.

LeBlanc, C. G., Bourbonniere, R. A., Schwarcz, H. P. & Risk, M. J. (1989). Carbon isotopes and fatty acids analysis of the sediments of Negro Harbor, Nova Scotia, Canada. *Estuarine, Coastal and Shelf Science,* **28**, 261–76.

Leckie, D. A., Singh, C., Goodarzi, F. & Wall, J. H. (1990). Organic-rich, radioactive marine shale: a case study of a shallow-water condensed section, Cretaceous Shaftsbury Formation, Alberta, Canada. *Journal of Sedimentary Petrology,* **60**, 101–17.

Leggett, J. K. (1980). British Lower Palaeozoic black shales and their palaeo-oceanographic significance. *Journal of the Geological Society of London,* **137**, 139–56.

Leithold, E. L. & Bourgeois, J. (1989). Sedimentation, sea-level change, and tectonics on an early Pleistocene continental shelf, northern California. *Geological Society of America Bulletin,* **101**, 1209–24.

LeNoir, E. A. (1987). Palynology of an offshore Miocene section, Gulf Coast, Louisiana. M.S. thesis, Louisiana State University.

LeNoir, E. A. & Hart, G. F. (1987). (abstract) Burdigalian (Early Miocene) dinocysts from offshore Louisiana. *Palynology,* **11**, 239.

LeNoir, E. A. & Hart, G. F. (1988). Palynofacies of some Miocene sands from the Gulf of Mexico, offshore Louisiana, U.S.A. *Palynology,* **12**, 151–65.

Lipson-Benitah, S., Flexer, A., Rosenfeld, A., Honigstein, A., Conway, B. & Eris, H. (1990). Dysoxic sedimentation in the Cenomanian-Turonian Daliyya Formation, Israel. In

Deposition of Organic Facies, ed. A. Y. Huc. *American Association of Petroleum Geologists Studies in Geology* **30**, 27–39.

Loutit, T. S., Baum, G. R. & Wright, R. C. (1983). (abstract) Eocene-Oligocene sea level changes as reflected in Alabama outcrop sections. *Bulletin of the American Association of Petroleum Geologists*, **67**, 506.

Loutit, T. S., Hardenbol, J., Vail, P. R. & Baum, G. R. (1988). Condensed sections: the key to age determination and correlation of continental margin sequences. In *Sea-level Changes: an Integrated Approach*, ed. C. K. Wilgus, B. H. Hastings, C. G. Kendall, H. W. Posamentier, C. A. Ross & J. C. Van Wagoner. *Society of Economic Paleontologists and Mineralogists Special Publication* **42**, 184–213.

Lucotte, M., Hillaire-Marcel, C. & Louchouarn, P. (1991). First-order organic carbon budget in the St. Lawrence lower estuary from ^{13}C data. *Estuarine, Coastal and Shelf Science*, **32**, 297–312.

MacNeil, F. S. (1944). Oligocene stratigraphy of southeastern United States. *American Association of Petroleum Geologists Bulletin*, **28**, 1313–54.

Mancini, E. A. & Tew, B. H. (1988). *Paleogene Stratigraphy and Biostratigraphy of Southern Alabama*. Field Trip Guidebook for the Gulf Coast Association of Geological Societies – Gulf Coast Section/Society of Economic Paleontologists and Mineralogists 38th Annual Convention.

Mancini, E. A. & Tew, B. H. (1990). (abstract) Relationships of Paleogene stage boundaries to group and unconformity-bounded depositional sequence contacts in Alabama and Mississippi. In *Sequence Stratigraphy as an Exploration Tool – Concepts and Practices in the Gulf Coast*, ed. J. M. Armentrout, J.M. Coleman, W.E. Galloway & P.R. Vail. Eleventh Annual Research Conference, Gulf Coast Section Society of Economic Paleontologists and Mineralogists Foundation, Program and Extended Abstracts, 221–8.

Mancini, E. A., Tew, B. H. & Waters, L. A. (1987). Eocene-Oligocene boundary in southeastern Mississippi and southwestern Alabama: a stratigraphically condensed section of a type 2 depositional sequence. In *Timing and Depositional History of Eustatic Sequences: Constraints on Seismic Stratigraphy*, ed. C. A. Ross & D. Haman. *Cushman Foundation for Foraminiferal Research Special Publication* **24**, 41–50.

McCave, I. N. (1973). Mud in the North Sea. In *North Sea Science*, ed. E. D. Goldberg. Cambridge, MA: Massachusetts Institute of Technology Press, 75–100.

McIver, R. D. (1967). Composition of kerogen – clue to its role in the origin of petroleum. *Seventh World Petroleum Congress Proceedings*, **2**, 25–36.

Miall, A. D. (1984). *Principles of Sedimentary Basin Analysis*. New York: Springer-Verlag.

Molenaar, C. M. (1983). Major depositional cycles and regional correlations of Upper Cretaceous rocks, southern Colorado Plateau and adjacent areas. In *Mesozoic Paleogeography of the West-central United States*, ed. M. W. Reynolds & E. D. Dolly. Denver, CO: Rocky Mountain Section, Society of Economic Paleontologists and Mineralogists, 201–24.

Mukhopadhyay, P. K., Hagemann, H. W. & Gormly, J. R. (1985). Characterization of kerogens as seen under the aspect of maturation and hydrocarbon generation. *Erdöl und Kohle*, **38**, 7–18.

Murray, G. E. (1961). *Geology of the Atlantic and Gulf Coastal Province of North America*. New York: Harper.

Nichols, M. M. & Biggs, R. B. (1985). Estuaries. In *Coastal Sedimentary Environments*, ed R. A. Davis. 2nd ed. New York: Springer-Verlag, 77–186.

Nummedal, D. (1990). Sequence stratigraphic analysis of Upper Turonian and Coniacian strata in the San Juan Basin, New Mexico, U.S.A. In *Cretaceous Resources, Events and Rhythms*, ed. R. N. Ginsburg & B. Beaudoin. Boston: Kluwer Academic Publications, 33–46.

Nummedal, D. & Swift, D. J. P. (1987). Transgressive stratigraphy at sequence-bounding unconformities: some principles derived from Holocene and Cretaceous examples. In *Sea Level Fluctuation and Coastal Evolution*, ed. D. Nummedal, O. H. Pilkey & J. D. Howard. *Society of Economic Paleontologists and Mineralogists Special Publication* **41**, 242–60.

Nummedal, D., Swift, D. J. P. & Kofron, B. M. (1989a). Sequence stratigraphic interpretation of Coniacian strata in the San Juan Basin, New Mexico. *Gulf Coast Section, Society of Economic Paleontologists and Mineralogists Foundation 7th Annual Research Conference Proceedings*, 175–202.

Nummedal, D., Wright, R., Cole, R. D. & Remy, R. R. (1989b). Cretaceous Shelf sandstones and shelf depositional sequences, Western Interior Basin, Utah, Colorado, and New Mexico. In *Sedimentation and Basin Analysis in Siliciclastic Rock Sequences*. Washington, DC: American Geophysical Union, vol. 1, section T119.

Nummedal, D., Wright, R. & Swift, D. J. P. (1988). Sequence Stratigraphy of Upper Cretaceous strata of the San Juan Basin, New Mexico. *American Association of Petroleum Geologists/Society of Economic Paleontologists and Mineralogists 73rd Annual Meeting, Field Trip Guidebook*, Houston, TX.

Pasley, M. A. (1991). (abstract) Organic matter variations in a depositional sequence: implications for use of source rock data in sequence stratigraphy. *Bulletin of the American Association of Petroleum Geologists*, **75**, 650.

Pasley, M. A., Gregory, W. A. & Hart G. F. (1991). Organic matter variations in transgressive and regressive shales. *Organic Geochemistry*, **17**, 483–509.

Pasley, M. A. & Hazel, J. E. (1990). Use of organic petrology and graphic correlation of biostratigraphic data in sequence stratigraphic interpretations: example from the Eocene-Oligocene boundary section, St. Stephens Quarry, Alabama. *Gulf Coast Association of Geological Societies Transactions*, **40**, 661–83.

Pavlik, T. S. (1981). Palynology and biostratigraphy of some lignite bearing sediments from Dolet Hills and Naborton Formations (lower Tertiary system) Desoto Parish, northwest Louisiana. M.S. thesis, Louisiana State University.

Pearson, D. L. (1984). *Pollen/spore color standard, version 2*. Bartlesville, Oklahoma: Phillips Petroleum Company, Exploration Projects Section.

Peters, K. E., Ishiwatari, R. & Kaplan, I. R. (1977). Color of kerogen as index of organic maturity. *Bulletin of the American Association of Petroleum Geologists*, **61**, 504–10.

Piel, K. M. (1965). Palynology of some Recent sediments from the Mississippi River Delta. M.S. thesis, Tulane University, New Orleans, LA.

Pierce, R. W. (1975). Coccoliths and related calcareous nannofossils from surficial bottom sediments of the Gulf of Mexico. Ph.D. dissertation, Louisiana State University.

Pierce, R. W. & Hart, G. F. (1978). (abstract) Preferential

preservation among calcareous nannofossils from the bottom of the Gulf of Mexico. *Palynology*, **2**, 228–9.

Pierce, R. W. & Hart, G. F. (1979). *Phytoplankton of the Gulf of Mexico: Taxonomy of the Calcareous Nannoplankton.* Baton Rouge: Louisiana State University School of Geoscience. *Geoscience and Man*, **20**.

Pike, W. S. (1947). Intertonguing marine and nonmarine Upper Cretaceous deposits of New Mexico, Arizona, and southwestern Colorado. *Geological Society of America Memoir* **24**.

Posamentier, H. W. & Chamberlain, C. J. (in press) Sequence stratigraphic analysis of Viking Formation lowstand beach deposition at Joarcam Field, Alberta, Canada. In International Association of Sedimentologists, Special Publication, ed. H. W. Posamentier, B. U. Haq & C. P. Summerhayes.

Posamentier, H. W. & James, D. H. (1991). (abstract) Variations of the sequence stratigraphic model: past concepts, present understandings, and future directions. *Bulletin of the American Association of Petroleum Geologists*, **75**, 655–6.

Posamentier, H. W., Jervey, M. T. & Vail, P. R. (1988). Eustatic controls on clastic deposition I – conceptual framework. In *Sealevel Changes: an Integrated Approach*, ed. C. K. Wilgus, B. H. Hastings, C. G. Kendall, H. W. Posamentier, C. A. Ross & J. C. Van Wagoner. *Society of Economic Paleontologists and Mineralogists Special Publication* **42**, 109–24.

Posamentier, H. W. & Vail P. R. (1988). Eustatic controls on clastic deposition II – sequence and system tract models. In *Sea-level Changes: an Integrated Approach*, ed. C. K. Wilgus, B. H. Hastings, C. G. Kendall, H. W. Posamentier, C. A. Ross & J. C. Van Wagoner. *Society of Economic Paleontologists and Mineralogists Special Publication* **42**, 125–54.

Pratt, L. M. (1984). Influence of paleoenvironmental factors on preservation of organic matter in Middle Cretaceous Greenhorn Formation, Pueblo, Colorado. *Bulletin of the American Association of Petroleum Geologists*, **68**, 1146–59.

Rice, D. D. (1983). Relation of natural gas composition to thermal maturity and source rock type in San Juan Basin, northwestern New Mexico and Southwestern Colorado. *Bulletin of the American Association of Petroleum Geologists*, **67**, 1199–218.

Rice, D. D. & Gautier, D. L. (1983). Patterns of sedimentation, diagenesis, and hydrocarbon accumulation in Cretaceous rocks of the Rocky Mountains. *Society of Economic Paleontologists and Mineralogists Short Course* **11**, 359.

Riley, G. W. & Nummedal, D. (1989). (abstract) Facies architecture of tidal shelf sandstone ridge: Tocito Sandstone. *Bulletin of the American Association of Petroleum Geologists*, **73**, 1172.

Riley, G. W. & Nummedal, D. (1991). (abstract) Sequence boundary modification by submarine erosion, Upper Cretaceous, San Juan Basin. *Bulletin of the American Association of Petroleum Geologists*, **75**, 662.

Robert, P. (1981). Classification of organic matter by means of fluorescence; application to hydrocarbon source rocks. *International Journal of Coal Geology*, **1**, 101–37.

Roberts, H. H. & Coleman, J. M. (1988). Lithofacies characteristics of shallow expanded and condensed sections of the Louisiana distal shelf and upper slope. *Gulf Coast Association of Geological Societies Transactions*, **38**, 291–301.

Ross, L. M. (1980). Geochemical correlation of San Juan Basin oils – a study. *Oil and Gas Journal*, **78**:44, 102–10.

Rullkötter, J. & Mukhopadhyay, P. K. (1986). Comparison of Mesozoic carbonaceous claystones in the western and eastern North Atlantic (DSDP Legs 76, 79, and 93). In *North Atlantic Palaeoceanography*, ed C. P. Summerhayes & N. J. Shackleton. Oxford: Blackwell Scientific, *Geological Society Special Publication* **21**, 377–87.

Savrda, C. E. & Bottjer, D. J. (1986). Trace-fossil model for reconstruction of paleo-oxygenation in bottom waters. *Geology*, **14**, 3–6.

Saxby, J. D. (1982). A reassessment of the range of kerogen maturities in which hydrocarbons are generated. *Journal of Petroleum Geology*, **5**, 117–28.

Schilling, P. E. & Hart, G. F. (1973). Statistical techniques and their application in palynology. *Journal of the International Association for Mathematical Geology*, **5**(3), 297–311.

Schlanger, S. O. & Jenkyns, H. C. (1976). Cretaceous oceanic anoxic events: causes and consequences. *Geologie en Mijnbouw*, **55**, 179–84.

Senftle, J. T., Brown, J. H. & Larter, S. R. (1987). Refinement of organic petrographic methods for kerogen characterization. *International Journal of Coal Geology*, **7**, 105–17.

Showers, W. J. & Angle, D. G. (1986). Stable isotopic characterization of organic carbon accumulation on the Amazon continental shelf. *Continental Shelf Research*, **6**, 227–44.

Smith, G. A. (1981). Evaluation of palynomorph and maceral distribution in sedimentary environments of the modern Mississippi and St. Bernard Deltas. M.S. thesis, Louisiana State University.

Stach, E., Mackowsky, M.-Th., Teichmüller, M., Taylor, G. H., Chandra, D. & Teichmüller, R. (1982). *Stach's Textbook of Coal Petrology*. 3rd ed. Berlin: Gebrüder Borntraeger.

Staplin, F. L. (1969). Sedimentary organic matter, organic metamorphism, and oil and gas occurrence. *Bulletin of Canadian Petroleum Geology*, **17**, 47–66.

Stefani, M. & Burchell, M. (1990). Upper Triassic (Rhaetic) argillaceous sequences in northern Italy: depositional dynamics and source potential. In *Deposition of Organic Facies*, ed. A. Y. Huc. *American Association of Petroleum Geologists Studies in Geology* **30**, 93–106.

Stein, R. & Littke, R. (1990). Organic-carbon-rich sediments and palaeoenvironment: results from Baffin Bay (ODP–Leg 105) and the upwelling area off northwest Africa (ODP–Leg 108). In *Deposition of Organic Facies*, ed. A. Y. Huc. *American Association of Petroleum Geologists Studies in Geology* **30**, 41–56.

Stopes, M. (1935). On the petrology of banded bituminous coal. *Fuel in Science and Practice*, **14**, 4–13.

Summerhayes, C. P. (1981). Organic facies of middle Cretaceous black shales in deep North Atlantic. *American Association of Petroleum Geologists Bulletin*, **65**, 2364–80.

Summerhayes, C. P. (1987). Organic-rich Cretaceous sediments from the North Atlantic. In *Marine Petroleum Source Rocks*, ed. J. Brooks & A. J. Fleet. Oxford: Blackwell Scientific, *Geological Society Special Publication* **26**, 301–16.

Swift, D. J. P. (1976). Continental shelf sedimentation. In *Marine Sediment Transport and Environmental Management*, ed. D. J. Stanley & D. J. P. Swift. New York: John Wiley & Sons, 311–50.

Teichmüller, M. & Wolf, M. (1977). Application of

fluorescence microscopy in coal petrology and oil exploration. *Journal of Microscopy*, **109**, 49–73.

Tissot, B. P. & Welte, D. H. (1984). *Petroleum Formation and Occurrence*. 2nd ed. Berlin: Springer-Verlag.

Tyson, R. V. (1987). The genesis and palynofacies characteristics of marine petroleum source rocks. In, *Marine Petroleum Source Rocks*, ed. J. Brooks & A. J. Fleet. Oxford: Blackwell Scientific, *Geological Society Special Publication* **26**, 47–68.

Vail, P. R. (1987). Seismic stratigraphy interpretation using sequence stratigraphy, part one: seismic stratigraphy interpretation procedure. In *Atlas of Seismic Stratigraphy*, ed. A. W. Bally. *American Association of Petroleum Geologists, Studies in Geology* **27**, 1–10.

Vail, P. R. (1990). Fundamentals of sequence stratigraphy. Sequence Stratigraphy Workbook, Part 4, Short Course Notes – 1990 Annual Meeting Gulf Coast Association of Geological Societies, Lafayette, LA, unpaginated.

Vail, P. R. & Wornardt, W. W. (1990). Well log-seismic stratigraphy: an integrated tool for the 90's. In *Sequence Stratigraphy as as Exploration Tool – Concepts and Practices in the Gulf Coast*, ed. J. M. Armentrout, J. M. Coleman, W. E. Galloway & P. R. Vail. *Eleventh Annual Research Conference, Gulf Coast Section Society of Economic Paleontologists and Mineralogists*, 379–88.

Valasek, D. W. (1991). (abstract) Sequence stratigraphic framework of the Cretaceous Gallup and Tocito Sandstones, San Juan Basin, New Mexico. *Bulletin of the American Association of Petroleum Geologists*, **75**, 687.

Van Pelt, R. & Habib, D. (1988). (abstract) Palynology of the Jurassic Twin Creek Limestone. *Palynology*, **12**, 248.

Van Wagoner, J. C., Mitchum, R. M., Campion, K. M. & Rahmanian, V. (1990). Siliciclastic sequence stratigraphy in well logs, cores, and outcrops: concepts for high-resolution correlation of time and facies. *American Association of Petroleum Geologists Methods in Exploration Series* **7**.

Van Wagoner, J. C., Mitchum, R. M., Posamentier, H. W. & Vail, P. R. (1987). Seismic stratigraphy interpretation using sequence stratigraphy, part 2: key definitions of sequence stratigraphy. In *Atlas of Seismic Stratigraphy*, ed. A. W. Bally. *American Association of Petroleum Geologists, Studies in Geology* **27**, 11–14.

Van Wagoner, J. C., Posamentier, H. W., Mitchum, R. M., Vail, P. R., Sarg, J. F., Loutit, T. S. & Hardenbol, J. (1988). An overview of the fundamentals of sequence stratigraphy and key definitions. In *Sea-level Changes: an Integrated Approach*, ed. C. K. Wilgus, B. H. Hastings, C. G. Kendall, H. W. Posamentier, C. A. Ross & J. C. Van Wagoner. *Society of Economic Paleontologists and Mineralogists Special Publication* **42**, 39–45.

Walker, R. G., ed. (1984). *Facies Models*. 2nd Edition. Geoscience Canada, Reprint Series 1, Geological Association of Canada.

Wanless, H. R. & Weller, J. M. (1932). Correlation and extent of Pennsylvanian cyclothems. *Geological Society of America Bulletin*, **43**, 1003–16.

Waples, D. W. (1985). *Geochemistry in Petroleum Exploration*. Boston: International Human Resources Development Corp.

Weeks, L. G. (1958). Habitat of oil and some factors that control it. In *Habitat of Oil*, ed. L. G. W. Tulsa, OK: American Association of Petroleum Geologists, 1–61.

Wignall, P. B. (1991). Model for transgressive black shales? *Geology*, **19**, 167–70.

Wrenn, J. (1982). Dinocyst biostratigraphy of Seymour Island, Palmer Peninsula, Antarctica. Ph.D. dissertation, Louisiana State University.

Wrenn, J. & Beckman, S. W. (1981). Maceral and total organic carbon analyses of DVDP drill core 11. In *Dry Valley Drilling Project*, ed. L. D. McGinnis. *American Geophysical Union, Antarctic Research Series*, **33**, 391–402.

Wrenn, J. & Beckman, S. W. (1982). Maceral, total organic carbon and palynological analysis of Ross Ice Shelf Project, Site J9 cores. *Science*, **216**, 187–9.

C Quantitative methods and applications thereof

18 Association of palynomorphs and palynodebris with depositional environments: quantitative approaches

WARREN L. KOVACH and DAVID J. BATTEN

Introduction

During the past few years there has been considerable interest in the way in which the composition of the acid resistant organic content of sedimentary rocks can be used to aid the interpretation of depositional environments. However, much of the published work has not involved fully comprehensive analyses. There has been a tendency to concentrate either on the distribution of palynomorphs at the expense of less well characterized phytoclasts, or on the shape and composition of the particulate detritus recovered, with scant attention being paid to the environmental implications of the presence of individual species of microfossils. As a result, much information of potential value to the aims of palynofacies analysis has been lost. In addition, comparatively little use has been made of numerical methods in analyzing and interpreting the data. Multivariate methods, which reduce the dimensionality of the data, can bring the most important trends to the fore and provide a quantitative basis for assessing these in the light of lithology and depositional environment. They also provide a graphical means of presenting the results much more effectively than through written descriptions alone.

In this chapter we review a variety of approaches that may be taken to analyze palynofacies and depositional environments, concentrating on the use of quantitative and multivariate methods. In order to demonstrate the usefulness of these techniques, we re-analyze a number of previously published sets of palynological and palyno-facies data with principal components and cluster analyses, using the computer program MVSP (Kovach, 1990). We also discuss approaches other authors have taken in applying numerical methods to palynofacies studies in order to demonstrate the range of techniques available. The names of the taxa mentioned below are as they appear in the original publications and hence are not necessarily consistent with current practice.

Numerical methods

Principal components analysis

There are many types of multivariate analytical methods available, but in this project we used mainly two techniques: R-mode principal components analysis (PCA) and minimum variance cluster analysis. PCA consists of a series of linear transformations of the original variables (a process called 'eigenanalysis') to extract simplified variables. Ideally, most of the variance in the data set will be accounted for by just two or three of these new variables, so that the data can be described in reduced dimensions. Graphically, it is a rotation of a swarm of data points in multidimensional space, so that the longest axis (that with the greatest variance) becomes the first PCA axis, the second longest perpendicular to the first becomes the second PCA axis, and so on. Thus the first few axes represent the greatest variation in the data set and any major trends should be apparent in these.

In R-mode PCA, the variables are the descriptors (the phytoclast categories or palynomorph taxa in palynofacies analysis). First, a covariance matrix is calculated for these variables. A correlation matrix may also be used if standardization is necessary for variables measured on a different scale or to give greater weight to rare taxa. An eigenanalysis of this matrix is then performed. The resulting eigenvalues indicate the amount of variance accounted for by each axis; the percentage of the total variance can be calculated to give an indication of the relative importance of each axis. Also produced are sets of component loadings (eigenvectors) which indicate the effect of each variable on each extracted PCA axis. Variables which have loadings greater than a certain cutoff point can be taken as the characteristic variables for that axis. The selection of this cutoff point can be arbitrary, but is usually chosen so that just a few (no more than 10%) of the variables are taken as most characteristic; 0.20 or 0.25 are commonly used. The loadings may be either positive or negative, as their signs are arbitrary, but variables that have opposite responses to the trend reflected by a particular PCA axis (such as an aridity-humidity gradient) will have opposite signs.

Component scores for the samples are then derived by multiplying the original data by the component loadings, thus giving a configuration of points that can be plotted to examine the relationships between the samples. Those that have high or low values for a variable with a high loading on a particular axis will be placed at the ends of that axis. Biplots may also be constructed, with the variables being represented by vectors emanating from the origin of a scattergram on which the samples have been plotted (Figs. 18.1, 18.3, 18.5, 18.9). These allow the investigator to see which variables have similar responses to the underlying trends as well as how they are affecting the placement of the samples.

Minimum variance clustering

Ordination methods such as PCA present the major trends in the first few axes of a scatterplot. Hierarchical cluster analysis, on the other hand, represents these trends in two dimensions by grouping together similar objects (samples or variables) in a tree-like form called a dendrogram. There are numerous forms of cluster analysis, but in this study we have mainly used minimum variance (= sum-of-squares or Ward's) clustering. This is generally recognized as being a very effective method that gives distinct clusters (Birks & Gordon, 1985; Pielou, 1984). It is also closely related mathematically to PCA in that both are based on covariance structure (Orloci, 1978), so that the results of the two techniques will be similar but presented in a different manner.

In all types of agglomerative cluster analyses, clusters are formed by fusing objects, one at a time, to other similar objects or groups. Different criteria are used to determine the similarity between them. In minimum variance clustering the criterion is within-group dispersion, defined as the sum of squared distances between each object and the centroid. The object that will increase this dispersion the least when added to a group is the one chosen for inclusion (Pielou, 1984). The results of the analysis are presented in a dendrogram which shows the linkages between the different objects and the distances between each group.

Constrained cluster analysis

It is important to remember that in most forms of cluster analysis the order of objects in the dendrogram is not significant, only the hierarchy of branching which connects the objects. However, there is a special type called constrained cluster analysis in which the objects are clustered so that they remain in their original order

(Birks & Gordon, 1985). This is valuable for analyzing sequential data, such as those collected in biostratigraphy. Distinct zones will be clearly indicated by distinct clusters in the dendrogram. However, if neighboring samples are very different, distortion will be introduced into the dendrogram in the form of reversals (Fig. 18.8). A constrained form of minimum variance clustering has been used in some of the studies discussed below.

Other methods

Many other types of multivariate analyses are available (Davis, 1986; Kovach, 1989; Pielou, 1984). The methods discussed above work well for the analyses presented here, as the data used are close to being normally distributed, particularly after log transformation. In cases where they depart drastically from normality, other non-metric techniques may be more effective (Kovach, 1989). It is best to experiment initially with a variety of methods, observing how well each performs with the available data. The results obtained, along with knowledge of the assumptions made by each method, may then guide the investigator in choosing the best one for a particular study.

Data transformations

Many types of geological data are recorded as proportions. Analysis of proportional data is problematic because of closure, which is the restriction that all variables must sum to a constant value, such as 1.0 or 100.0, or the number of miospores that are counted from each slide. This leads to a negative bias in correlations between variables. For example, if the absolute abundance of one taxon increases and another remains the same, a zero correlation would be expected. However, the increase in the proportion of one necessitates the decrease of the other, thus giving a negative correlation. Many techniques have been developed to deal with these problems. One is the centered logratio transformation (Aitchison, 1986), in which the original proportions are replaced by the log of the ratio between the proportion and the geometric mean of the sample. In mathematical terms, this is:

$$x'_{i,j} = \log(x_{i,j}/g_i)$$

where $x_{i,j}$ = proportion of taxon j in the ith sample; $x'_{i,j}$ = transformed value; $g_i = (x_{i,1} \ldots x_{i,n})^{1/n}$ = geometric mean of the ith sample; n = number of taxa in the sample. Problems arise when some of the proportions are zeros, since calculating the log of zero produces an error. This can be remedied by replacing them with a very small value, such as 0.000001, and then adjusting all other proportions so that the total is 1.0. Aitchison (1986) provided sophisticated techniques for choosing the correct replacement value; this is important particularly when hypothesis testing procedures are being carried out.

An alternative to logratio transformation of compositional data is to use a logarithmic abundance class scheme, such as the octave scale (Gauch, 1982). In this case, percentage data are converted into ten abundance classes, with the range of percentages for each class doubling at each step after the second (i.e., $0 = 0\%$, $1 = >0\%–0.5\%$, $2 = >0.5\%–1\%$, $3 = >1\%–2\%$, $4 = >2\%–4\%$, $5 = >4\%–8\%$, $6 = >8–16\%$, $7 = >16–32\%$, $8 = >32–64\%$, $9 = >64–100\%$). This not only eliminates some of the adverse effects of closure, but also simplifies the data set by reducing minor variations caused by local fluctuations and sampling error, leaving only the basic 'signal' of major variability in the data set.

Abundance class data can be analyzed using many of the standard multivariate analytical methods as long as they are intended only for descriptive purposes and do not include assessments of probability. Indeed, even presence-absence data can be analyzed by PCA and other similar methods (Jolliffe, 1986). The use of abundance classes also opens up the possibility of more rapid data collection by relying on visual estimations of abundances rather than on laborious counting and measuring. Sampling error is likely to be higher with these estimates, but many studies in community ecology (summarized in Orloci, 1978, and Gauch, 1982) show that the inherent variability (or noise) in a community is often even greater. In such cases, nothing is to be gained from increased sampling accuracy. The same situation may apply to paleontological data.

In order to compare the effects of the logratio transformation and the octave class scale, we applied both to a set of data on miospore occurrences in the Wealden of southern England (Batten, 1973) and performed two PCAs using covariance matrices. Figure 18.1 shows biplots of the first two axes from each analysis. The direction and length of the vectors for the dominant miospore taxa are very similar in the two graphs. In the octave analysis (Fig. 18.1B) three more taxa have loadings greater than the 0.2 cutoff and thus are plotted. The configuration of the samples is very similar. A PCA of untransformed percentage data is also shown (Fig. 18.1C). The results for this analysis are completely different from the others in that they are dominated by the four morphological groups with the highest overall proportions (*Classopollis*, *Pilasporites*, bisaccates, and smooth triletes) rather than by taxa whose distributions show the greatest differences between samples (such as *Pilasporites* and *Ischyosporites*, which dominate many Wadhurst Clay samples, and *Trilobosporites* and *Verrucosisporites*, which are most common in samples from the Ashdown Sands and the Fairlight Clay). The amount of variance accounted for by the first two axes of the analysis of the non-transformed data is

larger than for the other analyses, because the total variance is much greater, with the considerable differences in abundance being the main contributor. Comparisons using other data sets also show close similarity between the results of the two transformation methods but great differences for untransformed data (Kovach & Batten, unpublished). Thus, for descriptive purposes, the use of the ten-class octave scale can be a valuable and effective alternative to the logratio transformation.

Re-examination of published palynofacies data

Cretaceous palynofacies from the Lower Saxony Basin

Marshall and Batten (1988) investigated the distribution of dinoflagellate cysts and phytoclasts in the Cenomanian-Turonian 'black shale' sequences from Northern Europe in order to test a depositional model for these sediments. This model proposes that, because of the expansion of shallow shelf seas during the mid-Cretaceous transgressions, warm climate, sluggish oceanic circulation, and increased plankton productivity, large areas of the water column became anoxic, thus allowing much organic material to be deposited rather than biodegraded and destroyed. Our re-examination of their work focuses on two sections in the Lower Saxony Basin at Misburg and Wunstorf. The dinoflagellate and phytoclast data were presented using a five point abundance scale ($0 = $ absent, $1 = $ trace, $2 = $ present, $3 = $ common, $4 = $ abundant), and these were analyzed using minimum variance clustering and PCA.

The cluster analysis (Fig. 18.2) shows four main groups of samples. Group D consists of those that contain very few dinoflagellate cysts; most are from chalk or marly chalk deposits. The rest are divided into two large groups with distinct assemblages and a smaller one in which the palynological characters overlap. Group A contains the *Spiniferites* spp. association identified by Marshall and Batten (1988), which also includes *Pterodinium cingulatum*, *Palaeohystrichophora infusorioides*, and small amounts of organic matter. Group C contains the *Cyclonephelium compactum–C. membraniphorum*/*Eurydinium saxoniense* association and high abundances of organic matter, miospores, and amorphous algal detritus. Group B samples also contain large amounts of organic matter, but *Cyclonephelium* and *Eurydinium* are absent. Instead, *Spiniferites* is present along with the highest abundances of *Litosphaeridium siphoniphorum* and *Palaeohystrichophora infusorioides*.

The first axis of a covariance based PCA of the full data set is dominated by the low diversity samples which form Group D of the cluster analysis. When these are omitted from the analysis (Fig. 18.3A), the first axis clearly

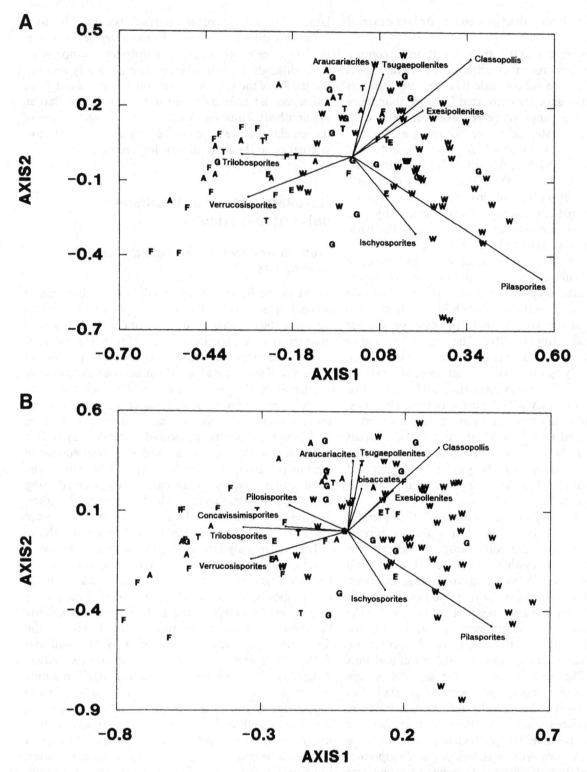

Figure 18.1. Biplots from PCAs of data from Batten (1973). (A) logratio transformation; (B) octave abundance scale; (C) raw percentage data. The letters plotted for each sample indicate geological formation: F – Fairlight Clay (now Fairlight Clays facies of the Ashdown Formation), A – Ashdown Sand (now Ashdown Formation), W – Wadhurst Clay, T – Tunbridge Wells Sand, G – Grinstead Clay, E – Weald Clay. Amount of total variance accounted for by first two axes of each plot: A – 22.72%, B – 22.91%, C – 50.64%.

separates the *Spiniferites* association from that characterized by an abundance of *Cyclonephelium* and *Eurydinium*. The samples in Group B, which contain large numbers of *Litosphaeridium siphoniphorum*, are placed high on the second axis with most being at the *Spiniferites* end of the first. Axis three (Fig. 18.3B) is characterized by some of the less common taxa, including *Canningia ringnesiorum*

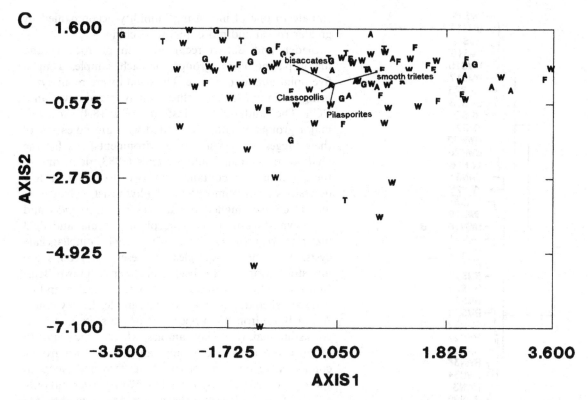

Figure 18.1 *continued*

and *Xiphophoridium alatum*, which occur mainly in the Misburg samples. The fourth axis largely accounts for the differences between samples that are dominated by either *Eurydinium* or *Cyclonephelium*.

All of the samples containing the *Cyclonephelium/ Eurydinium* association and most included in the *Litosphaeridium* group (Group B) are from bituminous marls, which are interpreted as having been deposited in anoxic conditions. The majority of the samples that yielded the *Spiniferites* association are from marls and marly chalk, although a few are from bituminous horizons. Marshall and Batten (1988) interpreted the marls containing this association as having accumulated in open marine, well-oxygenated waters, whereas the few bituminous samples in which it occurs were probably deposited at sites where conditions were locally anoxic but aerated above. *Eurydinium* and *Cyclonephelium* are most common in bituminous samples containing high amounts of bacterially degraded amorphous organic matter which is at least partially of algal origin. It was suggested that the anoxic areas were more widespread at these times, but the continuing presence of *Spiniferites* indicates that the upper portions of the water column were still oxygenated.

The Group B samples are unusual in that *Eurydinium saxoniense* and the *Cyclonephelium compactum–C. membraniphorum* complex are absent, and *Litosphaeridium siphoniphorum* is abundant. This may, in part, be a

reflection of stratigraphic control, as the samples are from the basal portions of both sections. Also, *L. siphoniphorum* is typically present in Cenomanian assemblages, and is cited by Marshall and Batten as evidence of a Cenomanian age for the oldest beds examined. The absence of the *C. compactum–C. membraniphorum* complex, which has a known range of Albian-Santonian (Lentin & Williams, 1989), indicates that there is also a facies difference between these samples and other bituminous deposits in the succession.

In our re-analysis of these data, the use of multivariate techniques has served to confirm that the trends identified by Marshall and Batten (1988) are indeed the major source of variation. The graphic presentation of the results provides a means of clearly illustrating these trends. The analyses have also enabled us to show that the differences between the Group B samples and all of the others are more important than was recognized previously. In the next set of analyses we go one step further and use the numerical results to correlate palynofacies trends with other environmental data.

Palynofacies from the Jurassic and Cretaceous of southern England

Sladen and Batten (1984) described the clay mineralogy and palynofacies of Upper Jurassic and Lower Cretaceous deposits in southern England in order to assess the climate and environments of the sediment source area.

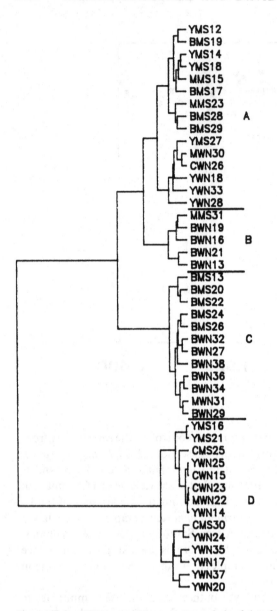

Figure 18.2. Minimum variance cluster analysis of data from Marshall and Batten (1988). The first letter preceding each sample indicates its lithology: B – bituminous marl, M – marl, Y – marly chalk, C – chalk. The second and third letters indicate the locality: MS – Misburg, WN – Wunstorf.

climate to one of increasing humidity accompanied by greater runoff and more intense leaching.

Sladen and Batten recorded both phytoclasts and major palynomorph groups for each sample, using a five-point abundance scale. These data were re-analyzed using minimum variance cluster analysis and covariance PCA. The results of the clustering (Fig. 18.4) show five major groups of samples (plotted with an indication of their suggested depositional environments). As for the analyses of Marshall and Batten's (1988) data, one of these (Group E) contains samples that have a low diversity of palynomorphs and phytoclasts. Groups C and D contain higher abundances of *Classopollis* and other gymnosperm pollen, amorphous vascular and algal material, *Botryococcus*, and, in a few samples, dinoflagellate cysts. Most of the samples in these two groups are mudstones from the Purbeck and lower Ashdown Beds. Many are also calcareous and contain bivalve and/or ostracod shell debris. Most of the samples from Groups A and B are from the upper part of the section, with B consisting mainly of clay samples, whereas the majority of those in A are siltstones and sandstones. Both groups contain large numbers of triradiate (fern and lycopod) spores, such as *Trilobosporites* and *Pilosisporites*, but only a few of the palynomorphs and phytoclasts that are common in Groups C and D.

In the PCA, the first axis (Fig. 18.5a) is dominated by the low diversity samples (Group E in the cluster analysis). The second axis (Fig. 18.5b) reflects the trend from samples dominated by *Classopollis*, other gymnosperm pollen, and algal material (Groups C and D) to samples dominated by triradiate spores (Groups A and B). Most of the Purbeck Beds samples are plotted on the positive end of this axis, while the upper Ashdown Beds samples are at the negative end. The third axis is dominated by amorphous vascular material at the negative end where most of the sandstone and siltstone samples are plotted. Smooth trilete spores, bisaccates, and *Gleicheniidites* prevail at the positive end.

The high abundance of *Classopollis* and other gymnosperm pollen in the Purbeck Beds is indicative of arid or semi-arid conditions (Vakhrameev, 1970), whereas the larger numbers of fern and lycopod spores higher up the section indicate an increasingly humid climate (Sladen & Batten, 1984). The presence of dinoflagellate cysts and amorphous algal material in some of the Purbeck and lower Ashdown Beds samples, and *Botryococcus* in many others, reflect a fluctuation between brackish and fresh water conditions of deposition. However, this feature does not become important in the PCA until the sixth axis (not illustrated), which accounts for only 3.79% of the variance. Although the fluctuations are environmentally interesting, the scarcity of dinoflagellates in most samples renders this of less numerical importance, thus emphasizing the need to note the dominant taxa in more than just the first few axes of a PCA.

We have focused our re-examination on their samples from that part of the Fairlight Borehole which penetrated the Purbeck and Ashdown Beds. The Purbeck Beds include shallow lagoonal and evaporitic deposits near the base of the succession, and sediments that towards the top indicate increasing freshwater influence on the depositional environments. The lower portion of the Ashdown Beds probably accumulated on a mudplain traversed by meandering sandy channels, whereas the upper portion reflects deposition on a braided alluvial sandplain. Clay mineralogical and palynological evidence suggests a significant change from an arid or semi-arid

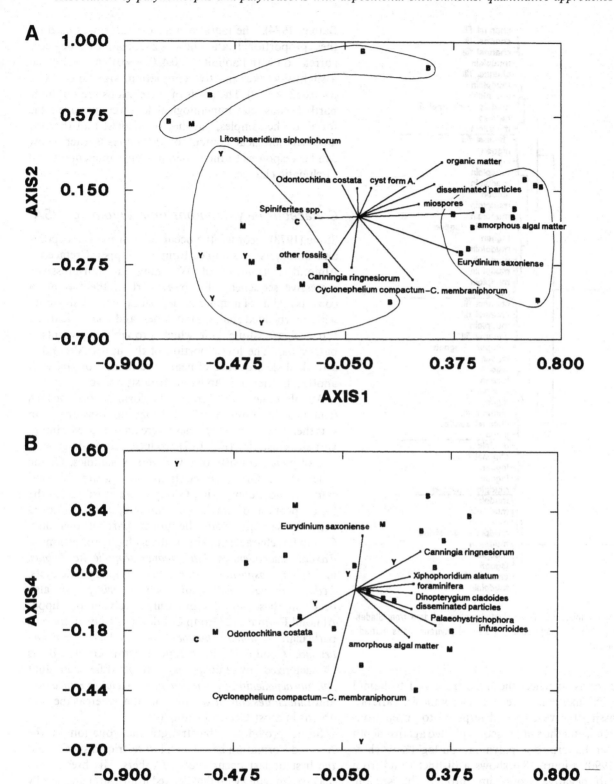

Figure 18.3. Biplot from a PCA of data from Marshall and Batten (1988), omitting taxa of Group D in Figure 18.2. (A) PCA axes 1 and 2; (B) PCA axes 3 and 4. See legend of Figure 18.2 for an explanation of the plotted letters. Amount of total variance accounted for by axes 1 & 2: 34.83%, 3 & 4: 14.30%.

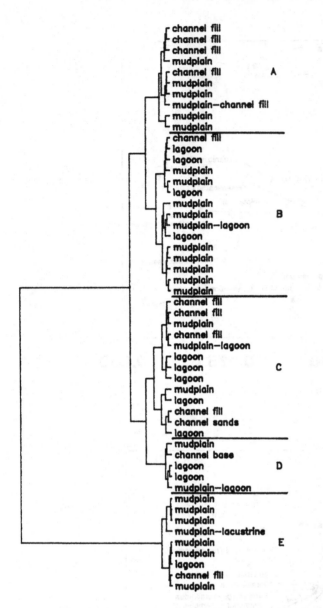

Figure 18.4. Minimum variance cluster analysis of data from Sladen and Batten (1984). The suggested depositional environment is plotted for each sample.

Batten, 1984). The logratio transformation was used on these proportional data. The axis 2 scores show a negative correlation with kaolinite (-0.430, $p = 0.046$, $n = 22$) and a slightly weaker positive correlation with illite (0.420, $p = 0.052$, $n = 22$). The levels of correlations are not high, partly because clay mineralogical data were not available for all of the samples, but they do suggest that the clay and palynological content of the deposits each show similar responses to the underlying environmental trend of climatic change.

Cretaceous palynomorphs from Wyoming, USA

Stone (1973) recorded the occurrence of both miospores and dinocysts in samples from the Upper Cretaceous Almond Formation of Wyoming, a transgressive–regressive sequence. The lower part of the formation consists mainly of nonmarine deposits, primarily siltstones with interbedded shales, sandstones, and coals. Near the top is a sandstone unit which probably represents a barrier bar. The lower portion of the upper Almond is a thick shale sequence of nearshore marine origin, with another barrier bar sandstone deposit above.

All of Stone's data are in the form of proportions (counts of 200 grains per slide). A logratio transformation was therefore used before these were analyzed. Minimum variance clustering (Fig. 18.7) produces four main groups, one of which consists of low diversity samples. Of the other three, Group B contains only lower Almond samples, those comprising Group C are mostly from the lower portion of the upper Almond, and Group D contains samples from the upper part of this unit. Group B is characterized by relatively high proportions of *Taxodiaceaepollenites hiatus*, *Inaperturopollenites dubius*, and *Laevigatosporites ovatus* but few dinoflagellate cysts. *Pediastrum paleogenites* and *Schizosporis parvus* also occur in these samples but only rarely in the upper Almond Formation. Group C is dominated by dinocysts, particularly *Deflandrea microgranulata* and *Trithyrodinium druggii*. Group D, in common with Group B, is characterized by miospores, but it differs in that *Abietineaepollenites foveoreticulatus*, *Arecipites reticulatus*, and *Liliacidites complexus* are common, whereas they are absent in most Group B samples.

Stone provided a biostratigraphic zonation of the Almond Formation based on a few restricted species and the first or last occurrences of others. His breakdown may be compared with that derived from a stratigraphically constrained minimum variance cluster analysis (Fig. 18.8). The separation of the lower and upper parts of the formation is very clearly indicated in the dendrogram, as most of the dinocysts are restricted to the younger unit. Stone's division of the lower Almond into two subzones correlates with the two main subgroups in the dendrogram. However, although he divided the upper Almond into

Axis 2 seems to reflect the trend from arid to humid climates. This axis can be treated as a composite variable, encompassing the vegetational response to an environmental trend, and the sample scores plotted against depth in what can be termed a 'palynomorph log' (Kovach & Batten, 1990). Figure 18.6 shows a distinct trend from positive to negative scores up through the section, reflecting the change from *Classopollis*- to spore-dominated samples. The correlation between these scores and the clay mineralogical data (presented in Fig. 2 of Sladen and Batten, 1984) can also be calculated, particularly for kaolinite, the presence of which indicates high levels of leaching, and for illite, which is commonly formed in alkaline profiles in arid climates (Sladen &

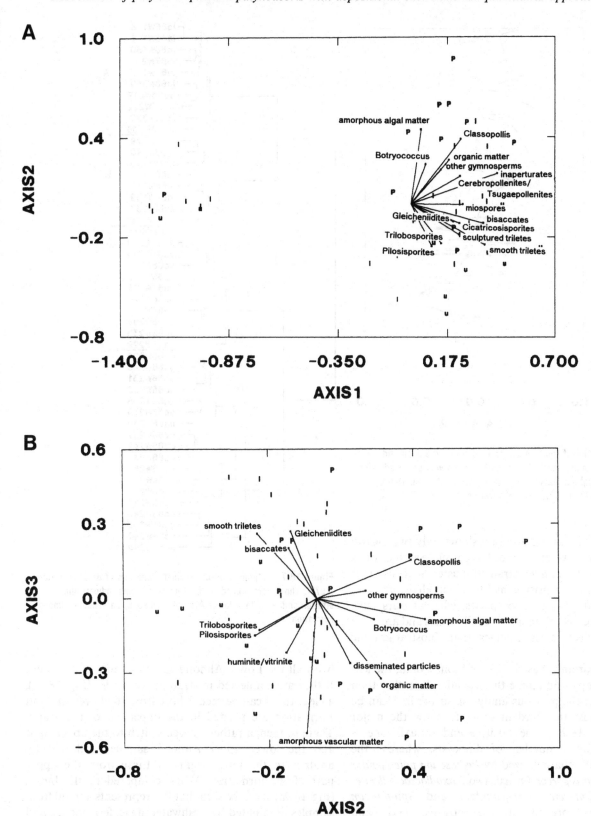

Figure 18.5. PCA of data from Sladen and Batten (1984). The
letters plotted for each sample indicate the stratigraphic unit:
P – Purbeck Beds, I – lower Ashdown Beds, u – upper Ashdown Beds.
Amount of total variance accounted for by axes 1 & 2: 48.47%, 2 & 3:
21.20%.

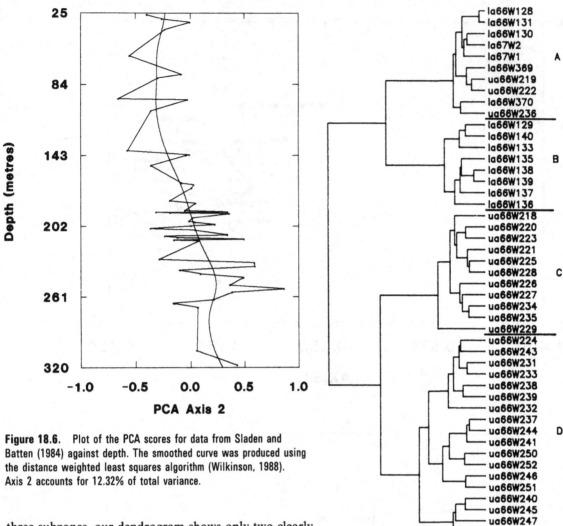

Figure 18.6.   Plot of the PCA scores for data from Sladen and Batten (1984) against depth. The smoothed curve was produced using the distance weighted least squares algorithm (Wilkinson, 1988). Axis 2 accounts for 12.32% of total variance.

Figure 18.7.   Minimum variance cluster analysis of data from Stone (1973). The letters preceding the sample numbers indicate the stratigraphic unit: la – lower Almond Formation, ua – upper Almond Formation.

three subzones, our dendrogram shows only two clearly defined clusters, which more or less correlate with Groups C and D of the non-constrained cluster analysis. This difference is not surprising, as Stone's subdivisions were based on only one or two species, whereas ours have taken into account abundances of all species and are more likely to reflect facies changes than biostratigraphic occurrences.

In the covariance based PCA of Stone's data, the low diversity samples dominate the second axis (rather than the first as in the previous analyses), so the first can be plotted against the third in order to show the major trends (Fig. 18.9a). The positive end of axis one is dominated by a number of dinocysts, whereas the negative end is characterized by *Pediastrum paleogenites* and *Schizosporis parvus*. On axis two, *Taxodiaceaepollenites hiatus*, *Deflandrea microgranulata*, and *Spinidinium densispinatum* prevail at the positive end, and *Abietineaepollenites foveoreticulatus*, *Arecipites reticulatus*, and *Liliacidites complexus* are most important at the negative end.

The samples on the biplot appear to fall into three main groups corresponding to those in the cluster analysis

(with all the Lower Almond samples being one cluster). If they are connected in stratigraphic order (Fig. 18.9b), a clear trend can be seen. Those from the lower Almond Formation are plotted in the upper left of the graph. There is, then, a rather sudden switch to the upper right and the lower upper Almond samples, followed by another to the lower right for those from the upper part of the formation. When compared to the biplot (Fig. 18.9a), it can be seen that this represents a trend from samples dominated by freshwater algae, fern spores, and angiosperms, and (to a lesser extent) gymnosperm pollen, to those dominated by dinoflagellate cysts, then to samples with a suite of gymnosperm and angiosperm pollen types different from that found in the oldest strata.

This trend very clearly represents the vegetational

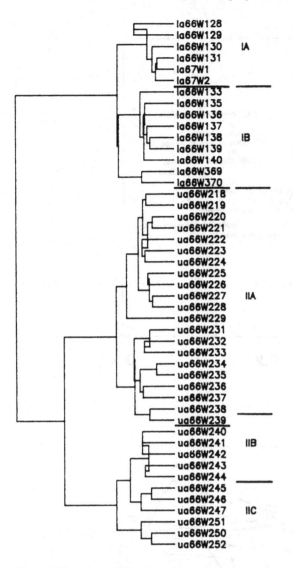

Figure 18.8. Stratigraphically constrained minimum variance cluster analysis of the data from Stone (1973). The letters preceding the sample numbers indicate the stratigraphic unit: la – lower Almond Formation, ua – upper Almond Formation. The subunits described by Stone are indicated to the right of the dendrogram. Note that, because of the restriction that neighboring samples must cluster together, the level of clustering of some taxa is higher than the level of their connection to the next cluster (e.g., samples ua66W243 and ua66W244). This feature is called a reversal.

response to a transgressive-regressive sequence of events, but with different assemblages of land plants dominating in the two nonmarine phases. The presence of *Taxodiaceae-pollenites* as a major component of the assemblages in the lower part of the section suggests that the coals which occur within it were formed in taxodiaceous swamps. The absence of this pollen type and the deposition of pollen grains of other species of angiosperms and gymnosperms during the second nonmarine phase implies that these stable swamp communities had not yet regenerated after the transgression, and that, instead, the newly exposed

land area was colonized by an early successional flora, in which monocots and gymnosperms were important elements.

Other approaches to quantitative palynofacies analysis

The above examples show the usefulness of PCA and minimum variance clustering in analyzing quantitative palynofacies data. These methods provide good results with data that are close to being normally distributed. Their mathematical basis is also relatively simple, which we think is important; nothing is to be gained from using a technique with a complex mathematical basis when a more basic one will provide equally satisfactory results. However, there is a wide variety of other methods which can be used for analyzing multivariate data. We now review some of the approaches that have been taken by other investigators in applying quantitative methods to palynofacies data (summarized in Table 18.1).

Miocene palynofacies from offshore Louisiana

LeNoir and Hart (1988) used several multivariate techniques in their investigation of the distribution of palynomorphs in a Miocene borehole from offshore Louisiana, USA. Principal components analysis was used first to extract the axes which account for the greatest amount of variance. These were then used as the variables in a factor analysis, which is similar to PCA but focuses on the correlation between variables rather than on the amount of variance. The resulting factor loadings were subsequently examined to identify groups of samples with similar values, and palynological zones were erected on this basis. LeNoir and Hart also performed a regression analysis and studied the residuals of some of the more environmentally sensitive species, in order to determine which zones are characterized by these taxa.

The basis on which some of the groups of factor loadings outlined in LeNoir and Hart's (1988) text-figure 5 were chosen is not readily apparent to us. However, their subsequent use of discriminant analysis to test the chosen palynofacies zones for coherence confirms that they are valid. A constrained cluster analysis of the raw data would provide a comparison with the results from the factor analysis. It would also be interesting to see the results of the initial PCA, as it alone might provide a useful ordination of these data without the extra complication of the factor analysis.

On the basis of their analyses, LeNoir and Hart defined seven palynofacies zones. Zones 7 and 1 (the bottom and top zones) show much 'continental' influence, in that they are dominated by many types of angiosperm and gymnosperm pollen. The others are dominated by

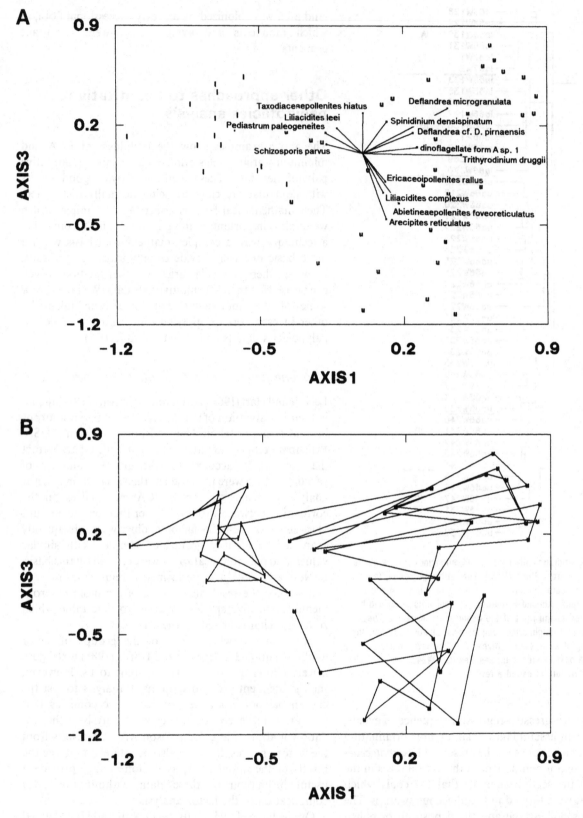

Figure 18.9. PCA of data from Stone (1973). (A) biplot; (B) same configuration of samples as (A) connected in stratigraphic order. The letters plotted for each sample indicate the stratigraphic unit: l – lower Almond Formation, u – upper Almond Formation. Axes 1 & 3 account for 21.14% of total variance.

Table 18.1. *Summary of selected publications demonstrating the use of quantitative methods in palynofacies studies*

The publications are listed in the same order as discussed in the text.

Publication	Type of data	Analyses	Results/comments
Lenoir & Hart, 1988	Counts/300 grains, no transformation	Regression analysis; PCA (matrix type unspecified) with subsequent factor analysis; discriminant function analysis	Seven palynomorph zones distinguished
Boulter & Riddick, 1986	Percentages, no transformation	Average linkage cluster analysis (strategy and distance measure unspecified); PCA (matrix type unspecified)	Identification of types of palynodebris having strong association with certain depositional environments
Lorente, 1986	Shape and size parameters, percentages of palynodebris based on point counting; both forms of data standardized	Minimum variance (Ward's) cluster analysis	Different depositional environments characterized by distribution of palynodebris particles of certain shapes and sizes, and by taxonomic and morphological content
Kovach, 1988	Counts per standard sized sediment sample; data ranked for analysis	UPGMA clustering and non-metric multidimensional scaling; both based on Spearman's Rank Order Correlation Coefficient	Depositional environments characterized by content of megaspores and other plant fragments; hypotheses of habitats of producing plants developed
Farley & Dilcher, 1986	Counts/750 grains; data ranked for clustering	Multinomial homogeneity test; UPGMA clustering based on Spearman's Rank Order Correlation Coefficient; species richness and Simpson's diversity index	Depositional environments characterized by miospore content
Medus *et al.*, 1988	Percentages, log transformed	PCA using centered covariance matrix	Trend identified in miospore distribution in relation to water level fluctuation (may be misinterpreted; see text)
Tyson, 1989	Counts/500 particles or grains, no transformation	Data plotted on scattergrams and ternary diagrams	Environmentally distinct palynofacies identified; hypotheses of sediment transport pathways developed

different assemblages of dinocysts. Zones 6, 5, and 3 are characterized by forms that are common in estuaries, while 4 and 2 contain species usually found in neritic environments. Lithological data suggest that these deposits were laid down in neritic environments, probably on the outer continental shelf. Therefore, the influx of continental and estuarine palynomorphs can be taken to indicate periods of high-energy deposition which

punctuated the otherwise lower energy conditions normally prevailing in the neritic zone.

Palynodebris from the Paleocene of the North Sea

An analysis of the palynodebris in Paleocene deposits from the Forties Field of the North Sea was performed by Boulter and Riddick (1986). Data were originally collected for 36 different classes of palynodebris as percentages. These were then analyzed using PCA and an unspecified form of average linkage cluster analysis. The results were used as a guide for determining which palynodebris classes had similar distributions and thus could usefully be lumped. It is advisable for such a first step to be performed in all studies, particularly in situations where the variables number 100 or more. This initial reduction of the data matrix allows for easier interpretation at later stages and may also enable the identification of groups of variables that need not be distinguished in the data collection phase of subsequent studies.

Boulter and Riddick then re-examined the samples, and data were collected for the resulting 17 variables using an abundance scale accurate to the nearest 5–10%. These were again analyzed using PCA and clustering in order to identify associations of palynodebris categories, the sedimentological value of which was then discussed. This study emphasized the correlations between the different types of palynodebris rather than their variability in distribution. For purposes such as this, factor analysis, which emphasizes correlations between variables rather than variance, may be more appropriate than PCA.

The phrase 'statistically significant' occurs in Boulter and Riddick's paper in the description of the six principal components with eigenvalues greater than 1.0 and in connection with the important variables. However, it must be pointed out that this commonly misused term has a very specific meaning in statistical studies. To avoid confusion and misinterpretation, it should only be used when a significant probability level has been demonstrated with hypothesis testing procedures such as a nested analysis of variance (ANOVA) or regression (Sokal & Rohlf, 1981). PCA as used here is a simple descriptive technique, and there is no element of probability involved. Some statistical tests have been developed for PCA (Morrison, 1976; Jolliffe, 1986), but their use is limited to special circumstances. In fact, PCA and other ordination and clustering methods would be better referred to as 'numerical' or 'exploratory' rather than 'statistical', to emphasize their differences from hypothesis testing techniques.

Tertiary palynofacies from Venezuela

Lorente (1986) used a number of interesting quantitative methods in a study of palynofacies in three cores from the upper Tertiary of Venezuela. The depositional environments of these samples were determined on the basis of sedimentological and paleontological criteria. The different facies were then characterized by their palynomorph and phytoclast content.

Data describing the shape and size of phytoclasts were collected using an image analysis system, and the resulting distributions for each depositional environment were compared using histograms and cumulative curves. Minimum variance (Ward's) cluster analysis was then applied to various combinations of these data, and the relationships between samples from different environments were studied.

Cluster analysis was also used to study the distribution of palynomorph data in each of the cores. Separate analyses were performed on the pollen and spores only, the other palynomorphs, the palynomacerals, the lithological characteristics, and all of the data combined. The placement and grouping of samples from different depositional environments were then examined in an attempt to determine which type of data best characterized each facies. A single dendrogram for the lithological data from all three cores was also constructed which clearly shows clusters corresponding to the major facies. Similar dendrograms for the palynomorph and phytoclast data, with the facies for each sample identified, would also have been useful in determining how well the palynofacies correlate with depositional environments, but these were not produced.

Megaspores and other plant remains from the Cretaceous of Kansas

Kovach (1988) investigated the distribution of megaspores and other small plant fragments in nonmarine and marginal marine deposits of the Cenomanian Dakota Formation of Kansas, USA. Most palynological data were collected as proportions. However, in this study the large size of the fossils allowed the total number in a sediment sample of a standard weight to be recorded. This brought an element of absolute abundance into the analysis, but, as a result, many of the species proved to have non-normal distributions, as the organic matter in some samples is very abundant while in others it is sparse. Non-metric techniques must be used in these circumstances to obtain useful results (Kovach, 1989). Cluster analysis was performed using unweighted pair group average linkage clustering (UPGMA), whereas non-metric multidimensional scaling was employed for ordination. The Spearman Rank Order Correlation Coefficient was used as the similarity measure for both analyses.

Kovach placed emphasis on determining the habitats of the plants which produced these fossils, as well as on characterizing the local depositional environments based on their fossil content. A number of megaspores were found to be common in both estuarine and lagoonal deposits as well as in some fluvial overbank sediments. Since many of these spores are thought to be botanically related to aquatic ferns and isoetalean lycopods, it was concluded that their producing plants were growing in wet areas on the floodplain and that their spores were transported downstream to the marginal marine sites of deposition. Other species of megaspores and seed cuticles were found only in lacustrine or fluvial deposits. Their producing plants probably grew near where they were deposited, but were far enough from the major river system in the area to have prevented long distance transport of the spores.

Miospores from the Cretaceous of Kansas

Farley and Dilcher (1986) have also worked on the palynology of the Dakota Formation, including some of the same localities studied by Kovach (1988). They emphasized correlations between miospore distribution and depositional environment. In order to determine the degree of sampling error, they counted the palynomorphs in three slides from each facies and then compared these replicate samples using a chi-square multinomial homogeneity test, with the result that the replicate counts of samples from all but one of the environments were statistically indistinguishable. This test was also used for comparisons between facies, showing that assemblages from each were significantly different from all of the others.

UPGMA clustering with the Spearman Rank Order Correlation Coefficient was also used in their study. The results obtained, along with an analysis of the species diversity of various miospore groups in the different environments, allowed the four facies sampled to be characterized palynologically. For instance, the swamp deposit was distinct in being of low diversity, lacking bisaccate and inaperturate pollen and containing higher proportions of certain other taxa. A laterally equivalent levee swale deposit at the same locality differed greatly from the swamp deposit in containing high proportions of bisaccate pollen and a greater diversity of miospores.

Miocene miospores from Nigeria

Medus *et al.* (1988) investigated the miospores, diatoms, and sedimentology of Miocene lake deposits from Nigeria. Their data show an environmental transition from an open lake through a boggy forest to poorly drained grasslands. Their use of PCA formed only a minor part of the study. They interpreted the first PCA axis as reflecting fluctuations in water level, as the Poaceae, Cyperaceae, and some types of spores dominate one end, while more aquatic taxa are placed at the other. However, it should also be noted that the grasses, sedges, and spores are by far the most abundant taxa, and that these are greatly separated from the other taxa on the first PCA axis. As shown by Kovach (1989) and in some of the examples discussed above, the first or second PCA axis is often dominated by the most abundant species or the

samples with fewest species. It is clear that the dominance of these four taxa on the first axis is related to their abundance and not to their distribution. In cases like this, it is best to ignore the suspect axis, plot the other major axes (2 and 3 in this case), and examine these for environmental trends.

Kerogen from the Jurassic of the North Sea and the UK

Tyson (1989) used quantitative techniques to analyze trends in Late Jurassic palynofacies from the North Sea and onshore UK. Ten kerogen categories were recognized and measured. The resulting data were analyzed using bi- and trivariate methods such as crossplots and ternary diagrams rather than the more powerful multivariate methods. Thus each analysis was limited to two or three dimensions, corresponding to two or three variables. Therefore all of the variables had to be either plotted separately, which gives a large number of graphs to study in order to discover the most important trends, or combined, thus losing information which could be of value. In either case, a subjective choice must be made as to which trends are most important or which sets of variables may be usefully combined. Inevitably, some important trends may be overlooked, or a poor choice may be made as to which characters to combine. Multivariate techniques allow all variables to be analyzed together so that data are distilled and the most important trends and patterns are more easily discovered. The subjective choices inherent in bi- and trivariate analyses are diminished, because the multivariate techniques allow for the numerical assessment of the importance of each trend in the data.

Despite the disadvantages of bi- and trivariate analyses, they can still provide useful results. Tyson combined the ten kerogen types into three categories (amorphous matter, phytoclasts, and exinitic material), which he then plotted on ternary diagrams. These were used to define ten paleoenvironmentally distinct palynofacies which were subsequently evaluated for their petroleum source potential. Hypotheses about the sediment transport pathways were also developed based on these diagrams. Crossplots of the phytoclast components were used in an attempt to distinguish distal from proximal facies in the basin.

Conclusions

The above examples have shown some of the many approaches that can be taken in analyzing palynofacies data using multivariate methods. These methods not only serve to simplify the data, making interpretation easier, but also provide both a graphical means of effectively demonstrating trends and a numerical basis for these that allows their correlation with other environmental parameters.

Multivariate techniques are not limited to acting on quantitative data. Many of the similarity measures that can be used in cluster analysis and some ordination techniques were specifically developed for dealing with presence-absence, abundance class, or other ordinal data. These types of data may also be analyzed using other forms of ordination, such as principal components analysis, as long as no statistical or probabilistic assumptions are made. The use of the ten-class octave abundance scale can be recommended as a means of simplifying data and removing the 'noise' of minor fluctuations.

Many of the studies reviewed in this chapter have been based on proportional data (including those in which counts out of a standard number of grains are recorded), but none has used transformations designed to eliminate problems caused by closure. As shown in Figure 18.1, the results from raw data can be dramatically different from those obtained after transformation. Thus the results of any analyses of proportional data that do not use the appropriate transformations, such as logratio, should be viewed with caution.

Most multivariate techniques have a variety of different options which can affect the results. For instance, PCA can be calculated using either a covariance or correlation matrix. The standardization implicit in the latter places greater weight on the rarer taxa, thus giving very different results (Kovach, 1989). Unfortunately, few authors specify which technique they have used. There are dozens of similarity measures applicable to cluster analysis, as well as many alternative clustering procedures. Different combinations of these will give varying results. The reader cannot evaluate the results properly if the precise method used in a particular study is not indicated.

Multivariate analyses are based on all available variables; no *a priori* decisions need to be made as to which to choose or combine, a prerequisite of bi- and trivariate analyses. However, it is still necessary to decide which variables to measure. Some palynofacies studies have included abundance data for each species present as well as of phytoclasts. Others are based on higher taxonomic and morphological groups which appear to have had similar environmental responses.

The inclusion of more specific miospore and plankton data usually aids the interpretation of environments of deposition, even when abundances are generally low. However, if occurrences of 100 or more taxa are recorded, interpretation can become difficult and the computer analyses more time consuming (or impossible because of limitations of the hardware). Grouping taxa that reflect similar habitat preferences and depositional conditions can simplify analyses with minimal loss of information.

Known environmental indicators such as *Classopollis* may be used at the generic level, whereas the majority of uncommon elements may be combined into morphological or taxonomic groups. In the early stages of a study, cluster analyses or ordination of all taxa may be performed initially as a guide to their combination into larger groups (Boulter & Riddick, 1986). Use of rank order similarity coefficients at this stage may be preferable so that the dominance of the most abundant taxa is reduced (Kovach, 1989).

Acknowledgments

The work for this chapter was initiated while WLK was the recipient of a NATO Postdoctoral Fellowship in Science, awarded in 1987. It was completed with the financial support of a research grant from BP International Ltd. to DJB for which we are most grateful. BP International gave permission for these results to be published. We would like to thank R. J. Howarth for useful suggestions during the course of this study, C. A. Duigan for comments on the manuscript and assistance with the graphs, and M. B. Farley and A. Traverse for valuable suggestions on a later version of this chapter.

References

Aitchison, J. (1986). *The Statistical Analysis of Compositional Data*. London: Chapman and Hall.

Batten, D. J. (1973). Use of palynologic assemblage-types in Wealden correlation. *Palaeontology*, **16**, 1–40.

Birks, H. J. B. & Gordon, A. D. (1985). *Numerical Methods in Quaternary Pollen Analysis*. London: Academic Press.

Boulter, M. C. & Riddick, A. (1986). Classification and analysis of palynodebris from the Palaeocene sediments of the Forties Field. *Sedimentology*, **33**, 871–86.

Davis, J. C. (1986). *Statistics and Data Analysis in Geology*, 2nd ed. New York: John Wiley & Sons.

Farley, M. B. & Dilcher, D. L. (1986). Correlation between miospores and depositional environments of the Dakota Formation (mid-Cretaceous) of north-central Kansas and adjacent Nebraska, U.S.A. *Palynology*, **10**, 117–33.

Gauch, H. G. (1982). *Multivariate Analysis in Community Ecology*. New York: Cambridge University Press.

Jolliffe, I. T. (1986). *Principal Components Analysis*. New York: Springer-Verlag.

Kovach, W. L. (1988). Quantitative palaeoecology of megaspores and other dispersed plant remains from the Cenomanian of Kansas, USA. *Cretaceous Research*, **9**, 265–83.

Kovach, W. L. (1989). Comparisons of multivariate analytical techniques for use in pre-Quaternary plant paleoecology. *Review of Palaeobotany and Palynology*, **60**, 255–82.

Kovach, W. L. (1990). MVSP: A MultiVariate Statistical Package, ver. 2. *INQUA – Commission for the Study of the Holocene, Working Group on Data-Handling Methods Newsletter*, **4**, 1–3.

Kovach, W. L. & Batten, D. J. (1990). Multivariate analysis

of palynofacies and other micropalaeontological data. *North Sea '90*, Nottingham, England, UK.

LeNoir, E. A. & Hart, G. F. (1988). Palynofacies of some Miocene sands from the Gulf of Mexico, offshore Louisiana, USA. *Palynology*, **12**, 151–66.

Lentin, J. K. & Williams, G. L. (1989). Fossil Dinoflagellates: Index to Genera and Species, 1989 Edition. *American Association of Stratigraphic Palynologists Contribution Series* **20**.

Lorente, M. A. (1986). *Palynology and Palinofacies of the Upper Tertiary of Venezuela*. Berlin: Gebrüder Borntraeger.

Marshall, K. L. & Batten, D. J. (1988). Dinoflagellate cyst associations in Cenomanian-Turonian 'black shale' sequences of northern Europe. *Review of Palaeobotany and Palynology*, **54**, 85–103.

Médus, J., Popoff, M., Fourtanier, E. & Sowunmi, M. A. (1988). Sedimentology, pollen, spores and diatoms of a 148m deep Miocene drill hole from Oku Lake, east central Nigeria. *Palaeogeography, Palaeoclimatology, Palaeoecology*, **68**, 79–94.

Morrison, D. F. (1976). *Multivariate Statistical Methods*, 2nd edition. New York: McGraw-Hill.

Orloci, L. (1978). *Multivariate Analysis in Vegetation Research*. The Hague: Dr W. Junk B.V.

Pielou, E. C. (1984). *The Interpretation of Ecological Data*. New York: John Wiley & Sons.

Sladen, C. P. & Batten, D. J. (1984). Source-area environments of Late Jurassic and Early Cretaceous sediments in Southeast England. *Proceedings of the Geologists' Association*, **95**, 149–63.

Sokal, R. R. & Rohlf, F. J. (1981). *Biometry*, 2nd edition. San Francisco: W. H. Freeman.

Stone, J. F. (1973). Palynology of the Almond Formation (Upper Cretaceous), Rock Springs Uplift, Wyoming. *Bulletins of American Paleontology* **64**, #278.

Tyson, R. V. (1989). Late Jurassic palynofacies trends, Piper and Kimmeridge Clay Formations, UK onshore and northern North Sea. In *Northwest European Micropalaeontology and Palynology*, ed. D. J. Batten & M. C. Keen. Chichester, UK: Ellis Horwood, 135–72.

Vakhrameev, V. A. (1970). Range and paleoecology of Mesozoic conifers, the Cheirolepidiaceae. *Paleontological Journal*, **1970/1**, 12–25.

Wilkinson, L. (1988). *Sygraph*. Evanston, IL: SYSTAT, Inc.

19 A quantitative approach to Triassic palynology: the Lettenkeuper of the Germanic Basin as an example

W. A. BRUGMAN, P. F. VAN BERGEN and J. H. F. KERP

Introduction

The Triassic in its type area, the Germanic Basin, is characterized by a threefold lithologic subdivision: (1) the Buntsandstein, (2) the Muschelkalk, and (3) the Keuper. This facies development can be recognized in large parts of Europe, namely, Germany, Poland, Luxembourg, France, The Netherlands, the North Sea Basin and Switzerland. Muschelkalk carbonates are not represented in Great Britain. In other areas, towards the basin margins, the Muschelkalk is developed in a more clastic facies.

Although the marine standard stages of the Triassic were established in the Alpine Tethys area, the traditional lithologic units of the Germanic Basin are still often incorrectly employed as chronostratigraphic units. These lithologic units are at least partly diachronous (Kelber, 1990). The Buntsandstein and the Keuper are developed in a terrestrial facies. The marine Muschelkalk faunas are largely endemic. Hence, correlations between the Germanic and Tethyan Triassic are extremely difficult. Palynology is one of the very few methods which may enable interregional correlations of marine and nonmarine deposits.

In the last 15 years one of the major research projects of the Laboratory of Palaeobotany and Palynology (Utrecht) has focused on the palynological characterization of the Alpine Triassic standard stages and substages (e.g., Schuurman, 1976, 1977, 1979; Visscher & Krystyn, 1978; Visscher & Brugman, 1981; Van der Eem, 1983; Brugman, 1986). After palynological characterization of the marine standard stages was attempted, research concentrated on the Triassic in its classical area.

The availability of a number of continuously cored wells offered excellent opportunities for detailed palynological investigations. Studies of these well and outcrop sections in the Germanic Basin were carried out in cooperation with the Department of Geology of the Ruhr University Bochum (Germany). Very closely spaced sampling and quantitative analysis enabled recognition

of developments in local vegetation and/or phytoplankton communities.

This study is based on a cored section of the Obernsees well in Franken, Bavaria, Germany (Fig. 19.1). The cored interval includes the basement, Rotliegend and Zechstein (Permian), Buntsandstein, Muschelkalk and Keuper (Triassic), and Lower Jurassic. A 40 m thick, densely sampled interval, covering the uppermost Muschelkalk, the Lettenkeuper and the lowermost Gipskeuper (Lower Myophorienschichten) was studied palynologically. Quantitative analysis enabled the differentiation of locally controlled environmental from large scale regional changes. As a sequel to the stratigraphical interpretation of palynological samples, attention was given to palynofacies analysis. The latter method, which focuses on the total acid-resistant organic matter content, may provide valuable environmental information (e.g., Lorente, 1986; Pocock et al., 1987–1988; Whitaker, 1984). The first results of the palynofacies analysis of Obernsees samples are presented in Van Bergen and Kerp (1990). The present study, combining the results of quantitative palynostratigraphic and palynofacies analyses, shows that such an approach contributes to a more accurate and detailed stratigraphical, depositional and environmental interpretation.

Stratigraphy

This chapter concentrates on the lower Keuper and the base of the middle Keuper, the uppermost unit of the Triassic in its classical subdivision. In addition, six samples from the underlying Muschelkalk were analysed.

In the German Triassic no formations, members and beds have been defined according to stratigraphic guidelines given by Hedberg (1976). Instead, subdivision in 'Schichten' and 'Bänke' is used. Most of these lithologic units bear German names only.

The name Keuper is derived from the old German term 'Kiper' or 'Kipper,' denoting weathered marly

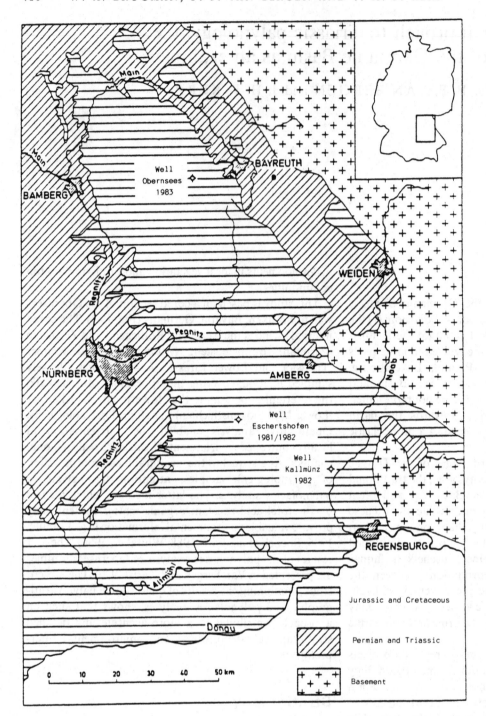

Figure 19.1. Geographical and geological setting of the Obernsees borehole, Germany. (Modified after Gudden & Schmid, 1985.)

sediments. In contrast to the primarily carbonate development of the underlying Muschelkalk, the Keuper consists predominantly of terrigenous clastic sediments. A large number of facies, varying from shallow marine to fluviatile and eolian, are represented in this lithostratigraphical unit. In the studied area, the Keuper is subdivided into: (1) the Lettenkeuper or Lettenkohle (=Lower Keuper), (2) the Gipskeuper and Sandsteinkeuper (=Middle Keuper), and (3) the Rätkeuper (=Upper Keuper). Brief characterizations of the first and second units are given below. The lithostratigraphic subdivision

of the studied interval is presented in Tables 19.1–19.3.

Lettenkeuper

The base of this lithological unit is defined at the base of the Vitriolschiefer. The top of the Grenzdolomit forms the upper limit of the Lettenkeuper.

The Lettenkeuper generally consists of multicolored

Table 19.1. *Quantitative distribution of principal palynomorph categories in the Upper Muschelkalk, Lettenkeuper (= Lower Keuper) and Lower Myophorienschichten of the Obernsees Well. For legend see Table 19.2*

The ecophases and phases are listed at the right. The first three curves at left are based on the second count (= relative frequencies of the land-derived palynomorphs). The other curves are based on the first count and depict the relative frequencies of all palynomorphs, marine elements as well as land-derived pollen, spores and algae.

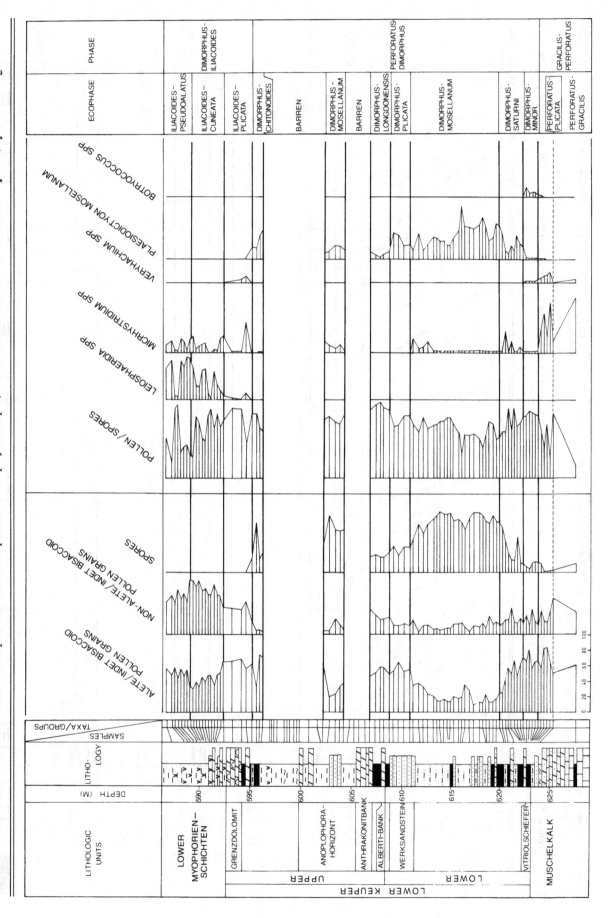

Table 19.2. *Quantitative distribution of relevant land-derived palynomorphs other than alete/indet. bisaccoid pollen grains in the Upper Muschelkalk, Lettenkeuper (= Lower Keuper) and Lower Myophorienschichten of the Obernsees Well*

This diagram is based on counts of all pollen and spores, except the group of alete/indet. bisaccoid pollen grains (= third count; see text). The individual palynomorph taxa distinguished are discussed in Appendix 19.1.

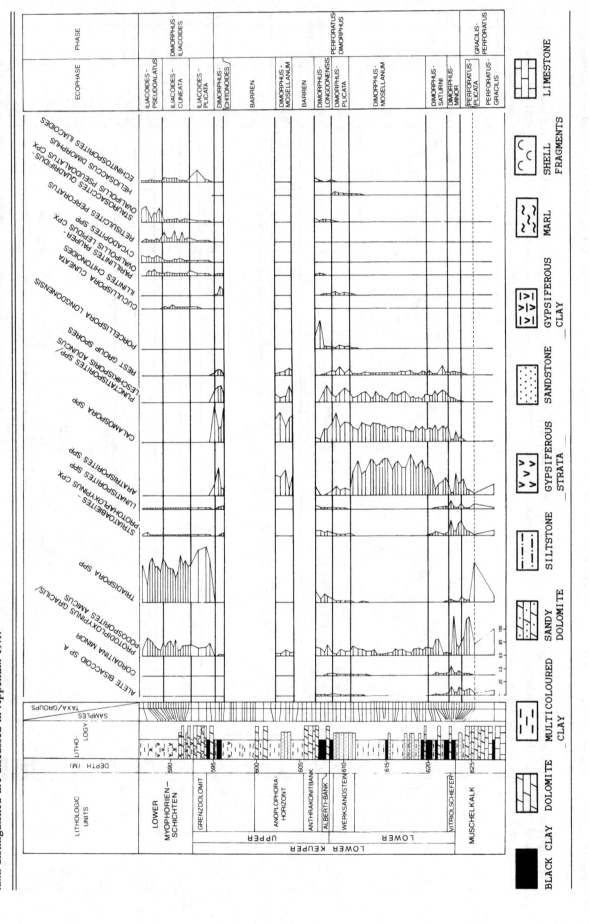

Table 19.3. *Quantitative distribution of organic matter in the Upper Muschelkalk, Lettenkeuper and Lower Myophorienschichten of the Obernsees Well*

General organization of table as for Table 19.1.

AGE			AMMONOID ZONE	PHASE
LADINIAN	UPPER LADINIAN	Upper Langobardian	Protrachyceras regoledanus	dimorphus-iliacoides
			Protrachyceras archelaus	
		Lower Langobardian		perforatus-dimorphus
			Protrachyceras gredleri	
	LOWER LADINIAN	Fassanian	Eoprotrachyceras curianii	gracilis-perforatus
			Nevadites sp.	

Figure 19.2. A comparison between ammonoid zones (modified after Krystyn in Visscher, 1983) and palynological phases in the Ladinian, upper Middle Triassic.

clays and siltstones, alternating with sandstones and dolomites. Thin coal layers are usually present. The Lettenkeuper in this area is subdivided into two major units, Lower and Upper.

THE LOWER LETTENKEUPER

This unit is considered to range from the base of the Vitriolschiefer to the base of the Alberti-Bank, a dolomitic limestone marker bed. The unit consists of alternating black, brown, gray to green clays, and uniformly coloured sandstones and siltstones, with a few dolomite intercalations. The Werksandstein is the main sandstone present in this interval. Plant remains, although generally fragmented, are quite common.

THE UPPER LETTENKEUPER

The Alberti-Bank is the base of the unit while the top of the Grenzdolomit is defined as the top of the Upper Lettenkeuper. In the Obernsees well a two-fold subdivision may be observed within this unit:

(1) The lower part consists of alternations of sandstones, siltstones, clays and dolomites. The main limestone unit is known as the Anthrakonitbank. In contrast to the Lower Lettenkeuper, clastics of the Upper Lettenkeuper are often reddish to violet; dolomites are generally thicker and more frequent.

(2) The upper part, the Grenzdolomit, consists of yellowish, cellular dolomites, and dolomitic limestones with clayey intercalations containing numerous pelecypod fragments.

The Gipskeuper

In Franken, Bavaria, the Gipskeuper is further subdivided into three lithological units, from the base to the top:

(1) the Myophorienschichten, (2) the Estherienschichten, and (3) the Schilfsandstein. Only the lower part of the Myophorienschichten has been studied. The base of this unit, named after the pelecypod, *Myophoria kefersteini*, is defined above the Grenzdolomit, whereas the top is traditionally placed at the base of the Bleiglanzbank (not included in this study). The lower part of the unit consists of evaporite layers (gypsum and/or anhydrite) with green-gray clayey intercalations, and occasionally dolomites. In the upper part evaporite beds are thinner, while clays are variegated, varying from green to red and/or violet. More detailed lithologic subdivisions of the studied interval are given in Emmert *et al.* (1985) and Haunschild (1985).

Chronostratigraphic interpretations of the Triassic of the Germanic Basin and the Alpine Tethys were given by Kozur (1972, 1974, 1975) and Dockter *et al.* (1980). The succession of ammonoid zones of the Ladinian is presented here in Figure 19.2 (after Krystyn, in Visscher, 1983). However, it should be realized that the Triassic of the Germanic Basin is characterized by a primarily endemic fauna. Hence, correlations between the Germanic Basin and the Alpine Tethys based on faunal elements are extremely difficult and controversial. Furthermore, as a result of provincialism, the palynological record from the Germanic Basin is by no means identical to that of the Tethys (see Brugman, 1986).

So far, only very few attempts have been made to correlate the Triassic of the Germanic Basin with the marine standard stages. No detailed biostratigraphical investigations have been undertaken of the interval studied here. However, on the basis of the overall composition of the palynological assemblages, the studied interval may be dated as Ladinian (upper Middle Triassic).

Material and methods

The study is an evaluation of 98 analyzed core samples from the Obernsees well in Franken, Bavaria, Germany (Fig. 19.1). The samples were provided by the Geological Survey of Bavaria. Eighty-two samples were selected for palynological analysis, and 75 for palynofacies analysis.

The investigated samples were prepared for palynological and palynofacies studies according to standard methods, using HCl (35%) and HF (40%) and sieved over a 10 μm sieve (Van Bergen *et al.*, 1990). Preparations were mounted in Elvacite. Permanent slides are stored in the collections of the Laboratory of Palaeobotany and Palynology, Utrecht. The slides were studied in normal transmitted light and under blue incident light.

Terminology

With respect to a palynostratigraphical evaluation of palynomorph assemblages from the Obernsees well, the phase concept propagated by the Laboratory of Palaeobotany and Palynology is followed (cf. Schuurman, 1977, 1979; Van der Zwan & Van Veen, 1978; Van der Zwan, 1980a, 1980b, 1981; Besems, 1981a, 1981b, 1982, 1983; Van Veen, 1981; Van der Eem, 1983; Brugman, 1986; Van Bergen & Kerp, 1990). The application of phases was first introduced by Schuurman (1977). Subsequently, Van der Zwan (1980b, p.192) defined a phase as 'any recognizable step in the (local, regional or interregional) gradual compositional development of successive (palynological) assemblages.'

In Triassic palynostratigraphy, the differential characters of phases have become primarily based on the qualitative composition of the land-derived fraction of palynomorph assemblages, i.e., spores and pollen grains (Brugman, 1986). Phases are considered to reflect gradual interregional time-significant floral developments and are independent of local environmental gradients.

With regard to the characterization of ecophases, emphasis is laid upon the quantitative distribution of palynomorph taxa (Brugman, 1986). Ecophases are considered to reflect developments in the local vegetation or phytoplankton communities. On a local scale they may have stratigraphical value. On a regional scale, however, ecophases do not necessarily represent successive developments of palynological assemblages in time. They may also reflect coeval lateral developments (Brugman, 1986). Therefore, an ecophase may be encountered more than once within a single phase. In addition, an ecophase can be used in ecological considerations (Brugman, 1986). This justifies the replacement of the term 'subphase' of earlier authors, e.g. Brugman (1986) and Van Bergen and Kerp (1990), by the term 'ecophase.'

Phases and ecophases are named after a combination of two species. Phases are preferably named after an element already present in an earlier phase (first part of phase name) and a first appearing or characteristic element (second part of phase name). Ecophases are named after the second part of a phase name (first part of ecophase name) and a characteristic element (second part of ecophase name).

In addition to phases and ecophases, palynofacies units are applied for interpretations of depositional environments. Palynofacies (sub)units are defined on the quantitative total organic matter data. Although ecophases based on palynomorph taxa may allow interpretations of depositional environments, some intervals are difficult to interpret due to the lack of significant palynomorphs. In addition to palynomorphs, taxonomically unassignable organic remains may reveal important information for paleoenvironmental interpretations (Lorente, 1986;

Pocock *et al.*, 1987–1988; Van Bergen & Kerp, 1990; Whitaker, 1984).

Palynofacies units have stratigraphical value on a local scale for the recognition of specific marker beds (Whitaker, 1984). However, on a large scale these palynofacies units have no stratigraphical value. Identifiable palynomorphs do not necessarily have to form part of the total assemblage. They can be absent. This means that samples definitely lacking identifiable palynomorphs can be assigned to (sub)units. These (sub)units indicate an environment of deposition and not the vegetation itself. Hence, such a unit based on unidentifiable organic matter only, might be distinguished in various post-Ordovician sediments. Palynofacies units do not necessarily correlate with ecophases (Fig. 19.3, Subunit IIIa).

Qualitative and quantitative analyses

Qualitative analysis

The slides are analyzed qualitatively and quantitatively with regard to palynomorphs and all other acid–resistant organic matter. All together 125 spore, pollen, alga and acritarch taxa were recognized. These are listed in Appendix 19.1 (annotated list of palynomorph species and categories). Botanical affinities and paleoecological significance of a selection of encountered taxa used for paleoenvironmental interpretations are given in Appendix 19.2.

Ten easily recognizable organic matter types were distinguished for palynofacies analysis. These types can be categorized in three major groups. Since the classification of dispersed organic matter is still a major point of discussion, and a uniformly accepted scheme has not been developed yet, the types applied here are briefly characterized below. In this chapter the types recognized in Van Bergen *et al.* (1990) and Van Bergen and Kerp (1990) are applied. However, it should be realized that some of these categories may have been defined and named differently by other authors. The numbers in parentheses refer to organic matter types recognized by Van Bergen *et al.* (1990).

(I)   Palynomorphs
This category includes:
   (I.1)   *Pollen grains*
   (I.2)   *Spores*
   (I.3)   *Algae*
      (I.3a)   *Unicellular algae* (acritarchs and prasinophytes)
      (I.3b)   *Multicellular algae*

(II)   Structured palynodebris
This category includes all organic remains clearly showing their nature by structure but usually not

LITHOLOGIC UNITS	(SUB) UNITS	ECOPHASE	PHASE	AGE	
Lower Myophorien- schichten	Vc	iliacoides- pseudoalatus	dimorphus- iliacoides	Upper Langobardian	UPPER MIDDLE TRIASSIC
	Vb	iliacoides- cuneata			
Lettenkeuper / Upper	Va	iliacoides- plicata	perforatus- dimorphus	UPPER LADINIAN	
	IIIa	dimorphus- chitonoides			
	-	-			
	IV				
	IIIa	dimorphus- mosellanum			
	IV	-			
	IIIc	dimorphus- longdonensis			
Lower	IIIb	dimorphus- plicata		Lower Langobardian	
	IIIa	dimorphus- mosellanum			
	IIb	dimorphus- saturni			
	IIa	dimorphus- minor			
Upper Muschel- kalk	Ia	perforatus- gracilis	gracilis- perforatus	Fassanian	LOWER LADINIAN
	Ib	perforatus- plicata			
	Ia	perforatus- gracilis			

Figure 19.3. Correlation of ecophases with palynofacies (sub)units for the upper Middle Triassic. Note that ecophases do not necessarily correlate one-to-one with palynofacies (sub)units. Subunit IIIa correlates with the *dimorphus-mosellanum* and with the *dimorphus-chitonoides* ecophases. Palynofacies Unit IV, which is based on assemblages lacking identifiable palynomorphs, cannot be correlated with an ecophase.

taxonomically assignable.

(II.1) *Wood remains*, includes, e.g., tracheids.

(II.2) *Cuticles*, can be distinguished from epidermal material since they display distinct fluorescence, whereas epidermal material does not.

(II.3) *Plant tissue*, includes all plant remains, not recognizable as wood remains or as cuticles. Cell outlines are characteristically visible.

(III) Structureless palynodebris

Organic remains of uncertain biological affinity, usually not showing well defined structure are included in the SOM (=structureless organic matter) category. The major groups were defined on the differences in color. It is known that color is ambiguous since it may depend on thickness, texture and thermal alteration of the material. Nevertheless, in practice the groups distinguished appear to be workable.

(III.1) *Transparent SOM* This group is thought to comprise mainly degradation products of (bisaccoid)[1] pollen grains and algae.

(III.2) *Yellow/Brown SOM* In the studied samples this group presumably represents mainly degradation products of plant tissue; in addition, pollen grains, spores and cuticles may in minor quantities contribute to the group.

(III.3 + III.4). *Black/Brown SOM* The origin of this group is considered to be mostly terrigenous, since degradation products of wood remains and plant tissues are among its precursors. This was verified by bleaching the material.

[1] In the Laboratory of Palaeobotany and Palynology, Utrecht, the term 'bisaccoid' is used for all bisaccate and proto(bi)saccate pollen (*sensu* Scheuring, 1974), because of the difficulty in distinguishing these forms in conventional microscopy.

Quantitative palynomorph analysis

For each slide the palynomorph content was counted in three different stages, in order to determine phases and ecophases.

(1) The general composition of the palynological assemblages (palynodebris excluded) was determined by counting a minimum of 100 palynomorphs per slide. The counts included both marine (e.g., acritarchs and prasinophytes) palynomorphs, as well as land-derived pollen, spores and algae (Table 19.1).

(2) Additional pollen grains and spores were then counted until 100 pollen and spores were identified, to study the compositional changes of the land-derived portion. The relative frequencies of alete/indet. bisaccoid pollen grains and of the land-derived palynomorph fraction exclusive of alete/indet. bisaccoid pollen grains and spores are also presented in Table 19.1.

(3) The third count consists of 100 spores and pollen after the alete/indet. bisaccoid pollen grains were excluded. Alete/indet. bisaccoid pollen grains frequently dominate land-derived palynological assemblages which may result in a squeezing effect. The data of the third count are presented as quantitative relative frequency diagrams in which the effect of dilution or masking is minimized (Table 19.2).

Quantitative palynofacies analysis

The general composition of the dispersed organic matter in each sample was determined in order to establish palynofacies (sub)units. At least 200 particles per slide were counted. The distribution and relative frequencies of the distinguished organic matter types are presented in Table 19.3.

Phases and ecophases and their interpretation

Because (eco)phases do not necessarily correlate with palynofacies (sub)units they are treated separately. The data used to define the ecophases can be found in Tables 19.1 and 19.2.

(1) *gracilis-perforatus* phase
Derivatio nominis: *Protodiploxypinus gracilis* and *Retisulcites perforatus*.

The base of the phase is marked by the appearance of *Retisulcites perforatus*, while the end is defined by the first appearance of *Echinitosporites iliacoides* and/or *Heliosaccus dimorphus*.

Chronostratigraphy: In the Obernsees well this phase can be distinguished in the uppermost part of the Muschelkalk. Based on a comparison with palynological assemblages from the Tethyan realm (Van der Eem, 1983), this phase is considered to indicate a Fassanian (Lower Ladinian) age (Fig. 19.3). However, according to the concepts of Dockter *et al.* (1980) and Kozur (1972, 1974, 1975) this interval can be dated as Fassanian to Lower Langobardian (Upper Ladinian).

(1a) *perforatus-gracilis* ecophase
Derivatio nominis: *Retisulcites perforatus* and *Proto-diploxypinus gracilis*

The dominance of acritarchs (*Micrhystridium* and *Veryhachium* spp.) characterizes this ecophase. Alete bisaccoid pollen grains are the dominant land-derived elements. Spores are extremely rare. Among the land-derived palynomorph fraction exclusive of alete/indet. bisaccoid pollen *Protodiploxypinus gracilis* is abundant. *Aratrisporites, Triadispora plicata*, taeniate bisaccoid pollen and/or 'alete bisaccoid sp. A' occur regularly to frequently.

Paleoenvironment: The alete/indet. and taeniate bi-saccoid pollen reflect an extremely high influx of hinterland elements. *Aratrisporites* spp. are thought to reflect mangrove-like, lycopod vegetation. *Protodiploxypinus gracilis/Podosporites amicus*-producing plants may represent a xerophytic coastal pioneer vegetation growing in front of the mangroves. A swamp or marsh vegetation, represented by spores of equisetophytic and pterophytic affinity, is just barely reflected. Based on the aquatic elements the marine environment is thought to be open to restricted.

(1b) *perforatus-plicata* ecophase
Derivatio nominis: *Retisulcites perforatus* and *Triadispora plicata*

Although only a single sample can be assigned to this ecophase, its striking difference from other assemblages justifies a division of phase 1 into two ecophases. Ecophase 1b is characterized by the dominance of pollen, and the regular presence of acritarchs assignable to *Micrhystridium* spp. Alete bisaccoid pollen grains are dominant among the land-derived elements. Spores are extremely rare. After exclusion of the alete/indet. bisaccoid pollen grains and algae, the ecophase is characterized by the dominance of *Triadispora plicata*. Taeniate bisaccoid pollen grains and *Protodiploxypinus gracilis* occur regularly.

Paleoenvironment: *Triadispora plicata* is here considered a xerophytic element which, judging from its absolute dominance, apparently grew in monospecific stands. It probably grew around saline mudflats. In addition, hinterland plant associations are represented by coniferalean elements which were dominant in the ecophase 1a. Although not so prominent, coastal elements are still present. The marine environment is more restricted than in ecophase 1a.

(2) *perforatus-dimorphus* phase

Derivatio nominis: Retisulcites perforatus and *Heliosaccus dimorphus*

The base of the phase is marked by the first appearance of *Heliosaccus dimorphus* and/or *Echinitosporites iliacoides*; the last occurrence of *Heliosaccus dimorphus* defines the upper limit. *Aulisporites astigmosus, Assereto-spora gyrata, Infernopollenites* spp., *Eucommiidites microgranulatus, Protodiploxypinus decus* and members of the *Partitisporites novimundanus* and *Duplicisporites granulatus* morphons have their first appearance. *Keuperisporites baculatus* is restricted to this phase.

Chronostratigraphy: This phase has been observed to extend approximately from the traditional Muschel-kalk/Lettenkeuper boundary to the base of the Grenzdolomit. The characteristic pollen forms, *Heliosaccus dimorphus* and *Echinitosporites iliacoides*, have not been observed in sediments below the *Protra-chyceras gredleri* and above the *Protrachyceras archelaus* Zones (see Scheuring, 1978; Van der Eem, 1983, for the Alpine/Tethyan Realm). In the Germanic Basin this phase may be correlated with the *Protrachyceras gredleri* Zone and the lower part of the *Protrachyceras archelaus* Zone, which would indicate a Lower Langobardian age. However, according to the concepts of Kozur (1972, 1974, 1975) and Dockter *et al.* (1980) the *perforatus-dimorphus* phase would correlate with the upper part of the *Protrachyceras archelaus* Zone.

(2a) *dimorphus-minor* ecophase

Derivatio nominis: Heliosaccus dimorphus and *Cordaitina minor*

Pollen grains, particularly alete/indet. bisaccoid forms, are the dominant group of palynomorphs. This ecophase is further characterized by the rare, but consistent occurrence of *Botryococcus* spp., *Plaesiodictyon mosellanum, Veryhachium* and *Micrhystridium* spp. (Table 19.1). In the land-derived fraction exclusive of alete/indet. bisaccoid pollen *Protodiploxypinus gracilis* are abundant to dominant while taeniate bisaccoid pollen are abundant. *Punctatisporites, Calamospora, Aratrisporites* spp., 'alete bisaccoid sp. A.' and/or *Cordaitina minor* occur regularly to frequently. *Echinitosporites iliacoides* and *Heliosaccus dimorphus* are extremely rare elements.

The *dimorphus-minor* ecophase correlates with the Vitriolschiefer.

Paleoenvironment: This ecophase represents the beginning of a regressive trend. *Botryococcus* spp. and *Plaesiodictyon mosellanum* are both regarded as fresh-brackish to freshwater elements (Hauschke & Heunisch, 1990). Their appearance indicates the first freshwater influx. The vegetation was apparently similar to that represented by ecophase 1a. However, marsh and swamp elements like *Punctatisporites* and *Calamospora* are present. The sediments were apparently deposited in a brackish to freshwater lagoon, while the surrounding vegetation types consisted predominantly of xerophytic elements although at least some hygrophytes were present.

(2b) *dimorphus-saturnii* ecophase

Derivatio nominis: Heliosaccus dimorphus and *Aratrisporites saturnii*

Alete/indet. bisaccoid pollen grains are the dominant palynomorphs. This ecophase is further characterized by the consistent to common occurrence of *Plaesio-dictyon mosellanum. Micrhystridium* spp. may be common to abundant. Apart from alete bisaccoid pollen grains, *Protodiploxypinus gracilis, Punctati-sporites, Calamospora* and/or *Aratrisporites* spp. are abundant to dominant. Taeniate bisaccoid pollen grains, *Cordaitina minor* and *Podosporites amicus* may occur consistently. *Echinitosporites iliacoides* and *Heliosaccus dimorphus* are extremely rare.

This ecophase is recognized in that part of the Lower Lettenkeuper which immediately overlies the Vitriolschiefer.

Paleoenvironment: The land-derived fraction of this assemblage is comparable to that of the previous ecophase. Although the hinterland vegetation is still strongly represented, an increased influence of freshwater input, here interpreted as a prograding delta, may be observed.

(2c) *dimorphus-mosellanum* ecophase

Derivatio nominis: Heliosaccus dimorphus and *Plaesio-dictyon mosellanum*

Spores are much more common than pollen. *Plaesiodictyon mosellanum* is abundant to dominant, whereas *Micrhystridium* spp. are consistent con-stituents. In the land-derived palynomorph fraction exclusive of alete/indet. bisaccoid pollen grains *Punctatisporites, Calamospora* and *Aratrisporites* spp. are all well represented. The latter genus is most abundant. *Protodiploxypinus gracilis* is a consistent to common element of the ecophase. *Echinitosporites iliacoides* and *Heliosaccus dimorphus* are relatively rare but occur more consistently than within ecophases 2a and 2b.

This ecophase correlates with the sandstone/

siltstone dominated interval of the Lower Letten-keuper below the Werksandstein and the sandstone (−604 m) above the Anthrakonitbank of the Upper Lettenkeuper.

Paleoenvironment: The relative scarcity of *Micrhy-stridium* spp. is probably related to the strong influx of clastic material. Such an influx of clastic material may have made the water turbid. Therefore, the photic zone may have been relatively thin. With regard to the land-derived component, the ecophase is characterized by the dominance of spores. They reflect the extreme proximity of coastal lycopodiophyte dominated swamp/marsh vegetation. The freshwater influence is indicated by the abundance of *Plaesiodictyon mosellanum*.

(2d) *dimorphus-plicata* ecophase
Derivatio nominis: Heliosaccus dimorphus and *Tria-dispora plicata*

Pollen grains, primarily alete/indet. bisaccoid pollen, are dominant among the palynomorphs. *Plaesio-dictyon mosellanum* occurs in significant amounts (Table I), while acritarchs are absent. The consistent occurrence of *Illinites chitonoides, Heliosaccus dimorphus, Triadispora plicata* and *Porcellispora longdonensis* in combination with the dominance of the combined occurrences of *Punctatisporites, Aratrisporites* and *Calamospora* spp. are characteristic for this ecophase. Among the spores, *Calamospora* is most abundant. *Protodiploxypinus gracilis* is still a consistent to common element.

Palynological assemblages, assignable to this ecophase, are restricted to the Werksandstein.

Paleoenvironment: marine influence cannot be demonstrated; acritarchs are absent. Pollen grains assignable to hinterland vegetation types are dominant among the palynomorphs. As in all of the designated ecophases these pollen grains are considered to be primarily water transported. Among the spores, *Calamospora*, reflecting equisetophyte stands growing along rivers, becomes dominant. *Punctatisporites* spp. and other fern spores are also well represented. These indicate the proximity of a hygrophytic fern- and equisetophyte-dominated swamp or marsh vegetation. The consistent occurrence of *Illinites chitonoides*, which most likely reflects reed-like vegetation on the delta plain, is of particular interest.

(2e) *dimorphus-longdonensis* ecophase
Derivatio nominis: Heliosaccus dimorphus and *Porcelli-spora longdonensis*

Alete/indet. bisaccoid pollen grains are the dominant palynomorphs. *Plaesiodictyon mosellanum* occurs in small numbers, while acritarchs are extremely scarce. The land-derived palynomorph fraction of this ecophase exclusive of alete/indet. bisaccoid pollen is characterized by the consistent and generally common occurrence of *Porcellispora longdonensis*, taeniate bisaccoid pollen grains and *Protodiploxypinus gracilis*. *Triadispora* spp., members of the *Parillinites pauper-Ovalipollis lepidus* and *Echinitosporites iliacoides* increase, whereas *Punctatisporites, Calamospora* and *Aratrisporites* spp. show a marked decrease towards the end of the ecophase. It may be noted that *Aratrisporites tenuispinosus* is the dominant species of *Aratrisporites* within this ecophase. *Illinites chitonoides* is consistently present. *Heliosaccus dimorphus* is scarce.

Assemblages assignable to this ecophase have been recognized in the dolomite dominated interval of the Alberti-Bank. This lithological unit, which marks the beginning of the Upper Lettenkeuper, represents a sharp change of environment in comparison with the Lower Lettenkeuper.

Paleoenvironment: This ecophase shows a combination of elements derived from coastal vegetation surrounding the evaporitic tidal flats and xerophytic hinterland vegetation. These replaced the hygrophytic vegetation types of the previous ecophases encountered in the underlying sediments. However, reed-like vegetation, represented by *Illinites chitonoides* pollen grains, is still reflected.

(2f) *dimorphus-chitonoides* ecophase
Derivatio nominis: Heliosaccus dimorphus and *Illinites chitonoides*

Spores and *Plaesiodictyon mosellanum* are the dominant palynomorphs, whereas acritarchs occur in small numbers only (Table 19.1). In the land-derived palynomorph fraction exclusive of alete/indet. bisaccoid pollen, *Illinites chitonoides* is common to abundant. The spores *Punctatisporites, Aratrisporites* and *Calamospora* spp. are all well represented to dominant. Taeniate bisaccoid pollen and *Proto-diploxypinus gracilis* are common.

Assemblages assignable to this ecophase are recognized in the silty/dolomitic interval just below the base of the Grenzdolomit.

Paleoenvironment: This ecophase shows the last reflection of the coastal lycopodiophyte-dominated swamp/marsh vegetation. The prominent freshwater influence is reflected by the abundance of *Plaesio-dictyon mosellanum*. Furthermore, there is evidence for well-developed reed-like vegetation.

(3) *dimorphus-iliacoides* phase
Derivatio nominis: Heliosaccus dimorphus and *Echinitosporites iliacoides*

The beginning of the phase is marked by the last occurrence of *Heliosaccus dimorphus* and/or *Keuperisporites baculatus*, whereas the first occurrence of *Camerosporites secatus* marks the end.

Chronostratigraphy: This phase extends from just below the base of the Grenzdolomit to the top of the studied interval. According to the concepts of Kozur (1972, 1974, 1975) and Dockter *et al.* (1980) this interval may be correlated with the *Frankites sutherlandi* Zone, the North American equivalent of the Alpine/Tethyan *Protrachy-ceras regoledanus* Zone and succeeding *Trachyceras aon* (ammonoid) Zones. Based on the absence of *Echinitosporites iliacoides* in fauna-controlled sediments belonging to the *Trachyceras aon* Zone, we hold the opinion that this phase may well represent the upper part of the *Protrachyceras archelaus* and possibly the *Protrachyceras regoledanus* (ammonoid) Zones. Controversies, however, exist whether the *Frankites sutherlandi* and/or the *Protrachyceras regoledanus* Zones should be regarded as Ladinian or Karnian (see Kozur, 1976). At present, the *dimorphus-iliacoides* phase is considered to represent the Upper Langobardian in the Germanic Basin.

(3a) *iliacoides-plicata* ecophase
Derivatio nominis: Echinitosporites iliacoides and *Triadispora plicata*

Pollen grains, primarily alete/indet. bisaccoid pollen grains, are the dominant group of palynomorphs. Spores are extremely rare, whereas acritarchs, including *Veryhachium* spp., are consistent to abundantly present. *Leiosphaeridia* spp. are relatively rare. The land-derived palynomorph fraction other than alete/indet. bisaccoid pollen is characterized by the dominance of *Triadispora* spp. (including *T. crassa*, *T. plicata* and *T. suspecta*), in combination with the abundance of *Protodiploxypinus gracilis*. *Echinitosporites iliacoides* increases towards the end of the ecophase, whereas *Retisulcites perforatus* becomes a rare but consistent constituent. *Heliosaccus dimorphus* is characteristically absent.

Assemblages assignable to this ecophase include the Grenzdolomit and a few samples from the immediately overlying and underlying beds.

Paleoenvironment: Within the land-derived fraction a distinct change can be observed at the transition of the underlying part of the Lettenkeuper to the Grenzdolomit. Spores and *Plaesiodictyon mosellanum* are extremely rare, whereas conifer and cycadophyte pollen grains become dominant. This association is

here considered to reflect xerophytic vegetation, surrounding evaporitic tidal flats. The conifers probably grew more inland. The presence of algae assignable to *Veryhachium* spp. indicates a more open marine environment in comparison to the other ecophases of the *dimorphus-iliacoides* phase.

(3b) *iliacoides-cuneata* ecophase
Derivatio nominis: Echinitosporites iliacoides and *Cucullispora cuneata*

As in ecophase 3a, pollen grains (alete/indet. bisaccoid pollen grains and *Triadispora* spp.) are the dominant palynomorphs. *Leiosphaeridia* spp. become common to abundant, whereas acritarchs are still common. However, only a single genus of acritarchs, namely *Micrhystridium*, has been encountered. After exclusion of the alete/indet. bisaccoid pollen grains and algae, the ecophase is characterized by the dominance of *Triadispora* spp. in combination with the abundance and occasional dominance of *Protodiploxypinus gracilis*. *Echinitosporites iliacoides* and *Retisulcites perforatus* are consistently to commonly present. The latter is, in addition to *Cucullispora cuneata*, *Cycadopites* spp., and members of the *Staurosaccites quadrifidus-Ovalipollis pseudoalatus* complex, a more common element of this ecophase.

Assemblages, assignable to this ecophase are restricted to the lower part of the Lower Myophorienschichten.

Paleoenvironment: Only the slight increase of *Cucullispora cuneata* is remarkable within the land-derived fraction. This in combination with the disappearance of *Veryhachium* and increase of *Leiosphaeridia* spp. indicates a shift in ecosystems. These aquatic elements reflect a change from a more open marine to a restricted, stressed hypersaline setting.

(3c) *iliacoides-pseudoalatus* ecophase
Derivatio nominis: Echinitosporites iliacoides and *Ovalipollis pseudoalatus*

Pollen grains (alete/indet. bisaccoid pollen grains and *Triadispora* spp.) and/or *Leiosphaeridia* spp. are the dominant groups of palynomorphs (Table 19.1). *Micrhystridium* spp. are generally common. The land-derived palynomorph fraction other than alete/indet. bisaccoid pollen and spores is characterized by the dominance of *Triadispora* spp. in combination with the common occurrence or abundance of *Cycadopites* spp., *Protodiploxypinus gracilis* and members of the *Staurosaccites quadrifidus-Ovalipollis pseudoalatus* complex. *Echinitosporites iliacoides*, *Retisulcites perforatus* and *Cucullispora cuneata* decrease relatively compared to the previous ecophase.

(SUB) UNITS	DEPOSITIONAL ENVIRONMENT	ECOPHASE
Vc	Like Vb, with increased oxidation level	*iliacoides-pseudoalatus*
Vb	Sub- to supratidal, restricted and stressed hypersaline	*iliacoides-cuneata*
Va	subtidal, restricted and stressed marine setting	*iliacoides-plicata*
IV	Inland sabkhas and, or saline mudflats	-
IIIc	Inter- to supratidal part of evaporitic tidalflats	*dimorphus-longdonensis*
IIIb	Fluviatile channels on a deltaic plain	*dimorphus-plicata*
IIIa	Nearshore restricted marine with a higher energy level than IIb	*dimorphus-chitonoides* & *dimorphus-mosellanum*
IIb	Like Ib, but more nearshore	*dimorphus-saturni*
IIa	Brackish lagoonal	*dimorphus-minor*
Ib	Low energy, restricted shallow marine	*perforatus-plicata*
Ia	Low energy, open to restricted, shallow marine	*perforatus-gracilis*

Figure 19.4. Depositional environments of (sub)units and ecophases recognized in the Obernsees well. (Modified after Van Bergen and Kerp, 1990.)

Paleoenvironment: With regard to the land-derived component the only significant change compared to the previous ecophase is the decrease in numbers of *Cucullispora cuneata* and an increase of the *Staurosaccites quadrifidus-Ovalipollis pseudoalatus* complex. This is probably indicative of a more nearshore environment, close to the habitat of the plants that produced *Staurosaccites quadrifidus-Ovalipollis pseudoalatus* pollen.

Palynofacies units

On the basis of relative frequencies of the different organic matter types five units and ten subunits can be recognized in the studied interval (Table 19.3). These (sub)units reflect different depositional environments. Interpretations of the respective depositional environments based on the total organic matter assemblages and lithology are given in Van Bergen and Kerp (1990; see Fig. 19.4). The data used to characterize the (sub)units are given in Tables 19.1 and 19.3. A correlation of (eco)phases with palynofacies (sub)units for the Obernsees well is presented on Figure 19.3.

Unit I

Unicellular algae are characteristic for this unit. In addition, abundant pollen grains, transparent SOM and black-brown SOM are present. This unit is subdivided into two subunits:

(1) Subunit Ia: Algae assignable to *Micrhystridium* and *Veryhachium* are characteristic for this subunit. Furthermore, transparent SOM is dominant while pollen grains, especially bisaccoid pollen and black-brown SOM are abundantly present. Black wood remains are common constituents.

(2) Subunit Ib: Although only a single sample can be assigned to this subunit, its striking difference in organic matter content with those assigned to subunit Ia necessitates a subdivision of Unit I into two subunits. Subunit Ib is characterized by the presence of *Micrhystridium* spp., the dominance of pollen grains, particularly coniferalean, and the absence of *Veryhachium* spp. Transparent SOM is abundant.

Unit II

This unit is characterized by the abundance of pollen grains and transparent SOM. Unit II is subdivided into two subunits:

(1) Subunit IIa: This subunit is characterized by the consistent occurrence of the multicellular alga *Botryococcus* sp. and unicellular algae, namely *Micrhystridium* spp. and *Veryhachium* spp.

(2) Subunit IIb: The common to consistent occurrence of unicellular algae assignable to *Micrhystridium*, is characteristic for subunit IIb.

Unit III

Black-brown SOM is characteristically the dominant category. Pollen grains, spores, multicellular coenobial algae (*Plaesiodictyon mosellanum*), wood remains, plant tissue, transparent and yellow-brown SOM are common to abundant. Three subunits are distinguished:

(1) Subunit IIIa: The subunit is characterized by the common to abundant presence of multicellular algae, *Plaesiodictyon mosellanum*, in combination with the dominance of black-brown SOM, and the common to abundant occurrence of spores (Table 19.3).

   Remark: Plant tissue is more abundant in subunit IIIa than in subunit IIb, whereas pollen grains and transparent SOM are less common in IIIa. This subunit is recognized at three intervals, the upper two being relatively thin.

(2) Subunit IIIb: This subunit is characterized by the absence of unicellular algae, in combination with the dominance of black-brown SOM.

   Remark: Spores, multicellular algae and transparent SOM are less common in subunit IIIb than in IIIa (Table 19.3).

(3) Subunit IIIc: The dominance of black-brown SOM, in combination with rare occurrences of plant tissue and common presence of pollen grains and spores, is characteristic for this subunit.

Unit IV

This unit is characterized by the dominance of black-brown SOM. Black wood remains are common while all the other organic matter types are rare or absent.

Unit V

The dominance of pollen grains, in combination with the occurrence of *Leiosphaeridia* spp., is characteristic for this unit. Wood remains, black-brown and transparent SOM are abundant. Three subunits are distinguished:

(1) Subunit Va: The unicellular algae, *Leiosphaeridia*, *Micrhystridium* and *Veryhachium* spp. are characteristic. Spores, multicellular algae and plant tissue are rare.

   Remark: *Leiosphaeridia* spp. are less common in comparison to the overlying subunits of Unit V.

(2) Subunit Vb: The common to abundant occurrence of *Leiosphaeridia* and *Micrhystridium* spp. are characteristic for this subunit. Spores, multicellular algae and plant tissue are absent.

(3) Subunit Vc: This subunit is characterized by the presence of *Leiosphaeridia* and *Micrhystridium* spp. together with abundant black-brown SOM.

   Remark: The amount of transparent SOM is lower than in the underlying subunit Vb.

Integrated palynological, palynofacies and sedimentological interpretation

The studied section can roughly be divided into two major parts: (1) a mainly carbonate to clayey/sandy lower part, and (2) a partly evaporitic/multicolored clayey upper part which starts above the Werksandstein (Table 19.1).

   The Germanic Basin did not show much difference in relief. Carbonate sediments were deposited in the relatively shallow Muschelkalk sea. Regarding the thickness of the sequence and the uniformity of the lithological composition, the subsidence rate must have been relatively high and stable. Sediments deposited in this rapidly subsiding basin mainly originated from the uplifted Fennoscandian Shield in the north (Kelber, 1990; Mader, 1990). Also the nearby Bohemian-Vindelician Massif should be considered a source of sediment supply. Large deltaic systems started to develop when the sedimentation rate began to exceed the subsidence rate. The deltaic system of the Lower Lettenkeuper prograded from northeast to southwest. The youngest of these diachronous delta deposits occurs in the southwestern part of the basin (Kelber, 1990).

   The main regressive trend of the lower part of the sequence is recognizable in the lithology as well as in the sedimentary organic matter content. The percentage of open-marine phytoplankton decreases upwards, whereas freshwater algae become more common. This increase in freshwater algae is correlated with a higher influx of land-derived organic matter (e.g., spores, plant tissue and brown wood remains). These latter elements are considered not to have been transported very far (Van Bergen & Kerp, 1990).

   The depositional environment changed from open/restricted shallow marine, via brackish lagoonal to deltaic/fluviatile. This latter environment is represented by channel sands (Hauschke & Heunisch, 1990; Van Bergen & Kerp, 1990). The impact of this regression on the local ecosystem development is obvious.

   The basal part of the sequence, deposited in an open marine environment, shows a high abundance of alete bisaccoid pollen, *Protodiploxypinus gracilis* and *Podosporites amicus*. The alete bisaccoid pollen are thought to be produced by conifers forming part of the hinterland plant associations. *Protodiploxypinus gracilis* and *Podosporites amicus* apparently represent xerophytic coastal pioneers. The lycopod spore *Aratrisporites*, which reflects mangrove vegetation, is also present, but in minor quantities. A general upward relative decrease of hinterland elements can be noticed, while mangrove, marsh and swamp elements become more common. The top of the lower part of the sequence is characterized by a high abundance of the equisetophyte spore *Calamospora*. As a result of the persisting regression, coastal vegetation was slowly replaced by marsh and swamp vegetation types growing on the encroaching deltaic system.

   Whereas the lower part of the studied interval shows a clearly recognizable regressive trend from open marine to deltaic, the upper part is much more varied. The upper part of the sequence mainly consists of evaporitic tidal flat, saline mudflat and inland sabkha deposits (Eugster & Hardie, 1975; Handford, 1982; Kendall, 1979; Aigner & Bachmann, 1989). Large-scale trends cannot be recognized in this interval. However, regarding the basin development, even minor sea level fluctuations had a major impact on such a mosaic pattern of lowland and marginally marine environments. Palynological assemblages show the presence of a number of, at least partly coexisting vegetation types. Because of the existence of a mosaic pattern of sedimentary environments no general trends in the floral succession have been observed. The most important vegetation types apparently were xerophytic coastal vegetation surrounding evaporitic tidal flats (ecophase 3a) and two different types of halophytic coastal vegetation (ecophases 3b and 3c). The upper part of the sequence includes two barren intervals which are interpreted as saline mudflat and inland sabkha deposits. These deposits yielded some black SOM and black woody remains, indicating high oxidation levels (Hauschke & Heunisch, 1990; Van Bergen & Kerp, 1990).

In the upper part of the sequence several short transgressive phases can be recognized, among which the Grenzdolomit represents the major one. This sudden event had a considerable impact over a large area and resulted in the formation of a stressed, marginal marine environment (Huth, 1954; Beutler & Schüler, 1979; Duchrow, 1984). This large transgressive pulse at the end of the Lettenkohle created large saline mudflats destroying the last hygrophytic vegetation represented by elements of ecophase 2f.

In addition, two small channels are present in the upper part of the studied sequence. One of them, encountered at −596 m, is even too small to be indicated on Tables 19.1–19.3. Such small channels are considered to be local phenomena (Mader, 1990). The palynomorph assemblages from the channel deposits show an abundance of hygrophytic elements which apparently reflect the local vegetation. These associations are similar to those from the sediments underlying the Werksandstein (Tables 19.2 and 19.3).

Mader (1990) interpreted floral changes observed in the studied stratigraphic interval in terms of paleoclimatic fluctuations. According to his view, the climate would have changed from semi-arid in the Muschelkalk to humid in the Lettenkeuper to semi-humid again in the lower Myophorienschichten. His interpretation of the Letten-keuper as humid is founded on the high abundance of hygrophytic elements. In addition he also mentioned the presence of xerophytes. However, it should be noted that these conclusions are exclusively based on the analysis of megaflora assemblages and on sedimentological criteria. Hygrophytic taxa which occur locally in wetland areas generally have a much higher preservation potential than hinterland elements. Therefore they are usually overrepresented in taphofloras. Even in palynofloral associations which normally provide a more complete picture, regional developments are easily masked by environmentally controlled local changes. Although they are usually present in minor quantities only, hinterland elements normally give a much better indication of climatic changes (Visscher & Van der Zwan, 1981). In order to avoid such biased views, stepwise counting procedures as applied in this study should be used. Only then does it become clear that xerophytic elements occur

persistently throughout the interval studied here. Therefore, the paleofloristic succession in the Middle Triassic as noticed by Mader (1990), in our opinion, represents a rather local, environmentally controlled floral change rather than a paleoclimatic fluctuation. During the deposition of the Muschelkalk, the Lettenkeuper and the Lower Myophorienschichten the climate remained semi-arid. The high abundance of hygrophytic elements in the Lettenkeuper can be ascribed to environmental changes which temporarily resulted in the flourishing of freshwater influenced ecosystems. Similarly, Visscher and Van der Zwan (1981) considered palynological assemblages from the circum-Mediterranean Upper Triassic (Karnian), which are dominated by hygrophytic elements to indicate a humid environment rather than a humid climate.

Concluding remarks

There is every indication that quantitative palynological/palynofacies analysis may contribute considerably to a better understanding of the stratigraphical, depositional and environmental interpretation of acid-resistant organic matter groups and their source sediments. The recognition of (eco)phases and palynofacies units may provide a distinct link between the depositional environment and the source areas of the organic material. Detailed quantitative palynological analyses can prove the co-existence of different ecosystems, as is demonstrated here for the Middle Triassic. Changes in overall palynological composition do not necessarily indicate climatic changes.

The quantitative approach followed here should be the basis for an adequate palynological characterization of the Triassic of the Germanic Basin. Eventually, when more information on other lithostratigraphic units of the Germanic Basin becomes available, a correlation with the marine standard stages of the Triassic can be attempted.

Appendix 19.1: Annotated list of palynomorph species and categories

SPORES
Laevigati
Calamospora spp. (Fig. 19.5.3)
Punctatisporites spp. (Fig. 19.5.4)
Aulisporites astigmosus Klaus, 1960
Apiculati
Keuperisporites baculatus Schulz, 1965 (Fig. 19.5.6)
Cyclogranisporites-Verrucosisporites complex
   Remarks: This complex includes species assignable to such form-genera as *Cyclogranisporites*, Mädler, 1964, *Cyclotriletes* Mädler, 1964 and *Verrucosisporites* Ibrahim, 1933 emend. Smith, 1971.
Porcellispora longdonensis (Clarke, 1965) Scheuring, 1970 (Fig. 19.5.7)

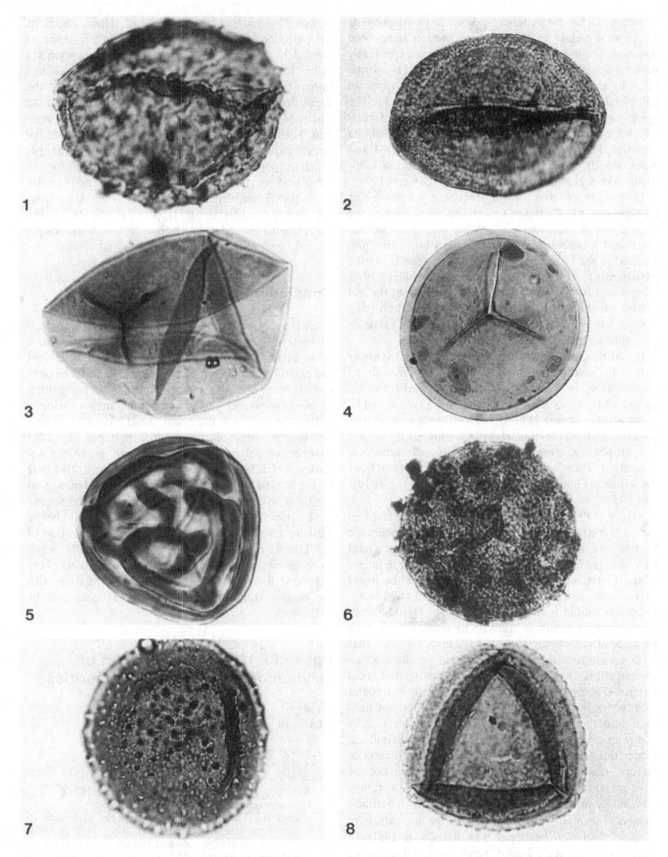

Figure 19.5. (1) *Aratrisporites tenuispinosus* Playford, 1965, 59 μm; (2) *Aratrisporites saturnii* (Thiergart, 1949) Mädler, 1964, 56 μm; (3) *Calamospora* sp., 47 μm; (4) *Punctatisporites* sp., 52 μm; (5) *Assertospora gyrata* (Playford et Dettman, 1965) Schuurman, 1977, 41 μm; (6) *Keuperisporites baculatus* Schulz, 1965, 94 μm; (7) *Porcellispora longdonensis* (Clarke, 1965) Scheuring, 1970, 79 μm; (8) *Duplicisporites verrucosus* Leschik, 1956, 37 μm.

Cingulati
Assertospora gyrata (Playford et Dettmann, 1965) Schuurman, 1977 (Fig. 19.5.5)
Cavatomonoletes
Aratrisporites saturnii (Thiergart, 1949) Mädler, 1964 (Fig. 19.5.2)
Aratrisporites pilosus (Leschik, 1956) Mädler, 1964
Aratrisporites centratus Leschik, 1956
Aratrisporites tenuispinosus Playford, 1965 (Fig. 19.5.1)
Aratrisporites paenulatus Playford et Dettmann, 1965
Acavatomonoletes
Leschikisporis aduncus (Leschik, 1956) Potonié, 1958

POLLEN GRAINS
Monosaccoid pollen grains
Cordaitina minor (Pautsch, 1971) Pautsch, 1973
Heliosaccus dimorphus Mädler, 1964 (Fig. 19.7.2)
Alete bisaccoid pollen grains
Alete/indet. bisaccoid pollen grains
Remark: Large numbers of alete and badly preserved bisaccoid pollen grains have been observed in the investigated material. Except for the distinct species presented below, we have applied the category Alete/indet. bisaccoid pollen grains. The use of form-genera and -species would suggest an unjustified precision within a group comprising so many taxonomic uncertainties.
Alete bisaccoid sp. A (Fig. 19.6.1)
Characteristic features: Alete bisaccoid pollen grains. The nexine constitutes a central body characterized by a circular to elliptical outline in polar view. The proximal sexine is differentially thickened and may be smooth, but is generally differentiated into closely spaced verrucae. Laterally the sexine is differentiated as two saccoid expansions. In plan view the maximum width of the expansions generally exceeds the maximum width of the central body. Distally a sexine-free area is present.
Cucullispora cuneata Scheuring, 1970
Protodiploxypinus gracilis Scheuring, 1970 (Fig. 19.6.6)
Protodiploxypinus decus Scheuring, 1970
Podosporites amicus Scheuring, 1970
Taeniate bisaccoid pollen grains
Infernopollenites sulcatus (Pautsch, 1958) Scheuring, 1970
Infernopollenites parvus Scheuring, 1970
Lunatisporites spp. (Fig. 19.6.8)
Protohaploxypinus-Striatoabieites complex (Fig. 19.6.7)
Remark: This complex contains forms with more than five taeniae, which are assignable to the form-genera *Protohaploxypinus* Samoilovich, 1953 and *Striatoabieites* Sedova, 1956.
Monolete, dilete and/or trilete bisaccoid pollen
Triadispora crassa Klaus, 1964
Triadispora plicata Klaus, 1964 (Fig. 19.6.2)
Triadispora suspecta Scheuring, 1970
Illinites chitonoides Klaus, 1964 (Fig. 19.6.4)
Parillinites pauper-Ovalipollis lepidus complex (Fig. 19.6.3)
Remark: This complex contains forms assignable to *Parillinites pauper* Scheuring, 1970 emend. Scheuring, 1978, *Parillinites eiectus* Scheuring, 1970, *Parillinites*

vanus Scheuring, 1970, *Ovalipollis lepidus* Scheuring, 1970 and *Ovalipollis ludens* Scheuring, 1970.
Staurosaccites quadrifidus. Ovalipollis pseudoalatus complex (Fig. 19.6.5)
Remark: This complex contains forms assignable to *Staurosaccites quadrifidus* Dolby, 1976, *Ovalipollis notabilis* Scheuring, 1970, *Ovalipollis brutus* Scheuring, 1970 emend. Scheuring, 1978, *Ovalipollis septimus* Scheuring, 1970 emend. Scheuring, 1978, *Ovalipollis ovalis* Scheuring, 1970 emend. Scheuring, 1978 and *Ovalipollis pseudoalatus* (Thiergart, 1949) Schuurman, 1976
Circumpolles
Partitisporites novimundanus morphon
Remark: See Van der Eem (1983). The morphon concept was introduced by Van der Zwan (1979) to denote a group of palynological species united by continuous variation of morphological characteristics. A morphon may be named informally after one or more of its constituent species and can be diagnosed by enumerations of both the common and the variable morphological characteristics.
Duplicisporites granulatus morphon (Fig. 19.5.8)
Remark: See Van der Eem (1983).
Sulcate pollen grains
Echinitosporites iliacoides Schulz & Krutzsch, 1961 (Fig. 19.7.1)
Retisulcites perforatus Mädler, 1964 Scheuring, 1970 (Fig. 19.7.4)
Cycadopites spp. (Fig. 19.7.3)
Eucommiidites microgranulatus Scheuring, 1970

ACRITARCHS
Micrhystridium spp. (Fig. 19.7.8)
Veryhachium spp. (Fig. 19.7.7)

PRASINOPHYTA
Leiosphaeridia spp.

CHLOROPHYTA
Plaesiodictyon mosellanum Wille, 1970 (Fig. 19.7.5,6)
Botryococcus spp.

Appendix 19.2 Botanical affinity and paleoecological significance of dominant taxa or categories in the Middle Triassic of the Germanic Basin

Calamospora spp.: Spores of equisetophytic affinity (e.g., Grauvogel-Stamm, 1973, 1978; Klaus, 1960).
Punctatisporites spp.: Spores of pterophytic affinity.
Cyclogranisporites-Verrucosisporites complex: Spores of pterophytic affinity.
Aratrisporites spp.: Spores of lycopodiophytic affinity (e.g., Helby & Martin, 1965; Boureau, 1967; Ash, 1979; Grauvogel-Stamm & Düringer, 1983).
Alete bisaccoid pollen grains: In Europe this category of pollen grains is known from Triassic conifers (e.g., Grauvogel-Stamm, 1969, 1978).
Triadispora spp.: Pollen grains assignable to this taxon have been encountered in male cones of conifers, i.e., *Sertostrobus laxus, Darneya peltata* and *Darneya mougeoutii* (Grauvogel-Stamm, 1978). These conifers probably formed xerophytic vegetation surrounding evaporitic tidal flats (Brugman, 1986).

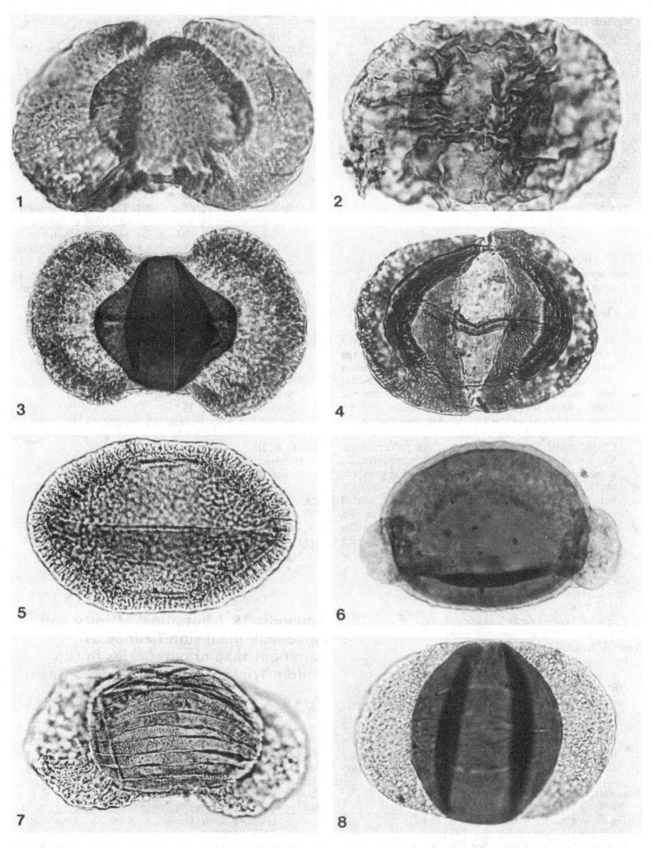

Figure 19.6. (1) alete bisaccoid pollen grain sp A., 105 μm; (2) *Triadispora plicata* Klaus, 1964, 72 μm; (3) *Parillinites* sp., 117 μm; (4) *Illinites chitonoides* Klaus, 1964, 85 μm; (5) *Ovalipollis pseudoalatus* (Thiergart, 1949) Schuurman, 1976, 76 μm; (6) *Protodiploxypinus gracilis* Scheuring, 1970, 41 μm; (7) *Striatoabieites aytugii* Visscher, 1966, 97 μm; (8) *Lunatisporites acutus* Leschik, 1956, 66 μm.

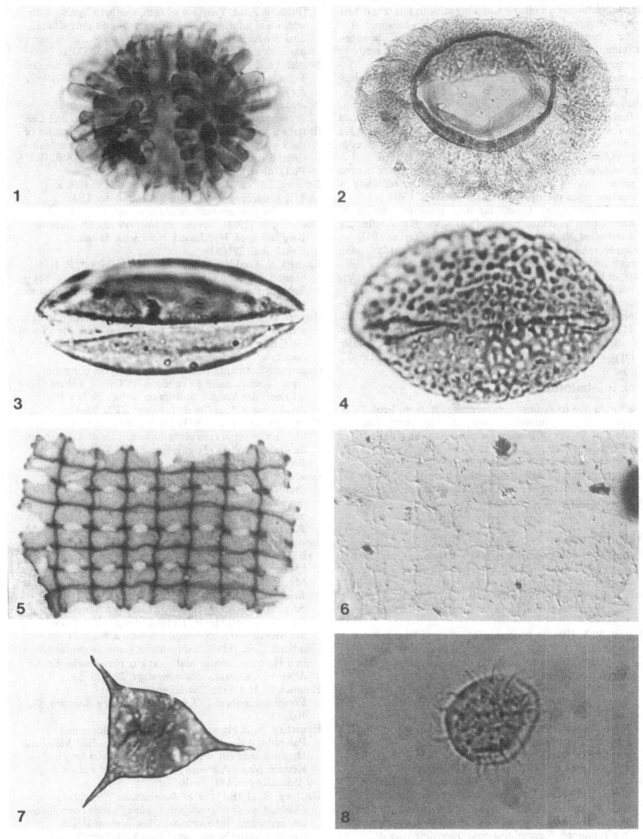

Figure 19.7. (1) *Echinitosporites iliacoides* Schulz & Krutsch, 1961, 51 μm; (2) *Heliosaccus dimorphus* Mädler, 1964, 185 μm; (3) *Cycadopites* sp., 35 μm; (4) *Retisulcites perforatus* Mädler 1964, 32 μm; (5) *Plaesiodictyon mosellanum* Wille, 1970, 152 μm × 81 μm; (6) *Plaesiodictyon mosellanum* Wille, 1970, 152 μm × 81 μm; (7) *Veryhachium* sp., 43 μm; (8) *Micrhystridium* sp., 18 μm.

Illinites chitonoides: Pollen grains assignable to this taxon have been observed in male cones (*Willsiostrobus acuminatus*) of the Triassic conifer *Aethophyllum stipulare* (Grauvogel-Stamm, 1978). This conifer presumably formed reed-like vegetation growing on the delta plain.

Taeniate bisaccoid pollen grains: Pollen grains of coniferalean or pteridospermous affinity.

Monosulcate asaccate pollen grains (e.g., *Cycadopites* spp., *Retisulcites perforatus* and *Echinitosporites iliacoides*): Pollen grains either of cycadophytic or ginkgophytic affinity; we believe that the Triassic forms generally represent cycadophytes.

Veryhachium spp.: Unicellular acid-resistant remains of marine phytoplankton groups of unknown affinity, reflecting a shallow, open marine environment (Staplin, 1961).

Micrhystridium spp.: Unicellular acid resistant remains of marine phytoplankton groups of unknown affinity, reflecting a restricted, shallow marine environment (Staplin, 1961).

Leiosphaeridia spp.: Prasinophyta, reflecting an intertidal, restricted shallow marine environment.

Botryococcus spp.: Chlorophyta, indicative of freshwater, brackish (coastal) lakes and lagoons (e.g., Tappan, 1980).

Plaesiodictyon mosellanum: This species belongs to the Chlorophyta and is related to the extant *Pediastrum* (see Wille, 1970). The taxon is restricted to freshwater environments; it is therefore considered that *Plaesiodictyon* reflects a freshwater input in marine environments (Hauschke & Heunisch, 1990).

Acknowledgments

We would like to express our sincere gratitude to Prof. Dr B. Schröder (Ruhr University, Bochum, Germany) and the Geological Survey of Bavaria for providing the samples from the Obernsees well. We want to thank Mrs Malcherek, H. Rijpkema and H. A. Elsendoorn for the preparation of the illustrations and photomicrographs. We highly appreciate the constructive remarks of Prof. Dr H. Visscher and the reviewers of an earlier draft of this chapter.

References

Aigner, T. & Bachmann, G. H. (1989). Dynamic stratigraphy of an evaporite-to-red bed sequence, Gipskeuper (Triassic), Southwest Germanic Basin. *Sedimentary Geology*, **62**, 5–25.

Ash, S. R. (1979). *Skilliostrobus* gen. nov., a new lycopsid cone from the early Triassic of Australia. *Alcheringa*, **3**, 73–89.

Besems, R. E. (1981a). Aspects of Middle and Late Triassic palynology. 1. Palynostratigraphical data from the Chiclana de Segura Formation of the Linares-Alcaraz region (southeastern Spain) and correlation with palynological assemblages from the Iberian Peninsula. *Review of Palaeobotany and Palynology*, **32**, 257–73.

Besems, R. E. (1981b). Aspects of Middle and Late Triassic palynology. 2. Preliminary palynological data from the Hornos-Siles Formation of the Prebetic Zone, NE province of Jaén (southeastern Spain). *Review of Palaeobotany and Palynology*, **32**, 389–400.

Besems, R. E. (1982). Aspects of Middle and Late Triassic palynology. 4. On the Triassic of the external zone of the Betic Cordilleras in the Province of Jaén, southern Spain (with a note on the presence of Cretaceous palynomorphs in a presumed 'Keuper' section). *Proceedings van de Koninklijke Nederlandse Akademie der Wetenschappen*, **B 85**, 29–52.

Besems, R. E. (1983). Aspects of Middle and Late Triassic palynology. 3. Palynology of the Hornos-Siles Formation

(Prebetic Zone, Province of Jaén, southern Spain), with additional information on the macro- and microfaunas. *Österreichische Akademie der Wissenschaften, Schriftenreihe der Erdwissenschaftlichen Kommissionen*, **5**, 37–56.

Beutler, G. & Schüler, F. (1979). Über Vorkommen salinarer Bildungen in der Trias im Norden der DDR. *Zeitschrift für geologische Wissenschaften*, **7**, 903–12.

Boureau, E. (1967). *Traité de paléobotanique, Tome II, Bryophyta, Psilophyta, Lycophyta*. Paris: Masson et Cie.

Brugman, W. A. (1986). A palynological characterization of the Upper Scythian and Anisian of the Transdanubian Central Range (Hungary) and the Vicentinian Alps (Italy). Ph.D. dissertation, State University of Utrecht.

Dockter, J., Puff, P., Seidel, G. & Kozur, H. (1980). Zur Triasgliederung und Symbolgebung in der DDR. *Zeitschrift für geologische Wissenschaften*, **8**, 951–63.

Duchrow, H. (1984). Keuper. In *Geologie des Osnabrücker Berglandes*, ed. H. Klassen. Naturwiss. Museum, Osnabrück, 221–333.

Emmert, E., Gudden, H., Haunschild, H., Meyer, R. K. F, Schmid, H., Schuh, H., Stettner, G. & Risch, H. (1985). Bohrgut-Beschreibung der Forschungsbohrung Obernsees. *Geologica Bavarica*, **88**, 23–47.

Eugster, H. P. & Hardie L. A. (1975). Sedimentation in an ancient playa-lake complex: the Wilkins Peak Member of the Green River Formation of Wyoming. *Geological Society of America Bulletin*, **86**, 319–34.

Grauvogel-Stamm, L. (1969). Nouveaux types d'organes reproducteurs males de conifères du Grès à Voltzia (Trias inférieur) des Vosges. *Bulletin du Service de la Carte Géologique d'Alsace et de Lorraine*, **22**(2), 93–120.

Grauvogel-Stamm, L. (1973). *Macrostrobus acuminatus* nom. nov., un novel organe reproducteur male de Gynosperme du Grès à Voltzia (Trias inférieur) des Vosges (France). *Géobios*, **6**(2), 143–6.

Grauvogel-Stamm, L. (1978). La flore du Grès à Voltzia (Buntsandstein supérieur) des Vosges du Nord (France) – morphologie, anatomie, interprétations phylogénique et paléogéographique. *Sciences Géologiques, Université Louis Pasteur de Strasbourg, Institut de Géologie, Mémoire*, **50**.

Grauvogel-Stamm, L. & Düringer, P. (1983). *Annalepis zeilleri* FLICHE 1910 emend., un organe reproducteur de Lycophyte de la Lettenkohle de l'Est de la France. Morphologie, spores in situ et paléoécologie. *Geologische Rundschau*, **72**, 23–51.

Gudden, H. & Schmid, H. (1985). Die Forschungsbohrung Obernsees-Konzeption, Durchführung und Untersuchung der Metallführung. *Geologica Bavarica*, **88**, 5–21.

Handford, C. R. (1982). Sedimentology and evaporite genesis in a Holocene continental – sabkha playa basin; Bristol dry lake, California. *Sedimentology*, **29**, 239–53.

Haunschild, H. (1985). Der Keuper in der Forschungsbohrung Obernsees. *Geologica Bavarica*, **88**, 103–30.

Hauschke, N. & Heunisch, C. (1990). Lithologie und Palynologie der Bohrung USB 3 (Horn – Bad Meinberg, Ostwestfalen): ein Beitrag zur Faziesentwicklung im Keuper. *Neues Jahrbuch für Geologie und Paläontologie, Abhandlungen*, **181**, 79–105.

Hedberg, H. D. (Ed.) (1976). *International Stratigraphic Guide. A guide to stratigraphic classification, terminology, and procedure*. International Classification IUGS, Commission on Stratigraphy. New York: John Wiley & Sons.

Helby, R. & Martin, A. R. H. (1965). *Cyclostrobus* n. g., cones of lycopsidean plants from the Narrabeen group (Triassic) of New South Wales. *Australian Journal of Botany*, **13**, 389–404.

Huth, R. (1956). Zur Geologie der Steinmergelbänke, insbesondere der Bleiglanzbank, im Gipskeuper. *Zeitschrift für angewandte Geologie*, **1**, 15–17.

Kelber, K.-P. (1990). Die versunkene Pflanzenwelt aus den Deltasümpfen Mainfrankens vor 230 Millionen Jahren. *Beringeria*, **1**.

Kendall, A. C. (1979). Continental and supratidal (sabkha) evaporites. *Geoscience Canada*, **5**(2), 66–78. (reprinted in: Walker, R. G., ed. (1981). Facies models. *Geoscience Canada, Reprint Series*, **1**, 145–57.)

Klaus, W. (1960). Sporen der Karnischen Stufe der ostalpinen Trias. *Jahrbuch der geologischen Bundesanstalt Wien, Sonderband*, **5**, 107–84.

Kozur, H. (1972). Vorläufige Mitteilung zur Paralleliserung der germanischen und tethyalen Trias sowie einige Bemerkungen zur Stufen- und Unterstufengliederung der Trias. *Mitteilungen der Gesellschaft der Geologie- und Bergbaustudenten in Österreich*, **21**, 363–412.

Kozur, H. (1974). Probleme der Triasgliederung und Paralleliserung der germanischen und tethyalen Trias. Teil I. Abgrenzung und Gliederung der germanischen Mitteltrias. *Freiberger Forschungshefte*, C **289**, 139–97.

Kozur, H. (1975). Probleme der Triasgliederung und Paralleliserung der germanischen und tethyalen Trias. Teil II. Anschluss der germanischen Trias und die internationale Triasgliederung. *Freiberger Forschungshefte*, C **304**, 51–77.

Kozur, H. (1976). Die stratigraphische Stellung der *Frankites sutherlandi*-Zone in der tethyalen Trias. *Geologisch-Paläontogische Mitteilungen Innsbruck*, **6**(4), 1–18.

Lorente, M. A. (1986). Palynology and palynofacies of the Upper Tertiary in Venezuela. *Dissertationes Botanicae*, **99**, Berlin: Stuttgart.

Mader, D. (1990). *Palaeoecology of the Flora in Buntsandstein and Keuper in the Triassic of Middle Europe*. Vol. 2, Stuttgart: Gustav Fisher Verlag.

Pocock, S. A. J., Vasanthy, G. & Venkatachala, B. S. (1987–1988). Introduction to the study of particulate organic materials and ecological perspectives. *Journal of Palynology*, **23–24**, 167–88.

Scheuring, B. W. (1974). 'Protosaccate' Strukturen, ein weitverbreitetes Pollenmerkmal zur frühen und mittleren Gymnospermenzeit. *Geologisch-Paläontologische Mitteilungen Innsbruck*, **4**:2, 1–30.

Scheuring, B. W. (1978). Mikrofloren aus den Meridekalken des Monte San Giorgio (Kanton Tessin). *Schweizerische paläontologische Abhandlungen*, **100**.

Schuurman, W. M. L. (1976). Aspects of Late Triassic palynology. 1. On the morphology, taxonomy and stratigraphical/geographical distribution of the formgenus *Ovalipollis*. *Review of Palaeobotany and Palynology*, **21**, 241–66.

Schuurman, W. M. L. (1977). Aspects of Late Triassic palynology. 2. Palynology of the 'Grès et Schiste à *Avicula contorta*' and 'Argiles de Levallois' (Rhaetian) of northeastern France and southern Luxembourg. *Review of Palaeobotany and Palynology*, **23**, 159–253.

Schuurman, W. M. L. (1979). Aspects of Late Triassic palynology. 3. Palynology of the latest Triassic and earliest Jurassic deposits of the northern Limestone Alps in Austria and southern Germany, with special reference to a palynological characterization of the Rhaetian Stage in Europe. *Review of Palaeobotany and Palynology*, **27**, 53–75.

Staplin F. L. (1961). Reef-controlled distribution of Devonian microplankton in Alberta. *Palaeontology*, **4**, 392–424.

Tappan, H. (1980). *The Paleobiology of Plant Protists*. San Francisco: W. H. Freeman & Company.

Van Bergen, P. F., Janssen, N. M. M., Alferink, M. & Kerp, J. H. F. (1990). Recognition of organic matter types in standard palynological slides. In *Proceedings of the International Symposium Organic Petrology*, Zeist, The Netherlands, ed. W. J. J. Fermont & J. W. Weegink. *Mededelingen Rijks Geologische Dienst*, **45**, 9–21.

Van Bergen, P. F. & Kerp, J. H. F. (1990). Palynofacies and sedimentary environment of a Triassic section in southern Germany. In *Proceedings of the International Symposium Organic Petrology*, Zeist, The Netherlands, ed. W. J. J. Fermont & J. W. Weegink. *Mededelingen Rijks Geologische Dienst*, **45**, 23–37.

Van der Eem, J. G. L. A. (1983). Aspects of Middle and Late Triassic palynology. 6. Palynological investigations in the Ladinian and Karnian of the Western Dolomites, Italy. *Review of Palaeobotany and Palynology*, **39**, 189–300.

Van der Zwan, C. J. (1979). Aspects of Late Devonian and Early Carboniferous palynology of southern Ireland. I. The *Cyrtospora cristifer* morphon. *Review of Palaeobotany and Palynology*, **28**, 1–20.

Van der Zwan, C. J. (1980a). Aspects of Late Devonian and Early Carboniferous palynology of southern Ireland. II. The *Aurospora macra* morphon. *Review of Palaeobotany and Palynology*, **30**, 133–55.

Van der Zwan, C. J. (1980b). Aspects of Late Devonian and Early Carboniferous palynology of southern Ireland. III. Palynology of Devonian-Carboniferous transition sequences with special reference to the Bantry Bay area, Cork. *Review of Palaeobotany and Palynology*, **30**, 165–286.

Van der Zwan, C. J. (1981). Palynology, phytogeography and climate of the Lower Carboniferous. *Palaeogeography, Palaeoclimatology, Palaeoecology*, **33**, 279–310.

Van der Zwan, C. J. & Van Veen, P. M., (1978). The Devonian-Carboniferous transition sequence in southern Ireland: integration of paleogeography and palynology. *Palinologia*, Número Extraordinario, **1**, 469–79.

Van Veen, P. M. (1981). Aspects of Late Devonian and Early Carboniferous palynology of southern Ireland. V. The change in composition of palynological assemblages at the Devonian-Carboniferous boundary. *Review of Palaeobotany and Palynology*, **34**, 67–97.

Visscher, H. (1983). I.G.C.P. project No. 4: A major achievement in Triassic research – A challenge to the S.T.S. *Albertiana*, **1**, 3–6.

Visscher, H. and Brugman, W. A. (1981). Ranges of selected Palynomorphs in the Alpine Triassic of Europe. *Review of Palaeobotany and Palynology*, **34**, 115–28.

Visscher, H. & Krystyn, L. (1978). Aspects of Late Triassic Palynology. 4. A palynological assemblage from ammonoid-controlled Late Karnian (Tuvalian) sediments of Sicily. *Review of Palaeobotany and Palynology*, **26**, 93–112.

Visscher, H. & Van der Zwan, C. J. (1981). Palynology of the circum-mediterranean Triassic: Phytogeographical and palaeoclimatological implications. *Geologische Rundschau*, **70**, 625–36.

Whitaker, M. F. (1984). The usage of palynostratigraphy and palynofacies in definition of Troll Field geology. In: *Offshore Northern Seas – Reduction of uncertainties by innovative reservoir geomodelling*. Norsk Petroleumsforening Article G6.

Wille, W. (1970). *Plaesiodictyon mosellanum* n.g., n.sp., eine mehrzellige Grünalge aus dem Unteren Keuper von Luxemburg. *Neues Jahrbuch für Geologie und Paläontologie, Monatshefte*, **5**, 283–310.

20 Palynomorph concentration in studies of Paleogene nonmarine depositional environments of Wyoming

MARTIN B. FARLEY

Introduction

Most quantitative palynologic work uses counts or percentages of taxa (i.e., relative frequency) as the basic measure of occurrence for analysis. Relative frequency has proved satisfactory for most purposes, but it has some undesirable properties. Most notable among these is the closed sum, that is, percentages always sum to 100 and counts sum to a fixed total (basically 100 times a constant). If one taxon goes up in frequency, the remaining taxa as a whole must go down. Percentages, therefore, have an artificial, imposed negative correlation that can distort multivariate ordinations and classifications, and can render common confirmatory statistical analysis such as analysis of variance invalid. Transformations to solve this problem for compositional data in general have been devised (Aitchison, 1986), but involve the logistic normal transformation which uses logarithms of the raw data. Because paleontological data matrices usually contain numerous zeroes, Aitchison's method has to be altered to change zeroes to small positive values. This alteration complicates matters, and the logistic normal transformation has not been used much in paleontology. Use of the logistic normal is treated more fully in Kovach and Batten (Chap. 18 this volume). The complications in using even transformed relative frequency data suggest that other possibilities for measuring quantitative abundance should be examined.

Use of palynomorph concentration (number of grains per unit volume or weight of sediment analyzed) is not common in pre-Quaternary palynology, but has theoretical advantages. Analyses of concentration can be based on total concentration of all palynomorphs or on concentrations of individual taxa in a sample. Concentrations are useful because they may distinguish depositional environments and because concentrations have favorable statistical properties. Palynomorph concentration per unit weight of sediment has a theoretical maximum limit which varies depending on exactly how it is calculated but is certainly in the tens of millions of grains per gram or higher (Brush & Brush, 1972, p. 365). The highest concentration so far found in sediments, however, is about 5 000 000/g from the Quaternary of the Black Sea (Traverse, 1988), at least an order of magnitude lower. Thus concentration values are not limited in practice to a theoretical limit, and the closed-sum problem does not exist. Further, each taxon's concentration is independent of the concentration of other taxa, so that variation in one taxon need not affect the others. As I will show below, concentration is commonly best analyzed after logarithmic transformation, which introduces both favorable statistical and graphical characteristics.

In this chapter, I analyze palynomorph concentration in depositional environments from the nonmarine Paleogene sediments of the central Bighorn Basin in Wyoming (Fig. 20.1). In particular, I explore the usefulness of palynomorph concentration for differentiating the depositional environments, using both the total concentration of all palynomorphs and the concentration of individual palynomorph taxa.

Terminology and general methods

Concentration values are sometimes referred to as absolute frequency (e.g., Darrell & Hart, 1970; Birks & Birks, 1980). However, absolute frequency has also been used to refer to pollen influx (also known as pollen depositional rate or accumulation rate; grains per unit area per unit time). Therefore, to prevent confusion, I avoid the term 'absolute.'

Concentration can use either unit weight or unit volume of sediment or water. Volume has invariably been used for water (e.g., Traverse & Ginsburg, 1966; Peck, 1973; Starling & Crowder, 1981; Farley, 1987b). Sediment can be analyzed either wet or dry. Choices here seem to have been made on the basis of what is most convenient for the analyst. Quaternary palynologists generally use unit volume, while pre-Quaternary palynologists (including people analyzing Quaternary sediments as

431

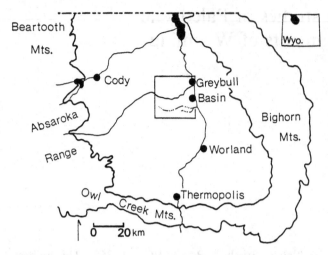

Figure 20.1. The Bighorn Basin in northwest Wyoming. Rectangle outlines area of sampling shown in greater detail in Figure 20.3.

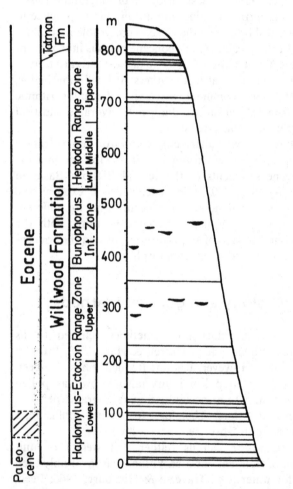

Figure 20.2. Stratigraphic section of Willwood Formation showing vertebrate biostratigraphic zones. Uncertainty of the Paleocene-Eocene boundary is indicated. Thicknesses are from the Wing-Schankler measured section (Wing, 1980). Lenses show zone of predominance of lenticular carbonaceous systems (ponds), and horizontal lines the occurrence of tabular systems (remaining environments). Thickness of deposits not to scale. Modified from Schankler (1980) and Wing (1984a).

pre-Quaternary analogs) tend to use unit weight. Authors differ over which method produces the best results (e.g., Bonny, 1972; Fletcher & Clapham, 1974). It has been easier for pre-Quaternary palynologists to weigh dry sediment than to measure its volume precisely (especially if more than a few cubic centimeters must be used), so they have tended to use dry weight sediment as the denominator.

There is also an extensive literature on laboratory methods for determining concentrations. These methods either use estimates for the volume or weight of the residue on a microscope slide as a proportion of the total residue produced by processing (e.g., Traverse & Ginsburg, 1966; Bonny, 1972), or use the addition of a known number of exotic pollen grains as a way of estimating what proportion of the total processed residue has been counted. Summaries of the techniques are given in Birks and Birks (1980) and Traverse (1988).

Setting

The sedimentary environments analyzed come from the Fort Union and Willwood Formations (Fig. 20.2; uppermost Paleocene to lower Eocene) of the central Bighorn Basin in Wyoming, USA. Two types of thin (about 2–5 m) organic-rich systems of sediment occur in the predominantly red-bedded sequence. Complete discussions of their sedimentology may be found in Wing (1984a) or Farley (1987a). What follows here is a summary of the important features.

The first type, the lenticular system, has an erosional base, is of limited lateral extent (less than 300 m long), and contains multiple packets of fining-up sediments. These units are interpreted as pondfill deposits formed in abandoned channels, where the fining-up sequences represent sediments introduced to the pond by episodic floods spilling over from the nearby active channel. I consider these units to be one lithofacies, because individual packets have the same origin and are too thin to consider separately.

The second type, the tabular system, has much greater lateral extent than lenticular systems (up to tens of kilometers), grades up from the underlying red or variegated beds, and can be divided into several distinct, recurring lithofacies. In a typical vertical sequence, these lithofacies are the variegated mudstone (normal oxidized deposition), clayey subunit, dark brown clay, carbonaceous shale, drab gray mudstone, interbedded silt and sand, and the sequence is sometimes capped by an erosively-based trough crossbedded sandstone. These are interpreted respectively as oxidized floodplain, subswamp soil, swamp margin, clastic swamp, reduced floodplain, levee/crevasse splay, and (subsequent) erosive channel.

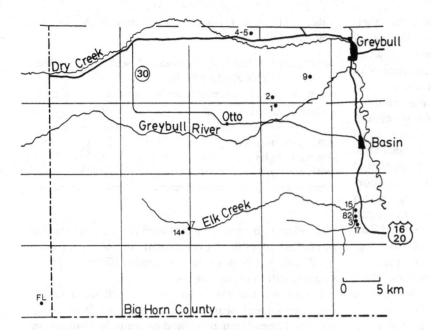

Figure 20.3. Localities of the sample profiles used in this study in the Bighorn Basin. Squares are Federal Land Survey townships. For locality descriptions, see Appendix 20.1.

I took a set of samples (a profile) vertically through the facies present in a tabular system at a locality or through the sediments as a whole in lenticular systems. I took four laterally separated vertical profiles of samples from one tabular system (Localities 3, 15, 17, and 82), single vertical profiles from three other tabular systems (Localities 2, 9, FL), and single vertical profiles from four lenticular systems (Localities 1, 4, 7, and 14). Localities are shown in Figure 20.3 and described in Appendix 20.1. Seventy-six samples were obtained from the 11 localities.

The majority of these samples (all samples from Localities 1–3, 15–17, 82) come from between 112 m and 150 m in the measured section of the Fort Union and Willwood Formations measured by Wing (1980) and Schankler (1980). Samples from Locality 9 occur just below the base of this measured section, and Locality 4 is known to be in the Lower *Haplomylus–Ectocion* Zone (LHEZ, vertebrate biostratigraphy, see Schankler, 1980; 50–380 m in the measured section). Thus most of the samples span only as much as the 1–1.5 Ma duration of the LHEZ (Wing & Bown, 1985), and the samples from the 112–150 m level span a fraction of this time. Wing's (1981) lowest megafloral biostratigraphic zone (the *Metasequoia-Cnemidaria* Concurrent Range Zone) spans 0–353 m in the measured section. The consistent nature of the megaflora over the interval of most of my samples supports the inference of limited effects of stratigraphic position on differences among my samples. Therefore, I conclude stratigraphic effects on palynofloral composition (e.g., the effects of time such as evolution, taxon migration, climate change) are of subordinate importance to lithofacies differences in this study.

Total concentration

Methods

I calculated concentration values for each sample using the method of Traverse and Ginsburg (1966, p. 426–8). In this method,

$$concentration = BD/CA,$$

where A = dry weight of the sediment processed, B = total weight of maceration residue plus mounting medium, C = weight of residue plus mounting medium on slide, and D = the number of palynomorphs on the microscope slide. Two estimates for each sample's concentration were generated by counting palynomorphs on two slides. In cases where palynomorphs occurred in numbers so large as to make counting the whole slide impractical (that is, more than about 1000 grains per slide), I divided the slide into halves or quarters by ruling straight lines from corner to corner of the coverslip with an India ink pen. Selection of which half or quarter to count was then made by flipping a coin. The fractional counts were then multiplied by two or four as appropriate to yield a count for the entire slide.

Examination of the raw concentration estimates revealed that the means for each sample and each depositional environment are proportional to their variances. This is consistent with other studies of palynomorph concentration in sediments or water (Farley, 1987b; Schuyler, 1987). Such a correlation of mean and variance implies that concentration is not distributed normally and makes statistical analyses dependent on normally distributed variables unreliable, particularly if the statistical method requires equal variances. Sokal and Rohlf (1981) suggest that logarithmic transformation of the data will sever the relationship of mean and variance and produce a normal distribution. Analysis of the transformed concentrations suggests that independence of means and variances, and equal variances within sampling variation, can be assumed. This indicates that palynomorph concentrations are log-normally distributed. In addition there are practical reasons supporting logarithmic transformation. These include the convenience of having values that can be graphed together reasonably and having differences of equal proportion be of equal magnitude (e.g., so that the distance from 10 to 50 is the same as the distance from 1000 to 5000).

The sampling design of two estimates within each sample and of samples within depositional environments makes possible a nested analysis of variance (ANOVA) of total concentration. In nested (or hierarchical) ANOVA, the samples are classified into the environments of the tabular system and the lenticular system. The lenticular samples are not subdivided because this carbonaceous system is considered a single depositional setting rather than separate, laterally adjacent, environments as in the tabular systems. The oxidized floodplain and channel environments of the tabular systems were omitted from the ANOVA because the number of samples in each ($n = 1$ and $n = 2$, respectively) is too small. The variability in the two estimates of concentration for each sample is used to test for differences among samples, and then the variance in the samples within environments is used to test for differences among environments.

Results

The ANOVA (Table 20.1) shows that differences among samples within environments and differences among environments themselves are statistically significant at less than the 0.1% level. The differences between the replicate estimates for each sample average about 0.18 log units. This amount (about 50% on an arithmetic scale) is small relative to the difference between many samples (see Fig. 20.4). This explains the highly significant difference found among samples. The small difference in the replicates suggests that concentration in samples can be estimated accurately by counting a single slide. This coincides with Darrell's (1973) analysis of modern Mississippi Delta sediments, which calculated concentration

Table 20.1. *Nested analysis of variance table for total concentration*

df = degrees of freedom; SS = sum of squares; MS = mean square; *(a)* = probability < 0.1%.

Source of variation	df	SS	MS	F-ratio
Among environments	5	104.4	20.88	13.8 (a)
Among samples	62	94.1	1.52	35.2 (a)
Within samples	68	2.93	0.043	

by a different method from that used here. Using one-slide estimates will result in a substantial time saving although it will not then be possible to evaluate differences among samples within environments.

I used two techniques to examine individual sample and environment means: graphical reference distributions and the Tukey-Kramer method of multiple comparisons (for environments only). Box *et al.* (1978, p. 190–3) provide a useful graphical technique for comparing variable means if the null hypothesis that all means are equal is rejected. Box *et al* point out that rejection of the null hypothesis of equal means is equivalent to saying that all the samples *cannot* be considered as randomly taken from a single normal distribution. Box *et al* approximate this normal distribution (assumed by the model of the ANOVA) by a Student's *t* distribution, and use this reference distribution to examine which samples fail to fit it as the distribution slides along the graph of samples (i.e., as the grand mean is allowed to vary). The samples that cannot fit are likely the ones whose means are different.

Figure 20.4 shows the reference distribution derived for comparison of samples. This method implies that samples that cannot be reasonably seen as random samples from the reference distribution have significantly different means. Every environment shown on Figure 20.4 contains samples which cover a larger range than that of the reference distribution.

The differences between environments can also be evaluated by a reference distribution (Fig. 20.5). From this figure, it is apparent that the oxidized floodplain, channel and subswamp soil environments have lower means and differ from the remaining environments. The reference distribution shows that the means of the remaining five environments can only be considered equal if the levee/crevasse splay, pond, and swamp are from the extreme tail ends of a distribution centered at 3.6 log units. This is unlikely, given the assumption of the method that these be *random* samples from this distribution.

Tukey-Kramer's method of multiple comparisons (Table 20.2; Sokal & Rohlf, 1981) indicates that pond and reduced floodplain concentrations can be considered

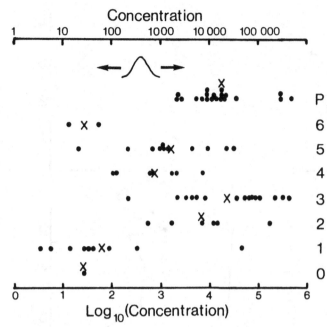

Figure 20.4. Mean palynomorph concentration of samples classified by environment. Environments: 0 = oxidized floodplain; 1 = subswamp soil; 2 = swamp margin; 3 = swamp; 4 = levee/crevasse splay; 5 = reduced floodplain; 6 = channel; P = pond. X's are at environment means. The curve (reference distribution) at the top of the graph shows the concentration range over which samples *within* environments can be considered as derived from a single population. This curve can be shifted laterally to take on any desired mean value for comparing samples within environments. See text for further discussion.

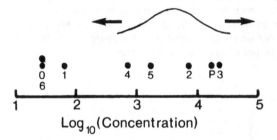

Figure 20.5. Graph of environment concentration means (the X's in Fig. 20.4). Environment abbreviations as in Figure 20.4. The curve (reference distribution) shows the reasonable range for concentrations if the environments have the same mean concentration. Environments that cannot be viewed as random samples from this reference distribution probably have statistically distinguishable means. The reference distribution is centered over the midpoint between the five organic-rich environments for convenience in assessing their differences; however, to compare any other set of environments the reference distribution can be shifted laterally to attempt to encompass that set. See text for further discussion.

different at the 5% level of significance. Pairs of environments in Figure 20.5 more widely separated than these are statistically separable at the 1% level. These data and the reference distributions suggest a continuum in which environments with adjacent means are statistically

Table 20.2. *Tukey–Kramer multiple comparisons of differences between environment means*

Each value is the difference between mean concentrations for the pair of environments indicated by the appropriate row and column. Statistical significance: ns = not significant; (a) significant at $P = 5\%$; (b) significant at $P = 1\%$.

	3	P	2	5	4	1
3	–					
P	0.151ns	–				
2	0.508ns	0.357ns	–			
5	1.175(b)	1.024(a)	0.667ns	–		
4	1.512(b)	1.362(b)	1.004ns	0.337ns	–	
1	2.582(b)	2.432(b)	2.074(b)	1.408(b)	1.070ns	–

indistinguishable, but that statistically significant differences exist over wider ranges of environments.

The vertical distribution of concentrations within sampling profiles (localities) does not show any consistent pattern, either in the tabular or lenticular systems (Figs. 20.6, 20.7). In part this reflects the variability in local superposition of environments (see Fig. 20.6).

Analysis of the residuals for each sample (sample concentration minus environment mean), however, reveals an interesting pattern in four of the vertical profiles taken in tabular systems (Fig. 20.8). Residuals averaged across all occurrences of an environment are zero, and so within a vertical profile residuals could be expected to vary about the zero line. All samples from Localities 9 and FL have concentrations greater than the averages for each sub-environment, whereas Localities 3 and 17 have concentrations consistently less than the averages. The extreme values for the subswamp soil, swamp, and reduced floodplain environments belong to one of these aberrant profiles (the highest sample total concentration in the subswamp soil is in Locality 9, and the lowest sample concentrations in the swamp and reduced floodplain are both in Locality 17). This suggests that some local environmental factors (perhaps local water table variations) slightly enhanced or degraded palynomorph preservation in the area of each profile.

Residuals for the ponds do not appear anomalous (Fig. 20.9), probably because of the pond fill's lithogenetic homogeneity and lack of sub-environments.

Taxon concentration

Methods

Taxon concentrations can be calculated by multiplying the percentage frequencies of each palynomorph taxon for a sample by that sample's total palynomorph

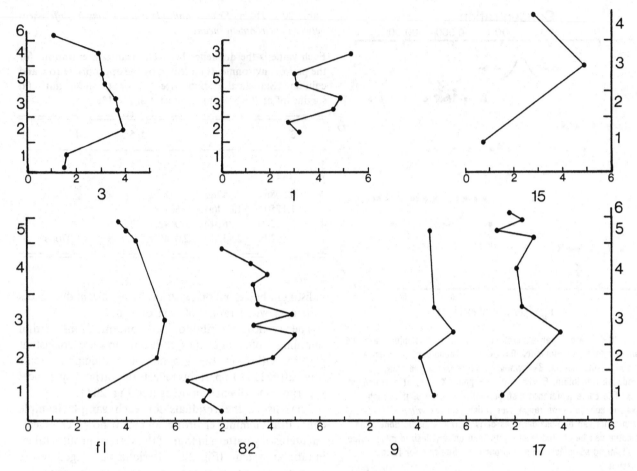

Figure 20.6. Mean sample concentration in vertical profiles (localities) for tabular systems. Vertical axis: Numbers are facies abbreviations as in Figure 20.4. Vertical scale is arbitrary. Horizontal axis: Scale numbers are log (concentration in grains/g). Numbers centered under plotted points are locality numbers.

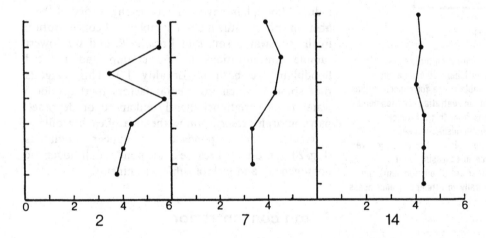

Figure 20.7. Sample concentration means in vertical profiles (localities) for lenticular units (ponds). Vertical axis: Scaled to equal, arbitrary length (not to sedimentary thickness). Tick marks represent samples. Horizontal axis: Scale numbers are log (concentration in grains/g). Large, centered numbers are locality numbers.

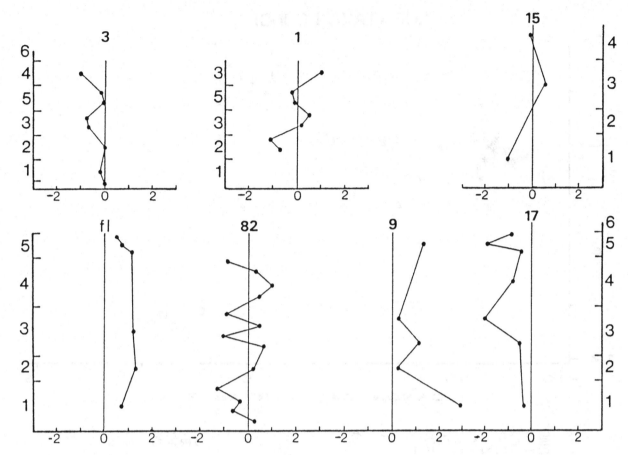

Figure 20.8. Residuals (difference between sample concentration and its environment's mean concentration) for each vertical profile (locality) in tabular systems (oxidized floodplain, subswamp soil, swamp margin, reduced floodplain, levee/crevasse splay, channel). Vertical scaling and labelling as for Figure 20.6. Numbers above each graph are localities. Horizontal scale is in log (concentration).

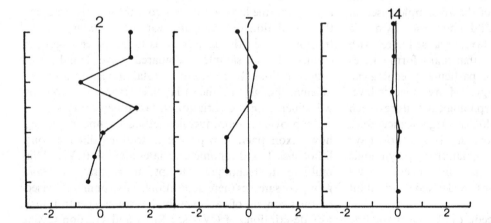

Figure 20.9. Residuals (differences between sample concentration and its environment's mean concentration) for each vertical profile (locality) in lenticular systems (ponds). Vertical scaling and labelling as for Figure 20.7. Large numbers at top of zero lines are locality numbers. Horizontal scale is log (concentration). Note that because lenticular units are not subdivided into subenvironments, the residual graphs are the same shape as the original concentration graphs, except that the origin is shifted.

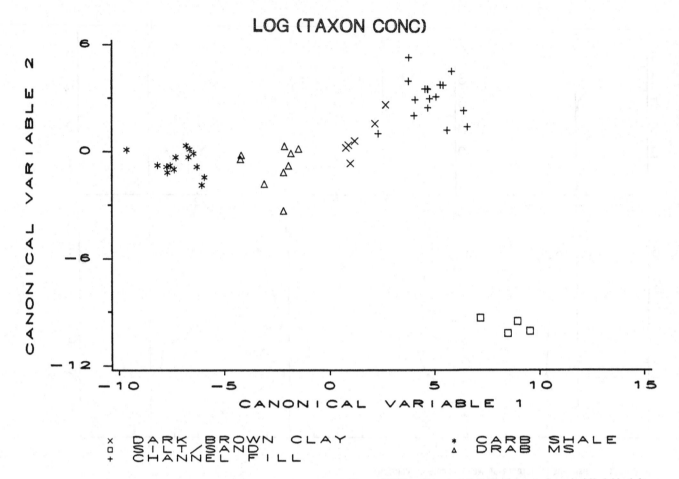

Figure 20.10. Canonical Variable 2 versus Canonical Variable 1 from canonical variates analysis of the logarithm of taxon concentrations. Dark brown clay = swamp margin; silt/sand = levee/crevasse splay; channel fill = pond; carb shale = swamp; drab ms = reduced floodplain.

concentration. I obtained taxon frequencies by counting 300 palynomorphs per sample excluding indeterminates for all samples which contained enough palynomorphs to count (at least 100 grains per slide). Complete 300-grain counts could be made for 50 of the 76 samples used in analysis of total concentration. I did not generally divide morphotypes into species-level taxa, because I agree with Boulter and Craig's (1979) view that many form-species do not have much biological, particularly ecological, value in Tertiary palynology. Many species-level distinctions are based on morphological features such as size that may have biological significance (e.g., ontogenetic) but which almost surely do not have phylogenetic significance. Most Quaternary palynologic work, for example, identifies pollen only to generic level even though these palynomorphs are derived from extant plant species.

In theory, the statistical procedures I follow to analyze total concentration (analysis of variance) could be extended to a multivariate analysis of variance of taxon concentrations, but the increase of statistical errors (uncertainty) in calculating percent taxon concentration from frequency and total concentration makes such an analysis of dubious value.

For multivariate analysis, the taxon concentrations are

log-transformed, as with total concentration, to reduce the association of mean and variance. Taxon concentrations vary in the same way as taxon percentages (or counts) if the samples compared have equal total concentrations. If the samples' total concentrations are different, though, then the variation in taxon concentrations will differ from the variation in taxon percentages.

To provide a reproducible method for understanding how taxon palynomorph concentrations differ among lithofacies, I used canonical variates analysis (CVA). This multivariate technique attempts to separate *a priori* groups of samples (here depositional environments) based on their content of variables (here taxon concentrations). For descriptions of CVA see Birks and Gordon (1985) or Pielou (1977). Use of all 127 morphotypes found is not possible or practical in the multivariate analysis. First, CVA requires fewer variables than samples. Further, most morphotypes occur with very low frequency in any particular sample; these rare types have variability that is high in proportion to their frequency. As a result, these taxa probably contribute mostly noise to the total

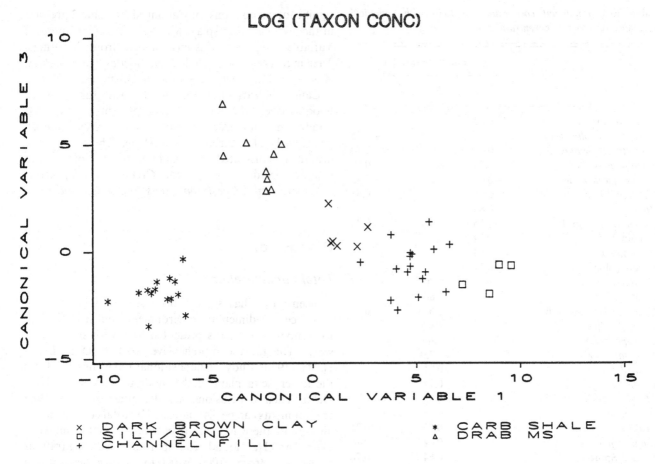

LOG (TAXON CONC)

x   DARK BROWN CLAY   *   CARB SHALE
□   SILT/SAND   △   DRAB MS
+   CHANNEL FILL

Figure 20.11.   Canonical Variable 3 versus Canonical Variable 1 from canonical variates analysis of the logarithm of taxon concentrations. Legend equivalents as in Figure 20.10. Note how the reduced floodplain, swamp margin, pond, and levee/crevasse splay environments occur in more or less homogeneous clouds along a continuum. On these axes, only the swamp samples are distinctly separated from the other environments.

variability of all taxa. This noise decreases the amount of variance explained by the first few eigenvectors (or equivalent vectors) and also tends to decrease the loading of the morphotypes on the eigenvectors. Quaternary palynologists commonly use only taxa with abundances greater than 5% in at least one sample (Birks & Gordon, 1985) in their multivariate analyses.

I included only taxa with total counts greater than 70 (approximately 0.4% averaged over all samples) in the multivariate analysis; this level represents a natural break in taxon abundances. Analysis of taxon abundance patterns shows furthermore that almost all taxa with total counts greater than 70 occur in at least one sample with a frequency greater than 5%. Taxa with total counts less than 70 have been omitted from multivariate analysis unless they can be lumped into a meaningful biological group (e.g., bisaccate pollen, fungal remains, all dino-flagellates). This produced 42 taxa for the multivariate analysis.

Results

Canonical variates analysis of the logarithms of taxon concentrations separates sets of lithofacies samples as shown in Figures 20.10 and 20.11. The facies occur approximately evenly spaced along Canonical Variable 1 (Fig. 20.10). Samples from the pond and the swamp margin environments overlap along Canonical Variable 1, but otherwise the samples from each environment occur in separate groups. Canonical Variable 2 separates the levee/crevasse splay from the others, which greatly overlap. The facies that overlap the most on Canonical Variables 1 and 2, the pond and swamp margin environments, are distinguishable on Canonical Variable 3 (Fig. 20.11). Canonical Variable 3 explains an additional 11% of total variance, bringing the cumulative total explained by three canonical variables to 94%.

Table 20.3 gives the significant correlations between the taxa and the new canonical variables (the CVA equivalent of loadings in principal components analysis). These correlations are mostly negative on Canonical Variable 1 and positive on Canonical Variable 2. This suggests that each canonical variable is measuring only taxon abundance and not contrasting taxa abundant in one environment (positively correlated taxa) with taxa rare in the same environment.

Table 20.3. *Important correlations of taxa with canonical variables log (taxon concentration) data*

	Canonical variable		
	1	2	3
Deltoidospora	−0.82		
Taxodiaceaepollenites	−0.60		−0.75
Inaperturopollenites	−0.69		−0.70
Tricolpites sp. A	−0.73		−0.68
Sparganiaceaepollenites	−0.68		−0.65
Pistillipollenites	−0.69		−0.61
Alnus	−0.85		
Gleicheniidites sp. B	0.65		0.71
Tricolpate 5	0.54		
Plicatopollis	0.63		
Dinoflagellates	0.77		
Tricolpites sp. C		0.70	−0.67
Pandaniidites		0.88	
Coryluspollenites sp. B		0.66	−0.63
Ulmipollenites		0.62	
Caryapollenites		0.87	
Laevigatosporites		0.72	
Leiotriletes		0.71	
Cyathidites		0.77	
Monosulcites		0.80	
Arecipites		0.75	
Multicellaesporites		0.62	−0.71
Inapertisporites		0.62	−0.76
Fungal Spore 6		0.64	−0.76
Misc. Fungal remains		0.63	−0.75
Tetraporina		0.63	
Cycadopites			−0.89
Cupuliferoipollenites			−0.79
Coryluspollenites sp. A			−0.81
Pluricellaesporites sp. A			−0.95
Platycaryapollenites			0.84

The taxa with high negative correlations on Canonical Variable 1 tend to be more common in the swamp and pond environments. The higher total concentrations that are characteristic of the swamp and pond create higher taxon concentrations in these two environments relative to the others regardless of a taxon's percentage frequency. Notable taxa with high negative correlations with Canonical Variable 1 are *Deltoidospora*, *Taxodiaceae-pollenites*, *Sparganiaceaepollenites*, *Pistillipollenites*, and *Alnus*.

The taxa with positive correlations on Canonical Variable 1 tend to be concentrated in the pond only. These include *Gleicheniidites*, *Plicatopollis*, and the dinoflagellates. The dinoflagellates are mostly reworked marine forms, and they indicate the highly allochthonous nature of the palynomorph assemblage in the ponds relative to the other environments. The taxa with

significant correlations on Canonical Variable 2 are rare in the levee/crevasse splay, which shows how Canonical Variable 2 separates this environment from the others. Taxa in this category include *Monosulcites*, *Pandaniidites*, *Arecipites*, *Caryapollenites*, and fungal spores.

Canonical Variable 3 separates the swamp and reduced floodplain on the basis of taxa rare in the reduced floodplain compared to most other environments, particularly the carbonaceous shale. These are taxa negatively correlated with Canonical Variable 3, and include taxodiaceous pollen, *Cycadopites*, fagaceous tricolpate forms, *Coryluspollenites*, and fungal spores.

Discussion

Total concentration

Concentration has seldom been studied in a detailed sequence of sedimentary environments, but it is possible to compare the results presented here with some other work. The most comprehensive work is Darrell and Hart's (1970) study of depositional environments from the lower delta plain of the modern Mississippi River Delta. Their conclusions parallel mine: at least some environments can be distinguished by total concentration alone, but there is overlap between the environments. Darrell and Hart found mean sporomorph concentrations in the range from 10 000–30 000 for environments which are not directly comparable with mine.

Muller (1959), in his study of the modern Orinoco Delta, found pollen concentrations on the terrestrial delta from 1000 grains/g on the natural levees and 100 000 grains/g in backswamp sediments. Muller concluded that total pollen concentration reaches its maximum where both detrital sedimentation and organic accumulation are at a minimum. Although Muller's lithofacies information is limited, his concentration values are quite similar to mine: the figures I have cited for Muller are maximum ranges; my average for the swamp environment is only about 25 000 grains/g, but the environment contains many samples with concentrations greater than 100 000 grains/g.

Schuyler (1987) studied palynomorph concentration in overbank, channel bar, and channel fill environments in the Upper Devonian Catskill Magnafacies in southern New York State. Overbank samples average four grains/g, channel bar samples 210 grains/g, and channel fill samples average 130 grains/g. These values are much lower than mine for comparable facies (channel fill and overbank), and may reflect the simpler plant ecosystems of the Devonian as compared to the early Tertiary.

Total palynomorph concentration among the samples and environments studied differs in a statistically significant way. Between sample differences reflect the heterogeneous nature of the terrestrial environments involved. Samples with highly disparate concentrations relative to other concentrations in the same carbonaceous system could reflect local conditions prevalent in the locality throughout the sedimentation history of this unit, or some transient environmental effect at one moment in the unit's history. For example, the sample with the highest concentration in the subswamp soil (see Fig. 20.4) occurs in a vertical profile (Locality 9) all of whose environments are characterized by higher than average palynomorph concentrations. This suggests a continuing local effect on palynomorph deposition or preservation at this site.

On the other hand, the much lower than average concentration in a sample in the pond profile at Locality 2 occurs between samples of higher than average concentration: this suggests that ephemeral factors (such as high sedimentation rate) caused this low concentration. The reference distribution for significant sample differences (Fig. 20.4) spans approximately 0.6 log units and suggests that differences between samples even smaller than those of the large residuals are statistically discernible. This results from the high reproducibility of concentration estimates for a single sample (0.2 log units) and from the general heterogeneity within depositional environments. Even the cluster of pond samples near the overall pond mean (Fig. 20.4) spans a wider range of concentration than can be considered as random samples from a population with a single mean. These results suggest that explaining concentration differences between samples from one environment will be difficult without either analysis of many more samples than used in this study, or additional information such as knowledge of anomalous local concentration values throughout a vertical profile of stacked sedimentary environments.

The geological significance of the differences between environmental means is difficult to assess. Clearly, a single sample cannot be assigned to an environment solely on the basis of its palynomorph concentration. The concentration of environments can be statistically distinguished, however, not only in the contrast between palynomorph-poor (oxidized floodplain, subswamp soil, and channel) and palynomorph-rich (pond, levee/crevasse splay, swamp, reduced floodplain) environments, but also in that the levee/crevasse splay and reduced floodplain environments can be distinguished from the pond and swamp environments (Table 20.2). Explaining these differences presents difficulties.

A large number of factors control palynomorph concentration in nonmarine sediments: (1) palynomorph production of the taxa in the environment; (2) density of vegetation in the environment; (3) access of wind and water, which carry palynomorphs to the depositional site; (4) grain size and sorting of transported sediments; (5) sedimentation rates; (6) preservation potential of organic matter in general; (7) differential destruction of some palynomorphs; (8) post-depositional changes such as leaching.

The variability of concentration within a single environment reflects these factors' effects. Ascertaining the cause of a particular environment's concentration relative to other environments can be problematic. The low concentrations in the subswamp soil could result from the leaching and other diagenetic alteration associated with the processes of soil formation below a swamp, but without knowing what its palynomorph content was initially, I am reluctant to conclude this. Because I consider the palynology of the subswamp soil to be mostly locally derived (see Farley, 1987a), I do not think that the low concentrations in this environment result from the non-hydraulic equivalence of most palynomorphs with clay-size particles.

The levee/crevasse splay contains a lower concentration of palynomorphs than the pond or swamp environments, despite containing considerable organic matter. A plausible explanation is that the higher sedimentation rates in the levee/crevasse splay implied by the high proportion of coarse clastic material and the levee/crevasse splay's position nearest the channel reduced the concentration of palynomorphs of the levee/crevasse splay relative to less detritally influenced facies. This interpretation is consistent with Muller's (1959) results from the Orinoco Delta where the natural levees had the lowest palynomorph concentration because of high clastic sediment supply. The levee/crevasse splay, which contains identifiable megafossils, has a lower concentration of palynomorphs than the reduced floodplain, which contains few if any recognizable megafossils (Wing, 1984a).

The oxidized floodplain contains few palynomorphs because it is oxidized; the channel because its grain size is much larger than the hydraulic equivalence of most palynomorphs.

Total concentration is amenable to analysis of variance; the residuals from the model for analysis of variance show that four of the tabular localities differ in having residuals for each sample in the profile always above or below zero. This means that all samples in Localities 9 and FL have total concentrations greater than their respective environments' averages, and all samples in Localities 3 and 17 have total concentrations less than their respective environments' averages. The anomalous residuals suggest that these localities differed from typical localities in their physical environment. Some physical feature promoted or degraded preservation of numbers of palynomorphs throughout the lifespan of tabular unit deposition at these spots. Features that could cause this might include differences in local elevation (a few decimeters would suffice) or in edaphic properties.

These anomalous profiles appear to differ in frequency of some palynomorph taxa as well. *Leiotriletes* is higher throughout the samples at Locality 3 than elsewhere, taxodiaceous pollen is lower in Locality 17, *Monosulcites* is higher in numbers and *Coryluspollenites* is lower in both Localities 9 and FL, and *Platycaryapollenites* is higher in numbers at Locality FL. Some of the differences may reflect other causes: the high abundance of *Platycaryapollenites* in Locality FL is a result of stratigraphic effects (Wing, 1984b). On the other hand, the occurrence of fewer pollen grains of *Taxodiaceae-pollenites* in localities with lower than average concentration (i.e., less swampy?) is consistent with what would be expected for abundance of taxodiaceous conifer trees in a drier location. *Monosulcites* also has contrasting abundance between the anomalously low and high concentration profiles that is not obviously a result of other causes. These differences in frequencies of taxa are not conclusive because of the small number of vertical profiles examined, but they are consistent with the suggestion of the concentration residuals, that these profiles show the effects of local physical differences, possibly edaphic in nature, in the tabular systems. These local variations in physical environment suggested by the residual anomalies would not be readily interpretable from the sedimentologic record alone. Palynomorph concentration therefore provides a new method for obtaining detailed data on the geomorphology of a depositional system.

Taxon concentration

The frequency of occurrence of palynomorph taxa in these environments depends on factors which can be classified into three major groups for each sample: depositional environment, geographic location, and stratigraphic position. My examination of the raw data arranged by locality suggests that effects of geographic and stratigraphic position are subordinate in importance to the effects related to depositional environment. The likelihood of stratigraphic factors confounding my facies differences is reduced by my use of primarily generic-level taxa. Such a taxonomic generalization tends to blur stratigraphic differences, but not paleoecologic differences (Boulter & Craig, 1979). This interpretation is supported by the separation of depositional environments in the canonical variates analysis. If geographic or stratigraphic position had a dominant role, then these effects would have resulted in a failure of the CVA to discriminate among environments. The success in general of the CVA indicates these effects are secondary in importance.

The CVA of taxon concentrations does not link the environments simply by their total palynomorph concentration, although total concentration plays an important role. For example, Canonical Variable 1, which accounts for 57% of the variability in the data, produces the lowest score for the environment with the highest concentration (swamp) and the highest score for the environment with the lowest concentration for which taxon frequency data are available (levee/crevasse splay), but the environments in between are not arranged in order of descending concentration (see Fig. 20.10). The remaining canonical variables for the concentration CVA produce sample scores that place the swamp and levee/crevasse splay samples together at the negative end of the scale (Fig. 20.11). The proximity of facies groups in these analyses reflects both the proximity of environments to each other in the original depositional setting and the similarity of the physical environment of some environments independent of their physical proximity or lack of it.

The CVA of taxon concentrations separates the different facies reasonably successfully, but not as distinctly as does CVA of taxon counts. Analysis of taxon counts (Farley, 1989) yields distinct, non-overlapping groups for each facies.

I suggest that there are two major explanations. First, determining taxon concentrations (proportion of taxon in count multiplied by total palynomorph concentration) leads to large uncertainties in each value by combining the uncertainties of total concentration in samples (approximately 0.2 log units or 50%) and of counts (confidence intervals as discussed by Maher, 1972: rare taxa have large 95% confidence intervals; for example, a taxon occurring at 0.5%, has an approximate confidence interval of 0.1–2.1%). The results of the calculations of taxon concentrations, therefore, have a great deal of noise that can disguise the structure of the data. This explanation is supported by the proportion of the total variance explained by the first two canonical variables in each analysis. The first two canonical variables in the taxon count analysis explain 95% of the total variance in the data, in contrast to only 83% explained by the first two canonical variables of taxon concentration data. The amount of noise in the data is apparently smaller in the taxon count CVA because two canonical variables are sufficient to separate the facies (Farley, 1989), whereas taxon concentration analysis requires three canonical variables to explain the same proportion of variance, and these are not as efficient in separating the facies.

Second, high total concentrations mean that few taxa present will have low concentrations, compared to their concentration in depositional environments with lower total concentrations. The smallest proportion in a 300 grain count is 0.33% (i.e., 1/300), so a sample with high total concentration (say 70 000 grains/g) has a minimum non-zero taxon concentration of 233 grains/g. The range of variation for rare taxa, therefore, breaks between zero (taxon does not occur in the 300 grain count) and 1/300 of the sample concentration. This problem could be reduced by basing taxon concentration on larger counts or on omission of rarer taxa. The former requires more work, and the latter drops information for the analysis. Analysis of taxon concentrations would perhaps be more effective in comparisons of differences within a single depositional environment or vertically at one locality through multiple environments.

Conclusion

Total palynomorph concentration among the samples within environments and among environments differs in a statistically significant way. The nested analysis of variance shows that even considering the large variability in concentration among individual samples that some of the environments have detectably different concentrations. The environments can be divided easily into palynomorph-rich environments (pond, levee/crevasse splay, swamp, reduced floodplain) with concentrations ranging from about 2.8 to 4.4 log units (630–25 000 grains/g), and palynomorph-poor environments (subswamp soil, oxidized floodplain, channel) with concentration about 1.6 log units (40 grains/g). Among the palynomorph-rich environments, multiple comparison methods suggest that the richest, the pond and swamp, are distinguishable from the levee/crevasse splay and reduced floodplain. Differences of less than about 0.8 log units of concentration between environments are probably not very important, so other distinctions between environments are not possible. It is possible that larger sets of samples could make smaller differences between samples statistically significant. While individual samples cannot be assigned to a particular environment on the basis of their total concentration, the pattern of residuals at the localities of the channel margin systems reveal that two localities have concentrations consistently greater than average and two consistently less than average. This result indicates that local differences in the physical environment occur at these localities and cause their palynomorph concentration to differ from other 'typical' profiles. Taxonomic composition at these localities compared to 'typical' localities is consistent with the hypothesis of an edaphic effect, although the compositional data are not conclusive.

Canonical variates analysis of taxon concentrations reveals that the depositional environments contain distinguishable assemblages of palynomorphs. These assemblages generally differ in abundance of taxa present rather than in their presence or absence. The assemblages from different environments are contiguous rather than clearly separated groups, because taxon concentrations, which incorporate the uncertainties of both taxon counts and total concentration, are relatively imprecise.

Thus, total concentration is easy to work with and suggests further avenues of investigation. Taxon concentration, on the other hand, produces results that are more difficult to interpret but may be useful for specifically defined problems.

Appendix 20.1

Localities

In this list, quadrangles are USGS 7.5′ topographic quadrangles in the Bighorn Basin, Wyoming, USA. Meter measurements indicated after locality numbers are positions in the measured section of Wing (1980) and Schankler (1980). See text for further details.

Locality 1 (Tabular)   150 m
    Gould Butte Quadrangle; SE1/4 NW1/4 sec. 5, T. (=township) 51 N., R. (=range) 94 W. Samples 8401–2 (swamp margin), 8403–4 (swamp), 8405–6 (reduced floodplain), 8407 (swamp)

Locality 2 (Lenticular)   140 m
    Gould Butte Quadrangle; SE1/4 NE1/4 sec. 31, T. 52 N., R. 94 W. Samples 8408–16 (pond)

Locality 3 (Tabular)   112 m
    Orchard Bench Quadrangle; SE1/4 SE1/4 sec. 20, T. 50 N., R. 93 W. Samples 8417 (oxidized floodplain), 8418 (subswamp soil), 8419 (swamp margin), 8420–1 (swamp), 8422–3 (reduced floodplain), 8424 (levee/crevasse splay), 8425 (channel)

Locality 4 (Lenticular)   LHEZ (see text)
    Sheep Canyon Quadrangle; sec. 2, T. 52 N., R. 95 W. Samples 8426–7 (pond)

Locality 7 (Lenticular)   420 m
    Wardel Reservoir Quadrangle; NE1/4 SE1/4 sec. 25, T. 50 N., R. 96 W. Samples 8430–4 (pond)

Locality 9 (Tabular)   just below 0 m
    Gould Butte Quadrangle; SW1/4 NE1/4 sec. 22, T. 52 N., R. 94 W. Samples 8437 (subswamp soil), 8438 (swamp margin), 8439–40 (swamp), 8441 (reduced floodplain)

Locality 14 (Lenticular)   530 m
    Wardel Reservoir Quadrangle; SE1/4 SW1/4 sec. 25, T. 50 N., R. 96 W. Samples 8448–54 (pond)

Locality 15 (Tabular)   112 m
    Orchard Bench Quadrangle; NE1/4 NE1/4 sec. 20, T. 50 N., R. 93 W. Samples 8455 (subswamp soil), 8456 (swamp), 8457 (levee/crevasse splay)

Locality 17 (Tabular)   112 m
    Orchard Bench Quadrangle; NE1/4 NE1/4 sec. 29, T. 50 N., R. 93 W. Samples 8460 (channel), 8461–3 (reduced floodplain), 8464 (levee/crevasse splay), 8465–6 (swamp), 8467 (subswamp soil)

Locality 82 (Tabular)   112 m
    Orchard Bench Quadrangle; SE1/4 NE1/4 sec. 20, T. 50 N., R. 93 W; samples collected by Scott Wing. Samples 8205–8 (subswamp soil), 8209 (swamp margin), 8210–13 (swamp), 8214-17 (levee/crevasse splay)

Locality FL (Tabular)   700 m
    NE1/4 NE1/4 NE1/4 sec. 35, T. 49 N., R. 98 W; samples collected by Scott Wing. Samples FL1 (subswamp soil), FL6 (swamp margin), FL7 (swamp), FL8–FL10 (reduced floodplain)

Acknowledgments

This work was supported by a National Science Foundation Graduate Fellowship, the Geological Society of America, and the Department of Geosciences, Pennsylvania State University. Scott L. Wing and Alfred Traverse provided invaluable assistance with the research and manuscript. W. L. Kovach and D. J. Batten are thanked for helpful comments on the manuscript. Manuscript revision was undertaken during the tenure of a Smithsonian Institution Postdoctoral Fellowship.

References

Aitchison, J. (1986). *The Statistical Analysis of Compositional Data*. London: Chapman & Hall.

Birks, H. J. B. & Birks, H. H. (1980). *Quaternary Palaeoecology*. London: Edward Arnold.

Birks, H. J. B. & Gordon, A. D. (1985). *Numerical Methods in Quaternary Pollen Analysis*. London: Academic Press.

Bonny, A. P. (1972). A method for determining absolute pollen frequencies in lake sediments. *New Phytologist*, **71**, 393–405.

Boulter, M. C. & Craig, D. L. (1979). A middle Oligocene pollen and spore assemblage from the Bristol Channel. *Review of Palaeobotany and Palynology*, **28**, 259–72.

Box, G. E. P., Hunter, W. G., & Hunter, J. S. (1978). *Statistics for Experimenters*. New York: John Wiley & Sons.

Brush, G. S. & Brush, L. M. Jr. (1972). Transport of pollen in a sediment-laden channel: a laboratory study. *American Journal of Science*, **272**, 359–81.

Darrell, J. H., II (1973). Statistical evaluation of palynomorph distribution in the sedimentary environments of the modern Mississippi River Delta. Ph.D. dissertation, Louisiana State University.

Darrell, J. H., II & Hart, G. F. (1970). Environmental determinations using absolute miospore frequency, Mississippi River Delta. *Geological Society of America Bulletin*, **81**, 2513–8.

Farley, M. B. (1987a). Sedimentologic and paleoecologic importance of palynomorphs in Paleogene nonmarine depositional environments, Bighorn Basin, Wyoming. Ph.D. dissertation, Pennsylvania State University.

Farley, M. B. (1987b). Palynomorphs from surface waters of the eastern and central Caribbean Sea. *Micropaleontology*, **33**, 254–62.

Farley, M. B. (1989). Palynological facies fossils in nonmarine environments in the Paleogene of the Bighorn Basin. *Palaios*, **4**(6), 565–73.

Fletcher, M. R. & Clapham, W. B. Jr. (1974). Sediment density and the limits to the repeatability of absolute pollen frequency determinations. *Geoscience and Man*, **9**, 27–35.

Maher, L. J. (1972). Nomograms for computing 0.95 confidence limits of pollen data. *Review of Palaeobotany and Palynology*, **13**, 85–93.

Muller, J. (1959). Palynology of Recent Orinoco Delta and shelf sediments. *Micropaleontology*, **5**, 1–32.

Peck, R. M. (1973). Pollen budget studies in a small Yorkshire catchment. In *Quaternary Plant Ecology*, ed. H. J. B. Birks & R. G. West. Oxford: Blackwell Scientific, 43–60.

Pielou, E. C. (1977). *Mathematical Ecology*. New York: Wiley-Interscience.

Schankler, D. M. (1980). Faunal zonation of the Willwood Formation in the central Bighorn Basin, Wyoming. *University of Michigan, Papers on Paleontology*, **24**, 99–114.

Schuyler, A. (1987). Sedimentology and paleoecology of miospores from the Middle to Upper Devonian Oneonta Formation: part of the Catskill Magnafacies, New York State. M.S. thesis. Pennsylvania State University.

Sokal, R. R. & Rohlf, F. J. (1981). *Biometry*. San Francisco: W.H. Freeman.

Starling, R. N. & Crowder, A. (1981). Pollen in the Salmon River system, Ontario, Canada. *Review of Palaeobotany and Palynology*, **31**, 311–34.

Traverse, A. (1988). *Paleopalynology*. London: Unwin Hyman.

Traverse, A. & Ginsburg, R. N. (1966). Palynology of the surface sediments of Great Bahama Bank, as related to water movement and sedimentation. *Marine Geology*, **4**, 417–59.

Wing, S. L. (1980). Fossil floras and plant-bearing beds of the central Bighorn Basin. *University of Michigan, Papers on Paleontology*, **24**, 119–25.

Wing, S. L. (1981). A study of paleoecology and paleobotany in the Willwood Formation (Early Eocene, Wyoming). Ph.D. dissertation, Yale University.

Wing, S. L. (1984a). Relation of paleovegetation to geometry and cyclicity of some fluvial carbonaceous deposits. *Journal of Sedimentary Petrology*, **54**, 52–66.

Wing, S. L. (1984b). A new basis for recognizing the Paleocene/Eocene boundary in western interior North America. *Science*, **226**, 439–41.

Wing, S. L. & Bown, T. M. (1985). Fine scale reconstruction of late Paleocene–early Eocene paleogeography in the Bighorn Basin of northern Wyoming. In *Cenozoic Paleogeography of West-central United States*, ed. R. M. Flores & S. S. Kaplan. Denver, Rocky Mountain Section – Society of Economic Paleontologists and Mineralogists, 93–105.

D Specific examples of applications

21 Multivariate analyses of palynomorph data as a key to depositional environments of Upper Cretaceous and Paleogene coal-bearing rocks of the western United States

R. FARLEY FLEMING and DOUGLAS J. NICHOLS

Introduction

In the reconstruction of ancient environments of deposition in nonmarine rocks on the basis of palynologic data, pollen and spores produced by the local flora (plants living near the area in which sediments were deposited) can provide important information, especially on the nature of local environments. This is most true for coal beds, which are usually autochthonous, and which among all ancient deposits have the best chance for preserving a record of the local flora from its pollen and spores. Knowledge of the nature of a coal-forming environment can contribute much to understanding the origin of the coal, its quality, and its significance in paleoclimatology and geologic history.

Disregarding rare occurrences of allochthonous or transported deposits of coal, coal beds represent mires in which vegetation growing on site or in the immediately surrounding area produces the macerals, often including abundant palynomorphs, that compose the peat that is eventually coalified. In this chapter we use the term mire to refer to wetland depositional environments characterized by accumulation of peat (Gore, 1983; see also detailed discussion of 'mire' in Nichols & Pocknall, Chap. 12 this volume); we use the term marsh to refer to wetland environments having aquatic vegetation and organic matter in the sediment, but without accumulation of peat (see Bates & Jackson, 1987). Studies of mires by Quaternary palynologists (e.g., Faegri & Iversen, 1975; Janssen, 1973) demonstrate that palynofloras from such depositional environments more accurately reflect the nature of the vegetation in the area than do those from most other types of sedimentary deposits. These studies also demonstrate, however, that because of differential productivity and preservation of palynomorphs, the relationship between plant microfossil assemblages and parent vegetation is not a simple one, and that reconstruction of ancient depositional environments can be fraught with uncertainties. This generalization becomes increasingly true the greater the geologic age, due to the greater proportion of extinct to extant plants represented and the greater importance of plants of uncertain or unknown ecology. The paleopalynologist must recognize these complications when endeavoring to reconstruct coal-depositional environments.

Another major aspect of paleopalynological investigations, perhaps more common than reconstruction of ancient depositional environments, is age determination and correlation on the basis of key taxa or assemblages of palynomorphs. Knowledge of which taxa are useful for biostratigraphy is critical to these investigations. The biostratigraphically most useful palynomorphs are usually those that are relatively independent of depositional environment; typically they are produced by regional rather than local vegetation. The ability to distinguish locally derived, facies-dependent palynomorphs from those more widespread in occurrence can permit refinement of palynological age determinations and correlations, especially in nonmarine rocks, including coal beds. Methods useful in reconstruction of coal-depositional environments on the basis of palynology may at the same time permit recognition of the regionally derived component of palynomorph assemblages in coals; these taxa are of potential value in palynostratigraphy.

Here we apply differing methods, including multivariate analyses, to reconstruct the environments of deposition of some coal beds and associated clastic rocks from two areas in the western United States. Some of these methods also make it possible to recognize palynostratigraphically useful taxa within the assemblages from these deposits. The primary purpose of our investigations was to examine the relationship between palynofacies and depositional environments as recorded in deposits that are largely autochthonous in origin. Our results bear on the more general problem of the use of palynomorphs in interpreting nonmarine sedimentary environments. Secondarily, our results demonstrate an approach to distinguishing facies-dependent plant microfossils from others more useful in palynostratigraphy. It is not our intent to compare different multivariate techniques, which has been done more thoroughly elsewhere (e.g., Gauch, 1982; Gauch & Whittaker, 1972; Gauch et al., 1977; Hill & Gauch, 1980; Kovach, 1988; Kovach, 1989). Nonetheless, the apparent utility, advantages, and disadvantages of the techniques we applied are evident in our results. In the Discussion section (see end of chapter) the role of the empirical method of palynofacies analysis is also considered.

Materials and methods

The material studied is the Vermillion Creek coal bed and associated rocks, which are deposits of local extent in the Niland Tongue of the Wasatch Formation (lower Eocene) in south-central Wyoming and northwestern Colorado, and several coal beds and associated rocks of greater regional extent in the lower part of the Raton Formation (uppermost Cretaceous and lower Paleocene) in southern Colorado and northern New Mexico (Fig. 21.1). In addition to differences in lateral extent, the rocks in these areas differ in geologic age. The Vermillion Creek beds date from about 51 Ma; Raton Formation rocks are distributed through a thicker stratigraphic interval that we estimate to range in age from about 67 to 64 Ma. The Vermillion Creek data set allows evaluation (using several methods) of samples all of about the same age but of varied depositional environment. The Raton data set also provides for evaluation of depositional environments, but additionally allows evaluation of the use of a multivariate technique on samples having a 'temporal gradient.' Tuckey and Anstey (1989) conducted a similar study on samples of Ordovician age containing marine invertebrate fossils; our Raton samples are of Cretaceous and Tertiary age and contain nonmarine plant microfossils, and the succession includes a celebrated systemic boundary.

As part of a comprehensive, multidisciplinary investigation of the Vermillion Creek coal bed, Nichols (1987) studied the palynology of the coal and its associated clastic facies and made an interpretation of the nature of the depositional environment based on the palynomorphs. The method used in that study may be described as empirical because it involved observation of kinds, relative abundances, and distributions by lithofacies of the palynomorphs. The interpretation relied on geological and botanical reasoning, and concepts originally developed for Quaternary palynological studies were adapted as a framework for the interpretation.

A method for deriving paleoecological interpretations of large palynomorph data bases using multivariate techniques (cluster analysis and principal components analysis) was described by Boulter and Hubbard (1982). We have applied the method of Boulter and Hubbard, slightly modified, to the Vermillion Creek data to see if

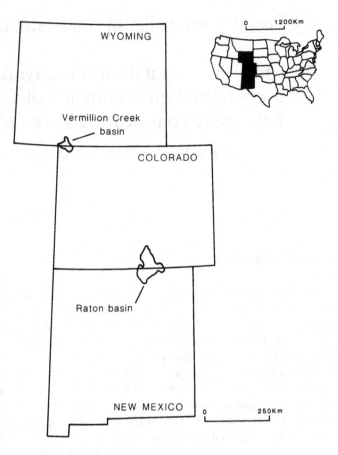

Figure 21.1. General location of study areas. Map of the United States shows positions of Colorado, New Mexico, and Wyoming.

different results would be obtained, and as a way of evaluating the results of the empirical method of analysis. The technique of Boulter and Hubbard and our modifications of it are outlined under principal components analysis below. Finally, we used an alternative multivariate technique (detrended correspondence analysis) on the same Vermillion Creek data base. This method showed much promise because its assumptions are ecologically more realistic than those of other multivariate techniques (Gauch, 1982).

Detrended correspondence analysis was applied also to a larger and more varied data base, the distribution and relative abundance of palynomorphs in coal and associated fine-grained clastic rocks from the Upper Cretaceous and lower Paleocene Raton Formation (Fleming, 1990). The Raton Formation data base includes almost twice as many taxa as the Vermillion Creek data base, a wider range of depositional environments, and also involves significant variation in the ages of samples analyzed.

Vermillion Creek coal bed and associated rocks

The lower Eocene Vermillion Creek coal bed is in the upper part of the Niland Tongue of the Wasatch

Table 21.1. *Vermillion Creek samples*

Samples listed in stratigraphic order within each of three cores; uppermost samples at top of each list.

Sample No.	Lithology	Stratigraphic Position
	Core VC-5 (locality D6310), sampled interval 3.3 m	
D6310-A	oil shale	just above coal seam
D6310-B	coal	top of seam
D6310-D	clay shale	upper parting
D6310-F	coal	high mid-seam
D6310-H	coal	above thick parting
D6310-I	clay shale	top of thick parting
D6310-J	clay shale	base of thick parting
D6310-K	coal	below thick parting
D6310-L	carbonaceous shale	lower parting
D6310-M	coal	near bottom of seam
D6310-N	carbonaceous shale	lowermost parting
D6310-O	coal	base of seam
	Core VC-7 (locality D6311), sampled interval 3.4 m	
D6311-A	carbonaceous shale	just above coal seam
D6311-B	coal	top of seam
D6311-C	carbonaceous shale	parting
D6311-D	coal	high mid-seam
D6311-E	coal	mid-seam
D6311-F	coal	mid-seam (below D6311-E)
D6311-G	coal	near bottom of seam
D6311-H	carbonaceous shale	just below coal seam
D6311-I	carbonaceous shale	well below coal seam
	Core VC-8 (locality D6312), sampled interval 3.8 m	
D6312-A	oil shale	well above coal seam
D6312-B	siltstone	below oil shale, above seam
D6312-C	carbonaceous shale	just above coal seam
D6312-D	coal	top of seam
D6312-E	coal	just above parting
D6312-F	carbonaceous shale	parting
D6312-G	coal	just below parting
D6312-I	coal	base of seam
D6312-J	carbonaceous shale	just below coal seam
D6312-K	carbonaceous shale	well below seam

Formation in the Vermillion Creek basin, Sweetwater County, Wyoming. Complete descriptions of various aspects of the deposit, summarized here, are included in chapters by several authors in Roehler and Martin (1987). The Niland Tongue is composed of about 50 to 130 m of coal, carbonaceous shale, clay shale, siltstone, and sandstone deposited in floodplain and paludal environments bordering ancient Lake Gosiute. The Niland Tongue is underlain and overlain by lacustrine deposits: the Luman Tongue and the Tipton Shale Member of the Green River Formation, respectively. The Luman Tongue and Tipton Shale Member are composed predominantly of oil shale. In the area sampled, the Vermillion Creek

coal bed ranges in thickness from 2.5 to 3.3 m and includes several partings up to 1 m thick, which are composed of clay shale or carbonaceous shale. Within the study area the stratigraphic sequence varied with respect to number, position, and thickness of coal splits and shale partings.

Samples discussed here and in Nichols (1987) came from three cores that included the coal bed and its partings and the overlying and underlying carbonaceous shale, oil shale, and siltstone. Samples were selected to represent the coal itself and the non-coal rock above, below, and within the coal. Table 21.1 is a list of the 31 samples analyzed, their stratigraphic positions in each of the three cores, and their lithological composition. Coal samples were taken at intervals within the bed. Samples of the coal were assumed to represent an ancient mire environment; the non-coal samples were assumed to represent other depositional environments, including those adjacent to the mire and those associated with Lake Gosiute. All samples yielded well-preserved palynomorphs, and recovery from many was excellent; coal samples were especially productive. The total assemblage from the Vermillion Creek beds included 54 species of pollen, spores, fungi, and algae, among which angiosperm pollen dominated (Nichols, 1987).

Raton Formation

The Upper Cretaceous and lower Paleocene Raton Formation is in the Raton basin of southeastern Colorado and northeastern New Mexico. The lithostratigraphy and sedimentology of the unit are discussed by Flores (1987) and Pillmore and Flores (1987) and summarized here. The Raton Formation is composed of 335 to more than 600 m of coal, carbonaceous shale, mudstone, siltstone, and sandstone. Conglomerate is also present, particularly at the base of the unit. The Raton is divided into three informal members: the lower coal zone, the barren series, and the upper coal zone. The two coal zones contain coal beds up to 2 m thick and the barren series contains thin coal beds, up to 15 cm thick. The K/T boundary is near the top of the lower coal zone. Sediments composing the Raton were deposited in a fluvial flood plain. The Raton Formation is underlain by coastal and delta plain deposits of the Vermejo Formation; it intertongues with and is overlain by conglomerates of the Poison Canyon Formation.

Samples were collected from a composite measured section of the entire thickness of the Raton Formation. Coal, carbonaceous shale, shale, mudstone, silty mudstone, and silty shale were sampled. Of 110 samples originally collected for this study, 66 samples yielded assemblages suitable for multivariate analysis. Table 21.2 is a list of the 66 samples analyzed, their stratigraphic position with respect to the K/T (Cretaceous/Tertiary) boundary, and their lithologies. Analyses were based on counts of species of pollen, spores, algae and fungi (93 species in all).

Table 21.2. *Raton Formation samples*

Samples listed in stratigraphic order. Stratigraphic position in meters above (positive values) or below (negative values) the K/T boundary.

Sample number	Lithology	Stratigraphic position
Upper coal zone		
D6831-F	mudstone	+370.1
D6831-B	mudstone	+369.1
D6831-A	carbonaceous shale	+366.3
D6830-G	carbonaceous shale	+319.0
D6830-B	carbonaceous shale	+318.3
D6829-C	mudstone	+267.0
D6829-B	coal	+266.2
D6829-A	mudstone	+265.7
D6828-E	mudstone	+204.9
D6828-C	carbonaceous shale	+204.5
D6828-B	coal	+204.4
D6828-A	mudstone	+204.1
D6827-F	coal	+187.1
D6827-E	coal	+186.7
D6827-D	coal	+186.6
D6826-A	mudstone	+162.8
D6825-I	carbonaceous shale	+124.4
D6825-H	coal	+124.0
D6825-G	coal	+123.8
D6825-F	coal	+123.7
D6825-E	coal	+123.6
D6825-B	coal	+123.4
D6825-A	coal	+123.2
D6824-K	mudstone	+107.5
D6824-J	coal	+107.2
D6824-I	carbonaceous shale	+105.0
D6824-H	coal	+104.6
D6824-G	carbonaceous shale	+104.4
D6824-F	carbonaceous shale	+103.9
D6824-E	carbonaceous shale	+103.5
D6824-D	carbonaceous shale	+102.9
D6824-B	mudstone	+102.5
D6824-A	mudstone	+102.2
Barren series		
D6823-H	siltstone	+52.8
D6823-G	carbonaceous shale	+52.4
D6823-F	carbonaceous shale	+51.3
D6823-E	mudstone	+51.1
D6823-D	mudstone	+51.0
D6823-C	mudstone	+50.9
D6823-B	mudstone	+48.4
D6823-A	mudstone	+48.2
D6822-E	mudstone	+34.7
D6822-C	siltstone	+32.4
D6821-I	shale	+27.4
D6821-H	mudstone	+27.3
D6820-E	silty mudstone	+19.6
D6820-A	mudstone	+17.4

Table 21.2. *Continued*

Sample number	Lithology	Stratigraphic position
Lower coal zone		
D6819-H	carbonaceous shale	+0.7
D6819-G	carbonaceous shale	+0.6
D6819-F	carbonaceous shale	+0.5
D6819-E	mudstone	+0.4
D6819-D	mudstone	+0.3
D6819-C	mudstone	+0.2
D6453-I	mudstone	−0.01
D6453-H	carbonaceous shale	−0.03
D6453-G	coal	−0.05
D6453-F	coal	−0.07
D6453-E	coal	−0.09
D6453-D	coal	−0.13
D6453-C	coal	−0.15
D6453-B	coal	−0.2
D6453-A	coal	−0.3
D6727-E	coal	−0.8
D6727-D	coal	−1.2
D7021-G	carbonaceous shale	−1.3
D7021-E	mudstone	−2.1

Generation of data bases

All samples discussed here were macerated following standard procedures for shales and post-Paleozoic coals described by Doher (1980). After taxonomic studies were completed, relative abundances of taxa were determined on the basis of counts of 200 or more specimens per slide. Relative abundances were calculated as percentages.

Empirical method of analysis

Some definitions of 'empirical' place emphasis on observation without reference to scientific principles, but our use of this word does not imply that no systems or theories lay behind our method. The empirical method of analysis of palynomorph data we employed was guided by geological, ecological, and botanical principles. Geological reasoning was based on parallel observations of the stratigraphy and sedimentology of samples analyzed. Ecological reasoning was uniformitarian in approach (cf. Ager, 1963; Dodd & Stanton, 1981), that is, based on comparison with living representatives of taxa having a microfossil record in our material, wherever possible. Botanical reasoning involved cognizance of differential productivity and preservation of plant microfossils and knowledge of the habitats of plants represented in our material. Of course, geological, ecological, and botanical reasoning also guided interpretations of results of multivariate analyses of both data bases discussed later. In its most direct application, however, the empirical method was applied to the initial analysis of the Vermillion Creek beds. In the empirical method of analysis of these beds, the significance of each palynomorph taxon was determined by consideration of its pattern of distribution, relative abundance, lithologic association, and botanical affinity.

Empirical analysis of palynomorph data from the Vermillion Creek beds was strongly influenced by the work of Janssen (1973) on Pleistocene and modern pollen deposition. Janssen defined four kinds of pollen deposition based on production and dispersion of pollen by parent plants: local (from plants at the sample site), extra-local (from plants within a few hundred meters of the sample site), regional (from major types of background vegetation), and extra-regional (from sources outside the area of consideration). Nichols (1987) altered these concepts somewhat to apply them to the Eocene deposits of the Vermillion Creek basin. He adopted Janssen's concept of the local, extra-local, and regional pollen deposition and designated them as components of the total palynoflora, although he did not recognize extra-regional as distinct from regional. Farley (1988) similarly modified Janssen's concepts and discussed three modes of palynomorph transport: local, waterborne (=extra-local), and aerial (=regional). Nichols defined another component, reworked palynomorphs, which might be thought of as analogous to extra-regional but with the added dimension of geologic time. He also expanded the concepts to include bryophyte and pteridophyte spores and algal cysts. Janssen had noted that his concepts were difficult to apply to the fossil record, by which he meant the Quaternary record, because of the effects of over-representation or under-representation of pollen due to differential productivity and the distribution of parent plant communities. These problems are even more serious in Tertiary or older rocks, and to them must be added the unknown role of plants that lived then but are now extinct. It may be that the ancestors of some modern plants had different ecologic requirements from those of their descendants, but unless that is known to be true, that potential complication is discounted or ignored in uniformitarian paleoecology, at least as a first assumption. Janssen noted that in Quaternary studies, the stratigraphic arrangement of assemblages of locally derived pollen provides important data, and this principle clearly is also valid for any pre-Quaternary study, including those discussed here.

Multivariate analyses

We used several multivariate analytical techniques to summarize patterns in the Vermillion Creek and Raton data sets. These techniques are cluster analysis, principal components analysis, and detrended correspondence analysis. Because there is a variety of options with each

of these techniques, the configurations we used are presented in this section.

CLUSTER ANALYSIS

All of the cluster analyses used the unweighted pair group method with average link clustering. *R*-mode clustering was run with MVSP version 1.3 (MultiVariate Statistical Package, written by Warren Kovach). Cluster analyses using the Pearson product-moment correlation coefficient and the Spearman rank order coefficient gave similar results; only the cluster dendrogram using the Pearson coefficient is presented. Cluster analyses were run on the Vermillion Creek data set only.

PRINCIPAL COMPONENTS ANALYSIS

Principal components analysis (PCA) was run using SPSS version 8.3 (Statistical Package for the Social Sciences). The PA1 factoring routine of SPSS was used with the option that the main diagonal of the correlation matrix was not altered. The Varimax option for rotation of the axes was employed. This configuration results in principal components analysis of a Pearson product-moment correlation matrix (Nie *et al.*, 1975).

PCA was used on the Vermillion Creek data set in conjunction with *R*-mode cluster analysis in the manner described by Boulter and Hubbard (1982). As they demonstrated, the use of *R*-mode cluster analysis in concert with PCA is a useful approach to interpreting multivariate analysis. They analyzed relative abundance data on 85 form-generic taxa in 200 samples from the Eocene of southern England that represented broad geographic, stratigraphic, and paleoclimatic regimes. They calculated a matrix of Pearson product-moment correlation coefficients from the relative abundances (percentage frequencies) of species in their samples. This matrix was used as a basis for PCA and *R*-mode cluster analysis. The cluster analysis dendrograms served as a guide to interpreting their principal components data.

We applied this approach to the Vermillion Creek data with some modification. Instead of analyzing all 54 species in the Vermillion Creek palynoflora, we selected 17 species to be analyzed, and their percentage relative abundances in each sample were recalculated on the basis of 100%. This procedure defined a group of taxa that is equivalent to the pollen sum of Faegri and Iversen (1975). Our 'pollen sum' includes not only pollen species but spores and other palynomorphs, however. The pollen sum technique was used because inspection of the raw relative abundance data suggested that some taxa were likely to represent significant members of local plant communities but others were not. Selection of taxa to be analyzed can preclude some possible correlations from being revealed, but it can eliminate noise and outliers in the data matrix that can hamper interpretation of the results (Gauch, 1982).

In compiling the list of taxa to be analyzed quantitatively, several criteria were used. The 17 taxa selected include the four most abundant species. Others selected are less abundant but appear to occur preferentially in certain rock types. Taxa whose paleoecological significance is thought to be established were given priority, although two species of algae and a species of water fern were not included, despite their presumed ecological significance, because together they amount to less than 0.1% of the total palynomorph assemblage. Other taxa excluded are so poorly represented in the counts that their importance in interpretation of paleoecological communities is highly dubious. The 17 taxa constitute 86.2% of the total palynomorph assemblage in the samples analyzed.

DETRENDED CORRESPONDENCE ANALYSIS

Detrended correspondence analysis (DCA) is a relatively new ordination technique that has proven useful in ecological studies (Gauch, 1982). DCA was run using DECORANA, which is program CEP-40 in the Cornell Ecology Program series (Hill, 1979). We used the IBM-PC version of DECORANA that was developed by Christopher Clampitt (Department of Botany, University of Washington). DCA was used on both the Vermillion Creek and Raton data sets. Although DCA permits plotting of both species and sample ordinations on the same scatterplot, for clarity we plotted each ordination separately.

Results

Vermillion Creek coal bed and associated rocks

LOCAL, EXTRA-LOCAL, REGIONAL, AND REWORKED PALYNOFLORAS

In the empirical analysis of Vermillion Creek data (Nichols, 1987; this study), the local, extra-local, regional, and reworked components of the total palynoflora are characterized respectively by the following attributes. The local palynoflora tends to be best represented in coal samples; some species tend to have high relative abundances (indicating dominance); certain rarely occurring diagnostic species may be present; masses of pollen or spores from individual anthers or sporangia deposited in place may be present; the association of species appears to represent a single plant community.

The extra-local palynoflora can occur in either coal or non-coal samples; relative abundances of certain species are greater than is normal for those species in the regional palynoflora; the association of species appears to represent one or more plant communities. We note that the depositional environments represented, marsh and lake, are each local environments in themselves; they are 'extra-local' only with respect to the mire (which was the focus of the study by Nichols, 1987).

The regional palynoflora tends to be best represented in rocks other than coal, including but not limited to rocks of fluvial or lacustrine origin. Relative abundances of individual species tend to be low, but occurrence is consistent. Occurrences and relative abundances are unrelated to lithology (facies-independent). Species included are common ones known from contemporaneous deposits of the region. The reworked palynoflora would more likely be found in clastic facies incorporating material eroded from older rocks than in autochthonous deposits. It includes species representative of plants not known to have been living in the region during the time of deposition of the sediments, and their occurrence is sporadic and relative abundance very low.

For the reconstruction of depositional environments, the local and extra-local palynofloras are the most important. For biostratigraphy the most useful components of the total palynoflora are regional, and the least useful are reworked.

The local palynoflora includes one species of palm pollen, three species of fern spores, three species of dicot angiosperm pollen, and one species of fungal spore; they are listed by name in Table 21.3. Palm pollen (*Arecipites*) was present in all Vermillion Creek samples, but it tended to be more abundant in coal, amounting to more than 25% of the assemblage in two-thirds of the coal samples analyzed. Masses of undispersed palm pollen were present in some samples, indicating local deposition. The fern spore species were clearly more abundant in coal than in non-coal facies, and two of the three were represented by masses of undispersed spores in the coal. The dicot pollen species of the local palynoflora were less common than the palm pollen or the fern spores, but were relatively more abundant in coal than in non-coal facies, and one was represented by undispersed masses. They are not species of known paleoecologic significance, but are included in the local palynoflora because of their pattern of occurrence. The fungal spores tended to be more abundant at the top of the Vermillion Creek coal bed than elsewhere in the sampled interval; they may be an indicator of a deterioration of the environmental conditions conducive to preservation of organic matter. The environment in which the peat that formed the Vermillion Creek coal accumulated is interpreted as a palm mire with subordinate ferns and with certain angiosperms as minor components. Palms are the dominant plants in some modern tropical mires (Richards, 1952).

The extra-local palynoflora includes three species of angiosperm pollen, two species of fern spores, and two species of algal cysts or coenobia (Table 21.3). The extra-local palynoflora was prevalent in shale that represented either clastic deposition at the mire margins, crevasse splay deposits that periodically invaded the mire, or freshwater lacustrine deposits. Two plant communities

Table 21.3. *Paleoecologically and biostratigraphically important palynomorph species from the Vermillion Creek coal bed and associated rocks and some commonly occurring species from the regional palynoflora as determined by the empirical method*

Local palynoflora

Arecipites tenuiexinous Leffingwell 1971

Laevigatosporites haardtii (Potonié & Venitz 1934) Thomson & Pflug 1953

Lygodiumsporites adriennis (Potonié & Gelletich 1933) Potonié 1956

Verrucatosporites prosecundus Elsik 1968

Intratiporopollenites sp.

Ranunculacidites sp.

Cupuliferoidaepollenites sp.

Pluricellaesporites sp.

Extralocal palynoflora

Platycarya platycaryoides (Roche 1969) Frederiksen & Christopher 1978

Pandaniidites radicus Leffingwell 1971

Sparganiaceaepollenites sp. cf. *S. polygonalis* Thiergart 1938

Deltoidospora sp.

Azolla cretacea Stanley 1965

Pediastrum paleogeneites Wilson & Hoffmeister 1953

Sigmopollis sp.

Regional palynoflora (part)

Aesculiidites circumstriatus (Fairchild in Stover *et al.* 1966) Elsik 1968

Alnus speciipites Wodehouse 1933

Tilia vescipites Wodehouse 1933

Tilia tetraforaminipites Wodehouse 1933

Tricolpites spp.

Ulmipollenites undulosus Wolff 1934

Lower Eocene congregation

Ailanthipites berryi Wodehouse 1933

Boehlensipollis sp.

Bombacacidites sp.

Eucommia sp.

Intratriporopollenites sp.

Momipites triradiatus Nichols 1973

Pistillipollenites mcgregorii Rouse 1962

Platycarya platycaryoides (Roche 1969) Frederiksen & Christopher 1978

Sparganiaceaepollenites sp. cf. *S. polygonalis* Thiergart 1938

Striatopollis sp.

appear to be represented, which inhabited areas adjacent to the palm mire. The most abundant angiosperm was a species of a modern genus (*Platycarya platycaryoides*) believed to have been produced by an early successional species in Eocene time (Wing, 1984); the producing plants probably were able to establish themselves quickly on crevasse splays and along the shifting margins of the mire.

The other two angiosperm pollen species (*Pandaniidites radicus* and *Sparganiaceaepollenites* sp.) are interpreted to have been produced by plants whose modern relatives are aquatic or inhabit coastal or marshy areas (Heywood, 1978). The algae include a fossil species of a modern freshwater genus (*Pediastrum*) and a fossil genus and species previously found in lacustrine deposits (*Sigmopollis* sp.). The spore species are thought to have represented ferns growing at the edge of the mire (extra-local) rather than in it (local) because of their more common occurrence in shale than in coal.

The components of the regional palynoflora need not be itemized here because they are not important to determination of local depositional environments; a few of the more commonly occurring species are listed in Table 21.3. The regional palynoflora included 29 species of angiosperm pollen, two species of gymnosperm pollen, and two species of fern spores. They were interpreted as having been produced by plants that were components of the regional flora and that lived at some distance from the site of deposition of the Vermillion Creek beds. Nichols (1987) also listed two other groups of palynomorph species within the Vermillion Creek palynoflora, which had biostratigraphic rather than paleoecologic significance. Four species were identified as reworked, and so had negative significance in biostratigraphy; that is, they should not be used for age determination (or for paleoecologic interpretation). Ten species were identified as potentially useful in establishing the age of the Vermillion Creek beds and in correlating them regionally; these species are listed in Table 21.3. This last group of species, which occur together but are only a part of the total Vermillion Creek assemblage, and which characterize these lower Eocene rocks, are appropriately referred to as a 'congregation' (cf. Bates & Jackson, 1987).

PCA AND *R*-MODE CLUSTER ANALYSIS

The dendrogram for the 17 species included in the *R*-mode cluster analysis of the Vermillion Creek samples is shown in Figure 21.2. By visual inspection of the dendrogram, the species were divided into four groups with one outlying species. Table 21.4 summarizes the results of both cluster analysis and PCA by listing the species in groups identified by cluster analysis and highlighting only the loadings on the principal components that are 0.25 or higher, except as noted.

All of the species in group 1 had their highest loadings on principal component 1. Inspection of the data matrix reveals that these species are most abundant in shale, especially carbonaceous shale. They occur only occasionally, in low numbers, in coal samples. Because this group includes *Pandaniidites radicus* and *Spargania-ceaepollenites* sp., it is interpreted to represent a marsh palynofacies, based on the previously mentioned affinities of these species with modern plants. This interpretation

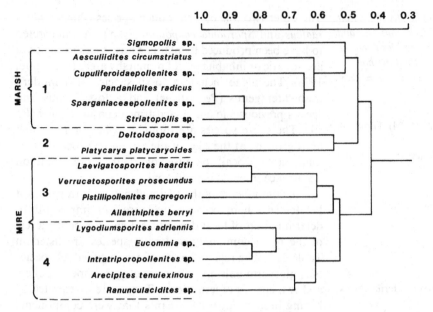

Figure 21.2. Dendrogram for cluster analysis of Vermillion Creek species. Scale is Pearson product-moment correlation coefficient. Species groups 1–4 determined by visual inspection of dendrogram.

suggests that the plants that produced the other angiosperm pollen in group 1 (Table 21.4) also lived in or very near a marsh, although their botanical affinities are uncertain.

Sigmopollis sp. was most closely linked to group 1 in the cluster analysis but is considered to be an outlier. It is weakly linked with the other species in the cluster dendrogram and loads highly on a different principal component than that of the group 1 species. In addition, inspection of the data matrix revealed that it has a unique distribution among the samples. *Sigmopollis* is an alga that is most abundant in one of the oil shale samples.

Group 2 includes only two species, *Platycarya platycaryoides* and *Deltoidospora* sp., which were linked together weakly by the cluster analysis. The principal component loadings for these species were very low compared to other taxa in the data matrix. The highest loading for *P. platycaryoides* is −0.06 on principal component 6. Inspection of the matrix of correlation coefficients reveals that this species always has a negative or very low coefficient of correlation with all other species. This lack of correlation probably is due to the pattern of relative abundance shown by this species in the Vermillion Creek samples. *Platycarya platycaryoides* is the most abundant species and occurs in all of the samples. Its relative abundance averages 30% and in some samples reaches more than 70%. No other species in the Vermillion Creek samples is as consistently abundant. *Deltoidospora* sp. was weakly linked with *P. platycaryoides* by the cluster analysis but has its highest loading (0.03) on principal component 1. This fern spore

occurs mostly in low numbers in shale samples but reaches as much as 40% in one sample, which also has a relatively high abundance of *P. platycaryoides* (54%). The two species in this cluster are not interpreted to represent a palynofacies.

Group 3 of the dendrogram (Fig. 21.2) contains two pterophyte and two angiosperm palynomorphs. The fern spores *Laevigatosporites haardtii* and *Verrucatosporites prosecundus* are closely linked by the cluster analysis and both loaded highly on principal component 3. The distribution and relative abundance patterns of group 3 species in the data matrix indicate that they are linked due to their co-occurrence in coal samples. Thus group 3 is interpreted to represent a mire palynofacies. Group 4 contains one pterophyte spore (*Lygodiumsporites adriennis*), the palm pollen (*Arecipites*), and pollen of three other angiosperms. Species of this group display patterns of distribution and relative abundance similar to that of group 3, indicating that both of these groups are part of the mire palynofacies.

Although groups 3 and 4 are both composed of taxa that have their highest abundances in coals, they were differentiated (only weakly linked) by the cluster analysis and by PCA. Table 21.4 shows that *Arecipites tenuiexinous* (group 4) is highly loaded on principal component 4 but is also moderately loaded on components 5 and 6. Other group 4 taxa have moderate to high loadings on principal component 5, but group 3 taxa have moderate to high loadings on principal components 4 and 6. Because *A. tenuiexinous* loads on principal components of both groups 3 and 4, it is evident that this species is important in both of these groups. Inspection of the data matrix revealed that this species is consistently present and most abundant in coal samples. Other species in groups 3 and

Table 21.4. *Paleoecologically important groups of palynomorph species from the Vermillion Creek coal bed and associated rocks as determined by R-mode cluster analysis and principal components analysis*

Species loadings for principal components (PC) 1 through 6 are listed; species loadings of 0.25 or more are highlighted (in bold) for the principal components of each group except for group 2, for which the highest loading for the species is highlighted. *Sigmopollis* sp. had a high loading (0.96) on PC 7, but was not included in any of the cluster groups and is not listed.

	PC 1	PC 2	PC 3	PC 4	PC 5	PC 6
Group 1						
Aesculiidites circumstriatus	**0.40**	0.04	−0.06	−0.51	0.01	**0.26**
Cupuliferoidaepollenites sp.	**0.36**	−0.34	−0.27	0.17	−0.33	**0.28**
Pandaniidites radicus	**0.94**	0.01	−0.06	0.01	−0.02	−0.09
Sparganiaceaepollenites sp.	**0.96**	−0.002	−0.03	0.06	−0.03	−0.01
Striatopollis sp.	**0.81**	−0.14	−0.19	−0.10	−0.17	0.13
Group 2						
Deltoidospora sp.	**0.03**	−0.15	−0.05	−0.01	−0.16	−0.89
Platycarya platycaryoides	−0.27	−0.32	−0.43	−0.64	−0.26	**−0.06**
Group 3						
Ailanthipites berryi	−0.09	−0.21	**0.35**	−0.12	−0.35	**0.44**
Laevigatosporites haardtii	−0.17	0.24	**0.82**	0.20	0.23	0.07
Pistillipollenites mcgregorii	0.08	0.05	0.15	**0.81**	−0.19	−0.03
Verrucatosporites prosecundus	−0.10	−0.08	**0.87**	0.17	−0.16	0.05
Group 4						
Arecipites tenuiexinous	−0.30	0.23	0.08	**0.60**	**0.49**	**0.25**
Eucommia sp.	−0.21	**0.84**	−0.10	0.18	−0.11	0.04
Intratriporopollenites sp.	−0.17	**0.84**	0.11	0.04	**0.29**	0.01
Lygodiumsporites adriennis	−0.40	**0.78**	0.12	0.02	0.04	0.07
Ranunculacidites sp.	−0.10	0.04	−0.003	−0.07	**0.88**	0.08
Cumulative percentage of variance	23.9	40.4	51.8	60.0	66.9	73.4

4 have their greatest abundances in coals, but are not consistently present in all coal samples. From this we interpret that groups 3 and 4 represent two subgroups within the mire palynofacies. Both are characterized by the presence of palms (represented by *A. tenuiexinous*) but they are distinguished by their different species of ferns and angiosperms.

DCA SPECIES ORDINATION

Two detrended correspondence analyses were done on palynomorph data from the Vermillion Creek samples. We used the same 17 species in the first of these analyses that were used in cluster analysis and PCA. The four DCA axes in this analysis had eigenvalues of 0.21, 0.10, 0.05, and 0.02. Axes having eigenvalues much less than the highest are unlikely to be significant (Hill, 1979), and only axes 1 and 2 were used to generate the scatterplot shown in Figure 21.3. This scatterplot shows a pattern that expresses an ecological gradient.

The scatterplot of the species ordination of 17 species (Fig. 21.3) is divided into three fields (labeled A, B, and C) representing a gradient along DCA axis 1. We drew the dashed lines between fields to separate what appear to be the major groups of plotted points. The species that occur within these fields are listed in Table 21.5 (names followed by asterisks). The second axis does not appear to have any ecological significance. Field A includes eight of the nine species that belong to the mire palynofacies as determined by PCA and cluster analysis. These include *Arecipites tenuiexinous* and the several fern spore species that dominate the mire palynofacies. Field A also includes the rare species *Eucommia* sp. The position of *Eucommia* in the scatterplot suggests that this species is restricted to the mire palynofacies. In the Vermillion Creek samples, it is present only in coal samples except for one shale sample immediately above a coal. Because this fossil is unequivocally referable to the Eucommiaceae, we interpret that *Eucommia* included species that inhabited mires in

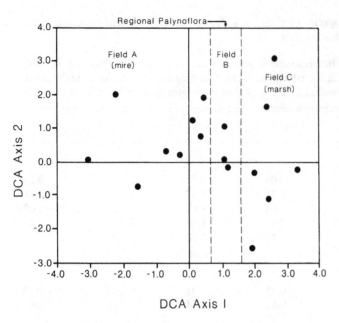

Figure 21.3. Detrended correspondence analysis ordination of 17 Vermillion Creek species. Fields delimited by dashed lines are: field A – species of the mire palynofacies; field B – regional palynoflora; and field C – species of the marsh palynofacies. See text for discussion of important species in each field.

Table 21.5. Paleoecologically important palynomorph species from the Vermillion Creek coal bed and associated rocks and some commonly occurring species from the regional palynoflora as determined by detrended correspondence analysis

Names followed by asterisks are the only ones included in the 17-species data base.

Field A (mire palynofacies)

Arecipites tenuiexinous *
Cycadopites sp.
Eucommia sp. *
Intratriporopollenites sp. *
Laevigatosporites haardtii *
Lygodiumsporites adriennis *
Pistillipollenites mcgregorii *
Ranunculacidites sp. *
Verrucatosporites prosecundus *

Field B (regional palynoflora)

Ailanthipites berryi *
Alnus speciipites
Cupuliferoidaepollenites sp. *
Momipites triradiatus
Platycarya platycaryoides *
Tilia tetraforaminipites
Tilia vescipites
Ulmipollenites undulosus

Field C (marsh palynofacies)

Aesculiidites circumstriatus *
Azolla cretacea
Deltoidospora sp. *
Pandaniidites radicus *
Sigmopollis sp. *
Sparganiaceaepollenites sp. *
Striatopollis sp. *

the Eocene. However, the only living species of *Eucommia* apparently does not inhabit mires (cf. Heywood, 1978; Leopold & MacGinitie, 1972).

Species in field B of Figure 21.3 show a pattern of distribution that is largely independent of lithology. Field B includes *Platycarya platycaryoides*, which is the most abundant species in the Vermillion Creek palynoflora and is present in all samples, and *Cupuliferoidaepollenites* sp. (which was included in the marsh palynofacies by PCA and cluster analysis), which also is present in all Vermillion Creek samples. The third species, *Ailanthipites berryi*, (which was included in the mire palynofacies by PCA and cluster analysis) is present in only about a third of the samples, but its occurrence also is independent of lithology. Independence of lithology is an important attribute of biostratigraphically useful taxa; therefore, these species are potentially useful in palynostratigraphy. *Platycarya platycaryoides* and *Ailanthipites berryi*, which are morphologically distinctive species, are members of the lower Eocene congregation (Table 21.3). *Cupuliferoidaepollenites* sp., which is not morphologically distinctive, is not regarded as having palynostratigraphic utility.

Field C contains four of the five species that belong to the marsh palynofacies as determined by PCA and cluster analysis. These include the environmentally sensitive species *Pandaniidites radicus* and *Sparganiaceaepollenites* sp. The fern spore species *Deltoidospora* sp. also is placed in Field C by DCA. This species was grouped with *Platycarya platycaryoides* by PCA and cluster

analysis, and these two species were not interpreted to represent a palynofacies on the basis of those methods. The implication that *Deltoidospora* sp. is part of the marsh palynofacies as indicated by DCA seems more reasonable because it is present almost exclusively in samples interpreted to represent marsh deposits.

The second analysis of Vermillion Creek data using DCA evaluated the robustness of the data set. The second analysis employed a matrix of 45 species (the total palynoflora excluding reworked specimens and trace occurrences) and 31 samples. Results were closely similar to those obtained using the 17-species data set. In the second analysis, the four DCA axes had eigenvalues of 0.18, 0.09, 0.05, and 0.04. As before, only the first and second were used to generate scatterplots. Figure 21.4 is the scatterplot of the species ordination. The original 17

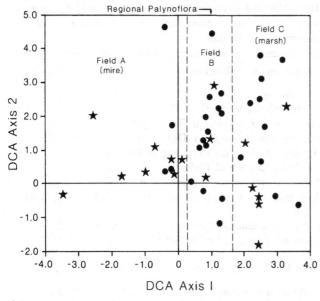

★ Original 17 Species
● Additional Species

Figure 21.4. Detrended correspondence analysis ordination of 45 Vermillion Creek species. Fields delimited by dashed lines are: field A – species of the mire palynofacies; field B – regional palynoflora; and field C – species of the marsh palynofacies. Stars indicate the original 17 species shown in Fig. 21.3; dots indicate additional species.

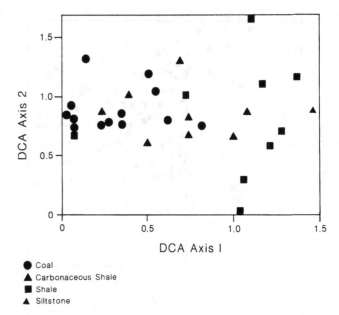

● Coal
▲ Carbonaceous Shale
■ Shale
▲ Siltstone

Figure 21.5. Detrended correspondence analysis ordination of Vermillion Creek samples. Circles represent coal, solid triangles represent carbonaceous shale, squares represent shale, and the small triangle represents siltstone. From left to right, the first axis expresses a lithologic gradient from coal to carbonaceous shale, shale, and siltstone.

taxa are shown as stars and the additional species are shown as dots. In both analyses, the original 17 species have the same relative distribution with respect to the gradient along the first DCA axis. Positions of some points (species) changed along axis 2 in the second analysis, but as noted, this axis has no ecologic significance. Dashed lines separating fields along the axis 1 gradient are in approximately the same positions in both scatterplots (Figs. 21.3 and 21.4). The second analysis demonstrates that the Vermillion Creek data set is robust, and that addition or elimination of rarely occurring species had no important effect on the results.

The significance of some of the additional species is worth noting. Although *Azolla cretacea* has known paleoecological significance, it was omitted from the original 'pollen sum' because it is extremely rare in the Vermillion Creek samples. *Azolla cretacea*, which is the megaspore of a heterosporous water fern, plotted in the field C, and its presence is consistent with the interpretation that field C represents the marsh palynofacies. As noted, field B includes species that are potentially biostrati-graphically useful. One of the additional taxa in field B is a species of *Momipites*, which, although it is not numerically prominent in Vermillion Creek samples, has palynostratigraphic significance and is a member of the lower Eocene congregation.

DCA SAMPLE ORDINATION

The scatterplot of the sample ordination based on both data sets arranged the samples along a lithologic gradient that is generally parallel to DCA axis 1. Here again the relative positions of points with respect to axis 1 are the same, irrespective of whether 17 or 45 species are included in the data sets. The only difference noted between sample ordination plots was relative movement of points along axis 2, which has no apparent ecological significance. Results based on the larger data set are shown in Figure 21.5.

In Figure 21.5, most coal samples are on the left and most shale samples are on the right, carbonaceous shale is distributed between these two lithologies, and there is one siltstone sample on the extreme right. The gradient is from most carbonaceous to least carbonaceous lithology. Although this distribution of samples seems obvious in terms of lithology, it is actually based on palynologic content of the samples. The distribution shows that there is a relationship between the lithology of the samples and their palynologic assemblages, because the distribution of samples in the scatterplot is based on the species ordination. Thus it is evident that both the palynomorph assemblage and the lithology of each Vermillion Creek sample are controlled by the depositional environment.

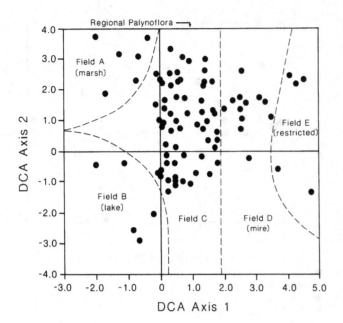

Figure 21.6. Detrended correspondence analysis ordination of Raton Formation species. Fields delimited by dotted lines are: field A – species of the marsh palynofacies; field B – species of the lake palynofacies; field C – species of the regional palynoflora; field D – species of the mire palynofacies; and field E – stratigraphically restricted taxa. See text for discussion of important species in each field.

Raton Formation

The Raton Formation palynomorph data were analyzed using DCA (PCA and cluster analysis also were used; results were comparable to those of DCA and only DCA results are discussed). Four DCA axes were ranked and scaled with species and sample scores provided for each DCA axis. DCA axes 1 through 4 had eigenvalues of 0.32, 0.15, 0.12, and 0.09, respectively. As mentioned, axes with eigenvalues much less than the highest eigenvalue are unlikely to be significant, and DCA axis 4 (eigenvalue = 0.09) was disregarded. Scatterplots of DCA axis 1 versus DCA axis 2 provided the most useful information, while those using DCA axis 3 did not show any significant pattern and are not discussed further.

DCA SPECIES ORDINATION

The species scores for the first two DCA axes were plotted against each other and the resulting scatterplot is shown in Figure 21.6. Visual inspection of the scatterplot revealed five fields of species. These were outlined by hand and labeled A, B, C, D, and E on the scatterplot. Species of field E include some taxa restricted to the Cretaceous (e.g., *Proteacidites* spp.). The other taxa have uncertain paleoecological or biostratigraphic significance. Because species of field E are not interpreted to have any paleoecological significance, they are not discussed further. Important species that occur in fields A–D are listed in Table 21.6.

Table 21.6. *Paleoecologically and biostratigraphically important palynomorph species from the Raton Formation (detrended correspondence analysis)*

Field A (marsh palynofacies)

Ovoidites sp.
Isoëtes subengelmanni Elsik 1968
Minerisporites succrassulus Tschudy 1976
Minerisporites glossoferus (Dijkstra 1961) Tschudy 1976
Azolla velus (Dijkstra 1961) Jain & Hall 1969
Azolla barbata Snead 1969
Kurtzipites trispissatus Anderson 1960
Kurtzipites circularis (Norton in Norton & Hall 1969) Srivastava 1981
Kurtzipites annulatus Norton in Norton & Hall 1969
Quercoidites spissus Leffingwell 1971

Field B (lake palynofacies)

Pediastrum paleogeneites Wilson & Hoffmeister 1953
Scenedesmus tschudyi Fleming 1989
Scenedesmus hanleyi Fleming 1989
Periporopollenites sp.

Field C (lower Paleocene congregation)

Momipites inaequalis Anderson 1960
Momipites waltmanensis Nichols & Ott 1978
Momipites wyomingensis Nichols & Ott 1978
Momipites tenuipolus Anderson 1960
Momipites leffingwellii Nichols & Ott 1978
Momipites anellus Nichols & Ott 1978
Momipites ventifluminis Nichols & Ott 1978
Caryapollenites prodromus Nichols & Ott 1978

Field D (mire palynofacies)

Stereisporites sp.
Reticuloidosporites pseudomurii Elsik 1968
Cycadopites sp.
Cupuliferoidaepollenites spp.
Tricolpites bathyreticulatus Stanley 1965
Fraxinoipollenites variabilis Stanley 1965
Triatriopollenites granilabratus (Stanley 1965) Norton in Norton & Hall 1969
Thomsonipollis magnificus (Pflug & Thomson in Thomson & Pflug 1953) Krutzsch 1960

Field A contains six palynomorph taxa: four angiosperms, a pteridophyte, and an alga. Samples that contain this array are commonly dominated by *Isoëtes subengelmanni* and always have one or more species of *Kurtzipites*. These samples also contain megaspores of *Minerisporites* and *Azolla* (they were not tabulated in the data matrix because they were recovered separately and therefore could not be included in the palynomorph counts). *Isoëtes* and *Minerisporites* are unequivocally referable to the Isoëtaceae, the modern quillwort family.

Members of this lycopsid family typically are found inhabiting the margins of lakes, ponds, or ephemeral pools (Tryon & Tryon, 1982). *Azolla* is a genus of aquatic ferns that inhabit freshwater marshes and mires. We interpret the species of field A to be the marsh palynofacies. The botanical affinity of the three species of *Kurtzipites* is uncertain. However, their inclusion in field A by DCA suggests that the plants that produced these pollen grains lived in or near a marsh.

Field B contains three species of algae and two angiosperm palynomorphs. The species of field B are *Pediastrum paleogeneites*, *Scenedesmus tschudyi*, *S. hanleyi*, *Periporopollenites* sp., and a simple tricolpate form (unnamed). Fleming (1989) discussed the paleoecological significance of fossil species of the modern genus *Scenedesmus*; today *Scenedesmus* occurs as phytoplankton restricted to freshwater habitats (Hutchinson, 1967; Round, 1973; 1981). *Pediastrum*, another modern genus of freshwater phytoplankton, is also known from Cretaceous and Tertiary rocks (Cookson, 1953; Wilson & Hoffmeister, 1953). On the basis of the presence of these freshwater phytoplankton, we interpret the species of field B to be the lake palynofacies.

The inclusion of angiosperm pollen such as *Periporopollenites* sp. with planktonic algae as part of the lake palynofacies at first appears to present a problem because few angiosperms inhabit lakes. However, plants fringing lakes would be expected to contribute greater amounts of pollen to lake sediments than plants living at some distance from the lake. A parallel high abundance of the morphologically similar, possibly congeneric, pollen of *Chenopodipollis* was reported in pond facies in the lower Eocene of the Bighorn basin by Farley (1988). Presumably based on the ecology of modern Chenopodiaceae, Farley suggested that his *Chenopodipollis* pollen was likely to have been transported from outside the immediate surroundings of the sampling site. Farley's *Chenopodipollis* and *Periporopollenites* from the Raton Formation may have been produced by the same kinds of plants. If this is true, then the distribution of the fossils in both the Bighorn and Raton basins suggests that the plants that produced the pollen lived around the margins of freshwater lakes, although modern chenopods do not occupy such habitats. This suggests that the autecology of the plants that produced this pollen, if they were chenopods, has changed. The possibility that the autecology of plants can change through time was considered by Farley, but in this case was overlooked.

Field C contains the largest number of species. It is located in the middle of the scatterplot and cannot be reasonably subdivided. Inspection of the species list and comparison with the raw count data reveal that these taxa are consistently present in the samples. In other words, Raton Formation samples have a high probability of containing taxa in this field. This group of species represents the regional palynoflora of the Raton Formation. Widespread distribution is an important attribute of biostratigraphically useful fossils, and therefore species of field C are potentially useful in palynostratigraphy. Some of these species, such as *Laevigatosporites*, are not useful because they are long-ranging forms that display no evolutionary change. Others, such as the morphologically distinctive juglandaceous species, exhibit a relatively high evolutionary rate that makes them valuable fossils in palynostratigraphy; they are members of the lower Paleocene congregation (Table 21.6).

Field D contains twelve taxa, all of which have their highest abundances in coal. Except for *Stereisporites* sp. and *Cycadopites* sp., the species of field D have unclear botanical affinities. The most abundant species are *Cupuliferoidaepollenites* spp. and *Fraxinoipollenites variabilis*, which are pollen of dicotyledonous angiosperms of uncertain affinity. Spores of *Stereisporites* were produced by *Sphagnum* moss, which is a common inhabitant of mires and bogs. The occurrence of all these species in coal indicates that field D is the mire palynofacies in the Raton Formation. The occurrence of *Sphagnum* supports this interpretation, although the other plants represented are not otherwise known to be mire plants. *Cycadopites* pollen was produced by cycads, which grow in a variety of tropical habitats in the modern world. The implication is that the plants that produced our fossil *Cycadopites* pollen inhabited the Raton mires. The same is true of pollen of another species in field D, the morphologically distinctive *Thomsonipollis magnificus*.

Ordination of Raton species shows a gradient similar to that seen in the Vermillion Creek species ordination. Figure 21.6 shows the marsh (field A) and lake (field B) palynofacies plotted on the left end of the scatterplot and the mire palynofacies plotted on the right. The marsh and lake palynofacies are separated along a gradient with respect to DCA axis 2. Field C species (the regional palynoflora) plot in the middle of the scatterplot. In the Vermillion Creek species ordination the marsh and mire palynofacies are also separated by a middle group that consists of ubiquitous taxa (e.g., *Platycarya platycaryoides*). The only difference is that the sequence along the gradient is reversed in Figure 21.3 with the marsh palynofacies on the right and the mire palynofacies on the left. These results suggest that detrended correspondence analysis of palynological data not only can identify palynofacies that are useful in interpretation of depositional environments, but also groups of palynomorphs that may include palynostratigraphically useful species.

DCA SAMPLE ORDINATION

The scatterplot of the Raton Formation sample ordination is shown in Figure 21.7. Initial inspection of the scatterplot did not reveal an easily interpretable pattern;

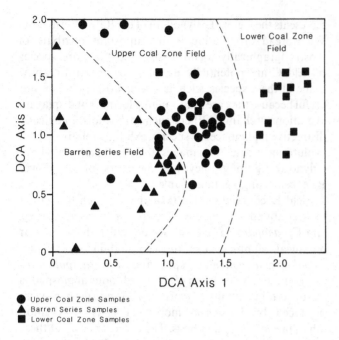

Figure 21.7. Detrended correspondence analysis ordination of Raton Formation samples. In ascending stratigraphic order, squares represent assemblages from lower coal zone (Upper Cretaceous) samples, triangles represent assemblages from barren series (Paleocene) samples, and circles represent assemblages from upper coal zone (Paleocene) samples. The first axis expresses a lithostratigraphic gradient that reflects division of the Raton Formation into three informal members. Fields of these members are delimited by dotted lines on the scatterplot.

the Raton samples do not plot along a lithologic gradient similar to that observed in the Vermillion Creek analysis. Coding the points to indicate the informal members of the Raton Formation represented, however, revealed three fairly distinct groups. These groups are outlined on the scatterplot and correspond to the lower coal zone, the barren series, and the upper coal zone of the Raton Formation. This distribution of samples is stratigraphically controlled. It distinguishes the members of the Raton Formation, even though two of the members (which are of different geologic age) are coal-bearing. Assemblages in both coal zones are characterized by the mire, marsh, and lake palynofacies; as discussed below, what distinguishes the coal zones from one another is that the are separated by the K/T boundary. Assemblages from the barren series generally lack the mire palynofacies and are characterized by species that are widely distributed.

It should be emphasized that the lower coal zone is Late Cretaceous in age, whereas the barren series and the upper coal zone are Paleocene. The Raton Formation data span the K/T boundary, a major event horizon. One purpose of our analysis of the Raton palynologic data is to evaluate changes across the K/T boundary as they might be revealed by DCA. Wolfe and Upchurch (1987) interpreted that a drastic change in the flora,

extinction of 75% of plant species, took place across the boundary. Palynologic data (Fleming, 1985; Fleming, 1990; Nichols et al., 1990; Nichols & Fleming, 1990) indicate that, regionally, the level of extinction was only about 45%, and that in the Raton basin it may have been as low as 15%. Wolfe and Upchurch attributed this discrepancy to preservational and taxonomic differences between fossil leaves and palynomorphs. Fleming (1990) noted that at least some of the discrepancy can be accounted for by sampling bias in the paleobotanical data. Raton palynologic data show that an extinction event indeed marks the K/T boundary, that part of the assemblage in Upper Cretaceous samples is absent in Paleocene samples, and that Paleocene samples contain juglandaceous species that are not present in the Upper Cretaceous. These differences clearly affected the ordination of the samples. DCA results and inspection of the data matrix indicate, however, that the change across the boundary does not involve a complete reorganization of the flora.

DCA axis 2 expresses a gradient along the scatterplot from the coal-bearing members on the right to the barren series on the left. This 'sequence' is out of stratigraphic order for two reasons. First, the palynologic similarities among assemblages from the two coal zones cause them to plot near each other. Second, because barren series and upper coal zone assemblages both lack species restricted to the Upper Cretaceous, they plot next to each other. A few upper coal zone assemblages (circles) plot in the barren series region. These assemblages are from a stratigraphic interval that is transitional between the barren series and the upper coal zone. Two assemblages from the lower coal zone (squares) plot in the upper coal zone field. They plot here because these assemblages happen to have few or none of the biostratigraphically restricted species that otherwise link assemblages from the lower coal zone.

Discussion and conclusions

All of the methods employed enabled us to distinguish depositional environments of some Upper Cretaceous and Paleogene coal-bearing rocks. By various means, patterns in the data were interpreted to represent palynofacies. By interpreting the palynofacies in the light of available sedimentologic and botanical data, we were able to draw inferences about the nature of the depositional environment. Once the paleoecological significance of the palynofacies was established, paleoecological inferences were made for some species of uncertain botanical affinity that were present in our samples.

In the Vermillion Creek beds, the empirical and multivariate methods revealed distinct palynofacies that represent mire and marsh depositional environments. There is a general agreement between the species compositions of the mire and marsh palynofacies as revealed by the empirical method, PCA and cluster analysis, and DCA (compare Tables 21.3, 21.4, and 21.5). The earlier results based on empiricism are thus corroborated. All three approaches indicate that the early Eocene coal-forming mire was dominated by palms and ferns with subordinate angiosperms. The associated Eocene marsh was characterized by plants whose modern analogs inhabit marshes. There is a clear correspondence between the palynofacies we defined and the environments of deposition determined from sedimentologic analysis (Roehler, 1987).

Results of the DCA of Raton Formation samples indicate that three palynofacies are present. They represent marsh, lake, and mire depositional environments of latest Cretaceous and early Paleocene age (Table 21.6). The marsh environment was characterized by aquatic pteridophytes and extinct plants of uncertain affinity. The lake environment was inhabited by freshwater phytoplankton. The flora of the Raton mire environments included ferns, *Sphagnum* moss, and cycads, but apparently was dominated by dicotyledonous angiosperms of uncertain affinity. In general, there is a close correspondence between Raton palynofacies and environments of deposition determined sedimentologically (Flores, 1987; Pillmore & Flores, 1987), in which coals contain the mire palynofacies and carbonaceous shales contain the marsh palynofacies. The third palynofacies, that of the lake, permitted a refinement in interpretation of Raton depositional environments. The lake palynofacies is present in certain mudstone samples that previously have not been recognized as lacustrine deposits.

All the methods revealed closely similar palynofacies, corroborating our interpretations of the depositional environments. The combination of cluster analysis and principal components analysis disclosed a subtle pattern not evident from the other methods. The Vermillion Creek mire flora was dominated by palms, but apparently had two different communities within it, characterized by different species of ferns and angiosperms.

The multivariate methods all showed closely comparable results. However, only detrended correspondence analysis revealed paleoecological gradients. Similar paleoecological gradients were demonstrated for the Vermillion Creek and Raton data sets. In both species ordinations by DCA, a marsh to mire gradient was expressed along one DCA axis, and a marsh to lake gradient was expressed along another axis in the Raton analysis. Our results suggest that multivariate techniques were useful in analyzing data such as ours, but if gradients are important to an analysis, DCA is to be preferred.

DCA also permitted interpretation of the paleoecology of some palynomorph species that have no known affinity with living plants. *Thomsonipollis magnificus* from the Raton Formation evidently was produced by an extinct plant that inhabited mires. *Eucommia* sp. from the Vermillion Creek beds apparently was produced by a plant that had similar ecological requirements, although modern *Eucommia* does not inhabit mires. These are ecological hypotheses that can be tested by additional observations of coal beds of comparable age.

DCA was found to provide a basis for recognizing congregations of potential value in palynostratigraphy. The method grouped widely occurring, facies-independent species in the center of the scatterplots. Among these species were several that are known to be important for age determination and correlation in rocks of Late Cretaceous and Paleogene age (lower Eocene and lower Paleocene congregations; Tables 21.3 and 21.6, respectively).

Our results demonstrate the practicality of multivariate techniques in analysis of paleopalynological data. However, no mathematical or statistical method can supplant geological and botanical reasoning and the necessity of detailed knowledge of the data set. In fact, such reasoning and knowledge enable one to interpret the results of multivariate analyses.

References

Ager, D. V. (1963). *Principles of Paleoecology*. New York: McGraw-Hill.

Bates, R. L., & Jackson, J. A. (1987). *Glossary of Geology* (3rd ed.). Alexandria, VA: American Geological Institute.

Boulter, M. C. & Hubbard, R. N. L. B. (1982). Objective paleoecologic and biostratigraphic interpretation of Tertiary palynological data by multivariate statistical analysis. *Palynology*, **6**, 55–68.

Cookson, I. C. (1953). Records of the occurrence of *Botryococcus braunii*, *Pediastrum* and the Hystrichosphaerideae in Cainozoic deposits of Australia. *Memoirs of the National Museum of Victoria, Melbourne*, **18**, 107–23.

Dodd, J. R. & Stanton, R. J., Jr. (1981). *Paleoecology, Concepts and Applications*. New York: John Wiley & Sons.

Doher, L. I. (1980). Palynomorph preparation procedures currently used in the paleontology and stratigraphy laboratories, U.S. Geological Survey. *US Geological Survey Circular*, **830**, 1–29.

Faegri, K. & Iversen, J. (1975). *Textbook of Pollen Analysis*. 3rd ed. New York: Hafner.

Farley, M. B. (1988). Environmental variation, palynofloras, and paleoecological interpretation. In *Methods and Applications of Plant Paleoecology*, ed. W. A. DiMichele & S. L. Wing. *Paleontological Society Special Publication*, **3**, 126–46.

Fleming, R. F. (1985). Palynological observations of the Cretaceous-Tertiary boundary in the Raton Formation, New Mexico. *Palynology*, **9**, 242.

Fleming, R. F. (1989). Fossil *Scenedesmus* (Chlorococcales) from the Raton Formation, Colorado and New Mexico, U.S.A. *Review of Palaeobotany and Palynology*, **59**, 1–6.

Fleming, R. F. (1990). Palynology of the Cretaceous-Tertiary boundary interval and Paleocene part of the Raton Formation, Colorado and New Mexico. Ph.D. dissertation, University of Colorado.

Flores, R. M. (1987). Sedimentology of Upper Cretaceous and Tertiary siliciclastics and coals in the Raton basin, New Mexico and Colorado. *New Mexico Geological Society Guidebook*, 38th Annual Field Conference, 255–64.

Gauch, H. G., Jr. (1982). *Multivariate Analysis in Community Ecology*. New York: Cambridge University Press.

Gauch, H. G., Jr., & Whittaker, R. H. (1972). Comparison of ordination techniques. *Ecology*, **53**, 868–75.

Gauch, H. G., Jr., Whittaker, R. H., & Wentworth, T. R. (1977). A comparative study of reciprocal averaging and other ordination techniques. *Journal of Ecology*, **65**, 157–74.

Gore, A. J. P. (1983). *Ecosystems of the World*, Volume 4A, *Swamp, Bog, Fen, and Moor*. Amsterdam: Elsevier.

Heywood, V. H. (1978). *Flowering Plants of the World*. Englewood Cliffs, NJ: Prentice Hall.

Hill, M. O. (1979). *DECORANA – a Fortran Program for Detrended Correspondence Analysis and Reciprocal Averaging*. Ithaca, NY: Cornell University.

Hill, M. O. & Gauch, H. G., Jr. (1980). Detrended correspondence analysis: an improved ordination technique. *Vegetatio*, **42**, 47–58.

Hutchinson, G. E. (1967). *A Treatise on Limnology*, Volume II – *Introduction to Lake Biology and the Limnoplankton*. New York: John Wiley & Sons.

Janssen, C. R. (1973). Local and regional pollen deposition. In *Quaternary Plant Ecology*, ed. H. J. B. Birks & R. G. West. New York: John Wiley & Sons.

Kovach, W. L. (1988). Multivariate methods of analyzing paleoecological data. In *Methods and Applications of Plant Paleoecology*, ed. W. A. DiMichele & S. L. Wing. *Paleontological Society Special Publication*, **3**, 72–104.

Kovach, W. L. (1989). Comparisons of multivariate analytical techniques for use in pre-Quaternary plant paleoecology. *Review of Palaeobotany and Palynology*, **60**, 255–82.

Leopold, E. B., & MacGinitie, H. D. (1972). Development and affinities of Tertiary floras in the Rocky Mountains. In *Floristics and Paleofloristics of Asia and Eastern North America*, ed. A. Graham. Amsterdam: Elsevier, 147–200.

Nichols, D. J. (1987). Palynology of the Vermillion Creek coal bed and associated strata. In *Geological Investigations of the Vermillion Creek Coal Bed in the Eocene Niland Tongue of the Wasatch Formation, Sweetwater County, Wyoming*, ed. H. W. Roehler & P. L. Martin. *US Geological Survey Professional Paper*, **1314-D**, 47–73.

Nichols, D. J. & Fleming, R. F. (1990) [1991]. Plant microfossil record of the terminal Cretaceous event in the western United States and Canada. In *Global Catastrophes in Earth History; an Interdisciplinary Conference on Impacts, Volcanism, and Mass Mortality*, ed. V. L. Sharpton & P. D. Ward. *Geological Society of America Special Paper*, **247**, 445–55.

Nichols, D. J., Fleming, R. F. & Frederiksen, N. O. (1990). Palynological evidence of effects of the terminal Cretaceous event on terrestrial floras in western North America. In *Extinction Events in Earth History*, ed. E. G. Kauffman & O. H. Walliser. *Lecture Notes in Earth Sciences*, **30**, 351–64.

Nie, N. H., Hull, C. H., Jenkins, J. G., Steinbrenner, K., & Bent, D. H. (1975). *Statistical Package for the Social Sciences*. New York: McGraw-Hill.

Pillmore, C. L., & Flores, R. M. (1987). Stratigraphy and depositional environments of the Cretaceous-Tertiary boundary clay and associated rocks, Raton basin, New Mexico and Colorado. In *The Cretaceous-Tertiary Boundary in the San Juan and Raton Basins*, ed. J. E. Fassett & J. K. Rigby, Jr. *Geological Society of America Special Paper*, **209**, 111–30.

Richards, P. W. (1952). *The Tropical Rain Forest*. Cambridge: Cambridge University Press.

Roehler, H. W. (1987). Paleoenvironments and sedimentology. In *Geological Investigations of the Vermillion Creek Coal Bed in the Eocene Niland Tongue of the Wasatch Formation, Sweetwater County, Wyoming*, ed. H. W. Roehler & P. L. Martin. *US Geological Survey Professional Paper*, **1314-C**, 25–45.

Roehler, H. W. & Martin, P. L. (1987). *Geological Investigations of the Vermillion Creek Coal Bed in the Eocene Niland Tongue of the Wasatch Formation, Sweetwater County, Wyoming*. *US Geological Survey Professional Paper*, **1314**.

Round, F. E. (1973). *The Biology of the Algae*. Cambridge: Cambridge University Press.

Round, F. E. (1981). *The Ecology of the Algae*. Cambridge: Cambridge University Press.

Tryon, R. M. & Tryon, A. F. (1982). *Ferns and Allied Plants*. New York: Springer-Verlag.

Tuckey, M. E. & Anstey, R. L. (1989). Gradient analysis: a quantitative technique for biostratigraphic correlation. *Palaios*, **4**, 475–79.

Wilson, L. R. & Hoffmeister, W. S. (1953). Four new species of fossil *Pediastrum*. *American Journal of Science*, **231**, 753–60.

Wing, S. L. (1984). A new basis for recognizing the Paleocene/Eocene boundary in western interior North America. *Science*, **226**, 439–41.

Wolfe, J. A. & Upchurch, G. R., Jr. (1987). Leaf assemblages across the Cretaceous-Tertiary boundary in the Raton basin, New Mexico and Colorado. *Proceedings of the National Academy of Science*, **84**, 5096–100.

22 Relationships between depositional environments and changes in palynofloras across the K/T boundary interval

ARTHUR R. SWEET

Introduction

Interest in understanding the reaction of plant communities to the Cretaceous-Tertiary (K/T) boundary event has focused increased attention on late Maastrichtian and Early Paleocene palynofloras. Both the relative abundances of major groups of palynomorphs and range truncations have been used to infer changes in terrestrial environments coincident with the K/T boundary in mid-continental North America (Orth *et al.*, 1981; Tschudy *et al.*, 1984; Nichols *et al.*, 1986; Bohor *et al.*, 1987; Lerbekmo *et al.*, 1987; Fleming & Nichols, 1990; Nichols, 1990; Sweet *et al.*, 1990; Nichols & Fleming, 1991; Sweet & Braman, 1992). These criteria sometimes are used to emphasize the more catastrophic aspects of floral change at the K/T boundary (Orth *et al.*, 1981; Tschudy *et al.*, 1984; Nichols *et al.*, 1986; Bohor *et al.*, 1987; Fleming & Nichols, 1990). In other publications the relative abundance spikes and range truncations are considered to be part of a continuum of change across the boundary (Tschudy & Tschudy, 1986; Lerbekmo *et al.*, 1987; Sweet *et al.*, 1990; Sweet & Braman, 1992). It now is generally accepted that many morphologically complex species of angiosperm pollen became extinct at or within an interval immediately contiguous to the K/T boundary, and that there is a causal connection between these extinctions and the K/T event.

The dominant late Maastrichtian miospore species often represent flowering plants (Tschudy & Tschudy, 1986; Lerbekmo *et al.*, 1987; Sweet *et al.*, 1990). A dramatic shift occurs in the relative abundances of spores and pollen immediately after the K/T extinction event in most North American terrestrial sections. In Alberta and southwestern Saskatchewan morphologically simple species of angiosperm pollen frequently dominate the post-boundary flora (Lerbekmo *et al.*, 1987; Sweet *et al.*, 1990; Sweet & Braman, 1992). However, in the mid-continental United States and in south-central Saskatchewan, fern spores usually overwhelmingly dominate in an interval up to 15 cm thick immediately overlying samples containing characteristic late Maastrichtian pollen taxa (Orth *et al.*, 1981; Tschudy *et al.*, 1984; Nichols *et al.*, 1986; Fleming & Nichols, 1990; Nichols & Fleming, 1991; Sweet & Braman, 1992). Tschudy *et al.* (1984) considered this 'fern spore spike' to represent the colonization of a denuded landscape by ferns following a catastrophic destruction of vegetation as the consequence of the K/T event. The species that increased in relative abundance subsequent to the extinction event are referred to here as opportunistic. This term was applied by Levinton (1969) to species whose abundance was controlled by the physical environment rather than by interspecific competition. The use of the alternate term 'pioneer' would imply an acceptance of Tschudy's concept of a denuded landscape, whereas the term opportunistic allows for a scenario in which there was only a selective dying of plant species.

The late Maastrichtian to early Paleocene floras vary in composition throughout mid-continental North America as a reflection of intra-continental (Wolfe & Upchurch, 1986; Nichols *et al.*, 1990) and regional (Jerzykiewicz & Sweet, 1988; Nichols, 1990; Sweet & Braman, 1992) differences in the paleoenvironment. Sequential changes occur within each of these regionally distinctive palynofloras through the K/T boundary interval (Sweet *et al.*, 1990; Sweet & Braman, 1992).

The purpose of this chapter is to relate palynofloral change to regional facies associations and local lithofacies and hence to latest Maastrichtian and earliest Paleocene, paleogeographic, climatic, and tectonic regimes. In this way the component of change that is independent from lithofacies and attributable to the K/T event will be isolated.

The following discussions focus on the upper Maastrichtian to lower Paleocene interval spanned by polarity chron 29R, which is variously estimated to be between 440 000 and 920 000 years long (Alvarez *et al.*, 1990; Rocchia *et al.*, 1990). During this time the only break in terrestrial sedimentation in mid-continental North America was the early Paleocene incursion of the

461

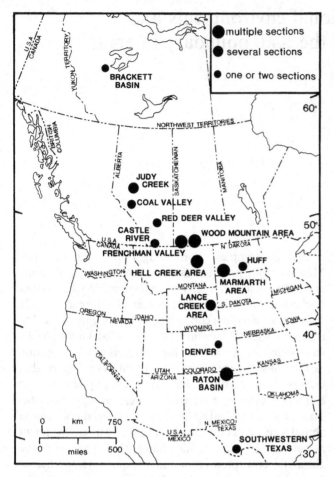

Figure 22.1. Map of mid-continental North America showing areas with terrestrial K/T boundary localities.

Cannonball seaway into the Dakotas and Manitoba (Keefer, 1965; Erickson, 1978; Cherven & Jacob, 1985).

Figured specimens will be permanently stored in the type collection of the Geological Survey of Canada, Ottawa, Ontario. These illustrations are placed at the end of this chapter for ease of reference (Figs. 22.13 and 22.14), since species illustrated are referred to throughout the chapter. Figure 22.1 gives the location of outcrop regions and reference sections used in this chapter. In the text the word 'contact' is substituted for 'boundary' in the discussion of those sections which lack a 'boundary claystone' and hence a complete record of the K/T event.

Significance of palynological data

Many fossils signify the death of an organism, but the presence of pollen or spores in a particular stratigraphic interval only indicates the former presence of a reproductively mature plant population. Hence, palynology provides no direct, positive evidence of catastrophic mass killings, but records subsequent changes in plant communities which may reflect ecological perturbations.

It is the premise here that the changes observed in the structure of plant communities at the K/T boundary were magnified or subdued by regional and local differences in environments of deposition. All environmentally related factors must therefore be addressed, before the effect and magnitude of the causal event can be inferred. Approaching the sequence of changes at the end of the Cretaceous from the perspective of changing depositional environments within a latitudinally and longitudinally graded physical system results in alternate explanations for some seemingly disparate paleontological trends.

Provincialism in fossil floras contiguous to the K/T boundary, reflecting physical factors with a latitudinal gradient (such as temperature), is a likely interpretation of the floral differences observed in samples from a north–south transect (Nichols *et al.*, 1990; Sweet & Braman, 1992; Nichols & Sweet, in press). Under these circumstances, each plant community would have had an optimum zone of habitation bounded by sub-optimum zones of tolerance. In the palynological record, a shift from sub-optimum to optimum conditions, either through time or geographically, predictably might be reflected in a shift from sporadic and low frequencies of any one species to consistent and frequent records. Alternately, if there was a perturbation in the environment exceeding the tolerance of a local community of plants, range truncations would occur. A true extinction event would occur only if this perturbation exceeds the tolerance of a community of plants throughout its entire range. From the above perspective, if the data base used in any specific study is derived from a geographically restricted region, palynological changes will be most evident upsection if the sections are located within the transition zone between two adjacent plant communities. If the sections are within the optimum area for a community of plants, conspicuous changes in the palynoflora will be recorded only by an extreme perturbation in the environment (Sweet & Braman, 1992).

Assessment of the dispersal vectors for angiosperm pollen also affects perception of the magnitude of change. There are no universal criteria for differentiating between pollen of zoophilous (animal, mostly insect, pollinated) and anemophilous (wind pollinated) plants. However, large pollen with a relatively conspicuous sculpture and a thick and structurally complex wall are likely to be from zoophilous plants. It is therefore probable that those late Maastrichtian species with morphologically complex pollen (e.g., most species of triprojectate pollen) having the above characteristics were zoophilous (Sweet *et al.*, 1990). In contrast, medium-sized (between about 18 and 28 μm) pollen with an oblate to spheroidal shape, relatively thin wall and subdued sculpture are more likely to be anemophilous (Whitehead, 1969; Sweet & Jerzykiewicz, 1987; Frederiksen, 1989; McIver *et al.*, 1991). Examples of the anemophilous type of pollen include *Kurtzipites trispissatus* and *Syncolporites minimus*, which occur in peak abundances immediately above the K/T boundary.

The presence of both zoophilous and anemophilous plants in the floras contiguous to the boundary affects the interpretation of relative abundance and range-truncation data. In one instance the pollen production of a wind-pollinated angiosperm weed species was calculated to be 27000 times greater than that of a comparable insect-pollinated weed by Payne (1981). Such a difference suggests that:

(1) zoophilous pollen often may occur at relative abundances of less than one percent in sediment, given a contributing plant community of mixed pollinating strategy;
(2) zoophilous plants are likely to be under-represented in pollen assemblages (Whitehead, 1969).

The former point implies that the quantification of trends of decreasing diversity (gradual extinctions), such as may occur in the latest Maastrichtian, will be difficult if the affected species in the fossil assemblage are zoophilous. Hotton (1988) attempted to resolve this technically difficult problem of quantifying the relative abundance of rarer species (zoophilous pollen) by proportionally increasing their count, while estimating the numbers of the abundant (anemophilous) species.

Regional environmental factors affecting the distribution of facies and composition of palynofloral assemblages

It has usually been argued that the late Maastrichtian was a time of falling sea level (Obradovich & Cobban, 1975; Jeletzky, 1978; Jones *et al.*, 1987; Hallam, 1988). Within this large-scale regressive trend Brinkhuis and Zachariasse (1988) inferred a rapid sea level fall in the final 17000 years of the Maastrichtian, followed by a rise in sea level at the beginning of the Tertiary, on the basis of the relative abundance of terrestrial miospores, dinoflagellates and land-derived organic matter in the section at El Haria, northwest Tunisia. However, Greenlee and Moore (1988) espoused a rapid rise in sea level in the latest Maastrichtian and earliest Paleocene on the basis of a regional erosional unconformity overlain by a major downlap surface identified in seismic sections along the continental margin of the southern United States. This is corroborated by a regional disconformity at the top of the upper Maastrichtian, Prairie Bluff Formation, in outcrop sections in central Alabama (Donovan *et al.*, 1988). The basal part of the overlying Clayton Formation, which spans the K/T contact, is interpreted as a transgressive unit. These latter observations

parallel those of Loutit *et al.* (1988) who described a transgressive system tract, representing an eustatic rise in sea level, as beginning in the latest Maastrichtian about 1 m below the K/T contact in the Braggs section, central Alabama.

The most exacting control on relative sea level position in the mid-continental area during the latest Maastrichtian and earliest Paleocene comes from the exposure at Huff, North Dakota (Figs. 22.1, 22.2). Lerbekmo and Coulter (1984) have integrated the magnetostratigraphy and lithostratigraphy to achieve the best approximation of the K/T contact at this site. They found that the 29R–29N polarity reversal occurs within the basal beds of the Cannonball Formation, about six meters above the most probable position for the K/T contact at the base of a coal. If the post-boundary coal at Huff represents the leading facies of the Cannonball transgression, then the K/T contact occurs within a transgressive phase as found in the marine sections discussed above. This implies that the withdrawal of the epicontinental seaway did not directly contribute to perturbations in the biota at the K/T boundary, although longer term climatic changes and the initiation of coal swamps in the latest Maastrichtian (discussed below) may be related to changing sea levels, especially on the eastern margin of the foreland basin.

Climatic factors

HUMIDITY

Both the physiognomy of plant megafossils and facies associations have been used to infer the wetness of the climate during the late Maastrichtian and early Paleocene. Plant characteristics used for determining humidity levels include leaf size, the abundance of pubescence (leaf hairs) on leaf surfaces, the presence or absence of leaf drip tips, and the thickness of the leaf cuticle, as well as the distribution of wood vessel elements and the development of tracheary annual growth rings (Wolfe & Upchurch, 1986, 1987a). Using these criteria, Wolfe and Upchurch have inferred a subhumid climate with low seasonality during the Late Cretaceous in mid-continental North America. 'In the south [Raton Basin] small leaf size, paucity of drip tips and thick cuticles with cutinized hairs on both surfaces indicate dry, open-canopy forest....' (Wolfe & Upchurch, 1986, p. 148). The similarity in the character of dispersed-cuticle assemblage through the interval from the base of the Raton Formation to the K/T boundary led them to conclude that the annual precipitation was consistent throughout the late Maastrichtian. They also concluded that there was a northward increase in humidity attributable to lower temperatures and therefore decreased evaporation, rather than to increased precipitation. Larger leaf size and an increased proportion of species with drip tips in the Paleocene suggested to them an increase in precipitation

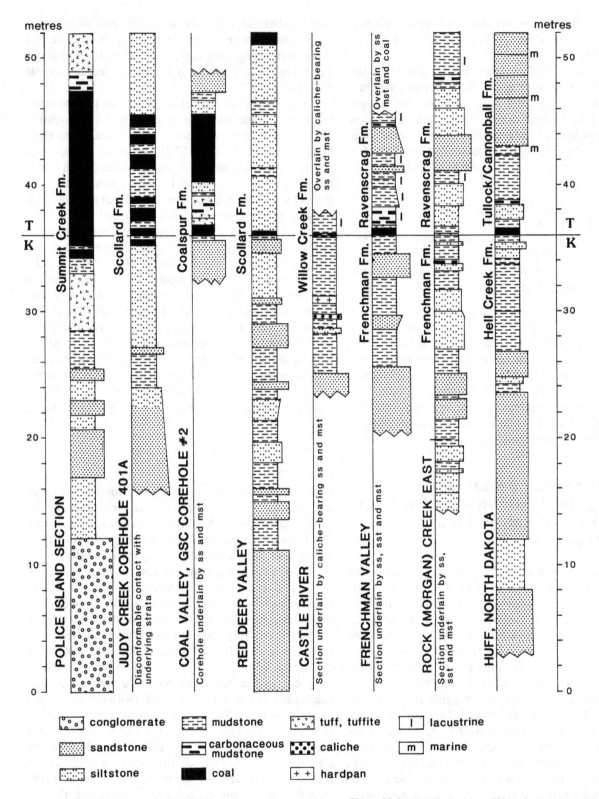

Figure 22.2. Lithological logs illustrating variations in facies associations in the interval contiguous to the K/T boundary based on sections from western and northern Canada and Huff, ND.

at the K/T boundary throughout the mid-continental region, including the Raton Basin (Wolfe & Upchurch, 1986, 1987b; Wolfe, 1990). Wolfe and Upchurch (1986) linked this to the initiation of peat deposition coeval with the K/T event in the central mid-continental region.

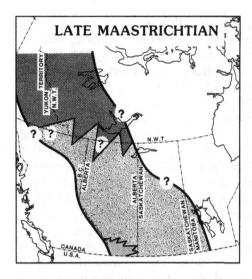

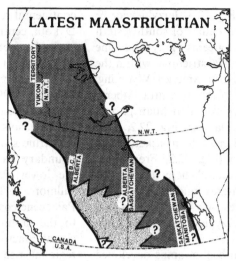

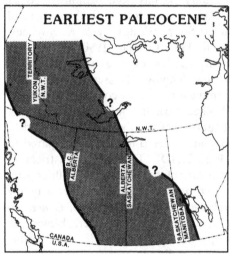

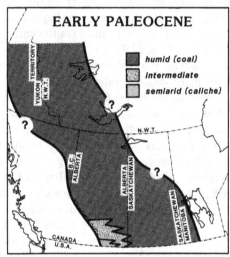

Figure 22.3. Geographic distribution of facies associations in western and northern Canada during the late Maastrichtian and early Paleocene. (From Sweet & Braman, 1992, *Cretaceous Research* 13.1, p. 35, fig. 2; used by permission of Academic Press.)

The presence of persistent, relatively thick coals also can be taken as indicating consistently humid conditions, or those in which precipitation generally exceeds evaporation (Hallam, 1984), assuming other factors related to basin subsidence and the supply of clastics remain more or less constant. The accumulation of organic matter requires an overall rate of vegetative production exceeding the rate of decomposition. Continuity of precipitation is therefore a major factor in both forest growth and the preservation of organic material (Ziegler et al., 1987), as decomposition is retarded in water-saturated environments (Moore, 1987). Alternatively, well-developed caliche in paleosols indicates seasonally dry (semi-arid) conditions, in which the amount of evaporation periodically exceeds precipitation (Jerzykiewicz & Sweet, 1988; Lehman 1989; 1990). However, it is noted by Blodgett (1988) that calcium supply, soil texture, and the influence of plants also can affect the development of caliche.

Both varicoloured, mudstone-containing mature caliche paleosol horizons, and coal-bearing floodplain facies associations are present in upper Maastrichtian and lower Paleocene rocks (Figs. 22.2, 22.3; Nichols, 1990, Fig. 2) of mid-continental North America. A third floodplain facies association, characterized by well-rooted, gray to greenish-gray mudstones and the absence of coal or caliche, is prevalent in upper Maastrichtian rocks. The greenish-gray mudstone facies association is interpreted as forming within a climatic setting intermediate between a persistently humid climate and a more seasonal, semi-arid climate (Sweet & Jerzykiewicz, 1987; Jerzykiewicz & Sweet, 1988).

These three facies associations are restricted regionally and stratigraphically (Fig. 22.3). The caliche-bearing facies occurs in the upper Maastrichtian and lower Paleocene Willow Creek Formation in southern Alberta

(Figs. 22.2, Castle River and Fig. 22.3; Jerzykiewicz & Sweet, 1988) and in the Maastrichtian of southwestern Texas (Lehman, 1989; 1990). The greenish-gray mudstone facies is prevalent in the upper Maastrichtian within the central part of mid-continental North America (Wyoming, Montana, Saskatchewan and Alberta). In central Alberta it is well-developed in the upper Maastrichtian part of the Coalspur and Scollard formations (Fig. 22.2, Coal Valley and Red Deer Valley) and in southern Saskatchewan in the Frenchman Formation (Fig. 22.2, Frenchman Valley and Rock Creek East). Where the upper Maastrichtian is in the greenish-gray mudstone facies the K/T boundary usually coincides with the base of a coal (Brown, 1962; Furnival, 1950; Jerzykiewicz & Sweet, 1988; Sweet & Braman, 1992).

Extensive late Maastrichtian coal swamps developed in two geographically separate regions in the mid-continental area. In the Brackett Basin (Northwest Territories) the K/T contact occurs within a major 12 m thick coal within the Police Island Section (Fig. 22.2) and 50 km to the southwest, in the proximal part of the depositional basin, within a 30 m thick conglomeratic unit, which is underlain by coal-bearing strata (Sweet et al., 1989). Likewise, in the Raton Basin of Colorado and New Mexico, the K/T boundary occurs near the top of an upper Maastrichtian coal-bearing interval, either immediately below a thin coal or, in the Sugarite section, near the top of a 1.8 m thick coal (Flores, 1984; Pillmore & Flores, 1987, fig. 1; Nichols, 1990; note that the coeval occurrence of upper Maastrichtian coal and the dry, open canopy forest of Wolfe and Upchurch appears anomalous and in need of rationalization).

Beginning in the latest Maastrichtian the total area covered by swamps apparently expanded. Coal occurs sporadically in the upper 2–3 m of the Maastrichtian in the Judy Creek coal field of north-central Alberta (Fig. 22.2) and in the Wood Mountain area (Twelve Mile Lake) in south-central Saskatchewan (Sweet & Braman, 1992). Coal also occurs in the uppermost Maastrichtian in the Marmarth region of southwestern North Dakota where at Pyramid Butte the K/T contact is at the top of an up to 0.9 m thick coal (Johnson et al. 1989; Nichols, 1990; Johnson & Hickey, 1991). Thin coals and coaly shale also occur sporadically in the uppermost Maastrichtian in Wyoming (Nichols, 1990), and in central Alberta and south-central Saskatchewan (Sweet & Braman, 1992). In southwestern Alberta a thin (2–3 cm) carbonaceous mudstone occurs at the K/T contact in the otherwise caliche-bearing floodplain facies (Jerzykiewicz & Sweet, 1988). This carbonaceous layer presumably records the same change to a higher water table regime that occurs concurrent with the end of the Cretaceous elsewhere.

In the earliest Paleocene, coal swamps were nearly ubiquitous through the central and northern mid-continental regions of North America (Wyoming, North Dakota, central and eastern Montana, and throughout most of western and northern Canada; Fig. 22.2; Nichols, 1990). Exceptions to this occur in southwestern Alberta (Sweet & Jerzykiewicz, 1988) and in southwestern Texas (Lehman, 1990), where a shift to wetter conditions (without coal formation) is recorded in the character of the paleosols. In the Raton Basin, although coal and lacustrine sediments occur immediately above the K/T boundary, an interval generally barren of coal occurs in the lower Paleocene (Fleming, 1989; Nichols, 1990; Pillmore & Flores, 1987). In this instance the lower Paleocene coal-barren interval is probably more related to the dominance of fluviatile environments than to climatic factors. The above observations corroborate a regional rise in the water table during the interval spanning the boundary, as has been demonstrated by pedogenic studies based on outcrops in northwestern and south-central United States (Fastovsky & McSweeney, 1987; Retallack et al., 1987; Lehman, 1990).

Climatically controlled facies also are reflected in the composition of palynological assemblages. A late Maastrichtian example is the nearly complete absence of triprojectate and allied taxa in the semi-arid, caliche-bearing strata of southwestern Alberta, as compared to more humid floodplain facies (Fig. 22.4A, compared to Fig. 22.4C & Fig. 22.4D). The late Maastrichtian palynomorph assemblage of the caliche-bearing floodplain facies is characterized by low abundance, extremely low diversity and a relatively high proportion of Concentricystes and Classopollis (= Corollina) (Jerzykiewicz & Sweet, 1988). The presence of Classopollis is compatible with semi-arid to arid conditions, because it is associated with aridity in Jurassic and Early Cretaceous strata (Vakhrameev, 1987) and with frenelopsids having xeromorphic characteristics in the Late Cretaceous (Pons et al., 1980). The angiosperm pollen component of the assemblage mostly consists of Ulmoideipites and simple tricol(por)ate pollen (Fig. 22.4A).

In the upper Maastrichtian, greenish-gray mudstone facies, nearly all fine-grained clastic rocks, yield a highly diverse and species-rich assemblage of mostly angiosperm pollen. In the Red Deer Valley this type of assemblage is typified by the presence of about 20 species of triprojectate and allied angiosperm pollen and the consistent presence of Wodehouseia spinata. The above taxa, in addition to species such as Liliacidites complexus, Leptopecopites pocockii and Siberiapollis spp., are mostly relatively large, thick-walled and distinctly sculptured. These morphologically complex taxa commonly represent over 5% of the total assemblage (Fig. 22.4B, triprojectates, Wodehouseia and others) with usually only one to several specimens of any one species being found in a single preparation. Penetetrapites inconspicuus occasionally comprises over 10% of the total angiosperm count [also included in

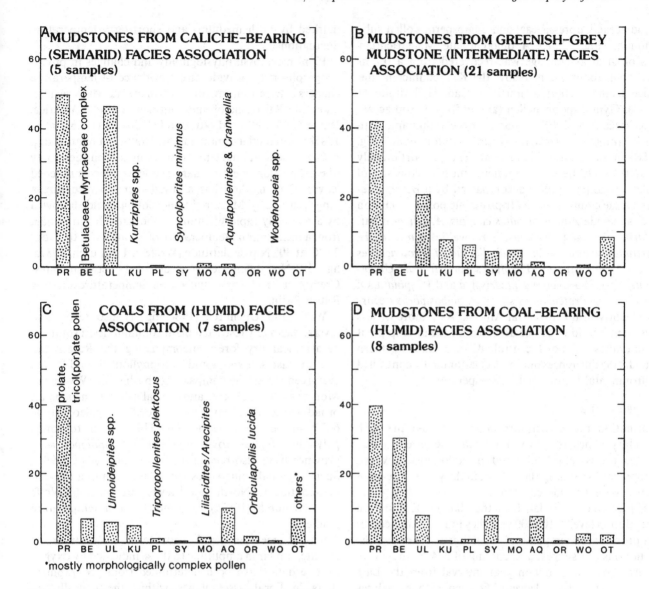

Figure 22.4. Histograms showing the average relative abundance of angiosperm pollen taxa (percent of the angiosperm component of the total assemblage) in: (A) the upper Maastrichtian caliche (semi-arid); (B) greenish-grey mudstone (intermediate); (C & D) coal (humid) facies associations. (In part after Sweet *et al.*, 1990.)

others (OT) in Fig. 22.4B]. The smaller, oblate, colpate or porate pollen with subdued sculpturing, such as *Kurtzipites trispissatus*, *Syncolporites minimus*, *Triporopollenites plektosus* and *Ulmoideipites*, are usually abundant if not dominant (Fig. 22.4B) within the angiosperm component of the assemblage.

The angiosperm component of upper Maastrichtian coal and clastic rock samples from the humid floodplain facies in north-central Alberta (latitude of 55° N) is characterized by the abundance of triprojectate pollen (averaging nearly 10%), triporate pollen of the Betulaceae-Myricaceae complex (averaging between 8 and 30%), frequent *Orbiculapollis*, and a comparatively low relative

abundance of *Ulmoideipites* (Figs. 22.4C, 22.4D and 22.5A). *Myrtipites scabratus* and *Penetetrapites inconspicuus* sometimes reach relative abundances of over 25% in mudstones (included in others in Fig. 22.4D). Both morphologically complex and simple angiosperm pollen are abundant as accessory species in coal samples. *Aquilapollenites conatus*, *A. delicatus* var. *collaris*, *A. quadricretaeus*, *A. reticulatus*, *Striatellipollis radiata*, *S. striatella*, *Polotricolporites rotundus* and *Wodehouseia octospina* are conspicuously present, with *W. spinata* occurring rarely. The smaller, morphologically simple species commonly present include *Orbiculapollis lucida*, *Syncolporites minimus*, *Tricolpites parvistriatus*, *Triporopollenites plektosus*, *Ulmoideipites hebridicus*, and *U. krempii*.

The upper Maastrichtian coal in the humid facies of north-central Alberta is dominated by spores (mostly *Laevigatosporites*), with gymnosperm (nearly exclusively

Taxodiaceae-Cupressaceae) and angiosperm pollen sub-dominant (Fig. 22.5A). A different late Maastrichtian palynological assemblage was recovered from a 0.7 m thick coal about 1.2 m below the K/T contact in the Police Island section of northern Canada (latitude of 64° N). Gymnosperm pollen (about 50% Taxodiaceae-Cupressaceae and 10% bisaccate) are dominant, with spores (mostly *Laevigatosporites*) subdominant (Fig. 22.5B) in the Police Island coal. The proportionately reduced angiosperm component mostly consists of prolate, tricolpate pollen accompanied by a Betulaceae-Myricaceae complex and by triprojectate pollen. Common species include *Aquilapollenites conatus*, *A. delicatus* var. *collaris*, *A. quadricretaeus*, *Cranwellia rumseyensis*, *Myrtipites scabratus*, *Orbiculapollis lucida*, *Proteacidites* sp. cf. *P. globosiporus*, *Tricolpites parvistriatus*, *Ulmoi-deipites* spp., *Wodehouseia quadrispina* and *W. spinata*. Of these species, *Proteacidites* sp. cf. *P. globosiporus* occurs infrequently in north-central Alberta. As a more or less consistent humidity can be assumed for the swamp plant communities of north-central Alberta and northern Canada, the differences must reflect latitudinally controlled environmental factors such as temperature.

TEMPERATURE

Information on temperature changes across the K/T boundary comes from two sources: marine-based oxygen isotope ratios and land-based paleofloristics. Oxygen isotope data spanning the K/T boundary from pre-1988 studies were considered inconclusive by Zachos *et al.* (1989). Based on their data from the Alabama K/T section (a resolution level > 100 000 yrs), they found no indication of a pronounced temperature excursion at the end of the Cretaceous but suggest a '... cooling of 3–4 °C may have occurred over a 3 million year interval from the Late Cretaceous to early Paleocene.' Oxygen isotope analyses of two deep sea K/T sequences from the Maud Rise, Weddell Sea, Antarctica, (65° S) by Stott and Kennett (1988) indicated a late Maastrichtian warming trend culminating in surface water temperatures of about 16 °C approximately 3 m below the top of the Cretaceous. In the 3 m interval below the K/T contact a cooling trend was established, which resulted in a 4–5 °C drop in surface temperatures by the end of the Cretaceous. Stable surface temperatures were indicated for the Paleocene. Keller and Lindinger (1989) also found evidence for a cooling of surface waters, possibly during a time of shallowing seas, in the upper 25 cm of the uppermost Maastrichtian, based on the disappearance of tropical foraminifera and an oxygen isotope excursion. Smit (1990) has accepted oxygen isotope data from the Agost section in Spain as indicating a rise of 8 °C in oceanic surface waters for about the first 5000 years of the Tertiary, followed by a cooling of surface temperatures. These results indicate that the K/T boundary interval most probably was

marked by both medium- and short-term temperature fluctuations.

Plant megafossil physiognomy and changes in phyto-geography may provide direct evidence of temperature changes in terrestrial environments in the late Maastrichtian and early Paleocene of mid-continental North America. Wolfe (1990; 1991) and Wolfe and Upchurch (1986, 1987a, 1987b) used modern analogues to establish a relationship between mean annual temperature and leaf physiognomy (the proportion of entire margined leaves versus notched leaves). This method has allowed them to recognize a long-term, early Maastrichtian cooling trend, followed by a relatively rapid warming in the mid-Maastrichtian, from a mean annual temperature of about 22 °C to about 27 °C at 30° N paleolatitude (Wolfe & Upchurch, 1987a, fig. 5). No difference was detected between latest Cretaceous and early Paleocene temperatures in the Raton Basin.

Wolfe and Upchurch (1986, fig. 1) also recognized several regionally restricted plant communities: a southern paratropical dry forest encompassing the Raton and Denver basins, a subhumid notophyllous, broad-leaved evergreen forest encompassing basins in Wyoming, Montana, North Dakota and central Alberta, and a polar, broad-leaved deciduous forest north of paleolatitude 66° N in the late Maastrichtian. Three (south to north) palynofloristic subprovinces within the *Aquilapollenites* province were also recognized by Nichols *et al.* (1990). Such regional differences in plant communities in mid-continental North America were attributed by Wolfe and Upchurch (1987a) to a north-south temperature gradient.

Several medium- and short-term fluctuations in climate, similar in amplitude to those indicated by oxygen isotope data, have been determined from geographical shifts in floral assemblages within the overall late Maastrichtian–Early Paleocene temperature optimum described by Wolfe and Upchurch. At Marmarth, North Dakota, a change in the type of plant macrofossils present occurs about 20 m below the K/T contact (Johnson *et al.*, 1989). This was taken to represent a warming event, as many of the taxa first appearing in this 20 m, pre-boundary interval previously were known from the upper Maastrichtian of the Raton basin, 1000 km to the south.

The apparent replacement of evergreen by deciduous forests in the early Paleocene led Wolfe and Upchurch (1986, 1987b) to conclude that an abrupt, low-temperature excursion occurred at the K/T boundary and that deciduousness had a selective advantage for survival. Wolfe (1990) apparently further refined the application of plant physiognomy for determining paleoclimates by using correspondence analysis. He proposed that a 10 °C increase in mean annual temperature and a four-fold increase in precipitation occurred after the K/T

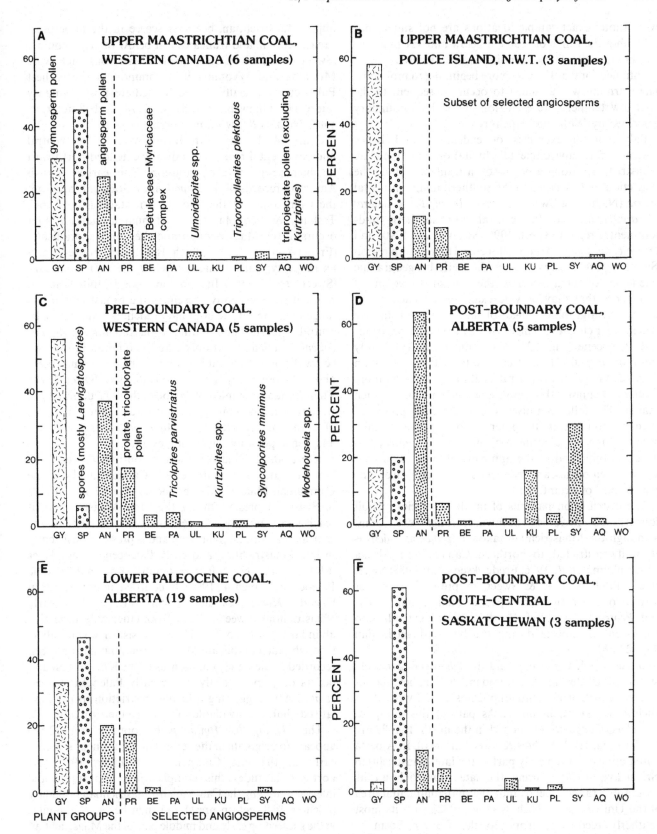

Figure 22.5. Relative abundance histograms based on an average percentage of selected groups of the total assemblage, excluding algae and *Sphagnum*. The histograms provide a comparison of assemblages from: (A & B) organic-rich sediments from below the immediate boundary interval (late Maastrichtian); (C) immediately below and (D & F) immediately above the boundary claystone; (E) from stratigraphically higher (lower Paleocene) coals. (Modified from Sweet & Braman, 1992.)

event (note: exact sample locations are not stated, but fall within the range of <30 cm below and <1 m above the boundary). These conditions were stated to be maintained for 0.5–1.0 Ma before beginning to moderate. The warming was assumed to occur subsequent to the brief low temperature excursion at the K/T boundary described by Wolfe and Upchurch.

Palynological examples of endemism, and hence evidence for environmental limitations in the late Maastrichtian along a north-south transect include the restriction of *Trisectoris* to the southern mid-continental region (Nichols & Sweet, in press), *Tricolpites* (*Gunnera*) *microreticulatus* to the central and southern mid-continental region (Sweet, 1990; Sweet et al., 1990) and *Triprojectus unicus* to the north-central to northern region (Sweet, 1990). However, it is the periodic geographic extension of taxa outside their normal region of occurrence that may provide examples of diachronism in the stratigraphic range of taxa and substantive evidence for climatic change during the latest Cretaceous and Paleocene (Fig. 22.6). The sometimes apparently contradictory direction of change from taxon to taxon (Fig. 22.6) requires further rationalization through more precisely defining the amplitude and timing of each change. The following discussion is therefore presented primarily to illustrate the potential for a palynologically based source of paleoclimatic information, recorded in the geographic and stratigraphic distribution of palynological assemblages, and independent of analogs from the distribution of extant plants.

The distribution patterns of northern species exhibit conspicuous periodic southward range extensions. Although the Maastrichtian taxon *Triprojectus unicus* is generally restricted to northern Canada and Alaska (Doerenkamp et al., 1976; Frederiksen et al., 1988; Sweet et al., 1989; Nichols & Sweet, in press), it (together with two other northern species, *Jiangsupollis major* and *Proteacidites* sp. cf. *P. globosiporus*) extends into north-central Alberta during the latest Maastrichtian (Fig. 22.6A).

Oculate pollen (*Azonia* and *Wodehouseia*) generally reaches its greatest relative abundance and highest species diversity north of 60° latitude (Wiggins, 1976; Sweet, 1990; Nichols & Sweet, in press). As part of this complex, *Wodehouseia spinata* originated in the north (Fig. 22.6B; Frederiksen, 1989; Nichols & Sweet, in press). In Alberta it first appears in the early part of the late Maastrichtian (Srivastava, 1970). During the late Maastrichtian its geographic range extends into the mid-continental area of the United States (Nichols et al., 1982), with its most southerly record occurring in the Denver Basin of Colorado (Nichols, 1985). The upward truncation in the stratigraphic range and/or the maximum abundance of *W. spinata* is also diachronous. In northern Canada *W. spinata* is abundant in the lower part of the

upper Maastrichtian, becomes scarce in the uppermost Maastrichtian, and truncates just below the K/T contact (Sweet et al., 1990). In central Alberta, Saskatchewan, Montana and Wyoming it is common in the earliest Paleocene, especially in the immediate post-boundary interval (Leffingwell, 1971; Sweet, 1978; Lerbekmo et al., 1987; Nichols & Sweet, in press).

Some of the taxa which originate in the central mid-continental region periodically extended northward in their geographic distribution. The earliest North American records of *Aquilapollenites reticulatus* are from the Campanian of the mid-western states (Fig. 22.6C; Tschudy & Leopold, 1971). At high latitudes it first occurs in the Maastrichtian and then only sporadically (Frederiksen et al., 1988; Nichols & Sweet, in press), and its range typically truncates below the K/T boundary (Sweet et al., 1990). In mid-continental United States, where it is consistently present in the Maastrichtian, its range truncates at the boundary (Nichols et al., 1982). In central Alberta and southern Saskatchewan *Aquilapollenites reticulatus* is only rare to common in samples below the boundary but is abundant in coals immediately above the boundary (Lerbekmo et al. 1987; Sweet, 1978). This abundance immediately above the K/T boundary in western Canada represents a northward shift in the area of optimum occurrence by *Aquilapollenites reticulatus* and hence possibly a warming event.

Kurtzipites originates in the mid-continental area of the United States and southwestern Canada at about the Campanian-Maastrichtian boundary (Fig. 22.6D; Nichols & Sweet, in press). During the early Maastrichtian it reaches relative abundances in excess of 60% in the central mid-continental region and is present consistently in late Maastrichtian and early Paleocene assemblages (McIver et al., 1991) before becoming extinct in the middle Paleocene (Demchuk, 1990). In contrast, in northwestern Canada *Kurtzipites* first occurs in the late early Maastrichtian (Sweet et al., 1989) and then only at relative abundances of 1% or less. Here it persists in low numbers through much of the late Maastrichtian but has not been recorded immediately contiguous to the K/T contact and occurs only sporadically in the early Paleocene (Sweet et al., 1990), suggesting a further restriction in the high latitude habitats available for the producing plant.

The *Siberiapollis-Montanapollis* complex (Fig. 22.6E) appears to originate in the central mid-continental region early in the late Campanian (Tschudy, 1971). At northern latitudes this complex first appears in the interval bounding the Campanian-Maastrichtian boundary (Nichols & Sweet, in press) and then becomes common in the succeeding early and middle part of the Maastrichtian (McIntyre, 1974; Sweet, et al. 1989) before becoming rare in the latest Maastrichtian (Sweet et al., 1990).

Mancicorpus follows a pattern similar to the *Siberiapollis-Montanapollis* complex. It originates in the lower

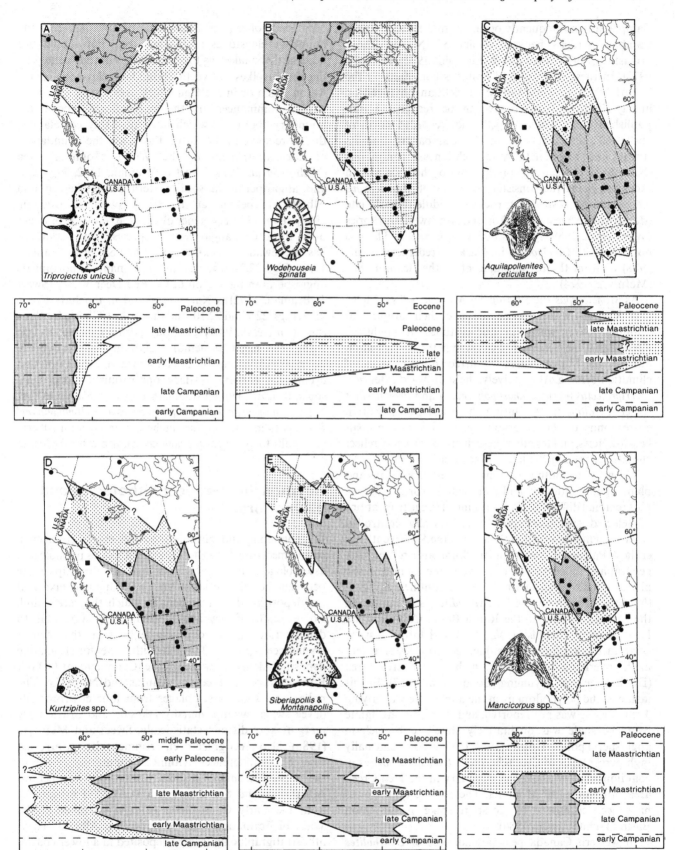

Figure 22.6. Geographic and stratigraphic distribution in mid-continental North America. (A) *Triprojectus unicus*; (B) *Wodehouseia spinata*; (C) *Aquilapollenites reticulatus*; (D) *Kurtzipites*; (E) *Siberiapollis-Montanapollis* complex; (F) *Mancicorpus* spp. Control points indicated by black dots (published) and squares (unpublished). Dark-shaded areas define regions of most frequent occurrence; light-shaded areas indicate more sporadic occurrence. (Modified in part from McIver *et al.*, 1991; Nichols & Sweet, in press; Sweet, 1990.)

Campanian and frequently occurs in mid-latitude upper Campanian rocks, between 40 and 55° N (Fig. 22.6F; Srivastava, 1970; Tschudy & Leopold, 1971; Tschudy, 1973). In these more central latitudes continuity in the record of Campanian and Maastrichtian phylogenetic lineages within *Mancicorpus* can be recognized (unpublished data). *Mancicorpus* is not recorded north of about 60° until the end of the Campanian and its stratigraphic record in the Maastrichtian is discontinuous (Nichols & Sweet, in press; unpublished data), an exception being the consistent presence of *Mancicorpus gibbus* and *M. rostratus* in rocks of middle to early late Maastrichtian age in the Yukon and Northwest Territories (Sweet *et al.*, 1989; Nichols & Sweet, in press). There is no record of *Mancicorpus* in Alaska (Frederiksen *et al.*, 1988) or from the region adjacent to the Beaufort Sea (McIntyre, 1974).

As *Mancicorpus*, the *Siberiapollis-Montanapollis* complex and *Kurtzipites* all appear to have their origin and optimum development in mid-latitudes, it might be inferred that the northward extension in their distribution during the Maastrichtian occurred during times of climatic warming. Alternatively, the southward extension of *Triprojectus unicus*, *Jiangsupollis major* and *Proteacidites* sp. cf. *P. globosiporus* during the latest Maastrichtian corresponds to the geographic expansion of the coal-bearing facies, and together these distributions may reflect climatic change in the latest Maastrichtian.

The north to south variations in facies associations and plant communities of mid-continental North America plus regional differences in the timing of range truncations of selected species support a climatically controlled environmental gradient at the time of the K/T event. The earliest Paleocene opportunistic floras also reflect such geographic differences. Ferns were the opportunistic plants in the southern part of the mid-continental region (Nichols & Fleming, 1991). *Cyathidites* is dominant in the earliest Tertiary in the Raton Basin, Colorado and New Mexico (Nichols *et al.*, 1985) and forms a spore spike in the kaolinitic claystone and, to a lesser extent in the overlying 'magic layer' in the Lance Creek area (Dogie Creek) of Wyoming (Bohor *et al.*, 1987). In this instance, the palynoflora from the immediately overlying 4 cm of shale was not reported, and that from the lignite over the shale was qualitatively stated to be rich in angiosperm pollen. Hotton (1984, 1988) reports unusually high percentages of *Cyathidites* and *Laevigatosporites* in coal immediately above the boundary in the Hell Creek area of east-central Montana. These spore dominances have been correlated to the spore spike in the Raton Basin (Tschudy *et al.*, 1984).

In western Canada peak abundances of *Cyathidites* occur in the latest Maastrichtian and sometimes in the boundary claystone (Sweet & Braman, 1992; Fig. 22.7). These occurrences of *Cyathidites* are comparable to the occurrence of a peak abundance of *Cyathidites* in the boundary claystones at Dogie Creek, but should not necessarily be taken as the same bioevent represented by the spore spikes above the boundary claystone in the Raton Basin or in the Hell Creek area.

The dominances in the palynofloras above the boundary claystone, which most probably correlate to the spore spike in the Raton Basin, are the abundances of *Laevigatosporites* in the Wood Mountain area (Nichols *et al.*, 1986; Sweet & Braman, 1992; Fig. 22.7) and angiosperms in southwestern Saskatchewan and Alberta (Lerbekmo *et al.*, 1987; Sweet & Braman, 1992). *Ulmoideipites* generally is the most prominent opportunisitic angiosperm species in southern Saskatchewan (Lerbekmo *et al.* 1987; Sweet & Braman, 1992; Fig. 22.7) whereas the dominant opportunistic angiosperm in the region of the Red Deer Valley (Sweet & Braman, 1992) is *Kurtzipites*, and in the Judy Creek area (Figs. 22.9 and 22.10) and the central Alberta foothills (Lerbekmo *et al.* 1987) it is *Syncolporites minimus*. Hence, within the mid-continental region in beds above the boundary claystone (or in its absence and the extinction event), the earliest Tertiary opportunistic flora shifts south to north from a dominance of *Cyathidites* to that of *Laevigatosporites* or *Laevigatosporites* and *Ulmoideipites*/*Kurtzipites* in southern Saskatchewan and central Alberta and finally to *Syncolporites minimus* in north-central Alberta.

Relationship between lithofacies and palynofloras

It is commonly thought that most low-energy continental environments are dominated by autochthonous palynological assemblages, and that a positive relationship exists between the relative abundance of individual taxa and the depositional environment in which they are found. For example, Farley and Dilcher (1986) were able to demonstrate an interdependence between the relative abundance of groups of taxa and lithofacies for nonmarine depositional environments in the (Cenomanian) Dakota Formation of north-central Kansas and Nebraska. The pollen and spore assemblages from four environments (levee swale, swamp, marshy lakeside and distributary margin) were distinct statistically. Sweet and McIntyre (1988) also recognized environmentally controlled assemblage differences within a nonmarine phase of the upper Turonian Cardium Formation (west-central Alberta), as did Sweet and Braman (1989) in samples from the Early Campanian, Chungo Member of the Wapiabi Formation from southwestern Alberta. These lithostratigraphic units were deposited in a lower coastal plain, marginal marine setting similar to that of the Dakota Formation described by Farley and Dilcher (1986). These three examples document palynologically distinctive plant communities during the initial phases of the Late Cretaceous radiation of the angiosperms. The evolution of morphologically complex Maastrichtian angiosperm taxa, including the triprojectates, and other taxa critical to the perception of floral change at the K/T boundary, occurred within these environmentally restricted plant communities. It is therefore considered necessary to examine the relationship between palynological assemblages and lithofacies within the K/T boundary interval to determine the degree of change across lithofacies contacts that can be attributed to the local depositional environment.

Characterization of the palynological assemblage of organic-rich sediments

As coals reflect a more or less consistent depositional environment, differences in coal derived palynological assemblages from below and above the K/T boundary within a restricted geographic area can be mostly attributed to the effect of the K/T event. Coal and carbonaceous mudstones occurring below the interval directly contiguous to the boundary in the Judy Creek coal field, north-central Alberta and at Twelve Mile Lake, south-central Saskatchewan, contain similar assemblages, and these have been used to calculate the percentage of selected palynomorphs in upper Maastrichtian organic-rich sediments (Fig. 22.5A; Sweet & Braman, 1992). The assemblages are characterized by dominant spores (mostly *Laevigatosporites*), with gymnosperm (Taxodiaceae-Cupressaceae) and angiosperm pollen subdominant. The angiosperms are dominated by simple prolate tricolpate pollen (commonly *Fraxinoipollenites variabilis*) and triporate pollen of the Betulaceae-Myricaceae complex, within a relatively diverse assemblage formed by abundant *Ulmoideipites, Triporopollenites plektosus, Syncolporites minimus*, triprojectate pollen (*Aquilapollenites conatus, A. quadricretaeus* and *Orbiculapollis lucida*), *Tricolpites parvistriatus, Wodehouseia quadrispina* and *W. spinata*.

Sweet and Braman (1992) report thin coals (4 to 14 cm thick) or coaly shales immediately below the boundary claystone at three localities: one from the Red Deer Valley and two in the Wood Mountain area. All palynological assemblages from these pre-boundary organic-rich sediments were found to be similar and hence the percentages were averaged to characterize their palynoflora (Fig. 22.5C). These organic-rich sediments are distinctive in being dominated by gymnosperm pollen, having a subdominance of angiosperms and a relatively low abundance of spores. Their angiosperm component is still relatively diverse and includes, in addition to those species shown on Figure 22.5, a number of morphologically complex pollen species (*Aquilapollenites conatus, A. delicatus* var. *collaris, A. quadricretaeus, A. quadrilobus*,

A. reticulatus, A. sp. cf. *A. vulsus, Cranwellia rumseyensis, Liliacidites complexus, Myrtipites granulatus, M. scabratus, Orbiculapollis lucida, Polotricolporites rotundus, Striatellipollis striatella, Wodehouseia octospina, W. quadrispina* and *W. spinata*). However, the relative abundance of the aquiloid triprojectates is one-third the abundance found in stratigraphically lower, upper Maastrichtian coals (Fig. 22.5A and 22.5C).

Assemblages from the coal above the level of the K/T extinction event are marked by peak abundances of opportunistic species (either taxa of angiosperm pollen: Fig. 22.5D, or of spores: Fig. 22.5F), as has been discussed previously. Excepting for localized *Sphagnum*-rich assemblages in the Wood Mountain area (see below), palynofloras from above the level of opportunistic floras appear to be regionally consistent in the proportion of the major plant groups and in the restricted number of species present (Sweet & Braman, 1992; Fig. 22.5E; based on samples from Judy Creek coreholes 83-313A and 83-401A). The percentage of gymnosperm pollen, spores and angiosperm pollen in these post-opportunistic, lower Paleocene palynomorph assemblages is closely comparable to that of assemblages from upper Maastrichtian coals away from the interval immediately contiguous to the K/T boundary.

As both the late Maastrichtian and early Paleocene swamp floras are dominated by fern spores (*Laevigatosporites*), with subdominant gymnosperm (Taxodiaceae-Cupressaceae) and angiosperm pollen, this must represent a stable forested-swamp complex in which the gymnosperms most likely formed the overstory. The close similarity in the overall structure of swamp communities in the late Maastrichtian and early Paleocene occurs notwithstanding the intervening interval of opportunism and the range truncations of such accessory species as *Aquilapollenites conatus, A. delicatus* var. *collaris, A. quadricretaeus, Wodehouseia quadrispina* and *Liliacidites complexus* near the K/T boundary.

The consistent and strong dominance by gymnosperm pollen of the assemblages from the uppermost Maastrichtian coaly shale and coals (Fig. 22.5C; Sweet & Braman, 1992) perhaps suggests that immediately prior to the boundary the swamps and surrounding wetlands were more extensively forested than underlying upper Maastrichtian and overlying lower Paleocene swamps. A depression in the abundance of gymnosperms coincides with maxima of opportunistic species (Figs. 22.5D and 22.5F). This serves to emphasize the short term, but apparently regional removal of the overstory (gymnosperms) by the K/T event, as the opportunistic palynofloras were dominated by probably herbaceous (fern) and/or herbaceous/shrubby (angiosperm) taxa. Such shifts as the change to an overwhelming dominance of the palynoflora in pre-boundary coals by gymnosperm pollen and to opportunistic fern and angiosperm taxa in

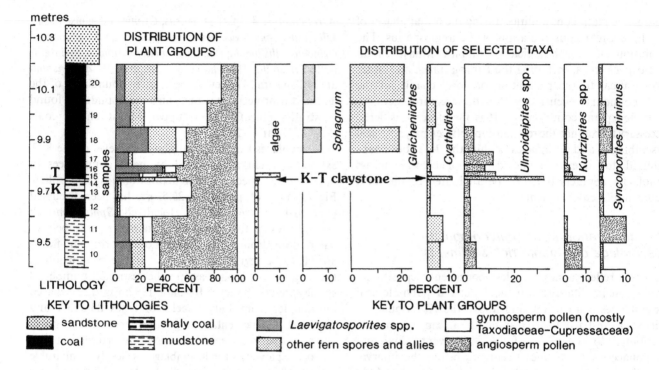

Figure 22.7. Relative abundance histograms showing the vertical distribution of groups and taxa of palynomorphs in the Wood Mountain Creek section, south-central Saskatchewan. (Modified from Sweet & Cameron, 1991, 'Palynofacies, coal petrographic facies and depositional environments: Amphitheatre Formation....', *International Journal of Coal Geology*, **19**, 121–44, fig. 6; used by permission of Elsevier Science Publishers.)

post-boundary coals imply that an environmental perturbation occurred contiguous to the K/T boundary interval which temporarily changed factors controlling the composition of swamp floras.

Comparisons of palynological and lithologic profiles

From the examination of coal swamp palynofloras across the K/T boundary it is apparent that significant changes in the palynoflora occurred independent of major changes in the depositional environment. In the following sections, changes in palynological profiles spanning several different rock lithologies are presented as a further framework against which to compare change within the immediate boundary interval.

K/T BOUNDARY SECTION, WOOD MOUNTAIN AREA (FIGS. 22.7 AND 22.8)

The Wood Mountain Creek K/T boundary section of south-central Saskatchewan illustrates compositional shifts in palynological assemblages through an interval of mudstone, coal and coaly shale (Sweet & Braman, 1992; Sweet & Cameron, 1991). The upper Maastrichtian medium-brown mudstones underlying the boundary coal (Fig. 22.7, samples 10 and 11) are dominated by angiosperm pollen. The upper mudstone surface is in sharp contact with 7 cm of blocky coal (sample 12) overlain by 7 cm of laminated, amber-rich, shaly coal

(sample 13). Both the blocky and shaly coals are relatively rich in liptinite (Fig. 22.8). The associated abundance of Taxodiaceae-Cupressaceae pollen (Fig. 22.7) and the conspicuous presence of macroscopic amber (also reflected in the high liptinite abundance), is interpreted to represent extensive forestation of the 'pre-boundary' swamps and surrounding wetlands. However, these forested-swamps must have included open areas of free-standing water, as the algae *Ovoidites*, *Sigmopollis* and *Tetraporina* also occur within the assemblage.

The 14 cm coal is overlain by a discrete (about 1 cm thick), pale, pinkish-tan K/T boundary claystone (sample 14). This claystone contains relatively little gymnosperm pollen and an abundance of spores, including a peak abundance of *Cyathidites* (Fig. 22.7). Angiosperm pollen present include the late Maastrichtian taxa *Orbiculapollis lucida* and *Wodehouseia quadrispina*. Betulaceae-Myricaceae pollen is abundant. Algae such as *Ovoidites*, *Sigmopollis* and *Tetraporina* sp. occur sporadically in these uppermost Maastrichtian sediments.

A sharply defined contact separates the claystone from the overlying 50 cm thick coal (samples 15 to 20). The early Paleocene assemblage in the lowest 2 cm of this coal (Fig. 22.7, sample 15) contains a species-rich and

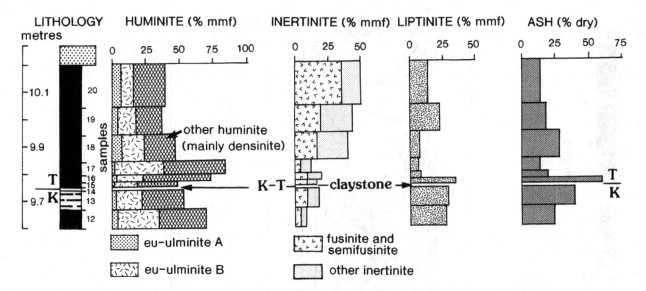

Figure 22.8. Relative abundance histograms showing the vertical distribution of coal macerals in the Wood Mountain Creek section, south-central Saskatchewan. (Modified from Sweet & Cameron, 1991, 'Palynofacies, coal petrographic facies and depositional environments: Amphitheatre Formation....', *International Journal of Coal Geology*, **19**, 121–44, fig. 7; used by permission of Elsevier Science Publishers.)

abundant suite of algae including *Pediastrum*, *Ovoidites*, *Sigmopollis*, *Tetranguladinium* and *Tetraporina* (Sweet & Braman, 1992). Angiosperm pollen also peaks opportunistically in this interval, especially *Ulmoideipites* and *Kurtzipites*. The greatest percentage of *Laevigatosporites* occurs in the next higher, 3 cm interval, of coal. The local depositional environment immediately above the claystone therefore likely included open water (as indicated by the high diversity and abundance of algae), bordered by fern and angiosperm dominated vegetation. Although the palynofloras immediately below and above the K/T boundary are substantially different, only relatively minor differences in the petrography of the coal occur contiguous to the boundary. These differences include a lower liptinite fraction and a higher inertinite fraction above the boundary (Fig. 22.8).

The most conspicuous aspect of the coal petrographic profile above the boundary claystone (Fig. 22.8) is the shift from a high percentage of huminite to abundant inertinite between sample intervals 17 and 18. At this level *Sphagnum* and *Gleicheniidites* make their appearance, with *Gleicheniidites* comprising approximately 20% of the total miospore assemblage. The abundance of *Sphagnum* and *Gleicheniidites* and the increase in inertinite in the upper part of the boundary coal is taken as evidence of a swamp to raised-bog succession similar to that illustrated by McCabe (1987). Extant species of the Gleicheniaceae prefer 'open habitats, often growing in sterile soil....' (Tryon & Tryon, 1982, p. 88) which corroborates this interpretation. Bog development was

probably localized, because species diversity is similar both in samples with abundant *Sphagnum* and *Gleicheniidites* and in *Laevigatosporites*-dominated samples. This interpretation also is compatible with the regional scarcity of *Sphagnum* within the immediate, boundary interval (Sweet & Braman, 1992). The parallel increase in Taxodiaceae-Cupressaceae pollen abundance and inertinite in the upper part of the coal suggests that the surrounding swamps or bogs were increasingly forested and subject to oxidation. Periodic oxidation of the bogs attests to times of dryness (Sweet & Cameron, 1991), even though their existence requires overall a relatively high rainfall evenly distributed throughout the year (Cecil *et al.*, 1985; Moore, 1987).

The shift from an angiosperm-dominated to a gymnosperm-dominated palynoflora between the mudstones and the coals below the boundary claystone corresponds to a change in lithofacies and hence the depositional environment prior to the K/T event. However, the contrast between gymnosperm-dominated swamp assemblage below the boundary, the *Cyathidites* fern spike in the boundary claystone, and the angiosperm and fern (*Laevigatosporites*) dominance above the claystone strongly supports a major shift in the ecology of plant communities subsequent to and during the K/T extinction event. The independence of the cause of ecological changes at the boundary from changes in the depositional environment is further emphasized by the similarity of the petrography of coals immediately contiguous to the K/T boundary, compared to those from the upper parts of the boundary coal. As the maceral composition of coals remain similar across the boundary, the environmental perturbations during the K/T event did not greatly affect the degradation of organic matter or were of too short a duration to be recorded by samples taken at intervals of two to several centimeters.

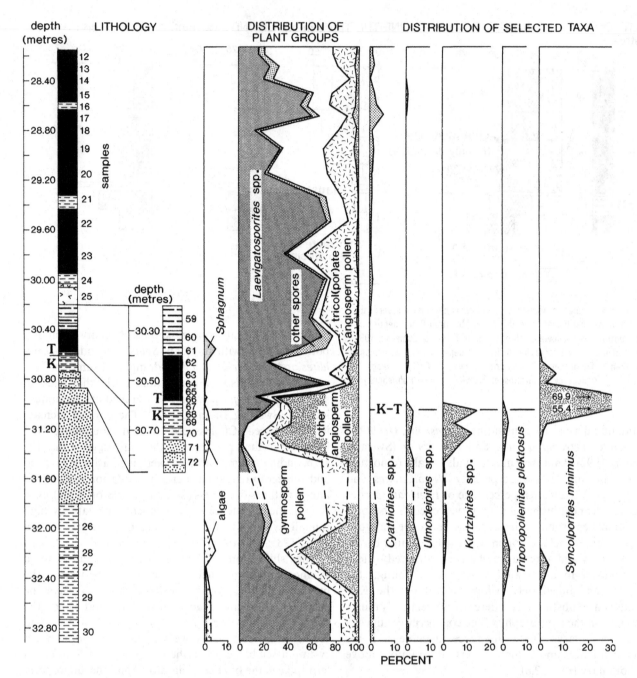

Figure 22.9. Relative abundance profiles for a 4.98 m interval contiguous to the K/T boundary in the Judy Creek North Corehole 83-313A. The numbers to the right of the lithology logs are sample numbers. A change in the vertical scale occurs between samples 26 and 72 and between 25 and 59. *Sphagnum* and algae percents are based on the entire assemblage. The remaining relative abundances of spores, gymnosperm pollen and angiosperm pollen are based on counts excluding *Sphagnum* and algae. Note the change in percent scale between the block of histograms representing the major groups and those for selected subgroups or species. The lithologies discussed in the text have been grouped into: sandstone and siltstone (dots, with the relative coarseness indicated by the width of the column), mudstone and shale (fine dashes), coaly mudstone or shale (heavy dashes), coal (black) and bentonite (random v's). (Modified from Sweet & Braman, 1992, *Cretaceous Research* 13.1, fig. 3; used by permission of Academic Press.)

K/T BOUNDARY SECTIONS, JUDY CREEK COREHOLES
83-313A AND 401A, NORTH-CENTRAL ALBERTA

In sections in the mid-continental region where the K/T contact occurs at a coal-clastic interface it is difficult to separate the component of change in plant communities attributable to the depositional environment from that generated by the K/T event. Cores from the Judy Creek coal field of north-central Alberta provide the opportunity to characterize palynological assemblages indigenous to clastic rocks and upper Maastrichtian and lower Paleocene coals, and to examine palynological profiles across mudstone-coal interfaces in both the upper Maastrichtian and at the K/T contact (Figs. 22.2 and 22.9 to 22.11; Sweet *et al.*, 1990; Sweet & Braman, 1992).

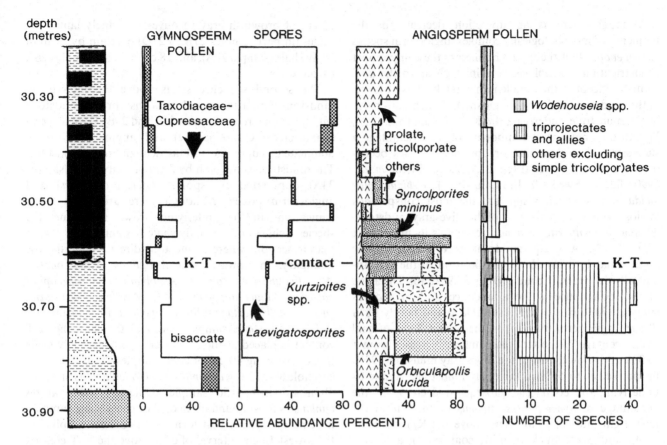

Figure 22.10. Relative abundance of spores, gymnosperm pollen, and angiosperm pollen, and the number of species of triprojectate and allied pollen over the interval spanning the K/T boundary in the Judy Creek North Corehole 313A. (Modified from Sweet *et al.*, 1990.) Lithology indicated to the left by same sorts of patterns as in Fig. 22.9.

Corehole 83-313A (Figs. 22.9 and 22.10) The upper Maastrichtian fining-upwards cycles in the lower part of corehole 83-313A, culminate in gray to yellowish-gray mudstones. The gray mudstones yield assemblages averaging 64% angiosperm pollen, 24% gymnosperm pollen, 10% fern spores and 5% *Sigmopollis*. *Penetetrapites inconspicuus* and prolate tricolpate pollen are the dominant forms of angiosperm pollen, within an overall low diversity assemblage, in which *Aquilapollenites delicatus* var. *collaris* and *A. conatus* are particularly abundant. Yellowish-gray mudstones (sample 30), forming the stratigraphically lowest floodplain sediments shown on Figure 22.9, yield little organic material and are barren of palynomorphs. The overlying 35 cm of pale yellowish-brown mudstone (sample 29), also with well-developed roots, contains an impoverished, *Laevigatosporites*-dominated assemblage. Their light color, the presence of roots, and the low yield of organics, suggests these mudstones were well-leached paleosols. The overlying 50 cm of light-olive to medium-gray shale with silty

laminations (samples 26 to 28) has a 2 cm thick coaly shale interbed near its base (sample 27). The coaly shale is characterized by a rich assemblage, including *Orbiculapollis lucida*, *Syncolporites minimus*, *Triporopollenites plektosus*, *Tricolpites parvistriatus*, *Wodehouseia quadrispina* and *Aquilapollenites quadricretaeus* (listed in order of decreasing abundance). This interval presumably is correlative with the lower coal of the 83-401A corehole (Fig. 22.11). The increased amounts of organics in the sediments and their olive-gray color suggests formation under wetter conditions than the underlying mudstones.

Above the top of the sandstone at 30.8 m, the mudstones (samples 68 to 72) are olive-gray and vertically rooted. The lower surface of the overlying 3 cm of coaly mudstone (containing distinct, elliptical clay inclusions; sample 67) is undulating but sharp. The position of the K/T contact is interpreted as occurring between the olive-gray and coaly mudstones, as the upward range of *Aquilapollenites attenuatus*, *Cranwellia rumseyensis*, *Myrtipites scabratus*, *Orbiculapollis lucida*, *Striatellipollis striatella* and *Wodehouseia octospina* truncate at this contact (one specimen of *Aquilapollenites conatus* was recorded from the coaly mudstone). The coaly mudstone is overlain by 3.5 m of coal with coaly mudstone and bentonitic interbeds, of which a 2.66 m interval (samples 59 to 66 and 12 to 25) is shown in Figure 22.9.

Although there is an upsection decrease in the frequency of morphologically complex angiosperm species (Sweet *et al.*, 1990), the species richness of these uppermost Maastrichtian assemblages remains high up to the K/T contact. Except in the sample immediately above the top of the sandstone at 30.8 m, angiosperm pollen dominates in the mudstone underlying the K/T contact (Fig. 22.9). Within the angiosperm component of the assemblage a succession of co-dominant or dominant species occurs near the K/T contact (Figs. 22.9, 22.10; Sweet *et al.*, 1990, fig. 7; Sweet & Braman, 1992). *Orbiculapollis lucida* and *Kurtzipites* spp. are the most conspicuous angiosperms in the lower part of the olive-gray mudstone, whereas *Syncolporites minimus* dominates its upper part (Fig. 22.10). Angiosperm dominance continues for approximately 8 cm above the position of the palynologically defined K/T contact (Figs. 22.9, 22.10). The acme of this dominance is the abundance of *Syncolporites minimus* (70% of the total assemblage and nearly 96% of the angiosperm fraction) in the lowest 2 cm of coal (its percentage of the total assemblage increases from 1 to 3% in the lower part of the olive-gray mudstone, 23% in sample 68, to 55% in the coaly mudstone sample 67 underlying the coal). The subsequent decrease in the percentage of *Syncolporites minimus* occurs within the next 3 cm interval (5 to 8 cm above the K/T contact), within an 18 cm thick vitrinitic coal having a macroscopically consistent lithology. *Fraxinoipollenites variabilis* is the most frequent species of angiosperm pollen above the interval of *Syncolporites* dominance. *Laevigatosporites* is most abundant in association with clastic partings within the Paleocene coal-bearing part of the sequence (Fig. 22.9). An exception to this is its high abundance in sample 64 from a coal 8 to 11 cm above the K/T contact.

The lithological sequence in this corehole apparently reflects a change from periodically wet to more consistently wet conditions. It starts with organic-poor, light-colored mudstones and terminates in well-laminated, possibly lacustrine shales, olive-gray mudstones (gleyed) and coal. The changes in the composition of the angiosperm pollen assemblage occur both coincident with and independent of the mudstone-coaly shale-coal interfaces (Fig. 22.10). As a palynofloral succession occurs below the K/T contact within a mudstone of macroscopically consistent lithology and above the contact in an apparently homogeneous coal, changes in ecological conditions both below and above the contact were independent of conspicuous changes in the depositional environment.

Corehole 83-401A (Fig. 22.11) The 3.71-m portion of the core illustrated in Figure 22.11 encompasses two mudstone to coal sequences, one of late Maastrichtian age and one which spans the palynologically defined K/T contact. The upper Maastrichtian sequence includes

33 cm of brownish-gray to olive-gray, finely laminated shales (in part samples 27B to 27D) overlain by 5 cm of coaly shale (sample 27A), and 28 cm of coal (samples 26A to 26C).

The second sequence starts above the intervening sandstone with a 16 cm, pale yellowish-brown mudstone with horizontal roots (samples 25A and 25B). Gymnosperm (Taxodiaceae-Cupressaceae) and angiosperm pollen dominate this uppermost Maastrichtian brown mudstone. The mudstone is overlain by 3 cm of coaly shale (sample 24A) dominated by spores (*Laevigatosporites*) and angiosperm pollen. Although there are some range truncations in the underlying brown mudstone, the species richness of the coaly shale is significantly higher than in sections where a mudstone directly underlies the boundary. The upward range of *Aquilapollenites conatus*, *A. delicatus* var. *collaris*, *A. oblatus*, *Leptopecopites pocockii*, *Liliacidites complexus*, *Myrtipites scabratus*, *Orbiculapollis lucida* and *Striatellipollis striatella* truncates in the coaly shale (sample 24A), and therefore the K/T contact is placed at the abrupt change from coaly shale to the overlying 40 cm of coal (samples 20 to 23). As in corehole 83-313A, *Kurtzipites* and *Orbiculapollis* are most abundant in clastic samples, whereas *Syncolporites minimus* is most abundant in coal or coaly shale samples, it being the species that forms an angiosperm 'spike' in the lowest 10 cm interval of coal above the K/T contact (Fig. 22.11).

Corehole 83-401A provides an opportunity to compare the palynological profiles from an upper Maastrichtian and a lowermost Paleocene coal. Within both coals the peak abundance of *Laevigatosporites*, the dominant type of spore, occurs well above the base of the coals (Fig. 22.11). Similarly the maximum abundance of angiosperm pollen and of *Syncolporites minimus* is in the stratigraphically lowest coal sample (Fig. 22.11), although the percentage of angiosperm pollen and of *S. minimus* is significantly higher in the coal immediately above the K/T contact than it is in the upper Maastrichtian coal. A similar succession was reported in the base of the Mynheer coal in the central Alberta foothills (Lerbekmo *et al.*, 1987). This suggests that the earliest Paleocene succession, i.e., an abundance of angiosperms followed by a dominance of *Laevigatosporites*, reflects a normal plant succession within coal swamps in north- and west-central Alberta. There are, however, major differences. The abundance of angiosperm pollen in the coal immediately above the K/T contact is 10 times greater than that in the base of the upper Maastrichtian coal. Another difference is the relatively high frequency (up to 2%) of *Aquilapollenites conatus*, *A. quadricretaeus*, and *Wodehouseia quadrispina* in the coal of the lower cycle and their absence in the upper cycle coal. These differences are attributable to intervening range truncations and ecological perturbations related to the K/T event.

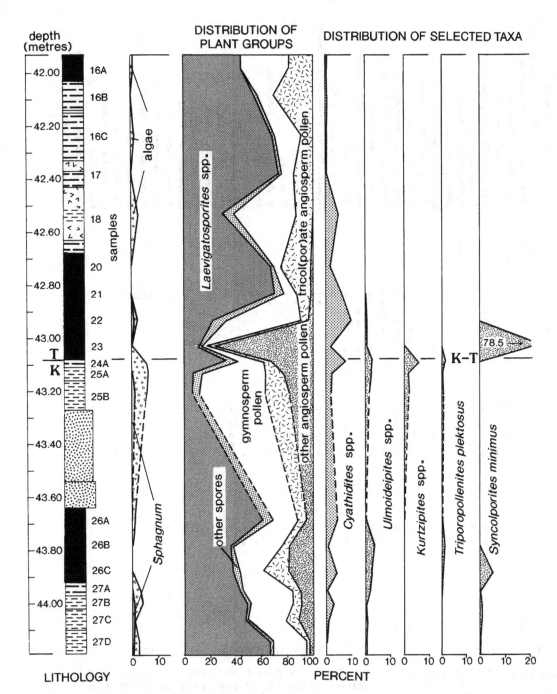

Figure 22.11. Relative abundance profiles for a 3.71 m interval contiguous to the K/T boundary in Judy Creek North Corehole 83-401A. See Fig. 22.3 for further information. (Modified from Sweet & Braman, 1992, *Cretaceous Research* 13.1, fig. 5; used by permission of Academic Press.) Lithology indicated by same patterns as in Fig. 22.9.

Discussion of relationships between regional facies associations, local lithofacies and floristic changes

Palynological changes occurring contiguous to the K/T boundary in western and northern Canada (Sweet *et al.*,

1990; Sweet & Braman, 1992) illustrate differences attributable to geographical, stratigraphical and environmental factors as well as to short term catastrophic events. In the Police Island section, below the K/T extinction event, Sweet *et al.* (1990) recognized two compositional shifts in the angiosperm component of the assemblage delimiting three distinctive palynological assemblages (Fig. 22.12). The lowest assemblage is distinguished by the persistent presence of several species more characteristic of the early Maastrichtian. The upward range of such species, including the morphologically complex (presumably zoophilous) ubiquitous northern species *Triprojectus*

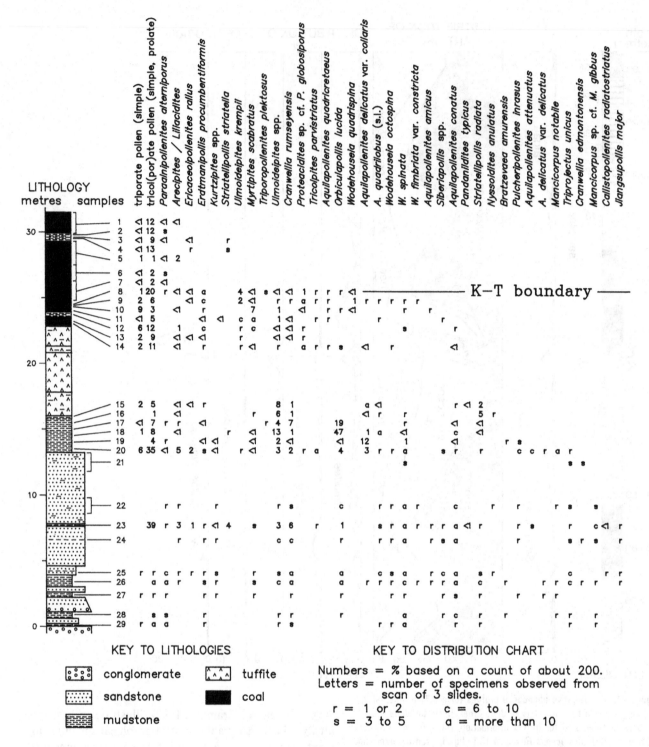

Figure 22.12. Distribution and abundance of angiosperm pollen species in the uppermost Maastrichtian and lowermost Paleocene of the Police Island section. (Modified from Sweet *et al.*, 1990.)

unicus, *Callistopollenites*, *Jiangsupollis*, and *Pulcheripollenites*, is truncated in the interval 10 to 16 m below the K/T contact. These truncations culminate within a homogeneous mudstone and therefore are not lithologically controlled.

The Red Deer Valley locality, which has stratigraphic control below the K/T boundary comparable to that in the Police Island section, has not produced a similar

interval of clustered range truncations in the late Maastrichtian. Those taxa which truncate in the Police Island section either were not present as far south as the Red Deer Valley (*Triprojectus unicus* and *Jiangsupollis*), or were restricted to the lower Maastrichtian in this more southern latitude (*Callistopollenites* and *Pulcheripollenites*). Nevertheless, through the interval 2.5 to 25 m below the boundary in the Red Deer Valley section, some sporadically occurring species, such as those of *Siberiapollis*, truncate within otherwise palynologically consistent assemblages typical of the *Wodehouseia spinata* zone of Srivastava (1970). These late Maastrichtian range truncations mostly seem to represent extinctions involving species originating and reaching acmes in abundance in the late Campanian or early Maastrichtian and probably reflect late Maastrichtian, medium- and long-term climatic changes. The greater vulnerability of higher latitude, late Maastrichtian paleofloras to climatic changes (Stott & Kennett, 1988; Wolfe & Upchurch, 1987a) may explain the more numerous range truncations in the Police Island section.

The late Maastrichtian truncations in the Police Island section result in an overall drop in the number of species in the 10 m interval below the K/T contact, which occurs about 1.9 m above the base of a 12 m coal (Fig. 22.12). The interval below the K/T contact is therefore characterized by a reduction in the number of taxa present, a factor enhanced, on a sample to sample basis, by the low frequency and erratic occurrence of the remaining morphologically complex (such as triprojectate) taxa.

A second apparent reduction in the number of species present occurs in the 0.25 m interval of coal immediately below the K/T contact in the Police Island section. Only 12 species typical of the Maastrichtian were recorded from immediately below the contact, compared to 36 species in the underlying mudstones and coal. Of the 12 species, five are represented by only one or two specimens, and only four occur in relative abundances of 0.5% or more (Sweet *et al.*, 1990).

Lerbekmo *et al.* (1987) and Sweet *et al.* (1990) emphasized this pattern of a reduced number and frequency of morphologically complex late Maastrichtian species within about 0.2 to 2 m of the boundary. They found that single samples in the immediate proximity of the boundary usually contained fewer specimens and fewer species of morphologically complex angiosperm pollen than did assemblages from underlying mudstones. Nevertheless, when species records from several localities were combined, most late Maastrichtian species were found to range into the boundary claystone. The western Canadian localities used in Lerbekmo *et al.* (1987) and Sweet *et al.* (1990) (Coal Valley; Judy Creek corehole 83-313A; Red Deer Valley, Knudsen's Farm; and Frenchman Valley) all have mudstone rather than coal or coaly shale underlying the K/T boundary or contact. Such pre-boundary mudstones are often well-rooted, show ped development, contain a low amount of organic material, and therefore probably were paleosols. Similar pedogenic upper Maastrichtian mudstones in Judy Creek

cores are notable for their apparent impoverishment in terms of the number of morphologically complex species present (see above), even though selected species such as *Aquilapollenites delicatus* var. *collaris* and *A. conatus* may be relatively abundant. Additionally, most sections used by Lerbekmo *et al.* (1987) and Sweet *et al.* (1990) occur in regions with an upper Maastrichtian intermediate facies association, which is accompanied by a comparatively low abundance of morphologically complex species, especially aquiloid taxa (Fig. 22.4).

It now is apparent that other factors had to be considered regarding the degree of diversity reduction contiguous to the K/T boundary. The species richness in the assemblage from the 3 cm coaly shale immediately below the K/T contact in Judy Creek corehole 83-401A was significantly higher than that found in other sections at the same stratigraphic level (Sweet & Braman, 1992). Indeed, all sections in western Canada with coaly shale or coal underlying the boundary claystone (Knudsen's Coulee, Wood Mountain Creek and Rock Creek West B) have comparatively higher abundances of morphologically complex pollen. This corresponds with the highest relative abundance of aquiloid pollen occurring in the humid facies association (Fig. 22.4) and with the relatively great species richness of upper Maastrichtian coals. Nevertheless, the abundance of morphologically complex pollen that occurs in the coals immediately below the K/T boundary is lower than in coals from lower in the upper Maastrichtian (Fig. 22.5A and 22.5C). The conclusion from these observations is that the apparent reduction in species richness below the K/T boundary relates in part but not completely to local and regional depositional environments.

Additionally, Sweet and Braman (1992) found that uppermost Maastrichtian coal swamp and swamp margin palynofloras of central Alberta and south-central Saskatchewan contain abundant Taxodiaceae-Cupressaceae-pollen. This was not predictable from data previously available on late Maastrichtian swamp floras. This dominance of gymnosperm pollen, together with the apparent decrease in species richness and the palynofloral succession in Judy Creek 83-313A immediately below the K/T contact, provides sufficient palynological data to conclude that plant ecosystems changed in the uppermost Maastrichtian. These changes correlate with the evidence for wetter depositional environments near the boundary. It also seems that local environmental factors may have been more critical in governing the occurrence of individual species of morphologically complex pollen because of their more sporadic occurrence in the latest Maastrichtian than earlier in the Maastrichtian. This suggests environmental constraints which exceeded the adaptive capacity of some species, narrowing the range of niches available to individual species in the latest Cretaceous. Many of the remaining morphologically complex late Maastrichtian species become extinct during the time the boundary claystone was deposited. The high relative abundance of zoophilous pollen in upper Maastrichtian coals makes their near absence from lower Paleocene coals even more exceptional and hence attributable to the K/T event.

Whitehead (1969) discussed pollination method versus vegetation type and suggested that for efficient anemophily the vegetation should be open in structure or go through a deciduous season (implying a seasonal climate) to minimize obstructions to wind transport. The floral diversity also should be relatively low to decrease the spacing between individuals of the same species. Zoophily tends to be associated with high diversity and closed vegetation. Wolfe and Upchurch (1986) interpreted the late Maastrichtian forests north of 60° paleolatitude to be deciduous. Wolfe (1987) suggested that deciduousness was advantageous, and that the evergreen habit was disadvantageous, to the survival of plants at the K/T boundary, on the assumption that the extinctions were the direct result of a sudden cooling or a short period of darkness. The linkage between anemophily and deciduousness might explain the contention of Wolfe and Upchurch that deciduousness had a selective advantage but does not address the physical cause of the extinctions.

Many triprojectate and other morphologically complex pollen taxa were part of the northern late Maastrichtian flora and were circumpolar in their distribution (Batten, 1984). It is unlikely that the extinction of the producing plants resulted from evergreen habit, as deciduousness was prevalent in northern forest. More likely, extinctions within the *Aquilapollenites* floral province resulted from ecological changes that affected the reproductive capacity of zoophilous plants having morphologically complex pollen. This suggests that the disruption of pollinating vectors indirectly led to the extinction of most zoophilous plants (McIver *et al.*, 1991; Sweet & Jerzykiewicz, 1987).

Assemblages from the boundary claystone record the next major change in the relative abundance of palynomorphs. Although there is considerable variation in the composition of these claystone assemblages, Sweet and Braman (1992) found that in their Rock Creek West B, Wood Mountain area, *Cyathidites minor* shifts from being a minor component of the palynoflora below the claystone to being dominant in the claystone. Higher percentages of *Myrtipites granulatus*, a late Maastrichtian species, also were found in the claystone than in underlying strata. In contrast, *Laevigatosporites* occurs more abundantly in strata below and above the claystone at all three Rock Creek localities described by Sweet and Braman (1992) than it does in the claystone. In the Knudsen's Farm section the percentage of angiosperm pollen in the claystone is nearly double that in underlying

strata, largely because of a major increase in the percentage of *Kurtzipites* (Sweet & Braman, 1992). In all cases the percentage of gymnosperm pollen is suppressed, a trend that starts in most sections with the boundary claystone (Sweet & Braman, 1992). These immediately post-boundary shifts in abundance were not predictable either from the lithofacies or from regional palynological patterns and hence must be related to the K/T event. Nevertheless, as seen in the Judy Creek corehole 83-401A, the palynofloral succession that occurs in the boundary coal has analogs in upper Maastrichtian coals, suggesting that the perturbation imposed by the K/T event was selective rather than catastrophic for the total community of plants. The similarity in the maceral composition of coals spanning the boundary helps corroborate this sense of continuity.

The last floral shift is an abrupt one, from opportunistic assemblages to those in which the more general late Maastrichtian style coal swamp flora is reestablished. *Laevigatosporites* and angiosperm pollen initially dominate these floras, albeit with a reduced diversity of morphologically complex pollen. An upsection increase in the percentage of Taxodiaceae-Cupressaceae pollen is apparent in most sections (Sweet & Braman, 1992; Sweet *et al.*, 1990). Short term fluctuations in relative abundances of palynological taxa in the organic-rich sediments consistently are most pronounced in the approximately 0.2 m interval spanning the boundary. However, fluctuations in the percentage of major groups of palynomorphs occur throughout a more extended interval. This is well-illustrated in Judy Creek corehole 83-313A, where most peak abundances of *Laevigatosporites* within the upper coal-bearing interval probably correlate with higher clastic input into the swamp and therefore possibly to the increased availability of nutrients. This observation also was made by Demchuk and Strobl (1989) in Paleocene coals of central Alberta. It is therefore not surprising that a second peak abundance of *Laevigatosporites* was recorded by Nichols *et al.* (1986) 3.65 m above the boundary claystone in their Rock Creek section. Such peak abundances mimic those that occur at the boundary, as each record a shift in ecological conditions. However, only at the boundary is there a direct association between the occurrence of peak abundances and a number of range truncations.

At a horizon 3 to 21 m above the K/T boundary, most relict late Maastrichtian species disappear (e.g. *Aquilapollenites reticulatus*), or become quite rare (e.g. *Wodehouseia spinata*), and *W. fimbriata* enters the assemblage. This entry level of *W. fimbriata* approximately coincides with the 29R–29N polarity reversal (Lerbekmo, 1985). The successive early Paleocene pollen assemblages consist of morphologically simple (anemophilous) angiosperm pollen in assemblages that continue to be dominated by miospores and gymnosperm pollen.

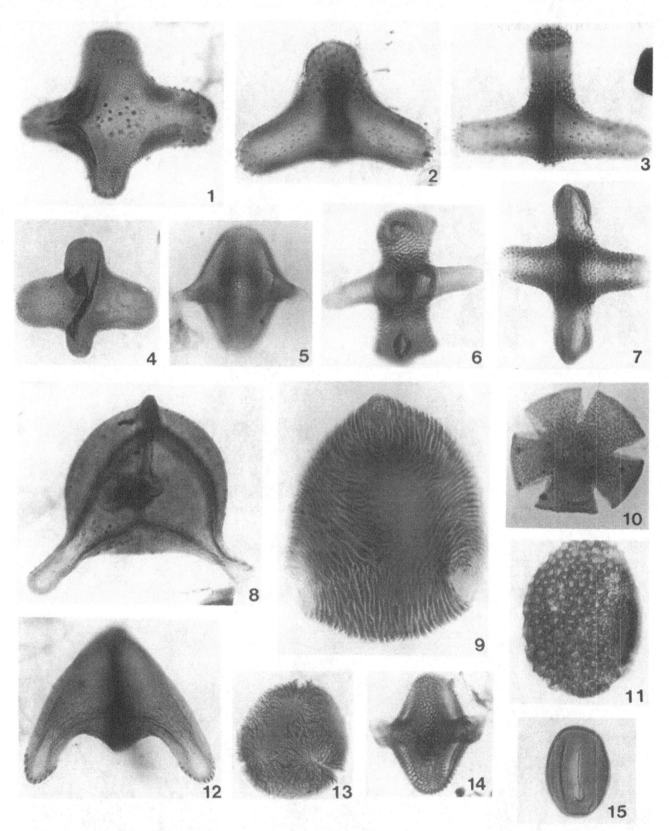

Figure 22.13.1–15. (1) *Aquilapollenites attenuatus* Funkhouser 1961, GSC 100717; (2) *A. quadrilobus* Rouse 1957, GSC 100718; (3) *A. delicatus* var. *collaris* Tschudy & Leopold 1971, GSC 100719; (4) *A.* sp. cf. *A. vulsus* Sweet 1986, GSC 100720; (5) *A. reticulatus* (Mchedlishvili) Tschudy & Leopold 1971, GSC 100721; (6) *A. conatus* Norton 1965, GSC 100722; (7) *A. quadricretaeus* Chlonova 1961, GSC 100723; (8) *Mancicorpus gibbus* Srivastava 1968, GSC 100724; (9) *Jiangsupollis major* Song *in* Song Zhi-chen, Zheng Ya-hui *et al.* 1980, GSC 100725; (10) *Leptopecopites pocockii* (Srivastava) Srivastava 1978, GSC 100726. (11) *Liliacidites complexus* (Leffingwell) Stanley 1965, GSC 100727; (12) *Mancicorpus rostratus* Srivastava 1968, GSC 100728; (13) *Callistopollenites radiatostriatus* (Mchedlishvili) Srivastava 1969, GSC 100729; (14) *Aquilapollenites oblatus* Srivastava 1968, GSC 100730; (15) *Fraxinoipollenites variabilis* Stanley 1965, GSC 100731. (All photographs are reproduced × 1000.)

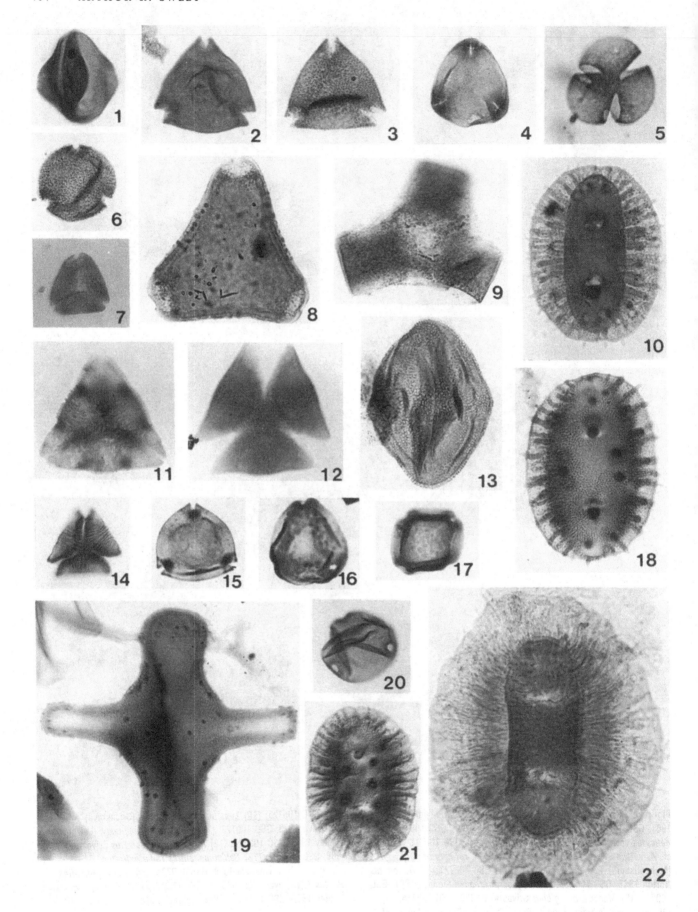

Summary and conclusions

Regional scenarios of plant extinctions and changes in plant assemblages at the K/T boundary are not simple. In western Canada, variations in palynological assemblages during the late Maastrichtian can be related to latitude, depositional environments, and successional changes. The degree of wetness is reflected both lithologically by differences in the overall associations of regional facies and palynologically in the diversity of the palynoflora. Successive changes in the palynoflora may or may not be accompanied by macroscopically different lithofacies, although changes in lithofacies generally are predictable with changes in the palynofloral assemblage. Together these data suggest a dynamic latest Maastrichtian biotic system, with changes in environmental and ecological conditions predating and postdating the K/T boundary.

A series of exceptional changes occurs in the interval immediately contiguous to the K/T boundary, causing increased range truncations and rapid changes in relative abundances within the palynofloral assemblages. Here, the floras appear to be affected by ecological changes independent of the usual fluvial and geographic controls on sedimentation (and lithofacies), even though some palynofloral changes coincide with those in lithofacies. These perturbations were continental in scope, even though their expression varied as a result of regional environmental factors, facies associations, and the local depositional regime.

Figure 22.14.1–22. (1, 2) *Orbiculapollis lucida* Chlonova 1961, GSC 100732; (3) *Myrtipites granulatus* Norton *in* Norton & Hall 1969, GSC 100733; (4) *M. scabratus* Norton *in* Norton & Hall 1969, GSC 100734; (5) *Tricolpites microreticulatus* Belsky, Boltenhagen & Potonié 1965, GSC 100735; (6) *Tricolpites parvistriatus* Norton 1967, GSC 100736; (7) *Syncolporites minimus* Leffingwell 1971, GSC 100737; (8) *Proteacidites* sp. cf. *P. globosiporus* Samoilovich 1961, GSC 100738; (9) *Penetetrapites inconspicuus* Sweet 1986, GSC 100739; (10) *Wodehouseia quadrispina* Wiggins 1976, GSC 100740; (11) `Striatellipollis striatella` (Mchedlishvili) Krutzsch 1969, GSC 100741; (12) *S. radiata* Sweet 1986, GSC 100742; (13) *Polotricolporites rotundus* Sweet 1986, GSC 100743; (14) *Cranwellia rumseyensis* Srivastava 1967, GSC 100744; (15) *Kurtzipites trispissatus* Anderson 1960, GSC 100745; (16) *Ulmoideipites krempii* Anderson 1960, GSC 100746. (17) *U. hebridicus* (Simpson) Sweet 1986, GSC 100747; (18) *Wodehouseia octospina* Wiggins 1976, GSC 100748; (19) *Triprojectus unicus* (Chlonova) Mchedlishvili 1961, GSC 100749; (20) *Triporopollenites plektosus* Anderson 1960, GSC 100750; (21) *Wodehouseia spinata* Stanley 1961, GSC 100751; (22) *W. fimbriata* Stanley 1961, GSC 100752.(All photographs are reproduced × 1000.)

Appendix 22.1 Location of illustrated sections

Wood Mountain Creek, Wood Mountain area, south-central Saskatchewan. Section located on the north facing valley wall of Wood Mountain Creek, 6 km north of Wood Mountain at 49° 25′ 20″ N, 106° 19′ 50″ E.

Judy Creek North coreholes, Judy Creek coalfield, north-central Alberta about 150 km northwest of Edmonton. Of the coreholes discussed in detail here, 83-313A (54° 33.7′ N, 115° 26.4′ W) is located farthest north, being 12 km west-northwest of 83-401A (54° 31.4′ N, 115° 16.2′ W).

Acknowledgments

Without the forbearance, gentle nudges and critiques of Al Traverse, this chapter would not have been written. Discussions with Elisabeth McIver and James White have led substantially to the development of the ideas expressed in the chapter and to some critical revisions. The continuing and valued influence of Jack Lerbekmo, Dennis Braman, Farley Fleming and Doug Nichols on the interpretation of K/T events is also gratefully acknowledged. However, the writer accepts full responsibility for the final content of the chapter. This chapter is GSC Contribution 29691.

References

Alvarez, W., Asaro, F. & Montanari, A. (1990). Iridium profile for 10 million years across the Cretaceous-Tertiary boundary at Gubbio (Italy). *Science*, **250**, 1700–2.

Batten, D. J. (1984). Palynology, climate and the development of Late Cretaceous floral provinces in the Northern Hemisphere; a review. In *Fossils and Climate*, ed. P. Brenchly. London: John Wiley & Sons, 127–64.

Blodgett, R. H. (1988). Calcareous paleosols in the Triassic Dolores Formation, southwestern Colorado. In *Paleosols and Weathering Through Geologic Time: Principles and Applications*, ed. J. Reinhardt & W. R. Sigleo. *Geological Society of America Special Paper*, **216**, 103–21.

Bohor, B. F., Triplehorn, D. M., Nichols, D. J. & Millard, H. T., Jr. (1987). Dinosaurs, spherules, and the 'magic' layer: a new K/T boundary clay site in Wyoming. *Geology*, **15**, 896–9.

Brinkhuis, H. & Zachariasse, W. J. (1988). Dinoflagellate cysts, sea level changes and planktonic foraminifers across the Cretaceous-Tertiary boundary at El Haria, northwest Tunisia. *Marine Micropaleontology*, **13**, 153–91.

Brown, R. W. (1962). Paleocene flora of the Rocky Mountains and Great Plains. *US Geological Survey Professional Paper*, **375**.

Cecil, C. B., Stanton, R. W., Neuzil, S. G., Dulong, F. T., Ruppert, L. F. & Pierce, B. S. (1985). Paleoclimate controls on late Paleozoic sedimentation and peat formation in the central Appalachian basin (U.S.A.). *International Journal of Coal Geology*, **5**, 195–230.

Cherven, V. B. & Jacob, A. F. (1985). Evolution of Paleogene depositional systems, Williston Basin, in response to global sea level changes. In *Cenozoic Paleogeography of the West-central United States*, ed. R. M. Flores & S. S.

Kaplan. The Rocky Mountain Section, Society of Economic Paleontologists and Mineralogists, *Rocky Mountain Paleogeography Symposium* **3**, 127–70.

Demchuk, T. D. (1990). Palynostratigraphic zonation of Paleocene strata in the central and south-central Alberta Plains. *Canadian Journal of Earth Sciences*, **27**, 1263–9.

Demchuck, T. D. & Strobl, R. (1989). Coal facies and in-seam profiling, Highvale No. 2 Seam, Highvale Alberta. In *Advances in Western Canadian Coal Geoscience*, ed. W. Langenberg. *Alberta Research Council Information Series*, **103**, 201–11.

Doerenkamp, A., Jardine, S. & Moreau, P. (1976). Cretaceous and Tertiary palynomorph assemblages from Banks Island and adjacent areas (N.W.T). *Bulletin Canadian Petroleum Geologists*, **24**, 372–417.

Donovan, A. D., Baum, G. R., Blechschmidt, C. L., Loutit, T. S., Pflum, C. E. & Vail, P. R. (1988). Sequence stratigraphic setting of the Cretaceous-Tertiary boundary in central Alabama. In *Sea-Level Changes – An Integrated Approach*, ed. C. K. Wilgus, B. S. Hastings, C. G. St. C. Kendall, H. W. Posamentier, C. A. Ross & J. C. Van Wagoner. *Society of Economic Paleontologists and Mineralogists Special Publication*, **42**, 299–307.

Erickson, J. (1978). Bivalve mollusk range extensions in the Fox Hills Formation (Maastrichtian) of North and South Dakota and their implications for the Late Cretaceous geologic history of the Williston Basin. *Annual Proceedings of the North Dakota Academy of Science*, **32**, 78–89.

Farley, M. B. & Dilcher, D. L. (1986). Correlation between miospores and depositional environments of the Dakota Formation (mid-Cretaceous) of north-central Kansas and adjacent Nebraska, U.S.A. *Palynology*, **10**, 117–33.

Fastovsky, D. E. & McSweeney, K. (1987). Paleosols spanning the Cretaceous-Paleogene transition eastern Montana and western North Dakota. *Geological Society of America Bulletin*, **99**, 66–77.

Fleming, R. F. (1989). Fossil *Scenedesmus* (Chlorococcales) from the Raton Formation, Colorado and New Mexico, U.S.A. *Review of Palaeobotany and Palynology*, **59**, 1–6.

Fleming, R. F. & Nichols, D. J. (1990). The fern-spore abundance anomaly at the Cretaceous-Tertiary boundary: a regional bioevent in western North America. In *Extinction Events in Earth History*, ed. E. G. Kauffman & O. H. Walliser. *Lecture Notes in Earth Sciences*, **30**. New York: Springer-Verlag, 351–64.

Flores, R. M. (1984). Comparative analysis of coal accumulation in Cretaceous alluvial deposits, southern United States Rocky Mountain Basins. In *The Mesozoic of Middle North America*, ed. D. F. Stott & D. J. Glass. *Canadian Society of Petroleum Geologists Memoir* **9**, 373–85.

Frederiksen, N. O. (1989). Changes in floral diversities, floral turnover rates, and climates in Campanian and Maastrichtian time, North Slope of Alaska. *Cretaceous Research*, **10**, 249–66.

Frederiksen, N. O., Ager, T. A. & Edwards, L. E. (1988). Palynology of Maastrichtian and Paleocene rocks, lower Colville River region, North Slope of Alaska. *Canadian Journal of Earth Sciences*, **25**, 512–27.

Furnival, G. M. (1950). Cypress Lake map-area, Saskatchewan. *Geological Survey of Canada Memoir*, **242**.

Greenlee, S. M. & Moore, T. C. (1988). Recognition and interpretation of depositional sequences and calculation of sea-level changes from stratigraphic data: offshore New Jersey and Alabama Tertiary. In *Sea-Level Changes – An Integrated Approach*, ed. C. K. Wilgus, B. S. Hastings,

C. G. St. C. Kendall, H. W. Posamentier, C. A. Ross & J. C. Van Wagoner. *Society of Economic Paleontologists and Mineralogists Special Publication* **42**, 329–53.

Hallam, A. (1984). Continental humid and arid zones during the Jurassic and Cretaceous. *Palaeogeography, Palaeoclimatology, Palaeoecology*, **47**, 195–223.

Hallam, A. (1988). A compound scenario for the end-Cretaceous mass extinctions. In *Palaeontology and Evolution: Extinction Events*, ed. M. A. Lamolda, E. G. Kauffman & O. H. Walliser. Revista Española de Paleontologia n. Extraordinario, 7–20.

Hotton, C. (1984). (abstract) Palynofloral changes across the Cretaceous-Tertiary boundary in east central Montana, U.S.A. *Abstracts, 6th International Palynological Conference, Calgary*, 66.

Hotton, C. (1988). Palynology of the Cretaceous-Tertiary boundary in central Montana, U.S.A. and its implication for extraterrestrial impact. Ph. D. dissertation, University of California, Davis.

Jeletzky, J. A. (1978). Causes of Cretaceous oscillations of sea level in western and arctic Canada and some general geotectonic implications. *Geological Survey of Canada Paper* **77–18**.

Jerzykiewicz, T. & Sweet, A. R. (1988). Sedimentological and palynological evidence of regional climatic changes in the Campanian to Paleocene sediments of the Rocky Mountain Foothills, Canada. *Sedimentary Geology*, **59**, 29–76.

Johnson, K. R. & Hickey, L. J. (1989). Megafloral change across the Cretaceous/Tertiary boundary in the northern Great Plains and Rocky Mountains, U.S.A. In *Global Catastrophes in Earth History; An Interdisciplinary Conference on Impacts, Volcanism, and Mass Mortality*, ed. V. L. Sharpton & P. D. Ward, *Geological Society of America Special Paper* **247**, 433–44.

Johnson, K. R., Nichols, D. J., Attrep, M., Jr. & Orth, C. J. (1989). High-resolution leaf-fossil record spanning the Cretaceous/Tertiary boundary. *Nature*, **340**, 708–11.

Jones, D. S., Mueller, P. A., Bryan, J. R., Dobson, J. P., Channell, J. E. T, Zachos, J. C. & Arthur, M. A. (1987). Biotic, geochemical, and paleomagnetic changes across the Cretaceous/Tertiary boundary at Braggs, Alabama. *Geology*, **15**, 311–15.

Keefer, W. R. (1965). Stratigraphy and geologic history of the uppermost Cretaceous, Paleocene, and Lower Eocene rocks in the Wind River Basin, Wyoming. *US Geological Survey Professional Paper* **495-A**.

Keller, G. & Lindinger, M. (1989). Stable isotope, TOC and $CaCO_3$ record across the Cretaceous-Tertiary boundary at El Kef, Tunisia. *Palaeogeography, Palaeoclimatology, Palaeoecology*, **73**, 243–65.

Leffingwell, H. A. (1971). Palynology of the Lance (Late Cretaceous) and Fort Union (Paleocene) Formations of the Type Lance Area, Wyoming. In *Symposium of Palynology of the Late Cretaceous and Early Tertiary*, ed. R. M. Kosanke & A. T. Cross. *Geological Society of America Special Paper* **127**, 1–64.

Lehman, T. M. (1989). Upper Cretaceous (Maastrichtian) paleosols in Trans-Pecos Texas. *Geological Society of America Bulletin*, **101**, 188–203.

Lehman, T. M. (1990). Paleosols and the Cretaceous/Tertiary transition in the Big Bend region of Texas. *Geology*, **18**, 362–4.

Lerbekmo, J. F. (1985). Magnetostratigraphic and biostratigraphic correlations of Maastrichtian to Early Paleocene strata between south-central Alberta and southwestern Saskatchewan. *Bulletin of Canadian Petroleum Geology*, **33**, 213–26.

Lerbekmo, J. F. & Coulter, K. C. (1984). Magnetostratigraphic and biostratigraphic correlations of Late Cretaceous to Early Paleocene strata between Alberta and North Dakota. In *The Mesozoic of Middle North America*, ed. D. F. Stott & D. J. Glass. *Canadian Society of Petroleum Geologists Memoir*, **9**, 313–17.

Lerbekmo, J. F., Sweet, A. R. & St. Louis, R. M. (1987). The relationship between the iridium anomaly and palynological floral events at three Cretaceous-Tertiary boundary localities in western Canada. *Geological Society of America Bulletin*, **99**, 325–30.

Levinton, J. S. (1969). The paleoecological significance of opportunistic species. *Lethaia*, **3**, 69–78.

Loutit, T. S., Hardenbol, J., & Vail, P. R. (1988). Condensed sections: the key to age determination and correlation of continental margin sequences. In *Sea-Level Changes – An Integrated Approach*, ed. C. K. Wilgus, B. S. Hastings, C. G. St. C. Kendall, H. W. Posamentier, C. A. Ross & J. C. Van Wagoner. *Society of Economic Paleontologists and Mineralogists Special Publication* **42**, 183–213.

McCabe, P.J. (1987). Facies studies of coal and coal-bearing strata. In *Coal and Coal-bearing Strata: Recent Advances*, ed. A. C. Scott. *Geological Society Special Publication* **32**, 51–66.

McIntyre, D. J. (1974). Palynology of an Upper Cretaceous section, Horton River, District of Mackenzie, N.W.T. *Geological Survey of Canada Paper* **74–14**.

McIver, E. E., Sweet, A. R. & Basinger, J. F. (1991). Sixty-five-million-year-old flowers bearing pollen of the extinct triprojectate complex – a Cretaceous-Tertiary boundary survivor. *Review of Palaeobotany and Palynology*, **70**, 77–88.

Moore, P. D. (1987). Ecological and hydrological aspects of peat formation. In *Coal and Coal-bearing Strata: Recent Advances*, ed. A. C. Scott. *Geological Society Special Publication* **32**, 7–15.

Nichols, D. J. (1985). (abstract) Palynomorph assemblages from uppermost Cretaceous deposits, Denver Basin, Colorado. *Palynology*, **9**, 249–50.

Nichols, D. J. (1990). Geologic and biostratigraphic framework of the non-marine Cretaceous-Tertiary boundary interval in western North America. *Review of Palaeobotany and Palynology*, **65**, 75–84.

Nichols, D. J. & Fleming, R. F. (1991). Plant microfossil record of the terminal Cretaceous event in the western United States and Canada. In *Global Catastrophes in Earth History; An Interdisciplinary Conference on Impacts, Volcanism, and Mass Mortality*, eds. V. L. Sharpton & P. D. Ward, *Geological Society of America Special Paper* **247**, 445–55.

Nichols, D. J., Fleming, R. F. & Frederiksen, N. O. (1990). Palynological evidence of effects of the terminal Cretaceous event on terrestrial floras in western North America. In *Extinction Events in Earth History*, ed. E. G. Kauffman & O. H. Walliser, *Lecture Notes in Earth Sciences*, **30**. New York: Springer-Verlag, 351–64.

Nichols, D. J., Fleming, R. F., Upchurch, G. R., Jr., Tschudy, R. H. & Pillmore, C. L. (1985). (abstract) Paleobotanical changes across the Cretaceous-Tertiary boundary at Sugarite, New Mexico: new data and interpretations. *Society of Economic Paleontologists and Mineralogists Annual Midyear Meeting, Abstracts* **2**, 68.

Nichols, D. J., Jacobson, S. R. & Tschudy, R. H. (1982). Cretaceous palynomorph biozones for the central and northern Rocky Mountain region of the United States. In *Geologic Studies of the Cordilleran Thrust Evolution Belt*, ed. R. P. Powers. Denver, CO: *Rocky Mountain Association of Geologists*, **2**, 721–33.

Nichols, D. J., Jarzen, D. M., Orth, C. J. & Oliver, P. Q. (1986). Palynological and iridium anomalies at Cretaceous-Tertiary boundary, south-central Saskatchewan. *Science*, **231**, 714–17.

Nichols, D. J. & Sweet, A. R. (in press). Biostratigraphy of Upper Cretaceous nonmarine palynofloras in a north-south transect of the Western Interior Basin. In *Evolution of the Western Interior Basin*, ed. W. G. E. Caldwell & E. G. Kauffman. *Geological Society of Canada Special Paper* **39**.

Obradovich, J. D. & Cobban, W. A. (1975). A time-scale for the Late Cretaceous of the western interior of North America. In *The Cretaceous System in the Western Interior of North America*, ed. W. G. E. Caldwell. *Geological Association of Canada Special Paper* **13**, 31–54.

Orth, C. J., Gilmore, J. S., Knight, J. D., Pillmore, C. L., Tschudy, R. H. & Fassett, J. E. (1981). An iridium abundance anomaly at the palynological Cretaceous-Tertiary boundary in northern New Mexico. *Science*, **214**, 1341–3.

Payne, W. W. (1981) Structure and function in angiosperm pollen wall evolution. *Review of Palaeobotany and Palynology*, **35**, 39–59.

Pillmore, C. L. & Flores, R. M. (1987). Stratigraphy and depositional environments of the Cretaceous-Tertiary boundary clay and associated rocks, Raton basin, New Mexico and Colorado. In *The Cretaceous-Tertiary Boundary in the San Juan and Raton Basins, New Mexico and Colorado*. ed. J. E. Fassett & J. K. Rigby, Jr. *Geological Society of America Special Paper* **209**, 111–30.

Pons, D., Lauverjat, J. & Broutin, J. (1980) Paléoclimatologie comparée de deux gisements du Cretacé supérieur d'Europe occidentale. *Mémoires Société Géologique de France*, **139**, 151–8.

Retallack, G. J., Leahy, G. D. & Spoon, M. D. (1987). Evidence from paleosols for ecosystem changes across the Cretaceous/Tertiary boundary in eastern Montana. *Geology*, **15**, 1090–3.

Rocchia, R., Boclet, D., Bontè, P., Jèhanno, C., Chen, Y., Courtillot, V., Mary, C. & Wezel, F. (1990). The Cretaceous-Tertiary boundary at Gubbio revisited: vertical extent of the Ir anomaly. *Earth and Planetary Science Letters*, **99**, 206–19.

Smit, J. (1990). Meteorite impact, extinctions and the Cretaceous-Tertiary boundary. *Geologie en Mijnbouw*, **69**, 187–204.

Srivastava, S. K. (1970). Pollen biostratigraphy and paleoecology of the Edmonton Formation (Maastrichtian), Alberta, Canada. *Palaeogeography, Palaeoclimatology, Palaeoecology*, **7**, 221–76.

Stott, L. D. & Kennett, J. P. (1988). Cretaceous-Tertiary boundary in the Antarctic: climatic cooling precedes biotic crisis. In *Global Catastrophes in Earth History: an Interdisciplinary Conference on Impacts, Volcanism, and Mass Mortality. Lunar and Planetary Institute Contribution* **673**, 184–5.

Sweet, A. R. (1978). Palynology of the Ravenscrag and Frenchman formations. In *Coal Resources of Southern Saskatchewan: A Model for Evaluation Methodology. Geological Survey of Canada Economic Geology Report* **30**, 29–39.

Sweet, A. R. (1990). Palynofloras and climates of the

Santonian to Paleocene of midcontinental North America. In *Proceedings of the Symposium Paleofloristic and Paleoclimatic Changes in the Cretaceous and Tertiary.* ed. E. Knobloch & Z. Kvacek. Prague, Czechoslovakia: Geological Survey Publisher, 99–103.

Sweet, A. R. & Braman, D. R. (1989). A distinctive terrestrial palynofloral assemblage from the lower Campanian Chungo Member, Wapiabi Formation, southwestern Alberta: a key to regional correlations. *Contributions to Canadian Coal Geoscience. Geological Survey of Canada Paper* **89-8**, 32–40.

Sweet, A. R. & Braman, D. R. (1992). The K-T boundary and contiguous strata in western Canada: interactions between paleoenvironments and palynological assemblages. *Cretaceous Research*, **13**, 31–79.

Sweet, A. R., Braman, D. R. & Lerbekmo, J. F. (1990). Palynofloral response to K-T boundary events; a transitory interruption within a dynamic system. *Geological Society of America Special Paper*, **247**, 457–69.

Sweet, A. R. & Cameron, A. R. (1991). Palynofacies, coal petrographic facies and depositional environments: Amphitheatre Formation (Eocene to Oligocene) and Ravenscrag Formation (Maastrichtian to Paleocene) Canada. *International Journal of Coal Geology*, **19**, 121–44.

Sweet, A. R. & Jerzykiewicz, T. (1987). Sedimentary facies and environmentally controlled palynological assemblages: their relevance to floral changes at the Cretaceous-Tertiary boundary. In *Fourth Symposium on Mesozoic Terrestrial Ecosystems, Short Papers*, ed. P. M. Currie & E. H. Koster. *Occasional Papers of the Tyrrell Museum of Palaeontology*, **3**, 208–13.

Sweet, A. R. & McIntyre, D. J. (1988). Late Turonian marine and nonmarine palynomorphs from the Cardium Formation, north-central Alberta foothills, Canada, In *Sequences, Stratigraphy, Sedimentology: Surface and Subsurface*, ed. D. P. James & D. A. Leckie. *Canadian Society of Petroleum Geologists Memoir* **15**, 499–516.

Sweet, A. R., Ricketts, B. D., Cameron, A. R. & Norris, D. K. (1989). An integrated analysis of the Brackett Coal Basin, Northwest Territories. *Current Research, Part G, Geological Survey of Canada Paper* **89-1G**, 85–99.

Tryon, R. M. & Tryon, A. F. (1982). *Ferns and Allied Plants with Special Reference to Tropical America.* New York: Springer-Verlag.

Tschudy, B. D. (1971). Two new fossil genera from upper Campanian (Cretaceous) rocks of Montana. *US Geological Survey Professional Paper*, **750-B**, B53–61.

Tschudy, B. D. (1973). Palynology of the Upper Campanian (Cretaceous) Judith River Formation, north-central Montana. *US Geological Survey Professional Paper* **770**.

Tschudy, B. D. & Leopold, E. B. (1971). *Aquilapollenites* (Rouse) Funkhouser – selected Rocky Mountain taxa and their stratigraphic ranges. In *Symposium on Palynology of the Late Cretaceous and Early Tertiary*, eds. R. M. Kosanke & A. T. Cross. *Geological Society of America Special Paper*, **127**, 113–67.

Tschudy, R. H., Pillmore, C. L., Orth, C. J., Gilmore, J. S. & Knight, J. D. (1984). Disruption of the terrestrial plant ecosystem at the Cretaceous-Tertiary boundary, Western Interior. *Science*, **225**, 1030–2.

Tschudy, R. H. & Tschudy, B. D. (1986). Extinction and survival of plant life following the Cretaceous/Tertiary boundary event, Western Interior, North America. *Geology*, **14**, 667–70.

Vakhrameev, V. A. (1987). Climates and the distribution of some gymnosperms in Asia during the Jurassic and Cretaceous. *Review of Palaeobotany and Palynology*, **51**, 205–12.

Whitehead, D. R. (1969). Wind pollination in the angiosperms: evolutionary and environmental considerations. *Evolution*, **23**, 28–35.

Wiggins, V. D. (1976). Fossil oculata pollen from Alaska. *Geoscience and Man*, **15**, 51–76.

Wolfe, J. A. (1987). Late Cretaceous-Cenozoic history of deciduousness and the terminal Cretaceous event. *Paleobiology*, **13**, 215–26.

Wolfe, J. A. (1990). Palaeobotanical evidence for a marked temperature increase following the Cretaceous/Tertiary boundary. *Nature*, **343**, 153–6.

Wolfe, J. A. (1991). Palaebotanical evidence for a June 'impact winter' at the Cretaceous-Tertiary boundary. *Nature*, **352**, 420–3.

Wolfe, J. A. & Upchurch, G. R., Jr. (1986). Vegetation, climatic and floral changes at the Cretaceous-Tertiary boundary. *Nature*, **324**, 148–52.

Wolfe, J. A. & Upchurch, G. R., Jr. (1987a). North American nonmarine climates and vegetation during the Late Cretaceous. *Palaeogeography, Palaeoclimatology, Palaeoecology*, **61**, 33–77.

Wolfe, J. A. & Upchurch, G. R., Jr. (1987b). Leaf assemblages across the Cretaceous-Tertiary boundary in the Raton Basin, New Mexico and Colorado. *Proceedings of the National Academy of Science, USA*, **84**, 5096–100.

Zachos, J. C., Arthur, M. A. & Dean, W. E. (1989). Geochemical and paleoenvironmental variations across the Cretaceous/Tertiary boundary at Braggs, Alabama. *Palaeogeography, Palaeoclimatology, Palaeoecology*, **69**, 245–66.

Ziegler, A. M., Raymond, A. L., Gierlowski, T. C., Horrell, M. A., Rowley, D. B. & Lottes, A. L. (1987). Coal, climate and terrestrial productivity: the present and early Cretaceous compared. In *Coal and Coal-bearing Strata: Recent Advances*, ed. A. C. Scott. *Geological Society Special Publication* **32**, 25–49.

23 Sedimentation of palynomorphs in rocks of pre-Devonian age

PAUL K. STROTHER

Introduction

Paleopalynology is the study of microscopic structurally preserved organic matter which can be extracted from siliciclastic rocks by acid maceration. Traditionally, paleopalynology includes the study of various algal remains, microscopic parts of some protists and animals, spores of cryptogamic plants and fungi, and the pollen grains of seed-bearing plants, all of which are thought to possess a resistant organic outer wall composed of sporopollenin, chitin, or a similar derivative. This definition restricts the palynological record prior to the evolution of the terrestrial embryophytic plants to the resistant remains of algal phytoplankton, chitinous scolecodonts, and possible zooplankton such as chitinozoans. However, in this chapter I expand the view of paleopalynology to include some other structurally intact, microscopic organic remains which are thought to be of planktonic origin. This means that phytoplankton preserved in both siliciclastic (Acritarcha) and in cherty carbonate facies (Cryptarcha, *sensu* Diver & Peat, 1979)) is included.

In broadest terms, the study of palynomorphs as sedimentary particles concerns the distribution and dynamics of biological carbon through geologic time. The stratigraphic record of microplankton is of great importance in modeling the dynamics of former marine and terrestrial ecosystems. The changes in the flux of carbon brought about by the rise of a standing biomass in terrestrial habitats represents a fundamental shift in the distribution of biogenic carbon on the earth. This shift occurred during the Silurian Period and is marked by the fossil record of the cryptospores, which represent the palynological evidence of a transition to an embryophyte-dominated terrestrial ecosystem (Strother, 1991). The lack of a well-developed and recognizable terrestrial flora prior to the Devonian adds to the difficulty of assessing the provenance of palynomorphs as sedimentary particles because terrestrial deposits can be difficult to recognize solely on the basis of sedimentological criteria. In part

this is because the life strategies of cyst forming phytoplankton are not fundamentally different in freshwater and marine habitats (Harris, 1986); therefore there are few morphological clues that allow us to distinguish freshwater from marine algal cysts. Surprisingly, most studies on the paleoecology of pre-Devonian acritarchs do not even consider the possibility of terrestrial (freshwater) origin for any taxa (e.g. Smith & Saunders, 1970; Vidal & Knoll, 1983), even though acritarchs have occasionally been reported from pre-Devonian strata thought to be terrestrial on sedimentological grounds (Diver, 1980; Downie, 1962; Zhang, 1982).

In order to utilize fully the record of microplankton from pre-Devonian time, we must stretch the concept of palynomorph from robust-walled microfossils extracted from shales to include organic-walled structures preserved in cherts. Many of Diver and Peat's (1979) group, Cryptarcha, fall into this category. Fossilization in chert can preserve three-dimensional spatial fabric which can be useful in paleoecological interpretation. The process of preservation in cherty lithofacies is fundamentally different from that in siliciclastic rocks, however, and it is important to realize that 'palynomorph' taxa described from Precambrian cherts are not necessarily directly comparable to palynomorphs found in shales. This is particularly true for small unicells such as *Myxococcoides* Schopf (Knoll, 1984). Additionally, there are fundamental changes in the distribution and formation of cherts of Precambrian versus Phanerozoic age, resulting in a more diverse sampling of depositional settings from Precambrian rocks (Malvia *et al.*, 1989).

Palynomorphs from pre-Devonian rocks can be used to address some fundamental problems in sedimentology and the distribution of organic matter. The accumulation of organic matter in sediments can be used to assess sedimentation rates (Brush, 1989), although this has not been attempted for rocks of pre-Devonian age. There is a relation between the accumulation of organic matter and local oxygen availability, and this, in turn, is important in the reconstruction of former ocean

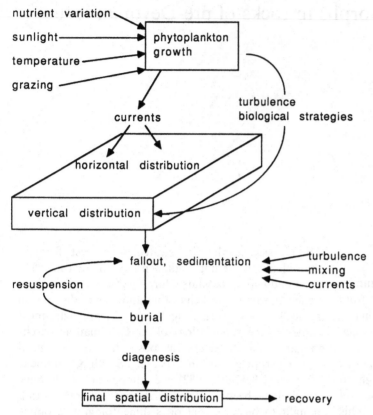

Figure 23.1. Factors affecting the distribution of palynomorphs in sediments.

ecosystems. The very presence of phytoplankton in ancient sediments is evidence of an active surficial ecosystem based on the light-harvesting abilities of phytoplankton. In some instances, we can characterize depositional environments on the basis of their palynomorph content (e.g., Knoll, 1984). This is of critical importance for the Precambrian, because the lack of other fossils from this time limits greatly the traditional uses of the fossil record in the reconstruction of paleo-environments.

By examining palynomorph diversity, morphology and spatial distribution in sediment, we expect to work backwards to reconstruct the biology of once-living phytoplankton communities. This is the classic problem in taphonomy: can one use the distribution of fossils in sedimentary rock to map backwards through the various dynamic taphonomic processes to the original system state? Because of information loss during sedimentation, it is often not possible to produce unique solutions to reverse mappings. Instead, the information available in the fossilized assemblage is used to establish a range of possible original states.

The use of palynomorphs in the reconstruction of pre-Devonian environments is undoubtedly complex (Fig. 23.1). The reason is that different factors act in

consort to produce a final distribution of palynomorphs in the sedimentary rock. Different combinations of these biological and physical factors can produce similar distributions of palynomorphs. For example, monospecific acritarch assemblages might reflect cyst accumulation from a geologically instantaneous algal bloom, but such assemblages could also be indicative of non-bloom-forming species from restrictive (low diversity) environments which accumulated over longer spans of time. In general, one must use palynomorph data in conjunction with lithologic information to create a composite picture of the depositional environment (e.g. Vidal, 1979, 1981). Vidal and Knoll conclude that low diversity assemblages composed of simple taxa with long strati-graphic ranges characterize 'inshore environments,' and that '..Proterozoic plankton diversity was highest in offshore shelf or platform environments.' (Vidal & Knoll 1983, p. 268).

Assumptions relating assemblage diversity and environment of deposition must be examined in light of the extant ecology and population dynamics of phytoplankton. There is initial temporal and spatial heterogeneity in phytoplankton diversity in the euphotic zone. For example, the number of cyst-forming dinoflagellate species in living communities does not correlate with the total taxonomic diversity of dinoflagellates found in the water column (Evitt, 1985). The production of cysts does not necessarily act as a measure of actual phytoplankton biomass. Mixing occurs over time as cysts fall to the sediment bottom. I show later in this chapter that water depth and cyst size are significant in this regard. Palynomorph diversity is lithofacies sensitive. Thus sediment size, sediment accumulation rate, and 'energy' of depositional environment all combine to modify the basic biological 'signal' which is represented in the fallout of cysts to the bottom sediment.

Several attempts have been made to use Paleozoic acritarch distributions to define paleoenvironments based on water depth, proximity to shoreline, and lithofacies (Dorning, 1981; Jacobson, 1979; Richardson & Rasul, 1990; Smith & Saunders, 1970). However, none of these authors has addressed the fundamental relationship between palynomorph distribution in sediment and the spatial/temporal distribution of extant cyst-forming phytoplankton in modern oceans and lakes. Likewise, Precambrian workers have considered 'random' spatial distributions of cells in cherty biofacies to indicate phytoplankton affinity of specific taxa (Knoll *et al.*, 1978; Vidal & Knoll, 1983) without addressing modern dynamics of the sedimentation of cyst-forming phytoplankton.

In the discussion that follows, I try to establish some of the limits on what we can and cannot do vis-à-vis the reconstruction of phytoplankton ecology from palynomorph distributions in pre-Devonian rocks. I focus on criteria for recognizing benthic versus planktic microfossils, recognition of non-marine versus marine palynomorphs, ecological aspects of phytoplankton cyst morphology, and the theoretical aspects of the relationship between palynomorph assemblage diversity and spatial distribution to temporal heterogeneity of cyst formation. These topics represent merely a small part of the much more general problems of biostratigraphy, taphonomy, the evolution of phytoplankton, and the history of the global carbon budget.

The dynamics of cyst sedimentation

Theoretically, the distribution of phytoplankton in sedimentary rock should bear some reflection of the ecology of living phytoplankton in the overlying water column. Many biological and physical factors interact in nature to produce a final spatial distribution of palynomorphs in bottom sediment and, ultimately, sedimentary rock (Fig. 23.1). The production of cells (whether vegetative or cyst) which make their way into bottom sediments varies over time and space as a function of the dynamics of phytoplankton population growth and vertical and horizontal distribution of cells during growth periods. The spatial distribution of living populations of phytoplankton is clearly governed by water turbulence (Harris, 1986). In particular, the horizontal component

of water movement in both marine and freshwater bodies is far greater than the speeds at which flagellated cells can propel themselves. Thus, the horizontal distribution of phytoplankton is controlled by physical factors related to energy inputs into the physical medium (Harris, 1986). Many individual phytoplankton species exhibit opportunistic strategies characterized by rapid growth during periods of nutrient availability, followed by equally rapid decline in growth. Studies of temporal variation in total phytoplankton biomass for various bodies of water show considerable variation in biomass at a given site over an annual period (Harris, 1983). At Hamilton Harbour, Ontario, Harris (1983) has demonstrated that patterns in water mixing are strongly correlated with phytoplankton diversity. This variation ranges from almost monospecific blooms during constant mixing to high diversity during intermittent mixing. Harris's work confirms the conclusions of Margalef (1978) and Reynolds (1980), who demonstrated that changes in the physical conditions of the water column determined changes in community diversity.

In the North Sea, the location of peak phytoplankton biomass shifts over time without resulting in a clear onshore-offshore gradient (Colebrook & Robinson, 1965). In the North Atlantic, there is a seasonal shift in peak chlorophyll production away from the shelf, but the duration of this peak decreases in deeper water (Robinson 1970). The importance of this work is, that the physical regime can determine phytoplankton diversity, and that this diversity may change at a specific location over an annual cycle. In the fossil record, therefore, we cannot assume that diversity of sedimented marine cysts is necessarily correlated with an offshore-onshore gradient. The notion that outer shelf environments are more nutrient-rich and tend to produce high diversity assemblages, whereas nearshore environments are stressed and produce monospecific blooms, is probably too simple a picture of the community dynamics of phytoplankton. This is especially true, if one considers the population dynamics of freshwater habitats and the possibility that freshwater sources can contribute to paralic deposition sites (DeSimone, 1988; Muller, 1959).

For all populations of phytoplankton there is fallout due to sinking, representing a continual kinetic loss of biomass to the population. The fallout consists of viable vegetative cells, physiologically dead vegetative cells and cysts. The proportion of these three components varies over the life of an individual phytoplankton bloom with physiologically dead cells and cysts increasing toward the end of a bloom. (Some chrysophytes, however, produce cysts at the height of population growth (Sandgren, 1988), so not all ecological strategies call for maximal cyst production near the end of a blooming period.) The distinction between low-level stochastic fallout and wholesale conversion to encystment is important for predicting the distribution of palynomorphs in bottom sediments. Stochastic loss should produce a continuous fallout of cells, the quantitative distribution of which should track the living population density. In non-bloom-forming species, the final spatial distribution of cells in sediment should be uniform within time scales associated with clastic sedimentation. With the onset of encystment, cell populations produce increasingly higher numbers of cysts which sink out of the euphotic zone. Since the onset of encystment can occur rapidly (at the scale of hours to days for many species), the production of spikes of sedimented cysts and physiologically dead cells should result in a punctuated sedimentary distribution of cells derived from these populations.

Cyst formation serves as a mechanism for survival during periods when growth conditions (nutrient availability and temperature) are too poor to sustain continued population growth (Sandgren 1988). In freshwater and shallow marine environments, encystment works only when the vegetative excysting cell can re-enter the euphotic zone at a later date. In waters where the sediment water interface is well below the euphotic zone, cysts which become entrapped in bottom sediment may be unable to repopulate surface waters. Because of this physical limitation, the life strategies of pelagic phytoplankton must be fundamentally different from those in shallow-water environments. Deep-water phytoplankton which form cysts must guarantee that cysts will not sink below a recovery depth before excystment occurs. This may be why dinoflagellate populations in shallow-marine temperate waters contain a higher percentage of cyst-forming species than populations from open tropical ocean waters (Wall et al., 1977). Deep-water cysts which are viable will excyst before sinking out of the upper portion of the water column. After excystment, the cyst wall will continue to sink and act as a particulate organic particle. The recovery of deep-water dinocysts with opened archeopyles is evidence of this particular strategy.

In general, Precambrian palynomorphs show little evidence of excystment structures. In cherty iron formation biofacies, *Huroniospora* is the only taxon that has preserved excystment-like features (Strother & Tobin, 1987). None of the taxa yet described from the Precambrian silicified carbonate lithofacies shows any sign of an excystment structure.

Precambrian palynomorphs with preserved excystment features are rare. For example, in Timofeev's monumental work on lower Paleozoic and upper Precambrian palynomorphs (Timofeev, 1959), of a total of 651 figured specimens, only one shows a possible excystment opening, consisting of a simple partial slit in the wall. Vidal (1976) records two aperturate species out of 28 taxa from the late Precambrian Visingsö Beds of southern Sweden. Vidal and Knoll (1983) review aperturate taxa in the Precambrian palynoflora and their discussion mentions

only five species that have distinct openings. They comment that median splitting is occasionally found in species of *Kildinosphaera* and 'other Proterozoic acritarchs' (Vidal & Knoll, 1983, p. 266). In general, the low number of figured specimens in the literature with excystment features indicates that most preserved Precambrian palynomorphs are failed cysts in which excystment did not occur.

Al-Ameri (1986) distinguishes five types of excystment structures found in lower Paleozoic acritarchs, but he does not give an indication of how common excystment structures are in lower Paleozoic taxa. My impression is that excystment structures are more prevalent in lower Paleozoic taxa than in the Precambrian. An increase in the abundance of preserved excystment structures could reflect an adaptive response to evolution in the euphotic zone.

Shallow-marine and freshwater phytoplankton will produce cysts which may remain viable in the upper layers of sediment prior to excystment. Evitt (1985, p. 11) cites reports of dinocyst viability in lake sediments for up to 16.5 years. The cysts of freshwater chlorophytes depend upon water turbulence to resuspend their cells in the euphotic zone (Happey-Wood, 1988), and the same is probably true for most freshwater and shallow-marine phytoplankton. Even algae with flagellated vegetative cells require cysts to be resuspended at or above the sediment interface in order for excystment to be successful. As with deep-water forms, successful excystment should leave behind the vacated cyst which may be preserved in the sedimentary record as organic particulate matter. The probability of permanent burial of unopened cysts must be higher in shallow-water species however, since their normal function includes shallow deposition in bottom sediment.

From the above discussion, it seems clear that the distribution of phytoplankton cysts in an accreting sedimentary environment is not readily predictable for a variety of reasons. Cysts may be formed continually, throughout the growth of a population, or they may form rapidly, generating a short-lived spike at the end of a bloom. Encystment rates expressed as cyst/thecae ratios in living marine dinoflagellates vary from 1:2 to 1:500 (Dale, 1976). Such a wide range makes it extremely difficult to map sedimented cyst density back to living population diversity. In addition to the differing encystment rates, there is a continual stochastic fallout from the euphotic zone into bottom sediments which should map the size of the total living phytoplankton community. Even though this continual cell loss will trace fluctuations in population size, the signal will degrade increasingly through time as mixing occurs during sedimentation. In order to determine whether the final distribution of cyst fallout can trace temporal variation in phytoplankton numbers in living populations, we need to

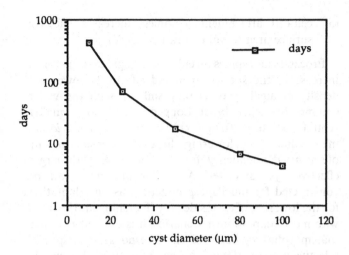

Figure 23.2. Fallout time for a hypothetical spherical cyst in 100 m water depth.

Table 23.1. *Calculated fallout time differences (δ_t) for Precambrian sphaeromorph acritarch taxa described in Vidal (1976)*

Taxon	Diameter range (μm)	δ_t (days)
Favosphaeridium favosum	33–144	16
Kildinella hyperboreica	40–77	8
K. cf. *sinica*	35–64	10
Protosphaeridium cf. *flexosum*	22–30	17
P. laccatum	20–68	41
P. papyraceum	46–68	5
cf. *Stictosphaeridium* sp.	30–150	19
S. cf. *sinapticuliferum*	50–80	4
S. verrucatum	18–32	38
Trachysphaeridium apertum	15–42	65
T. laminaritum	41–57	5
T. levis (Lop.) Vidal	10–100	176
T. laufeldii	42–50	3
T. timofeevii	25–115	27
Trematosphaeridium holtedahlii	28–64	18

examine the end points of a possible spectrum of variation in the timing of cyst fallout.

The first such end point is the case in which cyst-forming phytoplankton form a temporally continuous series of populations which overlap to maintain a constant population density. In this case, stochastic fallout could produce a uniform rain of cysts, and the end result will be a spatially random distribution of cysts in bottom sediment. Sediment cyst density would actually record the inverse of the mineralogic sedimentation rate.

On the other extreme, imagine two population spikes separated in time by a value, T_s. Each spike represents a signal which theoretically can be tracked by cyst density recorded in bottom sediment in which T_s is reflected in a thickness of sediment separating the depositional records of the two cyst-forming events, t_s. If the two spikes are to be recorded as separate events, clastic sedimentation rates must be high enough to accumulate lamination between the two events. In addition, variation in diameter may affect cyst settling rates, causing the two spikes to overlap in time.

In order to see the effect of variation in cyst diameter upon signal overlap, we can calculate settling times for cysts of different diameters and compare these values with realistic times between phytoplankton blooms, T_s. Assume that cysts fall with a terminal velocity determined by Stokes' Law:

$$\hat{v} = 2/9 g r^2 (\rho' - \rho) \mu^{-1}$$

where $\hat{v}$ is the terminal velocity, g is the acceleration due to gravity, r is the radius of a falling sphere, $(\rho' - \rho)$ is the excess density, and μ is the viscosity of the medium (Hutchinson, 1967). Given a system where $(\rho' - \rho)$, g and μ are constant, $\hat{v}$ will vary as a function of r^2:

$$\hat{v} = K r^2,$$

where $K = 2/9 g (\rho' - \rho) \mu^{-1}$. For a sphere with a diameter of 100 μm, in water with viscosity of 0.01 poise and excess density of 0.05:

$$\hat{v} = 2.7 \times 10^{-1} \text{ mm s}^{-1}$$

(based on data from Hutchinson, 1967; Reynolds, 1984, table 8). Under equivalent physical conditions, a spherical cyst with a diameter of 10 μm has a terminal velocity:

$$\hat{v} = 2.7 \times 10^{-3} \text{ mm s}^{-1}$$

These values are comparable to measured diatom sinking velocities of 1.27×10^{-3} to 0.35 mm s^{-1} reported by Smayda (1970) and summarized in Harris (1986).

For a water depth of 100 m, a reasonable figure for average water depth over continental shelf, fallout times can now be calculated as a function of cyst diameter (Fig. 23.2). Since the calculation assumes no turbulence, the fallout times in Figure 23.2 represent minimum values. We can distinguish between two separate phytoplankton blooms, only if the difference in fallout time for two populations is less than T_s. A 10 μm diameter cyst has a fallout time of over 400 days, thus in shelf environments, even under conditions of extremely low turbulence, small cyst fallout may overlap annual production cycles of larger cyst formers.

Even though many common Precambrian sphaeromorph species have a wide range in diameter, the resultant difference in fallout times is usually not very long. Table 23.1 shows diameter ranges and differences ($\delta_{t,100}$) in 100 m fallout times for species level taxa of sphaeromorphs compiled from the systematic description of Vidal (1976). The mode for $\delta_{t,100}$ is only 17 days, the mean is 30 days (although the same data set with the *Trachysphaeridium levis* value of 176 removed is 20 days which is a better estimate of a general figure). These numbers decrease if water depth is decreased, so deposition in nearshore deposits is potentially even more conducive to signal preservation.

In conclusion, the deposition of sphaeromorph acritarchs is theoretically capable of recording seasonal phytoplankton blooms in cyst-forming biological species. This assumes low turbulence, depths of 100 m or less (shelf), and a matrix sedimentation rate high enough to preserve signal separation. Bioturbation is also a potential factor in the preservation of a final palynomorph distribution in shale and siltstone. In modern oceans, bioturbation causes signal degradation at a scale of 10^3 to 10^5 years (Schiffelbein, 1984). However, prior to the origin of the Metazoa, individual laminae are retained in shales (Byers, 1976). Before the late Proterozoic, the degree to which palynomorph distribution reflected temporal heterogeneity of phytoplankton in the water column would have been principally a function of current mixing, turbulence, and sedimentation rate.

Vidal (1976) notes that the distribution of species such as *Octoedryxium truncatum* Rudavskaja, *Trachysphaeridium apertum* Vidal, *T. levis* (Lopukhin) Vidal, and *Pterospermosimorpha ? densicoronata* Vidal is 'anomalous' in that they tend to appear and disappear suddenly and are not related to sedimentary texture. He postulates that their distributions may track algal blooms in shallow marine environments (Vidal, 1976, p. 43). *T. apertum* is particularly significant in this regard because it retains a large circular to elliptical aperture which Vidal suggests is an excystment structure. In the one sample where *T. apertum* is common, it comprises 99% of the sample. Based on cyst diameter, this species has a 100 m fallout difference of $\delta_{t,100} = 65$ days, well within the normal temporal spacing in a biannual bloom-forming species of modern phytoplankton.

Adaptive constraints upon phytoplankton cyst morphology

The vegetative cells of photosynthetic planktonic organisms live under the adaptive limitation of having to remain in the photic zone. Vegetative cells have five adaptations to this constraint:

(1) undulapodia (flagella) to provide motility;
(2) intercellular gas-filled spaces to provide buoyancy (in cyanobacteria);
(3) spines (especially in dinoflagellates);
(4) extracellular mucilage (in diatoms, rhodophytes and cyanobacteria);

(5) cell-cell attachment (colonies) increases effective surface area to volume ratios (SA/V).

Processes and spines on eukaryotic algal cysts cause an increase in the surface area and effectively lower the density of algal cysts as they sink through the water column (Margalef, 1978). Long processes on a smaller central body (e.g., *Tunisphaeridium* and *Baltisphaeridium*) may reduce ϑ by lowering the excess density $(\rho' - \rho)$, offsetting the tendency for ϑ to increase with larger effective cyst diameter. A similar argument can be constructed for mucilage production, as an adaptation for the lowering of excess density (Hutchinson, 1967). As with most adaptationist scenarios, it is difficult to prove that morphology has evolved in response to a specific selective pressure (Gould & Lewontin, 1978). The nearly universal presence of spines and elaborate processes in unrelated marine and freshwater phytoplankton cysts from mid-Proterozic to Recent time attests to the functional value of this morphotype.

The excystment phase of the phytoplankton life cycle represents a fundamental adaptation to life in the photic zone. The physical constraints upon cyst morphology are felt across phylogentic boundaries, consequently phytoplankton cysts can be recognized in the fossil record even though the taxonomic affinities of various cysts remain unknown. The overall shape of phytoplankton cysts is quite variable, ranging from spherical (Sphaeromorphitae) to axial with a single mirror plane (Diacrodiaceae *sensu* Timofeev) to leiofusid form, to irregular. Thus the imposition of symmetry in cyst shape is not a fundamental constraint of the planktonic habitat, and shape alone cannot be used as a criterion for distinguishing benthic from planktic species of microorganisms.

The recognition of plankton in pre-Devonian rocks

In order to assess the impact of phytoplankton during pre-Devonian time, we must first establish a set of criteria for the recognition of phytoplankton in both shale and chert lithofacies. Shales pose less of a problem in this regard because palynomorphs are traditionally associated with argillaceous sediments. The acritarchs macerated from Precambrian shales are typically spherical to sub-spherical in overall shape. They occur with or without processes, and they occur with or without a surrounding organic membrane. The general similarity in form between lower Paleozoic acritarchs and late Precambrian species is strong evidence for the general planktonic habit of the Precambrian forms. The evidence for their planktonic affinity is summarized by Vidal and Knoll (1983), who add that the distribution of phytoplankton in cherts tends to be random and thus different from that of benthic microorganisms, which are localized along bedding planes.

Cyanobacteria present a special problem with regard to recognizing planktonic forms in pre-Devonian time. Cyanobacteria are traditionally considered to be a rare component of extant marine plankton, and the assumption that fossil cyanobacteria were also rare in marine deposits has prejudiced their interpretation as sedimentary particles. Even though only three genera of planktonic marine cyanobacteria are known (*Trichodesmium* = *Oscillatoria*, *Gloeocapsa*-like cells and *Synechococcus*), the small coccoids alone constitute as much as 6% of the total microbial plankton in a biomass in the open ocean (Johnson & Sieburth, 1979). Direct counts of coccoid cyanobacteria in sea water range from 6.0×10^2 to 1.4×10^4 cells per cm^3 (Johnson & Sieburth, 1979). It is remarkable that this same density range is reflected in counts of cyanobacteria preserved in late Precambrian cherts which range from 2.0×10^2 for planktonic spheroids to 1.6×10^4 for *Eosynechococcus thuleënsis* (Strother *et al.*, 1983). In spite of the figures cited above, which indicate the possibility of finding low diversity assemblages of marine cyanobacterial plankton in modern sediments, the cyanobacterial component of Phanerozoic rocks is essentially nil.

The situation is quite different in palynological macerations from upper Proterozoic sections where numerous authors have described oscillatoriacean filaments accompanying acritarchs in palynological macerations (Butterfield *et al.*, 1988; Damassa & Knoll, 1986; Hofmann & Aitken, 1979; Peat *et al.*, 1978; Tynni & Donner, 1980; and many others). The basic question surrounding this observation is whether the occurrence of cyanobacteria in palynological macerations represents (1) a component of normal marine phytoplankton, (2) an allochthonous benthic component which is washed into deeper water deposits, or (3) autochthonous benthos formed in paralic deposits. If the first scenario is true, it would indicate that cyanobacteria were ecologically more diverse during late Proterozoic time than they are today. If Precambrian cyanobacteria can be shown to be fundamentally benthic (littoral) or freshwater in distribution (as they are today), then their presence in palynological assemblages is of great value in determining depositional environments during the late Proterozoic. Essentially, their presence in an assemblage would indicate either proximity to freshwater or coastal habitats or, if cyanobacteria can be shown to be mat-forming by their spatial distribution within sedimentary laminae, they would indicate paralic depositional habitats.

Knoll and Swett (1985) used the presence of cyanobacterial filaments in palynological preparations of Upper Riphean rocks from Spitsbergen as indicators of very nearshore to coastal lagoonal affinity. They note that samples in which filaments are abundant (comprising up to 40% of the assemblage) are generally lacking in larger sphaeromorph acritarchs. In samples where filaments are rare, larger acritarchs constitute a more diverse and more dominant component of the total assemblage. They use this association to conclude that diverse assemblages with rare filaments probably represent the 'most off-shore or "normal" marine setting' within the sequence (Knoll & Swett, 1985, p. 458). Their data show the following correlation between filament abundance (indicated as rare, common, or abundant) and total diversity (number of taxa, D):

filaments abundant ... $D = 5.7$ taxa
filaments common ... $D = 6.8$ taxa
filaments rare ... $D = 7.5$ taxa.

Clearly the correlation between filament abundance and overall phytoplankton diversity is not strong, although it is consistent.

In a study of the upper Proterozoic and Lower Cambrian sequence in East Finnmark, Vidal (1981) notes the presence of cyanobacterial filaments. The average taxonomic diversity for samples with filaments present is 10.2 taxa, whereas for samples without filaments, the mean diversity is 5.0. For this sequence, filament presence is correlated with *higher* acritarch diversity. Although Vidal does not comment on this correlation, he cites Ivanovskaya & Timofeev (1971) and Vidal (1976), who indicate that acritarch diversity is salinity controlled, and that consequently, more diverse assemblages are perhaps derived from normal marine environments. This conclusion, with respect to the preservation of cyanobacterial filaments, is the opposite of that reached by Knoll and Swett (1985).

Several upper Precambrian deposits which on lithologic grounds are thought to have been deposited in nearshore environments contain diverse assemblages of cyanobacterial filaments in addition to leiospherid acritarchs. Zhang (1982) describes from Torridonian strata in Scotland a low diversity assemblage dominated by filamentous taxa. The Torridonian is generally thought to be fluvial, although Selley (1965) considered the grey shale from which the palynomorphs are described to be lacustrine. Germs *et al.* (1986) describe a filament-rich palynological assemblage from apparently quiet subtidal to fluvial settings in the uppermost Proterozoic Nama Group (Namibia). Cyanobacterial filaments, vendotaenids and simple leiospheres dominate these nearshore deposits. There is thus a body of evidence which suggests that cyanobacteria occupied planktic niches in marine and paralic habitats during the Proterozoic, but that their abundance in these habitats has since declined dramatically.

The earliest known fossil plankton with possible cyanobacterial affinity are *Leptoteichos golubicii* Knoll, Barghoorn & Awramik and *Eosphera tyleri* Barghoorn, which are found in the Gunflint Iron Formation and other cherts associated with iron formation at about 2 Ga (Barghoorn & Tyler, 1965; Knoll *et al.*, 1978). The spherical morphology of these organisms constitutes evidence of their planktonic habit (Knoll *et al.*, 1978). Both *Eosphera* and *Leptoteichos* are classified as *incertae sedis*, reflecting the possibility that they might not be cyanobacteria. Certainly, *Eosphera tyleri*, with its outer ring of ellipsoidal cells, is unlike any known organism. Nevertheless their sphericity and spatial distribution attest to their planktonic affinity.

Strother *et al.* (1983) were able to recognize a variety of planktonic colonial cyanobacteria in the intertidal to supratidal deposits of a late Proterozoic sequence in northwestern Greenland. They described two genera, *Gyalosphaera* and *Coleogleba* which show strong morphological affinities with the extant freshwater chrococcalean genera, *Gomphospheria* and *Microcystis*. The Precambrian forms, however, were deposited in coastal habitats which preserved a variety of microbial laminates and stromatolites, indicating a wider ecological niche for chrococcalean phytoplankton during the late Proterozoic.

The long stratigraphic record of benthic cyanobacteria found in mat-forming freshwater, littoral and shallow subtidal habitats is well established. Thus, filamentous cyanobacteria, particularly oscillatoriacean species, have persisted in these habitats throughout most of earth history. Their recovery in palynological preparations of Precambrian rocks constitutes, in many instances, evidence of autochthonous or allochthonous input from these habitats. The common occurrence of cyanobacterial filaments mixed with eukaryotic phytoplankton in what are clearly marine sequences of Proterozoic age, however, leads to the conclusion that oscillatoriacean species may also have occupied marine planktonic niches during the Proterozoic. This conclusion is strengthened by the present-day occurrence of *Trichodesmium* in the Sargasso Sea. But it has the consequence of diluting the strength of arguments which use the presence of filaments as indicators of nearshore deposition. Chrococcalean cyanobacteria appear to have become more restricted in their habitat preferences over time, with few remaining in the planktonic marine realm after the Early Cambrian decline of *Bavlinella*.

Paleoecological interpretations based on pre-Devonian phytoplankton

The recovery of algal cysts and cyanobacteria from sediments should in theory be useful in the reconstruction of former environments. If the environmental source of palynomorphs can be identified with some degree of certainty, their presence in a deposit can then be used for the interpretation of other depositional settings. Algal cysts which act as sedimentary particles originate from three basic source regimes: (1) terrestrial runoff, in which subaerial and freshwater cysts are brought to ocean margins via rivers which empty into oceans; (2) airborne cysts which are derived from resuspended marine forms and subaerial (terrestrial algae); (3) marine phytoplankton, in which cysts represent fallout from a horizontally dispersed source in the upper ocean layer (the euphotic zone).

The presence of airborne algae in coastal air samples has been most recently reviewed by Lee and Eggleston (1989), who identified ten genera of cyanobacteria, diatoms and chlorophyta in their samples. ZoBell (1942), citing the work of Erdtman (1937), estimated that the contribution of terrestrial pollen grains to the surface marine air was 0.7 to 18 grains per m^3 of air as compared to 18 000 grains per m^3 over land. If we assume that these figures also represent the relative contribution of terrestrial airborne algae over land and sea, then the fraction of airborne cyst fallout into the open oceans would be 1.4×10^{-5} to 1×10^{-3} times the terrestrial airborne component. Thus, far fewer airborne algal spores fall into open oceans than over land. In addition, much of the airborne fraction consists of bacterial and fungal spores, which are not likely to be found in pre-Devonian sedimentary deposits. Although the cases of rare aerial transport of terrestrially derived algal cysts may be of great significance in the ecology of algal dispersal, their mixing with marine phytoplankton should be insignificant, representing, at maximum, a potential density of 2×10^{-5} cysts per cm^3, based on Erdtman's figures for pollen grains. This is nine orders of magnitude less than typical phytoplankton densities of 10^4 cells per cm^3. Thus, we are left with freshwater runoff and marine surface waters as the principal sources of pre-Devonian cysts.

In principle, the factors which control the depositional distribution of terrestrially derived pre-Devonian algal cysts should follow those determined for more recent examples of pollen grains and spores. These factors have been reviewed by Traverse (1988) and need not be repeated here. It is important to recognize, however, that the correlation between percentage of terrestrial forms and distance from shore is not a simple linear function.

Palynomorphs from marine sources vary in relative concentration according to ecological factors governing the growth and diversity of living communities, the nature of fallout, and mixing during bottom sediment transport in both paralic and open shelf areas (Fig. 23.1). Dale (1976), in a comparative study of living marine dinoflagellates and their counterparts (cysts) in Recent sediments, found that only about 20% of living species were represented by sedimented cysts. He also noticed a correlation between sediment size and cyst density which was greatest in silts and generally increased with water depth. He also discussed rates of cyst formation in living populations noting that cyst/thecae ratios can vary from 1:500 to 1:2. This wide range in variation severely inhibits the ability to reconstruct true taxonomic diversity from palynological assemblages of marine phytoplankton. It does imply, however, that true phytoplankton diversity may be much higher than is indicated by the number of species preserved in the fossil record. Theoretically, it may be possible to estimate this factor by observing phytoplankton preserved in cherty deposits which allow for the preservation of both thecae and cysts. However, no such calculation has yet been made.

Other than their application in biostratigraphy, pre-Devonian palynomorphs have been most frequently used in the interpretation of depositional environments (Knoll, 1984; Jacobson, 1979; Vidal, 1976; and many others). The most common assumptions used in paleoecological interpretations are that cysts have a marine origin, and that low diversity assemblages occur in nearshore waters with restricted marine access, whereas higher diversity assemblages occur in offshore shelf environments with normal marine salinities and higher nutrient levels. This correlation between cyst diversity and proximity to land is most strongly presented by Knoll (1984) in his study of depositional environments in the upper Proterozoic Hunnberg Formation in Spitsbergen, in which sedimentological and stratigraphic evidence support the ecological interpretations based on palynomorph diversity. Both Vidal (1976) and Knoll (1984) cite the observations of Ivanovskaya and Timofeev (1971), who relate lower palynomorph diversity to reduced salinity based on boron content of their samples.

Richardson and Rasul (1990) use palynomorph assemblages from the upper Silurian of England and Wales to construct a series of microplankton phases which characterize different biotic associations. The changes in these phases are correlated with sedimentological evidence showing changes in sea level and storm-based sedimentation events. Recognition of terrestrial derived organic material in some of their samples strengthens their environmental interpretations.

Dorning (1981), in a study of acritarchs from the Silurian Welsh Basin, showed that acritarch diversity and general morphology can be divided into three regimes: (1) a low diversity nearshore assemblage dominated by sphaeromorphs but including *Veryhachium*, *Micrhystridium* and *Diexallophasis* as common genera; (2) a high diversity, offshore shelf assemblage dominated by spinose forms; (3) a low diversity, deep-water assemblage with a similar composition to the nearshore assemblage.

His generic diversity values range from 15 taxa (nearshore) to 17 taxa (shelf) and seven taxa (deep water) (Dorning, 1981, fig. 3.4). These numbers show that in the

absence of additional geologic information, acritarch diversity is a poor indicator of position from shoreline. Dorning's scheme has been widely cited, but Colbath (1985) warns that diversity expressed in terms of simply counting taxa is inadequate for distinguishing between assemblages of 'low' and 'moderate' diversity, because this method does not accommodate variance in sample size. In addition, he demonstrates that variations in processing procedure can affect diversity. Colbath (1985) recommends using species richness, d (Margalef, 1951), or information content, *H*, as better indicators of diversity in palynomorph assemblages.

Williams and Sarjeant (1966) conclude that acritarchs, by themselves, cannot be used to determine distance from shoreline or water depth. In combination with pollen and spores which come from known terrestrial sources, however, quantitative analysis of palynomorph assemblages can be used to trace ancient shorelines. Their results imply that in the absence of known terrestrial indicators, acritarchs may not be useful for determining water depth. Jacobson (1979) was able to distinguish three marine depositional environments (nearshore shallow water, shoal, and open-sea), each with characteristic acritarch species from Ordovician strata in the eastern USA. He first used sedimentological and stratigraphic data to identify these environments. He then confirmed the studies of Wall (1965) and Staplin (1961), who concluded that simple, smooth-walled leiospherids characterize nearshore environments. Leiospherids are found throughout the entire range of depositional settings, however, so, in effect, it is the absence of a more diverse acanthomorphic assemblage which typifies nearshore deposition. Given the return of leiosphaerid dominated assemblages in Dorning's deep-water Silurian sections, the use of leiospheres as depth indicators is questionable in the absence of additional sedimentological data.

In a study of acritarch presence and absence from clastic deposits of lower Silurian age in central Pennsylvania, Smith and Saunders (1970) claimed that acritarch abundance increased offshore. The percentage of palyniferous samples increased from 0% in fluvial deposits (Shawangunk and Tuscarora formations), to 32% in marginal marine strata (Clinton Beds), to 65–84% recovery in the open marine Rose Hill Shale. That acritarchs are indicators of marine deposition is not in question, but more recent interpretations of presumed fluvial deposits of the Tuscarora Formation now consider them to be nearshore marine (Cotter, 1983). Sedimentologic and petrographic evidence in combination with ichnofacies analysis (Freile, 1988) indicates that these deposits were laid down in a marine setting. The presence of *Rusophycus didymus* (a trilobite resting trace) in sandstones of the Tuscarora strongly reinforces Cotter's interpretation. Clinton beds and shale interbeds in the Shawangunk Conglomerate become more indurated eastward from the

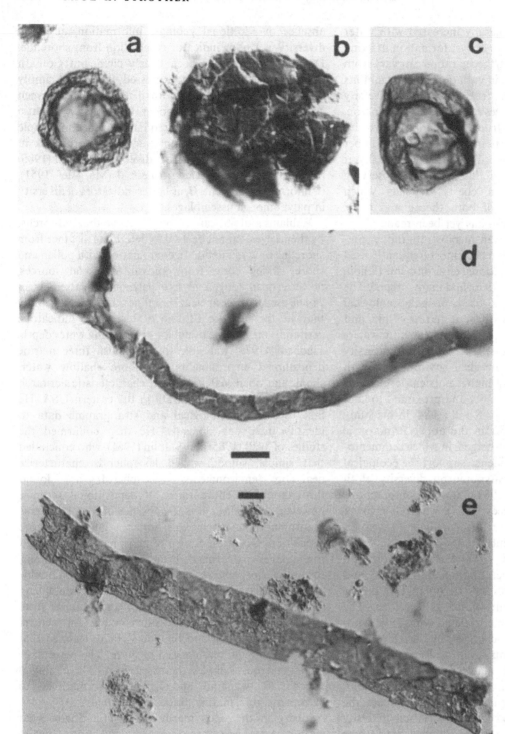

Figure 23.3. Palynomorphs from the Nonesuch Shale (1 Ga) which are thought to be of lacustrine origin. (a) *Trachysphaeridium* sp.; (b) *Kildinosphaera*? *chagrinata* Vidal; (c) *Leiosphaeridia* sp.; (d) smaller cyanobacterial (?) sheath; (e) large sheath attributable to *Siphonophycus*. Scale bars = 10 μm. Bar in d is for a–d. All specimens are from Manido Falls locality (Presque Isle River), northern Michigan (PKS collection number, MF-2, 1977). All photographs were taken under Nomarski Interference.

Susquehanna River, and this is reflected in increasing carbonization of organic fragments which are common in these deposits. Acritarch preservation in the first three (out of nine) localities sampled by Smith and Saunders is therefore poor, based on diagenetic loss of organic matter (Strother, unpublished observation). The inference, based on negative results, that environment of deposition is controlling acritarch distribution in this example is, therefore, suspect.

Ironically, some of the deposits which Smith and Saunders cite as barren, are now known to contain the best preserved assemblages of terrestrially derived, early land plant spore and cryptospore remains (Johnson 1985; Strother & Traverse, 1979). Within the Tuscarora Formation at Mill Hall in central Pennsylvania, arenaceous facies contain abundant indicators of marine deposition (petrographic analysis, ichnofacies analysis) but the shales with which they are interbedded contain a terrestrial palynological signature (DeSimone, 1988; Johnson, 1985). Based on the work of Cotter (1983), Johnson (1985), DeSimone (1988) and Freile (1988), I conclude that the Tuscarora represents a nearshore setting in which subtidal coastal sands are intermixed with fluvially derived muds. At the top of this barrier bar, coastal sequence, the shales gradually take on a marine signature, preserving a marine macrofauna (Swartz, 1923) and marine acritarchs (Cramer 1969).

Thus, it is quite clear, in the case of the Tuscarora shales, that nonmarine palynomorphs can be preserved in shales within deposits that are clearly nearshore marine based upon the study of primary depositional features preserved in the interbedded sandstones. The mixing of marine and terrestrial elements is not unusual in nearshore settings, however, as stated earlier, prior to the advent of a terrestrial flora, the recognition of terrestrially derived palynomorphs is quite difficult. The example from the Tuscarora Formation works, because we can recognize a terrestrial signal from the cryptospore component which dominates the palynological assemblage (Strother & Traverse, 1979; Johnson, 1985).

Since cryptospores and trilete spores first occur in the upper Ordovician (Gray *et al.*, 1982; Nøhr-Hansen & Koppelhus, 1988), prior to this time the terrestrial palynological signature in marine deposits cannot be easily assessed on palynomorphs alone. A low diversity palynomorph assemblage was recovered by Strother (1980) from the 1 Ga Nonesuch Formation of northern Michigan, an organic-rich, silt dominated formation thought to represent lacustrine deposition (Elmore *et al.*, 1989). It includes the sphaeromorphs, *Leiosphaeridia* spp. (Figs. 23.3c), *Kildinosphaera ? chagrinata* Vidal (Fig. 23.3b), *Trachysphaeridium* spp. (Fig. 23.3a), and rare sheaths (Fig. 23.3d,e). The assemblage is somewhat similar to one described from the Edicarian-age Tent Hill Formation from the Stuart Shelf in South Australia by Damassa and Knoll (1986). The Tent Hill assemblage consists entirely of simple sphaeromorph acritarchs mixed with 'sporadic' non-septate filaments (sheaths). Damassa and Knoll acknowledge that a nearshore paleoenvironmental interpretation of this assemblage cannot be excluded, but they seem to favor

an open marine habitat with an allochthonous filamentous component.

The sedimentology and regional geological settings of the Nonesuch and Tent Hill formations seem to carry more weight in the determination of depositional environments than does assemblage composition. Certainly, on the basis of palynomorph composition alone, both the Nonesuch and the Tent Hill could be construed as either freshwater or lagoonal. If the sheaths of filamentous cyanobacteria are derived from terrestrial or littoral mats, then, their presence in a low diversity sphaeromorph assemblage would be consistent with this interpretation.

Zhang (1982) described another low diversity acritarch assemblage from the fluvial, upper Riphean Torridonian Formation in Scotland. His treatment of the palynomorph taxonomy in this assemblage involves considerable lumping to produce a complex, pleiomorphic life cycle for a single hypothetical cyanobacterium, *Torridoniphycus lepidus*. Given the statement by Diver (1980), that the cryptarch assemblage from the Torridonian is both diverse and includes an autochthonous benthic component, consideration of the Torridonian as a model of a fluvial Precambrian assemblage remains inconclusive.

Conclusions

Although the factors which determine the distribution of living cyst-forming phytoplankton in the water column are complex, it appears that the depositional fallout of algal cysts into bottom sediments falls into two categories: a stochastic background fallout of mixed populations; and the rapid fallout of cysts from algal blooms. It is significant that, theoretically, based on Stokes' Law, spherical algal cysts can descend in the water column rapidly enough, even in waters of average shelf depth, to produce monospecific assemblages of buried cysts derived from a single bloom. The likelihood of preservation of such biotic events is increased as sedimentation rates go up and as water depth decreases. Also, the lack of bioturbation throughout most of the Precambrian increases the probability of preserving transient biotic events in the stratigraphic record. The interpretation of monospecific palynomorph assemblages must rely on physical data in addition to biological parameters, because extant blooms can occur in a wide range of freshwater and marine settings, not just in the inner shelf areas. Also, palynomorph assemblages of mixed diversity may record both stochastic fallout and bloom-forming taxa, especially in deeper shelf areas and with smaller diameter cysts which may take over an annual cycle to settle out. Sedimentation times increase with increasing turbulence.

The above conclusions support the conjecture of Vidal (1976) that some of his monospecific assemblages from the late Precambrian Visingsö Group could represent the sedimented remains of cyst-forming algal blooms in nearshore marine settings.

Few Precambrian palynomorph forms retain excystment features. This is especially true for assemblages preserved in chert facies. Two possible explanations exist for this paradox: most palynological assemblages were deposited under exceptional ecological conditions, and they therefore represent catastrophic losses of both vegetative and encysting cells; or the normal process of continuous cyst formation includes high losses, represented by failed cysts in the sedimentary record. A one percent survival rate is sufficient for recovery in *Melosira* (a diatom) (Jewson *et al.*, 1981). *Ceratium* (dinoflagellate) counts have been shown to fluctuate by as much as eight orders of magnitude in vegetative cells per cm^3 (Heaney *et al.*, 1983). If the above figures are normal, one survivor cell per cm^3 of sediment is capable of generating a subsequent population density of 10^4 cells per cm^3 in a 100 m overlying column of water within a growth season. Therefore, it seems likely that the burial of failed cysts is, in fact, the more common circumstance, and lack of preserved excystment features cannot be used to justify hypotheses of exceptional sedimentary conditions.

Many authors have assumed that the diversity of palynological assemblages is related to water depth or distance from shoreline. In situations where a terrestrial component can be identified, this supposition is probably valid. In situations where terrestrial components cannot be identified, however, such ecological arguments cannot be affirmed without additional evidence from sedimentological and regional geologic data. Even evidence from sedimentology can be misleading, as in the Tuscarora Formation (lower Silurian), where sedimentological data from shaly and sandy facies are at variance. Assemblage diversity must be treated in a more quantitative manner if palynology alone is to be used effectively in discerning paleoenvironments (Colbath, 1985).

With regard to Precambrian sediments, the ecological role of cyanobacterial sheaths must be clarified before they can be used to discern depositional environments. It is clear that the chroococcalean phytoplankton once occupied a wider variety of habitats (Strother *et al.*, 1983). However, the same is not necessarily true for filamentous cyanobacteria. The taxonomy of cyanobacterial sheaths found in palynological preparations must be clarified before the distribution and diversity of filamentous taxa can be used to help analyze depositional environments.

References

Al-Ameri, T. K. (1986). Observations on the wall structure and the excystment mechanism of acritarchs. *Journal of Micropalaeontology*, **5**(2), 27–35.

Barghoorn, E. S. & Tyler, S. (1965). Microorganisms from the Gunflint Chert. *Science*, **147**, 563–77.

Brush, G. (1989). Rates and patterns of estuarine sediment accumulation. *Limnology and Oceanography*, **34**(7), 1235–46.

Butterfield, N. J., Knoll, A. H. & Swett, K. (1988). Exceptional preservation of fossils in an Upper Proterozoic shale. *Nature*, **334**, 424–7.

Byers, C. W. (1976). Bioturbation and the origin of metazoans: Evidence from the Belt Supergroup, Montana. *Geology*, **4**, 565–7.

Colbath, K. G. (1985). A comparison of palynological extraction techniques using samples from the Silurian Bainbridge Formation, Missouri, U.S.A. *Review of Palaeobotany and Palynology*, **44**, 153–64.

Colebrook, J. M. & Robinson, G. A. (1965). Continuous plankton records: seasonal cycles of phytoplankton and copepods in the north-eastern Atlantic and the North Sea. *Bulletin of Marine Ecology*, **6**, 123–39.

Cotter, E. (1983). Shelf, paralic, and fluvial environments and eustatic sea-level fluctuations in the origin of the Tuscarora Formation (lower Silurian) of Central Pennsylvania. *Journal of Sedimentary Petrology*, **53**(1), 25–49.

Cramer, F. H. (1969). Possible implications for Silurian paleogeography from phytoplancton assemblages of the Rose Hill and Tuscarora Formations of Pennsylvania. *Journal of Paleontology*, **43**(2), 485–91.

Dale, B. (1976). Cyst formation, sedimentation, and preservation: Factors affecting dinoflagellate assemblages in Recent sediments from Trondheimsfjord, Norway. *Review of Palaeobotany and Palynology*, **22**, 39–60.

Damassa, S. P. & Knoll, A. H. (1986). Micropaleontology of the late Proterozoic Arcoona Quartzite Member of the Tent Hill Formation, Stuart Shelf, South Australia. *Alcheringa*, **10**, 417–30.

DeSimone, L. A. (1988). Paleoenvironmental interpretation of the lower Silurian Tuscarora Formation palynomorph suite in central Pennsylvania. M.S. thesis, Boston University, Boston, MA.

Diver, W. L. (1980). (abstract) Some factors controlling cryptarch distribution in the late Precambrian Torridon Group. *Fifth International Palynological Conference, Volume of Abstracts*, (Cambridge, 1980), 113.

Diver, W. L. & Peat, C. J. (1979). On the interpretation and classification of Precambrian organic-walled microfossils. *Geology*, **7**, 401–4.

Dorning, K. J. (1981). Silurian acritarch distribution in the Ludlovian shelf sea of South Wales and the Welsh Borderland. In *Microfossils from Recent and Fossil Shelf Seas*, ed. J. W. Neale & M. D. Brasier. Chichester, England: Ellis Horwood Ltd.

Downie, C. (1962). So-called spores from the Torridonian. *Proceedings of the Geological Society of London*, **1600**, 127–8.

Elmore, D. R., Milavec, G. J., Imbus, S. W. & Engel, M. H. (1989). The Precambrian Nonesuch Formation of the North American mid-continent rift, sedimentology and organic geochemical aspects of lacustrine deposition. *Precambrian Research*, **43**, 191–312.

Erdtman, G. (1937). Pollen grains recovered from the atmosphere over the Atlantic. *Meddelanden Göteborgs Botaniska Trädgård*, **12**, 185–96.

Evitt, W. R. (1985). *Sporopollenin Dinoflagellate Cysts*. Dallas, TX: American Association of Stratigraphic Palynologists Foundation.

Freile, D. (1988). Ichnofaunal and sedimentological analysis of the Tuscarora Formation in central Pennsylvania. M.S. thesis, Boston University, Boston, MA.

Germs, G. J. B., Knoll, A. H. & Vidal, G. (1986). Latest Proterozoic microfossils from the Nama Group, Namibia (South West Africa). *Precambrian Research*, **32**, 45–62.

Gould, S. J. & Lewontin, R. (1978). The spandrels of San Marco and the Panglossian Paradigm: A critique of the adaptationist programme. *Proceedings of the Royal Society of London*, **B205**, 581–98.

Gray, J., Massa, D. & Boucot, A. J. (1982). Carodocian land plant microfossils from Libya. *Geology*, **13**, 197–201.

Happey-Wood, C. M. (1988). Ecology of freshwater planktonic green algae. In *Growth and Reproductive Strategies of Freshwater Phytoplankton*, ed. C. D. Sandgren. Cambridge: Cambridge University Press, 175–226.

Harris, G. P. (1983). Mixed layer physics and phytoplankton populations; studies in equilibrium and non-equilibrium ecology. In *Progress in Phycological Research*, **2**, ed. F. E. Round & D. Chapman. Amsterdam: Elsevier, 1–52.

Harris, G. P. (1986). *Phytoplankton Ecology. Structure, Function and Fluctuation*. London: Chapman & Hall.

Heaney, S. I., Chapman, D. V. & Morison, H. R. (1983). The role of the cyst stage in the seasonal growth of the dinoflagellate *Ceratium hirundinella* within a small productive lake. *British Phycological Journal*, **18**, 47–59.

Hofmann, H. J., & Aitken, J. D. (1979). Precambrian biota from the Little Dal Group, Mackenzie Mountains, northwest Canada. *Canadian Journal of Earth Sciences*, **16**(1), 150–66.

Hutchinson, G. E. (1967). *A Treatise on Limnology*. New York: John Wiley & Sons.

Ivanovskaya, A. V. & Timofeev, B. V. (1971). Relations between salinity and the distribution of phytoplankton. *Geologiia i Geofysikiia*, **8**, 113–17. (In Russian).

Jacobson, S. R. (1979). Acritarchs as paleoenvironmental indicators in middle and upper Ordovician rocks from Kentucky, Ohio and New York. *Journal of Paleontology*, **53**(5), 1197–212.

Jewson, D. H., Rippy, B. H. & Gilmore, W. K. (1981). Loss rates from sedimentation, parasitism and grazing during the growth, nutrient limitation, and dormancy of a diatom crop. *Limnology and Oceanography*, **26**, 1045–56.

Johnson, N. J. (1985). Early Silurian palynomorphs from the Tuscarora Formation in central Pennsylvania and their paleobotanical and geological significance. *Review of Palaeobotany and Palynology*, **45**, 307–60.

Johnson, P. J. & Sieburth, J. McN. (1979). Chroococcid cyanobacteria in the sea: a ubiquitous and diverse phototrophic biomass. *Limnology and Oceanography*, **24**, 928–35.

Knoll, A. H. (1984). Microbiotas of the late Precambrian Hunnberg Formation, Nordaustlandet, Svalbard. *Journal of Paleontology*, **58**, 131–62.

Knoll, A. H., Barghoorn, E. S. & Awramik, S. M. (1978). New microorganisms from the Aphebian Gunflint Iron Formation, Ontario. *Journal of Paleontology*, **52**(5), 976–92.

Knoll, A. H. & Swett, K. (1985). Micropaleontology of the late Proterozoic Veteranen Group, Spitsbergen. *Palaeontology*, **28**(3), 451–73.

Lee, T. F. & Eggleston, P. M. (1989). Airborne algae and cyanobacteria. *Grana*, **28**, 63–6.

Malvia, R. G., Knoll, A. H. & Siever, R. (1989). Secular change in chert distribution: A reflection of evolving biological participation in the silica cycle. *Palaios*, **4**(6), 519–32.

Margalef, D. R. (1951). Diversidad de especies en los communidades naturales. *Publicaciones Instituto de Biologia Aplicado*, Barcelona, **9**, 5–27.

Margalef, D. R. (1978). Life-forms of phytoplankton as survival alternatives in an unstable environment. *Oceanologica Acta*, **1**, 493–509.

Muller, J. (1959). Palynology of Recent Orinoco Delta and shelf sediments. *Micropaleontology*, **5**, 1–32.

Nøhr-Hansen, H. & Koppelhus, E. B. (1988). Ordovician spores with trilete rays from Washington Land, North Greenland. *Review of Palaeobotany and Palynology*, **56**, 305–11.

Peat, C. J., Muir, M. D., Plumb, K. A., McKirdy, D. M., & Norvick, M. S. (1978). Proterozoic microfossils from the Roper Group, Northern Territory, Australia. *BMR Journal of Australian Geology & Geophysics*, **3**, 1–17.

Reynolds, C. S. (1980). Phytoplankton assemblages and their periodicity in stratifying lake systems. *Holocene Ecology*, **3**, 141–59.

Reynolds, C. S. (1984). *The Ecology of Freshwater Plankton*. Cambridge: Cambridge University Press.

Richardson, J. B. & Rasul, S. M. (1990). Palynofacies in a Late Silurian regressive sequence in the Welsh Borderland and Wales. *Journal of the Geological Society, London*, **147**, 675–86.

Robinson, G. A. (1970). Continuous plankton records: variation in the seasonal cycle of phytoplankton in the North Atlantic, *Bulletin of Marine Ecology*, **6**, 333–45.

Sandgren, C. D. (1988). The ecology of chrysophyte flagellates: their growth and perennation strategies as freshwater phytoplankton. In *Growth and Reproductive Strategies of Freshwater Phytoplankton*, ed. C. D. Sandgren. Cambridge: Cambridge University Press, 9–104.

Schiffelbein, P. (1984). Effect of benthic mixing on the information content of deep-sea stratigraphic signals. *Nature*, **311**, 651–3.

Selley, R. C. (1965). Diagnostic characters of fluviatile sediments of the Torridonian Formation (Precambrian) of northwest Scotland. *Journal of Sedimentary Petrology*, **35**, 366–80.

Smayda, T. J. (1970). The suspension and sinking of phytoplankton in the sea. *Annual Review of Oceanography and Marine Biology*, **8**, 353–414.

Smith, N. D. & Saunders, R. S. (1970). Paleoenvironments and their control of acritarch distribution: Silurian of east-central Pennsylvania. *Journal of Sedimentary Petrology*, **40**(1), 324–33.

Staplin, F. L. (1961). Reef-controlled distribution of Devonian microplankton in Alberta. *Palaeontology*, **4**(3), 392-424.

Strother, P. K. (1980). Microbial communities from Precambrian strata. Ph.D. dissertation, Harvard University, Cambridge, MA.

Strother, P. K. (1991). A classification schema for the cryptospores. *Palynology*, **15**, 219–36.

Strother, P. K, Knoll, A. H. & Barghoorn, E. S. (1983). Micro-organisms from the late Precambrian, Narssarssuk Formation, Northwest Greenland, *Palaeontology*, **35**, 1–25.

Strother, P. K. & Tobin, K. (1987). Observations on the genus *Huroniospora* Barghoorn: Implications for the paleoecology of the Gunflint microbiota. *Precambrian Research*, **36**, 323–33.

Strother, P. K. & Traverse, A. (1979). Plant microfossils from

Llandoverian and Wenlockian rocks of Pennsylvania. *Palynology*, **3**, 1–21.

Swartz, C. K. (1923). Stratigraphic and paleontologic relations of the Silurian strata of Maryland. In *Silurian*. Maryland Geological Survey, 25–52.

Timofeev, B. V. (1959). Ancient flora of the Baltic area and its stratigraphic significance. *Trudy Institute for the All-Union Scientific Investigation and Prospecting of Petroleum*, **129**, 1–320. (in Russian).

Traverse, A. (1988). *Paleopalynology*. Boston: Unwin Hyman.

Tynni, R. & Donner, J. (1980). A microfossil and sedimentation study of the Late Precambrian formation of Hailuoto, Finland. *Geological Survey of Finland, Bulletin*, **311**, 1–27.

Vidal, G. (1976). Late Precambrian microfossils from the Visingsö Beds in southern Sweden. *Fossils and Strata*, **9**, 1–56.

Vidal, G. (1979). Acritarchs from the Upper Proterozoic and Lower Cambrian of East Greenland. *Grønlands Geologiske Undersøgelse, Bulletin*, **134**, 1–40.

Vidal, G. (1981). Micropaleontology and biostratigraphy of the Upper Proterozoic and Lower Cambrian sequence in East Finnmark, northern Norway. *Norges Geologiske Undersøgelse, Bulletin*, **362**, 1–53.

Vidal, G. & Knoll, A. H. (1983). Proterozoic plankton. *Geological Society of America, Memoir*, **161**, 265–77.

Wall, D. (1965). Microplankton, pollen and spores from the Lower Jurassic in Britain. *Micropaleontology*, **11**, 151–90.

Wall, D., Dale, B., Lohman, G. P. & Smith, W. K. (1977). The environmental and climatic distribution of dinoflagellate cysts in modern marine sediments from regions in the North and South Atlantic Ocean and adjacent seas. *Marine Micropaleontology*, **2**, 121–200.

Williams, D. B. & Sarjeant, W. A. S. (1966). Organic-walled microfossils as depth and shoreline indicators. *Marine Geology*, **5**, 389–412.

Zang, W. L. & Walter, M. R. (1989). Latest Proterozoic plankton from the Amadeus Basin in central Australia. *Nature*, **337**, 642–5.

Zhang, Z. (1982). Upper Proterozoic microfossils from the Summer Isles, N. W. Scotland. *Palaeontology*, **25**(3), 443–60.

ZoBell, C. E. (1942). Microorganisms in marine air. In *Aerobiology*, ed. F. R. Moulton. Publication of the American Association for the Advancement of Science, **17**, 55–68.

Modern pollen transport and sedimentation: an annotated bibliography

MARTIN B. FARLEY

Introduction

This annotated bibliography contains entries on the modern transport and deposition of pollen. As a rule, studies that infer transport and deposition from older sediments are omitted. A few entries without annotations (e.g., Vronskiy 1981, 1984) are included for completeness. Some less well-known papers have longer summaries, to aid readers who might not have ready access to the originals.

Occasionally, secondary papers are cited within primary entries. If such a paper is treated in this chapter it is marked with an asterisk; if not treated here, its full bibliographic citation may be found by consulting the primary publication annotated here.

Acknowledgments

I have assembled this bibliography over a number of years. I would like to thank David L. Dilcher and Alfred Traverse for their encouragement, and to express much gratitude to the persevering Interlibrary Loan librarians, particularly in the Earth and Mineral Sciences Library at the Pennsylvania State University, for their help in obtaining many papers.

Annotated Bibliography

Andersen, S. Th. (1967). Tree-pollen rain in a mixed deciduous forest in South Jutland (Denmark). *Review of Palaeobotany and Palynology*, **3**, 267–75.
Transects through a deciduous forest show that 'pollen percentages decrease strongly 20–30 m from dense stands of the species in question, and the pollen spectra thus appear to represent highly local vegetation' (p. 269).

Andersen, S. Th. (1970). The relative pollen productivity and pollen representation of north European trees, and correction factors for tree pollen spectra. *Danmarks Geologiske Undersøgelse*, ser. II, **96**, 1–117.
Andersen's study in a forested upland (Jutland) concludes that tree pollen is wind-transported up to several hundred meters and that herbaceous pollen is transported only a few meters because the parent plants are not tall.

Andersen, S. Th. (1986). Palaeoecological studies of terrestrial soils. In *Handbook of Holocene Palaeoecology and Palaeohydrology*, ed. B. E. Berglund. New York: John Wiley & Sons, 165–77.
'Pollen assemblages deposited on the land surface mainly reflect the vegetation at or near the sampling site; pollen diagrams from the terrestrial soils therefore reflect strictly local changes in vegetation due to natural succession...' (p. 167).

Anderson, T. W. and Terasmae, J. (1966). Palynological study of bottom sediments in Georgian Bay, Lake Huron. *University of Michigan, Great Lakes Research Division, Publication No. 15*, 164–8.
A study of 18 surface bottom samples from central Georgian Bay (Lake Huron). Pine, oak, maple, elm, chenopods, ragweed, and *Artemisia* dominate the assemblages which generally reflect the lake's regional vegetation except for maple whose pollen's source is unclear.

Banks, H. H. and Nighswander, J. E. (1982). The application of a new method for the determination of particle deposition onto a lake. *Ecology*, **63**, 1254–8.
Gives a mathematical method of integrating pollen deposition at points on the lake surface to get total pollen deposition on the lake surface. Deposition of *Tsuga* and *Pinus* in 88 ha lake on the Canadian Shield best fits regression equations of the types:

$$y = a \log x + b$$

or

$$\log y = a \log x + b,$$

where y = pollen deposition (grains/m^2), x = distance from shore, a and b = constants. Both equations are exponential decreases with distance.

Berglund, B. E. (1973). Pollen dispersal and deposition in an area of southeastern Sweden – some preliminary results. In *Quaternary Plant Ecology*, ed. H. J. B. Birks and R. G. West. Oxford: Blackwell Scientific, 117–29.
A study of pollen deposition in three lakes of different sizes (1 ha–50 ha). Local vegetation contributes from 40–75% of the total pollen rain deposited in aerial traps on the lakes. Berglund shows that the lakes can be placed in a hierarchy: the two lakes with 1 ha and 5 ha area are dominated by local pollen and have an effective radius of pollen deposition of less than 500 m and 1000 m respectively. The larger lake (50 ha) has an effective radius of 10000 m and receives the highest proportion of

allochthonous pollen. These results match Tauber's (1965) theoretical calculations. Berglund shows that, for the largest lake, three-year averages of aerial trap results are similar to the results of sediment traps in the lake. Resuspension and redeposition in this relatively large, shallow (< 4 m), unstratified lake are higher than that found by Davis (1968*) for a deeper lake, that is, 10–20 times the pollen influx at the lake surface is found on the bottom. Bottom erosion affects layers more than three years old. Relative pollen values of the bottom mud are 'rather constant' over the whole lake.

Birks, H. J. B. (1970). Inwashed pollen spectra at Loch Fada, Isle of Skye. *New Phytologist*, **69**, 807–20.
Two classes of redeposited miospores exist in this lake: (1) reworked pollen from older sediments (ranging in age from Jurassic–?Pleistocene), and (2) inwashed pollen associated with inwashed bryophyte material. Types of pollen deterioration might allow differentiation between these and air transported grains.

Birks, H. J. B. (1981). Long-distance pollen in late Wisconsin sediments of Minnesota, U.S.A.: a quantitative analysis. *New Phytologist*, **87**, 630–61.
Very small amounts of pollen and spores (< 0.5%) occur in late Wisconsin sediments whose producing plants' local occurrence is extremely unlikely; these are therefore interpreted as present due to long-distance aerial transport.

Bonnefille, R. (1969). Analyse pollinique d'un sédiment récent: vases actuelles de la rivière Aouache (Éthiope). *Pollen et Spores*, **11**, 7–16.
Study of a modern river mud from the Awash River (East African Rift) near Addis Ababa. The spectrum is dominated by herb pollen and elements of the wooded savanna: Gramineae (56% of the total sporomorphs), Compositae, Amaranthaceae, Papilionaceae, *Polygala*. These are elements of the local vegetation. Trees of surrounding high areas are also present (10% of total sporomorphs): *Podocarpus* (9% of total sporomorphs), *Erica*, *Myrica*, *Hagenia*, *Teclea*, *Galiniera*.

Bonnefille, R. & Vincens, A. (1977). Représentation pollinique d'environnements arides a l'est du lac Turkana (Kenya). *Recherches françaises sur le Quaternaire (INQUA 1977), Supplèment au Bulletin AFEQ*, 1977-1, **50**, 235–47.
Study of 11 surface samples from three transects on the east side of Lake Turkana. The local vegetation is a subdesertic savanna with *Acacia*. One transect is marginal to the lake, one is on the edge of a river delta draining into the lake, and the third is fluvial (about 15 km above the mouth). Species richness increases from lake (7–15 spp/sample, 21 total) to delta (24–34/sample, 48 total) to fluvial (33–39/sample, 56 total). These totals reflect the floristic diversity of each environment. Important local taxa include Capparaceae, Amaranthaceae-Chenopodiaceae, Gramineae, and Compositae. Pollen of species not growing in the area amount to 0.04–2%. Of this, *Podocarpus* is most abundant (less than or equal to 1.3%). *Podocarpus* and the other non-local taxa (e.g., *Juniperus*, *Myrica*, *Olea*, *Erica*, *Artemisia*) are transported from the mountain forests 150 km to the east.

Bonny, A. P. (1976). Recruitment of pollen to the seston and sediment of some English Lake District lakes. *Journal of Ecology*, **64**, 859–87.
A study of five lakes in the English Lake District that attempts to relate pollen in lake sediments with the vegetation of surrounding areas. The two smallest lakes may receive 89% or 70% respectively of their pollen influx from streams. These figures are probably overestimates because of underestimation of airborne pollen deposition, and because placement of air traps created a bias against receiving maximum amounts of airborne deposition. Nonetheless, streamborne pollen is quantitatively important. Streamborne pollen also has a different taxonomic composition than airborne pollen: it increases the occurrence of taxa that are poorly dispersed in air (e.g., *Calluna*) and extra-local pollen (e.g., pine).

Bonny, A. P. (1978). The effect of pollen recruitment processes on pollen distribution over the sediment surface of a small lake in Cumbria. *Journal of Ecology*, **66**, 385–416.
A lake (Blelham Tarn, 10.2 ha, about 4.0 km² DB) from Bonny (1976*) is further investigated. The input of most major pollen taxa is significantly correlated with rainfall, thus indicating that substantial amounts of pollen received beyond the amount deposited by air are transported by runoff (streamborne) and not resuspended from the lake bottom. Total pollen deposited annually from the air was only about 15% of pollen caught in the lake, so that streamborne pollen may have amounted to up to 85% of the total annual pollen input. As mentioned in Bonny (1976*), however, airborne pollen was likely underestimated, that is, the aerial traps did not fully sample the airborne component. In addition, some pollen was resuspended from the lake bottom. This means that streamborne pollen was less than 85%, although how much less is unknown. In Bonny's results, no taxon plotted has a known airborne component that exceeds the sum of streamborne, resuspended, and airborne (but not found in the aerial traps) pollen. Most taxa are considerably less abundant in the known airborne pollen than in the total pollen from other sources.

Bonny, A. P. and Allen, P. V. (1984). Pollen recruitment to the sediments of an enclosed lake in Shropshire, England. In *Lake Sediments and Environmental History*, ed. E. Y. Haworth and J. W. G. Lund. Minneapolis: University of Minnesota Press, 231–59.
Analysis of pollen deposition to a 15 ha lake lacking streamborne input. Sampling used six aerial traps (one in *Alnus* forest at lake's edge, five in a transect on the lake's surface 5, 15, 25, 50, and 150 m from shore), one submerged sediment trap (5.5 m below the 150 m aerial trap), and 32 surface sediment samples. Aerial traps reflect flowering phenology. Pollen content of aerial traps is at a maximum in the forest trap (reflecting the strong local *Alnus* vegetation), and decreases sharply offshore to 50 m out (150 m trap total and composition is similar to those at 50 m). Gradient of decrease is greatest for taxa living in the area around the lake, whereas pollen of extra-local taxa occurs more evenly along the transect ('background' deposition). The submerged trap indicates substantial circulation of pollen in the water mass particularly from littoral areas to the center by the large amounts of *Alnus* it contains and the peaks in pollen catch coinciding with maximum water turbulence. The surface sediment results support this interpretation, as pollen composition is generally similar throughout the lake except in some littoral areas. The results of this investigation are consistent with Tauber's (1965*, 1977*) model: trunk space air currents deliver local pollen to the littoral region, while beyond the littoral region (about 30 m), lower amounts of pollen are deposited from air currents flowing above the surrounding vegetation and from pollen washed out of rain.

Bottema, S. & Van Straaten, L. M. J. U. (1966). Malacology and palynology of two cores from the Adriatic sea floor. *Marine Geology*, **4**, 553–64.

'An essential factor influencing the total pollen content of the investigated sediments seems to have been the distance to the coast (and notably to the mouths of large rivers) at the time of deposition'(p. 553). Statement is relatively unsupported except for noting that at times of lowered sea level the mouth of the Po River was less than 100 km from one of the core sites.

Bradshaw, R. H. W. (1988). Spatially-precise studies of forest
  dynamics. In *Vegetation History*, ed. B. Huntley and
  T. Webb, III. Dordrecht: Kluwer Academic, 725–51.
An excellent summary of work by Quaternary palynologists on closed-canopy forest sites. These are sites within the forest canopy (such as small hollows and mor humus sites) which receive the vast majority of their pollen from 20–30 m radius.

Brown, A. G. (1985). The potential use of pollen in the
  identification of suspended sediment sources. *Earth
  Surface Processes and Landforms*, **10**, 27–32.
An investigation of sporomorphs in stream water in southern England. The stream drains 11.4 km² above the sampling point. Samples were taken over a 10 hr period for one-day flood events in each of three months (June, September, January). Peak sporomorph concentration ranged from about 130 000 grains/l of water in the June event to about 240 000/l in January and September; average concentration was about 80 000 grains/l. Bulk samples between floods had a mean sporomorph concentration of 200 grains/l. Sporomorph concentration is basically correlated with discharge (1–6 m³/sec in September and 1–3 m³/sec in January and June) and suspended sediment concentration. January discharge and suspended sediment peaks reflect higher winter soil and bank erosion. Sporomorph concentrations are higher earlier in the day (during the rising limb of flow rate). Peak sporomorph concentrations are twice that Peck (1973*) found.

Brun, A. (1983). Étude palynologique des sédiments marins
  Holocènes de 5000 B.P. a l'actuel dans le Golfe de Gabès
  (mer pélagienne). *Pollen et Spores*, **25**, 437–60.
This paper discusses the palynology of 23 surface bottom samples near the islands of Kerkennah (off Tunisia). The pollen flora, in general, is dominated by *Olea* and *Pinus* with smaller amounts of *Pistacia* and *Quercus*. The taxa can be divided into five groups on the basis of ecological and geographic location; the most abundant group (averaging 55%) can be traced to sources in northern and central Tunisia. Pollen of some taxa (e.g., *Olea*) decrease in abundance as distance from Tunisia increases, allowing an increase in percentage of 'allochthonous' temperate pollen and Gramineae. The exact cause of this variation is unclear.

Brush, G. S. & Brush, L. M., Jr. (1972). Transport of pollen
  in a sediment-laden channel: a laboratory study. *American
  Journal of Science*, **272**, 359–81.
A flume study of pollen transport using 14 common pollen taxa. Samples of suspended water, the water surface, and the bed were used. The concentration of any pollen type with depth was relatively uniform. Pollen was preferentially deposited in the bottomsets of the silt to sand dunes, that is, in the finer sediment. About 38% of wetted pollen remained in suspension for some time in conditions equivalent to a shallow turbulent stream of moderate flow. They conclude that presence of large amounts of a pollen type in a fluvial deposit does not imply a large amount of source vegetation in the vicinity.

Brush, G. S. & DeFries, R. S. (1981). Spatial distributions of
  pollen in surface sediments of the Potomac Estuary.
  Limnology and Oceanography, **26**, 295–309.
Pollen of trees from the upper watershed of the Potomac River (about 60 km upstream from the upper limit of tidal influence) is not present in downstream estuarine deposits. The authors believe this is due to the deposition of large amounts of river sediments near the tidal-nontidal boundary in the river as hypothesized by Schubel (1972b, in Brush and DeFries' references).

Buller, R. E. (1951). Pollen shedding and pollen dispersal in
  corn. M.S. thesis. Pennsylvania State University,
  University Park.
A study designed to show minimum distance necessary to prevent outcrossing of seed corn by geographic isolation. Pollen deposited from a square source (32 m × 32 m) decreased to 0% (relative to maximum deposition in the plot center) within 427 m (85 rods) downwind of the plot edge.

Caratini, C. (1968). Analyse palynologique d'un sédiment
  actuel: vases de l'anse de l'Aiguillon (Charente maritime et
  Vendée). *Compte Rendu Sommaire des Séances de la
  Société Géologique de France*, 1968(**3**), 72–74.
An analysis of palynomorphs from shallow cores in Aiguillon Cove in the estuary of the Sèvre Niortaise (Atlantic coast of France). Pine occurs over a great range (6–70%), and is excluded from the pollen sum. Grass pollen ranges from 10–20% of the sum, Cyperaceae 5–15%. A wide variety of other taxa occur in amounts from 1–3% or less. Pollen originates from local vegetation (salt meadows, forests: pine, most of the grass pollen, Cyperaceae), river transport (hygrophilic plants), and wind (inland forests). The pollen of the sediments therefore has a mostly different source from the terrigenous sediments in which they are found.

Caratini, C., Blasco, F., & Thanikaimoni, G. (1973). Relation
  between the pollen spectra and the vegetation of a south
  Indian mangrove. *Pollen et Spores*, **15**, 281–92.
Autochthonous pollen of *Rhizophora* mangrove constitutes 75% of the pollen found in mangrove surface sediments. The highest value of allochthonous pollen occurs in channel sediments which implies water transport plays an important role in moving this pollen.

Caratini, C., Tastet, J-P., Tissot, C., & Fredoux, A. (1987).
  Sédimentation palynologique actuelle sur le plateau
  continental de Côte D'Ivoire. *Mémoires et Travaux, École
  Pratique des Hautes Études, Institut de Montpellier*, **17**,
  69–100.
A study of 29 samples of bottom sediments taken in four profiles perpendicular to the coast of the Ivory Coast west of Abidjan. Water depths range from 20–1050 m (shelf with some continental slope samples; bottom lithologies range from coarse sand to mud. Muds dominate, occur in two main patches, one of which is offshore from both the mouth of a lagoon and the only large river. Two winds influence the area: the southern trades (from the southwest, i.e., onshore) and the Harmattan, which blows from the northeast for only a few days and carries much dust (and therefore pollen) from the continent. Terrestrial palynomorphs come from five source vegetations: mangrove, marshy forest, humid forest, savanna, and anthropogenic. Mangrove plants (primarily *Rhizophora*) average 5%, are most abundant (10%) offshore from the lagoon mouth, and are over-represented relative to their proportion of the vegetation cover. The marshy, littoral forest sporomorphs are twice as

common as mangrove and vary the same way. The humid forest and savanna are inland and constitute most of the vegetation cover. The forest is about 20% of total sporomorphs; the savanna about 10%. Terrestrial sporomorphs are most abundant relative to dinoflagellates in the profiles offshore from lagoon mouths. Q-mode analysis of dinoflagellates group samples according to depth; Q-mode analysis of terrestrial palynomorphs is less clear: the three deepest samples occur together, the samples from the profile not opposite a lagoon mouth occur together, and the samples from water shallower than 200 m occur in a poorly defined group. This less clear grouping results from the influence of many factors (vegetation distribution irregularities, marine currents, variable winds) on the distribution of terrestrial sporomorphs.

Catto, N. R. (1985). Hydrodynamic distribution of palynomorphs in a fluvial succession, Yukon. *Canadian Journal of Earth Sciences*, **22**, 1552-6.
Analysis of Holocene braided stream deposits shows palynological differences between sand, coarse silt, and medium silt samples. Catto is able to exclude the possibility that the palynology of different grain sizes depends on their deposition when the producing plants reproduce by showing that the sands do not contain early season (spring) palynomorphs and the silts do not contain mid to late season (summer to fall) palynomorphs. *Lycopodium* and *Chenopodium* occur preferentially in the sands, and *Picea*, *Betula*, and *Alnus* in the silts. This illustrates that sedimentary sorting is the primary control on palynomorph occurrence. Contrast with Tippett (1964*).

Chmura, G. L. (1990). Palynological and carbon-isotopic techniques for reconstruction of paleomarsh salinity zones. Ph.D. diss. Louisiana State University, Baton Rouge.
Study of the Barataria Basin, a coastal marsh basin in the Mississippi River delta. The basin is characterized by trivial fluvial input. The dominant swamp arboreal vegetation is *Taxodium* and *Nyssa* with subsidiary *Fraxinus* and *Acer*; the swamps have many floating aquatics and herbs. There are four kinds of herbaceous marshes: (1) freshwater (dominated by *Panicum* with *Thelypteris*, *Osmunda*, *Vigna* and *Polygonum*); (2) intermediate marshes (*Spartina patens*, *Phragmites*, *Sagittaria*, *Bacopa*); (3) brackish (*Spartina patens*, *Distichlis*, *Juncus*, *Scirpus*); (4) salt (*Spartina alterniflora*, *S. patens*, *Juncus*, *Distichlis*). Sixty-two samples of simultaneously sampled vegetation and palynomorphs were ordinated by discriminant analysis, which separated 94% of the samples into their correct marsh type. The separation can be accomplished by use of only ten taxa (Tubuliflorae, Cyperaceae, Polypodiaceae, Gramineae, *Myrica*, *Vigna*, *Sagittaria*, *Osmunda*, *Typha latifolia*, and *Typha angustifolia*). Water drainage transport is seen in the occurrence of freshwater marsh types in the brackish marsh (Nymphaeaceae pollen, Polypodiaceae, and *Osmunda* spores) and salt marsh (*Myriophyllum*). Tidal transport of *Ambrosia*, *Nyssa*, *Quercus*, 'TCT' (= Taxodiaceae + Cupressaceae + Taxaceae), *Ulmus*, and *Pinus* is indicated by their higher abundance in the more distal (seaward) parts of the basin (i.e., if they were aerially transported, then the taxa should be more abundant in the proximal parts of the basin). In all, the samples' palynofloras best reflected the vegetation of the salinity zone in which they occurred, rather than the vegetation of the quadrat sampled.

Chmura, G. L. and Liu, K-B. (1990). Pollen in the lower Mississippi River: *Review of Palaeobotany and Palynology*, **64**, 253-61.
A study of 0.4-1.1 liter water samples taken on one day in each of seven months (February to August 1987) from a depth of one meter on both banks of the Mississippi River at Pointe a la Hache (about 50 km downstream from New Orleans). Sporomorph concentrations were lowest in August (657 grains/l) and highest in April and were highly correlated with suspended sediment concentration. The authors calculate that each gram of suspended sediment contains 14 000 grains on average; this is within the range of pollen concentration in bottom sediment of the Mississippi prodelta estimated by Darrell & Hart (1970*). The most common taxa include short-spined composites, *Quercus*, and Chenopodiaceae-Amaranthaceae. Less common types are Taxodiaceae-Cupressaceae-Taxaceae, Gramineae, *Salix*, and Cyperaceae. Concentration of indeterminate grains peaks in March whereas total pollen concentration (and the concentration of most identifiable taxa) peaks in April. Taxa from the Mississippi's headwaters are present in extremely small amounts. Peak occurrence of *Fraxinus* in May and of short-spined Compositae in early August suggests a northern US source because of the timing of these peaks relative to flowering. Reworked grains are less prominent than studies from Gulf of Mexico bottom sediments would suggest; this may result from reductions in river bank caving caused by artificial bank stabilization in the last 50 years. The authors estimate that the Mississippi delivers approximately 10^{19} grains/yr to the Gulf of Mexico (i.e., between 34 and 50×10^3 metric tons per year depending on the method of calculation).

Chowdhury, K. R. (1982). Distribution of Recent and fossil palynomorphs in the south-eastern North Sea (German Bay). *Senckenbergiana Maritima*, **14**, 79-145.
A study of modern surface sediment distribution of palynomorphs in river estuaries, tidal flats, and seafloor in the southeastern North Sea. Data are presented as percentages (no concentrations). Arboreal pollen (particularly *Betula*, *Alnus*, *Quercus*, *Pinus*) dominates, and is most abundant in German Bay and least abundant in the tidal flats. Shrubs (e.g., *Corylus*) occur in near and offshore areas. *Calluna* is abundant in tidal flats as are seashore plants (Chenopodiaceae, *Plantago maritima*, *Artemisia maritima*). Cyperaceae occur primarily in the river beds. Gramineae are relatively rare. Polypodiaceae and Sphagnaceae are very abundant in tidal flats (50% of all palynomorphs), but are somewhat less abundant elsewhere. Reworked pre-Tertiary, Tertiary, and Quaternary sporomorphs also occur. Other taxa present include *Pediastrum*, dinoflagellate cysts, and *Pachysphaera* (*Tasmanites*). Chowdhury presents results for each river, tidal flat area, and two subdivisions of German Bay. Winds during spring and summer are mostly onshore so water transport of sporomorphs appears most important. Local currents are especially effective. The offshore results show that only regional vegetation patterns onshore can be inferred from them.

Coetzee, J. A. (1976). A report on a pollen analytical investigation of Recent river mouth sediments on the southwest African coast: In *Palaeoecology of Africa*, **9**, ed. E. M. van Zinderen Bakker. Capetown: A. A. Balkema, 131-5.
River mouth sediments from 11 rivers draining 16 regional vegetation types were analyzed for the relationship of pollen content to vegetation. (The vegetation types include savannas, velds, steppes, karoo, and semi-desert.) In general, the pollen spectra reflect the regional vegetation.

Colwell, R. (1951). The use of radioactive isotopes in determining spore distribution patterns: American *Journal of Botany*, **38**, 511-23.
A study of pine pollen dispersal from an elevated source using

pollen labeled with P^{32}. The pollen (5×10^9 grains) was released from 4 m in height in a semi-open area and was collected by deposition into ground-based petri dishes. Deposition peaked approximately 7 m downwind and approached zero asymptotically with greater distance from the source (reaching 10% of the peak value at 9 m and 46 m for two traverses).

Cross, A. T., Thompson, G. G., & Zaitzeff, J. B. (1966). Source and distribution of palynomorphs in bottom sediments, southern part of Gulf of California: *Marine Geology*, **4**, 467–524.
Sporomorph distribution in the Gulf of California is related to patterns of terrigenous clastic sedimentation. The most important factors controlling distribution are bottom morphology, surface currents, tides, and fluvial transport. Water transport is especially important in distribution of *Quercus, Abies, Picea* pollen (all of which are normally thought of as wind transported). The authors believe hurricane force winds are of major significance, but present no supporting evidence.

Crowder, A. A. & Cuddy, D. G. (1973). Pollen in a small river basin: Wilton Creek, Ontario. In *Quaternary Plant Ecology*, ed. H. J. B. Birks and R. G. West. Oxford: Blackwell Scientific, 61–77.
Water transport of miospores in Wilton Creek must be a major contributor to Lake Ontario, because miospore proportions in the creek are nearly identical to those in the lake.

Currier, P. J. & Kapp, R. O. (1974). Local and regional pollen rain components at Davis Lake, Montcalm County, Michigan. *Michigan Academician*, **7**, 211–25.
A study of airborne pollen sedimentation above and into a 0.8 ha kettle. Airborne pollen was sampled during six periods (each about one week) chosen to match flowering periods of various local plants from May to August, 1973, from samplers suspended above the lake, in the woods surrounding the lake, and from jars submerged in the lake. Results suggest large amounts of pollen are carried to the lake in horizontal air currents just above the forest canopy. Pollen carried by air within the forest's trunk space, and at the canopy level itself occurs in lesser amounts.

Darrell, J. H., II (1973). Statistical evaluation of palynomorph distribution in the sedimentary environments of the modern Mississippi River delta. Ph.D. diss. Louisiana State University, Baton Rouge.
An analysis of the occurrence of 351 palynomorph groups (28 dinoflagellate, 302 pollen and spores, and 21 unknown affinity) in Mississippi River subenvironments (those of Darrell & Hart, 1970*). Eighty-seven percent of the taxa have a distribution independent of subenvironment; the rest are environment-dependent. Facies-dependent taxa are divisible into four groups: (1) reworked pollen concentrated in levees, channel, and offshore subenvironments; (2) bisaccate pollen concentrated offshore; (3) tree pollen (as opposed to grass and herb) concentrated in high-energy onshore (channel, levee) and offshore environments; (4) marsh vegetation pollen concentrated in marsh environments.

Darrell, J. H. & Hart, G. F. (1970). Environmental determinations using absolute miospore frequency, Mississippi River delta. *Geological Society of America Bulletin*, **81**, 2513–18.
The only published results to date of the only systematic study of a modern delta and its contained pollen and spores (see Hart, Chap. 9 this volume). (Muller's 1959* Orinoco study is at the reconnaisance level for the delta.) Darrell's dissertation (1973*) contains the full results. An analysis of the distribution of absolute pollen and spore frequency in four onshore subenvironments (marsh, interdistributary bay, channel, subaerial levee) and three offshore subenvironments (distributary mouth bar, distal bar, prodelta) of the Mississippi delta. Analysis of variance shows that certain combinations of subenvironments contain statistically significant differences in absolute pollen and spore abundance: onshore subenvironments vs. offshore subenvironments; marsh and bay subenvironments vs. levee and channel subenvironments; marsh vs. bay subenvironments. Darrell and Hart believe the marsh and bay abundances (which are highest) imply autochthonous pollen and spore deposition. The authors conclude that absolute sporomorph abundance is not very useful for distinguishing subenvironments, without additional data such as species content or grain size.

Davey, R. J. (1971). Palynology and palaeo-environmental studies, with special reference to the continental shelf sediments of South Africa. *Proceedings 2nd Planktonic Conference, Rome*, 331–47.
Miospore abundance decreases gradually offshore (from 8 km out, southern Africa) onto the shelf-slope break and then increases on the upper continental slope (see also Heusser, 1978b*; Heusser & Balsam, 1977*). The abundances appear to be correlated with sediment size. An abundance peak appears off the area of outflow of the two major local streams.

Davey, R. J. & Rogers, J. (1975). Palynomorph distribution in Recent offshore sediments along two traverses off South West Africa. *Marine Geology*, **18**, 213–25.
A study of nine samples, primarily for dinoflagellates, from the continental shelf and two from the continental slope off South West Africa between 25° and 30° S latitude. Local winds are strongly onshore, so sporomorph abundances are low except at the mouth of the Orange River.

Davis, M. B. (1968). Pollen grains in lake sediment: redeposition caused by seasonal water circulation. *Science*, **162**, 796–9.
Frains Lake (6.7 ha, thermally stratified, 9 m maximum depth, no permanent in- or out-flowing streams). Pollen deposition in bottom sediments ranged from 1200–21 000 grains/cm² per year averaged over the (then) 138 years since land settlement in the area. Concentrations were greatest in the deepest part of the lake. Sediment trap pollen deposition is 2–4 times greater than that calculated from the sediment. This excess is the result of lake bottom erosion and resuspension of previously deposited pollen which is then deposited again. This process occurs during fall and spring when the lake is not thermally stratified. The resuspension moves pollen preferentially from the lake edge to the deeper parts. It also mixes pollen deposited in different years creating a sort of running average of pollen deposition.

Davis, M. B. (1973). Redeposition of pollen grains in lake sediment. *Limnology and Oceanography*, **18**, 44–52.
A study of Frains Lake (see Davis, 1968*) and Sayles Lake (35 km west of Frains Lake, similar area and shape, no in- or out-flowing streams, but is mostly <1 m deep (maximum depth 3 m), never thermally stratified). Frains Lake has a great amount of resuspended pollen (Davis, 1968*) during periods lacking thermal stratification (fall and spring); Sayles Lake also has much resuspension but this process occurs throughout the ice-free season. Resuspension is the main factor distributing littoral sediment (and its pollen, especially ragweed) to the lake center. Pollen is not sorted by size or morphology during resuspension.

Davis, M. B. & Brubaker, L. B. (1973). Differential sedimentation of pollen grains in lakes. *Limnology and Oceanography*, **18**, 635–46.

Another study of Frains Lake. Deposition of forest tree pollen is uniform over the lake's surface. Pollen abundance is higher on the downwind side of the lake in the lake water. Uniform deposition of oak pollen on the lake floor during oak flowering was disturbed by resuspension which creates a concentration maximum in the deeper parts of the lake. Initial deposition of ragweed pollen is mainly littoral with resuspension creating a more even distribution across the lake. Thus, the ratio of the two taxa varies with water depth. More even deposition would be expected before stratification sets in.

Davis, M. B., Brubaker, L. B., & Beiswenger, J. M. (1971). Pollen grains in lake sediments: pollen percentages in surface sediments from southern Michigan. *Quaternary Research*, **1**, 450–67.

Results from sampling five lakes (especially Frains Lake) show differential pollen deposition occurs in the surface sediments. Pollen from plants immediately adjacent to or in the lake peaks in the lake sediments in the vicinity of the producing plants. Ragweed, pine, and some herbs with small pollen occur in highest abundance relative to oak pollen in shallow water sediment. Deciduous tree pollen, however, occurs in constant ratios (relative to oak) throughout the lake.

Davis, M. B., Moeller, R. E., & Ford, J. (1984). Sediment focusing and pollen influx. In *Lake Sediments and Environmental History*, ed. E. Y. Haworth and J. W. G. Lund. Leicester: Leicester University Press, 261–93.

This paper points out that the importance of streamborne input to an upland lake may depend in part on the ratio of watershed area to lake area. Lakes with ratios above 30:1 tend to have more important streamborne input relative to airborne input (they studied some New Hampshire lakes, plus Blelham Tarn, which was also investigated by Bonny, 1978*), whereas lakes with low ratios (10:1 and less) have relatively little streamborne input.

de Jekhowsky, B. (1963). Répartition quantitative des grands groupes de 'microorganontes' (spores, Hystrichosphéres, etc.) dans les sédiments marins du plateau continental. *Compte Rendu Sommaire des Séances de la Société de Biogéographie*, **349**, 29–47.

Primarily a theoretical discussion of palynomorphs as sedimentary particles in nearshore marine settings. Contains graphs and maps of ideal isopollen lines (both absolute and relative frequencies) of marginal marine systems. de Jekhowsky uses Muller's (1959) data from the Orinoco delta as an example. He concludes that while isopollen lines do show the general shape of the shoreline as W. S. Hoffmeister's US patent (1954: patent number 2 686 108) suggested, the data also can be used to obtain more detailed information on local sedimentologic conditions.

Delcourt, P. A., Delcourt, H. R., & Webb, T. III (1984). Atlas of mapped distributions of dominance and modern pollen percentages for important tree taxa of eastern North America. *American Association of Stratigraphic Palynologists, Contributions Series*, **14**.

Contains a series of maps of the correlation between a U.S. Forest Service measure of wood volume and percentages of 24 arboreal pollen taxa in modern surface sediments. Geometric-mean linear regression is used to calculate the relationship between the two variables. While this paper is not concerned directly with pollen transport and deposition, especially as to local circumstances, it does give a broad view of the relatively high correspondence between pollen occurrence and location of the producing plants (see also Huntley & Birks, 1983*).

Dreyfus, M., Miller, W. & Habib, D. (1971). (abstract) Marine sedimentation of spores and pollen grains, coastal western Puerto Rico. *Geoscience and Man*, **3**, 94.

Samples are from the narrow shelf near the Añasco River and the Cabo Rojo carbonate platform. On the shelf, concentrations range from 22 000 grains/g near the coast to 9000/g offshore. On the carbonate platform, concentrations range from less than 50 to 400/g.

Edmonds, R. L. (ed.) (1979). Aerobiology – The ecological systems approach. *U.S./I.B.P. Synthesis Series*, 10. Stroudsburg, PA: Dowden, Hutchinson & Ross.

A synthesis on all kinds of aerobiology with some review material on spores and pollen, mostly by A. M. Solomon (see Solomon & Harrington, 1979*).

Estlander, A. & Kulmala, A. (1973). Long distance dispersal of radioactive fall-out and natural pollen. In *Scandinavian Aerobiology. Ecological Research Committee Bulletin*, **18**, ed. S. Nilsson. Stockholm: Swedish Natural Resources Research Council, 61–70.

Weather patterns in central Europe during the winter and spring lead to transport of pollen from the Crimean steppes to Finland. Pollen deposition is caused by precipitation (rain or snow).

Faegri, K. & Iversen, J. (1975). *Textbook of Pollen Analysis*. 3rd ed. New York: Hafner.

Faegri, K., Kaland, P.E. & Krzywinski, K. (1989). *Textbook of Pollen Analysis by Knut Faegri and Johs. Iversen*. 4th ed. Chichester: John Wiley & Sons. (See also Faegri & Iversen, 1975, to which many people still refer.)

Considerable discussion of pollen transport, especially in the section, 'Where does the pollen go?', e.g. (p. 29): '...the distance 50–100 km forms a natural limit of pollen dispersal. It is self-evident that the greatest quantities are deposited long before this limit has been reached; on the other hand, there are, of course, those other grains which remain in the air for more than one day, and which can therefore be transported over great distances.'

Faegri, K. & Van der Pijl, L. (1979). *The Principles of Pollination Ecology*. Oxford, Pergamon Press.

Anemophilous pollen tends to be smaller than angiosperm pollen as a whole (20-30-[60] μm vs. 10–300 μm, respectively). Wind, of course, is necessary for wind pollination: in dense forests, Semerikov and Glotov (1971) estimate a maximum dispersal distance for *Quercus* pollen of 80 m and Koski (1970) has calculated a 'probable pollination distance' in a pine forest as 53 m. Transport distance can be great: see Hesselman (1919) or Rempe (1937) for a discussion of wholesale transport of pollen via large scale turbulence. Some coniferous pollen is known to have gone 150–1300 km. 'Once it is sufficiently high up in the air, a grain may go practically anywhere, but will probably be dead on arrival' (p. 40). Average maximum transport distance is 50 km, but generally will be less than 1 km.

Fall, P. L. (1987). Pollen taphonomy in a canyon stream. *Quaternary Research*, **28**, 393–406.

A comparison of alluvial vegetation and alluvial pollen in canyon streams in northeast Arizona. Surface soil palynological samples reflect the vegetation surrounding them reasonably well, but the pollen spectra of alluvial samples does not reflect floodplain vegetation or vegetation above canyon rims well (but see Hall, 1989*). The streams appear to homogenize the pollen of the multiple vegetation types through which the stream passes, and they lead to a dominance of alluvial spectra by *Pinus* and under-representation by *Juniperus* and nonarboreal types. Abundance of pollen taxa also depends on sediment grain size: Pinaceae and *Quercus* are best correlated with clay: Cheno-Am and *Artemisia* with sand; and *Juniperus* and *Ephedra* with silt.

Farley, M. B. (1987). Palynomorphs from surface water of the eastern and central Caribbean Sea. *Micropaleontology*, **33**, 254–62.
A study of 11 water samples taken along the Lesser Antilles and westward into the central Caribbean Sea. Sporomorph concentrations range from 1400 to 19 500 grains per 100 l of water. Analysis of variance of log-transformed concentrations and multiple comparison tests show that four of the samples could be considered to be higher in concentration than the others. The concentrations are primarily correlated with wind and water currents, and seem much less affected by distance from land. One sample, from east of St. Lucia, is downwater from Barbados but is not downwind from any land closer than the Cape Verde Islands (>4000 km). This sample has one of the highest concentrations (12 700 grains per 100 l); this suggests water transport is more important than wind transport. The most common taxa present include *Pinus-Podocarpus*, grass, Compositae, palms, and fungal spores. Some taxa are present whose sources are unclear (e.g., ?*Liquidambar*, ulmaceous pollen).

Farr, K. M. (1989). Palynomorph and palynodebris distributions in modern British and Irish estuarine sediments. In *Northwest European Micropaleontology and Palynology*, ed. D. J. Batten and M. C. Keen. Chichester, U.K.: Ellis Horwood, 265–85.
A study of palynomorphs in estuaries (primarily the Ythan) of the UK. Samples were taken in eight estuaries from their mouths to about two miles above the spring tidal limit (STL). In the Ythan, total sporomorphs occur as a reasonably uniform percentage of the total palynomorph assemblage (5–15%) upstream to the mean STL; beyond the STL sporomorphs reach at least 30%. Estimated concentrations of sporomorphs, however, do not increase much above the STL (but see Brush & Brush, Chap. 3 this volume). Most individual sporomorph taxa (including fungi) show trends similar to the total except *Pinus sylvestris* which has uniform percentage abundance. Phytocuticle also shows uniform abundance. Relative dominance of marine and terrestrial palynomorphs differs in some of the other estuaries; these are overwhelmingly dominated by sporomorphs. Most palynomorphs are transported by the tide and are deposited uniformly throughout the estuary. Sporomorph composition reflects drainage basin vegetation. Palynomorph abundance appears to be related to sediment grain size with fine silts (3–8 ϕ) producing most. Medium and long-range wind transport of sporomorphs to estuaries is insignificant. The sporomorph composition suggests fluvial transport from throughout the drainage basin. Saltmarsh sediments contain sporomorphs identical to those of intertidal flats except for increased halophyte pollen.

Fedorova, R. V. (1952). Dissemination of pollen and spores by running water. *Materialy po Geomorfologii i Paleogeografii SSSR. Raboty po Sporovo-pyltsevomu Analizu* 7. Trudy Instituta Geografii Moskva., Akademiya Nauk SSSR, **52**, 46–72. (In Russian).
Fedorova observed miospores in the Volga River that had been transported at least 1400 km.

Fisk, L. H. (1987). (abstract) Selective deposition of pollen and spores in the nearshore marine environment. *Palynology*, **11**, 226–7.
Study of terrestrial soils and marine mud from conifer forest to offshore in Puget Sound. Soil samples of pollen are closely related to the local vegetation, whereas marine samples contain more conifer pollen and fern spores and fewer angiosperm pollen. Pollen concentration increases offshore.

Fletcher, M. R. (1979). Distribution patterns of Holocene and reworked Paleozoic palynomorphs in sediments from southeastern Lake Michigan. Ph.D. diss. Case Western Reserve University, Cleveland.
A study of palynomorphs (frequency and concentration) from 28 surface bottom samples (8–66 km offshore) in southeastern Lake Michigan, ten river bottom sediment samples and nine river sediment trap samples from the four largest rivers draining southwestern Michigan into the lake, and eight soil samples from along these rivers. Rivers carry the majority of palynomorphs to the lake, and each carry a palynomorph assemblage related to the vegetational province of its drainage basin. Although the prevailing wind is from the west, pollen from the prairie (to the south and west of the lake) is only a secondary source. The most common taxa are pine, oak, cheno-ams, ragweed, and grass; 25–30 taxa occur in frequencies greater than 1%; 70 taxa occur in all. Diversity is inversely proportional to distance from shore. Pollen occurs mostly in the offshore fines with peak concentrations near river mouths. Lake bottom assemblages differ from one another especially in deep water, areas of steep bottom topography, and near major river mouths. Important controls on pollen distribution are location of source vegetation, differences in pollen density and decay resistance, lake circulation, phenology, location of fine silt deposition, and distance from shore. Reworked Paleozoic spores occur near shore and are apparently derived from Wisconsinan glacial deposits rather than directly from bedrock. Southeastern Lake Michigan has an average pollen spectrum that reflects regional vegetation around the lake. Minor pollen taxa in the lake are either the most common minor vegetation elements or are riparian. Patterns of pollen distribution in southeastern Lake Michigan are similar to those found for Lake Ontario by McAndrews and Power (1973*).

Fletcher, M. R. (1980a). (abstract) Principal components analysis of modern and Paleozoic palynomorph distribution patterns in southeastern Lake Michigan. *Palynology*, **4**, 240.
Lake circulation and sporomorph sources are primary controls on distribution in Lake Michigan (57 760 km²). Sediment sorting, at least in part, a function of lake circulation, exerts secondary control. Wind does not control ultimate deposition. Monitored pollen input and the distribution of reworked Paleozoic miospores implies four rivers are the lake's major sporomorph sources.

Fletcher, M. R. (1980b). (abstract) Fossil preservation and the interpretation of assemblages: a palynological example. *Geological Society of America, Abstracts with Programs*, **12**, 427.
Work on Holocene samples from Lake Michigan shows two mechanisms operate for destruction of sporomorphs: differential

destruction which removes susceptible taxa from the assemblage and transport destruction of all taxa in the system. Abundance of susceptible taxa decreases abruptly from rivers to the lake, and then decreases gradually in the lake itself. Damaged grains occur commonly in rivers and the lake center, but rarely in the intermediate portions of the lake. This rare occurrence of damaged taxa reflects a concentration of the most damage-resistant grains with complete absence of fragile types.

Fletcher, M. R. (1980c). (abstract) Microfossils as indicators of current patterns and sediment transport: seasonality and selective transport in Lake Michigan. *Geological Society of America, Abstracts with Programs*, 12, 427.
Palynomorphs are indicators of sediment sources, current patterns, and thermal structure of water in Lake Michigan, because the distribution of palynomorphs is controlled mostly by sediment transport (particularly from fluvial input) and water mass movements. Ragweed pollen is released in late summer when the lake is normally stratified, and counterclockwise currents predominate (as they do most of the year), so ragweed is preferentially deposited along the eastern shore. Similarly, less buoyant oak and reworked Paleozoic grains are deposited along the eastern shore by the counterclockwise current. Pine pollen, on the other hand, is released in the spring when clockwise currents dominate and thus is transported southward. This southward transport is preserved in the distribution of pine pollen because it is trapped in the thermal bar, a band of dense, sinking water.

Florschütz, F. & Jonker, F. (1939). A botanical analysis of a late Pleistocene and Recent profile in the Rhine Delta. *Recueil Travaux Botaniques Neerlandais*, 36, 686–96.
Substantial amounts (10%) of *Abies* pollen occur in sub-boreal sediments of the Rhine delta 400 km from the nearest areas of growth in the Black Forest.

Gregory, P. H. (1973). *The Microbiology of the Atmosphere*. Aylesbury, England: Leonard Hill.
This book contains a comprehensive account of dispersal of microorganisms through the atmosphere. Gregory covers theoretical, experimental, and field work on dispersal of such organic particles. Especially useful are Chapters 2, 6–8, 10–13, 16–18.

Grindrod, J. (1985). The palynology of mangroves on a prograded shore, Princess Charlotte Bay, North Queensland, Australia. *Journal of Biogeography*, 12, 323–48.
Includes analysis of six surface pollen transects across mangrove forest, tidal mud flat, and cheniers. Local winds are predominantly from the east. Most of the pollen at each sampling site is derived from local vegetation (< 50 m). This is particularly true for *Ceriops* and *Avicennia* in the mangrove forest; *Dodonea*, Cyperaceae and Chenopodiaceae on the tidal flats; and *Celtis* on the tidal flats near the chenier ridges (the chenier ridges themselves contain no pollen). Gramineae (coastal plain and uplands) and Myrtaceae (*Eucalyptus*; adjacent uplands) are transported greater distances as, to a certain extent, is *Rhizophora* (mangrove). Taxa present due to long-distance transport (in very low abundance) include *Casuarina*, *Podocarpus*, *Callitris*. In general, pollen reflects local vegetation, as shown by the rapid decrease of abundance of the mangrove forest pollen as distance from the forest increases. Grindrod uses these modern data to show progradation of the sedimentary environments over the last 2000 years.

Groot, J. (1966). Some observations on pollen grains in suspension in the estuary of the Delaware River. *Marine Geology*, 4, 409–16.
Sporomorph distribution in Delaware Bay represents the regional vegetation of the Delaware River drainage basin better than it does the local vegetation. *Tsuga* pollen, whose source is at least 200 km away, occurs in all samples. There is a correlation between the weight of water-suspended sediment and the number of contained pollen and spores.

Groot, J. & Groot, C. R. (1966). Marine palynology – possibilities, limitations, problems. *Marine Geology*, 4, 387–95.
Discusses the occurrence of sporomorphs in marine, especially deep-sea deposits, and their usefulness for regional climatic interpretation, provenance of fine-grained terrigenous clastics, etc. Groot and Groot estimate river transported pollen to the northern West Atlantic at about 15×10^{16} grains per year, and aerial deposition at about 15×10^{18}. This would yield about 100 grains/cm^2 per year in the area from the continental shelf edge to the Mid-Atlantic Ridge or about 5×10^5 grains/g. This is orders of magnitude higher than the actual average (20–40 grains/g). Therefore, much pollen is lost.

Habib, D., Donahue, J., & Krinsley, D. H. (1971). (abstract) Modern pollen distribution off southwestern Puerto Rico, a preliminary report. *Transactions of the Caribbean Geology Conference, Geological Bulletin*, Department of Geology, Queens College of the City University of New York, 5, 130.
Brief discussion of surface samples from the Cabo Rojo carbonate bank. The sporomorph assemblages reflect the vegetation of southwestern Puerto Rico (semiarid plants and mangroves), and not that of the humid central vegetation. Spores and pollen are concentrated in nearshore quiet waters. These results coincide with results from Mississippian carbonates of Indiana.

Hafsten, U. (1960). Pleistocene development of vegetation and vegetation and climate in Tristan da Cunha and Gough Island. *Arbøk for Universitetet I Bergen (Mat.-Naturv. Serie)*, 20, 1–48.
Some Tristan da Cunha samples contain *Nothofagus* which implies long distance air transport of 4000 km from South America.

Hall, S. A. (1989). Pollen analysis and paleoecology of alluvium. Quaternary Research, 31, 435–8.
A reinterpretation of the results of Fall (1987*), who concluded that pollen spectra of canyon alluvial samples do not reflect floodplain vegetation or the vegetation above canyon rims because the pollen sorts by grain size and morphology. Hall's reinterpretation is that the different character of pollen spectra from clay and sand results from their having different sources: higher percentages of *Pinus* occur in clays because the clays originate from the upstream drainage basin. Support for this comes from the similarity of pollen concentration and total *Pinus* percentage in dried mudcurls (Fall's modern analog to the alluvial clay) and the alluvial clays, the shale bedrock of the drainage basin, and the ponderosa/pinyon pine ratio in the mudcurls and the drainage basin's pine forest. Lower percentages of *Pinus* in sandy alluvium results from sand sources in the side canyons (sand bedrock in canyon walls). This is supported by *Pinus* frequencies in side canyon soils and the ponderosa/pinyon pine pollen ratios in the sandy alluvium and the canyon soils.

Heusser, L. E. (1978a). Pollen in Santa Barbara Basin, California: a 12,000 year record. *Geological Society of America Bulletin*, **89**, 673–8.
In the Santa Barbara Basin and the Saanich Basin (Vancouver Island), maximum pollen concentration coincides with low oxygen content of the overlying water; this implies pollen decomposition may be a factor in marine pollen concentration. Comparison of fluvial and marine samples shows a qualitative similarity. Eolian transport is probably not negligible.

Heusser, L. E. (1978b). Spores and pollen in the marine realm: In *Introduction to Marine Micropaleontology*, ed. B. U. Haq & A. Boersma. New York: Elsevier, 327-40.
Discusses pollen sedimentation off the Pacific Coast of North America. Pollen rain is 100–1000 grains/cm^2 per year. Pollen concentrations on the shelf are higher adjacent to river mouths and lower adjacent to areas without permanent drainage. The amount of pollen in more oceanward sediments is not directly proportional to distance from land, but is apparently related to lutites that bypass the shelf to reach the continental slope.

Heusser, L. E. (1983). Pollen distribution in the bottom sediments of the western North Atlantic Ocean. *Marine Micropaleontology*, **8**, 77–88.
Analysis of 130 cores (25–55° N latitude along coast of North America) from the continental slope and rise, areas with high sedimentation rates of fine-grained terrigenous clastics. Percentages of taxa were based on various pollen sums. These percentages were winnowed to the 15 most important types (pine, spruce, hemlock, fir, oak, willow, birch, alder, grass, sedge, composites, ericads, polypods, *Sphagnum*) which were then analyzed by Q-mode factor analysis. Pollen concentration ranged from 10 grains/g on the abyssal plain to 23 000 grains/g on the slope (her Fig. 1b is a contour map of concentrations), and generally speaking, concentrations are much less north of 45° N lat. Her Figure 2 contains onshore vegetation abundance isograds and offshore miospore abundance isograds of spruce, alder, birch, pine, oak, and hemlock. These taxa account for 80% of the total pollen data, and more than 50% of any sample. Heusser discusses abundance peaks of these taxa and their relation to onshore vegetation types. The first factor (of the Q-mode factor analysis) is dominated by pine and oak (67% of the total variance), the second factor by spruce and *Sphagnum* (19% of the total variance), and the third factor by birch and alder (6%). The areal distribution of the factor loadings coincides with the major vegetation formations onshore. 'Quantities of pollen in sediments on the slope of the western North Atlantic Ocean are proportionate to the quantities of suspended sediments of the major river systems of eastern North America' (p. 85). Some rivers are high in suspended sediment load (about 50 mg/l, e.g., Delaware, Susquehanna, Potomac), whereas some are low (< 1 mg/l, e.g., St. Lawrence), and pollen concentrations on the slope follow these trends. (Most suspended sediment is trapped in saltmarshes, estuaries, and lagoons.) Marine currents also affect pollen concentrations. Highest pollen concentrations are in olive-colored muds low in dissolved oxygen.

Heusser, L. E. (1988). Pollen distribution in marine sediments on the continental margin off northern California. *Marine Geology*, **80**, 131–47.
A study of 84 surface bottom samples between 38–39° N and 123–124° W, 46 fluvial sediment samples, and 23 depth-integrated water samples from the Russian River. Pollen concentrations (grains/cm^3 wet sediment) range from < 100 to 5000. Concentrations increase seawards from the shoreline to

5 km offshore. Beyond this (in depths > 50 m on the shelf and on the slope), concentrations show no trend related to distance from shore or water depth. High pollen concentrations are associated with the area west of the Russian River. Midslope sediment trap samples contain 100–2000 grains/cm^3 with a winter maximum and early spring minimum similar to the trend in the Russian River water samples. Most palynomorphs are of arboreal species (especially *Pinus*, *Sequoia*, *Quercus*, *Alnus*) with lesser numbers of shrubs and herbs (Gramineae, Compositae), and polypod fern spores. Reworked palynomorphs are rare. *Sequoia* is highest in percentage abundance on the shelf south of major rivers, *Pinus* rises rapidly offshore, *Alnus* and Gramineae decline rapidly offshore. Fluvial pollen assemblages are highly variable. In the midslope trap, pine peaks in late fall and winter, oak in spring and summer, alder in summer, grass in fall. Palynomorph concentration is comparable to those known from other continental shelves off North America, but is 1/10 the concentration at major river mouths, and 1/100 the estuarine concentration. The pattern of occurrence is consistent with fluvial transport modified by marine currents: highest palynomorph sedimentation in the fall and winter when fluvial clastic input is highest, high concentrations near river mouths, close relationship of pollen to sediment texture (i.e., high concentration in fine silt and lay, low in well-sorted coarse sizes such as nearshore sandy sediments).

Heusser, L. E. & Balsam, W. L. (1977). Pollen distribution in the northeast Pacific Ocean. *Quaternary Research*, **7**, 45–62.
An analysis of 61 cores from about 35° N to 60° N latitude along the western coast of North America. Maximum pollen concentrations occur off the mouth of the Columbia River and the outlet of Sacramento and San Joaquin Rivers (San Francisco Bay). Low concentrations occur north of 53° N lat. and > 250 km offshore. The coincidence of pollen and spore concentration maxima with areas of high sediment influx suggests fluvial transport is a prominent factor in pollen influx. Concentration does not decrease linearly offshore; it is lower on the shelf and in the basin, higher on the slope and rise. This suggests that pollen deposition is related to the deposition of lutites which bypass the shelf. Concentration of individual pollen taxa primarily reflects fluvial and marine sedimentation and secondarily vegetational distribution and abundance. Pine pollen appears to be extremely susceptible to marine transport.

Heusser, L. E. & Balsam, W. L. (1985). Pollen sedimentation in the northwest Atlantic: effects of the Western Boundary Undercurrent. *Marine Geology*, **69**, 149–53.
Study of 12 coretops in a transect perpendicular to shore in water from 2000–4800 m deep off the Chesapeake Peninsula. Some sediment is less than 145 years old, as shown by the presence of ragweed pollen. Pollen concentration increases downslope from 1000 grains/cm^3 (< 2400 m depth) to more than 2300 grains/cm^3 below 2900 m depth, and decreases precipitously on the lower rise to 225/cm^3 below 4400 m. Below 4400 m, reworked pollen (from Maritime Canada; see also Needham, *et al.*, 1969*) increases sharply from 7 to 44% of the total. These changes below 4400 m reflect the effects of the Western Boundary Undercurrent.

Hinga, K. R., Sieburth, J. M. & Heath, G. R. (1979). The supply and use of organic material at the deep-sea floor. *Journal of Marine Research*, **37**, 557–79.
A study involving three sediment traps in the North Atlantic off the east coast of North America. Pine pollen occurred in

all samples; other taxa (not identified) were rarely observed. In some samples, pollen was found clumped with other organics; these clumps are inferred to be the remains of fecal pellets. The water in the southernmost trap (77F–G, off Florida, just north of the Bahamas) contained 60 pine pollen/l (June, 1977) at 660 m depth. Pine pollen flux is $760/m^2$ per day in the trap south of Cape Hatteras (77D), $1200/m^2$ per day in the trap (77A) west of Delaware Bay (the pollen in the water from 77F–G prevented calculation of an accurate flux there). Trap 77A's pine pollen is about 4% of the flux of organic material, whereas 77D's is about 1%. Groot and Groot's (1966*) calculated supply of pollen to the North Atlantic is similar to the flux at 77A. Preservation of pollen in oceanic bottom sediments is not as high as the flux suggests; thus degradation (decay, fungi) must play an important role.

Hirst, J. M. (1973). Spore transport and vertical profiles. In *Scandinavian Aerobiology, Ecological Research Committee Bulletin*, **18**, ed. S. Nilsson. Stockholm: Swedish Natural Resources Research Council, 95–8.

Mostly about fungal spores (e.g., *Puccinia*), but concludes that (1) most spore dispersal is local, especially in vegetation canopies; (2) the proportion of spores not (locally) deposited depends on height of liberation, obstacles, and turbulence; (3) most pollen returns to the ground within two to five days, but small (fungal) spores can stay up at least twice as long, so long as they escape rain-forming clouds.

Hirst, J. M., Stedman, O. J. & Hogg, W. H. (1967). Long-distance spore transport: methods of measurement, vertical spore profiles and the detection of immigrant spores. *Journal of General Microbiology*, **48**, 329–55.

Theoretical calculations conservatively imply that 1–25% (day) and 10–90% (night) of spores (12–32 μm) are deposited within 40 m of the source. Airplane sampling led to an estimate of 10^4 spores/m^3 at 600 m altitude, and 100s/m^3 at 3000 m. ('Spore' here includes fungal spores and pollen.) Concentrations during the day differed from those at night.

Hirst, J. M., Stedman, O. J. & Hurst, G. W. (1967). Long-distance spore transport: vertical sections of spore clouds over the sea. *Journal of General Microbiology*, **48**, 357–77.

More airplane sampling (see Hirst, Stedman & Hogg, 1967*). Spores = Pollen + fungal spores. Spore concentration downwind of the English coast reached a maximum between 500 and 1500 m altitude, because lower levels were preferentially deposited. Pollen clouds were 300–600 m lower than those of *Cladosporium*, which has a much lower terminal settling velocity.

Holmes, P. L. (1990). Differential transport of spores and pollen: a laboratory study. *Review of Palaeobotany and Palynology*, **64**, 289–96.

Flume and field study of sporomorph transport and deposition. Water-saturated pollen of various taxa were used in a flume with and without sediment. Without sediment, pollen deposition ceased once flow velocity reached 25 cm/sec. Sediment increased the rate of pollen deposition at higher flow velocities by trapping spores within bedforms. Field investigation in a 1.2 ha lake (maximum depth 1.8 m, one inflowing stream) was based on 40 samples of the top 5 cm of bottom sediment. The largest sporomorph taxa (e.g, fern spores, some grass pollen) are primarily deposited in the stream delta, mid-sized taxa (some grasses, *Corylus*) are deposited more evenly in the lake, and small taxa (Urticaceae) are predominantly deposited in the

sediments farthest from the delta. Sorting is mainly controlled by grain size.

Hooghiemstra, H. (1988). (abstract) The eolian component in pollen transport and the interpretation of the pollen signal in marine sediments: exceptional conditions in the NW African Atlantic area? *Stuifmail*, **6**(2), 20.

Summarizes work in offshore NW Africa, showing that sporomorph distribution is directly related to the wind systems. Distribution of pollen in cored sediment therefore permits interpretation from the pollen data of paleo-tradewinds and onshore vegetation patterns in the late Pleistocene.

Hooghiemstra, H. (1988). Palynological records from northwest African marine sediments: a general outline of the interpretation of the pollen signal. *Philosophical Transactions of the Royal Society of London*, **B318**, 431–49.

Contains a summary of the data presented in Hooghiemstra, et al. (1986*).

Hooghiemstra, H. & Agwu, C. O. C. (1986). Distribution of palynomorphs in marine sediments: a record for seasonal wind patterns over NW Africa and adjacent Atlantic. *Geologische Rundschau*, **75**, 81–95.

Earlier work (Agwu & Beug, 1982, 1984) established seven pollen assemblages for the NW Africa area, five of which can be directly related to latitudinal vegetation zones. Most pollen and spores are transported by two wind belts (northeast trades and Africa Easterly Jet), fluvial transport being relatively unimportant. This paper gives results from more than 100 bottom surface samples collected between 4° and 34° N latitude. The authors give percentage abundance maps for *Pinus*, Mediterranean elements, Chenopodiaceae-Amaranthaceae, steppe and desert taxa, Gramineae, tropical forest-savanna taxa, trilete fern spores, and Cyperaceae. Most *Pinus* is apparently derived from Europe. Cheno-Am pollen occurrence reflects the two wind belts and the position of the intertropical weather front. Tropical forest-savanna taxa reflect Gulf of Guinea fluvial transport and the intertropical front. Trilete fern spores appear to reflect fluvial transport. Cyperaceae occurrence is apparently controlled by fluvial transport and the Canary Current. See also Hooghiemstra, et al. (1986*) and Melia (1984*).

Hooghiemstra, H., Agwu, C. O. C., and Beug, H-J. (1986). Pollen and spore distribution in recent marine sediments: a record of NW-African seasonal wind patterns and vegetation belts: '*Meteor*' Forschungsergebnisse, **C40**, 87–135.

Study of palynomorph percentages in 109 surface bottom samples from the Atlantic Ocean between 34 and 6° N and 11° to 23° W off Africa in water depths from 50–4870 m. Isopoll maps are presented for taxa in five source-related groups: (I) European and North African mountain trees and shrubs; (II) Mediterranean trees and shrubs; (III) steppe and desert plants; (IV) steppe and desert plants identifiable only to family; (V) tropical forest and savanna taxa. Because of the arid climate, fluvial transport is relatively slight; therefore, most pollen is transported by the NE trades (whose position varies seasonally depending on the ITCZ (Inter-Tropical Convergence Zone) position) or the African Easterly Jet (Saharan-derived). Both winds shift seasonally and vary by altitude. Ocean currents are from the NE along the NW African coast to 10° N (North Equatorial Current) and from the W from 10–0° N (Guinea Current). Samples contain 1–1400 grains; pollen was more abundant in the area south of 22° N. Isopoll maps are presented for 40 genera and unified groups of taxa (e.g., monolete spores).

Group I pollen (e.g., *Alnus, Betula, Pinus*) peaks (25–40% of the palynoflora) at the northern edge of the study area consistent with the March–April ITCZ position. Rainout removes pollen from the NE trades. Group II pollen (e.g., *Quercus, Erica, Olea*) occurs in a narrow offshore belt with maximum abundance (20–24%) as far south as 23° N reflecting ITCZ rainout in April–May. Group III pollen (e.g., *Ephedra, Calligonum*) occurs in a narrow offshore range from 26–35° N directly offshore from North African source areas. A second peak occurs from 15–20° N offshore from Saharan sources. The authors attribute this to tradewind transport as well. Group IV pollen (e.g., *Artemisia*, Gramineae) occur primarily from 29–31° N and 15–22° N (60–80% of the total palynoflora), and are closely related to the main sources at the northern and southern edges of the Sahara. Group V pollen (e.g., *Cassia, Combretum, Elaeis, Rhizophora*), dominant (≥80%) from 6–11° N, reflects fluvial transport. Cyperaceae, *Sparganium-Typha*, and *Isöetes* are also primarily fluvially transported. There is no evidence of any transport of palynomorphs by the southerly trades from Angola. Therefore, in this area, offshore from a large arid continent, eolian transport dominates with subordinate effects of oceanic currents and fluvial supply.

Hopkins, J. S. (1950). Differential flotation and deposition of coniferous and deciduous tree pollen. *Ecology*, **31**, 633–41.
Bladdered coniferous pollen remained afloat much longer than deciduous angiosperm pollen, but there was much taxon-dependent variation.

Horowitz, A. (1969). Recent pollen sedimentation in Lake Kinneret, Israel. *Pollen et Spores*, **11**, 353–84.
Pollen spectra of the surface muds are relatively uniform over the lake's area (total area 170 km²). Jordan River influx has a profound influence on pollen spectra, especially for hydrophilous species (e.g., *Thelypteris* and *Tamarix*) and to some extent for *Quercus*.

Huntley, B. & Birks, H. J. B. (1983). *An Atlas of Past and Present Pollen Maps for Europe: 0–13 000 Years Ago.* Cambridge: Cambridge University Press.
Contains a study of the correlation between pollen in surface samples (mud, moss polsters, traps) and vegetation on a broad ('vegetational formation') scale. Their Figures 4.44 and 4.45 compare reconstructed (from pollen) and actual vegetation maps. The maps are quite comparable on a fairly broad scale (1° × 1° grid). See also the somewhat similar study by Delcourt, *et al.*, 1984*.

Jackson, S. T. (1988). (abstract) Pollen-vegetation relationships in small lake basins: evidence for varying pollen source areas within and among taxa. *American Quaternary Association, Program and Abstracts, 10th Biennial Meeting*, 76.
A study of modern pollen in 20 ponds (0.1–0.5 ha) in New York and New England shows that pollen abundance depends on the taxon, its abundance in the vegetation, and its distance from the pond.

Jones, M. D. & Newell, L. C. (1946). Pollination cycles and pollen dispersal in relation to grass improvement. *Nebraska Agricultural Experiment Station, Research Bulletin*, **148**, 1–42.
A study similar to that of Buller (1951*). Plot size not given. Deposition (on glass slides) decreased to 0.8% (relative to amount in plot center) at 301 m (60 rods) from the plot center.

Kershaw, A. P. & Hyland, B. P. M. (1975). Pollen transfer and periodicity in a rain-forest situation. *Review of Palaeobotany and Palynology*, **19**, 129–38.
Study of aerial deposition of pollen in traps on a small lake in the marginal rainforest of northeastern Queensland. Absolute pollen deposition ranges from 90–250 grains/cm² per year. This value is comparable to values from tundra, and is much less than values from temperate deciduous forest.

Kershaw, A. P. & Strickland, K. M. (1990). A 10-year pollen trapping record from rainforest in northeastern Queensland, Australia. *Review of Palaeobotany and Palynology*, **64**, 281–8.
A study of modern pollen deposition within simple notophyll (leaf size between microphyll and mesophyll) vine rain forest (altitude 1000 m) in northeast Australia over a 10 year period. Pollen was trapped at ground level and at heights to 27 m (two traps within the trunk space, two within the canopy, and one above the canopy). Of 137 taxa recorded, 120 were derived from the rainforest (56 total taxa on average annually). Local plants (pollen from all woody plants within 10 m of the traps: *Acmena*, Sapotaceae, Elaeocarpaceae, Proteaceae, *Symplocos*) constituted about two-thirds of the total number of grains, extralocal (other woody taxa occurring within a 0.5 ha censused plot surrounding the sample site) constituted 19%, and the remainder was from taxa outside the plot. Yearly variation of pollen influx is high: fern spore values are relatively constant, whereas pollen taxa vary from year to year, thus suggesting that flowering is not annual for each tree (does not necessarily apply to the species as a whole). The canopy traps recorded the highest pollen influx, and also had the highest percentage of local pollen, probably because of the high concentration of zoopollinators in the canopy. These results suggest low pollen transport of most woody rainforest taxa; this is also supported by other studies showing rainforest taxa never reach more than 2% in pollen diagrams outside the forest edge.

Kondratyene, O., Blazhchishim, A. N. & Emelianov, E. M. (1970). Composition and distribution of pollen and spores in surface layers of sediment in the central and southeastern parts of the Baltic Sea. *Baltica*, **4**, 181–95. (In Russian).
A study of 100 samples from the central and southeastern Baltic. Pollen concentrations range from an average of 21 grains/g in sand to an average of 8000/g in pelites. These concentrations are 100–1000 times those found for the Mediterranean (Vronskiy and Panov, 1963*) or the coastal Pacific (Koreneva, 1964b*). Baltic pollen concentrations increase with distance from shore (unlike the other examples just cited) and reach a maximum in the Baltic's deep basins. Pollen concentration depends on lithology (especially pelitic content) and water depth. The increase in concentration with water depth occurs across all lithologic types: sand in depth <40 m, 12–19/g; sand 50–70 m, 17–105/g; pelitic muds <100 m, <6000/g; pelitic muds 100–160 m, 2800–17 400/g. Pollen of tree species dominates (generally >90%) with pine (*Pinus*: 65% of the tree pollen), birch (*Betula*: 8–16%), alder (*Alnus*: 7–14%), and spruce (*Picea*: 2–12%) most important. Present in small amounts is pollen of broadleaved trees (2–10%), herbs (mostly Gramineae, Compositae, and Chenopodiaceae: 2–6%). and pteridophytes (fern spores and *Sphagnum*: 2–6%). Pollen composition, as concentration, is correlated with water depth and not distance from shore: pine content is greatest in water <25 m (73.5%), and decreases to 57.8% at depths >100 m. Angiosperm pollen, on the other hand, increases toward deepwater (e.g., birch 8.7% in water <25 m deep, 18.2% at

depths > 100 m). Pine and spruce pollen, the authors conclude, is large and heavy so it is deposited in shallow water, whereas angiosperm pollen is small and light, so it is deposited in deeper water. (Traverse and Ginsburg, 1966*, found that pine pollen remains afloat easily and for long periods.) The pollen spectra generally reflect the overall surrounding forest composition, rather than the effects of nearby plants.

Koreneva, E. V. (1964a). Distribution and preservation of pollen in the western part of the Pacific Ocean. *Geological Bulletin* Department of Geology, Queens College, City University of New York, **2**, 1–17.
A study of sea bottom surface samples from about 25° N lat. to about 42° S latitude along a more or less north–south zone straddling the International Date Line. The following sediments were sampled: red clay, calcareous oozes, siliceous pelagic oozes, terrigenous silty clays, sand, gravel, and volcanic and organic sediments. Red clays were essentially barren of sporomorphs and they are rare in calcareous oozes. The largest numbers are found in samples near New Zealand. The concentration of sporomorphs is greater on the east side of New Zealand than on the west side, that is, on the downwind side of the prevailing winds. Coniferous pollen and trilete spores are dominant in these samples, with subsidiary angiosperm pollen. The assemblages are biased towards sporomorph morphological types transportable by wind or turbulent water for great distances. The concentration of pollen and spores in a sample is generally correlated with the amount of terrigenous sediment present. Large quantities of sporomorphs occur only within about 300 km of large islands (like New Zealand).

Koreneva, E. V. (1964b). Distribution of spores and pollen of terrestrial plants in bottom sediments of the Pacific Ocean. In *Ancient Pacific Floras – the Pollen Story*, ed. L. M. Cranwell. Honolulu: University of Hawaii Press, 31.
Sporomorphs occur in bottom sediments of the marginal northwestern Pacific and near islands that are offshore. High concentrations are found in submarine depressions. Samples from the Sea of Okhotsk contain pollen and spores of four 'provinces': northern tundra, Kamchatka grassy-deciduous forest, southwestern conifer-*Quercus* forest, and central mixed vegetation. Grass pollen is not important anywhere and less so away from the coast.

Koreneva, E. V. (1966). Marine palynological researches in the U.S.S.R. *Marine Geology*, **4**, 565–74.
Summary of several separate Soviet studies: In the Sea of Okhotsk, the concentration of sporomorphs is 'very closely related to the granulometric and mineralogical composition of the sediments, and also to the relief of the seafloor' (p. 566). In the sea of Azov, pollen concentrations were inversely proportional to sediment grain size. Arboreal pollen is supplied by both the wind and the Kuban River. Pollen concentrations in the Mediterranean Sea are generally low; little river water enters the sea.

Koreneva, E. V. (1971). Spores and pollen in Mediterranean bottom sediments. In *The Micropaleontology of Oceans*, ed. B. M. Funnell & W. R. Riedel. New York: Cambridge University Press, 361–71.
Includes results from about 74 surface bottom samples throughout the Mediterranean. Highest sporomorph concentrations (> 100 grains/g) occurred in the extreme western end of the Mediterranean and in the northwest Adriatic; lowest concentrations occurred in some samples off the desert coast of Africa. Vegetation of surrounding land is best reflected in coastal samples. *Pinus* dominates the arboreal pollen. Most diverse palynology occurs in the western end of the Mediterranean (including *Cedrus*, *Abies*, *Quercus*, *Castanea*, *Juglans*, Ericaceae, Euphorbiaceae, Leguminosae, *Pinus*) and is derived from the varied vegetation of southern Spain and northwest Africa.

Lézine, A.-M. & Edorh, T. M., 1991. Modern pollen deposition in west African Sudanian environments. *Review of Palaeobotany and Palynology*, **67**, 41–58.
A study of 79 land surface and fluvial mud samples from Senegal and Togo. The climate is summer-wet, winter-dry. Fluvial mud and nearby soil surface samples show great similarity; thus, fluvial pollen transport is slight in these areas. The samples generally represent the local vegetation, except that Gramineae and Cyperaceae are overrepresented. In particular, combretaceous forests and wetter forest with *Isoberlinia* (Leguminosae) are recognizable by their pollen assemblages. The samples proved to be sensitive records of local vegetation differences.

Lubliner-Mianowska, K. (1962). Pollen analysis of the surface samples of bottom sediments in the bay of Gdansk. *Acta Societatis Botanicorum Poloniae*, **31**, 305–12.
Study of 55 samples taken from five transects in the Bay of Gdansk. The transects are concentrated in areas away from the sandy sediments associated with Vistula River discharge. Samples from the deeper areas of the bay have palynomorph concentrations from 50 to some hundreds per gram. Coniferous pollen increases in percentage abundance offshore because of its high buoyancy. Forest spores (Filicales, *Lycopodium*) increase in the central, deeper part of the bay due to transport by marine currents from the edge of the bay.

Lundqvist, J. & Bengtsson, K. (1970). The red snow – A meteorological and pollen-analytical study of long transported material from snowfalls in Sweden. *Geologiska Föreningens i Stockholm Förhandlingar*, **92**, 288–301.
Colored snow is shown to contain exotic (to Sweden) pollen whose taxonomic makeup coincides fairly well with the vegetation of the South Russian steppes. Storms in the area north of the Black Sea and prevailing winds from there support the inference that the exotic pollen is derived from the steppes. The concentration of solids (pollen + dust + diatoms) was about 0.1 g/l of snow or about 0.5–1.0 g/m² of ground. This leads to an estimate of 1 g/m² per year. The authors conclude that this could explain the appearance of *Ephedra* pollen in the Pleistocene of Scandinavia.

Maher, L. J., Jr. (1963). Pollen analyses of surface materials from the southern San Juan Mountains, Colorado. *Geological Society of America Bulletin*, **74**, 1485–504.
Meltwater overflow in cirque lakes carries early season pollen out of the lakes.

Maher, L. J., Jr. (1964). *Ephedra* pollen in sediments of the Great Lakes region. *Ecology*, **45**, 391–5.
Ephedra and *Sarcobatus* pollen can be shown to be wind transported more than 700 km.

Maley, J. (1972). La sedimentation pollinique actuelle dans la zone du lac Tchad (Afrique centrale). *Pollen et Spores*, **14**, 263–307.
Study of 12 samples from recent sediments of Lake Chad, a river draining into it, and four small ponds adjoining it. These palynomorphs in these sediments correspond only partly to the basin's vegetation. The 'allochthonous southern group' occurs only in sediments influenced by the Chari River; the prevailing

winds generally oppose the river's direction of flow. Rare grains of *Artemisia* and *Erica* may have been transported by wind as much as 800 km (from the north).

Mal'gina, E. A. & Maev, E. G. (1966). Spore-pollen spectra of bottom sediments of the Caspian Sea. *Izvestiya, Akademii Nauk SSSR, Seriya Geograficheskaya,* 1966(2), 61–70. (In Russian).
This paper covers two separate sampling areas in and near the Caspian Sea. The first reports on four bottom samples from the northwestern arm of the Caspian (60–80 km offshore from the Volga delta), and four nearby surface soil samples. Average concentration is 10 grains/g. These samples contain predominantly 'steppe'-type pollen: chenopods, *Artemisia,* mosses. This is similar to the spectra of surface soils of nearby land areas. The tree pollen composition matches that of the lower reaches of the Volga as reported by Fedorova (1952*), but the total amount of tree pollen is relatively slight. Most sporomorphs in these samples are transported aerially and reflect the adjacent semi-desert vegetation. Marine currents are apparently unimportant in transporting sporomorphs. The second area is in the southern Caspian and comprises seven bottom samples, mostly on a profile along 39° N latitude, and three surface soils samples. The soil samples contain mostly chenopod pollen, and 20% exotic tree pollen (e.g., *Carpinus, Alnus*). The bottom samples' concentration ranges from about 1000/g in lime muds to 240/g in sandy samples. The highest percentage of tree pollen is in the middle of the basin, mostly alder and broad-leaved species. The broad-leaved species (*Carpinus,* oak, beech, elm) are mostly from forests of northern Iran. Chenopods and *Artemisia* dominate the herb and shrub fraction. The pattern of sporomorph occurrence is related to easterly winds and the resulting counterclockwise water circulation: pollen of trees is deposited more or less uniformly in the surface waters, while herb and shrub pollen is carried along the edge of the sea. As a result, tree pollen is more important in the sea's center.

Martin, P. S. & Gray, J. (1962). Pollen analysis and the Cenozoic: *Science,* **137,** 103–11.
Discusses dramatic examples of long-distance aerial pollen transport, but notes that local pollen generally drowns out 'foreign' pollen, especially in environments (e.g., forests) where the local vegetation is abundant.

Matsushita, M. (1981). Palynological researches of surface sediments in the Harima Nada, Seto Inland Sea – Comparison of the actual vegetation with the composition of pollen grains and spores. *Quaternary Research (Dai-yonki Kenkyu, Tokyo),* **20**(2), 89–100. (In Japanese).
Same stations sampled as in Matsushita (1982*). Pine, the most common type, is over-represented relative to source vegetation; Fagaceae is under-represented.

Matsushita, M. (1982). Palynological researches of surface sediments in the Harima Nada, Seto Inland Sea – Behavior of pollen grains and spores. *Quaternary Research (Dai-yonki Kenkyu, Tokyo),* **21**(1), 15–22.
Analysis of 38 surface samples from this arm of the sea (between Honshu and Hikoku Islands; about 50 km west of Kobe) indicates that the distribution of sporomorphs is coincident with high percentages of fine particles and (low?) specific gravity of the sediments. Matsushita found high correlation of the log (number of sporomorphs) and the percentage of particles 16 μm or smaller. The best correlation coefficients occur between 4–8 μm ($r = 0.82$). Sporomorph abundances ranged from 3100 to 50000 per gram of dry sediment. Common taxa present

include *Pinus, Abies, Tsuga, Cryptomeria,* Fagaceae, Gramineae, and fern spores. Contour maps of abundance for *Pinus, Cryptomeria,* Fagaceae, and Gramineae are similar to the overall trends.

Matsushita, M. (1985). The behavior of streamborne pollen in the Kako River, Hyogo Prefecture, western Japan: *Quaternary Research (Dai-yonki Kenkyu, Tokyo),* **24**(1), 57–61.
Matsushita shows that palynomorph concentration in the water of the Kako River (about 35 km west of Kobe) is highly correlated with suspended sediment concentration and river discharge (i.e., rainfall with a lag). Maximum palynomorph concentration reaches 10^4 grains/l (at river discharge of 200 m³/sec). Relative importance of taxa varies from sampling period to sampling period (Matsushita sampled daily over about a one week period in each of May, June, July, 1980), but individual taxa show the same correlation with discharge as does total pollen concentration regardless of month. Most pollen occurs with suspended sediment of 11–44 μm (6.5–5 ϕ) size. Matsushita (1982*) found pollen in Harima Sound sediments, into which the Kako River drains, associated with sediment particles finer than 16 μm. In this paper, Matsushita attributes this discrepancy to an artifact of processing the marine surface sediments. Frequency distributions for total pollen and four common taxa by sediment grain size reveals a different pattern for each taxon. Most *Pinus* grains occur with particles larger than 5 ϕ, very few with grains smaller than 5.5 ϕ. One-quarter of *Cryptomeria* pollen occurs with particles coarser than 5 ϕ, and half occur with 5–5.5 ϕ particles. Gramineae occurs dominantly in the 4.5–5 ϕ fraction with gradually decreasing amounts in finer sizes; this resulted from occurrence of many species of grasses and of both fresh and weathered pollen. *Platycarya* pollen occurs in more or less constant frequencies in grain sizes from 5–7 ϕ. Pollen grain behavior in this river, therefore, appears to be determined by size and not shape.

McAndrews, J. H. (1972). Pollen analyses of the sediments of Lake Ontario. *24th International Geological Congress (Montreal), Section 8,* 223–7.
Analysis of 91 surface sediment samples taken on approximately a 16 km grid for palynomorph percentages and concentrations for 61 taxa. Eighty samples contained at least 100 tree pollen. Highest average concentrations (~ 80000/g) occurred in the deepwater clays; concentrations ranged down to less than 40000/g in shallow nearshore silts and sands. The deepwater clays contained more uniform assemblages than nearshore sediments. Ragweed (*Ambrosia/Iva*), pine (*Pinus*), and oak (*Quercus*) pollen dominate the assemblages with subordinate cedar (*Thuja, Juniperus*), birch (*Betula*), elm (*Ulmus*), and grass (Gramineae). Pollen percentage variations reflect proximity of river mouths, of areas with large concentrations of particular producing plants, and of areas with late-glacial sediment outcrop. See McAndrews and Power (1973*) for trend surface analysis and further details.

McAndrews, J. H. & Power, D. M. (1973). Palynology of the Great Lakes: the surface sediments of Lake Ontario. *Canadian Journal of Earth Sciences,* **10,** 777–92.
A comprehensive study of the palynology of Lake Ontario (19000 km²). McAndrews & Power's major conclusions are: (1) pollen concentration is higher in silt-clay (offshore) than in sandy (inshore) sediments; (2) relative percentages of different pollen types are uniform over the lake and reflect the drainage basin's vegetation as a whole; (3) most pollen is modern;

(4) runoff and rivers are prime pollen sources (pollen type percentages offshore from river mouths reflect river basin vegetation); (5) air movement does not explain pattern of pollen percentages; (6) local pollen patterns offshore from river mouths are related to lake currents. McAndrews & Power's results are presented primarily as trend surface maps (p. 786–7) of the distribution of pollen types over the 80 sampling points in the lake. More specific results include: (1) samples commonly included up to 5% Paleozoic spores; (2) *Picea, Ulmus, Ambrosia, Quercus, Betula,* Gramineae, Chenopodiaceae are correlated best with points of river influx; (3) pollen concentration ranged up to 240 000 grains/g dry sediment and averaged about 130 000/g in productive samples.

McDonald, J. E. (1962). Collection and washout of airborne pollens and spores by raindrops. *Science,* **135,** 435–7.
Theoretical discussion on efficiency of rain washout of pollen and spores. Collection efficiency of rain depends on raindrop size and fall velocity, and the product of palynomorph density and diameter squared (computable from its terminal velocity by Stokes' Law). Calculation suggests small palynomorphs (3.5–7 μm, such as *Penicillium*) have generally low collection efficiencies; larger grains have larger efficiencies. Further analysis gives an equation for the probable fraction of palynomorphs removed from a column of air. A light shower (rainfall 0.1 cm) will remove almost all palynomorphs only if raindrops are small (0.2 mm, drizzle size), but a 1 cm rainfall will remove essentially all pollen larger than about 10 μm regardless of raindrop size. (See Muller, 1959*, for empirical support for this.)

Melia, M. B. (1980). Distribution and provenance of palynomorphs in northeast Atlantic aerosols and bottom sediments. Ph.D. diss. Michigan State University, East Lansing.
A study of eolian dust and bottom sediments off the coast of Northwest Africa. Pollen concentrations range from 2000/g off the Saharan coast in Mauritania to less than 50/g in deep ocean basins. Mediterranean aerosols may contain more than 40 grains/m^3 during the summer and range from 4–6/m^3 for tropical aerosols during the winter. Fungal spores are most abundant in tropical aerosols and bottom sediments. Distribution patterns of opal phytoliths and freshwater diatoms in both air and sediment indicate dust storms are the major transport agents for these from the interior of Africa. Some diatoms may have been transported in this fashion as much as 6000–7000 km from Chad or Niger. Some Mediterranean pollen may have been transported as much as 5000 km. 'Palynomorph distributions overall are related closely to both source vegetation and to atmospheric and oceanic transport mechanisms.'

Melia, M. B. (1984). The distribution and relationship between palynomorphs in aerosols and deep-sea sediments off the coast of northwest Africa. *Marine Geology,* **58,** 345–71.
Published version of Melia (1980*). Forty air samples and 71 bottom samples offshore from the coast between 0° and 32° N latitude were analyzed. Melia found three palynofloral zones in both the aerosols and bottom sediments: Equatorial-Tropical zone (south of 10°), Saharan zone (10°-25° N), Mediterranean zone (north of 25° N). Maps of concentration and percentage frequency for a number of important taxa (e.g., bisaccates, *Quercus, Betula,* cheno-ams, Gramineae, and Compositae) are given. The Mediterranean zone contains types such as *Olea, Quercus, Betula, Alnus, Ulmus, Plantago, Pinus, Abies,* etc. The Saharan zone is dominated by Chenopodiaceae-Amaranthaceae

with Gramineae and Compositae. The Equatorial-Tropical zone includes Palmae, mangroves, and Euphorbiaceae. These zones represent, in a general way, the major climatic-vegetational zones of northwest Africa. Aerosol samples reflect the pollen in the air at the moment of sampling so local times of flowering are important, whereas bottom samples integrate pollen deposition over a long period of time. Sporomorph distribution patterns primarily reflect wind transport (especially dust storms), and secondarily water transport (surface ocean currents, fluvial transport – especially locally – and deep ocean currents in descending order of importance). See also Hooghiemstra and Agwu (1986*).

Mudie, P. J. (1982). Pollen distribution in recent marine sediments eastern Canada. *Canadian Journal of Earth Sciences,* **19,** 729–47.
Examines pollen concentrations and taxonomic compositions in relation to onshore vegetation; fluvial discharge, wind, and ocean currents; and continental shelf morphology along the eastern coast of Canada from the Bay of Fundy north to 68° N lat. on Baffin Island. Quaternary pollen and spores are absent or very rare in marine sediments containing >40% sand, and concentrations decrease rapidly offshore. Estuaries of major rivers have more sporomorphs than other environments. Some types are probably wind-transported: *Picea* (spruce), *Pinus* (pine, in sediments north of Newfoundland), *Sphagnum* (moss). Others are fluvially transported to the sea: *Pinus* (south of Newfoundland), *Abies* (balsam fir), fern spores. Transport modes for others are unclear (e.g., *Quercus, Tsuga*). Offshore concentration decrease is apparently related to decrease in average sedimentation rate offshore. Greatest variability in concentration occurs nearshore. Mudie calculates that fluvial deposition accounts for only 10–25% of measured pollen deposition (2000 grains/cm^2 per year; Nova Scotia rivers may transport 230×10^{12} grains per year). Much of the fluvial pollen load from large rivers is deposited near river mouths. Mudie concludes that 'wind transport is the simplest explanation for the marine distribution of bisaccate pollen' (p. 743), although ocean currents probably also contribute.

Muller, J. (1959). Palynology of Recent Orinoco delta and shelf sediments. *Micropaleontology,* **5,** 1–32.
A classic study of miospore distribution in modern environments. Prevailing winds are onshore, so the final distribution of pollen and spores in marine environments of the delta is controlled by water currents. The offshore areas of greatest sporomorph concentration coincide with the major deltaic distributaries. Atmospheric concentrations of pollen and spores appear to be very low, perhaps owing to the cleansing effect of heavy daily rains. On the terrestrial part of the delta, levee and backswamp environments differed palynologically: levees contained reworked and transported sporomorphs; backswamps contained more or less autochthonous pollen and spores. In addition, Muller said he could recognize backswamp subprovinces that reflect local plants and sediment supply.

Needham, H. D., Habib, D. & Heezen, B. C. (1969). Upper Carboniferous palynomorphs as a tracer of red sediment dispersal patterns in the northwest Atlantic. *Journal of Geology,* **77,** 113–20.
Pennsylvanian age spores were found in a distinctive red sediment on the modern Atlantic coastal slope and the combination of these two features was used to infer: (1) the red sediments' provenance in the Canadian Maritime Provinces, and (2) the existence of contourites (currents that run along the

continental shelf parallel to bathymetric contours) along the continental slope.

Nilsson, S. (1973). Scandinavian aerobiology. *Swedish Natural Resources Research Council, Ecological Research Committee Bulletin*, **18**, 1–222.
This is mostly useful as a literature review, but some entries (e.g., Hirst, 1973*; Estlander & Kulmala, 1973*) are interesting as primary sources.

Okubo, A. and Levin, S. A., (1989). A theoretical framework for data analysis of wind dispersal of seeds and pollen. *Ecology*, **70**, 329–38.
An exposition of several models for the wind dispersal and deposition of pollen including inverse power, negative exponential (log-linear), Gaussian Plume (basically that used by Tauber, 1965*), and advection-diffusion-settling models. The last is then compared to data on seeds and sporomorphs (*Pinus*, *Plantago*, *Ambrosia* pollen; *Lycopodium* spores). Actual modal dispersal distances (from the literature) range from 0.8 m (*Plantago*) to 20 m (*Ambrosia*). Modal dispersal distances are approximated by the height of release times wind velocity divided by the turbulent mixing velocity.

Panov, D. G., Vronskiy, V. A. & Aleksandrov, A. N. (1964). Distribution and composition of pollen and spores in surface sediments of the Sea of Azov. *Doklady, Earth Science Sections (A.G.I. Translations), Academy of Sciences USSR*, **155**, 81–3.
A study of 25 surface sediment samples from throughout the Sea of Azov. Herbaceous pollen, mainly chenopods (46–68%) and *Artemisia* (12–20%), constitutes 74–89% of total sporomorphs. Also included are *Ephedra*, thistles (Compositae), Gramineae, Leguminosae. Arboreal pollen comprises 5–20%, mostly pine (*Pinus*, subgenus *Diploxylon*) with birch (*Betula*) and alder (*Alnus*). Reworked pollen occurs near the Kuban' estuary. Spores comprise 5–11% of the total. The spore-pollen spectra agree well with the vegetation of the coastal areas: pine pollen comes from the Caucasus, chenopods from the steppe, and spores are found off the southeastern shore where swampy coasts occur near the Kuban' River delta. There is no significant relation between amount of pollen and distance from the coast. The authors believe this results from the small size of the Sea of Azov, and the water currents which produce poorly sorted sediments. There is a relationship between clayey sediments and increased pollen content.

Peart, D. R. (1985). The quantitative representation of seed and pollen dispersal. *Ecology*, **66**, 1081–3.
A short theoretical note deriving a seed-fall (pollen-fall) density and a probability function of the deposition of a particular seed (or pollen grain) in the same area. Peart shows how these two functions can be equated and how dispersal curves for each function are related.

Peck, R. (1973). Pollen budget studies in a small Yorkshire catchment. In *Quaternary Plant Ecology*, ed. H. J. B. Birks & R. G. West. Oxford: Blackwell Scientific, 43–60.
A study of pollen influx into two artificial lakes, each about 1.25 ha in area. Peck found a direct correlation between water turbulence and number of (previously deposited) grains resuspended in the lakes' water. The upper, more open lake was much more turbulent than the lower, partially forest-enclosed lake. Stream transport accounts for 91% of miospore influx into the upper lake and 97% of miospore influx into the lower lake. Stream input occurred at a more constant rate than

aerial influx (which peaked at the time of peak flowering). Peck inferred streamborne pollen and spores as derived from direct fall from bankside species, bank erosion, and overland runoff. These last two are especially active during floods when sporomorph content of the stream water increases from 150–30 000 grains/l.

Pennington, W. (1979). The origin of pollen in lake sediments: an enclosed lake compared with one receiving inflow streams. *New Phytologist*, **83**, 189–213.
Comparison of a 5.5 ha 'enclosed' lake (that is, one without inflowing streams) with a 10 ha lake with stream influx (Blelham Tarn, studied by Bonny, 1976*, 1978*). Pollen percentages of individual taxa in both lakes are approximately the same, but estimated pollen influx is much higher for Blelham Tarn (about 30 000 grains/cm^2 per year versus about 5000 grains/cm^2 per year). The enclosed lake had a more or less constant pollen influx over the entire 7000 year period studied, which presumably is due solely to aerial transport. Using Bonny's (1978*) work on aerial pollen deposition, Pennington concludes 83% of the pollen influx to Blelham Tarn is water transported. The palynological record over the last 7000 years indicates that stream transported pollen was less than 50% of pollen influx when the region was forested (5000–1000 years BP) and only rose after deforestation of the area. Pennington attributes the increase in pollen influx after deforestation to soil erosion and consequent transport of the pollen deposited in the soil. Relative abundance of fern spores and grains unidentifiable due to crumpling or degradation is higher in Blelham Tarn, a result consistent with soil storage and stream transport.

Potter, L. D. (1967). Differential pollen accumulation in water-tank sediments and adjacent soils. *Ecology*, **48**, 1041–3.
A study of pollen distribution in sediment in six water tanks in New Mexico (each tank about 470 m^2 in area). These receive only aerial pollen deposition. Pollen composition in different areas of the tank varies in a statistically significant manner: greatest deviations from the overall composition were found in the south and west areas of the tanks. These deviations are not simply related to wind direction. Water-tank samples contained higher percentages of arboreal pollen than samples of soils surrounding the tanks, due to the local ground deposition of Cheno-Am and grass pollen.

Prentice, I. C. (1988). Records of vegetation in time and space: the principles of pollen analysis. In *Vegetation History*, ed. B. Huntley and T. Webb, III. Dordrecht: Kluwer Academic, 17–42.
Summary of Prentice's theoretical analysis of pollen source areas and deposition for Quaternary palynology. Includes information on how percentages of pollen in sediment must be adjusted for pollen productivity and transportability in order to derive the percentages of tree species in the source area.

Ratan, R. & Chandra, A. (1983). Palynological investigation of the Arabian Sea sediments: fungal spores. *Geophytology*, **13**, 195–201.
An analysis of 35 bottom samples from the Indian continental shelf west of Bombay and the Gulf of Kachchh (=Kutch). Fungal spore diversity is higher near the coast and decreases offshore. Because fungal spores are produced in vegetative mats, they are much less susceptible to wind transport and thus are mostly waterborne.

Raynor, G. S., Ogden, E. C. & Hayes, J. V. (1970).
Dispersion and deposition of ragweed pollen from experimental sources. *Journal of Applied Meteorology*, **9**, 885–95.
Study of the aerial transport of *Ambrosia* pollen from circular area sources and point sources. Aerial concentration decreased downwind to about 5% of the concentration at 1 m from the source by a downwind distance of about 80 m (over the test duration, <1 hr for point; >3 hr for area sources). This decrease is due to both diffusion and deposition. Deposition at 100 m is less than 2% of that measured 1 m from the source (deposition=grains/unit area). Air loss at each sampling distance appears to be roughly comparable to amount deposited. Extrapolations (subject to much uncertainty) indicate about 1% of the pollen would still be airborne 1 km from the source.

Raynor, G. S., Ogden, E. C. & Hayes, J. V. (1972).
Dispersion and deposition of corn pollen from experimental sources. *Agronomy Journal*, **64**, 420–7.
An analysis of the dispersion of *Zea mays* pollen (diameter 90–100 μm) similar to the one by Raynor, *et al.* (1970*) for *Ambrosia* (diameter 20 μm). The total amount of *Zea* pollen remaining airborne at 60 m transport is 5% of that at 1 m from the source. The large size of the pollen produces a relatively large settling rate. About 63% of the pollen settled inside the source area (18.3 m circles). By 60 m distance, deposition/unit area is 0.2% that at 1 m (compared to 1.4% for *Phleum* and 2.6% for *Ambrosia*) along the centerline of wind velocity.

Raynor, G. S., Ogden, E. C. & Hayes, J. V. (1976).
Dispersion of fern spores into and within a forest. *Rhodora*, **78**, 473–87.
Dryopteris and *Osmunda* spores compared with ragweed pollen dispersion (see also Raynor, *et al.*, 1970*, 1972*). Fern spores disperse in a manner qualitatively similar to smaller particles (i.e., ragweed pollen) but are deposited over shorter transport distances (most spores deposited within 50 m).

Rich, F. J. & Spackman, W. (1979). Modern and ancient pollen sedimentation around tree islands in the Okefenokee Swamp. *Palynology*, **3**, 219–26.
Palynology of tree islands in the Okefenokee Swamp reflects the vegetation (shrubs) of the tree island, whereas the open-marsh habitats contain little shrub pollen, relatively high concentrations of aquatic plant pollen, and very large concentrations of arboreal pollen.

Richelot, C. & Streel, M. (1985). Transport et sédimentation du pollen par les courants aériens, fluviatiles, et marins a Calvi (Corse). *Pollen et Spores*, **27**, 349–64.
Study of 75 surface samples from the drainage basin of a bay in Corsica including river bed, interfluves, and marine bay bottom samples. Nearshore samples were taken from the organic mats formed by *Posidonia oceanica* roots. Abundance of eight principal taxa is used to interpret transport patterns: *Pinus nigra*, *P. pinaster*, *Quercus*, *Alnus*, *Erica*, Oleaceae, Gramineae, and sum of Cistaceae and Compositae. Offshore bay samples are dominated by the bisaccates. Abundance of *Alnus* and Oleaceae decreases steadily towards the offshore. The fluvial and interfluve samples are dominated by local pollen. For example, *Alnus* is common adjacent to the river but is absent from interfluve samples less than 300 m away. On the coast, wind transport of pollen from the mountain slopes is subordinate in importance to local pollen. Low-water fluvial transport carries pollen such as *Pinus* downstream even in the absence of wind. Flood-stage fluvial transport carries large amounts of sediment and sporomorphs. These sporomorphs are a mixture of upstream and midstream spectra. This flood-stage assemblage matches the nearshore marine assemblage of the *Posidonia* mats (15% *Pinus*, 10% *Quercus*, 5% Oleaceae, 3% *Alnus*, 3% *Erica*, and 4% Cistaceae+Compositae); therefore the origin of these marine assemblages appears to be a result of flood transport to the sea of 'upland' taxa and then coastal current transport to the *Posidonia* mats. This provides an alternate interpretation for the 'Neves effect' of Chaloner and Muir (1968), who hypothesized that certain airborne, upland sporomorph taxa occur abundantly in Carboniferous marine sediments, because transgressions eliminated much of the lowland vegetation that would otherwise provide most of the spores and pollen to the offshore sediments. Richelot & Streel suggest instead that floods carry upland pollen across the lowlands to the sea.

Richerson, P. J., Moshiri, G. A. & Goldshalk, G. L. (1970). Certain ecological aspects of pollen deposition in Lake Tahoe (California-Nevada). *Limnology and Oceanography*, **15**, 149–52.
Discusses total mass of pollen deposited (presumably aerially) in Lake Tahoe in one month (June-July 1968):
3055 kg *Abies* (4.554×10^{12} grains);
1101 kg *Pinus* (2.616×10^{12} grains).

Riegel, W. L. (1965). Palynology of environments of peat formation in southwestern Florida. Ph.D. diss. Pennsylvania State University, University Park.
Modern palynology of mangrove and related environments, fresh and brackish water marshes, tree hammocks. Most pollen types have a restricted environmental distribution, thus making at least qualitative reconstruction of plant communities possible. Palynology of some environments varied considerably over distances of miles mostly due to changing hydrologic conditions. Variations in mangrove environments occur with distance from shore, coastline shape, and drainage pattern; fresh-water marsh assemblages change in relation to the drainage system. Inland bay assemblages are in part reworked, and do not reflect local vegetation. Fresh-water marsh palynomorphs are found in the mangrove, but the converse does not occur. Water transport of pollen explains the distribution patterns better than wind.

Ritchie, J. C. & Lichti-Federovich, S. (1967). Pollen dispersal phenomena in Arctic – Subarctic Canada. *Review of Palaeobotany and Palynology*, **3**, 255–66.
Air sampling (Hirst trap) at Churchill, Manitoba: Pollen from sources more than 30 km from the sampling point constituted about 5.4% of the 21 554 grains collected. Some of these exotic types must have been transported at least 2000 km. Types transported more than 1000 km constituted about 0.7% of the total. Petri dish sampling at 20 arctic and subarctic stations: high proportions of exotics (about 50%) occurred at stations with low collection amounts (i.e., less than 200 grains total) or low concentrations (less than 10% of the more concentrated traps).

Rossignol, M. (1961). Analyse pollinique de sédiments marins Quaternaires en Israel. I. Sédiments récents. *Pollen et Spores*, **3**, 303–24.
A study of 32 samples taken off the coast of Israel in 1960. Only the second lot of 14 samples contained numerous sporomorphs. Pollen per gram of sediment is low (4–43), reflecting the low density of onshore vegetation. Pollen

concentration is nil near the shore, increases to a maximum about 8 km offshore, and then decreases farther offshore. Sporomorphs can be divided into two types which are approximately equal in abundance: autochthonous, i.e., from Israel (e.g., *Asphodelus*, chenopods, composites, grasses) and allochthonous (e.g., *Podocarpus, Pinus*, Ericaceae, Combretaceae). These allochthonous types do not grow in Israel and are inferred to have been transported with mud from the Nile. This view is reinforced by analysis of two samples from the Nile delta which contain allochthonous types. Both source types have similar areal distributions suggesting that their deposition is controlled by local hydrodynamics, but the autochthonous pollen appears to have a northeastern source and the allochthonous pollen a southwestern source.

Rossignol, M. (1969). Sédimentation palynologique récente dans la Mer Morte. *Pollen et Spores*, **11**, 17–38.
Study of six surface samples, three from the North Basin and three from the South Basin of the Dead Sea. The most abundant taxa (in decreasing order) are Chenopodiaceae, *Artemisia*, Gramineae, Compositae, Cruciferae, *Plantago*, and Zygophyllaceae. Representatives of Irano-Touranian vegetation (e.g., *Artemisia*) are in general less well represented than the local (to the Dead Sea) Saharo-Arabic vegetation (e.g., Chenopodiaceae, Gramineae, Cruciferae, Compositae, Zygophyllaceae) or Mediterranean vegetation (e.g., *Pinus, Quercus, Olea, Cupressus*) transported to the sea by runoff, water currents, and wind. Because Mediterranean pollen is more abundant in the North Basin into which the Jordan and Arnon Rivers discharge, water transport of the pollen from the Trans-Jordan Mountains seems likely. (Wind transport from the west [Israel] should produce more or less equal proportions in both basins.)

Rossignol-Strick, M. (1973). Pollen analysis of some sapropel layers from the deep-sea floor of the eastern Mediterranean. *Initial Reports of the Deep Sea Drilling Project*, **13**, 971–91.
Four transport mechanisms apparently control miospore distribution in the Nile River and adjacent Mediterranean: long-distance fluvial transport; long-distance eolian transport (especially from the desert); transport to the sea by short-distance water or wind; and long-distance dispersal, mostly by surface currents, in the sea. Rossignol-Strick found, in the Nile River, miospores with the following probable sources: Sudanese marshes (Gramineae, Cyperaceae; Nile delta marshes also a possible source); Lake Victoria or Ethiopian Mesozoic strata (pteridophytes); East African savanna (*Podocarpus*, Acanthaceae, Euphorbiaceae, Capparaceae, Moraceae, *Ficus*); and equatorial mountains.

Rowley, J. R. & Walch, K. M. (1972). Recovery of introduced pollen from a mountain glacier stream. *Grana*, **12**, 146–52.
Pollen of seven exotic genera (*Ulmus, Juglans, Liquidambar, Onoclea, Typha, Zea, Carya*) was introduced into Lyell Fork (Lyell glacier area, Yosemite National Park). Subsequent samples downstream in ponds and the stream yielded introduced pollen, and it was also recovered from ponds in the area at higher elevation and having no water connection with the seeded stream. The authors infer that some introduced pollen was released into the air via spray bubbles and other means (see also Valencia, 1967*).

Saad, S. I. & Sami, S. (1967). Studies of pollen and spores content of Nile Delta deposits (Berenbal Region). *Pollen et Spores*, **9**, 467–503.
A study of 22 samples, mostly from a well extending 37 m into the northwest Nile delta. The resulting pollen and spore assemblages can be divided into allochthonous and autochthonous taxa. Allochthonous taxa include mainly trilete and monolete spores, with minor amounts of *Podocarpus*, Ericaceae, Combretaceae. Autochthonous taxa are mainly Gramineae and Cyperaceae with Chenopodiaceae, Umbelliferae, Typhaceae, Compositae and minor amounts of other angiosperms. No statistical evaluation of the data is given. The authors conclude that the overall Nile source vegetation has not changed much since deposition of the samples at 30 m depth (deeper samples were barren). On the other hand, at least once they infer the introduction of new drainage to the Nile basin by the appearance of new allochthonous grains.

Salas, M. R. (1983). Long-distance pollen transport over the southern Tasman Sea: evidence from Macquarie Island. *New Zealand Journal of Botany*, **21**, 285–92.
Discusses the occurrence of exotic sporomorphs in Holocene lake sediments of Macquarie Island (>1200 km southeast of Tasmania and >1000 km southwest of New Zealand). Exotic pollen constitutes up to 5% of total count in some lake core samples. Sources for exotics could include New Zealand, Australia, or Tasmania. *Acacia*, an entomophilous taxon, occurs; this suggests that a pollen type's potential for long-distance dispersal depends as much on source area abundance as on adaptations to anemophily.

Sanukida, S. & Matsushita, M. (1986). Studies on the modern sedimentary environment in Lake Hamana on the Pacific coast of central Japan. *Quaternary Research (Dai-yonki Kenkyu, Tokyo)*, **25**(1), 1–12.
Study of a lagoon system to evaluate distribution of palynomorphs in the bottom sediments, relating this distribution to bathymetry, grain size, and lagoonal circulation. A very high correlation was found between the log of the palynomorph concentration of the sediment and the proportion of sediment particles finer than 6ϕ. The deeper (more inland) parts of the lagoon contained greater concentrations of palynomorphs (up to $10^5/g$), as do the mouths of some of the rivers feeding the lagoonal arms. Gramineae pollen is useful for examining organic sources because *Oryza*, the principal grass found, is not widely dispersed by air. Gramineae pollen is transported out of the main river drainage and mouth by the large fluvial discharge and is deposited in the northern part of the main lagoon, where fluvial flow and the vertical salinity stratification is reduced. The southern part of the main lagoon contains few palynomorphs because of a topographic high in the lagoon bottom separating north from south. Also, because of the shallower water here, the higher energy sediments (e.g., sands) are not hydraulically equivalent for palynomorphs.

Schmidt, F. H. (1973). Some aspects of the transport process in the atmosphere. In *Scandinavian Aerobiology*. *Ecological Research Committee Bulletin*, **18**, ed. S. Nilsson. Stockholm: Swedish Natural Resources Research Council, 53–9.
Turbulence is necessary for small particles to reach the high altitudes required for long distance transport by air. Despite the washout effect of showers, the most effective such turbulent convection is found in thunderstorms.

Solomon, A. M. & Harrington, J. B. (1979). Palynology models. In *Aerobiology – The Ecological Systems Approach. United States/International Biological Programme (US/IBP) Synthesis Series*, **10**, ed. R. L.

Edmonds. Stroudsburg, PA: Dowden, Hutchinson & Ross, 338–61.
Palynologic models of airborne transport, especially as they apply to Quaternary palynology. Reviews system variables (e.g., time, data base, processes, etc.), simple models (Davis's R-value method, modern analog methods), nonparametric statistical models (simple use of Spearman's Rank Order Correlation Coefficient, canonical correlation analysis), and the IBP (International Biological Programme) model (application of atmospheric transport equations).

Somboon, J. R. P., (1990). Palynological study of mangrove and marine sediments of the Gulf of Thailand. *Journal of Southeast Asian Earth Sciences*, **4**(2), 85–97.
A study of modern surface sediments from 23 localities in a mangrove system on the northeast coast of the Gulf of Thailand and ten localities along the Gulf's main axis. Mangrove vegetation is zoned landward into beach ridge, *Avicennia*-dominated, mangrove (*Rhizophora*) proper, *Nypa*-dominated, and transitional swamp. Somboon groups all pollen taxa found into four assemblages: mangrove, brackish water, mainly freshwater, and unidentified pollen. The mangrove assemblage dominates both the *Avicennia* and mangrove zones. In the part of the mangrove swamp dominated by *Rhizophora*, *Avicennia*, and *Sonneratia*, their pollen is dominant. *Rhizophora* pollen dominates the *Avicennia*, mangrove, and *Nypa* zones with maximum absolute abundance in the mangrove. *Rhizophora*'s small size encourages transport into laterally adjacent zones. *Avicennia* pollen occurs mostly in the *Avicennia* and mangrove zones. *Sonneratia* is most common in tidal creeks and the *Avicennia* zone. *Lumnitzera*, *Diospyros*, *Melaleuca*, and *Acrostichum* occur mostly in the *Nypa* and transitional swamp zones. *Nypa* and *Oncosperma* pollen occur only very rarely. Upland plant pollen (e.g., *Podocarpus*, *Quercus*, Dipterocarpaceae) is also rare. Palmae and Gramineae occur in greatest abundance in the *Avicennia* zone and decrease landward. Marine samples were generally poorly productive of palynomorphs. Pollen in tidal flats and offshore areas is mainly water-transported, with larger grains preferentially deposited in tidal areas and smaller grains in shallow marine sediments.

Sowunmi, M. A. (1979). (abstract) Palynological study in the Niger Delta. *Palaeoecology of Africa*, **11**, 191.
A modern pollen spectrum on a forested deltaic island on the Niger delta was derived mainly from this island and vicinity (within a 2.3 km radius). *Rhizophora* and Pteropsida are over-represented, relative to the vegetation; allochthonous pollen comprises 17% of the pollen sum.

Stanley, E. A. (1965). Abundance of pollen and spores off the eastern coast of the U.S. *Southeastern Geology*, **7**, 25–33.
In Atlantic Coastal Shelf sediments, the highest concentration of pollen and spores (per gram of sediment) occurs off the mouths of major river systems (e.g., Chesapeake and New York Bays). Stanley also concluded that quartz of medium silt or finer grain size is hydraulically equivalent to pollen.

Stanley, E. A. (1966). The application of palynology to oceanology with reference to the northwestern Atlantic. *Deep-Sea Research*, **13**, 921–39.
A study of 250 shelf and basin samples from 16 cores. Percentages of reworked sporomorphs (Paleozoic-Pliocene) ranged above 50% in some samples. Practically all samples had at least 5% reworked grains. Although concentration generally increased offshore to the continental shelf, two cores taken from river mouths showed very high concentrations. The

variation in percent of reworked microfossils is related to Pleistocene climatic variation. One Gulf of Mexico core contained reworked *Wodehouseia* and *Aquilapollenites* whose North American source is at least 1500 km away. (Closer sources may be known now.) Samples off Nova Scotia contain Carboniferous as well as Mesozoic and Cenozoic types (see also Needham, et al., 1969*).

Stanley, E. A. (1969). The occurrence and distribution of pollen and spores in marine sediments. *Proceedings First International Conference on Planktonic Microfossils*. Leiden: E. J. Brill, 640–3.
Late Cretaceous miospores occur in Pleistocene sediments of the Gulf of Mexico. The nearest source for these reworked grains is in the northwest Great Plains over 2600 km away.

Starling, R. N. & Crowder, A. (1981). Pollen in the Salmon River system, Ontario, Canada. *Review of Palaeobotany and Palynology*, **31**, 311–34.
A study of aerial and fluvial transport of pollen along and in a river 140 km long which flows through lakes (drainage basin about 1000 km²). The Salmon river flows opposite the prevailing winds. Pollen was concentrated in the zone of maximum velocity (with the silt) of the river, but significant amounts were also carried in the bedload (this pollen was abraded). Peak pollen load coincided with peak river flow and maximum aerial concentration of pollen (early spring). Arboreal pollen, aquatics, *Ambrosia*, and Gramineae are the major fluvial components. Grass and herbs are best represented at the outflow of a lake along the river's path showing the lake surface's high trapping efficiency. *Pinus*, *Betula*, *Quercus* are over-represented relative to local vegetation; *Fraxinus*, *Ostrya/Carpinus*, and Cupressineae are under-represented. Pollen concentration in the river varied during 1976 from 370–21 000 grains/35 liters of water. Pollen deposited in the bed is transported during late fall and early winter by storm-induced floods, as indicated by types found, abrasion, and complete ice-cover of the river.

Stix, E. (1975). Pollen- und Sporengehalt der Luft im Herbst über dem Atlantik. *Oecologia*, **18**, 235–42.
Two late autumn collections of aerial palynomorphs: one on a leg from Hamburg to the Azores, the other from the Azores to the West Indies. Fungal spores dominate the collections (up to 2000/m³ of air). Pollen constitute 0–0.7 grains/m³ in each sample.

Sun, X. & Wu, Y. (1988). The distribution of pollen and algae in surface sediments of Dianchi, Yunnan Province, China: *Review of Palaeobotany and Palynology*, **55**, 193–206.
A study of 59 surface sediment samples from freshwater Lake Dianchi (310 km²) in southwest China. The lake is at 1890 m elevation, and is divided into two sub-basins: Caohai (about 8 km²) which is shallow (maximum depth 2.7 m), bounded by swamps, and has bottom sediment of organic-rich mud (16 samples); and Dianchi proper (about 300 km²) which is deeper (maximum depth 6.5 m), and has bottom sediment of clay (43 samples). Total discharge from inflowing streams is slight. The natural vegetation is seasonal (winter-dry, summer-wet) subtropical semihumid evergreen broad-leaved forest; however, agriculture has modified much of the original cover. Prevailing winds are southwesterly. The authors identified 72 sporomorph taxa. Pollen concentrations ranged from 2150–80 990 grains for Caohai (average 23 790) and 280–14 370 for Dianchi (average 4470). Highest concentrations occur near the middle of each basin, and lowest mostly near stream mouths where the input

is diluted by clastics. Arboreal pollen was 65% of total land plant pollen on average in the Caohai, 90% on average in the Dianchi. Herbaceous pollen in Dianchi peaked near stream mouths, while pteridophyte spores reached minima near Caohai stream mouths. Arboreal types were dominated by *Pinus* (average 80% in Dianchi, 59% in Caohai) with highest percentages near stream mouths on the western side of the lake. *Alnus* (12% average in both) and fagaceous types (5% average in both) were concentrated near Dianchi river mouths and at one end of Caohai. Rare arboreal pollen includes *Castanopsis*, *Betula*, and *Engelhardia*. Cyperaceae distribution was similar to *Pinus*. *Artemisia*, Gramineae, Gesneriaceae, and Chenopodiaceae were more abundant near the shores than in the lake center.

Tauber, H. (1965). Differential pollen dispersion and the interpretation of pollen diagrams. *Danmarks Geologiske Undersøgelse*, **II.89**, 1–69.
An application of theoretical meteorological equations to pollen dispersal in air. Tauber derives several variants on a standard equation and calculates some dispersal results for typical botanical conditions. For example, for an infinitely producing elevated (25 m) line source, most pollen is deposited within about 3 km of the source. These theoretical equations agree with reality within about a factor of two at least for pollutants. See also Gregory (1973*).

Tauber, H. (1977). Investigations of aerial pollen transport in a forested area. *Dansk Botanisk Arkiv*, **32**(1), 1–121.
Study of a lake (about 20 000 m²) and a surrounding forest (about 50 km²) in Zealand, Denmark. The lake is shallow (mean depth 80–90 cm) and has one intermittent stream flowing into and out of it. Intensive sampling on the lake surface, at the lake bottom, and in the surrounding forest was conducted over a five year period. Contains detailed data on total concentration and taxon distribution for all sampling sites. Tauber concludes that about one half the pollen grains deposited on the lake bottom are derived from the inflowing stream. Resuspension of previously deposited grains (see Davis, 1968*, 1973*) is also high. Considerable fractions (greater than or equal to 14%) of *Alnus*, *Corylus*, *Betula*, and Gramineae pollen show signs of abrasion; this is taken to imply stream transport. Stream samples are deficient in *Fagus* relative to the lake, which implies little stream transport of this genus. *Alnus* and *Corylus* are in surplus in the lake relative to air sampling; more evidence for their stream transport. Tauber concludes that pollen sedimented on the lake bottom should be intermediate in composition between that in the stream (i.e., from the plants bordering the stream according to Tauber) and that derived from aerial transport into the lake (dominantly local pollen with some from a more extended area). The waterborne input sufficiently distorts the pollen spectra that 'the surrounding vegetation can hardly be reconstructed [quantitatively]' (p. 72).

Tippett, R. (1964). An investigation into the nature of the layering of deepwater sediments in two eastern Ontario lakes. *Canadian Journal of Botany*, **42**, 1693–709.
A study of banded sediment in two lakes (one about 11 ha and 17 m maximum depth, the other 6 ha and 10.5 m maximum depth) in eastern Ontario. Sediment is banded in organic-rich and inorganic-rich layers approximately 1.2 mm thick. The inorganic-rich layers contain consistently higher percentages of bisaccate and *Tilia* pollen, while the organic-rich layers contain higher percentages of *Ulmus* and *Betula*. *Quercus* and *Tsuga* have no preferred occurrence. These results are interpreted to result from organic layer deposition during the spring when *Ulmus* and *Betula* flower, and inorganic-dominated deposition in summer when bisaccate-producing gymnosperm taxa and *Tilia* flower. *Quercus* and *Tsuga* flower during late spring and early summer when the sediment type is changing from organic to inorganic rich. Contrast with Catto (1985*).

Traverse, A. (1990). Studies of pollen and spores in rivers and other bodies of water in terms of source-vegetation and sedimentation with special reference to Trinity River and Bay, Texas: *Review of Palaeobotany and Palynology*, **64**, 297–303.
Study of water samples (20 1 each) from surface and midwater depths at three stations: Trinity Bay about 10 km from the mouth of the Trinity River, Trinity River delta, and about 15 km upstream from the river mouth. Sampling was conducted during 1961–62 and 1985–86. The bay samples had higher concentrations at low tide when the Trinity River floods over the bay. Surface water samples are generally lower in sporomorphs than midwater depths, but lower in fungal spores and algal palynomorphs (e.g., *Botryococcus*, *Pediastrum*). Concentrations ranged from $10^4 - 3 \times 10^5$ sporomorph grains per 100 liters of water and $2 \times 10^3 - 4 \times 10^5$ grains per 100 liters of both fungal spores and algal palynomorphs. Construction of a dam about 200 km upstream in the later 1960s resulted in a decrease in palynomorph concentration of an order of magnitude in downstream water.

Traverse, A. & Ginsburg, R. N. (1966). Palynology of the surface sediments of Great Bahama Bank, as related to water movement and sedimentation. *Marine Geology*, **4**, 417–59.
Shows the distribution of palynomorphs on a major carbonate platform that has no stream influx, no terrigenous sediments, and on which surface sediments have been deposited only since the last major sea level rise, a few thousand years ago. Water transport must be used to explain pollen distribution. Pine pollen distribution closely matches sediment-type distribution. Pine pollen occurs in sediments adjacent to Eleuthera Island, which has no pines, and in low concentration adjacent to New Providence Island, which has pines. Low pollen concentrations occur in sediments indicative of high water velocities (e.g., oolite sands, pelletoidal sands), and high pollen concentrations in sediments indicative of low velocities (e.g., lime muds).

Traverse, A. & Ginsburg, R. N. (1967). Pollen and associated microfossils in the marine surface sediments of the Great Bahama Bank. *Review of Palaeobotany and Palynology*, **3**, 243–54.
See Traverse & Ginsburg (1966*).

Tschudy, R. H. (1969). Relationship of palynomorphs to sedimentation. In *Aspects of Palynology*, ed. R. H. Tschudy & R. A. Scott. New York: Wiley-Interscience, 79–96.
A study of Lake Maracaibo, Venezuela (13 000 km²). Tschudy found two transport groups: pollen (<24 μm diameter) and trilete fern spores (>24 μm diameter). The former are wind-transported into the lake and subsequent water transport of them is relatively less than such transport in smaller lakes. The trilete fern spores, however, are carried by rivers into the lake and lake circulation patterns clearly control fern spore distribution.

Turon, J-L. (1980). Les pollens et les spores dans la sédimentation actuelle le long de la ride Reykjanes et dans la fracture de Gibbs. *Academie des Sciences Paris, Comptes Rendus, Série D*, **291**, 453–6.

A study of four samples taken near Iceland along the Reykjanes Ridge and in the Gibbs Fracture. Palynomorphs present include *Pinus, Picea, Sphagnum, Lycopodium, Selaginella,* and *Dryopteris* types. Spores dominate closest to Iceland, conifer pollen in the Gibbs Fracture. The spores come from Iceland (suggested by Icelandic vegetation and dominance of spores nearest Iceland). Conifer pollen comes from Europe or North America (conifers not present in Iceland). The relatively large proportion of *Picea* pollen relative to total conifer pollen (50%) and the occurrence of haploxylonoid *Pinus* grains (i.e., *Pinus strobus*) suggests a Canadian source. Having identified the sources for these palynomorphs, Turon shows how they are in accord with the known hydrologic dynamics of water masses in the North Atlantic, so that the pollen could be water-transported from Canada across the Labrador Sea and into the Gibbs Fracture.

Valencia, M. J. (1967). Recycling of pollen from an air-water interface. *American Journal of Science*, **265**, 843–7.

A simple experiment shows that pollen can be liberated from a water surface by enclosure in bubbles ejected from the water. The amount released is inversely related to salinity (artificial 'sea water' used, limited test results).

Van Campo, M. & Quet, L. (1982). Transport par les vents de pollen et de poussières rouges du Sud au Nord de la Méditerranée. *Comptes Rendus des Séances de l'Academie des Sciences, Série 2*, **295**, 289–92.

Gives results of sampling red dust storms with gauze coated with silicone oil at Odeillo (eastern Pyrenees) and Montpellier. The occurrence of *Casuarina, Acacia, Calligonum,* and Combretaceae prove that the dust originated in Africa. Note, however, that local pollen (e.g., *Taxus, Populus, Pinus,* Cupressaceae) constitutes more than 90% of the total counted and usually is more than 96% of the total.

Vincens, A. (1984). Environnement végétal et sédimentation pollinique lacustre actuelle dans la bassin du lac Turkana (Kenya). *Revue de Paléobiologie, Volume Special (Museum d'Histoire naturelle de Genève, Suisse)*, 235–42.

Study of 34 samples from modern bottom samples of Lake Turkana in the East African Rift. Vincens finds that pollen occurrence depends primarily on the position of river mouths and on the lake's internal currents. Aerial transport can account for no more than 10% of the total pollen contributed. The delta of the Omo, the only perennial river feeding the lake, contains mostly 'allochthonous taxa of altitude,' spores, and Typhaceae; riparian taxa are rare. Deltas of intermittent rivers with large drainage basins contain less 'allochthonous taxa of altitude' than in the Omo and these taxa are less diverse and dominated by *Podocarpus*. This reflects the lower discharge of these rivers. Riparian taxa occur in abundances of less than 15% and demonstrate the existence of populations of reeds upstream from the rivers' mouths, whereas in the Omo reeds occur as far down as the delta. Secondary intermittent rivers (<50 km long) carry primarily local steppe taxa. These rivers' contributions lack regional taxa of altitude, and taxa requiring permanent water (riparian forest taxa, Typhaceae, spores). The deltas of these rivers differ from the lake itself in that the deltas contain pollen poorly dispersed by the wind. The steppe pollen is more important in these deltas than the littoral vegetation compared to the typical lake edge. Currents in Lake Turkana move 'allochthonous taxa' away from the river deltas along the lake's edge. The distribution of *Myrica, Hyphaene,* and Typhaceae matches the known current patterns well, thus providing evidence for the importance of lake currents in transporting palynomorphs.

Vronskiy, V. A. (1970). Pollen and spores from bottom deposits of the Aral Sea. *Doklady, Earth Science Sections (A.G.I. Translations), Academy of Sciences of the USSR*, **195**, 79–81. (Translation of *Doklady Akademii Nauk SSSR*, **195**(5), 1163–66.)

A study of 24 samples of surface bottom sediment from the Aral Sea, which is entirely within a desert zone of the former USSR. The palynoflora is dominated by herbaceous pollen (90–96%), especially *Artemisia* (33–62%, particularly subgenus *Seriphidium*) and Chenopodiaceae (18–43%, particularly subfamily Spirolobeae). *Phragmites* occurs up to 28% near the Amu Dar'ya River delta. Arboreal pollen is dominated by pine (*Pinus*: up to 4%, subgenus *Diploxylon*), birch (*Betula*: up to 2.5%), and willow (*Salix*: up to 1.5%). Reworked pollen occurs in quantity only near river deltas. Bottom sediment palynology is consistent with the flora of the land surrounding the sea: the deserts create the predominance of herbaceous over arboreal pollen. There is no large variation in the pollen content with distance from shore because of the small size of the sea (slightly larger than Lake Michigan) and the presence of islands in it. The relative homogeneity of Aral palynology results from the action of water currents throughout the basin. Maximum concentrations of pollen and spores occur in clayey-calcareous mud, and concentration decreases as sand or biogenic (e.g., diatom) content increases.

Vronskiy, V. A. (1974). Composition and distribution of pollen and spores in the surface sediment layer of the Persian Gulf. *Doklady, Earth Science Sections (A.G.I. Translations), Academy of Sciences of the USSR*, **215**, 230–3. (Translation of *Doklady Akademii Nauk SSSR*, **215**(1), 200–3.)

Study of 25 samples from throughout the Persian Gulf. The spectra are dominated by grass pollen (80–92%). *Pinus* is 0.5–3.5%, Oleaceae 0.5–1.6%, *Quercus* 0.4–1.6%, and *Acacia* 0.5–1% (7% in one sample). Samples from the southern Persian Gulf contain mangrove (*Rhizophora* and *Avicennia*) pollen. Spores are 5–13%. The palynology reflects the surrounding land vegetation. Very low average sporomorph concentrations were found: 1 grain/g in coarse sand samples, 10 grains/g in sandy mud, and 30–40 grains/g in clayey mud. Maximum concentrations were found in the northern and northeastern parts of the Gulf where mud transported by rivers is common.

Vronskiy, V. A. (1975). On modern 'pollen rain' above the surface of the Aral Sea. *Doklady Akademii Nauk SSSR*, **222**(1), 167–70. (In Russian.)

A study of 24 aerial deposition samples taken by glycerin-jelly-coated slides on an oceanographic vessel in July 1971 and July–August 1972. Twenty samples were taken while the ship was in motion, and four when it was stopped. Fifteen samples were taken during the day, and nine at night. Day stations received much more pollen (100–700 grains; average about 300) than night stations (0–225 grains; average about 100). A station near the sea's center was barren. Wind velocity appears to be the primary controlling factor (pollen quantity peaked at stations with wind up to 6m/sec; night station's wind ranged from 0–2 m/sec). Herbaceous pollen comprises 96–99% of the palynoflora, mainly Chenopodiaceae (60–87%, mostly subfam. Cyclolobeae) and *Artemisia* (13–26%). Gramineae pollen (5–19%) occurs in highest numbers near the mouths of the Amu

Dar'ya and Syr Dar'ya rivers. *Pinus* and *Betula* are the main arboreal taxa. The pollen rain is similar to the coastal vegetation. Comparison with previously reported surface bottom samples (cf. Vronskiy, 1970*) shows pollen of grasses, aquatics and reworked forms that occur in sediments near the deltas of large rivers must be water-transported, as these are absent from the pollen rain. Vronskiy estimates pollen rain at about 10^{16} grains in summer over the Aral Sea. In turn, Vronskiy calculates about 167 grains/cm² are deposited aerially in summer, and about 21 grains/cm² fluvially.

Vronskiy, V. A. (1979). Some aspects of using marine palynology in oil geology. *Izvestiya, Akademii Nauk SSSR, Seriya Geologicheskaya*, **1979(10)**, 150–3. (In Russian.)
A summary of Soviet and foreign work that appears intended to present spores and pollen as a possible source for hydrocarbons. Includes a summary of work partly discussed elsewhere in this bibliography (cf. Vronskiy, 1975*) and partly not yet available to this compiler. Vronskiy says the Aral Sea receives 97% of its sporomorphs aerially (summer pollen rain 11×10^{16} grains), and the Sea of Azov 99% (summer rain 12×10^{16} grains). The North Caspian Sea receives summer pollen rain of 16×10^{16} grains. (By comparison, Groot & Groot, 1966*, calculated summer pollen rain to the western North Atlantic of 15×10^{18} grains.) Total weight of sporomorphs deposited annually for these three seas is 5460 tons by air, 191 tons by water. Sporomorph concentrations are inversely correlated with sediment grain size with the exception that fine-grained biogenic sediments (e.g., coccolith oozes) may contain low sporomorph concentrations. Pollen concentrations of 4500 grains/g are known from the southern Caspian Sea, 20 000/g from the Baltic, and up to 160 000 from the Black Sea. By using estimates of the weight of single pollen grains, Vronskiy calculates that up to 5 mg of pollen can occur in 1 g of bottom sediment in the Black Sea.

Vronskiy, V. A. (1981). The present-day 'pollen rain' above the surface of the Caspian Sea. *Izvestiya, Akademii Nauk SSSR, Seriya Geograficheskaya*, **1981(2)**, 67–74. (In Russian.)
Vronskiy, V. A. (1984). Methodological aspects of palynological research of the southern seas in the USSR. *Izvestiya, Akademii Nauk SSSR, Seriya Geograficheskaya*, **1984(4)**, 70–6. (In Russian.)
Vronskiy, V. A. & Chernousov, S. Ya. (1984). Sedimentation and distribution of pollen in modern sediments of the northwestern part of the Black Sea. *Izvestiya, Akademii Nauk SSSR, Seriya Geologicheskaya*, **1984(10)**, 112–17. (In Russian.)
A study of 28 surface bottom samples of the northwestern part of the Black Sea (near Odessa and the mouths of the Dneiper and Danube Rivers). Herb pollen predominates (53–77% of the total). Of these, Chenopodiaceae are most abundant (18–37%) followed by Gramineae (8–22%) and *Artemisia* (7–20%) with rarer Cyperaceae and Compositae. Arboreal pollen comprises 20–39%, mostly *Pinus* (12–33%) with *Betula* and *Alnus* (each 1–5%) and 12 other taxa. Spores are 4–8%. Reworked forms are usually 2–12%, and reach up to 37% near the Danube delta. The palynological spectra reflect the coastal vegetation well. For example, chenopod pollen reaches its maximum in a bay which is bordered by an alkaline desert and salt marshes. Maximum grass pollen occurs in the areas near the mouths of the Danube and Dneiper Rivers and Karkinitskiy estuary. Pollen concentration is highly correlated with the pelitic fraction content and organic matter content, and inversely related with carbonate content. Pollen is absent from clean coquina, occurs

in negligible quantities in sandy coquina, occurs from 2–10 grains/g in silty coquinas, and reaches a maximum in the clayey, slightly shelly muds (40–86 grains/g). This last lithology occurs in the areas near the Danube's mouth as a result of flocculation of fine-grained fluvial particles mixing with seawater and the reduction in water velocity at the river's mouth. The concentrations found in this study are much less than those known for the finely pelitic, deepwater sediments of the Black Sea.

Vronskiy, V. A. & Fedorova, R. V. (1981). Pollen and spore concentrations in recent continental and marine sediments. *Izvestiya, Akademii Nauk SSSR, Seriya Geologicheskaya*, **1981(12)**, 79–86. (In Russian.)
A summary of palynomorph concentration investigations from a variety of continental and marine sediments, mostly from Soviet research. The continental sediments include examples from soils of the forest, forest-steppe, steppe, semi-desert, and desert of the former USSR. Steppe soils contain the highest concentrations (average 51 000 grains/g), followed by semi-desert (20 000 grains/g), forest-steppe (17 000 grains/g), forest zone (6700 grains/g), and desert (generally barren, but concentrations of about 10 000 grains/g do occur in takyrs [clay-silt playas]). Other continental sediments include peats, lagoons, and lake deposits. Concentrations are highly variable in these deposits (peat 140 000 grains/g; lakes 400–99 000 grains/g; lagoons 4–61 800/g). The authors point out that many factors control preservation of palynomorphs, particularly the sporopollenin content of the spores or pollen. In the southern seas of the former USSR (Azov, Aral, Caspian), aerial transport predominates (>97%) over fluvial transport in carrying pollen and spores; in all 11 620 tons of pollen and spores are deposited after the summer by air, and only 211 tons by water. In general for marine environments, there is a high correlation between concentration and bottom lithology: maximum amounts of pollen occur in clayey-silts, minimum amounts in sands. There is also a high positive correlation between palynomorph concentration and organic matter content.

Vronskiy, V. A. & Panov, D. G. (1963). Pollen and spore composition in the surface layer of marine sediments in the Mediterranean. *Doklady, Oceanology (A.G.U. Translations), Academy of Sciences of the USSR*, **153**, 43–5. (Translation of *Doklady Akademii Nauk SSSR*, **153**(2), 447–9.)
A study of 34 samples of surface marine sediments from throughout the Mediterranean Sea; seven samples were barren, and seven other samples contained only ten or fewer pine pollen. Samples from the northern Mediterranean are dominated by pine (*Pinus*, subgenus *Diploxylon*); other trees are represented only by single grains. Herbaceous pollen includes chenopods and composites (especially *Artemisia*); spores of *Sphagnum*, mosses, and ferns also occur. Some spectra dominated by grasses occur in the southern Mediterranean. Most samples offshore from North Africa are barren; the authors attribute this to intense physical weathering of pollen in the desert climate. The amount of pollen found decreases with distance from shore and only single palynomorphs occur 600–700 km offshore.

Wang, K., Zhang, Y. & Sun, Y. (1982). The spore-pollen and algae assemblages from the surface layer sediments of the Yangtze River delta. *Acta Geographica Sinica*, **37**(3), 261–71. (In Chinese; information from English summary and figure captions.)
A study of 66 samples taken from the onshore and offshore Yangtze River delta between 119–124° E and 30–33° N. The

palynomorph assemblages are closely related to the vegetation of the deltaic plain and surrounding hills. Most palynomorphs of the deltaic plain are derived from local vegetation, and the rest are transported primarily by wind from neighboring hills. Delta front and prodelta palynomorphs are transported by wind and fluvial currents; marine currents have little effect. Different (litho?) facies of the delta have different palynomorph assemblages. Different geomorphic units of the delta have different densities (concentrations?) of palynomorphs: higher densities in the subaerial sediments, lower in the submarine delta, and lowest in the river mouth. Densities are higher in fine sediments and lower in coarse sediments. Hydrophyte pollen distribution in marine sediments corresponds approximately to the limit of recent sedimentation (about 123° 20′ E in the East China Sea).

Wolfenbarger, D. O. (1946). Dispersion of small organisms, distance dispersion rates of bacteria, spores, seeds, pollen, and insects; incidence rates of diseases and injuries. *American Midland Naturalist*, **35**, 1–152.
See Wolfenbarger (1959*).

Wolfenbarger, D. O. (1959). Dispersion of small organisms, incidence of virus and pollen; dispersion of fungus spores and insects. *Lloydia* (*Cincinnati*), **22**, 1–106.

Literature review; see for early references to decrease of pollen abundance with distance from the source (e.g., Jones & Newell, 1946*).

Xu, J. (1987). Action of spore pollen to the sedimentary study on radiating sandy ridges along the coast of northern Jiangsu Province. *Acta Sedimentologica Sinica*, **5**(4), 147–58. (In Chinese; information from the English summary and figure captions.)
A study of 65 samples (apparently 61 bottom samples, and four seawater samples) from the East China Sea between the Yangtze (Chiangjiang) Estuary and the estuary of the Old Yellow (Huanghe) River (31.5–34° N, 120.5–122.5° E). The sporomorph assemblage in the Yangtze Estuary is dominated by *Pinus*, *Pteridium*, Polypodiaceae, and *Quercus*, whereas that of the Old Yellow is dominated by *Pinus*, *Pteridium*, *Artemisia*, and Chenopodiaceae. The assemblage from the radial, i.e., perpendicular-to-shore, sandy ridges between the two estuaries is dominated by *Pinus*, Compositae, *Pteridium*, and *Artemisia*. The limit of Yangtze delta sediments, as indicated by palynomorphs, is 122° 30′ E and 32° 20′ N, but Old Yellow palynomorphs occur only in the Old Yellow delta. High concentrations of palynomorphs occur in fine-grained sediments, low in coarse to fine sand. Palynomorphs are poorly preserved in areas of stronger dynamics and older sediments, better preserved in areas of weaker dynamics and younger sediments.

Index

Names of organisms and people are included with all other terms in this combined index. Because the book is multi-authored, it has no general bibliography. Therefore, names of people from all chapters are indexed here, making it possible to locate their citations in individual chapters and thus the bibliographic references. Indexing of persons is limited to the first author in multi-author publications. Authors contributing to this book are referenced only once for their own chapters, e.g., 'Boulter, M. C., 199f.' For Farley's annotated bibliography, only the authors of main entries are indexed, not authors mentioned within individual annotations. F. = Figure, T. = Table. Numbers in bold indicate illustrations of cited taxa.

Aaby, B., 246
Aario, L., 256
Abies, 11, F. 14.3, F. 14.7
  balsamea, F. 14.6
  cones vs. basal area of trees, F. 14.6
  veitchii, 265
Abietineaepollenites foveoreticulatus, 398, 400
Absaroka Range, F. 20.1
absolute frequency, contrasted with concentration, 431
Acacia farnesiana, 74
Acanthaulax sp.1, F. 16.7
Acer, 2, 11, 24, 259, T. 3.6, T. 13.1
  negundo, 265
  rubrum, 246, 265, F. 14.15, T. 3.6
  saccharinum, 265
Acetabularia, 179
Acritarcha, 489
acritarchs, 1, 2, 6, 325, 331, 415, 492, 499, F. 17.4
  distribution vs. paleoenvironments, 491
  diversity vs. environments, 497
Actinomycetes, attack on pollen exines, 48
Adam, D. P., 50
Adams, C. C., 257
Adirondack Mountains, F. 14.3
Adovasio, J. M., 55
Adriatic Sea, pollen in sediment of, 504–5
aerial forest canopy parts, fate of, 104
aerial transport (*see* transport, atmospheric)
aerobiology vs. pollen, 508, 517
aerosol water droplets, pollen in, 24, 516, 522
Aesculiidites circumstriatus, F. 21.2, T. 21.3–21.5
Aethophyllum stipulare, 428
Africa, southwest coast, pollen in river sediment, 506–7
African easterly jet, in pollen transport, 512
Agathis australis, 121
Ager, T. A., 227, 302, 448
Agost section, Spain, 468
Agwu, C. O. C., 512
Ahlgren, C. E., 122
Aigner, T., 422
Ailanthipites berryi, 454, F. 21.2, T. 21.3–21.5
Aitchison, J., 392, 431
Aitken, J. D., 495
Alabama K/T section, 468
Alabama River, F. 17.35
Al-Ameri, T. K., 492
Albertibank, 414, 419, T. 19.1–19.3

Albian, 329, 330, F. 16.8
Albian–Cenomanian episode, rich marine amorphous debris, F. 16.9
Albian–Santonian, 395
alder, 59
Aldorfia aldorfensis, F. 16.7
Aleksandrov, A. N., 517
Alexopoulos, C. J., 105
Alfaroa, 229, T. 12.3
alga sp. A, F. 16.7
algae, 1, 2, 201, 415
alkaline soils of American Southwest, 47f.
alkalinity, effect on pollen, 47
All American Pipeline, 50, F. 4.2
Allen, G. P., 125
Allen, P. V., 269, 504
Alligator Reef, 183
Allison, P. A., 1
Allison, T. D., 279
alluvial valley maceral facies, 351
alluvium, palynology of, 510
Almond Formation, 398, 400, F. 18.7–18.9
Alnipollenites, **F. 12.6**
verus **T. 12.3**
Alnus, 11, 26, 60–4, 229, 440, F. 12.6, F. 15.8, T. 12.3, T. 20.3
  glutinosa, 11, 59
  pollen, importance of in pollen rain of Corsica, 63
  speciipites, T. 21.3, T. 21.5
  viridis, 11
alpine moor-heaths, 59, 60
Alpine/Tethyan realm, 418
Altamaha River, 289, F. 15.1–15.3
Alterbidinium acutulum, T. 16.1
  minus, T. 16.1
Amaranthaceae (*see also* cheno-ams), 76, 80, 81
Amaranthus, 51, 54, 55, 304
  cannabinus, 304
Amazon cone, 137
Amazon River, 130, 134, 137, 138, 383
Ambrosia, 2, 10, 11, 13, 260, T. 14.1
  artemisifolia, 36, 38
  experimental dispersion studies, 518
  trifida, 36, 37, 38, **F. 3.3**
Amelia Island, Florida, F. 15.1, F. 15.3
Ammonia, 179
ammonoid zones, Ladinian, 414, F. 19.2
amorphous debris, 312, F. 16.3
  facies, 312
amorphous indeterminate maceral, T. 10.10
  delta model, F. 17.13

amorphous macerals (*see* macerals, amorphous)
amorphous matter, 201
  palynodebris, sheet-like, F. 11.4
  palynodebris with specks, 202, F. 11.3
amorphous non-structured macerals, defined, 177, T. 10.1
amorphous organic matter (*see* organic matter, amorphous)
amorphous protistoclast, F. 17.10
  delta model, F. 17.13
amorphous structured macerals, defined, 177, T. 10.1
amorphous vascular matter, F. 18.5
Amphitheatre Formation, F. 22.7, F. 22.8
Amsterdam Palynological Organic Matter Classification, 3
Andalusiella
  polymorpha, T. 16.1
  sp. B, T. 16.1
Andersen, S. T., 20, 246, 257, 262–4, 266, 268, 270, 272, 503
Anderson coal seam, 224, 230
Anderson, T. W., 257, 503
Andropogon, 74
Andros Island, Bahamas, F. 10.1
anemophilous aquatic plants, 261
anemophilous pollination vs. entomophilous pollination, 259f.
Angle, D. G., 383
Anisian, F. 16.11, F. 16.12
Anoplophora-horizont, T. 19.1–19.3
ANOVA, 349, 404, 434
Anstey, R. L., 446
Antarctica, Maud Rise, Weddell Sea, 468
Anthrakonitbank, 419, T. 19.1–19.3
Aouache River, Ethiopia, 504
Appalachian physiographic province, 40
Apteodinium sp. 1, F. 16.7
Aptian, 328–30
Aquilapollenites, F. 22.4
  amicus, F. 22.12
  attenuatus, 477, F. 22.12, **F. 22.13**
  conatus, 467, 468, 473, 477, 478, 481, F. 22.12, F. 22.13
  delicatus var. *collaris*, 467, 468, 473, 477, 478, 481, F. 22.12, **F. 22.13**
  delicatus var. *delicatus*, F. 22.12
  floral province, 468, 482
  oblatus, 478, **F. 22.13**
  quadricretaeus, 467, 468, 473, 477, 478, F. 22.12, **F. 22.13**
  quadrilobus (s.l.), F.22.12, **F. 22.13**

525

Printed in the United States
By Bookmasters